Emerging Trends in Antibacterial Discovery

Answering the Call to Arms

Edited by

Alita A. Miller and Paul F. Miller

Antibacterials Research Unit
Pharmatherapeutics Division
Pfizer Worldwide Research & Development
Groton, CT
USA

Caister Academic Press

Caister Academic Press
Norfolk, UK

www.caister.com

British Library Cataloguing-in-Publication Data
A catalogue record for this book is available from the British Library

ISBN: 978-1-904455-89-9

Cover design by Alita A. Miller. Top section from left to right: Photographs of *Acinetobacter baumannii* grown on agar (courtesy of Andrew Tomaras), *Pseudomonas aeruginosa* at 100× from a light microscope (courtesy of Craig McPherson) and *Klebsiella pneumoniae* at 50,000× from a transmission electron microscope (courtesy of Germaine Boucher). Lower section: Pfizer Antibacterial senior scientist James Smith reading an MIC plate.

Printed and bound in Great Britain

Contents

Contributors

Annaliesa S. Anderson
Senior Director, Bacteriology
Pfizer Vaccine Research
Pearl River, NY
USA

annaliesa.anderson@pfizer.com

Thomas Bjarnsholt
Department of International Health, Immunology and Microbiology
Faculty of Health Sciences
University of Copenhagen;
Department of Clinical Microbiology
Copenhagen University Hospital
Copenhagen
Denmark

tbjarnsholt@sund.ku.dk

Kenneth F. Blount
BioRelix
New Haven, CT
USA

kblount@biorelix.com

Ronald R. Breaker
Department of Molecular, Cellular and Developmental Biology
Department of Molecular Biophysics and Biochemistry
Howard Hughes Medical Institute
Yale University
New Haven, CT
USA

ronald.breaker@yale.edu

Arturo Casadevall
Department of Microbiology and Immunology
Albert Einstein College of Medicine
Bronx, NY
USA

casadeva@aecom.yu.edu

Michael A. D'Elia
Antimicrobial Research Centre
Department of Biochemistry
McMaster University
Hamilton, ON
Canada

Michael.Delia@ucdconnect.ie

Robert G.K. Donald
Senior Principal Scientist, Bacteriology
Pfizer Vaccine Research
Pearl River, NY
USA

robert.donald@pfizer.com

Heather B. Felise
Department of Microbiology
University of Washington
Seattle, WA
USA

hrb@u.washington.edu

Jay Fitzgerald
Stanford University
Stanford, CA
USA

jayfitz@stanford.edu

Jason Gill
Center for Phage Technology
Department of Biochemistry and Biophysics
Texas A&M University
College Station, TX
USA

jason.gill@tamu.edu

Michael Givskov
Department of International Health, Immunology and Microbiology
Faculty of Health Sciences
University of Copenhagen
Denmark

mgivskov@sund.ku.dk

John D. F. Hale
Facility of Anti-infective Drug Development and Innovation
Department of Pharmacy Practice
Monash University
Parkville
Melbourne, VIC
Australia

jdfhale@gmail.com

Robert E.W. Hancock
Centre for Microbial Diseases and Immunity Research
University of British Columbia
Vancouver, BC
Canada

bob@cmdr.ubc.ca

Elizabeth J. Harry
The ithree institute
University of Technology, Sydney (UTS)
Sydney, NSW
Australia

liz.harry@uts.edu.au

Eric T. Harvill
Department of Veterinary and Biomedical Science
The Pennsylvania State University
University Park, PA
USA

harvill@psu.edu

Jeffrey E. Janso
Natural Products Research
Pharmatherapeutics Division
Pfizer Worldwide Research and Development
Groton, CT
USA

jeffrey.janso@pfizer.com

Chaitan Khosla
Stanford University
Stanford, CA
USA

khosla@stanford.edu

Toni Kline
Department of Microbiology
University of Washington
Seattle, WA
USA

klinet@u.washington.edu

Elaine R. Lee
Department of Molecular, Cellular and Developmental Biology
Yale University
New Haven, CT
USA

elaine.lee@yale.edu

Younjoo Lee
Stanford University
Stanford, CA
USA

lee.younjoo@gmail.com

Olga Lomovskaya
Mpex Pharmaceuticals, Inc.
San Diego, CA
USA

olomovskaya@mpexpharma.com

Michael J. McCluskie
Pfizer Vaccine Research
Ottawa, ON
Canada

michael.mccluskie@pfizer.com

Alita A. Miller
Senior Principal Scientist
Antibacterials Research Unit
Pharmatherapeutics Division
Pfizer Worldwide Research & Development
Groton, CT
USA

alita.a.miller@pfizer.com

Paul F. Miller
Vice President and Chief Scientific Officer
Antibacterials Research Unit
Pharmatherapeutics Division
Pfizer Worldwide Research & Development
Groton, CT
USA

paul.f.miller@pfizer.com

Samuel I. Miller
Departments of Microbiology, Immunology, Genome Sciences & Medicine
University of Washington
Seattle, WA
USA

millersi@u.washington.edu

Leigh G. Monahan
The ithree Institute
University of Technology, Sydney (UTS)
Sydney, NSW
Australia

leigh.monahan@uts.edu.au

L. Silvia Munoz-Price
University of Miami
Miller School of Medicine
Division of Infectious Diseases
Jackson Memorial Hospital
Miami, FL
USA

simunozprice@gmail.com

Adam M. Nelson
Departments of Internal Medicine/Division of Infectious Diseases and Internal Medicine/ Pulmonary and Critical Care Medicine
The University of Michigan
Ann Arbor, MI
USA

adnels@med.umich.edu

Alex J. O'Neill
Antimicrobial Research Centre and Institute of Molecular and Cellular Biology
University of Leeds
Leeds
UK

a.j.oneill@leeds.ac.uk

Olga M. Pena
Centre for Microbial Diseases and Immunity Research
University of British Columbia
Vancouver, BC
Canada

opena@cmdr.ubc.ca

John P. Quinn
Antibacterials Research Unit
Pharmatherapeutics Division
Pfizer Worldwide Research and Development
Groton, CT
USA

john.p.quinn@pfizer.com

Ronald J. Quinn
Eskitis Institute for Cell and Molecular Therapies
Griffith University
Brisbane, Queensland
Australia

r.quinn@griffith.edu.au

Audrey N. Schuetz
Department of Pathology & Laboratory Medicine
Weill Cornell Medical College/New York-Presbyterian Hospital
New York, NY
USA

ans9112@med.cornell.edu

Bret R. Sellman
Department of Infectious Disease
MedImmune, LLC
Gaithersburg, MD
USA

sellmanb@medimmune.com

C. Ken Stover
Department of Infectious Disease
MedImmune, LLC
Gaithersburg, MD
USA

stoverk@medimmune.com

Yi-Wei Tang
Departments of Pathology and Medicine
Vanderbilt University Medical Center
Nashville, TN
USA

yiwei.tang@vanderbilt.edu

Juilee Thakar
Department of Physics
The Pennsylvania State University
University Park, PA
USA

juilee.thakar@gmail.com

Tim Tolker-Nielsen
Department of International Health, Immunology and Microbiology
Faculty of Health Sciences
University of Copenhagen
Denmark

ttn@sund.ku.dk

Haruaki Tomioka
Department of Microbiology and Immunology
Faculty of Medicine
Shimane University
Izumo, Shimane
Japan

tomioka@med.shimane-u.ac.jp

Risini D. Weeratna
Pfizer Vaccine Research
Ottawa, ON
Canada

risini.weeratna@pfizer.com

Ryland F. Young, III
Center for Phage Technology
Department of Biochemistry and Biophysics
Texas A&M University
College Station, TX
USA

ryland@tamu.edu

Vincent B. Young
Departments of Internal Medicine/Division of Infectious Diseases and Microbiology and Immunology
The University of Michigan
Ann Arbor, MI
USA

youngvi@med.umich.edu

Helen I. Zgurskaya
University of Oklahoma
Department of Chemistry and Biochemistry
Stephenson Life Sciences Research Center
Norman, OK
USA

elenaz@ou.edu

Part I

Answering the Call to Arms

1 Answering the Call to Arms: Introduction and Overview

Alita A. Miller* and Paul F. Miller

Abstract

A global public health crisis due to antibiotic resistance may be imminent. Several organizations are working to mitigate the lack of new, effective drugs either in development or in the clinic by proposing strategies for re-investment in antibacterial research. Although it is imperative that regulatory issues be resolved and strategic policies be put in place, it is equally important to define the scientific path required to address this crisis. The goal of this textbook, therefore, is to offer new ways of thinking about antibiotics and technical solutions for the resistance problems we face. By summarizing innovative new concepts and approaches from leading experts around the world, we hope to enable the implementation of the re-investment strategies that are so urgently needed.

Introduction

Not so long ago, our society basked in a 'golden age of antibiotics', when a physician's ability to successfully treat bacterial infections was essentially taken for granted, at least in developed countries. However, as highlighted in the introduction to nearly every chapter in this book, the emergence and dissemination of antibiotic resistance throughout the world, exacerbated by dwindling pharmaceutical investment in antibacterial research, has threatened to close the door on that era. Some important classes of antibiotics, such as fluoroquinolones and cephalosporins, are gradually becoming obsolete against certain types of infections [most notably caused by multidrug resistant (MDR) Gram-negative strains] and new therapies to replace them simply do not exist.

Several organizations, such as the Infectious Diseases Society of America, have raised a 'call to arms' (Talbot *et al.*, 2006; Finch, 2007; Spellberg *et al.*, 2008; Boucher *et al.*, 2009; Gilbert *et al.*, 2010) to encourage researchers, physicians and policy-makers to collectively and effectively address this looming crisis which has obvious implications for modern medicine and public health on a global scale: in addition to keeping routine infections at bay, physicians rely on antibiotics to enable a variety of modern medical practices, ranging from anti-cancer and other chemotherapies to many important surgical procedures. The future of medical practice without effective antibiotics is unthinkable.

Fortunately, as mentioned below and described in Chapter 2 by Munoz-Price and Quinn, some movement towards a 'solution space' is starting to occur, at least with respect to possible regulatory reforms and financial incentives for re-investment in antibacterial research. However, these discussions and subsequent policy changes are only part of the response necessary to combat the problem of antibacterial resistance. Equally important are the scientific advances that must be made in order to discover novel antibacterial therapies and approaches. As scientists working in the private pharmaceutical sector whose aggregate investment has diminished significantly over the past decade, we feel that we are in a relatively unique

*Corresponding author

position to contribute to this significant undertaking. Although it is obviously our business to make and sell antibiotics in a competitive market, the current high medical need and probable future health care crisis compels us to proactively engage in the global fight against antibiotic resistance, in part by sharing both our own perspectives, as well as our knowledge of emerging trends in the field. This textbook serves as one manifestation of that commitment: with it, we hope to highlight the many opportunities that exist within antibacterial research in order to encourage investigators to rise up and answer this 'call to arms'. In addition, we intentionally selected authors from both academic and industrial settings from around the world, as we felt an international representation of this emerging science would best reflect the global nature not only of the problem but also of the most effective response.

We would like to emphasize that this book is not completely comprehensive in scope. We felt that the topics that we selected for this volume are representative of the many concepts that are currently under consideration in antibacterial research. We realize that in addition to the many ideas that are addressed here, there are a number of others in existence that are equally intriguing but are in their infancy; some (but not all) of these are touched upon in this introductory chapter. Finally, the many on-going efforts to either improve existing antibacterial chemical scaffolds or optimize combinations of antibiotics will not be covered here because they are beyond the scope of this book, which is meant to focus on novel approaches and ideas.

Filling the gaps in antibacterial discovery

Novel targets

Considering their enormous historical success, it is no surprise that pursuit of inhibitors of the relatively few 'traditional' antibacterial targets (components of either cell wall, protein, DNA, RNA or metabolic pathways) continues. Our understanding of how these compounds affect these pathways is still growing, as exemplified by recent structural studies that define how certain inhibitors bind to macromolecular targets such as RNA polymerase (Mukhopadhyay *et al.*, 2008), the ribosome (Ippolito *et al.*, 2008; Stanley *et al.*, 2010) or a DNA:Gyrase complex (Bax *et al.*, 2010). However, development of any novel inhibitor of these precedented targets must by necessity include strategies to mitigate the numerous resistance mechanisms that bacteria have evolved in response to 'classical' antibiotic therapy. Therefore it is very appealing to seriously consider 'non-traditional' targets for future antibacterial discovery, as pre-existing resistance mechanisms are theoretically much less likely to be present. Because it was not feasible to properly cover every compelling non-traditional target in this textbook, we elected to highlight recent advances in the understanding of several of those which are not only intriguing but also are at various stages of characterization, such that the corresponding science would be representative of the types of innovation currently in the works for others. For example, there is an extensive knowledge base around bacterial cell division, yet inhibitors of this essential function have yet to be brought to the clinic. To address this gap, Harry and co-workers not only provide an excellent, comprehensive overview of the current state of knowledge on the process of bacterial cell division in Chapter 3, but also include detailed descriptions of corresponding inhibitors and the overall future potential of this system for the discovery of novel antibacterial compounds. Another attractive non-traditional target which has been the subject of a great deal of interest and preclinical research is bacterial efflux, especially in Gram-negative species, which plays a significant role in antibiotic resistance as well as other important functions (Li and Nikaido, 2009). Because efflux pump inhibitors potentiate a number of antibiotics currently used in the clinic, it is logical to hypothesize that their co-administration would extend the clinical longevity or even restore the utility of some of these classes of drugs. In Chapter 4, Lomovskaya and Zgurskaya summarize the significant progress that has been made to further understand antibiotic resistance-related efflux transporters, the complexity of their regulatory systems and phenotypes as well as describe the discovery and evolution of some inhibitors which could eventually lead to the nomination of clinical candidates.

In contrast, researchers have only recently begun to investigate whether structured non-coding RNAs called riboswitches might be amenable to antibacterial drug discovery. These fascinating structures regulate expression of genes required for basic metabolism and other functions in response to binding of a specific small molecule ligand (Roth and Breaker, 2009). In Chapter 5, Breaker and colleagues summarize the research performed to date to validate riboswitches as drug targets, and then consider some of the important questions that will determine their ultimate success as marketed antibiotics. Finally, a preliminary discussion around virulence as an antibacterial target is presented in Chapter 6, in which Miller and coworkers at the University of Washington argue that this may be where the future of antibacterial therapy lies, using Gram-negative secretion systems as an example of this strategy. Accordingly, in their chapter they provide a synopsis of current research on the identification and development of secretion inhibitors and also discuss their feasibility as antimicrobial agents.

As intellectually appealing as it is to evaluate various novel pathways and targets for antibacterial therapy based solely on their role in bacterial physiology, it is essential not to overlook two key concepts realized by researchers in the field over the last couple of decades. First, almost all antibiotics that have been successfully employed for decades as monotherapies use much more complex mechanisms than mere inhibition of a single enzyme to arrest bacterial growth. As summarized in several reviews, these antibiotics either target multiple gene products (such as PBPs, topoisomerases or the ribosome) or a target an essential substrate which has to be processed by multiple enzymes (such as components of peptidoglycan) (Brötz-Oesterhelt and Brunner, 2008; Silver, 2007). Not surprisingly, marketed antibiotics which do target single enzymes (for example rifampicin or the sulfonamides) are used only in drug combinations or as topical agents owing to the extremely rapid rates of clinical resistance emergence against them. Although this may not prove to be the case for inhibitors of non-essential functions, definitive scientific proof of that hypothesis has yet to be published.

Second, the overall lack of success in the past decade or so of genomics-driven, target-based antibacterial drug discovery has taught us a difficult lesson industry-wide: that potent inhibitors of essential bacterial genes, regardless of the degree of target validation and availability of structural biology or other enabling drug discovery platforms, are not very easily turned into viable clinical candidates. This realization was summarized perfectly in a review by Payne and colleagues in 2007, in which they described how the antibacterial discovery group at GlaxoSmith-Kline (GSK) evaluated 300 genes and completed 70 high throughput screens over a seven-year period: this effort resulted in only five lead series and, unfortunately, nothing in the clinic (Payne *et al.*, 2007). Our experience at Pfizer was almost identical to theirs. We, like the GSK group, realized that 'it made little scientific or economic sense to simply keep screening more targets' and that 'experience suggests that it is easier to find the cellular target of an antibacterial compound than it is to engineer permeability into an enzyme inhibitor'. In retrospect, this concept seems almost intuitive, as all early antibiotics were first identified because of their antibacterial properties; their mechanisms of action (MOAs) were subsequently determined much later.

Our early discovery strategy now includes screens that interrogate various chemical libraries for compounds that have 'whole cell activity' (WCA), i.e. that are capable of inhibiting bacterial growth. Once hit compounds are triaged for promiscuity, toxicity and other liabilities, we subject them to our 'reverse genomics' platform, which uses a wide array of biochemical and genetic[1] tools to rapidly determine their MOA. Like GSK, our success rate using this approach is appreciably higher than with a purified target-based screening approach, as exemplified in recent publications (Miller *et al.*, 2008, 2009a). Researchers in academia also successfully apply this strategy, as shown by two recent examples. First, the Brown lab at McMaster University

1 Note that some antibacterial groups have added whole genome sequencing in their discovery 'toolbox' as a result of its increased performance and reduced cost in the last couple of years.

used this approach, which they call 'chemical genomics', to identify novel inhibitors of Gram-negative lipoprotein trafficking (Pathania *et al.*, 2009). Second, researchers at the University of Zurich showed that that a *Pseudomonas*-specific peptidomimetic compound inhibits bacterial growth in a non-lytic way, namely by targeting the outer membrane protein Imp (also known as LptA and OstA), which plays an important role in outer membrane biogenesis (Srinivas *et al.*, 2010).

Given the above, it may be reasonable to consider screening strategies that capture both target-specific and WCA phenotypes as having the highest potential for rapid identification of novel inhibitors with a desirable MOA. In 2006, researchers at Merck used this approach to identify platensimycin, a novel inhibitor of fatty acid biosynthesis (Wang *et al.*, 2006). Their screen consisted of interrogating a collection of 250,000 natural product extracts for WCA against a *Staphylococcus aureus* strain that contained a xylose-inducible antisense RNA of *fabF*, whose induction selectively sensitized the strain to FabF-specific inhibitors (Young *et al.*, 2006; Wang *et al.*, 2006). Hit specificity was subsequently confirmed both genetically and biochemically. The platform has since been automated and is robust, reproducible and relatively high throughput (Donald *et al.*, 2009). This 'fitness test profiling' has been used to discover a number of additional novel compounds, some of which have been published: platensin, a related but more potent inhibitor that targets both FabF and FabH (Wang *et al.*, 2007), novel inhibitors of SAV1754, an *S. aureus*-specific protein involved in peptidoglycan synthesis (Huber *et al.*, 2009) and coelomycin, whose MOA is still being elucidated but also is likely to involve inhibition of cell wall synthesis, as a variety of sensitized strains are hypersusceptible to it (Goetz *et al.*, 2010). Finally, another interesting variation on antibacterial WCA screens that has recently surfaced is to perform them in chemically defined media that represent environments bacteria encounter during infection, thereby potentially allowing for identification of inhibitors that target factors required for survival *in vivo* (Dunn *et al.*, 2009), a concept which is further discussed below.

In search of whole cell activity

Although quite a bit of knowledge has been generated over the years on the physicochemical properties of antibacterial compounds (Macielag, 2005), relatively little has been published despite some important distinctions from other pharmacophores, such as their divergence from the Lipinski 'Rule of Five'[2] (Lipinski *et al.*, 1997). Although we did not include a chapter on this topic, we felt it relevant enough to mention here. Recently, scientists at Achaogen compared the average property values of 147 antibacterial compounds, delineated into Gram-positive only or Gram-negative actives, to a list of commercially available drugs from the Comprehensive Medicinal Chemistry (CMC) database (O'Shea and Moser, 2008). Substantial differences were seen not only between the test and CMC sets, but also between those with Gram-positive-only versus Gram-negative activity. For example, Gram-negative active compounds have significantly higher polarity than the Gram-positive actives, which were closer in polarity to the CMC group. Greater polarity and a sharp molecular weight cut-off at 600 g/mol for Gram-negative actives suggest selective entry is dominated by physical properties of outer membrane porins. Other noteworthy reviews on the subject include those from Nikaido and colleagues, which focus on the properties of the Gram-negative outer membrane that allow it to exclude certain classes of compounds (for example see Nikaido, 2003). Some groups have used computational tools, including artificial neural networks, linear discriminant analysis and logistic regression analysis, to predict antibacterial activity with some success (for example see Cronin *et al.*, 2002). Despite this knowledge base, 'designing in' antibacterial activity to an otherwise promising target inhibitor remains a challenge for antibacterial medicinal chemists.

An interesting approach to circumvent this challenge, at least in some Gram-negative species, is to trick bacteria into taking up compounds

2 This 'rule of five' resulted from a landmark study that was the first to correlate specific physicochemical properties of compounds (namely molecular weight, lipophilicity, hydrogen bond donors and acceptors) with successful, orally available drug candidates.

using a 'Trojan horse' strategy. One of the main mechanisms by which these bacteria acquire iron is through the use of siderophores, which are low molecular weight iron chelators that are synthesized and secreted by bacteria into the environment when very low levels of iron are available, such as during infection (Guerinot, 1994). Once they scavenge the iron, the siderophores are then taken up by a number of specific outer membrane receptors that are up-regulated in response to low iron. The Miller lab at the University of Notre Dame and others have taken advantage of this system to synthesize siderophore–drug conjugates that have antibacterial activity (Miller *et al.*, 2009b; Möllmann *et al.*, 2009). One potential concern for this approach is facile resistance emergence; however, some studies have demonstrated that iron-uptake mutants are avirulent (Tsolis *et al.*, 1996; Wyckoff *et al.*, 2006).

A fairly radical way to give compounds whole cell activity that has received some recent attention is via photoactivation. Photodynamic therapy (PDT) involves the use of a non-toxic dye and low intensity visible light that combine to produce a cytotoxic species in the presence of oxygen (Hamblin and Hasan, 2004). This technology has been known for many years, and sometimes is used as a disinfectant strategy[3]. Photoactivation appears to contribute to the potency of the relatively new series of poryphorin-based, membrane-active antibacterial compounds (Maisch *et al.*, 2005). One of these agents, XF-73, is currently in clinical trials for nasal decolonization[4]. Although it comes with many disadvantages, such as the requirement for light (and obvious lack of selectivity) at the site of infection, the approach may have some utility in very serious, localized infections, and there is a recent report of its efficacy in a mouse model of *Acinetobacter* infection (Ragàs *et al.*, 2010).

Novel sources of chemical matter

In addition to changes in screening strategies and efforts to engineer bacterial cell-penetrant compounds, antibacterial researchers have also started seriously evaluating new sources of novel and diverse chemical matter. As discussed in the Payne review (Payne *et al.*, 2007), a significant fraction of the chemical libraries of large pharmaceutical companies consists of combinatorial chemistry-derived compounds which follow the Lipinski 'rule of five' mentioned above (Lipinski *et al.*, 1997). Therefore it is probable that focusing on those libraries, which were enriched for compounds with higher polarity and lower lipophilicity, as sources for antibacterial screens contributed to the lack of success of the genomics era described above. Accordingly, a number of recent reviews have encouraged revisiting natural products for antibacterial discovery (von Nussbaum *et al.*, 2006; Newman, 2008; Fischbach and Walsh, 2009; Li and Vederas, 2009), especially considering the progress made in both natural product chemistry and biology (Koehn, 2008; Nicolaou *et al.*, 2009). As summarized by Quinn and Janso in Chapter 7, many encouraging advances in the field have also been recently made in the discovery and exploitation of natural products from novel as well as 'traditional' sources. These experts also provide important details around the evolution of both culture and compound dereplication techniques in their chapter, reinforcing the potential for future success using this approach.

Advances in microbial genetics have helped accelerate this revitalization of natural product discovery. Thanks to the significant knowledge base that has been built around the biosynthesis of natural products in the last few decades, scientists have begun to genetically manipulate strains into producing novel derivatives of these complex structures, essentially conducting medicinal chemistry via bacterial genetics. This fascinating approach is referred to as 'biochemical engineering', 'metabolic engineering' or 'synthetic biology'. In Chapter 8, Khosla and colleagues illustrate the significant potential of this concept for optimization of current scaffolds not only to achieve improved potency and spectrum (including overcoming resistance) but also reduced toxicity and desirable pharmacokinetic properties. The related and emerging field of metagenomics, i.e. the extraction and

3 Ondine Biomedical, a biotech based in Vancouver, British Columbia, markets several disinfectant products based on this concept – see http://www.ondinebio.com/.

4 See http://www.destinypharma.com/ for details.

cloning of DNA from non-culturable organisms (which represent the vast majority on earth), has allowed scientists to discover a great number of novel cryptic genes, gene products and secondary metabolites, some of which, not surprisingly, have antimicrobial properties (Handelsman, 2004).

Know thine enemy, know thyself

The future success of antibacterial discovery does not depend solely on implementing new screening strategies and pursuing unique sources of matter. It must also include a comprehensive understanding of our 'enemies' on a molecular level: not only the 'weapons' they employ during infection (i.e. their components required for pathogenesis) but also the detailed mechanisms of their response to the battery of insults they constantly face. Recent research suggests that the bacterial response to antibiotic drug treatment is quite complex, involving many genetic and biochemical pathways (Kohanski *et al.*, 2010). Kohanski and colleagues showed in 2007 that hydroxyl radical formation plays an important role in the antibacterial activity of three bactericidal antibiotics (a β-lactam, an aminoglycoside and a fluoroquinolone) through a common 'oxidative damage cell death' pathway that involves the TCA cycle, a transient depletion of NADH, destabilization of Fe–S clusters and stimulation of the Fenton reaction; this was not observed for bacteriostatic drugs (Kohanski *et al.*, 2007). This finding suggests that all major bactericidal drugs can be potentiated by inhibiting the stress response network (or SOS response). Recent work by Lu and Collins (2009) supports this hypothesis. Their studies involved deployment of a bacteriophage engineered to overexpress the LexA protein, a negative regulator of the SOS response. They found that infecting *E. coli* with this designer phage not only resulted in potentiation of the activity of a quinolone both *in vitro* (including within biofilms) and *in vivo*, but also reduced the frequency of antibiotic resistance and restored activity against a *gyrA* mutant resistant strain. These findings have obvious implications for the ability of bacteria to survive lethal doses of antibiotics that go beyond traditional resistance mechanisms. In addition, as O'Neill discusses in Chapter 9, bacteria can become refractory to antibacterial agents in other ways that also do not fit the classical definition of 'resistance' because no heritable genetic determinants are involved. These phenomena, which include drug indifference, tolerance and persistence, are thought to play an important role in antibiotic treatment failures. In addition to employing the stress response mentioned above, bacteria can also resist antibiotic killing merely by halting active metabolism (yet remaining completely viable). In his chapter, O'Neill proposes some ways in which antibacterial discovery scientists might consider addressing these challenging phenotypes to improve the efficacy of both existing and future classes of antibacterial therapies.

Anti-virulence approaches

No review of potential antibacterial therapies would be complete without consideration of virulence as a target for intervention, which has been a subject of extensive research and debate for over a decade. Virulence determinants are those bacterial factors or mechanisms which either allow pathogens to cause damage to host tissues, such as toxins or proteases, and/or permit them to effectively adapt to their surroundings and colonize their host, such as adhesins and quorum-sensing systems. In contrast to antimicrobial agents which are designed to kill bacteria outright, inhibition of these factors or processes (or the regulatory networks that control them), would theoretically create an *in vivo* scenario that is similar to vaccination with a live, attenuated strain, whereby the 'disarmed' bacteria would be cleared by the host immune system with little or no impact on endogenous microflora. As described in a number of excellent reviews (see, for example, Cegelski *et al.*, 2008, and Rasko and Sperandio, 2010), other potential advantages to this approach include expanding the repertoire of potential bacterial targets for intervention and exerting less selective pressure, which may result in lower resistance rates (although, as mentioned above, that hypothesis has yet to be proven). We have included two representative chapters in this book that we hope will provide an appreciation for this subfield of antibacterial research. As mentioned above, Chapter 3 describes the current efforts and future prospects of targeting secretion systems in Gram-negative bacteria. In Chapter 10, Bjarnholt and colleagues present compelling data supporting the development of quorum sensing inhibitors, specifically with the aim of disrupting bacterial biofilms, which are, as described below, complex multicomponent structures that play an important role in the ability of certain pathogens to successfully establish infection and evade traditional antibiotic therapy (Flemming and Wingender, 2010).

With the exception of toxin-mediated diseases such as anthrax or cholera, it is clear that successful bacterial infections involve quite complex mechanisms of pathogenesis, often relying on a number of factors whose expression or activity is growth phase or site dependent. Although a number of studies have shown that inactivation of certain of these factors leads to attenuated virulence in preclinical animal models (as well-summarized in Clatworthy *et al.*, 2007), our own experience leads us to believe that many of these results depend on the relatively specialized model in which a given mutant was tested. For example, we recently generated a number of mutants in *P. aeruginosa* corresponding to inactivation of a wide variety of virulence factors (alone and in combination) that have been shown to be attenuated *in vivo* in various models. However, when we tested these mutant strains in our 'standard' mouse models of infection[5], we saw no significant difference in lethality versus the parent strain for any of the strains, contrary to what was reported in the literature. Whether this means that our infection models are not refined enough to capture the importance of these virulence factors for infection or that the lack of those factors themselves really does not significantly deter *P. aeruginosa* to establish infection is unresolved. However, this lack of translation to 'standard' preclinical models may be one reason why more virulence inhibitors haven't advanced to clinical trials. Perhaps this disconnect can be turned into an opportunity to improve collaborative interactions between academic and industrial antibacterial research, including a better definition of the criteria required for advancement of clinical candidates with 'non-traditional' mechanisms of action.

Targeting genes which are '*in vivo* essential' may be a good compromise between finding those inhibitors of targets which are essential both *in vitro* and *in vivo* (and which could be accompanied by higher rates of resistance) and those of virulence, which may involve a much smaller window for efficacy and narrower spectrum of activity. In the mid-1990s, two techniques were developed to identify such genes (Chiang *et al.*, 1999). Signature-tagged mutagenesis (STM) involves the use of a transposon (Tn) library in which each transposon has a unique DNA sequence or 'tag' that allows for its identification as follows: bacteria mutagenized by STM-Tn library are used to infect a given host and the *in vivo* essential genes are identified by subsequent selective hybridization of bacteria recovered from the infection to the entire Tn library arrayed in singleton. Those that come up absent in the hybridization are those that did not survive the infection, implying that the gene which was inactivated by the transposon is important for bacterial survival *in vivo*. *In vivo* expression technology (IVET) is a promoter-trap strategy designed to identify genes whose expression is induced in during infection. This strategy involves the use of a Tn library consisting of genomic DNA fragments fused to a selective marker. IVET-Tn-bearing bacteria are used to infect a given host; only those genes containing fusions which result in expression are recovered after infection. Because this approach only captures those genes that are very highly expressed, scientists refined the technology to be recombination-based (RIVET), changing the reporter gene to encode $\gamma\delta$ resolvase, which catalyses irreversible recombination between specific DNA sequences called *res* sites and engineering the construct so that the selective marker (usually tet^R) is flanked by *res* sites. Therefore any level of expression of the reporter would result in a permanent, heritable change (from tet^R to tet^S), which can be detected by replica plating after the bacteria are recovered from the infection. A large body of literature has resulted from the use of

5 We used both septicaemia and lung infection models, which we normally rely upon to demonstrate preclinical *in vivo* antibacterial efficacy.

these screens for numerous pathogens and host settings (for example see Hsiao and Zhu, 2009; Slauch and Camilli, 2000).

However, as is the case for the target-based screens described above, identification of attractive *in vivo* essential targets and subsequent *in vitro* screening of compound libraries for inhibitors of these factors might not necessarily translate into finding compounds that will truly inhibit their function during infection. Therefore, some researchers have chosen instead to focus on a 'reverse genomics' equivalent approach for the discovery of *in vivo* essential targets: screening for compounds which abrogate infection in whole animal models. Almost all of these models involve invertebrate hosts, most commonly the nematode *Caenorhabditis elegans* (Sifri *et al.*, 2005)[6], but also various insects such as fruit flies (*Drosophila melanogaster*; Apidianakis and Rahme 2009) or the larvae of wax moths (*Galleria mellonella*; Fedhila *et al.*, 2010), gypsy moths (*Lymantria dispar*; Broderick *et al.*, 2010) or silk moths (*Bomboyx mori*; Hamamoto *et al.*, 2004). Even much simpler eukaryotes such as *Tetrahymena pyriformis* (Benghezal *et al.*, 2007), *Dictyostelium discoideum* (Lima *et al.*, 2011) and *Candida albicans* (Hogan and Kolter, 2002) have been used to study bacterial virulence. Here, the challenge in advancing these hits preclinically would be in the deconvolution of their MOAs, which may not be as straightforward as for inhibitors of essential genes, and also, like the virulence inhibitors, the translation of the inhibitory activity seen in these primitive invertebrate models to efficacy in a more relevant (i.e. rodent) model of infection.

No germ is an island

Like the humans they infect, bacteria exist in complex communities. By utilizing sophisticated communication mechanisms such as quorum sensing, they are capable of extensive multicellular collaborations, demonstrating remarkable patterns of cooperation (Shapiro, 2007). Biofilms are the quintessential example of this societal behaviour. These surface-associated communities, composed of a protective extracellular polymeric matrix of polysaccharides, DNA and proteins, play a major role in chronic infections, especially because of their notable recalcitrance to antibacterial therapy (Lewis, 2010). As described in Chapter 9, a dormant subpopulation of persister cells is likely to be the main culprit responsible for this phenotype. Therefore it is not surprising that, as described in Chapter 10, a significant amount of research is currently devoted to the understanding of biofilm communities as well as various strategies to disarm them.

The success of these communities involves a remarkable exhibition of bacterial 'altruism' in which a given phenotype of a subpopulation results in great overall benefit to the community, even if the phenotype is detrimental to the subpopulation itself. Just one example of this behaviour is the role autolysis plays in biofilm formation. The DNA embedded within the extracellular matrix which helps protect the bacteria against the activity of host antimicrobial peptides (Mulcahy *et al.*, 2008), is a result of the autolysis of a small fraction of the resident bacteria (Bayles, 2007). This autolytic event is finely regulated during biofilm development (D'Argenio *et al.*, 2002; Webb *et al.*, 2003; Thomas *et al.*, 2008) and mutants defective in autolysis are unable to form biofilms (Karatan and Watnick, 2009; Ma *et al.*, 2009). However, whether this is indeed an 'altruistic' behaviour may be debatable, as some evidence exists to suggest that biofilm-associated autolysis is linked to both suicidal and fratricidal activities within bacterial societies (Thomas and Hancock, 2009). Either way, it is undeniable that one of the ways in which bacteria outsmart us is by working together.

Luckily, not all bacterial communities are our adversaries. In fact, we are colonized by vast ecosystems of beneficial microbes, collectively referred to as our 'microbiome', whose cellular counts exceed the number of cells in our bodies by an estimated factor of ten, with one hundred times as many genes as within our genome (Hooper and Gordon, 2001). As described in Chapter 11 by Nelson and Young, the study of these ecosystems has exploded in the last several years thanks to the metagenomics approaches mentioned above as well as new 'deep sequencing' techniques which

[6] Note that this screen can be conducted in a semi high throughput fashion for some (but not all) pathogens of interest (Moy *et al.*, 2009).

enable, for the first time, their detection, classification and characterization. Chapter 11 provides an excellent overview of the techniques being used to study these populations as well as descriptions of the deleterious effects observed when they are disrupted. Researchers have discovered that these communities are greatly beneficial to human health, playing important roles in nutrient utilization and immune regulation, as well as protection from pathogens (Fujimura *et al.*, 2010). We included this chapter because it has become apparent that researchers should take the microbiome into account during the antibacterial discovery process. Not only should we consider whether or not a particular compound of interest will be detrimental to the beneficial effects of our microbiome but also whether or not we can use what we are beginning to learn about them to our advantage, from modulating host defences to mining for novel matter we could potentially turn into new antibacterial therapies. Note that this exciting field is only in its infancy; we expect remarkable advances in our understanding of our own physiology to result from continued research in this area.

Bacterial hijackers

Another way in which pathogenic bacteria outsmart us is by hijacking the very cells whose job it is to protect us from them. Tuberculosis (TB) is a classic example of such a strategy: that one in three humans residing on our planet carries the bacteria responsible for this disease is testament to its success. *Mycobacterium tuberculosis* is a facultative, intracellular pathogen that causes infection by subverting the intracellular signal transduction pathways of host lymphocytes to suit its own designs. This, coupled with a very slow growth rate and its ability to become latent for long periods of time, makes TB quite a difficult disease to treat, especially considering that most current anti-TB drugs target metabolic reactions and proteins which are essential for bacterial growth in extracellular (i.e. less medically relevant) environments. The need for new treatments for tuberculosis is particularly high given the emergence of extensively drug-resistance strains, which is complicated by incomplete and delayed diagnoses, drug costs and challenging supply issues in areas where the disease is most rampant (Cegielski, 2010). In Chapter 12, Tomioka provides a very detailed description of the molecular mechanisms involved in *M. tuberculosis* pathogenesis, focusing on a variety of immunomodulatory events that ensue. He argues that future anti-TB research may be more successful if it included targeting the factors required for cross-talk with cellular signalling pathways as well as those related to mycobacterial dormancy.

Antibacterial biological warfare

Although some pathogens manipulate our immune system to their own advantage, we humans have also learned to harness its power in very effective ways. Indeed, the use of biological agents as a means to combat infection can be considered both a fairly old and a relatively new idea. The only effective antibacterial therapies available in the pre-antibiotic era were biological agents: vaccines to prevent infection, various antibody-containing sera or bacteriophage preparations to treat them. Today, our increased understanding of the mechanisms involved in immunomodulation, emerging technologies in the fields of peptide and protein chemistry and a renewed interest in phage therapy has led researchers to revisit these important approaches in the fight against MDR bacteria.

It is well known that successful vaccination against bacterial infections resulted in a dramatic improvement in human mortality rates in the last century and it remains a cornerstone of public health practices today. Indeed, significant research is currently under way to discover vaccines for use as precautionary measures against bioterrorism agents or in communities at risk for MRSA infections, such as those found in prisons or athletic organizations. As summarized by Donald and Anderson in Chapter 13, now that production of many of the early vaccines (which were based on attenuated or killed strains) has been discontinued owing to safety concerns, current antibacterial vaccine research is focused on bacterial antigens and their combinations. These authors provide an excellent overview of the recent science around several representative vaccines and conclude the chapter with a discussion of the future directions in this field, most notably in the challenges associated with developing successful vaccines against intracellular pathogens such as *M. tuberculosis* and *Chlamydia trachomatis*. In Chapter 14, Weeratna and McCluskie continue this line of thinking by describing the most recent advances in adjuvants, which play a key role in enhancing the immunogenicity of both acellular prophylactic and therapeutic vaccines. This chapter provides important details around recently characterized signalling pathways involved in innate immunity and the corresponding recognition and response to bacterial intruders. Like many of the other topics in this book, these recent learnings have also led to a corresponding increased investment in this promising field.

Selective modulation of innate immunity as an anti-infective strategy in and of itself is an emerging concept driven by the huge advances in our understanding of this crucial host defence system. Indeed, there are a number of potential targets whose modulation either mimic the activities of the natural immune response such as growth factors, chemokines and cytokines, or rescue various host signalling pathways from the bacterial hijacking described above. A number of these types of agents are currently in development (although many, but not all, are for anti-viral indications; Hamill *et al.*, 2008). This approach has several potential advantages. First, because the mechanisms at play are effective against a wide array of pathogens, immunomodulation could possibly be a broad spectrum therapy, potentially even including activity against fungal or viral pathogens. Second, since the pathogen itself is not targeted, there would be no selective pressure to develop resistance to treatment. Third, it is possible that manipulating innate immune responses may allow for the selection of an adaptive response that is appropriate and specific for the invading pathogen (i.e. Type 1 for intracellular and Type 2 for extracellular pathogens; Hamill *et al.*, 2008). As described in Chapter 15 by Hancock and colleagues, many host defence peptides have immunomodulatory activity; some have antimicrobial activity as well. The combination of our advanced knowledge in the field as well as recent advances in peptide design and synthesis makes their utilization as antibacterial agents particularly intriguing. However, there is always the risk that inappropriate interventions of the innate immune system would lead to undesirable inflammatory responses, resulting in either acute immunotoxicities such as septic shock or chronic ones, such as the induction of autoimmunity or chronic inflammation (Hamill *et al.*, 2008). Nonetheless, the prospect of manipulating the immune response to treat infections is an exciting one.

Major advances in antibody engineering over the last few decades have resulted in the generation of safe, specific and potent antibody therapeutics for a variety of diseases. Currently, over two dozen monoclonal antibodies have been approved by the FDA for clinical use and hundreds more are in clinical trials (Dimitrov and Marks, 2009), yet only one of these targets an infectious agent [Synagis, an anti-respiratory syncytial virus (RSV) mAb for a neonatal indication]. This is most probably because until the significant rise in antibiotic resistance made it clear that urgent measures must be taken, most pharmaceutical companies shied away from the development of pathogen-specific agents. Thanks to the emerging crisis, there is now significant interest and investment in the development of antibody-based therapeutics for the treatment of infectious diseases by a number of different research groups, including ours. In Chapter 16, Stover and Sellman discuss how most of the earlier monoclonal antibody efforts for antibacterial research failed because they occurred before the relatively recent technological revolution and then provide evidence to suggest that the field does indeed have significant potential for future success. Recent deal-making and promising clinical trial results for antibody-based antibacterial agents reinforce this idea (Extance, 2010). Although the approach implies a move away from empiric therapy with a reliance on pathogen-specific diagnostics, this may not necessarily be true, given the following trends. First, there appears to be an increasing acceptance by regulatory agents to consider combinations of antibodies as single therapies if the synergy is clearly established preclinically. For example, Symphogen, a biotech company whose platform is the generation of mixtures of recombinant antibodies, is currently testing their lead product candidate consisting of a mixture of 25 antibodies[7] in phase II clinical trials. Second, it is now economically feasible to develop and commercialize these antibody cocktails thanks to decreasing costs of antibody production which has resulted from vast improvements in their expression, purification and manufacture (Chartrain and Chu, 2008). Finally, advances in antibody engineering have led to the generation of 'designer antibodies', which have such features as bispecificity (Jain *et al.*, 2007) or enhanced effector function (Lazar *et al.*, 2006). This type of engineering may eventually lead to the generation of multiple desired activities in a single molecule.

Last, but certainly not least, of the biological agents included in this textbook are the remarkably clever bacteriophages. The recent and significant re-emergence of the field of phage therapy in response to the antibacterial resistance crisis comes as no surprise. These bacterial viruses have been historically quite valuable, not only as antibacterial agents (they were used around the world in the pre-antibiotic era and are still in use in some parts of Eastern Europe), but also as tools which played an important role in the early days of molecular biology, especially with respect to understanding the mechanisms of gene regulation. There are currently at least a dozen biotech companies devoted to discovering effective phage therapies for a variety of bacterial infections, several of which are in or nearing human clinical trials. In Chapter 17, Young and Gill lay the foundation for this approach by thoroughly describing the history of phage therapy, the molecular mechanisms by which phage and bacteria interact with each other and the overwhelming body of evidence demonstrating efficacy in preclinical animal models of infection. These phage experts then go on to discuss the key issues researchers should take into account when considering these agents for antibacterial therapy, from the selection of preclinical candidates, including ways in which to overcome facile resistance emergence, to important considerations for regulatory and IP concerns during the development process.

Enabling antibacterial discovery

This book would not be complete without inclusion of several emerging concepts that have the potential to be greatly enabling for future antibacterial discovery, as described below. Like other areas of focus in this book, we consider these to be representative of the future ways in which antibacterial discovery might be revolutionized: we are sure that many other innovative approaches will follow suit as the medical need continues to increase.

Understanding bacterial infections using mathematical modelling

Before discussing how mathematical modelling can enhance antibacterial discovery efforts, we felt it might be useful to provide a bit of background on the approach. It goes without saying that preclinical pharmacokinetic (PK) and pharmacodynamic (PD) studies[8] are critical for the successful development of novel antibacterial agents. Many years ago, it was observed *in vitro* that the mechanism of killing of antibiotics can differ with respect to the correlation between drug concentration and effects on bacteria: some classes of compounds such as aminoglycosides or quinolones show concentration-dependent killing (because the rate and extent of killing increases over a wide range of concentrations) whereas for others, such as β-lactams, killing is time-dependent (because it increases with concentration over a relatively narrow range of doses, so the amount of time the bacteria is exposed to the drug is more important than the total concentration of drug; Shah *et al.*, 1976). However tempting it is to predict dose and duration of therapy based on these *in vitro* observations, it is important to remember than drug levels fluctuate *in vivo*.[9] Preclinically, an

[7] Sym001 is a mixture of 25 unique anti-Rhesus D antibodies being developed to treat idiopathic thrombocytopenic purpura (ITP) and the prevention of hemolytic disease of newborns (HDN).

[8] PK refers to the absorption, distribution and elimination of drugs. PD describes the activity of a given drug on its target.

[9] To account for this, the concentration *in vivo* can be represented as the integral of drug concentration (C) over time ($\int C\,dt$) (or the area under the concentration–time curve) as the area under the curve (AUC) at a

empirical relationship between drug exposure, MIC and overall effect can be established using animal infection models which are designed to closely mimic *in vivo* conditions, thereby allowing for selection of frequency and amount of dose for human clinical trials (Owens and Ambrose, 2007). However, a major disadvantage to this approach is that, although similar growing conditions and infection endpoints (i.e. cure or death) seen in humans can be obtained in preclinical species, the pharmacokinetics of the antibiotic could significantly differ between the two. *In vitro* models, on the other hand, can accurately represent human PK and therefore can be used more extensively for dose selection, as well as determining susceptibility breakpoints and for evaluating the clinical meaning of antibiotic resistance. There are a number of different *in vitro* PK–PD models which have been optimized to capture certain parameters of infection, such site and distribution; these have been extensively reviewed in the literature (as summarized in Gloede *et al.*, 2010). A further step is to apply mechanism-based mathematical models[10] to better define the infection process itself. In addition to the above components described above, mathematical models allow the description of the number of microorganisms and the momentary effects of the antibiotics over the course of time. These models generally include at least a submodel of bacterial replication and a submodel of antibacterial effects (Czock *et al.*, 2009) and can therefore be used to potentially identify targets for intervention that may not have been selected based on their *in vitro* phenotypes. In Chapter 18, Thakar and Harvill take this modelling to the next level by describing their methods to include the more complex mechanisms that occur during infection, such as immune-mediated sequestration of bacteria or evasion strategies employed by the pathogen. Their mathematical modelling of these events uses information obtained from experimental data to construct the interactions and dependencies between the various components involved in the infection process. As proposed by the authors, adding components of this interplay to current models should make them more robust, and thereby help select antibacterial drug targets that will be effective *in vivo*.

Bacterial diagnostics

If it were possible for physicians to rapidly, accurately and cheaply determine the aetiological agent of any given infection at the time of clinical presentation, it would clearly change not only the way infectious diseases were treated clinically, but also antibacterial discovery itself. Indeed, many of the emerging trends described above imply a move from empiric to such 'evidence-based' therapies. Given the significant technological advances in other areas in recent decades, it is actually somewhat surprising that bacterial diagnostics have not yet evolved to such a point. The reason for this is almost certainly the same reason antibacterial discovery efforts have dwindled while resistance rears its ugly head: the return on investment for these types of diagnostics simply was not sufficient for the pharmaceutical industry to pursue them. Now that we are faced with this crisis, we antibacterial researchers find this to be a significant gap: not only in not having enabling diagnostic technologies available but also not having a comprehensive understanding of what types are currently available, in development or under consideration. Successful clinical trials for pathogen-specific therapies would almost be impossible without such a resource. Likewise, the success of clinical trials designed to differentiate on activity against certain types of resistance

given time (i.e. 24 hours would be AUC_{0-24}). Hence, from a PK–PD perspective, bacterial killing is a function of an antibiotic's AUC_{0-24} when compared with the minimal inhibitory concentration required to arrest bacterial growth or MIC (AUC_{0-24}:MIC ratio). The AUC_{0-24}:MIC ratio is the PK–PD measure correlated with the clinical efficacy of quinolones, tetracyclines and macrolides. For some agents, such as aminoglycosides, the ratio of the maximal drug concentration (Cmax) to the MIC of the antibacterial agent to the infecting bacteria is still the best correlate for clinical activity. The efficacy of others, such as β-lactams, is driven by the duration of time that drug concentrations remain above the MIC (T > MIC).

[10] These are not to be confused with mathematical modelling of the epidemiology of infectious diseases, which has long played a central role in our understanding of the spread of disease, even helping guide public health policies in a number of countries (see Grassly, N.C., and Fraser, C. (2008). Mathematical models of infectious disease transmission. Nat. Rev. Micro. *6*, 477-487.).

elements would also be highly enabled by diagnostics which accurately identify them. Schuetz and Tang help fill this gap in knowledge by providing an excellent and comprehensive overview of the subject in Chapter 19. They compare those techniques that are currently in use to a number of emerging, quite promising approaches. Although some of these are too complex and expensive to be useful in a clinical setting, others could be adapted to a 'point of care' type of device, which, as discussed in the chapter, could be transformative for the field.

Cultivating healthy pharmacoeconomics

While the focus of this book is on scientific opportunities, we can only translate these into therapeutic realities if other aspects of this disease area evolve as well. These include the processes used to evaluate the safety and efficacy of candidate agents in human clinical trials, the evolving requirements for demonstrating a benefit sufficient to merit approval for use by subscribing physicians, the role of payer organizations in determining the order in which new agents will be used in the treatment of disease, and the manner in which sponsoring companies make decisions on whether to traverse this path at all, relative to other disease areas in which they might choose to invest.

In order to understand contemporary pharmaceutical decision making and its impact on antibacterial investment, it is important to appreciate the concept of estimated net present value (eNPV). An eNPV for a project reflects the future value for the resulting product by considering costs, anticipated revenues and probabilities of success, and takes into account the future value of the money needed to advance a project. While it is clear that the high medical need for new antibacterial agents will assure that most new drugs that address a key resistance issue will enjoy a favourable return on investment, the high cost of drug development in general makes it difficult for drug companies to advance large portfolios in all disease areas of interest. Accordingly, eNPV calculations are frequently used to permit 'apples-to-apples' comparisons across disease areas. This assessment has a few worrisome implications for antibacterial projects in today's challenging financial environments. First, the significant number of approved antibacterial agents coupled with the acute treatment period for most bacterial infections, among other factors, creates a valuation for antibacterial projects that is less than that of many other disease areas that have few competing agents on the market and that are used for chronic disease treatment (Projan, 2003). Moreover, challenges in obtaining regulatory approval for many new agents further impacts value calculations in a negative manner by decreasing the probability of success, and causing attrition calculations to occur after development costs have been incurred. In the most extreme cases, this leads to a calculation in which investing in antibacterials appears less worthwhile than using that same money to acquire a stable financial security, an option that investing sponsors must consider.

Fortunately, as discussed in Chapter 2, there are potential solutions to this rather grim picture, but these require accurately defining where antibacterial projects fall short in eNPV calculations and tying these issues to achievable solutions. Particularly encouraging are recent meetings and analyses that have both highlighted the worsening medical need situation as well as addressed the underlying causes of the current weak antibacterial pipeline (Talbot *et al.*, 2006; Spellberg, 2008; Boucher, 2009). During its term of presidency of the European Union, the Swedish government selected the lack of antibacterial R&D activity and productivity as the key issue it sought to advance. This decision led to a multiparty meeting held in Stockholm in September 2009. Representatives from governments, regulatory bodies, academia and industry contributed to the meeting and produced a set of declarations that are binding on subsequent EU presidencies (Cars, 2009). A follow-up meeting was held in Uppsala, Sweden, in September 2010 to continue the discussion. Several published economic analyses helped pave the way for the Sweden meetings, and further evaluated some of the tools and options that might be considered as incentives to stimulate increased antibacterial R&D (Mossialos, 2009; Kesselheim, 2010). Among the incentive categories under consideration are prizes and awards for successful development of new antibiotics, early market commitments to reward successful innovations, patent extensions, vouchers, various tax incentives and frank conversations regarding the prices that could be charged for truly innovative new medicines. There is real hope that one or more of these will emerge from these discussions, but a truly meaningful outcome will require the continued engagement of all key stakeholders, acting with the dire medical need of patients in mind. The outcome must also support the key value of encouraging appropriate use of effective antibacterials, lest we fall into this hole again.

While financial incentives are critical as they will have the effect of helping the eNPV issue, regulatory issues must also be addressed (Projan, 2004). Opportunities for improvement could include the issuance of updated drug development guidance documents, simplification of phase III trial requirements through the more thorough integration of PK–PD modelling, and perhaps the limited licensing of urgently needed agents following phase II studies with controlled usage of the new product, permitting post-registration knowledge acquisition in a more 'real world' setting than is often encountered in controlled clinical trials. A more detailed discussion of these concepts is presented in Chapter 2.

It may also be necessary for the clinical community to go back to its history through the effective use of narrower spectrum agents, as discussed by Casadevall in Chapter 20. Such agents will certainly be easier to discover as the challenge of drug candidate discovery is often inversely related to the breadth of spectrum sought. This could be particularly helpful for intrinsically resistant species like *Pseudomonas* and *Acinetobacter*. Interestingly, the drug development path for such agents may be even more of an issue than any challenge associated with their clinical use. It is here, again, that rapid, accurate and cost-effective diagnostics may come to the rescue, and we should be encouraged by the array of technologies that may produce game-changing devices, as discussed in Chapter 19 by Schuetz and Tang.

The success of the industry has, in many ways, been its own curse. Decades of successful antibiotic discovery has led to significant use, and in some cases overuse or misuse, of these life-saving agents. This has led to the current era characterized not only by growing antibiotic resistance but also the transition of an entire therapeutic repertoire from patent-protected to generic status. It has also created the unique cultural expectation that antibiotics should be cheap since prices almost always drop as agents lose patent coverage (Laxminarayan, 2007). Unfortunately, this expectation belies the incredible medical benefit that new antibiotics which truly address a key medical need can offer – the ability to *save the life* of a patient infected with an MDR pathogen. Higher pricing, as a reward for innovation, is a generally unpopular concept, but can be objectively considered as reasonable when one compares them to the prices attainable for new cancer treatments, or for a lifetime of therapy for a chronic condition like diabetes. Higher prices will also discourage inappropriate use as low prices create an environment where high sales volumes are the only way to generate an adequate return on investment. Thus, educating payers and patients on the rationale for premium pricing for life-saving antibiotics is a worthwhile use of healthcare dollars.

Concluding remarks

In summary, we should look back on the past few decades of antibiotic discovery and development as a mostly successful chapter in pharmaceutical history, as these medicines have helped extend the life expectancy of the human race (Armstrong, 1999). They can also be characterized as a time in which we, as a society, thought we would be able to continue this success story indefinitely. We now understand that antibiotic use impacts both the patient as well as society as a whole (Laxminarayan, 2007). However, although we now realize how mistaken our previous perception was, we still have the opportunity to learn effectively from the many lessons of this rich chapter in our history, and to collectively plan how to achieve another golden age, one that incorporates scientific, medical and economic insights that will allow the next chapter to last far longer. Critical to our ability to recapture an age of effective antibacterial agents is our conversion of the exciting and promising scientific opportunities described in this book into new agents and strategies that will address the needs of patients throughout the world.

Acknowledgements

We would like to thank the following individuals for their comments on this chapter and overall organization of the book: Arturo Casadevall, Matthew Crawford, MaryKay Lescoe, Tom Magee, Phoebe Roberts, Jeremy Starr and Andrew Tomaras. We would also like to thank James Smith, Craig McPherson and Germaine Boucher for their artistic contributions to the book cover.

References

Apidianakis, Y., and Rahme, L.G. (2009). *Drosophila melanogaster* as a model host for studying *Pseudomonas aeruginosa* infection. Nat. Protoc. *4*, 1285–1294.

Armstrong, G.L., Conn, L.A., Pinner, R.W. (1999). Trends in infectious disease mortality in the United States during the 20th century. J. Am. Med. Assoc. *281*, 61–66.

Bax, B.D., Chan, P.F., Eggleston, D.S., Fosberry, A., Gentry, D.R., Gorrec, F., Giordano, I., Hann, M.M., Hennessy, A., Hibbs, M., *et al.* (2010). Type IIA topoisomerase inhibition by a new class of antibacterial agents. Nature *466*, 935–940.

Bayles, K.W. (2007). The biological role of death and lysis in biofilm development. Nat. Rev. Micro. *5*, 721–726.

Benghezal, M., Adam, E., Lucas, A., Burn, C., Orchard, M.G., Deuschel, C., Valentino, E., Braillard, S., Paccaud, J.-P., and Cosson, P. (2007). Inhibitors of bacterial virulence identified in a surrogate host model. Cell. Microbiol. *9*, 1336–1342.

Boucher H.W., T.G.H., Bradley J.S., Edwards J.E., Gilbert D., Rice L.B., Scheld M., Spellberg B., and Bartlett J. (2009). Bad bugs, no drugs: no ESKAPE! An update from the Infectious Diseases Society of America. Clin. Infect. Dis. *48*, 1–12.

Broderick, N., Raffa, K., and Handelsman, J. (2010). Chemical modulators of the innate immune response alter gypsy moth larval susceptibility to *Bacillus thuringiensis*. BMC Microbiol. *10*, 129.

Brötz-Oesterhelt, H., and Brunner, N.A. (2008). How many modes of action should an antibiotic have? Curr. Opin.Pharmacol. *8*, 564–573.

Cars, O. (2009). Innovative Incentives for Effective Antibacterials Conference (Stockholm), 17 September 2009. www.se2009.eu/en/meeting_news/

Cegelski, L., Marshall, G.R., Eldridge, G.R., and Hultgren, S.J. (2008). The biology and future prospects of antivirulence therapies. Nat. Rev. Micro. *6*, 17–27.

Cegielski, J.P. (2010). Extensively drug-resistant tuberculosis: 'There must be some kind of way out of here'. Clin. Infect. Dis. *50*, S195–S200.

Chartrain, M., and Chu, L. (2008). Development and production of commercial therapeutic monoclonal antibodies in mammalian cell expression systems: an overview of the current upstream technologies. Curr. Pharm. Biotechnol. *9*, 447–467.

Chiang, S.L., Mekalanos, J.J., and Holden, D.W. (1999). *In vivo* genetic analysis of bacterial virulence. Ann. Rev. Microbiol. *53*, 129–154.

Clatworthy, A.E., Pierson, E., and Hung, D.T. (2007). Targeting virulence: a new paradigm for antimicrobial therapy. Nat. Chem. Biol. *3*, 541–548.

Cronin, M.T.D., Aptula, A.O., Dearden, J.C., Duffy, J.C., Netzeva, T.I., Patel, H., Rowe, P.H., Schultz, T.W., Worth, A.P., Voutzoulidis, K., *et al.* (2002). Structure-based classification of antibacterial activity. J. Chem. Inform. Comput. Sci. *42*, 869–878.

Czock, D., Markert, C., Hartman, B., and Keller, F. (2009). Pharmacokinetics and pharmacodynamics of antimicrobial drugs. Expert Opin. Drug Metabol. Toxicol. *5*, 475–487.

D'Argenio, D.A., Calfee, M.W., Rainey, P.B., and Pesci, E.C. (2002). Autolysis and autoaggregation in *Pseudomonas aeruginosa* colony morphology mutants. J. Bacteriol. *184*, 6481–6489.

Dimitrov, D., and Marks, J. (2009). Therapeutic antibodies: current state and future trends – is a paradigm change coming soon? Meth. Mol. Biol. *525*, 1–27.

Donald, R., Skwish, S., Forsyth, R., Anderson, J., Zhong, T., Burns, C., Lee, S., Meng, X., LoCastro, L., Jarantow, L., *et al.* (2009). A *Staphylococcus aureus* fitness test platform for mechanism-based profiling of antibacterial compounds. Chem. Biol. *16*, 826–836.

Dunn, M., Ramírez-Trujillo, J., and Hernández-Lucas, I. (2009). Major roles of isocitrate lyase and malate synthase in bacterial and fungal pathogenesis. Microbiology *155*, 3166–3175.

Extance, A. (2010). Biologics target bad bugs. Nat. Rev. Drug Discov. *9*, 177–178.

Fedhila, S., Buisson, C., Dussurget, O., Serror, P., Glomski, I.J., Liehl, P., Lereclus, D., and Nielsen-LeRoux, C. (2010). Comparative analysis of the virulence of invertebrate and mammalian pathogenic bacteria in the oral insect infection model *Galleria mellonella*. J. Invert. Pathol. *103*, 24–29.

Finch, R. (2007). Innovation–drugs and diagnostics. J. Antimicrob. Chemother. *60*, i79–i82.

Fischbach, M.A., and Walsh, C.T. (2009). Antibiotics for emerging pathogens. Science *325*, 1089–1093.

Flemming, H., and Wingender, J. (2010). The biofilm matrix. Nat. Rev. Microbiol. *8*, 623–633.

Fujimura, K.E., Slusher, N.A., Cabana, M.D., and Lynch, S.V. (2010). Role of the gut microbiota in defining human health. Exp. Rev.Anti-infect. Ther. *8*, 435–454.

Gilbert, D., Guidos, R., Boucher, H., Talbot, G., Spellberg, B., Edwards, J.J., Scheld, W., Bradley, J., and Bartlett, J. (2010). The 10 × 20 initiative: pursuing a global commitment to develop 10 new antibacterial drugs by 2020. Clin. Infect. Dis. *50*, 1081–1083.

Gloede, J., Scheerans, C., Derendorf, H., and Kloft, C. (2010). *In vitro* pharmacodynamic models to determine the effect of antibacterial drugs. J. Antimicrob. Chemother. *65*, 186–201.

Goetz, M.A., Zhang, C., Zink, D.L., Arocho, M., Vicente, F., Bills, G.F., Polishook, J., Dorso, K., Onishi, R., Gill,

Tsolis, R., Baumler, A., Heffron, F., and Stojiljkovic, I. (1996). Contribution of TonB- and Feo-mediated iron uptake to growth of *Salmonella typhimurium* in the mouse. Infect. Immun. *64*, 4549–4556.

Wang, J., Soisson, S.M., Young, K., Shoop, W., Kodali, S., Galgoci, A., Painter, R., Parthasarathy, G., Tang, Y.S., Cummings, R., *et al.* (2006). Platensimycin is a selective FabF inhibitor with potent antibiotic properties. Nature *441*, 358–361.

Wang, J., Kodali, S., Lee, S.H., Galgoci, A., Painter, R., Dorso, K., Racine, F., Motyl, M., Hernandez, L., Tinney, E., *et al.* (2007). Discovery of platencin, a dual FabF and FabH inhibitor with *in vivo* antibiotic properties. Proc. Natl. Acad. Sci. *104*, 7612–7616.

Webb, J.S., Thompson, L.S., James, S., Charlton, T., Tolker-Nielsen, T., Koch, B., Givskov, M., and Kjelleberg, S. (2003). Cell death in *Pseudomonas aeruginosa* biofilm development. J. Bacteriol. *185*, 4585–4592.

Wyckoff, E.E., Mey, A.R., Leimbach, A., Fisher, C.F., and Payne, S.M. (2006). Characterization of ferric and ferrous iron transport systems in *Vibrio cholerae*. J. Bacteriol. *188*, 6515–6523.

Young, K., Jayasuriya, H., Ondeyka, J.G., Herath, K., Zhang, C., Kodali, S., Galgoci, A., Painter, R., Brown-Driver, V., Yamamoto, R., *et al.* (2006). Discovery of FabH/fabf inhibitors from natural products. Antimicrob. Agents Chemother. *50*, 519–526.

Renewing Investment in Antibacterial Research

2

L. Silvia Munoz-Price and John P. Quinn*

Abstract

We summarize the epidemiology, clinical presentation, and current treatment options for the most clinically relevant multidrug resistant Gram-positive and Gram-negative organisms. Additionally, we describe the challenges faced by pharmaceutical companies within the antimicrobial research and development field, especially the disproportion between the degree of investment (both monetary and time) required and the relatively small profit antimicrobial agents bring. Finally, some potential solutions for the lack of antimicrobial agents are discussed. These include more widespread use of the Orphan Drug Act, patent extensions, and the Biomedical Advanced Research and Development Authority (BARDA).

Introduction

During the past decade the world has experienced a significant increase in the number of multidrug resistant organisms, both among Gram-positive and Gram-negative species. At the same time, efforts by pharmaceutical companies to find novel antimicrobials have decreased. This chapter describes a clinician's view of the current status of multidrug resistant organisms, the reasons for the decreased production of anti-infectives, as well as potential solutions that are currently being considered to change this situation.

Clinical overview of common resistant organisms

The number of multidrug resistant organisms has been on a steep rise for the past decade. In an effort to summarize and publicize this urgent problem, the Infectious Diseases Society of America (IDSA) published several papers describing the desperate need for new antibiotics to target these resistant bacteria. In 2008, Louis Rice crafted the term ESKAPE (*Enterococcus faecium*, *Staphylococcus aureus*, *Klebsiella pneumoniae*, *Acinetobacter baumannii*, *Pseudomonas aeruginosa*, and *Enterobacter* species) to describe the most troublesome organisms (Boucher *et al.*, 2009). This section of the chapter summarizes the main clinical and microbiological characteristics for each of these pathogens in the USA and Europe, where the greatest amount of surveillance data exist. This is not to deny the extent of drug resistance in the developing world, which in some cases seems to exceed the high rates seen in the USA and European Union (EU). For example, rates of extended spectrum beta-lactamase (ESBL) production among enteric organisms in the Asia-Pacific region appear to exceed those in the west (Hawser *et al.*, 2009).

Klebsiella pneumoniae and other enterics

A carbapenemase (KPC)-producing *Klebsiella pneumoniae* strain was first isolated in a patient

*Corresponding author

from North Carolina in 1996, although the actual publication of this case by Jones and collaborators did not appear until 2001 (Yigit *et al.*, 2001). During the subsequent years, a worldwide dissemination of this enzyme occurred, with cases reported in Colombia, Greece, Israel, France, Turkey, and Asia (Nordmann, 2009). KPCs are currently considered endemic in New York and New Jersey. KPC enzymes have been isolated not only in *K. pneumoniae* but also in *Escherichia coli*, *Enterobacter cloacae*, *P. aeruginosa*, *P. fluorescens*, and *A. baumannii*. KPCs confer resistance to virtually all beta-lactams and carbapenems. Most KPC-producing isolates remain susceptible to tigecycline and polymyxins. Ertapenem resistance is frequently used as an initial screening tool (up to 40% of strains will test susceptible to imipenem), followed by confirmatory testing using PCR against the KPC-encoding genes.

Clinical presentations of *K. pneumoniae* vary from mere colonization to serious infections such as those involving the bloodstream and urinary tract, as well as ventilator-associated pneumonia. Mortality has been described in the range of 40–60% depending on the severity of the infection. Another important development is the intercontinental transfer of KPC-colonized patients. This phenomenon was described in several instances, including transfers from New York City area hospitals to Israel, and from Israel to Colombia and Miami (Navon-Venezia *et al.*, 2009). In some cases, these patient transfers resulted in hospital outbreaks by KPC-producing strains as well as changes in the flora of the accepting institutions.

ESBLs of the TEM and SHV family have been problematic in *Klebsiella* for 20 years and continue to expand worldwide. Considerable geographic variability exists. For example, in Europe there is a marked gradient of increasing resistance from north to south and west to east (Fig. 2.1A).

During the last decade, there has been a dramatic transfer of ESBLs from *Klebsiella* to *E. coli*, as well as a shift from the classic TEM and SHV variants to a newer family, CTX-M. Of special clinical concern, *E. coli* isolates resistant to advanced spectrum cephalosporins and quinolones have been identified as causing community-associated infections, especially urinary tract infections. ESBLs associated with these phenomena are predominantly the CTX-M-15 variant. Furthermore, this occurrence has been described not only in Europe, but also in the Americas and Asia (Canton, 2006). A recent study showed CTX-M carriage to be present in up to a third of patients transferred from Haiti to a referral hospital in south Florida (Munoz-Price, unpublished), which further emphasizes the worldwide dissemination of these genes. Another study by Robisceck and colleagues showed that one of the main risk factors for CTX-M-15 carriage was a recent trip to India (Jacoby *et al.*, 2006). This further highlights the degree of global interdependence that exists in terms of travel, exposures and acquisition of resistant organisms.

CTX-M-15 has also been isolated from animals, including migratory seagulls and poultry (Bortolaia *et al.*, 2010). Additionally, these genes have been identified in environmental sources, such as rivers (Chen *et al.*, 2010). These factors together with the detection of CTX-M-15 from patients without previous exposure to healthcare settings, make it tempting to speculate that the prevalence of these genes in the environment is immense and under-recognized. The observation that a high percentage of patients entering healthcare facilities from the community carry CTX-M-15 producing pathogens is particularly disturbing. These patients are at higher risk for treatment failure with oral antibiotics owing to cross resistance among the beta-lactam class. However, clinical outcomes of these patients, at this time, are still not well characterized.

A particular concern is the intercontinental spread of a new *E coli* clone designated MLST 131 (Nicolas-Chanoine *et al.*, 2008). This clone is virtually always quinolone-resistant, and often harbours CTX-M-15 as well. It appears to be more virulent than comparator strains and hence is more likely to cause pyelonephritis and bacteraemia.

Acinetobacter baumannii

Acinetobacter baumannii is another Gram-negative organism renowned for its ability to accumulate antimicrobial-resistant genes. It was previously considered an organism of questionable pathogenicity. However, *A. baumannii* is now

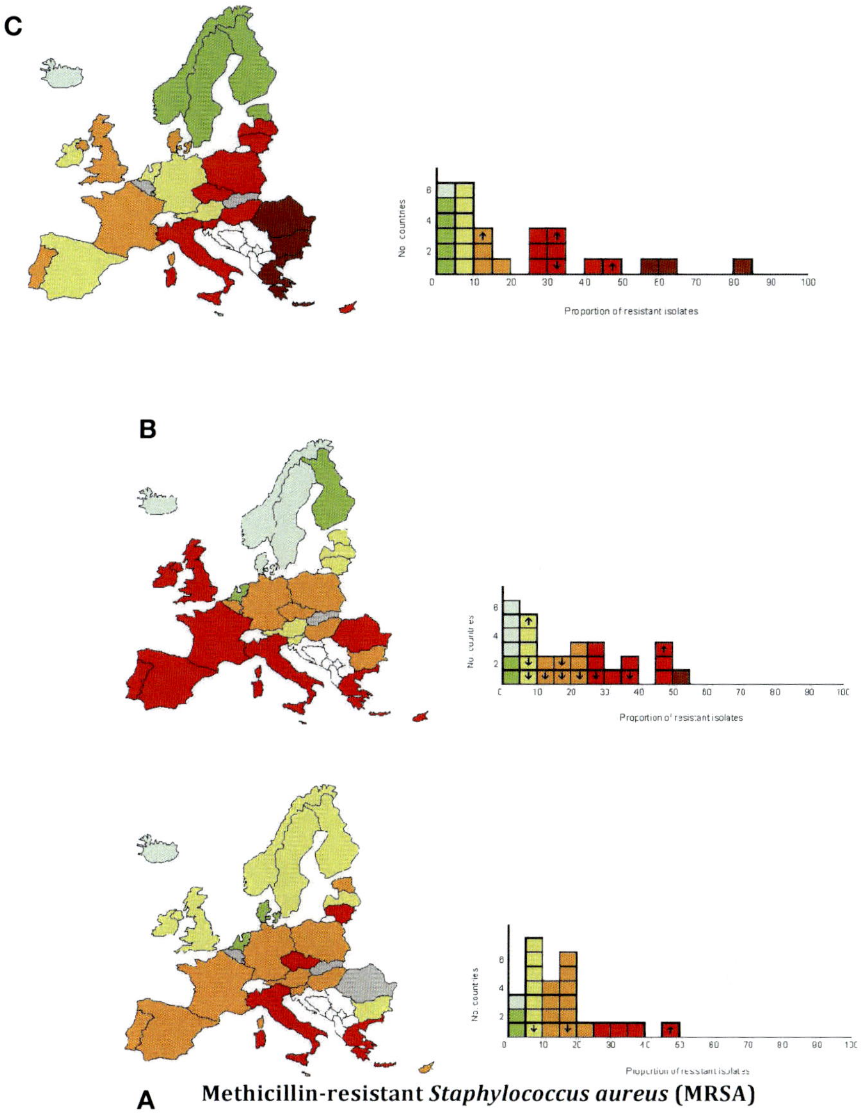

Figure 2.1 The proportion of resistant isolates in bloodstream infections in European Union (EU) member states, Iceland and Norway, in 2007 and trends for 2005–2007. (A) Third-generation cephalosporin-resistant *Klebsiella pneumoniae*. (B) Carbapenem-resistant *Pseudomonas aeruginosa* (C) Methicillin-resistant *Staphylococcus aureus* (MRSA). Reproduced with permission from Figure 4 of the ECDC/EMEA joint technical report (www.ecdc.europa/eu).

recognized as one of the preeminent pathogens in intensive care units across the world. Usually found in soil, one of its main characteristics is its resistance to desiccation which allows it to remain viable on inanimate objects for several months. Furthermore, this characteristic explains the period of endemicity that usually follows hospital outbreaks: high peaks of cases for a brief period followed by a long phase of lower number of cases over time (Villegas and Hartstein, 2003).

The most common clinical manifestations of *A. baumannii* are ventilator-associated pneumonia, and infections of the bloodstream and urinary tract. Patients colonized or infected with *A. baumannii* tend to be chronically ill and have multiple co-morbidities. Mortality with this organism varies depending on the case series, and recent evidence would tend to suggest that mortality among infected cases is due to the underlying conditions rather than to the actual pathogen. Treatment of resistant isolates is limited to polymyxins and in some cases tigecycline, although strains resistant to both of these antibiotics have been described.

During the past decade the percentage of imipenem-resistant *A. baumannii* increased from ~10% to ~20% in the USA (Rhomberg and Jones, 2009). The mechanisms of resistance most frequently identified are the carbapenemases, usually class D enzymes from the OXA family. These genes are frequently up-regulated by adjacent insertion sequences, especially *ISA*ba-1 (Turton *et al.*, 2006). OXA type enzymes vary in distribution depending on the region of the world. However, a recent study by Mugnier and collaborators showed that OXA-23 have a worldwide distribution among carbapenem-resistant *A. baumanni* (Mugnier *et al.*, 2010).

Community infections with *A. baumannii* are for the most part geographically limited to Australia and Asia, and are rarely described in the USA and Europe. These community infections are usually characterized by aggressive pneumonias seen among pharyngeal carriers and immunosuppressed patients (alcoholics or patients with malignancies). Another special population affected by *A. baumannii* is the military. Descriptions of military personnel infected with *A. baumannii* were first published during the Korean War and have recently been revived during the wars in Afghanistan and Iraq (Dallo and Weitao, 2010).

Pseudomonas aeruginosa

This Gram-negative organism can develop resistance to multiple antibiotics by accumulating mutations that affect the concentration of beta-lactams in the periplasmic space, including hyper-expression of efflux pumps (which increase excretion), porin mutations (which decrease permeability), or overexpression of chromosomal beta-lactamases (leading to increased enzymatic hydrolysis within the periplasmic space). Additionally, plasmid-mediated beta-lactamases, such as metallo-beta-lactamases, have been described among pseudomonads in different regions of the world. Moreover, these enzymes are usually carried in conjunction with aminoglycoside-modifying enzymes, making these isolates multidrug resistant or pan-resistant.

P. aeurginosa is a water-based organism and is associated with hospital acquired infections. This opportunistic pathogen causes infections especially among intensive care unit patients, immunocompromised hosts and burn patients. *Pseudomonas* spp. are typically associated with ventilator-associated pneumonia, bloodstream infections, and skin and soft-tissue infections.[1]

Treatment options for susceptible *Pseudomonas* isolates include anti-pseudomonal penicillins (i.e. piperacillin), anti-pseudomonal cephalosporins (i.e. cefepime), monobactams (i.e. aztreonam), carbapenems (i.e. imipenem), aminoglycosides, and fluoroquinolones (i.e. ciprofloxacin). Combination therapy of a β-lactam and an aminoglycosides is preferred, although solid data favouring this modality are not available.

Resistance to three or more antibiotic classes among *P. aeruginosa* varies depending on the locale. A recent study published by Giske and colleagues shows a prevalence of resistance among *P. aeruginosa* of up to 50% in some areas of Europe (Fig. 2.2; Miriagou *et al.*, 2010). Although no large multicentre studies are currently available, the data seem to indicate an increase in mortality among patients infected with multidrug resistant *P. aeruginosa* (Lautenbach *et al.*, 2010). Colistin is an older antibiotic class that has activity against pan-resistant isolates.

Methicillin-resistant *Staphylococcus aureus* (MRSA)

MRSA was first reported shortly after the introduction of methicillin in the 1960s. According to the National Nosocomial Infections Surveillance system, the percentage of methicillin resistance among American strains of *S. aureus* increased

[1] From the comprehensive clinical database www.uptodate.com.

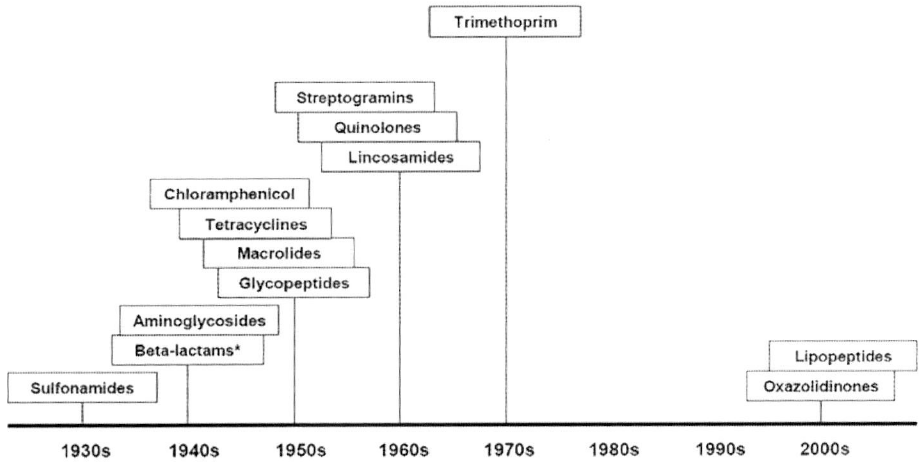

Figure 2.2 Discovery of new classes of antibiotics. Reproduced with permission from the ECDC/EMEA joint technical report (www.ecdc.europa/eu).

from 36% in 1992 to 64% in 2003 (Klevens *et al.*, 2007). Resistance is conferred by alterations in a penicillin-binding-protein (PBP-2A). This protein is encoded by *mecA* which is located within a staphylococcal chromosomal cassette, also known as SCC*mec*. There are two main types of MRSA, hospital acquired and community acquired, although distinction between the two strains by a general microbiology laboratory is quite challenging. Hospital-acquired MRSA has a clonal and worldwide distribution; however, its prevalence varies depending on the country. It tends to cause infections in hospitalized patients with multiple co-morbidities; thus, mortality associated with these infections is difficult to separate from the underlying conditions.

Since the late 1990s, a rapid spread of community acquired MRSA (CA-MRSA) strains has occurred across the USA, and to a lesser extent in Europe and Asia. CA-MRSA was first described in healthy patients without any exposure to healthcare systems (CDC, 1999). Infections caused by CA-MRSA are usually associated with the skin and soft tissues in the form of boils, deep abscesses, cellulitis and, less frequently, necrotizing pneumonia. Antibiotic treatment with either trimethoprim–sulfamethoxazole or other active oral agents can be used if there is large skin involvement.

Despite its name, CA-MRSA is no longer confined to the community but rather is becoming a healthcare-associated strain that might even become more prevalent than HA-MRSA among hospital settings (Popovich *et al.*, 2008). Therefore, it might be impossible to differentiate between CA- and HA-MRSA strains solely on epidemiological links.

Vancomycin-intermediate *S. aureus* (VISA) was first described in Japan in 1997 and believed to be caused by an increase in the thickness of its cell wall (CDC, 1997). More worrisome, vancomycin-resistant *S. aureus* (VRSA) was first reported in Michigan in 2002 (Sievert *et al.*, 2008), although only a handful of cases have been described since then. These strains appear to carry a *vanA* determinant that was originally described in enterococci. This mutation seems to make *S. aureus* strains 'unfit', which would partially explain their limited dissemination since 2002 (Perichon and Courvalin, 2009).

Vancomycin-resistant *Enterococcus* (VRE)

Enterococci have different penicillin-binding proteins (PBPs) than staphylococci or streptococci in that they have a lower affinity for beta-lactams, which explains the lack of activity of cephalosporins against this organism.

Vancomycin-resistant enterococcus (VRE) was first described in Europe in 1986, and rapidly spread within the USA in the late 1990s. In the

USA, the prevalence of VRE went from 0% in 1989 to 28% in 2003 (NNIS, 2004; Top *et al.*, 2008). Currently, over 50% of *E. faecium* in the USA are VRE. On the contrary, *E. faecalis* still remains susceptible to vancomycin. Mechanisms of resistance to vancomycin among VRE are usually encoded by *vanA* and *vanB* genes. Enzymes produced by these genes result in the production of an altered peptidoglycan structure (D-Ala-D-Lac as compared with the normal D-Ala-D-Ala; Top *et al.*, 2008), which decreases the affinity of vancomycin for its cell wall binding sites.

VRE has become one of the main multidrug resistant organisms within healthcare facilities across the USA. Risk factors for VRE acquisition include prolonged hospitalization, previous use of antibiotics, acuity of illness, and surgery. VRE is often present as bacterial colonization among hospitalized patients (i.e. skin, wounds, and rectum); however, true infections can be fatal. As with HA-MRSA, VRE can cause blood stream and surgical site infections. Treatment options for VRE include daptomycin, linezolid, and tigecycline.

The current crisis of a diminished antibiotic pipeline

According to the IDSA White Paper 'Bad Bugs, No Drugs', many of the large pharmaceutical companies [including Bristol-Myers Squibb, Procter and Gamble, Abbott Laboratories, Eli Lilly, Aventis, and Wyeth (acquired by Pfizer)] have significantly decreased their anti-infective research and development efforts (Boucher *et al.*, 2006). As a result, the number of new antibacterial agents approved since 1998 have progressively decreased throughout the years to only five drugs during the period from 2001 to 2004 (see Fig. 2.2). Furthermore, a 2008 review by Abbanat and colleagues showed that most antibiotics in development at that time targeted only Gram-positive organisms (Abbanat *et al.*, 2008). Compounds targeting resistant Gram-negative organisms or antibiotics with novel mechanisms of action were lacking.

The potential explanations for this change in the pipeline are of course multiple and start with the financial element: antibiotics have a lower return on investment compared with other disease areas. In addition to taking approximately 14 years to go from a laboratory idea to an actual drug launch, each successful antibiotic represents an exorbitant investment of money, in the range 400–800 million dollars (Payne *et al.*, 2007). Additionally, antibiotics are drugs that are used for short amounts of time, thus further limiting their profitability when compared with agents that are used chronically. A helpful comparison can be made between a cardiovascular drug, such as the anti-hypertensive drug enalapril (used by patients for decades or a lifetime) and an antibiotic with Gram-positive coverage, such as daptomycin. Even though daptomycin is the most frequently used antibiotic for hard-to-treat Gram-positive infections such as osteomyelitis, its use is limited to a maximum of 6 weeks, and most courses are much shorter, in the range of 7–10 days. From this perspective, it is exceedingly difficult for an antibiotic to compete financially with cardiovascular agents – or any other long-term use drug.

Additionally, not all infections are caused by recalcitrant organisms. Even though the percentage of infections caused by resistant organisms is increasing, most outpatient and inpatient infections are still caused by non-resistant bacteria or bacteria that can be treated with older, cheaper antibiotics. Thus, despite the fact that antibiotics frequently save lives and are desperately needed for some types of infections, market dynamics have kept antibiotic prices historically low.

It is important to remember that pharmaceutical companies are for-profit organizations that answer to shareholders. Therefore, it is hard to argue against research and development investment decisions that favour drugs that will offer the greatest revenue potential (long-term usage per patient), and investing in antibiotics may not be considered a wise business decision when R&D budgets are shrinking. As stated in the IDSA White Paper '... a musculoskeletal drug is worth about $1.150 billion, a neuroscience treatment is rated at $720 million, and a medicine for resistant Gram-positive cocci is worth only $100 million.' (Boucher *et al.*, 2006).

Additionally, there are unique characteristics of anti-infectives that are not shared by other drugs. First, the more that antibiotics are used, the more antibiotic resistance they provoke. This reality leads to the second feature, which is that infectious diseases physicians react by limiting their use, 'saving' them for when they are urgently needed. These two factors conspire to further decrease the already limited market for anti-infectives, and thus worsen the return on investment (Abbanat *et al.*, 2008).

Another important factor that has contributed to the drying pipeline is the lack of productive ways to identify novel chemical classes with bona fide antibiotic potential. After the entire genome of *Haemophilus influenzae* was completely sequenced in 1995, pharmaceutical companies expected to find an abundance of potential targets using the newly available information. One of those companies was GlaxoSmithKline, which spent seven years pursuing potential antibiotic targets from among 300 *H. influenzae* genes. Seventy high-throughput screening (HTS) tests were performed at a cost of $1 million each. By 2001, and after a huge capital investment, only five leads (potential antibiotics) were identified. After years of research and investment, GlaxoSmithKline was not able to develop any commercial products from this endeavour (Payne *et al.*, 2007).

Once 'leads' (chemical starting points for subsequent projects) are identified, they still need to go through years of chemistry refinement to produce an optimized compound suitable for experimentation in humans. Additionally, these antimicrobial molecules need to be able to penetrate different bacterial envelope structures, act at evolutionarily divergent versions of these bacterial targets in order to be active against phylogenetically diverse species, and overcome hydrolysing enzymes within resistant bacteria (Payne *et al.*, 2007). This overwhelming number of characteristics makes antibacterial lead discovery even more challenging than in most other therapeutic areas.

Clinical trials for approval of antibiotics add another layer of complexity to the process. Given that antibiotics are used to treat different infections (i.e. pneumonia, urinary tract infections, skin and sift tissue infections), regulatory agencies require separate clinical trials for each site of infection. This not only increases the monetary investment required, but also the time needed to complete clinical development and gain regulatory approval. Furthermore, there is a lack of standardized guidelines from regulatory entities, such as the Food and Drug Administration (FDA), stipulating the type of clinical trial required for approval. For many infection sites, the FDA has provided only provisional guidelines, and even these take years to develop and are infrequently updated to reflect new knowledge.

While these challenges have contributed to the current shortfall of private sector investment in antibacterial R&D, it is a misconception to believe that government can fill this gap. Thus far, the only drug created by the US government is artemisin, an antimalarial drug. However, small biotech companies seem to be stepping in. Initially, biotech companies brought forward compounds that were originally discovered (and subsequently abandoned) by big pharmaceutical companies, and continued the processes required to take them to final approval and commercialization (e.g. daptomycin by Cubist; Bradley *et al.*, 2007). More recently, biotechs are working on discovering new compounds from their own internal efforts with the help of experienced drug developers (most of them with big pharma experience).

In conclusion, there is significant risk and time involved in antibiotic research and development, leading many larger companies to question the potential return on investment. Thus, novel approaches outside of conventional business models are needed in order to restore much-needed commercial attractiveness to this area.

Potential solutions

Since the regulatory and political environments in the USA and EU are quite different, they will be discussed separately. For the US, the most comprehensive list of potential solutions to the current state of anti-infectives research and development are found in two sources: the IDSA White Paper 'Bad Bugs, No Drugs' (Boucher *et al.*, 2006) and Brad Spellberg's book *Rising Plague: the Global Threat from Deadly Bacteria and our Dwindling Arsenal to Fight Them* (Spellberg, 2009). For the EU, we will refer to proceedings from a meeting in Sweden sponsored in part by the European Medicines Evaluation Agency (EMEA) in 2009.[2]

2 See www.ecdc.europa.eu.

United States

The Orphan Drug Act (ODA) was approved in 1983 by the United States Congress with the goal of increasing the production of drugs targeting diseases with an incidence of less than 200,000 patients within the USA. This law gives incentives to manufacturers who invest in research and development of drugs with a small market. These incentives include seven years of marketing exclusivity, grant money from the FDA, smaller patient requirements for clinical trials, and tax incentives on clinical trials. There is a proposal to consider including anti-infectives targeting multidrug resistant bacteria as part of the ODA.

There is evidence showing that the ODA has worked for other pharmaceutical products. Before 1983, only ten drugs that met ODA criteria were approved to the market. After the ODA's conception in 1983 and through 2004, a total of 249 orphan drugs were approved. This underscores the important incentive that ODA constituted for drug developers.

The Biomedical Advanced Research and Development Authority (BARDA) is a branch of the Office of the Assistant Secretary for Preparedness and Response in the US Department of Health and Human Services. BARDA integrates in one office the development and purchase of treatments and diagnostic tools for public health emergencies. Additionally, BARDA manages Project Bioshield which is responsible for developing countermeasures for chemical, biological, radiological, and nuclear agents, as well as the advanced development and procurement of medical countermeasures for pandemic influenza and other emerging infectious diseases that fall outside the auspices of Project Bioshield.

According to BARDA's website, the Influenza and Emerging Infectious Diseases Division is in the planning phase for the Emerging Infectious Disease programme. This programme is currently envisaged to support the development of vaccines and drugs to address emerging disease threats. Including antimicrobial resistant pathogens as part of the definition of 'emerging disease threats' could be one additional incentive for R&D in this area.

Another potential solution is the 'wild-card patent extension'. This concept was originally proposed in 2003 by Senators Lieberman and Hatch as part of the Biological, Chemical, and Radiological Weapons Countermeasures Research Act (S.666). This incentive would allow a pharmaceutical company that develops an anti-infective targeting specific pathogens (e.g. pan-resistant *Acinetobacter* or pan-resistant *Pseudomonas*) to prolong the marketing exclusivity of *any of the pharmaceutical company's other drugs*.

This could certainly constitute a major monetary incentive for pharmaceutical companies to renew development of anti-infectives. S.666 originally stipulated a two-year extension of patents; however, this proposal has been quite controversial and might not even be considered feasible at this time. This legislation also provided tax incentives for companies willing to fund research for antibiotics against special pathogens, such as special tax credits and a zero capital gain tax rate on any stock gains. Even though S.666 was proposed in 2003, as of 2010 it is still a pending legislation.

Another major barrier for pharmaceutical companies within the antibiotic field is the degree of liability that companies have once the product is on the market. Rare side-effects may emerge during commercial use that are impossible to detect in clinical trials and therefore will only be discovered once the drug is on the market. An injury compensation fund, similar to the one currently in effect for vaccines, has been proposed to help offset this risk. This common fund would be made of a percentage of the antibiotic cost and would be used to pay settlements occurring as a result of unexpected side-effects of the antimicrobial products.

The FDA usually takes 18 months to review new antibiotic applications. Currently, this time counts as part of the marketing exclusivity period. A potential incentive would be to reinstate these months to the patent exclusivity period once the drug is on the market.

Other potential incentives include FDA approval of antibiotics for the treatment of pathogen-specific conditions rather than specific sites of infection (e.g. MRSA vs. pneumonia), and facilitating research and development by small biopharmaceutical companies. IDSA's proposal for biopharmaceutical companies includes providing tax incentives to these companies (retained earnings), increasing the amount of dollars spent on Small Business Innovation Research (SBIR) grants, and waiving the fees for supplemental FDA applications.

European Union

In 2007, the European Centre for Disease Prevention and Control (ECDC), the European Medicines Agency (EMEA) and the international network Action on Antibiotic Resistance (ReAct) entered into a discussion on the need to document the gap between the frequency of multidrug-resistant bacterial infections in the EU and the development of new antibiotics. As a result, an ECDC/EMEA Joint Working Group was established in 2008 to give an account of facts and figures that would allow reasonable predictions of the extent of the gap in the coming years.

The following antibiotic-resistant bacteria were selected because they are frequently responsible for bloodstream infections in the EU and their associated antibiotic resistance trait is, in most cases, a marker for multiresistance:

- *S. aureus*, methicillin resistance (MRSA);
- *S. aureus*, vancomycin intermediate resistance and vancomycin resistance (VISA/VRSA);
- *Enterococcus* spp. (e.g. *Enterococcus faecium*), vancomycin resistance (VRE);
- *S. pneumoniae*, penicillin resistance (PRSP);
- Enterobacteriaceae (e.g. *E. coli*, *K. pneumoniae*), third-generation cephalosporin resistance;
- Enterobacteriaceae (e.g. *K. pneumoniae*), carbapenem resistance;
- non-fermentative Gram-negative bacteria (e.g. *P. aeruginosa*), carbapenem resistance.

It is no surprise that these are the same problematic pathogens addressed by the IDSA in the USA. While the first part of this chapter summarized many of the major antibiotic resistance issues that are driving current medical need, differences exist in the prevalence of some of these pathogens between the USA and EU, as well as within the EU itself, and therefore a few of these will be summarized here as they may uniquely influence EU policy movements.

In 2007, the average proportion of *S. aureus* blood isolates that showed resistance to methicillin was the highest percentage of antibiotic-resistant isolates among the bacteria frequently responsible for bloodstream infections in the EU. However, this proportion has been decreasing in recent years (Fig. 2.1C). This is because of decreasing MRSA trends in several Member States, probably because of action plans at the national level as documented for France, Slovenia and the United Kingdom. The average proportion of MRSA in the EU has reached a level close to that of the selected antibiotic-resistant Gram-negative bacteria.

The proportion of *S. aureus* blood isolates that showed intermediate resistance to vancomycin (VISA) was very low (less than 0.1%) in EU Member States, Iceland and Norway. No vancomycin-resistant *S. aureus* isolates were reported to the EARSS surveillance programme in 2007.

In contrast, the average proportion of *E. coli* blood isolates – the most common Gram-negative bacterium responsible for infections in humans – showing resistance to third-generation cephalosporins has been rising steadily. As mentioned previously, this is largely attributable to the rise of the CTX-M class of ESBLs.

At the same time, there is no sign of decreasing resistance to third-generation cephalosporins in *K. pneumoniae* (Fig. 2.1A) or to carbapenems in *P. aeruginosa* (Fig. 2.1B). In 2007, the proportion of *K. pneumoniae* blood isolates from EU Member States, Iceland and Norway that showed resistance to carbapenems was, in general, very low (median = 0%) with the exception of Greece, where it reached 42%.

The human and economic burden of antibiotic-resistant bacteria could only be estimated for the following five species: MRSA, vancomycin-resistant *E. faecium*, third-generation cephalosporin-resistant *E. coli* and *K. pneumoniae* and carbapenem-resistant *P. aeruginosa*. Overall, it was estimated that in 2007 approximately 25,000 patients died from an infection due to any of the selected five antibiotic-resistant bacteria in the European Union, Iceland and Norway. In addition, infections due to any of the selected antibiotic-resistant bacteria resulted in approximately 2.5 million extra hospital days and additional in-hospital costs of more than EUR 900 million.

The Swedish Government has commissioned the European Observatory on Health Systems

Klevens, R.M., Morrison, M.A., Nadle, J., Petit, S., Gershman, K., Ray, S., Harrison, L.H., Lynfield, R., Dumyati, G., Townes, J.M., *et al.* (2007). Active Bacterial Core surveillance (ABCs) MRSA Investigators: invasive methicillin-resistant *Staphylococcus aureus* infections in the United States. JAMA *298*, 1763–1771.

Lautenbach, E., Synnestvedt, M., Weiner, M.G., Bilker, W.B., Vo, L., Schein, J., and Kim, M. (2010). Imipenem resistance in *Pseudomonas aeruginosa*: emergence, epidemiology, and impact on clinical and economic outcomes. Infect. Control Hosp. Epidemiol. *31*, 47–53.

Miriagou, V., Cornaglia, G., Edelstein, M., Galani, I., Giske, C.G., Gniadkowski, M., Malamou-Lada, E., Martinez-Martinez, L., Navarro, F., Nordmann, P., *et al.* (2010). Acquired carbapenemases in Gram-negative bacterial pathogens: detection and surveillance issues. Clin. Microbiol. Infect. *16*, 112–122.

Mugnier, P.D., Poirel, L., Naas, T., and Nordmann, P. (2010). Worldwide dissemination of the blaOXA-23 carbapenemase gene of *Acinetobacter baumannii*. Emerg. Infect. Dis. *16*, 35–40.

Nordmann, P. (2009). The real threat of KPC producing bacteria. Lancet Infect. Dis. 9, 228–236.

Navon-Venezia, S., Leavitt, A., Scwaber, M.J., Rasheed, J.K., Srinivasan, A., Patel, J.B., Carmell, Y; Israeli KPC Kpn Study Group. (2009). First report on a hyperepidemic clone of KPC-3-producing *K. pneumoniae* in Israel genetically related to a strain causing outbreaks in the United States. Antimicrob. Agents Chemother. *53*, 818–820.

National Nosocomial Infections Surveillance (NNIS) System Report, data summary from January 1992 through June 2004 (2004). Am. J. Infect. Control 32, 470–485.

Nicolas-Chanoine, M.H., Blanco, J., Leflon-Guibout, V., Demarty, R., Alonso, M.P. Caniça, M.M., Park, Y.J., Lavigne, J.P., Pitout, J., Johnson, J.R. (2008). Intercontinental emergence of *E. coli* clone O25:H4-ST131 producing CTX-M-15. J. Antimicrob. Chemother. *61*, 273–281.

Payne, D.J., Gwynn, M.N., Holmes, D.J. ,and Pompliano, D.L. (2007). Drugs for bad bugs: confronting the challenges of antibacterial discovery. Nat. Rev. Drug Discov. *6*, 29–40.

Périchon, B., and Courvalin, P. (2009). VanA-type vancomycin-resistant *Staphylococcus aureus*. Antimicrob. Agents Chemother. *53*, 4580–4587.

Popovich, K.J., Weinstein, R.A., and Hota, B. (2008) Are community-associated methicillin-resistant *Staphylococcus aureus* (MRSA) strains replacing traditional nosocomial MRSA strains? Clin. Infect. Dis. *46*, 787–794.

Rhomberg, P.R., and Jones, R.N. (2009). Summary trends for the Meropenem Yearly Susceptibility Test Information Collection Program: a 10-year experience in the United States (1999–2008). Diagn. Microbiol. Infect. Dis. *65*, 414–426.

Sievert, D.M., Rudrik, J.T., Patel, J.B., McDonald, L.C., Wilkins, M.J., and Hageman, J.C. (2008). Vancomycin-resistant *Staphylococcus aureus* in the United States, 2002–2006. Clin. Infect. Dis. 46. 668–674.

Spellberg B. (2009). Rising Plague: the Global Threat from Deadly Bacteria and our Dwindling Arsenal to Fight Them (Prometheus Books, Amherst, NY).

Top, J., Willems, R., and Bonten, M. (2008). Emergence of CC17 *Enterococcus faecium*: from commensal to hospital-adapted pathogen., FEMS Immunol. Med. Microbiol. *52*, 297–308.

Turton, J.F., Ward, M.E., Woodford, N., Kaufmann, M.E., Pike, R., Livermore, D.M., and Pitt. T.L. (2006). The role of ISAba1 in expression of OXA carbapenemase genes in *Acinetobacter baumannii*. FEMS Microbiol. Lett. *258*, 72–77.

Villegas, M.V. and Hartstein, A.I. (2003). *Acinetobacter* outbreaks, 1977–2000. Infect. Control Hosp. Epidemiol. *24*, 284–295.

Yigit, H., Queenan, A.M., Anderson, G.J., Domenech-Sanchez, A., Biddle, J.W., Steward, C.D., Alberti, S., Bush, K., and Tenover, F.C. (2001). Novel carbapenem-hydrolyzing β-lactamase, KPC-*1*, from a carbapenem-resistant strain of *Klebsiella pneumoniae*. Antimicrob. Agents Chemother. *45*, 1151–1161.

Part II

Novel Targets and Sources

Mining Bacterial Cell Division for New Antibacterial Drugs

3

Leigh G. Monahan, Michael A. D'Elia and Elizabeth J. Harry*

Abstract

The alarming rise of antibiotic resistant bacteria in hospitals and the community has exposed a critical need for new drugs that are not merely variants of older antibiotics, but target previously unexploited proteins and pathways. The wealth of available knowledge on the process of bacterial cell division implicates the division pathway as an excellent potential target, and has aided target-driven approaches to identify novel inhibitors. In this chapter we discuss the therapeutic potential of inhibiting bacterial division based on a strong foundation of basic research into the division mechanism and its regulation in model bacteria, and more recently, clinically relevant pathogens. In addition, we review the progress made towards identifying division inhibitors, describe new approaches for antibacterial drug development targeting division and discuss the potential challenges for the future of this exciting new area of antibacterial discovery.

Introduction

The serious and growing problem of antibiotic resistance has been greatly exacerbated by a reliance on new drugs that are simply variants of older overused antibiotics. As such, a desperate need for drugs with novel modes of action has been recognized by health organizations, industry and academia alike. Almost all traditional antibiotics target one of only four cellular processes: protein synthesis, nucleic acid synthesis, cell-wall synthesis or folate synthesis (Walsh, 2003). Antibiotic research has focused on these biosynthetic pathways largely because they are extremely well characterized and essential for bacterial viability. However, with a flood of new information from genomic and proteomic research, as well as the relatively recent development of powerful cell biology techniques for bacteria, several new potential antibiotic targets have emerged (Freiberg and Brotz-Oesterhelt, 2005).

One such target is bacterial cell division, the process by which a single progenitor cell splits into two independent daughter cells. Bacteria must divide in order to establish an infection, and cannot survive for more than a few generations when division is blocked. Bacterial division is driven by the coordinated action of several specific proteins, most of which are essential for viability and highly conserved among bacteria. Many of these proteins are well characterized, both structurally and biochemically, and represent excellent potential targets for antibacterial protein inhibitors. While bacterial division is not yet targeted clinically, drug discovery efforts to exploit this essential process are now well under way (Vollmer, 2006; Lock and Harry, 2008; Kapoor and Panda, 2009).

This chapter will review the current and future potential of cell division as an antibacterial target for the development of effective antibiotics with new modes of action. The molecular framework for division in the model organisms *Escherichia coli* and *Bacillus subtilis* will be described, identifying

*Corresponding author

the various proteins involved in septum formation, and their suitability as drug targets. Division in some of the key bacterial pathogens will then be described with emphasis on unique insights gained from studies on both the division process and its regulation in these organisms. An appraisal of current approaches used to identify division inhibitors, and new approaches for antibacterial drug discovery targeting division will also be discussed, followed by challenges and future directions.

Bacterial cell division: septum formation

Bacterial division involves the coordinated ingrowth of the cell envelope layers (the cell membrane(s) and cell wall) to form a septum or cross-wall between two newly replicated chromosomes (Fig. 3.1), effectively splitting the cell into two independent compartments. Peptidoglycan in the septum is hydrolysed by specific enzymes (peptidoglycan or murein hydrolases), enabling the two newborn cells to separate (Fig. 3.1). Research in the bacterial division field has focused most extensively on the rod-shaped bacteria *E. coli* and *B. subtilis* (for recent reviews see Goehring and Beckwith, 2005; Dajkovic and Lutkenhaus, 2006; Harry *et al.*, 2006; Adams and Errington, 2009). These organisms grow by elongation along the long axis of the rod-shaped cell (i.e. the rod progressively increases in length while maintaining the same diameter) and divide by forming a septum

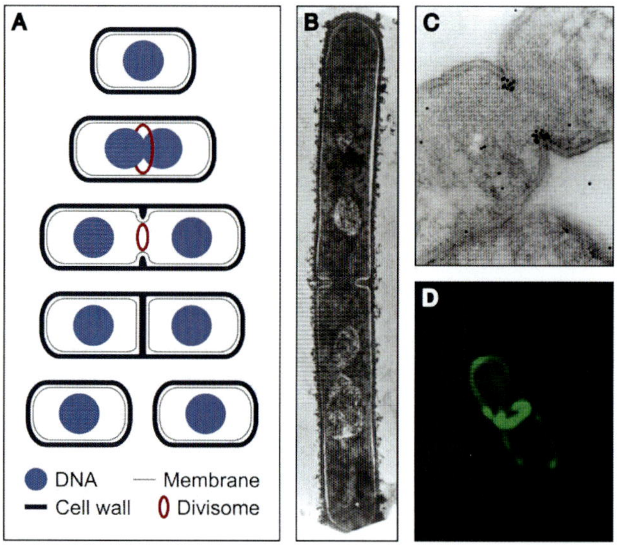

Figure 3.1 Bacterial cell division: septum formation and the divisome. (A) The division cycle of a model rod-shaped bacterium. Rod-shaped cells grow by elongation along their long axis. After reaching an appropriate length, the newly replicated chromosomes (blue circles) segregate to opposite sides of the cell, and a multimeric protein ring is formed precisely at the cell centre, composed of at least 12 division proteins collectively termed the divisome (red ring). At a later stage of the cell cycle, the division ring constricts. This is accompanied by ingrowth of the cell membrane and cell wall to produce a septum that splits the cell into two identical compartments. Peptidoglycan in the septum is then hydrolysed, enabling the two newborn cells to separate. This example shows a Gram-positive rod such as *Bacillus subtilis*, which has a single cell membrane. Gram-negative bacteria, such as *Escherichia coli*, contain an inner and outer membrane, both of which are split during the division process to produce separate daughter cells. (B) An electron micrograph of a *B. subtilis* cell undergoing division. A partially formed septum can be seen at the cell centre. (C) An *E. coli* cell captured during the process of septal invagination. In this cross section, the FtsZ protein has been labelled by immunogold staining. FtsZ is the first protein to form a ring at the division site, and as such the FtsZ signal can be seen at the leading edge of the developing septum (reproduced with permission from Lutkenhaus,1993). (D) An *E. coli* cell expressing a green fluorescent protein (GFP) fusion to FtsL, another member of the divisome. FtsL localizes into a ring at the cell centre. This figure is reproduced with permission from Ghigo *et al.* (1999).

precisely at the cell centre at the appropriate time, producing daughter cells that are identical in size (Fig. 3.1). It is important to note that with the wide array of different cell shapes and structures found throughout the bacteria, there are some fundamental differences in how division is orchestrated between species (see below). However, work using *E. coli* and *B. subtilis* as model systems has revealed a great deal about the proteins involved in bacterial division, how they coordinate septum formation and how they might be targeted by antibiotics.

The divisome: a multiprotein apparatus that drives septum formation

Because cell division is essential for viability, many of the proteins responsible for this process were identified in *E. coli* and *B. subtilis* by screening for conditional, temperature-sensitive division mutants (Goehring and Beckwith, 2005). These mutants grow and divide normally at low (permissive) temperatures, but are specifically defective in division at high (restrictive) temperatures. Under restrictive conditions, the mutants continue to grow (elongate) and undergo DNA replication normally, leading to the formation of long cells (or filaments) containing multiple chromosomes. For this reason, many division genes are denoted *fts* (filamentous temperature sensitive). In *B. subtilis*, some division genes are named *div*, for division. Once division has been inhibited at the restrictive temperature, *fts* mutants can only survive for a few mass doublings before the filaments eventually lyse. It is reasonable to predict that specific inhibitors of division proteins could occasion cell death in a similar way, at least in rod-shaped bacteria (Lock and Harry, 2008).

In total, there are at least 12 cell division proteins in *E. coli* and *B. subtilis* that are recruited to the division site at mid-cell (Fig. 3.2). These proteins form a ring-like complex known as the

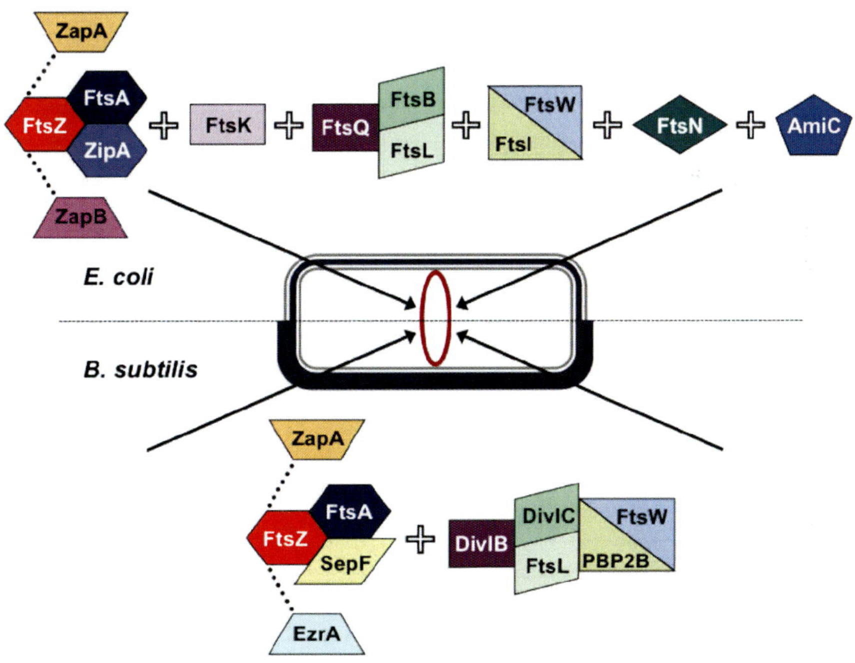

Figure 3.2 Recruitment of key components to the cell division ring in *Escherichia coli* and *Bacillus subtilis*. Bacterial septum formation requires the localization of the division proteins to a ring at the nascent division site. In *E. coli*, these proteins appear to be recruited in an essentially linear order (as individual proteins and protein subcomplexes), while in *B. subtilis* the septal recruitment pathway apparently occurs in two stages. This figure shows only the best-characterized members of the cell division pathway in *E. coli* and *B. subtilis*. Proteins depicted with the same shape and colour are homologous. Dotted lines indicate that a protein is not absolutely required (under standard laboratory conditions) for downstream recruitment but acts to form or stabilize the divisome. This figure is adapted with permission from Lock and Harry (2008).

division ring or divisome, which extends around the circumference of the cell and ultimately drives the formation of the septum (Figs. 3.1 and 3.2; Harry *et al.*, 2006). The first protein to localize to the division site is the bacterial tubulin homologue FtsZ (Errington *et al.*, 2003). FtsZ assembles into a polymeric ring structure (the Z ring) on the inner surface of the cytoplasmic membrane (Bi and Lutkenhaus, 1991), which serves as a scaffold upon which all other cell division proteins are assembled. The division proteins are recruited to the Z ring in a hierarchical order, which is slightly different in *E. coli* and *B. subtilis* (Fig. 3.2).

In *E. coli*, the FtsA, ZipA, ZapA and ZapB proteins interact directly with FtsZ and probably assemble at midcell concomitantly with the Z ring (Adams and Errington, 2009). The FtsEX complex is then recruited to the ring via direct interaction with FtsZ (Schmidt *et al.*, 2004; Corbin *et al.*, 2007), and is thought to aid in the subsequent recruitment of FtsK and a pre-formed complex of FtsL, FtsQ and FtsB (Goehring *et al.*, 2006). These proteins are required for the recruitment of an additional complex of FtsW and FtsI, and this is followed sequentially by FtsN, AmiC and EnvC (Fig. 3.2; Goehring *et al.*, 2006). The recently characterized YmgF protein is also believed to form part of the *E. coli* divisome, and is recruited to mid-cell in a manner dependent on the presence of at least FtsZ, FtsA, FtsQ and FtsN (Karimova *et al.*, 2009). Interestingly, studies have shown that there is a considerable time delay between the formation of the Z ring and the recruitment of proteins downstream of FtsK in *E. coli*, suggesting that the divisome matures in two distinct stages (Aarsman *et al.*, 2005).

Divisome assembly in *B. subtilis* also occurs in two stages (Errington *et al.*, 2003; Gamba *et al.*, 2009). First, the Z ring assembles at midcell concomitantly with the FtsZ-interacting proteins FtsA, ZapA, EzrA (and probably SepF; Jensen *et al.*, 2005; Gamba *et al.*, 2009). Second, the membrane-bound division proteins FtsL, DivIB (the *B. subtilis* homologue of *E. coli* FtsQ), DivIC (FtsB homologue), FtsW, PBP2B (FtsI homologue) and probably YpsB (Tavares *et al.*, 2008) are recruited in a concerted fashion (Fig. 3.2; Gamba *et al.*, 2009).

Cell division proteins as antibacterial targets

Bacterial cell division proteins have a number of properties that could make them excellent antibacterial targets (Table 3.1). First, many of the division proteins are essential for the propagation and viability of the organism, and are therefore likely to be essential in establishing an infection. Second, the division proteins are conserved among many bacterial pathogens, but mainly absent in eukaryotes. While there are exceptions to this rule (for example, FtsZ is a structural homologue of tubulin; Lowe and Amos, 1998; Nogales *et al.*, 1998a) and FtsA shares a similar fold with actin (van den Ent and Lowe 2000), inhibitors specific to the prokaryotic homologues have been identified (see below). A third advantage of targeting bacterial cell division proteins is that many contain extracellular domains that would be readily accessible to inhibitory compounds, avoiding the need to optimize hit compounds for entry into the cytoplasm.

Another potential benefit of developing inhibitors against cell division is that division is not simply an isolated event, but is intimately connected with other aspects of the bacterial cell cycle. For example, recent evidence shows that the division process is coordinated with carbon metabolism in *B. subtilis*, via a direct interaction between FtsZ and the enzyme UgtP, a component of the glucolipid biosynthesis pathway (Weart *et al.*, 2007; Wang and Levin, 2009). The *B. subtilis* cell utilizes this mechanism to regulate the timing of cell division in response to nutrient-dependent changes in growth rate. Furthermore, work in *E. coli* and *Caulobacter crescentus* demonstrates that FtsZ plays a general role in cell growth, independent of its function in septum formation, by helping to direct peptidoglycan synthesizing enzymes for lateral cell wall synthesis (Varma and Young, 2004, 2009; Aaron *et al.*, 2007; Varma *et al.*, 2007). Taking these findings into account, it is probable that specific inhibitors of cell division will not only block the septation process, but may have more far reaching effects on the regulation of the cell cycle, increasing the potential to efficiently kill the bacterium.

Many bacterial cell division proteins have now been well characterized both *in vitro* and *in vivo*,

Table 3.1 Cell division proteins as potential antibacterial targets[a]

Protein	Proposed function	Essential?[b]	Conservation	Human homologue?[c]	Interactions	Structural data available?
FtsZ	Forms a polymeric ring at the division site, required for the recruitment of all downstream division proteins	Yes (Bi *et al.*, 1991; Dai and Lutkenhaus, 1991)	Widely conserved across the bacteria	Yes (Lowe and Amos, 1998; Nogales *et al.*, 1998)	FtsA (Wang *et al.*, 1997), ZipA (Hale and de Boer, 1997), ZapA (Gueiros-Filho and Losick, 2002), ZapB (Ebersbach *et al.*, 2008), SepF (Hamoen *et al.*, 2006; Ishikawa *et al.*, 2006), EzrA (Haeusser *et al.*, 2004), FtsE (Corbin *et al.*, 2007), FtsK (Di Lallo *et al.*, 2003)	Yes (Lowe and Amos, 1998)
FtsA	Tethers FtsZ to the membrane, stabilizes Z rings, recruits downstream proteins	Yes[d] (Lutkenhaus and Donachie, 1979)	Widely conserved, often co-transcribed with *ftsZ*	Yes (Bork *et al.*, 1992; van den Ent *et al.*, 2008)	FtsZ (Wang *et al.*, 1997)	Yes (van den Ent and Lowe 2000)
ZipA	Stabilises Z rings at the membrane	Yes (Hale and de Boer, 1997)	γ-proteobacteria	No	FtsZ (Hale and de Boer, 1997)	Yes (Mosyak *et al.*, 2000)
ZapA	Promotes Z ring formation by inducing protofilament bundling	No (Gueiros-Filho and Losick, 2002)	Widely conserved	No	FtsZ (Gueiros-Filho and Losick, 2002)	Yes (Low *et al.*, 2004)
ZapB	Stimulates Z ring assembly	No (Ebersbach *et al.*, 2008)	Subset of γ-proteobacteria	No	FtsZ (Ebersbach *et al.*, 2008)	Yes (Ebersbach *et al.*, 2008)
SepF/YlmF (*B. subtilis* only, no *E. coli* homologue)	Promotes Z ring formation and proper septum development, function overlaps with FtsA	No (Hamoen *et al.*, 2006; Ishikawa *et al.*, 2006)	Gram-positives and cyanobacteria	No	FtsZ (Hamoen *et al.*, 2006; Ishikawa *et al.*, 2006)	No
EzrA (*B. subtilis* only, no *E. coli* homologue)	Negatively regulates FtsZ polymerization and Z ring assembly, contributes to the dynamic turnover of the Z ring, involved in coordinating cell elongation with division	No (Levin *et al.*, 1999)	Low GC Gram-positive bacteria	No	FtsZ (Haeusser *et al.*, 2004)	No
FtsE	Similar to ABC transporter, unknown function	No[e] (Schmidt *et al.*, 2004; Reddy, 2007)	Widely conserved	Yes	FtsZ (Corbin *et al.*, 2007), FtsX (Schmidt *et al.*, 2004)	No
FtsX	Similar to ABC transporter, unknown function	No[e] (Schmidt *et al.*, 2004; Reddy, 2007)	Widely conserved	No	FtsE (Schmidt *et al.*, 2004)	No

Table 3.1 continued

Protein	Proposed function	Essential?[b]	Conservation	Human homologue?[c]	Interactions	Structural data available?
FtsK	DNA translocase (ATPase), roles in chromosome partitioning and septum formation	Yes[d] (Wang and Lutkenhaus, 1998)	Widely conserved	No	FtsZ (Di Lallo *et al.*, 2003), FtsL (Di Lallo *et al.*, 2003), TopoIV (Espeli *et al.*, 2003)	Yes (Massey *et al.*, 2006)
FtsQ (DivIB)	Unknown, possible role in septal peptidoglycan synthesis	Yes[f] (Storts *et al.*, 1989)	Widely conserved, although conservation at the sequence level is poor	No	FtsB and FtsL (Buddelmeijer and Beckwith, 2004), FtsI (D'Ulisse *et al.*, 2007), FtsN (D'Ulisse *et al.*, 2007), FtsW (D'Ulisse *et al.*, 2007)	Yes (Robson and King, 2006; van den Ent *et al.*, 2008)
FtsB (DivIC)	Unknown, possible role in septal peptidoglycan synthesis	Yes (Levin and Losick, 1994; Buddelmeijer *et al.*, 2002)	Widely conserved, poor sequence conservation	No	FtsQ and FtsL (Buddelmeijer and Beckwith, 2004)	Yes (Masson *et al.*, 2009)
FtsL	Unknown, possible role in septal peptidoglycan synthesis	Yes (Guzman *et al.*, 1992; Daniel *et al.*, 1998)	Widely conserved, poor sequence conservation	No	FtsQ and FtsB (Buddelmeijer and Beckwith, 2004), FtsW (Di Lallo *et al.*, 2003), FtsK (Di Lallo *et al.*, 2003)	Yes (Masson *et al.*, 2009)
FtsW	SEDS protein required for septal peptidoglycan synthesis. Recruits FtsI to the division site	Yes (Boyle *et al.*, 1997)	Widely conserved	No	FtsI (Datta *et al.*, 2006), FtsZ (*M. tuberculosis* only; Datta *et al.*, 2006), FtsQ (D'Ulisse *et al.*, 2007), FtsL (Di Lallo *et al.*, 2003), FtsN (Di Lallo *et al.*, 2003)	No
FtsI/PBP3 (PBP2B)	Transpeptidase (class B PBP), introduces cross-links into septal peptidoglycan	Yes (Spratt 1977; Daniel *et al.*, 1996)	Widely conserved	No	FtsW (Datta *et al.*, 2006), FtsQ (D'Ulisse *et al.*, 2007), FtsN (Di Lallo *et al.*, 2003)	Yes (Dessen *et al.*, 2001)
FtsN	Unknown, contains peptidoglycan binding domain	Yes (Dai *et al.*, 1993)	Present throughout the proteobacteria, poor sequence conservation (Moll and Thanbichler, 2009)	No	FtsQ (D'Ulisse *et al.*, 2007), FtsI (Di Lallo *et al.*, 2003), FtsW (Di Lallo *et al.*, 2003)	Yes (Yang *et al.*, 2004)

AmiC	Amidase (peptidoglycan hydrolase), required for septal cell wall degradation/cell separation	Yes (Heidrich *et al.*, 2001; Kobayashi *et al.*, 2003)	Widely conserved	No	–	No
EnvC	Peptidoglycan hydrolase required for septal cell wall degradation/cell separation	Yes (Kobayashi *et al.*, 2003; Bernhardt and de Boer, 2004)	Widely conserved	No	–	No

[a]Only the major *E. coli* cell division proteins are shown, with the names of their *B. subtilis* homologues shown in brackets where the names differ. Properties of the proteins that implicate them as good antibacterial targets are underlined.

[b]Essentiality refers to *E. coli* and/or *B. subtilis*.

[c]Data on human homology to the bacterial division proteins come from a bioinformatic analysis by Lock and Harry (2008) in addition to the cited references.

[d]Essential in *E. coli*, but not in *B. subtilis* (Beall and Lutkenhaus, 1992; Jensen *et al.*, 2005; Biller and Burkholder, 2009; Kaimer *et al.*, 2009). Note that *B. subtilis* has two different FtsK homologues, SpoIIIE and SftA.

[e]Essential only under conditions of low ionic strength in *E. coli*.

[f]Essential in *E. coli*, but only essential at high temperatures in *B. subtilis* (Beall and Lutkenhaus, 1989; Katis *et al.*, 2000).

ABC, ATP-binding cassette; PBP, penicillin-binding protein; SEDS, shape, elongation, division, sporulation.

providing a good starting point for the target-driven discovery of inhibitory compounds. Crystal structures are available for most of the essential division proteins (Table 3.1), and catalytic activity has been established for some of these. Through these studies some division proteins have been shown to contain clearly identifiable druggable domains, i.e. domains likely to be amenable to drug development (Hopkins and Groom, 2002), including the GTP-binding domain of FtsZ and the ATP-binding site of FtsA (de Boer *et al.*, 1992; RayChaudhuri and Park, 1992; Mukherjee *et al.*, 1993; Sanchez *et al.*, 1994; van den Ent and Lowe 2000; Feucht *et al.*, 2001; Lara *et al.*, 2005). Active site binding domains represent excellent targets for small molecule inhibitors, given that they are required for protein function, and that in general most drugs act by competing with endogenous molecules for binding to such sites. However, targeting ATP- or GTPases does pose potential selectivity issues.

An alternative approach for developing anti-bacterial cell division inhibitors would be to target protein–protein interactions between members of the divisome (see below). Several essential division protein interactions have been identified in bacteria and at least partially characterized (Table 3.1; Goehring and Beckwith, 2005; Harry *et al.*, 2006). These are required for the assembly of the divisome and/or for the regulation of the division process, making them attractive potential targets.

Below we discuss some of the individual proteins of bacterial cell division in more detail, including the specific properties that may make them suitable as antibacterial targets. A summary of all major division proteins in *E. coli* and *B. subtilis* is provided in Table 3.1.

FtsZ as a target

The vast majority of studies aiming to develop lead compounds against the bacterial cell division pathway have focussed specifically on inhibiting the protein FtsZ (see below). This is not surprising, as FtsZ exhibits several properties that make it an obvious candidate for drug targeting. FtsZ is the first protein to assemble at the nascent division site, and is required for the localization and correct functioning of all other cell division proteins (Errington *et al.*, 2003). As described above, FtsZ polymerizes to form the Z ring on the inner surface of the cytoplasmic membrane (Bi and Lutkenhaus, 1991). The formation of this ring establishes both the timing and location of septum formation, and as such it is subject to multiple levels of regulation. Accordingly, FtsZ interacts with a variety of other proteins, and some of these interactions are essential for viability (Adams and Errington, 2009). The FtsZ protein itself is of course essential, and moreover, it is the most highly conserved cell division protein in bacteria. Only a very small number of bacterial species have been found to lack an FtsZ homologue, with the only pathogens of note being the chlamydiae, which are obligate intracellular parasites, and the mucosal pathogen *Ureaplasma urealyticum* (Glass *et al.*, 2000; Vaughan *et al.*, 2004; Adams and Errington, 2009). Finally, recent studies have shown that as well as providing a scaffold for the recruitment of the other division proteins, the Z ring plays an active role in septum formation by generating an inward pinching force on the cell membrane (Osawa *et al.*, 2008, 2009). For these reasons, it is probable that any drug that effectively inhibits FtsZ function will have potent broad-spectrum antibacterial activity.

The biggest potential obstacle in developing successful antibacterials against FtsZ is its homology with the eukaryotic cytoskeletal protein tubulin. Although tubulin and FtsZ exhibit only limited amino acid sequence identity (<10%; Erickson, 2007), the two proteins share a strikingly similar three-dimensional fold (Lowe and Amos, 1998; Nogales *et al.*, 1998a,b) and similar biochemical properties (Romberg and Levin, 2003). However, there are some subtle structural differences (Nogales *et al.*, 1998a), and with a wealth of information available on FtsZ and tubulin structure and function, researchers have been able to exploit these differences to identify specific compounds that inhibit FtsZ and not human tubulin (see below). Furthermore, it has been shown that none of the classical tubulin inhibitors (albendazole, colchicine, nocodazole, paclitaxel 3-methoxybenzamide or thiabendazole) exhibit any significant activity against FtsZ *in vitro* (Wang *et al.*, 2003). These preliminary data strengthen the notion that FtsZ specific inhibitors are possible despite the significant homology with tubulin.

Like tubulin, FtsZ is a GTPase (de Boer *et al.*, 1992; RayChaudhuri and Park, 1992; Mukherjee and Lutkenhaus, 1998), and upon binding to GTP it reversibly assembles into linear polymer strands (protofilaments) by the head-to-tail association of monomers (Fig. 3.3; Mukherjee and Lutkenhaus, 1994, 1998). As two subunits come together in the protofilament, a catalytic loop region termed the T7 loop (Nogales *et al.*, 1998a) is inserted from one subunit into the GTP-binding pocket of the next (Fig. 3.3A and B). This completes the active site and enables GTP hydrolysis to occur (Scheffers *et al.*, 2002; Oliva *et al.*, 2004). As a result, GTPase activity is completely dependent on polymerization. *In vitro* studies have shown that FtsZ protofilaments can exist in isolation in solution (Romberg *et al.*, 2001), or associate together via lateral interactions between filaments to form protofilament bundles and variety of other higher-order polymers (Fig. 3.3D; Bramhill and Thompson, 1994; Erickson *et al.*, 1996; Yu and Margolin, 1997; Lowe and Amos, 1999, 2000; Gonzalez *et al.*, 2003; Oliva *et al.*, 2003; Popp *et al.*, 2009). However, unlike tubulin, FtsZ protofilaments do not assemble into tubule structures, apparently because of a difference in the lateral surface conformation of the two proteins (Nogales *et al.*, 1998a). Critically, very little is known about the actual *in vivo* arrangement of FtsZ protofilaments within the Z ring (Fig. 3.3E). Nonetheless, recent evidence indicates that in addition to head-to-tail polymerization, lateral association of FtsZ

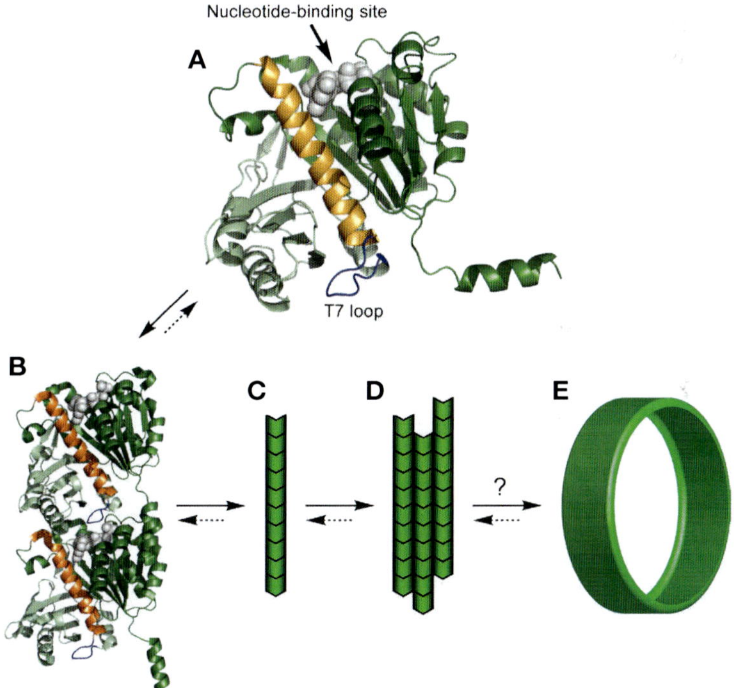

Figure 3.3 FtsZ and the formation of the Z ring. (A) Crystal structure of FtsZ bound to GDP (grey spheres) from *Methanococcus jannaschii*. The FtsZ monomer consists of an N-terminal nucleotide-binding domain (dark green) and a C-terminal globular domain (pale green) separated by a central helix (orange). The T7 loop, which is important for catalysis. (B) Crystal structure of an FtsZ dimer from *M. jannaschii*, illustrating the head-to-tail association of two monomers and the insertion of the T7 loop (blue) from one subunit into the nucleotide-binding pocket of the other. (C) The addition of further monomers to the growing chain produces a linear protofilament with distinct polarity. (D) FtsZ protofilaments can associate laterally to form higher-order structures. (E) The completed Z ring is thought to consist of some higher-order arrangement of FtsZ protofilaments, although the fine details of this structure are poorly understood. It is important to note that FtsZ assembly is a completely reversible process, as indicated by the reverse arrows. In the cell, this equilibrium is controlled by a network of FtsZ-binding proteins that either promote or inhibit FtsZ polymerization (see text).

protofilaments is essential for Z ring formation and thus cellular viability (Monahan *et al.*, 2009). It is possible therefore, that specific inhibitors of either GTP binding, head-to-tail subunit association or lateral FtsZ interactions could function as potent antibacterial drugs. However, while the mechanism of protofilament formation and GTP binding are intimately understood at the structural level (Oliva *et al.*, 2004, 2007), very little is currently known about how FtsZ filaments associate together laterally and where the lateral interaction sites are located.

The role of GTP hydrolysis in FtsZ function is also somewhat unclear at present. *In vitro* work indicates that while GTP binding stimulates the formation of FtsZ protofilaments, hydrolysis destabilizes the polymer, which eventually leads to its disassembly (Mukherjee and Lutkenhaus, 1994, 1998). GTP hydrolysis is required *in vivo* to maintain a state of dynamic turnover within the Z ring (the Z ring undergoes continual remodelling throughout its existence, through FtsZ exchange between the ring and the intracellular FtsZ pool; Stricker *et al.*, 2002; Anderson *et al.*, 2004). Interestingly, mutants that are severely impaired in GTP hydrolysis are still able to divide, despite exhibiting a slower Z ring turnover (Dai *et al.*, 1994; Mukherjee *et al.*, 2001; Stricker *et al.*, 2002). This brings into question the functional significance of FtsZ's GTPase activity and the role for a dynamic exchange of FtsZ within the Z ring. GTP hydrolysis has also been implicated in the generation of mechanical force during invagination of the septum, but again this is a matter of intense debate (Adams and Errington, 2009; Osawa *et al.*, 2009). Therefore, it is important to be aware that compounds that specifically interfere with the ability of FtsZ to hydrolyse GTP may not prove to be effective as antibacterial agents.

Targeting proteins that affect FtsZ assembly

FtsZ interacts directly with a large number of proteins that modulate its assembly properties *in vivo*. These include FtsA, ZipA, ZapA, ZapB and SepF (or YlmF), all of which promote FtsZ assembly or stabilize FtsZ polymers, as well as MinC, SlmA, Noc, EzrA, SulA and ClpX, which destabilize or inhibit FtsZ polymerization. These proteins are thought to act in a concerted manner to ensure that the Z ring forms only once per cell cycle and only at the correct time and place. They may also help to regulate the dynamics and constriction of the Z ring (reviewed by Harry *et al.*, 2006; Michie and Lowe 2006; Adams and Errington, 2009). For example, SlmA (in *E. coli*) and Noc (in *B. subtilis*) associate with the DNA to prevent Z rings from forming inappropriately over the chromosome (Wu and Errington, 2004; Bernhardt and de Boer, 2005). On the other hand, the ZapA protein acts to promote and stabilize Z ring formation at the division site by stimulating the lateral association of FtsZ protofilaments *in vivo* (Gueiros-Filho and Losick, 2002; Monahan *et al.*, 2009).

While some of the FtsZ-binding proteins, such FtsA, ZapA and MinC, are well conserved throughout bacteria, several others are present only in a small subset of bacterial genomes (see Table 3.1). As a result, the specific complement of FtsZ regulatory proteins varies significantly from species to species, and is probably tailored to meet the particular physiological requirements and environmental conditions of each different bacterium. Importantly, the vast majority of FtsZ-binding proteins have been shown to be individually dispensable for Z ring formation and cellular viability, at least in the model organisms *E. coli* and *B. subtilis*. However, these proteins often become essential in the absence of a second FtsZ modulator or when Z ring formation is perturbed by a reduction in the intracellular concentration of FtsZ, demonstrating that there is a large degree of redundancy/functional overlap within the FtsZ regulatory network (Levin *et al.*, 1999; Gueiros-Filho and Losick, 2002; Wu and Errington, 2004; Ishikawa *et al.*, 2006).

While this redundancy argues that FtsZ-binding proteins in general might make poor antibacterial targets, it is interesting to note that several proteins, including FtsA, ZipA, MinC and EzrA, compete for the same binding site on FtsZ: a short stretch of residues at the extreme C-terminus (Din *et al.*, 1998; Liu *et al.*, 1999; Ma and Margolin, 1999; Yan *et al.*, 2000; Haney *et al.*, 2001; Singh *et al.*, 2007; Shen and Lutkenhaus, 2009). This binding site contains a highly conserved motif of nine amino acids that is essential for FtsZ function (Vaughan *et al.*, 2004).

If a drug-like compound could be developed that binds to the C-terminus of FtsZ and simultaneously prevents access to several FtsZ modulators, it would almost certainly exhibit broad-spectrum antibacterial activity. By the same token, a compound that mimics the FtsZ C-terminus and can bind and inhibit multiple accessory proteins would be likely to kill bacterial cells. Nonetheless, research emerging from pathogenic bacteria suggests that some of FtsZ-binding proteins, despite being dispensable for viability in *E. coli* and *B. subtilis*, are in fact essential in certain pathogens. For example, homologues of SepF (YlmF), EzrA and Noc have all been shown to be essential in *Staphylococcus aureus* (see below). This suggests that the FtsZ-binding proteins may individually be good targets in their own right. As further information comes to hand, it will become clear which of the FtsZ modulators are attractive to pursue as broad- or narrow-spectrum antibacterial targets.

Two proteins that are essential even in model bacteria, and do appear to be good potential targets, are FtsA and ZipA. These proteins are discussed in turn below.

FtsA

Outside of FtsZ, FtsA is the most well conserved bacterial cell division protein, being found throughout the bacteria and in many clinically relevant pathogens (Table 3.1). FtsA has been demonstrated to be essential for cell division and viability in a number of organisms, including *E. coli* (Lutkenhaus and Donachie 1979), *S. aureus* (Chaudhuri *et al.*, 2009) and *Streptococcus pneumoniae* (Lara *et al.*, 2005), highlighting its potential as an antibacterial target. Although FtsA is not strictly essential in *B. subtilis*, cells are less viable and highly filamentous in its absence (Beall and Lutkenhaus, 1992; Jensen *et al.*, 2005).

FtsA is a multifunctional protein that is important not only for Z ring formation and stabilization, but for the recruitment of other proteins to the division site and potentially for the constriction of the Z ring and septal invagination (Pichoff and Lutkenhaus, 2002; Goehring *et al.*, 2005; Jensen *et al.*, 2005; Pichoff and Lutkenhaus, 2005). The three-dimensional fold of FtsA resembles that of eukaryotic actin, comprising two globular domains arranged around an ATP-binding pocket (van den Ent and Lowe 2000). Although this poses a potential problem for antibacterial development, FtsA lacks an entire actin subdomain, and instead possesses a novel subdomain termed 1C, which is unique among members of the actin family. FtsA has been shown to bind itself, ATP, FtsZ and the cytoplasmic membrane, and all of these interactions appear to be crucial for its function (Sanchez *et al.*, 1994; Wang *et al.*, 1997; Yim *et al.*, 2000; Feucht *et al.*, 2001; Di Lallo *et al.*, 2003; Rico *et al.*, 2004; Karimova *et al.*, 2005; Lara *et al.*, 2005; Pichoff and Lutkenhaus, 2005, 2007; Shiomi and Margolin, 2007; Beuria *et al.*, 2009a). FtsA has also been found to exhibit ATPase activity in *B. subtilis* (Feucht *et al.*, 2001), although the significance of this is unclear and complicated by the fact that only very limited ATP hydrolysis was observed in a study of *E. coli* FtsA (Beuria *et al.*, 2009a) and no nucleotide hydrolysis activity was seen for the *S. pneumoniae* protein (Lara *et al.*, 2005).

Significant insight into the molecular basis of FtsA function has been gained by dissecting the roles of specific regions of the protein. Such studies have revealed that subdomain 1C is required for the recruitment of downstream cell division proteins to the divisome (Corbin *et al.*, 2004; Rico *et al.*, 2004), while residues clustered on the surface of another subdomain (2B) are important for the interaction of FtsA with FtsZ (Pichoff and Lutkenhaus, 2007). As mentioned above, FtsA interacts with the conserved C-terminal tail of FtsZ, and this interaction is essential for viability. Studies have shown that FtsA binds FtsZ prior to Z ring formation, consistent with it playing a direct role in the assembly of the ring (Jensen *et al.*, 2005). As well as binding FtsZ, Pichoff and Lutkenhaus (2005) demonstrated that FtsA contains a conserved amphipathic helix at its C-terminus, which targets FtsA to the plasma membrane and is essential for function. FtsZ lacks a membrane targeting sequence, suggesting that at least part of FtsA's role as a promoter and stabilizer of Z ring formation is to tether FtsZ to the membrane. Recent evidence hints that FtsA may also function to directly modulate FtsZ assembly (Beuria *et al.*, 2009a). Although further work needs to be done to fully characterize the interactions of FtsA with itself and other proteins, these findings indicate the model organisms *E. coli* and *B. subtilis*. These include GpsB (Claessen *et al.*, 2008) and YpsB (Tavares *et al.*, 2008) in *B. subtilis*, and YmgF in *E. coli* (Karimova *et al.*, 2009). As yet there are no structural data available for these proteins, and only very limited information regarding their precise roles in division. As such, the suitability of these proteins as antibiotic targets remains unclear. However, this may change as new results come to hand.

Cell division in pathogens

It is becoming apparent that nearly all bacteria have a set of conserved cell division proteins. However, this set is variable depending on the species, and some proteins of the set diverge more than others. This probably reflects differences in cell envelope morphology and cell shape, which governs how bacteria grow and divide (Goehring and Beckwith, 2005). A detailed knowledge of cell division in important pathogens will facilitate the identification of division and division-related targets suitable for narrow- and broad-spectrum therapies. Furthermore, it will reveal targets that are relevant in the *in vivo* (host) situation. It is only very recently that cell division research has broadened to include clinically relevant bacteria, particularly the cocci and mycobacteria.

Pseudomonas and *Acinetobacter*

In general, Gram-negative pathogens have homologues of most of the division proteins found in *E. coli*, and presumably their function is the same. However, there are very few division studies involving important drug-resistant Gram-negative pathogens such as *P. aeruginosa* and closely related *Acinetobacter* spp. The prevalence of multidrug resistance in *Pseudomonas* and *Acinetobacter* is substantial in ICU patients worldwide (Opal 2009). Most alarmingly, *A. baumannii* strains have now been reported that are resistant to every antibiotic compound in clinical use, including recently approved drugs such as tigecyclin (Dijkshoorn *et al.*, 2007; Navon-Venezia *et al.*, 2007).

In the case of *Acinetobacter* many division proteins have diverged substantially from their *E. coli* counterparts, and a comprehensive review on this topic has been published recently (Robinson *et al.*, 2010). Notably, no homologues could be found for FtsE, FtsX or SulA, which are conserved in almost all other γ-proteobacteria. SulA is an inhibitor of FtsZ polymerization that is induced as part of the SOS response in *E. coli*, where it serves to temporarily halt cell division in the event of DNA damage (Huisman, 1981; Cordell, 2003). *Acinetobacter* also lacks the SOS regulator protein, LexA. It has been proposed that a unique stress response occurs in this group of organisms (Robinson *et al.*, 2010).

The cocci

Numerous important pathogens are cocci, and include staphylococci, streptococci, enterococci and *Neisseria*. Cocci can be divided further into the truly spherical cocci, including staphylococci and *Neisseria*, and the ellipsoid ovococci, such as streptococci and enterococci. Generally, cocci and ovococci have the same cohort of cell division proteins as *B. subtilis*. However, important differences in septum formation have been uncovered recently that are relevant to drug discovery.

Division in true cocci

Clearly the most successful antibiotics have been those that target PG synthesis, such as the β-lactams. This has been largely because of their high specificity for bacteria and their potency, since cell wall synthesis is essential for bacterial growth and viability. In cocci, PG synthesis and PBP function are also inextricably linked to cell division (Zapun *et al.*, 2008b). Unlike rod-shaped bacteria, which have two distinct modes of PG synthesis (lateral and septal), true cocci have only one mode that is specific to division (Fig. 3.4; Higgins, 1970; Giesbrecht, 1976; Pinho and Errington, 2003; Zapun *et al.*, 2008b). In other words, these organisms actually grow by division, and any antibiotic targeting PG synthesis in cocci is actually targeting the division process. In line with unimodal PG synthesis, there is a general absence of MreB and related proteins (required for elongation in rod-shaped bacteria) and fewer PBPs in cocci (Jones *et al.*, 2001; Scheffers and Pinho, 2005; Zapun *et al.*, 2008b). For example, while *E. coli* and *B. subtilis* have 12 and 16 PBPs, respectively, *S. aureus* and *Neisseria meningitidis* or *N. gonorhoeae* have only four (Foster and Popham,

2002; Zapun *et al.*, 2008b). The low number of PBPs and the single mode of PG synthesis (division) in cocci suggest a simpler PG synthesis mechanism and regulation which could facilitate the discovery of novel antibacterials that target cell division in these important pathogens.

In *S. aureus*, the division septum forms a circular plate, which is subsequently cleaved and remodelled to produce the hemispherical poles of the daughter cells (Fig. 3.4). *S. aureus* has two essential PBPs: PBP1 (HMW class B) and PBP2 (HMW class A), and two non-essential ones: PBP3 (HMW class B) and -4 (LMW class C; Pinho and Errington, 2005; Pereira *et al.*, 2007). PBP4 appears to be involved in secondary cross-linking of PG. The FtsI homologue in *S. aureus*, PBP1, has been shown to be essential for cell division, and growth (Pereira *et al.*, 2007). As expected PBP1, the product of the *pbpA* gene, localizes to the septum, as do probably all four PBPs in this organism (Pinho and Errington, 2005; Pereira *et al.*, 2007; Zapun *et al.*, 2008b).

S. aureus has homologues of all cell division proteins present in its Gram-positive relative, *B. subtilis*, with varying degrees of conservation. Importantly, all division genes identified so far in *S. aureus* are essential (Chaudhuri *et al.*, 2009). This includes *ftsA*, *sepF* (*ylmF*), *ezrA*, *yyaA* and *ypsB* which are all non-essential in *B. subtilis* (Table 3.1). This probably reflects the need for this organism to successfully divide in order to grow. In *B. subtilis* SepF (YlmF) can complement the activity of the FtsZ-binding protein FtsA, to the extent that overexpression of SepF complemented the otherwise filamentous *ftsA* null mutant (Ishikawa *et al.*, 2006). These overlapping roles may have diverged in *S. aureus*, since both genes are essential (Chaudhuri *et al.*, 2009). The YyaA protein, also called Noc for nucleoid occlusion protein in *B. subtilis*, is described later (Chaudhuri *et al.*, 2009).

Depletion of FtsZ in *S. aureus* cells is lethal resulting in inhibition of division, and an 8-fold increase in cell volume (Pinho and Errington, 2003). Interestingly, rather than preventing PG synthesis, FtsZ depletion results in improper localization of PG synthesis, so that it no longer occurs at the septum (Pinho and Errington, 2003). PBP2, the only PBP in *S. aureus* that contains both transglycosylase and transpeptidase domains (Goffin and Ghuysen, 1998), is mislocalized, but continues to synthesize cell wall at the periphery, resulting in accumulation of cell wall in confined spots (Pinho and Errington, 2003). Therefore, FtsZ is required for regulated PG synthesis in *S. aureus*. Adding back FtsZ does not revive FtsZ-depleted cells, suggesting that compounds inhibiting FtsZ are likely to be bactericidal (Pinho and Errington, 2003).

Division in ovococci

While cocci divide in either two or three alternating perpendicular planes, ovococci divide in the same plane, perpendicular to the longer axis of the cell like *E. coli* and *B. subtilis* (Fig. 3.4; Zapun *et al.*, 2008b). Ovococci cells synthesize PG mostly at the division site. Ovococci cells are surrounded in their middle by an annular outgrowth of PG (Fig. 3.4; Tomasz *et al.*, 1964; Higgins, 1970; Daniel and Errington, 2003; Ng *et al.*, 2004). The division septum forms immediately below this PG ring, splitting it into two rings that move apart leaving new cell wall to form in between (Fig. 3.4). Ovococci are said to have two types of division–specific PG synthesis: septal (centripetal) and peripheral (elongation between two old hemispheres; Zapun *et al.*, 2008a). Both are considered division-specific, with peripheral synthesis being distinct from elongation of rod-shaped bacteria (Zapun *et al.*, 2008a).

Like cocci, ovococci may have a common set of PBPs that are required for the two types of division-specific PG synthesis (Zapun *et al.*, 2008a). These include three class A HMW PBPs (PBP1a 1b, 2a), two class B HMW PBPs (PBP2b and 2x), and a LMW PBP (PBP3) with carboxypeptidase activity (Scheffers and Pinho, 2005). The FtsI homologue in *S. pneumonia* is PBP2x (Morlot *et al.*, 2003; Scheffers and Pinho, 2005). FtsZ-depletion studies have not been carried out in ovococci (Zapun *et al.*, 2008b). However, colocalization of all high-molecular-weight (HMW) PBPs with FtsZ at the onset of division is consistent with an FtsZ-triggered cell wall synthesis, as with *S. aureus* (Zapun *et al.*, 2008b). Also, zantrins, which are small molecules that perturb FtsZ function, inhibit division in *S. pneumoniae*, resulting in cell enlargement (Margalit *et al.*, 2004).

Table 3.2 Inhibitors of bacterial cell division

Compound	Example of structure[a]	Target organism tested	Target (mechanism of action)
2-Alkoxycarbonyl-aminopyridine/2-carbamoyl pteridine	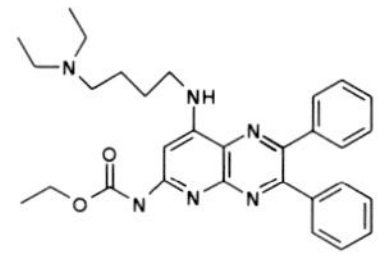	*M. tuberculosis*	FtsZ (inhibits polymerization and GTPase activity)
2-Phenyl indole/3-(2-indolyl) piperidine/ Carboxybiphenyl-indole	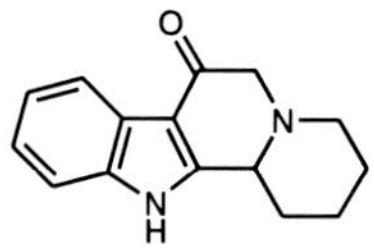	*S. aureus*, *S. pneumoniae*, *E. faecalis*, *E. coli* (imp mutant), *B. subtilis*, *H. influenzae*, *M. catarrhalis*, *C. albicans*	FtsZ–ZipA (inhibits protein–protein interaction)
534F6 derivatives	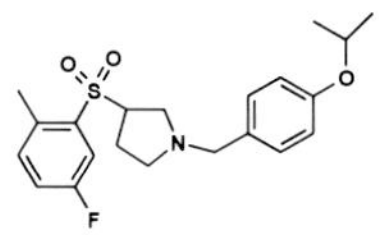	*E. coli* (efflux pump mutant)	FtsZ (inhibits GTPase activity)
A189	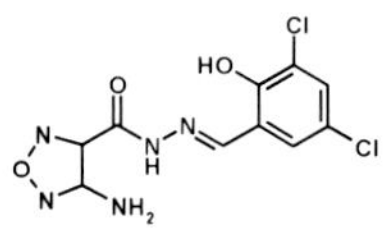	*E. coli* (efflux pump mutant), *S. aureus*	FtsZ (inhibits polymerization and GTPase activity)
Amikacin	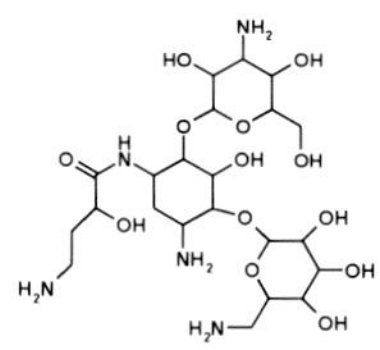	*E. coli*	FtsZ (inhibits proper Z-ring assembly)
Berberine	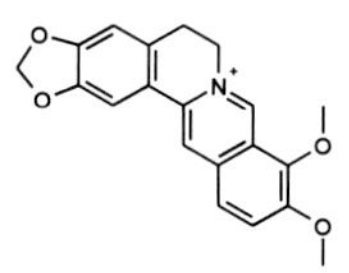	*E. coli*, *S. aureus*, *S. mutans*, *S. pyogenes*	FtsZ (inhibits polymerization and GTPase activity)
Cinnamaldehyde	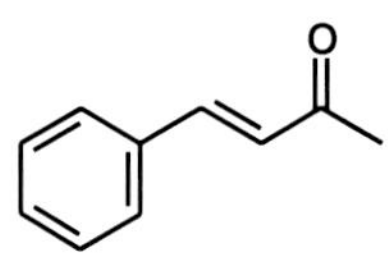	*E. coli*, *B. subtilis*, *S. aureus*	FtsZ (inhibits polymerization and GTPase activity)
Curcumin	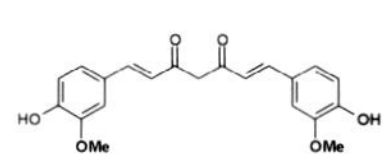	*B. subtilis*, *E. coli*	FtsZ (inhibits polymerization and increases GTPase activity)
Dichamanetin/2-hydroxy-5-benzylisouvarinol-B	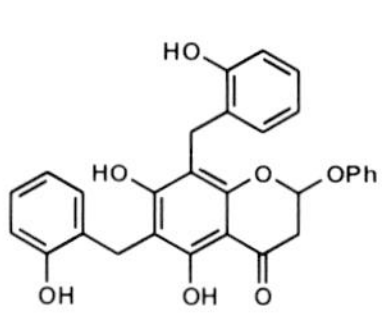	*S. aureus*, *B. subtilis*, *M. smegmatis*, *E. coli*, *P. aeruginosa*	FtsZ (inhibits GTPase activity)
GTP analogues – modification of nucleobase	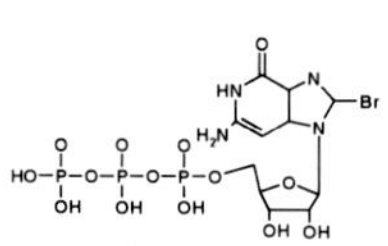	*E. coli (in vitro)*	FtsZ (inhibits polymerization and GTPase activity)

Target determination	Natural or synthetic/ semisynthetic	MIC[b]	Tubulin activity tested	Reference
In vitro polymerization and GTPase assay	Synthetic/ semisynthetic	0.15–6.25 μg/ml (0.3–12.5 μM)	Yes	White *et al.* (2002), Reynolds *et al.* (2004)
In vitro fluorescence polarization and NMR assays	Synthetic/ semisynthetic	4–64 μg/ml (8–128 μM)	No	Sutherland *et al.* (2003), Jennings *et al.* (2004)
In vitro GTPase	Synthetic/ semisynthetic	10–>80 μM	No	Mukherjee *et al.* (2007)
In vitro light scattering and GTPase assay	Synthetic/ semisynthetic	16–64 μg/ml (32–128 μM)	No	Ito *et al.* (2006)
In vivo Z-ring assembly	Synthetic/ semisynthetic	32 μg/ml (64 μM)	Not directly (Lietma, 1983)	Possoz *et al.* (2007)
In vitro light scattering and binding assays, *in vivo* Z-ring assembly	Natural (*Berberis* species)	31.3–125 μg/ml (62.6–250 μM)	Yes (Khan, 2010)	Villinski *et al.* (2003), Domadia *et al.* (2008)
In vitro light scattering, GTPase and binding assays, *in vivo* Z-ring assembly	Natural (*Cinnamonum cassia*)	250–1000 μg/ml (500–2000 μM)	No	Domadia *et al.* (2007)
In vitro light scattering, GTPase and binding assays, *in vivo* Z-ring assembly	Natural (*Curcuma longa*)	~100 μM	Yes (Thomas, 2008 #192)	Rai *et al.* (2008)
In vitro GTPase assay	Natural (*Uvaria. chamae* and *Xylopia afticana*)	1.7–15.4 μM	No	Urgaonkar *et al.* (2005)
In vitro light scattering and GTPase assay	Synthetic/semi-synthetic	N.D.	Yes	Lappchen *et al.* (2005, 2008)

Table 3.2 continued

Compound	Example of structure[a]	Target organism tested	Target (mechanism of action)
GTP analogues – GAL core (guanine moiety linked to Ala)	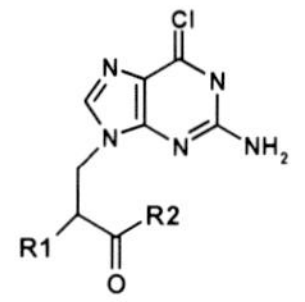	*E. coli*, *S. aureus*	FtsZ (inhibits GTPase activity)
MciZ	40 amino acid peptide	*B. subtilis*	FtsZ (inhibits polymerization and GTPase activity)
OTBA	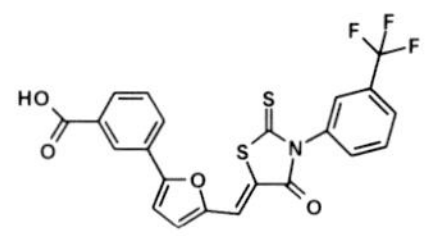	*B. subtilis*	FtsZ (enhanced stability of polymers and inhibits GTPase activity)
PC5538/PC17094	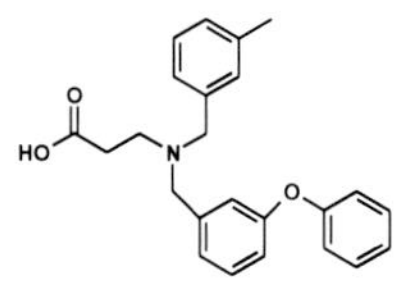	*B. subtilis*, *M. catarrhalis*, *S. aureus*	FtsZ (inhibits polymerization and GTPase activity)
PC19072/ alkoxybenzamide (PC19072 analogues)	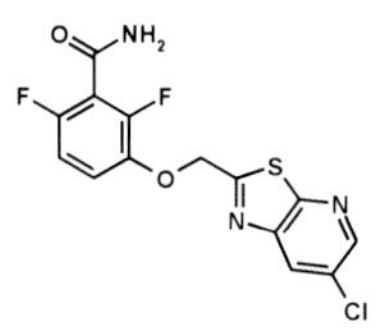	*B. subtilis*, *S. aureus*, other *Staphylococcus* sp.	FtsZ (inhibits polymerization and GTPase activity)
Sanguinarine	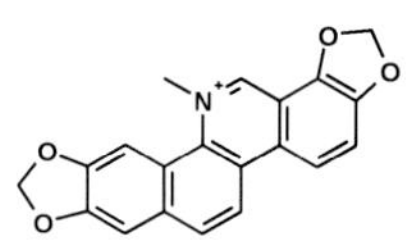	*B. subtilis*, *E. coli*	FtsZ (inhibits polymerization activity)
Taxanes	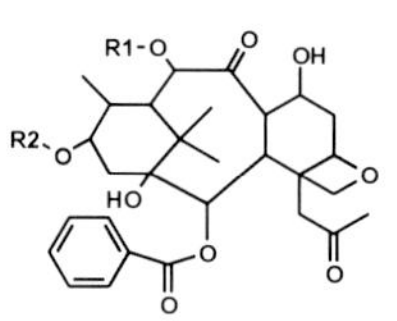	*M. tuberculosis*	FtsZ (stabilization against depolymerization)
Totarol	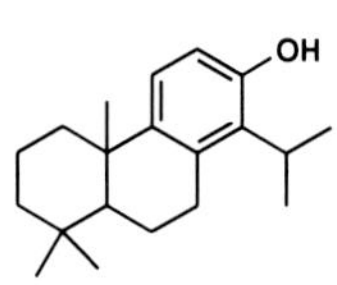	*B. subtilis*, *M. tuberculosis* (*in vitro*)	FtsZ (inhibits polymerization and GTPase activity)
Viriditoxin	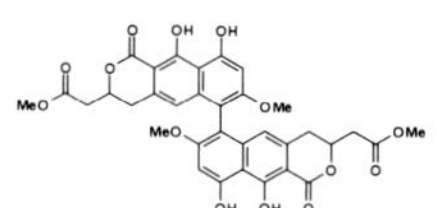	*S. aurues*, *E. faecium*, *E. faeclis*, *S. pneumoniae*, *E. coli* (hyperpermeable mutant)	FtsZ (inhibits polymerization and GTPase activity)
Zantrins	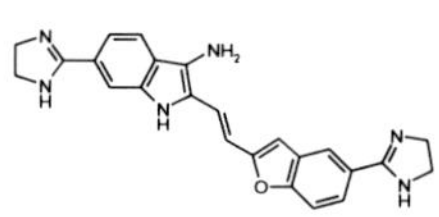	*E. coli*, *S. dysenteriae*, *B. subtilis*, *B. cereus*, *S. aureus*, *S. pneumoniae*, *C. perfringens*	FtsZ (some compounds destabilize, while others stabilize assembly)

[a]One example for each class of compound is shown. See references for additional structures.

[b]Note units. For minimal inhibitory concentrations (MICs) cited in μg/ml, the concentration was converted to μM by using a general molecular weight of 500 g/mol.

NA, not applicable; ND, not described.

Target determination	Natural or synthetic/ semisynthetic	MIC[b]	Tubulin activity tested	Reference
In vitro GTPase assay	Synthetic/semi-synthetic	N/A – Disc diffusion assay	No	Paradis-Bleau *et al.* (2007)
In vitro light scattering, GTPase and binding assays, *in vivo* Z-ring assembly	Natural (peptide expressed in *B. subtilis*)	N.D.	No	Handler *et al.* (2008)
In vitro light scattering, GTPase and binding assays, *in vivo* Z-ring assembly	Synthetic/semi-synthetic	2 μM	No	Beuria *et al.* (2009)
In vitro sedimentation and GTPase assays, *in vivo* FtsZ localization, MCS	Synthetic/semi-synthetic	8–64 μg/ml (16–128 μM)	No	Stokes *et al.* (2005)
In vitro GTPase assay, *in vivo* FtsZ localization	Synthetic/semi-synthetic	0.5–1 μg/ml (1–2 μM)	Yes	Haydon *et al.* (2008), Czaplewski *et al.* (2009)
In vitro light scattering, sedimentation and binding assays, *in vivo* Z-ring assembly	Natural (rhizomes of *Sanguinaria canadensis*)	10–75 μM	Yes (Wolff, 1993 #193)	Beuria *et al.* (2005)
In vitro light scattering	Synthetic/semi-synthetic	1.25–5 μg/ml (2.5–10 μM)	Yes (McGrogan, 2008)	Huang *et al.* (2006)
In vitro light scattering, GTPase and binding assays, *in vivo* Z-ring assembly	Natural (*Podocarpus totara*)	2 μM	No	Jaiswal *et al.* (2007)
In vitro polymerization and GTPase assay, MCS	Natural (*Aspergillius* species)	2–25 μg/ml (4–50 μM)	No	Wang *et al.* (2003)
In vitro polymerization assay, *in vivo* Z-ring assembly	Synthetic/semi-synthetic	0.321–20 μM	No	Margalit *et al.* (2004)

of FtsZ to eukaryotic tubulin potentially presents a significant hurdle to the advancement of FtsZ inhibitors. Given the structural similarities between the two proteins, parallel testing of FtsZ inhibitors for anti–tubulin activity and cytotoxicity to eukaryotic cells is vital for pursuing these compounds towards drug development. Accordingly, Table 3.2 highlights FtsZ inhibitors tested against tubulin. Given their potential utility as anti-cancer compounds, the molecules extracted from natural sources have been tested against eukaryotic tubulin. The same is not typically true of the synthetic compounds, although the reports of FtsZ inhibitors from White *et al.* (2002) and Reynolds *et al.* (2004) are exceptions to this. The structure activity relationship (SAR) studies presented in these reports provide interesting insights into the homology of FtsZ and tubulin from a drug development standpoint highlighting the potential to exploit FtsZ despite this homology. Compounds SRI-3072 and SRI-7614, despite having significant structural differences both appear to inhibit *M. tuberculosis* FtsZ to similar degrees. However, SRI-7614 inhibits bovine brain tubulin polymerization with an ID_{50} of 4 μM but SRI-3072 does not inhibit polymerization (up to 100 μM; White *et al.*, 2002). SAR data presented by Reynolds *et al.* (2004) are even more striking. This report compares two molecules that differ by only a single ring nitrogen, but the results are akin to those found with SRI-3072 and SRI-7614. Although many more studies are needed, this suggests that there is real promise in finding FtsZ specific inhibitors despite its structural similarity with tubulin.

There are several key parameters used to test the *in vitro* effectiveness of an antibacterial compound, and the minimal inhibitory concentration (MIC) for death is one typical parameter used. In this respect, the FtsZ inhibitors identified display a significant range in potency (Table 3.2). Although the MIC is not necessarily a useful determinant for a successful clinical drug, given the importance of other factors including pharmacokinetics, it provides a useful starting point to test drug effectiveness. Compounds with low MICs will require lower concentrations to kill the bacteria, consequently requiring a lower concentration of compound to reach the site of infection (comparatively, the typical MIC for ampicillin against penicillinase-sensitive *S. aureus* is 0.1–0.5 μg/ml; Harrison and Cropp, 1965; Zygmunt *et al.*, 1968; Wehrli *et al.*, 1983). In addition to the MIC, organism specificity for the compound is also a key feature. As discussed, FtsZ is broadly conserved, with high homology, among both Gram-positive and Gram-negative bacteria. Therefore, it is expected that inhibitors targeting FtsZ have the capacity for a broad spectrum of antibacterial activity. Surveying the organisms targeted in Table 3.2, the vast majority of the compounds were tested against Gram-positive species. Where these were tested against *E. coli*, a mutant with a compromised outer membrane was often used suggesting that the outer membrane is an important determinant for drug access. Outer membrane permeability is a significant challenge in designing broad spectrum antibiotics and highlights the innate capacity for resistance among many Gram-negative bacteria (Hancock, 1997). Thus, broad-spectrum inhibitors are often very difficult to find.

In vivo efficacy plays a strategic role in assaying lead compounds for potential anti-infectives. Only one FtsZ inhibitor has been reported as being tested in a model of infection, namely PC19072 (Haydon *et al.*, 2008). Using a murine model, Haydon *et al.*, showed that PC19072 is able to completely cure a lethal dose of *S. aureus* infection at 30 mg/kg body weight and allow 50% survival at 10 mg/kg. In this experiment, vancomycin at 3 mg/kg was used as a control with results comparable to PC19072 at a concentration 10-fold higher. Additional drug development may be necessary to reduce the effective concentration, but these infection models provide great promise for this compound as a new therapeutic.

With the exception of one, the molecules identified as inhibitors of FtsZ are an array of small chemical compounds. Recently, using a yeast two-hybrid screen with a *B. subtilis* genomic library, Handler *et al.* (2008) identified a small peptide, MciZ, which inhibits FtsZ. MciZ shares significant homology to a macrophage produced antimicrobial peptide CRAMP (Rosenberger *et al.*, 2004), and is expressed in the mother cell during sporulation, where it has been shown to inhibit Z ring formation. In addition, the recombinant 40-amino-acid peptide was shown *in vitro* to inhibit the GTPase activity and polymerization of FtsZ. This recent work raises a number of interesting questions, not the least of which is the potential for use of antimicrobial peptides as a therapeutic strategy to inhibit division.

The compounds identified as inhibitors of FtsZ (Table 3.2) reflect a diverse set of structures. The exploitation of bacterial division, specifically FtsZ, as a druggable target is still in its infancy with most of the studies arising in the last five to six years. Furthermore, the division machinery comprises many proteins that are extracellular, with several of them probably being promising targets. In recent years, tangible success has been seen in examining only one of these proteins. Thus there remains a large number of division proteins completely unexplored for chemical inhibition offering real promise for bacterial division as a new and emerging therapeutic alternative.

New approaches for antibacterial discovery targeting division

Inhibition of division as a pathogen survival strategy

While cell division is vital for viability, colonization and infection, evidence is now emerging that suggests the inhibition of division, also known as filamentation, actually plays a vital role in pathogenesis and virulence in many pathogenic species, both Gram-negative and Gram-positive (Justice *et al.*, 2008). Furthermore, a novel mechanism involving filamentation has been described to allow bacteria to survive antimicrobial lethality (Miller *et al.*, 2004). An understanding of these cell division control mechanisms, particularly in *in vivo* situations is likely to provide novel and improved anti-infective strategies.

In some cases this type of filamentation occurs via the SOS response, and division is arrested at the level of Z ring assembly. The SOS response is induced in response to DNA damage (and other stresses) and temporarily inhibits cell division until the damage is repaired (Huisman, 1981). In *E. coli* the SOS-induced protein, SulA, inhibits Z ring formation (Cordell, 2003). Uropathogenic *E. coli* (UPEC), the predominant cause of urinary tract infections, is a well-studied example. During acute infection of the mammalian bladder, filamentous bacteria form on the luminal surface of the bladder where they evade neutrophil phagocytosis (Justice *et al.*, 2004). Shorter, bacillary forms are killed. The protein that mediates UPEC filamentation, the SOS-response protein SulA, was found to be essential for pathogenesis in immunocompetent hosts (Justice *et al.*, 2006). Moreover, mouse hosts that are defective in innate immune system recognition do not trigger filamentation of UPEC (Justice *et al.*, 2004), and SulA-mediated filamentation is dispensable for the establishment of urinary tract infections in these mice (Justice *et al.*, 2006). These data demonstrate that filamentation is a vital part of UPEC survival during the infection process.

Various observations, both *in vivo* and *in vitro*, have identified several other examples of filamentation during pathogenesis to increase survival of the pathogen. Filamentation in *in vitro* models of infection include *M. tuberculosis* (see above; Chauhan *et al.*, 2006b), *S. typhimurium* (Rosenberger and Finlay, 2002; Henry *et al.*, 2005), *Legionella* spp. (aetiological agent of Legionnaires' disease; Piao *et al.*, 2006), *Shigella flexneri* (causing enteritis; Lucchini *et al.*, 2005) and *Burkholderia pseudomallei* (causing melioidosis; Jones *et al.*, 1996).

In vivo studies include (i) *Haemophilus influenzae* during otitis media, where filamentous forms have been suggested to provide protection from neutrophil phagocytosis (Leroy *et al.*, 2007); (ii) the opportunistic urinary pathogen, *Proteus mirabilis*, which forms elongated, filamentous cells in response to contact with solid surfaces, such as catheters (Allison *et al.*, 1992). This is associated with increased virulence and degradation of mammalian antibacterial peptides (Belas *et al.*, 2004); and (iii) *S. enterica* serovar Typhimurium growing in murine fibroblast cells (Martinez-Moya *et al.*, 1998). In this case the filamentous cells are blocked in division at a stage after Z ring formation, possibly at FtsI function, and a defect in the histidine biosynthetic pathway is correlated with the observed filamentation phenotype (Henry *et al.*, 2005).

Targeting division regulation

The earliest observable stage in cell division in bacteria is the formation of the Z ring precisely at the division site. It is this stage that commits the bacterial cell to divide. This process must therefore be tightly regulated in time and space to ensure equal partitioning of DNA into newborn cells, and thus viability. Another way to target cell division then is to inhibit proteins that regulate Z ring formation.

In rod-shaped bacteria, such as *E. coli* and *B. subtilis*, two factors known to be involved in the spatial control of division are nucleoid occlusion (NO) and the Min system (Goehring and Beckwith, 2005; Dajkovic and Lutkenhaus, 2006; Harry *et al.*, 2006; Lutkenhaus, 2007; Adams and Errington, 2009). These systems prevent Z rings forming in all positions in the cell except the division site at midcell. Although not essential on their own, absence of both is synthetically lethal because of the inability to form functional Z rings (Wu and Errington, 2004; Bernhardt and de Boer, 2005).

While Min is present in many bacteria, it is notably absent from some important pathogens. This includes most cocci and ovococci including staphylococci, enterococci and streptococci, and mycobacteria (Margolin, 2000; Hett and Rubin, 2008; Zapun *et al.*, 2008b). While DivIVA is present in *S. aureus*, *S. pneumoniae*, *Enterococcus faecalis*, *Streptomyces*, and also mycobacteria, current evidence and the apparent absence of MinC homologues in these organisms (the negative effector of the Min system) suggest a morphogenesis role for this protein, rather than a division site placement function (Allison *et al.*, 1992; Hermans *et al.*, 1995; Fadda *et al.*, 2003; Flardh, 2003; Sassetti *et al.*, 2003; Pinho and Errington, 2004).

In ovococci, nucleoid occlusion also does not seem to operate, since the division site is located precisely around the nucleoid (Zapun *et al.*, 2008b). However, a Noc homologue can be identified in *S. aureus* that shows 48% amino acid identity to the *B. subtilis* Noc protein. Importantly this protein is likely to be essential (Chaudhuri *et al.*, 2009). Determining the mechanism of action of Noc in this organism could identify a way to maintain nucleoid occlusion in this organism and prevent Z ring formation, leading to cell death.

How is Z ring formation regulated in time and space for cells that have no Min system or nucleoid occlusion? Indeed even when both the Min and NO systems are inactivated in *E. coli* and *B. subtilis*, Z ring positioning is still biased towards cell division sites (Wu and Errington, 2004; Bernhardt and de Boer, 2005). Furthermore, when FtsZ is overproduced in *minD noc* double mutants of *B. subtilis*, a substantial number of Z rings form at internucleoid positions (Wu and Errington, 2004). These observations strongly suggest that mechanisms in addition to nucleoid occlusion and Min regulate Z ring formation even in these rod-shaped bacteria.

Other more recently identified temporal and spatial regulators of Z ring formation in these model bacteria include metabolic enzymes (as growth rate regulators; Weart *et al.*, 2007; Wang and Levin, 2009) chromosome replication (Harry *et al.*, 1999; Regamey *et al.*, 2000; Moriya *et al.*, 2001), and localized membrane lipid domains including cardiolipin and phosphatidylethanolamine (Mileykovskaya and Dowhan, 2000; Koppelman *et al.*, 2001; Kawai *et al.*, 2004). It is probable that further investigation of cell division regulation in pathogens that do not have Min and/or NO systems will identify candidate antibacterial targets. Relatively recent intensive research in mycobacterium growth and division has highlighted some of these mechanisms, which may turn out to be well conserved in prokaryotes (Hett and Rubin, 2008), and are discussed below.

Regulation of growth and cell division in mycobacteria: a special case or not?

The importance of cell division in infection, persistence and reactivation of TB means that molecules that regulate division are potential antitubercular targets. The apparent absence of FtsZ accessory proteins, such as FtsA, ZapA and EzrA, and of nucleoid occlusion and the Min system in actinomycetes indicates the presence of other key division regulators, some of which may be conserved among a wide range of bacteria. These may include proteins that affect Z ring formation or septum formation, such as PG hydrolases and PknA, PknB and WhiB.

PG hydrolases as regulators of cell division

Several hydrolases are also present in *M. tuberculosis* some of which must play a role in septum splitting (Hett and Rubin, 2008). One of these, Rv2719c, has been implicated in division control at the level of Z ring assembly (Chauhan *et al.*, 2006a). It shows limited homology to *B. subtilis* YneA protein (the SOS response protein that leads to division inhibition; Chauhan *et al.*, 2006a). The up-regulation of Rv2719c that occurs during growth inside macrophages and in response to DNA damage or β-lactams results in aberrant Z ring formation. The effect of Rv2719c appears to be indirect, and is proposed to occur via a change in the balance of PG synthesis and hydrolase activities (Chauhan *et al.*, 2006a). It would be worthwhile investigating the molecular details of this balance as it is likely to be important for infection and perhaps persistence.

PG hydrolases are redundant and since bacteria can usually survive *in vitro* in chains, these proteins tend not to be essential. Surprisingly therefore RipA (Rpf-interacting protein) is an essential PG endopeptidase identified in *M. smegmatis* (Sassetti *et al.*, 2003). A *ripA* deletion also results in increased susceptibility to β-lactams (Hett *et al.*, 2008). RipA interacts with two resuscitation promotion proteins (Rpfs), RpfB and RpfE; and both RipA and RpfB are localized to the division septum in *M. smegmatis* (Hett *et al.*, 2008). Rpfs are a family of lytic transglycosylases that hydrolyse PG (Mukamolova *et al.*, 2002). They have been strongly implicated in the reactivation of dormant *M. tuberculosis* to resume division *in vivo* (Downing *et al.*, 2005; Tufariello *et al.*, 2006). The role of Rpfs in latent tuberculosis has yet to be confirmed, however it has been hypothesized hydrolases will be required to nick PG allowing growth out of its latent state (Hett and Rubin, 2008). RipA is a potential drug target, perhaps ideally suited to combination therapies with other cell wall drugs. Furthermore, identifying what cell wall changes need to occur for reactivation and resumption of cell division, and how RipA might enable this, will provide further insights for intervention of this key event in TB disease.

PknA and PknB

Both PknA and PknB appear to be regulators of cell division. These proteins are members of a family of serine/threonine protein kinases (STPKs), present in both eukaryotes and prokaryotes. STPKs are sensors of environmental signals important for bacterial pathogenesis and development (Narayan *et al.*, 2007). *pknA* and *pknB*, are part of an operon which also includes the cognate phosphatase, PstP, and the morphogenic genes, *rodA* (involved in cell shape) and *pbpA* (a PG-synthesizing enzyme which localizes to the septum, and is important for cell division; Boitel *et al.*, 2003; Chopra *et al.*, 2003). Increased or decreased levels of both PknA and PknB results in division defects (Kang *et al.*, 2005). *In vivo* targets identified for PknA and PknB, are PknB itself, DivIVA (Wag31) and Rv1422 (a conserved hypothetical protein of unknown function; Kang *et al.*, 2005). PknB also phosphorylates PBPA (Kang *et al.*, 2005), while PstP dephosphorylates PBPA (Dasgupta *et al.*, 2006). It has been proposed that PknB and PstP regulate the localization and activity of PBPA, thereby regulating septation (Kang *et al.*, 2005). Available evidence suggests that PknA regulates division by interfering with FtsZ function, including the overexpression of PknA in *E. coli* cells that causes them to elongate more than usual (Chaba *et al.*, 2002). PknA interacts with, and phosphorylates, FtsZ in *M. tuberculosis*, reducing its GTPase activity and its ability to polymerize *in vitro* (Thakur and Chakraborti, 2006). The details for this division regulation pathway will no doubt provide avenues for antibacterial discovery in TB and other prokaryotes where this signal transduction pathway is conserved. The extracellular kinase domain PknB would be more accessible to drugs than PknA (Jones and Dyson, 2006).

WhiB2

A growing body of evidence suggests that the WhiB-like proteins exclusive to the GC-rich actinomycete genera play significant roles in pathogenesis and cell division (Hett and Rubin,

2008). The function of these proteins is not well understood, but they are postulated to be transcription factors and to play a role in intracellular redox sensing (Davis and Chater, 1992; Raghunand and Bishai, 2006; Alam *et al.*, 2007). *S. coelicolor* WhiB mutants fail to assemble FtsZ rings in their aerial hyphae, leading to the arrest of hyphal development before cross-wall formation (Schwedock *et al.*, 1997). *M. tuberculosis* has seven WhiB proteins (WhiB1–7; Cole *et al.*, 1998; Soliveri *et al.*, 2000). Addition of cell wall inhibitors to *M. tuberculosis* results in a specific increase in WhiB2 suggesting a role for this protein in PG remodelling and cell division (Geiman *et al.*, 2006). The homologue of WhiB2 in *M. smegmatis*, WhmD, is essential and changes in intracellular levels of WhmD suggest an activation role for this protein specifically in the formation and maturation of the division septum (Gomez and Bishai, 2000). FtsZ levels were unaltered, so presumably Z ring formation is defective in this situation as is the case with *S. coelicolor* (Gomez and Bishai, 2000). It will be interesting to see what septum formation genes might be regulated by WhiB2 in *M. tuberculosis*.

Targeting protein–protein interactions

Despite recent advances in peptide mimetics, a significant limitation as a therapy is their limited utility in reaching intracellular targets (Torchilin, 2006, 2008). Intracellular protein–protein interactions are of course the largest pool to draw from for therapeutic intervention. For this reason, conventional small molecule inhibitors have found more application in targeting protein–protein interactions despite having their own particular challenges (see Cochran, 2001; Toogood, 2002; Berg, 2003; Gadek and Nicholas, 2003; Arkin and Wells, 2004; Fry, 2006; Wells and McClendon, 2007). However, protein–protein interactions represent an enormous, yet largely unexploited class of therapeutic targets. This interaction landscape is massive and represents crucial components in biological processes (Butland *et al.*, 2005; Krogan *et al.*, 2006). Significant success in targeting peptide–protein interactions has been made through the identification of peptide fragments mimicking natural peptide substrates. By and large, these studies have centred on the extracellular domains of membrane bound signalling receptors (Livnah *et al.*, 1996; Wrighton *et al.*, 1996; Yanofsky *et al.*, 1996; Cwirla *et al.*, 1997). Peptide mimetics could well be a promising avenue for inhibiting the myriad of protein interactions that occur in the divisome as many of the interacting domains are essential, and their cellular location makes targeting both accessible and unaffected by drug-efflux pumps. The conserved set of complex interactions between FtsL, FtsB(DivIC) and FtsQ(DivIB) would be a good starting point.

Although more challenging than drug targets that naturally bind small molecules, significant progress in the discovery and development of small-molecule inhibitors of protein–protein interactions has been made through several strategies, including fragment discovery approaches and phage display (Shuker *et al.*, 1996; Boehm *et al.*, 2000; Toogood, 2002; Sidhu, 2003; Arkin and Wells, 2004). These strategies could be applied to target interactions between FtsZ and the various proteins (e.g. FtsA, ZipA, MinC) that modulate its polymerization. Since many of these modulators interact with the widely conserved C-terminal domain of FtsZ, inhibiting interaction with one binding partner is likely to inhibit binding of others to FtsZ, which is lethal (see earlier). More importantly, mutagenesis experiments suggest that ZipA and FtsA interact differently with the C-terminal tail of FtsZ (Haney *et al.*, 2001). This means that inhibitors designed to prevent the essential FtsA–FtsZ interaction, may attain resistance at a much slower rate, since single mutations that suppress the FtsA–FtsZ interaction may interfere with other proteins interacting with the same (or overlapping) residues of FtsZ.

The methodology to identify inhibitors of protein–protein interactions is relatively well defined (Toogood, 2002). From the myriad of approaches, one that may be particularly valuable in the identification of division inhibitors is the use of fluorescent-based approaches. Fluorescent protein tagging has played a considerable role in the functional assignment of division machinery proteins (Harry *et al.*, 2006) and remains a promising method for high-throughput small molecule lead generation.

The key to success for targeting protein–protein interactions will depend on whether the chemical nature of the individual interacting sites enables specific and potent inhibition by small molecules or peptides. Whether targeting protein–protein interactions will, in some instances, be less susceptible to the attainment of resistance by mutation is hard to predict at this early stage.

Target-based approaches for new penicillin-binding protein inhibitors

Penicillin binding proteins (PBPs) have played a significant role as antimicrobials since the discovery of penicillin. The ability to inhibit PBP activity has drastic effects on the growth and division of nearly all bacterial species. However, with the rise in resistance to the therapeutic activity of penicillin and penicillin-like molecules, the utility of targeting PBPs with penicillin-like chemistry is in rapid decline. A renewed hope has emerged with the crystallization of a bifunctional PBP, PBP2 from *S. aureus*, presenting the first crystal structure of the transglycosylase domain of any member of this class of proteins (Lovering *et al.*, 2007). Until this study, several PBP structures were reported from a variety of organisms, however given its importance in binding β-lactam antibiotics, these studies were limited to the transpeptidase domain (Contreras-Martel *et al.*, 2006; Sauvage *et al.*, 2007; Urbach *et al.*, 2009). The advancement in this area to include the transglycosylase domain has opened the door to the possibilities for structure-based drug design to this unexploited domain. The structure of PBP2 in complex with moenomycin has also been determined (Lovering *et al.*, 2007). Moenomycin is currently used in animal health (Ostach and Walker, 2005) and is one of the very few antibiotics thus far that target the transglycosylase reaction in peptidoglycan biosynthesis (Wright 2007). In short order, bifunctional PBPs from two additional strains have been solved, PBP1A from *Aquifex aeolicus* (Yuan *et al.*, 2007) and PBP1B from *E. coli* (Sung *et al.*, 2009). These new crystal structures will provide new tools and prospects for the renewed inhibition of well-validated and successful targets with a new chemical class of antibiotics for the treatment of human infection.

Lipoteichoic acid: a potential new component of bacterial cell division

Lipoteichoic acid (LTA) polymers are lipid bound exopolysaccharides extending into the cell wall matrix. These polymers are highly conserved among Gram-positive organisms and are composed of repeating glycerol phosphate units covalently attached to a glycolipid moiety in the cell membrane (Fischer, 1988). Not until recently have the enzymes involved in polymer synthesis been identified (Grundling and Schneewind, 2007a,b). Allelic replacement studies in *S. aureus* (Grundling and Schneewind, 2007b), *B. subtilis* (Schirner *et al.*, 2009) and *Listeria monocytogenes* (Webb *et al.*, 2009) have demonstrated profound defects in septum formation, suggesting a role for LTA in cell division. Furthermore, using microscopic examination, Schirner *et al.* (2009) have shown that deletion of a homologue in *B. subtilis* impacts Z ring formation. LTA therefore represents an emerging target for Gram-positive division that awaits further understanding of its specific role and prospects for therapeutic exploitation.

Essential genes: *in vitro* versus *in vivo* dispensability

The most basic requirement for selecting bacterial protein targets for therapeutic intervention is that inhibition of the protein is lethal to the cell. This property is typically investigated through gene dispensability studies. These studies are usually limited to conditions requiring growth in the laboratory setting. Furthermore, as the studies of essential genes have developed to include multiple gene deletions in the same background, the genomic context under study has been found to be important in defining an essential gene. This is probably because of the complexity of biology and gene essentiality may not mean much without understanding what underlies the essential phenotype (D'Elia *et al.*, 2009). With respect to antimicrobial interest, these studies are disconnected from the desired outcome, given that pathogenesis requires survival in a host organism. Therefore, the important determinant is gene essentiality *in vivo*.

Within the protein complement that comprises the division machinery several proteins are considered dispensable for growth (Lock and Harry, 2008). However, studies concerning the virulence potential of division mutants are almost non-existent. In *M. tuberculosis* for example, several genes must be intact so that bacteria can survive during infection in a mouse but not for growth in defined medium (Wei JR, 2008). There is considerable understanding of otherwise dispensable genes being required for pathogenesis, especially amongst pathogenicity island genes (Censini *et al.*, 1996; Ochman *et al.*, 1996; Lindsay *et al.*, 1998). Therefore, although genes essential in the laboratory setting remain attractive targets for drug development, genes that have been described as non-essential in the context of laboratory conditions should not be overlooked in this respect.

Conclusions and challenges for the future

As a potential target for novel antibacterial therapies, the bacterial cell division pathway appears to tick all the right boxes. Not only is division essential for survival and infection, but it is controlled by a group of mostly well-characterized and essential proteins, some of which are clearly amenable to drug targeting. The appeal of cell division as an antibacterial target is not merely theoretical, however, as real inroads have now been made towards exploiting this pathway for drug development.

There is of course still a long way to go. At the basic research level, there is a great deal yet to uncover about the molecular mechanism of bacterial division. The precise functions of many of the division proteins are still unknown, and it is unclear exactly how they act in concert to synthesize the septum and regulate this process after assembling into a ring at the division site. These proteins may perform structural roles, or could perhaps exhibit enzymatic activity and thus provide a cleft for potent and specific inhibition by small molecules. An intimate understanding of how division occurs in model bacteria, and how it differs in important pathogens, will place us in an excellent position to develop new and effective cell division inhibitors using target-driven approaches.

Importantly, the vast majority of known division proteins are essential for the growth, and thus pathogenesis, of bacteria. In addition, some genes that are dispensable for vegetative cell division under standard laboratory conditions can become essential under specific conditions of stress encountered within a host. This appears to be the case for resuscitation-promoting factors, or Rpfs, present in a wide variety of bacteria including mycobacteria. Efforts to characterize the division process and the proteins required during pathogenesis could uncover many novel antibacterial targets that are not apparent from traditional *in vitro* studies.

Ultimately, cell division must be tightly regulated and intimately coordinated with other events in the bacterial cell cycle to ensure the continued viability of the organism. As research starts to move from studies on single proteins towards global systems biology approaches, we will begin to understand how different signals and pathways are integrated within the bacterium, and how this influences the cell's decision about when and where to divide. As well as uncovering information on the basic biology of bacteria, this will reveal the most effective points of intervention to stop them in their tracks.

Acknowledgements

We thank Eric Brown for critical reading of this chapter.

References

Aaron, M., Charbon, G., Lam, H., Schwarz, H., Vollmer, W., and Jacobs-Wagner, C. (2007). The tubulin homologue FtsZ contributes to cell elongation by guiding cell wall precursor synthesis in *Caulobacter crescentus*. Mol. Microbiol. *64*, 938–952.

Aarsman, M.E.G., Piette, A., Fraipont, C., Vinkenvleugel, T.M.F., Nguyen-Disteche, M., and den Blaauwen, T. (2005). Maturation of the *Escherichia coli* divisome occurs in two steps. Mol. Microbiol. *55*, 1631–1645.

Adams, D.W., and Errington, J. (2009). Bacterial cell division: assembly, maintenance and disassembly of the Z ring. Nat. Rev. Microbiol. *7*, 642–653.

Alam, M.S., Garg, S.K., and Agrawal, P. (2007). Molecular function of WhiB4/Rv3681c of *Mycobacterium tuberculosis* H37Rv: a [4Fe–4S] cluster co-ordinating protein disulphide reductase. Mol. Microbiol. *63*, 1414–1431.

Allison, C., Coleman, N., Jones, P.L., and Hughes, C. (1992). Ability of *Proteus mirabilis* to invade human urothelial cells is coupled to motility and swarming differentiation. Infect. Immun. *60*, 4740–4746.

Anderson, D.E., Gueiros-Filho, F.J., and Erickson, H.P. (2004). Assembly dynamics of FtsZ rings in *Bacillus subtilis* and *Escherichia coli* and effects of FtsZ-regulating proteins. J. Bacteriol. *186*, 5775–5781.

Arkin, M.R., and Wells, J.A. (2004). Small-molecule inhibitors of protein–protein interactions: progressing towards the dream. Nat. Rev. Drug. Discov. *3*, 301–317.

Beall, B., and Lutkenhaus, J. (1989). Nucleotide sequence and insertional inactivation of a *Bacillus subtilis* gene that affects cell division, sporulation, and temperature sensitivity. J. Bacteriol. *171*, 6821–6834.

Beall, B., and Lutkenhaus, J. (1992). Impaired cell division and sporulation of a *Bacillus subtilis* strain with the *ftsA* gene deleted. J. Bacteriol. *174*, 2398–2403.

Belas, R., Manos, J., and Suvanasuthi, R. (2004). *Proteus mirabilis* ZapA metalloprotease degrades a broad spectrum of substrates, including antimicrobial peptides. Infect. Immun. *72*, 5159–5167.

Ben-Yehuda, S., and Losick, R. (2002). Asymmetric cell division in *B. subtilis* involves a spiral-like intermediate of the cytokinetic protein FtsZ. Cell *109*, 257–266.

Berg, T. (2003). Modulation of protein–protein interactions with small organic molecules. Angew. Chem. Int. Ed. Engl. *42*, 2462–2481.

Bernhardt, T.G., and de Boer, P.A. (2004). Screening for synthetic lethal mutants in *Escherichia coli* and identification of EnvC (YibP) as a periplasmic septal ring factor with murein hydrolase activity. Mol. Microbiol. *52*, 1255–1269.

Bernhardt, T.G., and de Boer, P.A. (2005). SlmA, a nucleoid-associated, FtsZ binding protein required for blocking septal ring assembly over chromosomes in *E. coli*. Mol. Cell. *18*, 555–564.

Beuria, T.K., Santra, M.K., and Panda, D. (2005). Sanguinarine blocks cytokinesis in bacteria by inhibiting FtsZ assembly and bundling. Biochemistry *44*, 16584–16593.

Beuria, T.K., Mullapudi, S., Mileykovskaya, E., Sadasivam, M., Dowhan, W., and Margolin, W. (2009a). Adenine nucleotide-dependent regulation of assembly of bacterial tubulin-like FtsZ by a hypermorph ofbacterial actin-like FtsA. J. Biol. Chem. *284*, 14079–14086.

Beuria, T.K., Singh, P., Surolia, A., and Panda, D. (2009b). Promoting assembly and bundling of FtsZ as a strategy to inhibit bacterial cell division: a new approach for developing novel antibacterial drugs. Biochem. J. *423*, 61–69.

Bi, E.F., and Lutkenhaus, J. (1991). FtsZ ring structure associated with division in *Escherichia coli*. Nature *354*, 161–164.

Bi, E., Dai, K., Subbarao, S., Beall, B., and Lutkenhaus, J. (1991). FtsZ and cell division. Res. Microbiol. *142*, 249–252.

Biller, S.J., and Burkholder, W.F. (2009). The *Bacillus subtilis* SftA (YtpS) and SpoIIIE DNA translocases play distinct roles in growing cells to ensure faithful chromosome partitioning. Mol. Microbiol. *74*, 790–809.

den Blaauwen, T., de Pedro, M.A., Nguyen-Disteche, M., and Ayala, J.A. (2008). Morphogenesis of rod-shaped sacculi. FEMS Microbiol. Rev. *32*, 321–344.

Boehm, H.J., Boehringer, M., Bur, D., Gmuender, H., Huber, W., Klaus, W., Kostrewa, D., Kuehne, H., Luebbers, T., Meunier-Keller, N., *et al.* (2000). Novel inhibitors of DNA gyrase: 3-D structure based biased needle screening, hit validation by biophysical methods, and 3-D guided optimization. A promising alternative to random screening. J. Med. Chem. *43*, 2664–2674.

de Boer, P., Crossley, R., and Rothfield, L. (1992). The essential bacterial cell-division protein FtsZ is a GTPase. Nature *359*, 254–256.

Boitel, B., Ortiz-Lombardia, M., Duran, R., Pompeo, F., Cole, S.T., Cervenansky, C., and Alzari, P.M. (2003). PknB kinase activity is regulated by phosphorylation in two Thr residues and dephosphorylation by PstP, the cognate phospho-Ser/Thr phosphatase, in *Mycobacterium tuberculosis*. Mol. Microbiol. *49*, 1493–1508.

Bork, P., Sander, C., and Valencia, A. (1992). An ATPase domain common to prokaryotic cell cycle proteins, sugar kinases, actin, and hsp70 heat shock proteins. Proc. Natl. Acad. Sci. U.S.A. *89*, 7290–7294.

Botta, G.A., and Park, J.T. (1981). Evidence for involvement of penicillin-binding protein 3 in murein synthesis during septation but not during cell elongation. J. Bacteriol. *145*, 333–340.

Bowler, L.D., and Spratt, B.G. (1989). Membrane topology of penicillin-binding protein 3 of *Escherichia coli*. Mol. Microbiol. *3*, 1277–1286.

Boyle, D.S., Khattar, M.M., Addinall, S.G., Lutkenhaus, J., and Donachie, W.D. (1997). *ftsW* is an essential cell-division gene in *Escherichia coli*. Mol. Microbiol. *24*, 1263–1273.

Bramhill, D., and Thompson, C.M. (1994). GTP-dependent polymerization of *Escherichia coli* FtsZ protein to form tubules. Proc. Natl. Acad. Sci. U.S.A. *91*, 5813–5817.

Buddelmeijer, N., Judson, N., Boyd, D., Mekalanos, J.J., and Beckwith, J. (2002). YgbQ, a cell division protein in *Escherichia coli* and *Vibrio cholerae*, localizes in codependent fashion with FtsL to the division site. Proc. Natl. Acad. Sci. U.S.A. *99*, 6316–6321.

Buddelmeijer, N., and Beckwith, J. (2004). A complex of the *Escherichia coli* cell division proteins FtsL, FtsB and FtsQ forms independently of its localization to the septal region. Mol. Microbiol. *52*, 1315–1327.

Butland, G., Peregrin-Alvarez, J.M., Li, J., Yang, W., Yang, X., Canadien, V., Starostine, A., Richards, D., Beattie, B., Krogan, N., *et al.* (2005). Interaction network containing conserved and essential protein complexes in *Escherichia coli*. Nature *433*, 531–537.

Censini, S., Lange, C., Xiang, Z., Crabtree, J.E., Ghiara, P., Borodovsky, M., Rappuoli, R., and Covacci, A. (1996). *cag*, a pathogenicity island of *Helicobacter pylori*, encodes type I-specific and disease-associated virulence factors. Proc. Natl. Acad. Sci. U.S.A. 93, 14648–14653.

Chaba, R., Raje, M., and Chakraborti, P.K. (2002). Evidence that a eukaryotic-type serine/threonine protein kinase from *Mycobacterium tuberculosis* regulates morphological changes associated with cell division. Eur. J. Biochem. *269*, 1078–1085.

Chaudhuri, R.R., Allen, A.G., Owen, P.J., Shalom, G., Stone, K., Harrison, M., Burgis, T.A., Lockyer, M., Garcia-Lara, J., Foster, S.J., *et al.* (2009). Comprehensive identification of essential *Staphylococcus aureus* genes using transposon-mediated differential hybridisation (TMDH). BMC Genomics *10*, 291.

Chauhan, A., Lofton, H., Maloney, E., Moore, J., Fol, M., Madiraju, M.V., and Rajagopalan, M. (2006a). Interference of *Mycobacterium tuberculosis* cell division by Rv2719c, a cell wall hydrolase. Mol. Microbiol. *62*, 132–147.

Chauhan, A., Madiraju, M.V., Fol, M., Lofton, H., Maloney, E., Reynolds, R., and Rajagopalan, M. (2006b). *Mycobacterium tuberculosis* cells growing in macrophages are filamentous and deficient in FtsZ rings. J. Bacteriol. *188*, 1856–1865.

Chopra, P., Singh, B., Singh, R., Vohra, R., Koul, A., Meena, L.S., Koduri, H., Ghildiyal, M., Deol, P., Das, T.K., *et al.* (2003). Phosphoprotein phosphatase of *Mycobacterium tuberculosis* dephosphorylates serine-threonine kinases PknA and PknB. Biochem. Biophys. Res. Commun. *311*, 112–120.

Chung, H.S., Yao, Z., Goehring, N.W., Kishony, R., Beckwith, J., and Kahne, D. (2009). Rapid {beta}-lactam-induced lysis requires successful assembly of the cell division machinery. Proc. Natl. Acad. Sci. U.S.A. *106*, 21872–21877.

Claessen, D., Emmins, R., Hamoen, L.W., Daniel, R.A., Errington, J., and Edwards, D.H. (2008). Control of the cell elongation-division cycle by shuttling of PBP1 protein in *Bacillus subtilis*. Mol. Microbiol. *68*, 1029–1046.

Cochran, A.G. (2001). Protein–protein interfaces: mimics and inhibitors. Curr. Opin. Chem. Biol. *5*, 654–659.

Cole, S.T., Brosch, R., Parkhill, J., Garnier, T., Churcher, C., Harris, D., Gordon, S.V., Eiglmeier, K., Gas, S., Barry, C.E. 3rd, *et al.* (1998). Deciphering the biology of *Mycobacterium tuberculosis* from the complete genome sequence. Nature *393*, 537–544.

Contreras-Martel, C., Job, V., Di Guilmi, A.M., Vernet, T., Dideberg, O., and Dessen, A. (2006). Crystal structure of penicillin-binding protein 1a (PBP1a) reveals a mutational hotspot implicated in beta-lactam resistance in *Streptococcus pneumoniae*. J. Mol. Biol. *355*, 684–696.

Corbin, B.D., Geissler, B., Sadasivam, M., and Margolin, W. (2004). Z-ring-independent interaction between a subdomain of FtsA and late septation proteins as revealed by a polar recruitment assay. J. Bacteriol. *186*, 7736–7744.

Corbin, B.D., Wang, Y., Beuria, T.K., and Margolin, W. (2007). Interaction between cell division proteins FtsE and FtsZ. J. Bacteriol. *189*, 3026–3035.

Cordell, S.C., E. J. H. Robinson, and J. Löwe (2003). Crystal structure of the SOS cell division inhibitor SulA and in complex with FtsZ. Proc. Natl. Acad. Sci. U.S.A. *199*, 7889–7894.

Cwirla, S.E., Balasubramanian, P., Duffin, D.J., Wagstrom, C.R., Gates, C.M., Singer, S.C., Davis, A.M., Tansik, R.L., Mattheakis, L.C., Boytos, C.M., *et al.* (1997). Peptide agonist of the thrombopoietin receptor as potent as the natural cytokine. Science *276*, 1696–1699.

Czaplewski, L.G., Collins, I., Boyd, E.A., Brown, D., East, S.P., Gardiner, M., Fletcher, R., Haydon, D.J., Henstock, V., Ingram, P., *et al.* (2009). Antibacterial alkoxybenzamide inhibitors of the essential bacterial cell division protein FtsZ. Bioorg. Med. Chem. Lett. *19*, 524–527.

D'Elia, M.A., Pereira, M.P., and Brown, E.D. (2009). Are essential genes really essential? Trends Microbiol. *17*, 433–438.

D'Ulisse, V., Fagioli, M., Ghelardini, P., and Paolozzi, L. (2007). Three functional subdomains of the *Escherichia coli* FtsQ protein are involved in its interaction with the other division proteins. Microbiology *153*, 124–138.

Dai, K., and Lutkenhaus, J. (1991). *ftsZ* is an essential cell division gene in *Escherichia coli*. J. Bacteriol. *173*, 3500–3506.

Dai, K., Xu, Y., and Lutkenhaus, J. (1993). Cloning and characterization of *ftsN*, an essential cell division gene in *Escherichia coli* isolated as a multicopy suppressor of *ftsA12*(Ts). J. Bacteriol. *175*, 3790–3797.

Dai, K., Mukherjee, A., Xu, Y., and Lutkenhaus, J. (1994). Mutations in *ftsZ* that confer resistance to SulA affect the interaction of FtsZ with GTP. J. Bacteriol. *176*, 130–136.

Dajkovic, A., and Lutkenhaus, J. (2006). Z ring as executor of bacterial cell division. J. Mol. Microbiol. Biotechnol *11*, 140–151.

Daniel, R.A., and Errington, J. (2000). Intrinsic instability of the essential cell division protein FtsL of *Bacillus subtilis* and a role for DivIB protein in FtsL turnover. Mol. Microbiol. *36*, 278–289.

Daniel, R.A., and Errington, J. (2003). Control of cell morphogenesis in bacteria: two distinct ways to make a rod-shaped cell. Cell *113*, 767–776.

Daniel, R.A., Williams, A.M., and Errington, J. (1996). A complex four-gene operon containing essential cell division gene *pbpB* in *Bacillus subtilis*. J. Bacteriol. *178*, 2343–2350.

Daniel, R.A., Harry, E.J., Katis, V.L., Wake, R.G., and Errington, J. (1998). Characterization of the essential cell division gene *ftsL(yIID)* of *Bacillus subtilis* and its role in the assembly of the division apparatus. Mol. Microbiol. *29*, 593–604.

Daniel, R.A., Noirot-Gros, M.F., Noirot, P., and Errington, J. (2006). Multiple interactions between the transmembrane division proteins of *Bacillus subtilis* and the role of FtsL instability in divisome assembly. J. Bacteriol. *188*, 7396–7404.

Dasgupta, A., Datta, P., Kundu, M., and Basu, J. (2006). The serine/threonine kinase PknB of *Mycobacterium tuberculosis* phosphorylates PBPA, a penicillin-binding protein required for cell division. Microbiology *152*, 493–504.

Datta, P., Dasgupta, A., Bhakta, S., and Basu, J. (2002). Interaction between FtsZ and FtsW of *Mycobacterium tuberculosis*. J. Biol. Chem. *277*, 24983–24987.

Datta, P., Dasgupta, A., Singh, A.K., Mukherjee, P., Kundu, M., and Basu, J. (2006). Interaction between FtsW and penicillin-binding protein 3 (PBP3) directs PBP3 to mid-cell, controls cell septation and mediates the formation of a trimeric complex involving FtsZ, FtsW and PBP3 in mycobacteria. Mol. Microbiol. *62*, 1655–1673.

Davis, N.K., and Chater, K.F. (1992). The *Streptomyces coelicolor whiB* gene encodes a small transcription factor-like protein dispensable for growth but essential for sporulation. Mol. Gen. Genet. *232*, 351–358.

Dessen, A., Mouz, N., Gordon, E., Hopkins, J., and Dideberg, O. (2001). Crystal structure of PBP2x from a highly penicillin-resistant *Streptococcus pneumoniae* clinical isolate: a mosaic framework containing 83 mutations. J. Biol. Chem. *276*, 45106–45112.

Di Lallo, G., Fagioli, M., Barionovi, D., Ghelardini, P., and Paolozzi, L. (2003). Use of a two-hybrid assay to study the assembly of a complex multicomponent protein machinery: bacterial septosome differentiation. Microbiology *149*, 3353–3359.

Dijkshoorn, L., Nemec, A., and Seifert, H. (2007). An increasing threat in hospitals: multidrug-resistant *Acinetobacter baumannii*. Nat. Rev. Microbiol. *5*, 939–951.

Din, N., Quardokus, E.M., Sackett, M.J., and Brun, Y.V. (1998). Dominant C-terminal deletions of FtsZ that affect its ability to localize in *Caulobacter* and its interaction with FtsA. Mol. Microbiol. *27*, 1051–1063.

Domadia, P., Swarup, S., Bhunia, A., Sivaraman, J., and Dasgupta, D. (2007). Inhibition of bacterial cell division protein FtsZ by cinnamaldehyde. Biochem. Pharmacol. *74*, 831–840.

Domadia, P.N., Bhunia, A., Sivaraman, J., Swarup, S., and Dasgupta, D. (2008). Berberine targets assembly of *Escherichia coli* cell division protein FtsZ. Biochemistry *47*, 3225–3234.

Downing, K.J., Mischenko, V.V., Shleeva, M.O., Young, D.I., Young, M., Kaprelyants, A.S., Apt, A.S., and Mizrahi, V. (2005). Mutants of *Mycobacterium tuberculosis* lacking three of the five *rpf*-like genes are defective for growth *in vivo* and for resuscitation *in vitro*. Infect. Immun. *73*, 3038–3043.

Duncan, K., and Barry, C.E. 3rd (2004). Prospects for new antitubercular drugs. Curr. Opin. Microbiol. *7*, 460–465.

Dziadek, J., Madiraju, M.V., Rutherford, S.A., Atkinson, M.A., and Rajagopalan, M. (2002). Physiological consequences associated with overproduction of *Mycobacterium tuberculosis* FtsZ in mycobacterial hosts. Microbiology *148*, 961–971.

Ebersbach, G., Galli, E., Moller-Jensen, J., Lowe, J., and Gerdes, K. (2008). Novel coiled-coil cell division factor ZapB stimulates Z ring assembly and cell division. Mol. Microbiol. *68*, 720–735.

van den Ent, F., and Lowe, J. (2000). Crystal structure of the cell division protein FtsA from *Thermotoga maritima*. EMBO J. *19*, 5300–5307.

van den Ent, F., Vinkenvleugel, T.M., Ind, A., West, P., Veprintsev, D., Nanninga, N., den Blaauwen, T., and Lowe, J. (2008). Structural and mutational analysis of the cell division protein FtsQ. Mol. Microbiol. *68*, 110–123.

Erickson, H.P. (2007). Evolution of the cytoskeleton. Bioessays *29*, 668–677.

Erickson, H.P., Taylor, D.W., Taylor, K.A., and Bramhill, D. (1996). Bacterial cell division protein FtsZ assembles into protofilament sheets and minirings, structural homologs of tubulin polymers. Proc. Natl. Acad. Sci. U.S.A. *93*, 519–523.

Errington, J., Daniel, R.A., and Scheffers, D.J. (2003). Cytokinesis in bacteria. Microbiol. Mol. Biol. Rev. *67*, 52–65.

Espeli, O., Lee, C., and Marians, K.J. (2003). A physical and functional interaction between *Escherichia coli* FtsK and topoisomerase IV. J. Biol. Chem. *278*, 44639–44644.

Fadda, D., Pischedda, C., Caldara, F., Whalen, M.B., Anderluzzi, D., Domenici, E., and Massidda, O. (2003). Characterization of *divIVA* and other genes located in the chromosomal region downstream of the dcw cluster in *Streptococcus pneumoniae*. J. Bacteriol. *185*, 6209–6214.

Feucht, A., Lucet, I., Yudkin, M.D., and Errington, J. (2001). Cytological and biochemical characterization of the FtsA cell division protein of *Bacillus subtilis*. Mol. Microbiol. *40*, 115–125.

Fischer, W. (1988). Physiology of lipoteichoic acids in bacteria. Adv. Microb. Physiol. *29*, 233–302.

Flardh, K. (2003). Essential role of DivIVA in polar growth and morphogenesis in *Streptomyces coelicolor* A3(2). Mol. Microbiol. *49*, 1523–1536.

Foster, S.J., and Popham, D.L. (2002). Structure and synthesis of cell wall, spore cortex, teichoic acids, S-layers, and capsules. In *Bacillus subtilis* and its Closest Relatives: From Genes to Cells, Sonenshein, A.L., Hoch, J.A., and Losick, R., eds. (Washington DC, U.S.A, American Society for Microbiology Press), pp. 21–41.

Freiberg, C., and Brotz-Oesterhelt, H. (2005). Functional genomics in antibacterial drug discovery. Drug Discov. Today *10*, 927–935.

Fry, D.C. (2006). Protein–protein interactions as targets for small molecule drug discovery. Biopolymers *84*, 535–552.

Gadek, T.R., and Nicholas, J.B. (2003). Small molecule antagonists of proteins. Biochem. Pharmacol. *65*, 1–8.

Gamba, P., Veening, J.W., Saunders, N.J., Hamoen, L.W., and Daniel, R.A. (2009). Two-step assembly dynamics of the *Bacillus subtilis* divisome. J. Bacteriol. *191*, 4186–4194.

Geiman, D.E., Raghunand, T.R., Agarwal, N., and Bishai, W.R. (2006). Differential gene expression in response to exposure to antimycobacterial agents and other stress conditions among seven *Mycobacterium tuberculosis whiB*-like genes. Antimicrob. Agents. Chemother. *50*, 2836–2841.

Geissler, B., Elraheb, D., and Margolin, W. (2003). A gain-of-function mutation in *ftsA* bypasses the requirement for the essential cell division gene *zipA* in *Escherichia coli*. Proc. Natl. Acad. Sci. U.S.A. *100*, 4197–4202.

Gerard, P., Vernet, T., and Zapun, A. (2002). Membrane topology of the *Streptococcus pneumoniae* FtsW division protein. J. Bacteriol. *184*, 1925–1931.

Ghigo, J.M., Weiss, D.S., Chen, J.C., Yarrow, J.C., and Beckwith, J. (1999). Localization of FtsL to the *Escherichia coli* septal ring. Mol. Microbiol. *31*, 725–737.

Giesbrecht, P., Wecke, J., and Reinicke, B. (1976). On the morphogenesis of the cell wall of staphylococci. Int. Rev. Cytol. *44*, 225–318.

Glass, J.I., Lefkowitz, E.J., Glass, J.S., Heiner, C.R., Chen, E.Y., and Cassell, G.H. (2000). The complete sequence of the mucosal pathogen *Ureaplasma urealyticum*. Nature *407*, 757–762.

Goehring, N.W., and Beckwith, J. (2005). Diverse paths to midcell: assembly of the bacterial cell division machinery. Curr. Biol. *15*, R514–526.

Goehring, N.W., Gonzalez, M.D., and Beckwith, J. (2006). Premature targeting of cell division proteins to midcell reveals hierarchies of protein interactions involved in divisome assembly. Mol. Microbiol. *61*, 33–45.

Goehring, N.W., Gueiros-Filho, F., and Beckwith, J. (2005). Premature targeting of a cell division protein to midcell allows dissection of divisome assembly in *Escherichia coli*. Genes Dev *19*, 127–137.

Goffin, C., and Ghuysen, J.M. (1998). Multimodular penicillin-binding proteins: an enigmatic family of orthologs and paralogs. Microbiol. Mol. Biol. Rev. *62*, 1079–1093.

Gomez, J.E., and Bishai, W.R. (2000). *whmD* is an essential mycobacterial gene required for proper septation and cell division. Proc. Natl. Acad. Sci. U.S.A *97*, 8554–8559.

Gonzalez, J.M., Jimenez, M., Velez, M., Mingorance, J., Andreu, J.M., Vicente, M., and Rivas, G. (2003). Essential cell division protein FtsZ assembles into one monomer-thick ribbons under conditions resembling the crowded intracellular environment. J. Biol. Chem. *278*, 37664–37671.

Gonzalez, M.D., and Beckwith, J. (2009). Divisome under construction: distinct domains of the small membrane protein FtsB are necessary for interaction with multiple cell division proteins. J. Bacteriol. *191*, 2815–2825.

Grundling, A., and Schneewind, O. (2007a). Genes required for glycolipid synthesis and lipoteichoic acid anchoring in *Staphylococcus aureus*. J. Bacteriol. *189*, 2521–2530.

Grundling, A., and Schneewind, O. (2007b). Synthesis of glycerol phosphate lipoteichoic acid in *Staphylococcus aureus*. Proc. Natl. Acad. Sci. U.S.A *104*, 8478–8483.

Gueiros-Filho, F.J., and Losick, R. (2002). A widely conserved bacterial cell division protein that promotes assembly of the tubulin-like protein FtsZ. Genes Dev. *16*, 2544–2556.

Guzman, L.M., Barondess, J.J., and Beckwith, J. (1992). FtsL, an essential cytoplasmic membrane protein involved in cell division in *Escherichia coli*. J. Bacteriol. *174*, 7716–7728.

Haeusser, D.P., Schwartz, R.L., Smith, A.M., Oates, M.E., and Levin, P.A. (2004). EzrA prevents aberrant cell division by modulating assembly of the cytoskeletal protein FtsZ. Mol. Microbiol. *52*, 801–814.

Hale, C., Rhee, A., and de Boer, P. (2000). ZipA-Induced bundling of FtsZ polymers mediated by an interaction between C-terminal domains. J. Bacteriol. *182*, 5153–5166.

Hale, C.A., and de Boer, P.A. (1997). Direct binding of FtsZ to ZipA, an essential component of the septal ring structure that mediates cell division in *E. coli*. Cell *88*, 175–185.

Hale, C.A., and de Boer, P.A. (1999). Recruitment of ZipA to the septal ring of *Escherichia coli* is dependent on FtsZ and independent of FtsA. J. Bacteriol. *181*, 167–176.

Hale, C.A., and de Boer, P.A.J. (2002). ZipA is required for recruitment of FtsK, FtsQ, FtsL, and FtsN to the septal ring in *Escherichia coli*. J. Bacteriol. *184*, 2552–2556.

Hamoen, L.W., Meile, J.C., de Jong, W., Noirot, P., and Errington, J. (2006). SepF, a novel FtsZ-interacting protein required for a late step in cell division. Mol. Microbiol. *59*, 989–999.

Hancock, R.E. (1997). The bacterial outer membrane as a drug barrier. Trends Microbiol. *5*, 37–42.

Handler, A.A., Lim, J.E., and Losick, R. (2008). Peptide inhibitor of cytokinesis during sporulation in *Bacillus subtilis*. Mol. Microbiol. *68*, 588–599.

Haney, S.A., Glasfeld, E., Hale, C., Keeney, D., He, Z., and de Boer, P. (2001). Genetic analysis of the *Escherichia coli* FtsZ.ZipA interaction in the yeast two-hybrid system: characterization of FtsZ residues essential for the interactions with ZipA and with FtsA. J. Biol. Chem. *276*, 11980–11987.

Harrison, E.F., and Cropp, C.B. (1965). Comparative *in vitro* activities of lysostaphin and other antistaphylococcal antibiotics on clinical isolates of *Staphylococcus aureus*. Appl. Microbiol. *13*, 212–215.

Harry, E., Monahan, L., and Thompson, L. (2006). Bacterial cell division: the mechanism and its precision. Int. Rev. Cytol. *253*, 27–94.

Harry, E.J., Rodwell, J., and Wake, R.G. (1999). Co-ordinating DNA replication with cell division in bacteria: a link between the early stages of a round of replication and mid-cell Z ring assembly. Mol. Microbiol. *33*, 33–40.

Haydon, D.J., Stokes, N.R., Ure, R., Galbraith, G., Bennett, J.M., Brown, D.R., Baker, P.J., Barynin, V.V., Rice, D.W., Sedelnikova, S.E., *et al.* (2008). An inhibitor of FtsZ with potent and selective anti-staphylococcal activity. Science *321*, 1673–1675.

Heidrich, C., Templin, M.F., Ursinus, A., Merdanovic, M., Berger, J., Schwarz, H., de Pedro, M.A., and Holtje, J.V. (2001). Involvement of N-acetylmuramyl-L-alanine amidases in cell separation and antibiotic-induced autolysis of *Escherichia coli*. Mol. Microbiol. *41*, 167–178.

Henriques, A.O., Glaser, P., Piggot, P.J., and Moran, C.P., Jr. (1998). Control of cell shape and elongation by the *rodA* gene in *Bacillus subtilis*. Mol. Microbiol. *28*, 235–247.

Henry, T., Garcia-Del Portillo, F., and Gorvel, J.P. (2005). Identification of *Salmonella* functions critical for bacterial cell division within eukaryotic cells. Mol. Microbiol. *56*, 252–267.

Hermans, P.W., Abebe, F., Kuteyi, V.I., Kolk, A.H., Thole, J.E., and Harboe, M. (1995). Molecular and immunological characterization of the highly conserved antigen 84 from *Mycobacterium tuberculosis* and *Mycobacterium leprae*. Infect. Immun. *63*, 954–960.

Hett, E.C., Chao, M.C., Deng, L.L., and Rubin, E.J. (2008). A mycobacterial enzyme essential for cell division synergizes with resuscitation-promoting factor. PLoS Pathog. *4*, e100000.

Hett, E.C., and Rubin, E.J. (2008). Bacterial growth and cell division: a mycobacterial perspective. Microbiol. Mol. Biol. Rev. *72*, 126–156.

Higgins, M.L., and Shockman, G.D. (1970). Model for cell wall growth of *Streptococcus faecalis*. J. Bacteriol. *101*, 643–648.

Holtje, J.V. (1998). Growth of the stress-bearing and shape-maintaining murein sacculus of *Escherichia coli*. Microbiol. Mol. Biol. Rev. *62*, 181–203.

Hopkins, A.L., and Groom, C.R. (2002). The druggable genome. Nat. Rev. Drug. Discov. *1*, 727–730.

Huang, Q., Kirikae, F., Kirikae, T., Pepe, A., Amin, A., Respicio, L., Slayden, R.A., Tonge, P.J., and Ojima, I. (2006). Targeting FtsZ for antituberculosis drug discovery: noncytotoxic taxanes as novel antituberculosis agents. J. Med. Chem. *49*, 463–466.

Huisman, O., and D'Ari, R. (1981). An inducible DNA replication-cell division coupling mechanism in *E. coli*. Nature *290*, 797–799.

Ishikawa, S., Kawai, Y., Hiramatsu, K., Kuwano, M., and Ogasawara, N. (2006). A new FtsZ-interacting protein, YlmF, complements the activity of FtsA during progression of cell division in *Bacillus subtilis*. Mol. Microbiol. *60*, 1364–1380.

Ito, H., Ura, A., Oyamada, Y., Tanitame, A., Yoshida, H., Yamada, S., Wachi, M., and Yamagishi, J. (2006). A 4-aminofurazan derivative-A189-inhibits assembly of bacterial cell division protein FtsZ *in vitro* and *in vivo*. Microbiol. Immunol. *50*, 759–764.

Jaiswal, R., Beuria, T.K., Mohan, R., Mahajan, S.K., and Panda, D. (2007). Totarol inhibits bacterial cytokinesis by perturbing the assembly dynamics of FtsZ. Biochemistry *46*, 4211–4220.

Jennings, L.D., Foreman, K.W., Rush, T.S. 3rd, Tsao, D.H., Mosyak, L., Kincaid, S.L., Sukhdeo, M.N., Sutherland, A.G., Ding, W., Kenny, C.H., *et al.* (2004). Combinatorial synthesis of substituted 3-(2-indolyl) piperidines and 2-phenyl indoles as inhibitors of ZipA–FtsZ interaction. Bioorg. Med. Chem. *12*, 5115–5131.

Jensen, S.O., Thompson, L.S., and Harry, E.J. (2005). Cell division in *Bacillus subtilis*: FtsZ and FtsA association is Z-ring independent, and FtsA is required for efficient midcell Z-ring assembly. J. Bacteriol. *187*, 6536–6544.

Jones, A.L., Beveridge, T.J., and Woods, D.E. (1996). Intracellular survival of *Burkholderia pseudomallei*. Infect. Immun. *64*, 782–790.

Jones, G., and Dyson, P. (2006). Evolution of transmembrane protein kinases implicated in coordinating remodelling of Gram-positive peptidoglycan: inside versus outside. J. Bacteriol. *188*, 7470–7476.

Jones, L.J., Carballido-Lopez, R., and Errington, J. (2001). Control of cell shape in bacteria: helical, actin-like filaments in *Bacillus subtilis*. Cell *104*, 913–922.

Justice, S.S., Hung, C., Theriot, J.A., Fletcher, D.A., Anderson, G.G., Footer, M.J., and Hultgren, S.J. (2004). Differentiation and developmental pathways of uropathogenic *Escherichia coli* in urinary tract pathogenesis. Proc. Natl. Acad. Sci. U.S.A *101*, 1333–1338.

Justice, S.S., Hunstad, D.A., Cegelski, L., and Hultgren, S.J. (2008). Morphological plasticity as a bacterial survival strategy. Nat. Rev. Microbiol. *6*, 162–168.

Justice, S.S., Hunstad, D.A., Seed, P.C., and Hultgren, S.J. (2006). Filamentation by *Escherichia coli* subverts innate defenses during urinary tract infection. Proc. Natl. Acad. Sci. U.S.A *103*, 19884–19889.

Kaimer, C., Gonzalez-Pastor, J.E., and Graumann, P.L. (2009). SpoIIIE and a novel type of DNA translocase, SftA, couple chromosome segregation with cell division in *Bacillus subtilis*. Mol. Microbiol. *74*, 810–825.

Kang, C.M., Abbott, D.W., Park, S.T., Dascher, C.C., Cantley, L.C., and Husson, R.N. (2005). The *Mycobacterium tuberculosis* serine/threonine kinases PknA and PknB: substrate identification and regulation of cell shape. Genes Dev. *19*, 1692–1704.

Kapoor, S., and Panda, D. (2009). Targeting FtsZ for antibacterial therapy: a promising avenue. Expert Opin. Ther. Targets *13*, 1037–1051.

Karimova, G., Dautin, N., and Ladant, D. (2005). Interaction network among *Escherichia coli* membrane proteins involved in cell division as revealed by bacterial two-hybrid analysis. J. Bacteriol. *187*, 2233–2243.

Karimova, G., Robichon, C., and Ladant, D. (2009). Characterization of YmgF, a 72-residue inner membrane protein that associates with the *Escherichia coli* cell division machinery. J. Bacteriol. *191*, 333–346.

Katis, V.L., Wake, R.G., and Harry, E.J. (2000). Septal localization of the membrane-bound division proteins of *Bacillus subtilis* DivIB and DivIC is codependent only at high temperatures and requires FtsZ. J. Bacteriol. *182*, 3607–3611.

Kawai, F., Shoda, M., Harashima, R., Sadaie, Y., Hara, H., and Matsumoto, K. (2004). Cardiolipin domains in *Bacillus subtilis* marburg membranes. J. Bacteriol. *186*, 1475–1483.

Kell, C.M., Sharma, U.K., Dowson, C.G., Town, C., Balganesh, T.S., and Spratt, B.G. (1993). Deletion analysis of the essentiality of penicillin-binding proteins 1A, 2B and 2X of *Streptococcus pneumoniae*. FEMS Microbiol. Lett. *106*, 171–175.

Khan, M., Giessrigl, B., Vonach, C., Madlener, S., Prinz, S., Herbaceck, I., Holzl, C., Bauer, S., Viola, K., Mikulits, W., *et al.* (2010). Berberine and a *Berberis lycium* extract inactivate Cdc25A and induce alpha-tubulin acetylation that correlates with HL-60 cell cycle inhibition and apoptosis. Mutat. Res. *683*, 123–130.

Kobayashi, K., Ehrlich, S.D., Albertini, A., Amati, G., Andersen, K.K., Arnaud, M., Asai, K., Ashikaga, S., Aymerich, S., Bessieres, P., *et al.* (2003). Essential *Bacillus subtilis* genes. Proc. Natl. Acad. Sci. U.S.A *100*, 4678–4683.

Koppelman, C.M., Den Blaauwen, T., Duursma, M.C., Heeren, R.M., and Nanninga, N. (2001). *Escherichia coli* minicell membranes are enriched in cardiolipin. J. Bacteriol. *183*, 6144–6147.

Krogan, N.J., Cagney, G., Yu, H., Zhong, G., Guo, X., Ignatchenko, A., Li, J., Pu, S., Datta, N., Tikuisis, A.P., *et al.* (2006). Global landscape of protein complexes in the yeast *Saccharomyces cerevisiae.* Nature *440*, 637–643.

Lappchen, T., Hartog, A.F., Pinas, V.A., Koomen, G.J., and den Blaauwen, T. (2005). GTP analogue inhibits polymerization and GTPase activity of the bacterial protein FtsZ without affecting its eukaryotic homologue tubulin. Biochemistry *44*, 7879–7884.

Lappchen, T., Pinas, V.A., Hartog, A.F., Koomen, G.J., Schaffner-Barbero, C., Andreu, J.M., Trambaiolo, D., Lowe, J., Juhem, A., Popov, A.V., *et al.* (2008). Probing FtsZ and tubulin with C8-substituted GTP analogs reveals differences in their nucleotide binding sites. Chem. Biol. *15*, 189–199.

Lara, B., and Ayala, J.A. (2002). Topological characterization of the essential *Escherichia coli* cell division protein FtsW. FEMS Microbiol. Lett. *216*, 23–32.

Lara, B., Rico, A.I., Petruzzelli, S., Santona, A., Dumas, J., Biton, J., Vicente, M., Mingorance, J., and Massidda, O. (2005). Cell division in cocci: localization and properties of the *Streptococcus pneumoniae* FtsA protein. Mol. Microbiol. *55*, 699–711.

Le Gouellec, A., Roux, L., Fadda, D., Massidda, O., Vernet, T., and Zapun, A. (2008). Roles of pneumococcal DivIB in cell division. J. Bacteriol. *190*, 4501–4511.

Leroy, M., Cabral, H., Figueira, M., Bouchet, V., Huot, H., Ram, S., Pelton, S.I., and Goldstein, R. (2007). Multiple consecutive lavage samplings reveal greater burden of disease and provide direct access to the nontypeable *Haemophilus influenzae* biofilm in experimental otitis media. Infect. Immun. *75*, 4158–4172.

Levin, P.A., Kurtser, I.G., and Grossman, A.D. (1999). Identification and characterization of a negative regulator of FtsZ ring formation in *Bacillus subtilis.* Proc. Natl. Acad. Sci. U.S.A *96*, 9642–9647.

Levin, P.A., and Losick, R. (1994). Characterization of a cell division gene from *Bacillus subtilis* that is required for vegetative and sporulation septum formation. J. Bacteriol. *176*, 1451–1459.

Lietman, P.S., and Smith, C.R. (1983). Aminoglycoside nephrotoxicity in humans. Rev. Infect. Dis. *5*, S284-S293.

Lindsay, J.A., Ruzin, A., Ross, H.F., Kurepina, N., and Novick, R.P. (1998). The gene for toxic shock toxin is carried by a family of mobile pathogenicity islands in *Staphylococcus aureus.* Mol. Microbiol. *29*, 527–543.

Liu, Z., Mukherjee, A., and Lutkenhaus, J. (1999). Recruitment of ZipA to the division site by interaction with FtsZ. Mol. Microbiol. *31*, 1853–1861.

Livnah, O., Stura, E.A., Johnson, D.L., Middleton, S.A., Mulcahy, L.S., Wrighton, N.C., Dower, W.J., Jolliffe, L.K., and Wilson, I.A. (1996). Functional mimicry of a protein hormone by a peptide agonist: the EPO receptor complex at 2.8 A. Science *273*, 464–471.

Lock, R.L., and Harry, E.J. (2008). Cell-division inhibitors: new insights for future antibiotics. Nat. Rev. Drug. Discov. *7*, 324–338.

Lovering, A.L., de Castro, L.H., Lim, D., and Strynadka, N.C. (2007). Structural insight into the transglycosylation step of bacterial cell-wall biosynthesis. Science *315*, 1402–1405.

Low, H.H., Moncrieffe, M.C., and Lowe, J. (2004). The crystal structure of ZapA and its modulation of FtsZ polymerisation. J. Mol. Biol. *341*, 839–852.

Lowe, J., and Amos, L.A. (1998). Crystal structure of the bacterial cell-division protein FtsZ. Nature *391*, 203–206.

Lowe, J., and Amos, L.A. (1999). Tubulin-like protofilaments in Ca^{2+}-induced FtsZ sheets. EMBO J. *18*, 2364–2371.

Lowe, J., and Amos, L.A. (2000). Helical tubes of FtsZ from *Methanococcus jannaschii.* Biol. Chem. *381*, 993–999.

Lucchini, S., Liu, H., Jin, Q., Hinton, J.C., and Yu, J. (2005). Transcriptional adaptation of *Shigella flexneri* during infection of macrophages and epithelial cells: insights into the strategies of a cytosolic bacterial pathogen. Infect. Immun. *73*, 88–102.

Lutkenhaus, J. (1993). FtsZ ring in bacterial cytokinesis. Mol. Microbiol. *9*, 403–409.

Lutkenhaus, J. (2007). Assembly dynamics of the bacterial MinCDE system and spatial regulation of the Z ring. Annu. Rev. Biochem. *76*, 539–562.

Lutkenhaus, J.F., and Donachie, W.D. (1979). Identification of the *ftsA* gene product. J. Bacteriol. *137*, 1088–1094.

Ma, X., and Margolin, W. (1999). Genetic and functional analyses of the conserved C-terminal core domain of *Escherichia coli* FtsZ. J. Bacteriol. *181*, 7531–7544.

Maggi, S., Massidda, O., Luzi, G., Fadda, D., Paolozzi, L., and Ghelardini, P. (2008). Division protein interaction web: identification of a phylogenetically conserved common interactome between *Streptococcus pneumoniae* and *Escherichia coli.* Microbiology *154*, 3042–3052.

Margalit, D.N., Romberg, L., Mets, R.B., Hebert, A.M., Mitchison, T.J., Kirschner, M.W., and RayChaudhuri, D. (2004). Targeting cell division: small-molecule inhibitors of FtsZ GTPase perturb cytokinetic ring assembly and induce bacterial lethality. Proc. Natl. Acad. Sci. U.S.A *101*, 11821–11826.

Margolin, W. (2000). Themes and variations in prokaryotic cell division. FEMS Microbiol. Rev. *24*, 531–548.

Martinez-Moya, M., de Pedro, M.A., Schwarz, H., and Garcia-del Portillo, F. (1998). Inhibition of *Salmonella* intracellular proliferation by non-phagocytic eucaryotic cells. Res. Microbiol. *149*, 309–318.

Massey, T.H., Mercogliano, C.P., Yates, J., Sherratt, D.J., and Lowe, J. (2006). Double-stranded DNA translocation:

structure and mechanism of hexameric FtsK. Mol. Cell *23*, 457–469.

Masson, S., Kern, T., Le Gouellec, A., Giustini, C., Simorre, J.P., Callow, P., Vernet, T., Gabel, F., and Zapun, A. (2009). Central domain of DivIB caps the C-terminal regions of the FtsL/DivIC coiled-coil rod. J. Biol. Chem. *284*, 27687–27700.

McCormick, J.R. (2009). Cell division is dispensable but not irrelevant in Streptomyces. Curr. Opin. Microbiol. *12*, 689–698.

McGrogan, B.T., Gilmartin, B., Carney, D.N., and McCann, A. (2008). Taxanes, microtubules and chemoresistant breast cancer. Biochim. Biophys. Acta *1785*, 96–132.

Mercer, K.L., and Weiss, D.S. (2002). The *Escherichia coli* cell division protein FtsW is required to recruit its cognate transpeptidase, FtsI (PBP3), to the division site. J. Bacteriol. *184*, 904–912.

Michie, K.A., and Lowe, J. (2006). Dynamic filaments of the bacterial cytoskeleton. Annu. Rev. Biochem. *75*, 467–492.

Mileykovskaya, E., and Dowhan, W. (2000). Visualization of phospholipid domains in *Escherichia coli* by using the cardiolipin-specific fluorescent dye 10-N-nonyl acridine orange. J. Bacteriol. *182*, 1172–1175.

Miller, C., Thomsen, L.E., Gaggero, C., Mosseri, R., Ingmer, H., and Cohen, S.N. (2004). SOS response induction by beta-lactams and bacterial defense against antibiotic lethality. Science *305*, 1629–1631.

Moll, A., and Thanbichler, M. (2009). FtsN-like proteins are conserved components of the cell division machinery in proteobacteria. Mol. Microbiol. *72*, 1037–1053.

Monahan, L.G., Robinson, A., and Harry, E.J. (2009). Lateral FtsZ association and the assembly of the cytokinetic Z ring in bacteria. Mol. Microbiol. *74*, 1004–1017.

Moriya, S., Rashid, R., Andrade Rogrigues, C.D., and Harry, E. (2010). Influence of the nucleoid and early stages of DNA replication on positioning the division site in *Bacillus sibtilis*. Mol. Microbiol. *76*, 634–647.

Morlot, C., Zapun, A., Dideberg, O., and Vernet, T. (2003). Growth and division of *Streptococcus pneumoniae*: localization of the high molecular weight penicillin-binding proteins during the cell cycle. Mol. Microbiol. *50*, 845–855.

Mosyak, L., Zhang, Y., Glasfeld, E., Haney, S., Stahl, M., Seehra, J., and Somers, W.S. (2000). The bacterial cell-division protein ZipA and its interaction with an FtsZ fragment revealed by X-ray crystallography. EMBO J. *19*, 3179–3191.

Moy, F.J., Glasfeld, E., Mosyak, L., and Powers, R. (2000). Solution structure of ZipA, a crucial component of *Escherichia coli* cell division. Biochemistry *39*, 9146–9156.

Mukamolova, G.V., Turapov, O.A., Young, D.I., Kaprelyants, A.S., Kell, D.B., and Young, M. (2002). A family of autocrine growth factors in *Mycobacterium tuberculosis*. Mol. Microbiol. *46*, 623–635.

Mukherjee, A., and Lutkenhaus, J. (1994). Guanine nucleotide-dependent assembly of FtsZ into filaments. J. Bacteriol. *176*, 2754–2758.

Mukherjee, A., and Lutkenhaus, J. (1998). Dynamic assembly of FtsZ regulated by GTP hydrolysis. EMBO J. *17*, 462–469.

Mukherjee, A., Dai, K., and Lutkenhaus, J. (1993). *Escherichia coli* cell division protein FtsZ is a guanine nucleotide binding protein. Proc. Natl. Acad. Sci. U.S.A *90*, 1053–1057.

Mukherjee, A., Saez, C., and Lutkenhaus, J. (2001). Assembly of an FtsZ mutant deficient in GTPase activity has implications for FtsZ assembly and the role of the Z ring in cell division. J. Bacteriol. *183*, 7190–7197.

Mukherjee, S., Robinson, C.A., Howe, A.G., Mazor, T., Wood, P.A., Urgaonkar, S., Hebert, A.M., Raychaudhuri, D., and Shaw, J.T. (2007). N-Benzyl-3-sulfonamidopyrrolidines as novel inhibitors of cell division in *E. coli*. Bioorg. Med. Chem. Lett. *17*, 6651–6655.

Narayan, A., Sachdeva, P., Sharma, K., Saini, A.K., Tyagi, A.K., and Singh, Y. (2007). Serine threonine protein kinases of mycobacterial genus: phylogeny to function. Physiol. Genomics *29*, 66–75.

Navon-Venezia, S., Leavitt, A., and Carmeli, Y. (2007). High tigecycline resistance in multidrug-resistant *Acinetobacter baumannii*. J. Antimicrob. Chemother. *59*, 772–774.

Ng, W.L., Kazmierczak, K.M., and Winkler, M.E. (2004). Defective cell wall synthesis in *Streptococcus pneumoniae* R6 depleted for the essential PcsB putative murein hydrolase or the VicR (YycF) response regulator. Mol. Microbiol. *53*, 1161–1175.

Nguyen-Disteche, M., Fraipont, C., Buddelmeijer, N., and Nanninga, N. (1998). The structure and function of *Escherichia coli* penicillin-binding protein 3. Cell. Mol. Life. Sci. *54*, 309–316.

Nogales, E., Downing, K.H., Amos, L.A., and Lowe, J. (1998a). Tubulin and FtsZ form a distinct family of GTPases. Nat. Struct. Biol. *5*, 451–458.

Nogales, E., Wolf, S.G., and Downing, K.H. (1998b). Structure of the alpha/beta tubulin dimer by electron crystallography. Nature *391*, 199–203.

Noirclerc-Savoye, M., Le Gouellec, A., Morlot, C., Dideberg, O., Vernet, T., and Zapun, A. (2005). *In vitro* reconstitution of a trimeric complex of DivIB, DivIC and FtsL, and their transient co-localization at the division site in *Streptococcus pneumoniae*. Mol. Microbiol. *55*, 413–424.

Ochman, H., Soncini, F.C., Solomon, F., and Groisman, E.A. (1996). Identification of a pathogenicity island required for *Salmonella* survival in host cells. Proc. Natl. Acad. Sci. U.S.A *93*, 7800–7804.

Ohashi, T., Hale, C.A., de Boer, P.A., and Erickson, H.P. (2002). Structural evidence that the P/Q domain of ZipA is an unstructured, flexible tether between the membrane and the C-terminal FtsZ-binding domain. J. Bacteriol. *184*, 4313–4315.

Oliva, M.A., Huecas, S., Palacios, J.M., Martin-Benito, J., Valpuesta, J.M., and Andreu, J.M. (2003). Assembly of archaeal cell division protein FtsZ and a GTPase-inactive mutant into double-stranded filaments. J. Biol. Chem. *278*, 33562–33570.

Oliva, M.A., Cordell, S.C., and Lowe, J. (2004). Structural insights into FtsZ protofilament formation. Nat. Struct. Mol. Biol. *11*, 1243–1250.

Oliva, M.A., Trambaiolo, D., and Lowe, J. (2007). Structural insights into the conformational variability of FtsZ. J. Mol. Biol. 373, 1229–1242.

Opal, S.M., and Calandra, T. (2009). Antibiotic usage and resistance: gaining or losing ground on infections in critically ill patients. J. Am. Med. Assoc. *302*, 2367–2368.

Osawa, M., Anderson, D.E., and Erickson, H.P. (2008). Reconstitution of contractile FtsZ rings in liposomes. Science *320*, 792–794.

Osawa, M., Anderson, D.E., and Erickson, H.P. (2009). Curved FtsZ protofilaments generate bending forces on liposome membranes. EMBO J. *28*, 3476–3484.

Ostach, B., and Walker, S. (2005). Bacterial transglycosylase inhibitors. Curr. Opin. Chem. Biol. *9*, 459–466.

Paradis-Bleau, C., Beaumont, M., Sanschagrin, F., Voyer, N., and Levesque, R.C. (2007). Parallel solid synthesis of inhibitors of the essential cell division FtsZ enzyme as a new potential class of antibacterials. Bioorg. Med. Chem. *15*, 1330–1340.

Pereira, S.F., Henriques, A.O., Pinho, M.G., de Lencastre, H., and Tomasz, A. (2007). Role of PBP1 in cell division of *Staphylococcus aureus*. J. Bacteriol. *189*, 3525–3531.

Peters, P.C., Migocki, M.D., Thoni, C., and Harry, E.J. (2007). A new assembly pathway for the cytokinetic Z ring from a dynamic helical structure in vegetatively growing cells of *Bacillus subtilis*. Mol. Microbiol. *64*, 487–499.

Piao, Z., Sze, C.C., Barysheva, O., Iida, K., and Yoshida, S. (2006). Temperature-regulated formation of mycelial mat-like biofilms by *Legionella pneumophila*. Appl. Environ. Microbiol. 72, 1613–1622.

Pichoff, S., and Lutkenhaus, J. (2002). Unique and overlapping roles for ZipA and FtsA in septal ring assembly in *Escherichia coli*. EMBO J. *21*, 685–693.

Pichoff, S., and Lutkenhaus, J. (2005). Tethering the Z ring to the membrane through a conserved membrane targeting sequence in FtsA. Mol. Microbiol. *55*, 1722–1734.

Pichoff, S., and Lutkenhaus, J. (2007). Identification of a region of FtsA required for interaction with FtsZ. Mol. Microbiol. *64*, 1129–1138.

Pinho, M.G., and Errington, J. (2003). Dispersed mode of *Staphylococcus aureus* cell wall synthesis in the absence of the division machinery. Mol. Microbiol. *50*, 871–881.

Pinho, M.G., and Errington, J. (2004). A *divIVA* null mutant of *Staphylococcus aureus* undergoes normal cell division. FEMS Microbiol. Lett. *240*, 145–149.

Pinho, M.G., and Errington, J. (2005). Recruitment of penicillin-binding protein PBP2 to the division site of *Staphylococcus aureus* is dependent on its transpeptidation substrates. Mol. Microbiol. *55*, 799–807.

Popp, D., Iwasa, M., Narita, A., Erickson, H.P., and Maeda, Y. (2009). FtsZ condensates: an *in vitro* electron microscopy study. Biopolymers *91*, 340–350.

Possoz, C., Newmark, J., Sorto, N., Sherratt, D.J., and Tolmasky, M.E. (2007). Sublethal concentrations of the aminoglycoside amikacin interfere with cell division without affecting chromosome dynamics. Antimicrob. Agents Chemother. *51*, 252–256.

Raghunand, T.R., and Bishai, W.R. (2006). Mapping essential domains of *Mycobacterium smegmatis* WhmD: insights into WhiB structure and function. J. Bacteriol. *188*, 6966–6976.

Rai, D., Singh, J.K., Roy, N., and Panda, D. (2008). Curcumin inhibits FtsZ assembly: an attractive mechanism for its antibacterial activity. Biochem. J. *410*, 147–155.

Rajagopalan, M., Maloney, E., Dziadek, J., Poplawska, M., Lofton, H., Chauhan, A., and Madiraju, M.V. (2005). Genetic evidence that mycobacterial FtsZ and FtsW proteins interact, and colocalize to the division site in *Mycobacterium smegmatis*. FEMS Microbiol. Lett. *250*, 9–17.

RayChaudhuri, D. (1999). ZipA is a MAP-Tau homolog and is essential for structural integrity of the cytokinetic FtsZ ring during bacterial cell division. EMBO J. *18*, 2372–2383.

RayChaudhuri, D., and Park, J.T. (1992). *Escherichia coli* cell-division gene *ftsZ* encodes a novel GTP-binding protein. Nature 359, 251–254.

Reddy, M. (2007). Role of FtsEX in cell division of *Escherichia coli*: viability of *ftsEX* mutants is dependent on functional SufI or high osmotic strength. J. Bacteriol. *189*, 98–108.

Regamey, A., Harry, E.J., and Wake, R.G. (2000). Mid-cell Z ring assembly in the absence of entry into the elongation phase of the round of replication in bacteria: co-ordinating chromosome replication with cell division. Mol. Microbiol. 38, 423–434.

Reynolds, R.C., Srivastava, S., Ross, L.J., Suling, W.J., and White, E.L. (2004). A new 2-carbamoyl pteridine that inhibits mycobacterial FtsZ. Bioorg. Med. Chem. Lett. *14*, 3161–3164.

Rico, A.I., Garcia-Ovalle, M., Mingorance, J., and Vicente, M. (2004). Role of two essential domains of *Escherichia coli* FtsA in localization and progression of the division ring. Mol. Microbiol. 53, 1359–1371.

Risinger, A.L., Giles, F.J., and Mooberry, S.L. (2009). Microtubule dynamics as a target in oncology. Cancer Treat. Rev. 35, 255–261.

Robichon, C., King, G.F., Goehring, N.W., and Beckwith, J. (2008). Artificial septal targeting of *Bacillus subtilis* cell division proteins in *Escherichia coli*: an interspecies approach to the study of protein–protein interactions in multiprotein complexes. J. Bacteriol. *190*, 6048–6059.

Robinson, A., Brzoska, A.J., Turner, K.M., Withers, R., Harry, E.J., Lewis, P.J. and Dixon, N.E. (2010). Essential biological processes of an emerging pathogen: DNA replication, transcription and cell division in *Acinetobacter* spp. Microbiol. Mol. Biol. Rev. *74*, 273–297.

Robson, S.A., and King, G.F. (2006). Domain architecture and structure of the bacterial cell division protein DivIB. Proc. Natl. Acad. Sci. U.S.A. *103*, 6700–6705.

Romberg, L., and Levin, P.A. (2003). Assembly dynamics of the bacterial cell division protein FtsZ: poised at the edge of stability. Annu. Rev. Microbiol. *57*, 125–154.

Romberg, L., Simon, M., and Erickson, H.P. (2001). Polymerization of FtsZ, a bacterial homolog of tubulin: is assembly cooperative? J. Biol. Chem. *276*, 11743–11753.

Rosenberger, C.M., and Finlay, B.B. (2002). Macrophages inhibit *Salmonella typhimurium* replication through MEK/ERK kinase and phagocyte NADPH oxidase activities. J. Biol. Chem. *277*, 18753–18762.

Rosenberger, C.M., Gallo, R.L., and Finlay, B.B. (2004). Interplay between antibacterial effectors: a macrophage antimicrobial peptide impairs intracellular *Salmonella* replication. Proc. Natl. Acad. Sci. U.S.A. *101*, 2422–2427.

Sanchez, M., Valencia, A., Ferrandiz, M.J., Sander, C., and Vicente, M. (1994). Correlation between the structure and biochemical activities of FtsA, an essential cell division protein of the actin family. EMBO J. *13*, 4919–4925.

Sassetti, C.M., Boyd, D.H., and Rubin, E.J. (2003). Genes required for mycobacterial growth defined by high density mutagenesis. Mol. Microbiol. *48*, 77–84.

Sauvage, E., Duez, C., Herman, R., Kerff, F., Petrella, S., Anderson, J.W., Adediran, S.A., Pratt, R.F., Frere, J.M., and Charlier, P. (2007). Crystal structure of the *Bacillus subtilis* penicillin-binding protein 4a, and its complex with a peptidoglycan mimetic peptide. J. Mol. Biol. *371*, 528–539.

Sauvage, E., Kerff, F., Terrak, M., Ayala, J.A., and Charlier, P. (2008). The penicillin-binding proteins: structure and role in peptidoglycan biosynthesis. FEMS Microbiol. Rev. 32, 234–258.

Scheffers, D.J., de Wit, J.G., den Blaauwen, T., and Driessen, A.J. (2002). GTP hydrolysis of cell division protein FtsZ: evidence that the active site is formed by the association of monomers. Biochemistry *41*, 521–529.

Scheffers, D.J., and Pinho, M.G. (2005). Bacterial cell wall synthesis: new insights from localization studies. Microbiol. Mol. Biol. Rev. *69*, 585–607.

Schirner, K., Marles-Wright, J., Lewis, R.J., and Errington, J. (2009). Distinct and essential morphogenic functions for wall- and lipo-teichoic acids in *Bacillus subtilis*. EMBO J. *28*, 830–842.

Schmidt, K.L., Peterson, N.D., Kustusch, R.J., Wissel, M.C., Graham, B., Phillips, G.J., and Weiss, D.S. (2004). A predicted ABC transporter, FtsEX, is needed for cell division in *Escherichia coli*. J. Bacteriol. *186*, 785–793.

Schwedock, J., McCormick, J.R., Angert, E.R., Nodwell, J.R., and Losick, R. (1997). Assembly of the cell division protein FtsZ into ladder-like structures in the aerial hyphae of *Streptomyces coelicolor*. Mol. Microbiol. *25*, 847–858.

Shen, B., and Lutkenhaus, J. (2009). The conserved C-terminal tail of FtsZ is required for the septal localization and division inhibitory activity of MinC(C)/MinD. Mol. Microbiol. 72, 410–424.

Shiomi, D., and Margolin, W. (2007). Dimerization or oligomerization of the actin-like FtsA protein enhances the integrity of the cytokinetic Z ring. Mol. Microbiol. *66*, 1396–1415.

Shuker, S.B., Hajduk, P.J., Meadows, R.P., and Fesik, S.W. (1996). Discovering high-affinity ligands for proteins: SAR by NMR. Science *274*, 1531–1534.

Sidhu, S.S., Fairbrother, W.J. and Deshayes, K. (2003). Exploring protein–protein interactions with phage display. Chem. Biochem. *4*, 14–25.

Singh, J.K., Makde, R.D., Kumar, V., and Panda, D. (2007). A membrane protein, EzrA, regulates assembly dynamics of FtsZ by interacting with the C-terminal tail of FtsZ. Biochemistry *46*, 11013–11022.

Slayden, R.A., Knudson, D.L., and Belisle, J.T. (2006). Identification of cell cycle regulators in *Mycobacterium tuberculosis* by inhibition of septum formation and global transcriptional analysis. Microbiology *152*, 1789–1797.

Soliveri, J.A., Gomez, J., Bishai, W.R., and Chater, K.F. (2000). Multiple paralogous genes related to the *Streptomyces coelicolor* developmental regulatory gene *whiB* are present in *Streptomyces* and other actinomycetes. Microbiology *146*, 333–343.

Spratt, B.G. (1975). Distinct penicillin binding proteins involved in the division, elongation, and shape of *Escherichia coli* K12. Proc. Natl. Acad. Sci. U.S.A. *72*, 2999–3003.

Spratt, B.G. (1977a). Properties of the penicillin-binding proteins of *Escherichia coli* K12. Eur. J. Biochem. *72*, 341–352.

Spratt, B.G. (1977b). Temperature-sensitive cell division mutants of *Escherichia coli* with thermolabile penicillin-binding proteins. J. Bacteriol. *131*, 293–305.

Stokes, N.R., Sievers, J., Barker, S., Bennett, J.M., Brown, D.R., Collins, I., Errington, V.M., Foulger, D., Hall, M., Halsey, R., *et al.* (2005). Novel inhibitors of bacterial cytokinesis identified by a cell-based antibiotic screening assay. J. Biol. Chem. *280*, 39709–39715.

Storts, D.R., Aparicio, O.M., Schoemaker, J.M., and Markovitz, A. (1989). Overproduction and identification of the ftsQ gene product, an essential cell division protein in *Escherichia coli* K-12. J. Bacteriol. *171*, 4290–4297.

Stricker, J., Maddox, P., Salmon, E.D., and Erickson, H.P. (2002). Rapid assembly dynamics of the *Escherichia coli* FtsZ-ring demonstrated by fluorescence recovery after photobleaching. Proc. Natl. Acad. Sci. U.S.A. *99*, 3171–3175.

Sung, M.T., Lai, Y.T., Huang, C.Y., Chou, L.Y., Shih, H.W., Cheng, W.C., Wong, C.H., and Ma, C. (2009). Crystal structure of the membrane-bound bifunctional transglycosylase PBP1b from *Escherichia coli*. Proc. Natl. Acad. Sci. U.S.A. *106*, 8824–8829.

Sutherland, A.G., Alvarez, J., Ding, W., Foreman, K.W., Kenny, C.H., Labthavikul, P., Mosyak, L., Petersen, P.J., Rush, T.S. 3rd, Ruzin, A., *et al.* (2003). Structure-based design of carboxybiphenylindole inhibitors of the ZipA–FtsZ interaction. Org. Biomol. Chem. *1*, 4138–4140.

Tavares, J.R., de Souza, R.F., Meira, G.L., and Gueiros-Filho, F.J. (2008). Cytological characterization of YpsB,

Efflux Pumps from Gram-negative Bacteria: From Structure and Function to Inhibition

4

Olga Lomovskaya and Helen I. Zgurskaya*

Abstract

Multidrug efflux pumps adversely affect both the clinical effectiveness of existing antibiotics as well as the discovery process to find new ones. In this chapter, we summarize recent advances in structural and functional analyses of multicomponent efflux pumps from Gram-negative bacteria with the focus on transporters belonging to the resistance-nodulation cell division superfamily. The unquestionably significant impact of these pumps on the effectiveness of antibiotics in clinical settings and their emerging role in bacterial pathogenesis makes them attractive targets for inhibition. We discuss modes of inhibition and current efforts to develop effective inhibitors of multidrug efflux pumps.

Introduction

The problem of antibiotic resistance and superbugs

In most developed countries, antibiotics are the second most widely used class of drugs after simple analgesics. Introduction of antibiotics in the late 1930s resulted in a significant jump in human life expectancy. Antibiotics tamed *Streptococcus pyogenes* that caused half of all post-birth deaths and was a major cause of death from burns or *Staphylococcus aureus* that was fatal in 80% of infected wounds. Tuberculosis (TB) and bacterial pneumonia became much less successful killers. Nowadays, antibiotics are important components of pre- and post-surgical therapeutic regimens; they are invaluable for the maintenance of patients that become immunocompromised as a result of cancer, organ transplantation or contracting AIDS. As was first shown with stomach ulcers, more and more disorders are being found to have an infectious origin, providing new indications for antibiotics. In short, antibiotics are one of the most important components of human medicine.

However, with the end of the 'golden era of discovery' of new antibiotic classes some time in the 1960s, and the end of the 'golden age of antibiotic medicinal chemistry' in the 1980s, when many useful modifications to known antibiotics were devised in order to improve their efficacy and safety and to overcome growing resistance, we are left with limited choices to fight bacterial infections. This arsenal of drugs continues to shrink owing to the dramatic increase in the prevalence of untreatable pan-resistant pathogens (Projan and Shlaes, 2004; Wright 2007; Demain, 2009).

There are two types of multidrug resistant superbugs. Normally sensitive bacteria can acquire multiple distinct resistance genes, each responsible for a specific biochemical resistance mechanism. Resistance genes can be located on plasmids and transposons and the corresponding mechanisms are numerous and efficient. Antibiotics can be degraded, inactivated or sequestered, their targets can be modified so that binding affinity is reduced, or sensitive targets can be replaced altogether with in-sensitive ones (Woodford, 2005; Alekshun and Levy, 2006; Tenover, 2006).

*Corresponding author

This type of multidrug resistance is observed in the case of the notorious methicillin-resistant *S. aureus* (MRSA). Another example is *Acinetobacter baummanii* that was recently added to the 'hit-list' of important pathogens identified by the Infectious Diseases Society of America (Talbot *et al.*, 2006; Peleg *et al.*, 2008; Perez *et al.*, 2008). Natural competence, or the ability to easily take up foreign DNA, significantly facilitates acquisition of antibiotic resistance genes by this bacterium.

Superbugs of a different type possess powerful and broad defence mechanisms which are not specific towards particular antibiotics. One of them is *Pseudomonas aeruginosa*, a frequent and deadly nosocomial pathogen (Rice 2007; Talbot 2008; Boucher *et al.*, 2009; Peleg and Hooper, 2010). The natural ecological niches of this and similar bacteria are soil and water, diverse environments crammed with multiple chemical challenges to overcome. For these bacteria, multidrug resistance appears to be their natural state since it is essential for their survival in natural habitats (Wright 2007). It is achieved by the cooperation of two attributes: a low permeability cell envelope which provides a physical barrier to the entry of both lipophilic and hydrophilic molecules (Nikaido, 2003) and efflux systems consisting of membrane transporters with unusually broad substrate specificity (Piddock, 2006a; Lomovskaya *et al.*, 2007; Poole 2007; Li and Nikaido, 2009). This non-specific defence might be further fortified by acquisition of antibiotic-specific resistance genes.

Diversity of multidrug resistance transporters

Multidrug resistance, or polyspecific transporters, are present in all living systems, be they bacteria or man. Anticancer drugs that are aimed at fighting tumours, drugs that are meant to cure various brain or liver disorders, and drugs that are much more preferably taken orally than through a needle more often than not are intercepted by these versatile proteins. They provide a front-line, non-specific defence, independent of attacked specific target, which then might be further enhanced by much more toxin-specific mechanisms, such as toxin metabolism or desensitization of the toxin's target (Higgins, 2007).

Bacteria, which have been around for a couple of billion years longer than eukaryotic organisms, appear to be particularly endowed with polyspecific transporters. According to predictions from the bioinformatic analysis of more than 200 available bacterial genomes, putative multidrug resistance factors (MDRs) comprise 2–7% of the total bacterial protein complement (Paulsen *et al.*, 2001; Saier and Paulsen, 2001). Such putative MDRs are identified based on sequence similarity with experimentally confirmed transporters able to handle multiple 'drug-like' substrates. Most of these substrates are hydrophobic or amphipathic molecules frequently containing weakly basic moieties. Other substrates are organic cations whose permanent charge is distributed over a large hydrophobic surface (Zgurskaya and Nikaido, 2000, 2002; Zgurskaya, 2002). These general substrate characteristics parallel physiochemical, 'drug-like' features which are required for crossing biological membranes, and must be present in both natural toxins and man-made drugs (Higgins and Gottesman, 1992).

Obviously not all predicted transporters have been confirmed to be polyspecific, but quite a few of them are (Nishino and Yamaguchi, 2001), emphasizing the impressive scale of the bacterial 'resistome'. We should consider ourselves lucky that not all of these polyspecific transporters, frequently rushed to be dubbed MDRs, are in fact clinically relevant. While they effectively expel dyes, quaternary ammonium compounds, and other organic chemicals, efflux of therapeutically useful antibiotics either cannot be detected or often is not significant enough to confer resistance associated with treatment failures. However, enough of them do their job so well that they frequently leave us completely defenceless against some bacterial pathogens.

Functional studies and subsequent phylogenetic analysis demonstrated that bacterial MDR transporters can be organized into five evolutionarily distinct protein families that significantly differ in bioenergetics, structure and transport mechanism (Yen *et al.*, 2010). The important implication is that during evolution, polyspecific transporters appear to have arisen independently several times. This structural and functional diversity strengthens anti-toxin barriers. By the same token, however, it also significantly complicates the task of developing anti-microbial agents that could overcome these protective barriers.

Most MDRs are found in three large and diverse superfamilies: ABC (ATP-binding cassette; Locher, 2009), MF (major facilitator; Saier *et al.*, 1999; Fluman and Bibi, 2009) and RND (resistance-nodulation cell division; Tseng *et al.*, 1999; Nikaido and Takatsuka 2009). In addition, some MDRs form a core of smaller superfamilies: SMR (small multidrug resistance) family (now part of the DMT (drug/metabolite transporter) superfamily; Jack *et al.*, 2001; Bay and Turner, 2009) and MATE (multidrug and toxic extrusion) family [recently joined the MOP (multidrug/oligosaccharidyl-lipid/polysaccharide) superfamily; Hvorup *et al.*, 2003; Moriyama *et al.*, 2008]. A relational database (TransportDB) that classifies and predicts drug efflux systems from organisms whose genomes have been sequenced is available at http://membranetransport.org/.

Genomic studies indicate that the total number of MDR systems is approximately proportional to the total number of all of the transport systems identified in a given organism (Ren and Paulsen, 2007). The complement of MDR transporters in an organism appears to correlate to both the evolutionary history and overall physiology and lifestyles of the organism. Accordingly, intracellular pathogens seem to contain a somewhat limited repertoire of MDRs, presumably because of their stable environment. On the contrary, soil- and plant-associated bacteria generally appear to encode a larger variety and number of MDR transporters, perhaps reflecting much more variable conditions of their own ecological niches.

ABC MDRs (and all other members of this superfamily) are primary active transporters which couple substrate translocation with binding and hydrolysis of ATP. MDRs in all the other superfamilies are secondary transporters that utilize electro-chemical gradients of ions (most frequently protons but sometime sodium) to transport their diverse substrates. Both primary and secondary transporters are ubiquitous in bacteria; however, their relative presence seems to correlate with energy generation: fermentative bacteria tend to rely more on the primary transporters while aerobic bacteria contain somewhat more secondary transporters in their genomes (Paulsen *et al.*, 1998, 2001).

While being polyspecific, bacterial MDRs from different families are still vastly different in the level of promiscuity towards various substrates. Nevertheless, some substrates, such as hydrophobic cations (for example ethidium bromide) can be handled by the majority of them. Bacteria have apparently specifically evolved to ensure protection against such compounds, as they have the potential to be particularly harmful. If not intercepted they could accumulate intracellularly, driven by the inside negative membrane potential. Perhaps not by coincidence but exactly because of this extra precaution taken by bacteria long ago, hydrophobic cations are conspicuously absent from the list of clinically useful antibiotics (Paulsen and Lewis, 2001).

For some efflux transporters a protective role is their main, natural function, while others extrude drugs only when presented with a choice 'to efflux or to die' (Vazquez-Laslop *et al.*, 2000; Neyfakh, 2001). 'Dedicated guards' can be identified relatively easily by their broad substrate spectrum, and their basal level in cells is high and may become even higher in the presence of drugs.

Efflux and antibiotic resistance

Gram-positive bacteria

Multidrug efflux plays a significant role as a mechanism for antibiotic resistance in many important human pathogens. However, the prevalence of efflux-mediated antibiotic resistance varies significantly depending upon the specific antibiotic and bacterium. In general, efflux plays a more significant role in Gram-negative bacteria (see below) and it is Gram-negative transporters which are the main subject of this chapter. Still, several examples of efflux-mediated intrinsic or acquired resistance in Gram-positive bacteria are worth mentioning. *E. faecalis* that is included on the recent list of 'superbugs' developed by the Infectious Disease Society of America (Talbot *et al.*, 2006) has an unusually high level of intrinsic antibiotic resistance for a Gram-positive bacterium. It is not susceptible to lincosamides (lincomycin and clindamycin) and significantly less susceptible to aminoglycosides, macrolides and fluoroquinolones than most Gram-positives. *E. faecalis* is also intrinsically resistant to the recently developed antibiotic Synercid (the synergistic mixture of the two semisynthetic derivatives of the streptogramin class, pristinamycin, quinupristin and dalfopristin; Tendolkar *et al.*, 2003). Systematic genetic disruption of 23 genes encoding putative ABC transporters showed that two were involved in conferring high level intrinsic antibiotic resistance in this organism. Deletion of *abc16* resulted in a 10-fold increase in susceptibility to macrolides, while deletion of *abc23* made *E. faecalis* 50- to 200-fold more susceptible to lincosamides and Synercid (Davis *et al.*, 2001). No significant reduction in resistance to fluoroquinolones was observed. It is quite possible that other, non ABC-type MDRs contribute to fluoroquinolone resistance in this organism.

Staphylococcus aureus and *Streptococcus pneumoniae* are the leading causative agents for many community and hospital infections. Both contain constitutively expressed MDR efflux pumps defining the basal level of resistance to ciprofloxacin (Hassan *et al.*, 2007). The MFS-transporter NorA is a constitutive efflux pump in *S. aureus* (Kaatz *et al.*, 1993; Neyfakh *et al.*, 1993), whereas the recently discovered heterodimeric ABC-transporter SP2073/SP2075 (PatA/PatB) protects *S. pneumoniae*. Inactivation of either of these transporters results in 4-fold increased susceptibility to fluoroquinolones (Robertson *et al.*, 2005; Marrer *et al.*, 2006).

Efflux resistance to fluoroquinolones in some clinical isolates of *S. pneumoniae* was found to result from overproduction of its constitutive ABC-transporter as well as an additional MDR, PmrA, from the MFS family (Gill *et al.*, 1999; Piddock *et al.*, 2002; Avrain *et al.*, 2007). In *S. aureus*, efflux resistance in clinical isolates can result from overexpression of at least four different MFS MDR pumps: NorA, NorB, NorC, and MdeA (DeMarco *et al.*, 2007). Importantly, the newer generation fluoroquinolones, including levofloxacin, moxifloxacin, and gemifloxacin, are affected by the multidrug resistance pumps in *S. aureus* and *S. pneumoniae* to a much lesser degree than is ciprofloxacin (Gill *et al.*, 1999; Truong-Bolduc *et al.*, 2006; Avrain *et al.*, 2007). Among all of the mentioned transporters, it is only NorB from *S. aureus* which is capable to confer somewhat increased resistance to moxifloxacin (2-fold; Truong-Bolduc *et al.*, 2005).

Recent findings implicate the MATE transporter MepA from *S. aureus* in a 2- to 4-fold level of resistance to the antibiotic tigecycline (McAleese *et al.*, 2005), the most recent antibiotic approved by the Food and Drug Administration (FDA) for the treatment of some multidrug resistant Gram-negative and Gram-positive bacteria, including MRSA (Lentino *et al.*, 2008).

In the case of the intracellular Gram-positive pathogen *Mycobacterium tuberculosis*, resistance to various anti-TB drugs such as izoniazid, rifampicin, pyrazinamide, ethambutol, aminoglycosides and fluoroquinolones is mainly attributed to mutations in specific target genes. However, it was long ago realized that resistance in a proportion of clinical isolates cannot be explained by such mutations and there is a growing list of examples of relevant efflux pumps (Louw *et al.*, 2009). Efflux of a majority of anti-TB drugs was demonstrated using efflux inhibitors (Colangeli *et al.*, 2005; Gupta *et al.*, 2006; Rodrigues *et al.*, 2009). Recent microarray analysis of 25 efflux genes demonstrated that ten of them (mainly ABC and MF transporters) showed increased expression after exposure to common anti-TB drugs (Gupta *et al.*, 2010). The mmpL7 gene from *M. tuberculosis* encoding a RND transporter was shown to confer isoniazid resistance when cloned in *M. smegmatis* (Pasca *et al.*, 2005). Another RND efflux system, MmpS5–MmpL5, is apparently involved in azole resistance in *M. tuberculosis* (Milano *et al.*, 2009). Importantly, a recent study demonstrated that inhibition of efflux activity significantly reduced the frequency of resistance to ethambutol (Srivastava *et al.*, 2010). There is much to be done to determine the clinical relevance of these data such as the incidence of efflux-mediated resistance and the identity of the most relevant efflux pumps. The emerging picture, however, is that in *M. tuberculosis* as in other bacteria efflux may influence the level of intrinsic drug resistance and clinical resistance to anti-TB drugs.

Gram-negative bacteria

The evidence supporting the clinical significance of efflux in Gram-negative bacterial drug resistance is overwhelming. Most classes of currently available antibiotics (macrolides/ketolides, oxazolidinones, lipopeptides, glycopeptides, rifamycins, streptogramins, etc.) are considered to be 'Gram-positive only', since they lack utility against Gram-negative bacteria. Some bacteria are particularly notorious in this respect. Among them, as mentioned above, is *Pseudomonas aeruginosa*, a common nosocomial pathogen, the causative agent of many life-threatening infections and the major reason for the shortened life span of people with cystic fibrosis. *Pseudomonas* infections (Rello, 2005) can be successfully treated by only a few of some 160 approved antibiotics, in particular by members of the fluoroquinolone, β-lactam or aminoglycoside classes.

That intrinsic efflux is the cause of non-susceptibility to antibiotics is evident from the fact that deletion of major constitutively expressed MDR efflux pumps that belong to the RND family of transporters renders *P. aeruginosa* and many other Gram-negative bacteria as sensitive to these antibiotics as the much more susceptible Gram-positive pathogens. As might be expected and is well-documented (Piddock, 2006a; Poole 2005, 2007), enhanced efflux due to overexpression of RND MDR pumps plays a prominent role in acquired resistance to a limited number of useful antibiotics as well. While MDR efflux is typically not the only mechanism of resistance to a particular agent, numerous *in vitro* studies provide strong evidence that in the absence of efflux, the emergence of unrelated resistance mechanisms might be significantly suppressed (Markham, 1999; Oethinger *et al.*, 2000; Lomovskaya *et al.*, 2001; Ricci *et al.*, 2006; Fange *et al.*, 2009).

Those who are involved in antibiotic drug discovery are very well aware that the primary hit rate in random antibacterial whole cell screens using large libraries of synthetic compounds or natural products is up to 1000-fold lower against Gram-negative than Gram-positive bacteria. The hit rate is significantly increased when wild-type strains are replaced with mutants lacking efflux pumps (Hsieh *et al.*, 1998). Thus, it is clear that these MDRs can successfully handle a wide array of substrates. The result of this intrinsic efflux is an inadequate number of primary hits, which are the starting point for the discovery of new agents. In short, MDR pumps adversely affect both the clinical effectiveness of existing antibiotics as well as the discovery process to find new ones.

The extraordinary efficiency of efflux in Gram-negative bacteria is brought about by a combination of three major factors: (i) the existence of multicomponent MDR pumps with astounding breadth of substrate specificity which are able to capture their substrates in the periplasm and extrude them across the outer membrane (Zgurskaya and Nikaido, 2000) in a process that will be referred from now on as a trans-envelope efflux; (ii) the low permeability of the typical Gram-negative outer membrane which slows down uptake of extruded substrates back into bacterial cell (Nikaido, 2003) and (iii) cooperation between complex trans-envelope pumps and 'simple' inner membrane transporters (Lee *et al.*, 2000; Tal and Schuldiner, 2009).

A transenvelope efflux pump consists of an inner, an outer and periplasmic component, all assembled into a multiprotein complex that traverses the whole envelope of Gram-negative bacteria (see below). The inner membrane component is largely responsible for substrate capture and its subsequent transport is energized by proton antiport or ATP hydrolysis. Transporters from three different families, ABC, MF and RND, can form multicomponent efflux pumps (Lomovskaya *et al.*, 2002; Horiyama *et al.*, 2010). However, those that contain RND-transporters play the most significant role in multidrug resistance in Gram-negative bacteria owing to their unprecedented broad substrate specificity. RND pumps can recognize and extrude positive, negative or neutral charged molecules, substances as hydrophobic as organic solvents and lipids, and compounds as hydrophilic as aminoglycoside antibiotics. Importantly, a transporter captures its substrates in the periplasm or the outer leaflet of the inner membrane and extrudes them across the outer membrane of Gram-negative bacteria. Since this membrane provides a significant permeability barrier for both hydrophilic and hydrophobic compounds, it slows down the uptake of substrates back into cells, thus preventing a pump from being overwhelmed.

The final touch for the efficient 'Gram-negative' defence is provided by cooperation between multicomponent trans-envelope and 'simple' inner membrane pumps in situations where they share common substrates. For example, cells of *E. coli* or *P. aeruginosa* may acquire plasmid-encoded antibiotic transporters such as TetA or CmlA (Poole, 2005), even though they already encode in their genomes endogenous MDR pumps (AcrAB-TolC and MexAB-OprM, respectively) that also can efflux tetracycline and chloramphenicol. TetA and CmlA are examples of 'simple' single-component pumps that are located in the inner membrane and extrude their substrate into the periplasm of Gram-negative bacteria.

When efflux pumps of different structural types were combined in the same cell (Lee *et al.*, 2000), the observed antibiotic resistance was much higher than that conferred by each of the pumps expressed singly, and the fold-increase in drug resistance was close to the product of the fold-increases owing to the individual pumps. For example, when singly expressed, the AcrAB-TolC and the CmlA pumps conferred similar levels of resistance to chloramphenicol, raising the minimal inhibitory concentration (MIC) of the strain that lacked both pumps from 1 μg/ml to 16–32 μg/ml in *E. coli*. The strain expressing both pumps had MICs of >256 μg/ml. This multiplicative effect on drug resistance can be easily understood if one accepts that the two substrate gradients are maintained independently by two types of pumps: gradients across the inner and the outer membrane by a single and a multicomponent pump, respectively. In this scenario, a single compound pump extrudes its substrates into the periplasm, where they are 'picked' up by the trans-envelope pump and further transported across the outer membrane. In other words, these pumps work 'in series' as opposed to 'parallel' organization of transporters that capture their substrates from and extrude them into the same cellular compartment (Lee *et al.*, 2000).

The case of both series and parallel pumps appears to be relevant in clinical settings. In some strains of *Salmonella* that are highly resistant (MIC >128 μg/ml) to chloramphenicol and tetracycline, the resistance observed is the result of simultaneous overexpression of both the multicomponent AcrAB-TolC pump and the single-component specific efflux pump: Flo and Tet, respectively (Quinn *et al.*, 2006). Strains overexpressing pairs of Mex pumps have been reported among clinical isolates of *P. aeruginosa* (Pumbwe and Piddock, 2000; Piddock, 2006a).

Very recently, cooperation between endogenous transporters expressed at normal levels was shown to be the basis for high-level intrinsic resistance to lipophilic cations (such as ethidium bromide and acriflavin) in *E. coli* (Tal and Schuldiner, 2009). Inner membrane transporters MdfA (major facilitator family) and EmrE (SMR family) that share these common substrates with each other and with AcrAB-TolC were simultaneously inactivated in a strain expressing AcrAB-TolC. This resulted in a significant decrease in drug resistance. Inactivation of either MdfA or EmrE alone did not have any measurable effect. This result indicates that MdfA and EmrE work in parallel and can complement each other, and that both pumps work in series with AcrAB-TolC. It also means that AcrAB-TolC alone cannot confer the high level intrinsic resistance (to at least lipophilic cations) that is seen in the wild-type strain of *E. coli* and relies on help from the inner membrane transporters for cell detoxification. As was discussed above, lipophilic cations represent a particular danger to bacterial cells; so it is not by accident that bacteria adapted a robust efflux-based strategy to defend themselves against such molecules. It remains to be seen if endogenous single and multicomponent transporters might cooperate to confer intrinsic resistance to conventional antibiotics.

Role of multidrug resistance efflux pumps in pathogenesis

There is accumulating evidence that MDR-transporters that were initially recognized for their ability to confer multidrug resistance possess functions unrelated to antibiotics. This is not at all surprising given the multiplicity of efflux systems in bacterial genomes and their obviously ancient evolutionary origin. Polyspecific efflux pumps from various bacteria have recently become implicated in various stress responses such as oxidative and nitrosative stresses or the stress from compromised translation (reviewed in Poole 2008). They appear to be involved in the excretion of potentially toxic bacterial metabolites (Helling *et al.*, 2002; Rosner and Martin, 2009). More and more data demonstrate the importance of MDR pumps for bacterial pathogenesis (Blair and Piddock, 2009; Nishino *et al.*, 2009; Piddock, 2006b). To be pathogenic, an organism must be able to invade a host, multiply in the host, evade host defences, and harm the host in some way. Efflux pumps may have a role in all of these phenotypes and several suggestive examples implicating RND-driven pumps are given below.

Mutants of *S. typhimurium* that lack *acrB* are less able to adhere to and invade mouse macrophages (Buckley *et al.*, 2006). A recent study demonstrated that disruption of *acrAB-tolC* in *S. typhimurium* resulted in reduced expression of multiple genes including genes for proteins involved in pathogenicity (Webber *et al.*, 2009). Among them are chemotaxis and motility genes as well as genes encoding type III secretion system required for *Salmonella* invasion. The mechanism of efflux-dependent gene regulation is not yet understood, but these data might help to explain why efflux-deficient mutants of *S. typhimurium* are attenuated in animal models of infection (Nishino *et al.*, 2006). Mutants of *P. aeruginosa* that lack MexAB-OprM had reduced ability to invade Madin–Darby canine kidney (MDCK) cells (Hirakata *et al.*, 2002) and inhibitors of this pump were recently shown to affect invasiveness in the same system (Hirakata *et al.*, 2009). Importantly, the defect in invasion was restored by addition of culture supernatant from MDCK cells infected with the wild-type cells, indicating that decreased invasiveness of efflux mutants could be due to the reduced secretion of an 'invasion factor'. Its nature is not known at present time, but at least acyl-homoserine lactones, the auto-inducers involved in the regulation of cell-to-cell communication (quorum sensing) that plays an important role in pathogenesis (Parker and Sperandio, 2009), have been excluded from the list. The defect in cell invasion (lung epithelial and human macrophage cells) was also demonstrated with *Burkholderia pseudomallei* mutants lacking the MDR pump BpeAB-OprB that was also shown to export multiple drugs (Chan and Chua, 2005). This pump is essential for the secretion of acyl homoserine lactone auto-inducers and expression of diverse virulence factors (Chan *et al.*, 2007a). However, recent data indicate that this essentiality could be dependent on the genetic background of the strain (Mima and Schweizer, 2010), indicating that other pumps from *B. pseudomallei* could play similar role.

There are also several examples where an increased efflux has been shown to be detrimental for the expression of virulence genes and pathogenesis (Evans *et al.*, 1998; Alonso *et al.*, 2004; Juhas *et al.*, 2004; Linares *et al.*, 2005; Sanchez *et al.*, 2002). The emerging picture is that while for some MDR pumps, the export of intracellularly produced signal molecules that are involved in regulation of virulence and pathogenicity might be their true physiological function, others can affect the intracellular concentration of these molecules unintentionally, just by being polyspecific.

Perhaps the best documented and the least controversial feature is the importance of MDR efflux pumps in evading host defences. *Escherichia coli*, *Salmonella typhimurium*, *Campylobacter jejuni* and *Vibrio cholerae* are all inhabitants of the intestinal tract of vertebrates, an environment that is rich in bile salts that act as detergents to facilitate digestion of food fat. All these bacteria become hypersensitive to bile salts when major RND-transporters are deleted from their genomes. Moreover, bile salts appear to be preferred substrates of AcrAB-TolC from *E. coli* (Zgurskaya and Nikaido, 1999b) and are capable of inducing expression of several MDR transporters from *E. coli* and *S. typhimurium* (Rosenberg *et al.*, 2003; Nikaido *et al.*, 2008; Nishino *et al.*, 2009). The natural niche for the strict human pathogen *Neisseria gonorrhoeae* is the genitourinary mucosae which is rich in free fatty acids that have potent anti-gonococcal activity *in vitro* against mutants lacking the MDR efflux pump MtrCDE. The ability of these agents to induce *mtrCDE* expression indicates that they are its natural substrates (Hagman *et al.*, 1995). Mutants lacking MtrCDE are also hypersensitive to human cathelicidin peptide LL-37 found in genitourinary mucosa (Hagman *et al.*, 1995). *Klebsiella pneumoniae* is an important respiratory pathogen. A recent study demonstrated that mutants that lack the MDR pump AcrAB-TolC are hypersensitive to antimicrobial activity of human bronchoalveolar lavage fluid which is rich in anti-microbial peptides. Purified peptides were also shown to be substrates of this pump (Padilla *et al.*, 2010). Importantly, efflux-deficient mutants of pathogens mentioned above show various degree of attenuation in infection models (Buckley *et al.*, 2006; Quinn *et al.*, 2007; Padilla *et al.*, 2010). Likewise, mutants overexpressing RND efflux pumps can be readily isolated in animal models even in the absence of antibiotic pressure (Join-Lambert *et al.*, 2001).

The involvement of efflux pumps in antibiotic resistance and pathogenesis makes them attractive targets for inhibition. Understanding their structures and transport mechanisms is absolutely essential for the development of credible inhibitory strategies.

Structure and transport mechanism of RND transporters

Structure

An extraordinary amount of structural information on bacterial RND-transporters has been published in the past few years (Fig. 4.1). Several high-resolution structures of the AcrB efflux pump from *E. coli*, with and without co-crystallized substrates, as well several mutant AcrBs, have emerged from several laboratories across the world (Murakami *et al.*, 2002, 2006; Yu *et al.*, 2003b, 2005; Seeger *et al.*, 2006; Su *et al.*, 2006; Sennhauser *et al.*, 2007; Tornroth-Horsefield *et al.*, 2007; Drew *et al.*, 2008). Structure of the MexB protein from *P. aeruginosa* that shares 69% identity and 83% similarity with AcrB was also recently solved (Sennhauser *et al.*, 2009). As discussed below, RND-transporters from Gram-negative bacteria function with two other proteins (MFP and TolC) in a large complex. X-ray structures of the other proteins have also become available (Koronakis *et al.*, 2000; Akama *et al.*, 2004a,b; Higgins *et al.*, 2004a; Mikolosko *et al.*, 2006), and only the structure of the tripartite complex remains to be determined (see Fig. 4.2, which depicts the fitted model of the *E. coli* AcrAB-TolC tripartite efflux complex).

The high-resolution structure of *E. coli* AcrB highlights the structural features of RND transporters responsible for their ability to bind multiple drugs and associate with the accessory proteins (Fig. 4.3). RND transporters function as homo- or hetero-trimers consisting of protomers of about 1100 amino acid residues in length. Each protomer consists of the 12 transmembrane α-helices (TMS1 to TMS12) and a large periplasmic domain. The periplasmic domains assemble through multiple contacts between protomers into a 'mushroom'-like structure with a central channel, three cavities and three vestibules, which

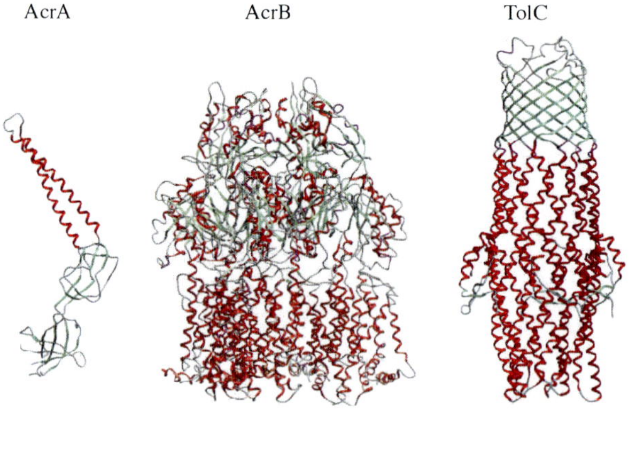

Figure 4.1 Structure of AcrA (PBD code 2f1m), AcrB (PBD code 1ek9) and TolC (PBD code 2dhh) proteins from *E. coli*. α-Helices and β-sheets are coloured in red and green, respectively. The figure shows a monomer of AcrA and trimers of AcrB and TolC. All proteins are shown to scale (relative to each other). Adapted from Lomovskaya *et al.* (2007) with permission.

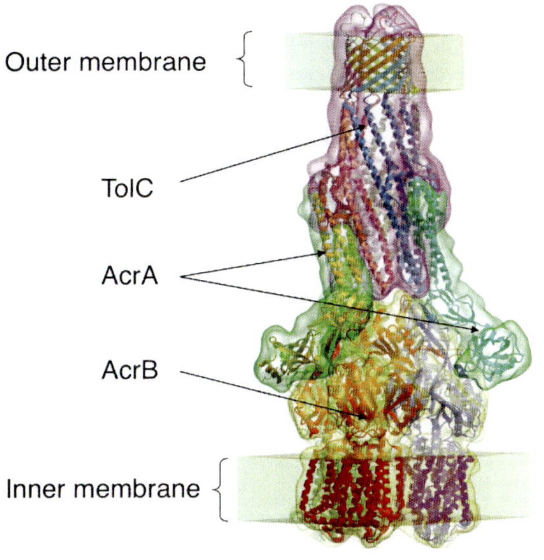

Figure 4.2 Fitted Model of the *E. coli* AcrAB-TolC tripartite efflux complex. AcrA, AcrB and TolC are coloured magenta, dark-yellow and sky blue, respectively. The simplified shape of the protein surface was generated in ICM (Molsoft) using a fast Fourier transfrom (FFT)-based smoothing algorithm. Most of the AcrA trimers are removed by cross-section. There is significant evidence that the 'crater' in the fully assembled pump is connected to TolC, either directly or via an adaptor composed of AcrA subunits. The tubular TolC structure serves as an 'exhaust pipe' that penetrates the outer membrane, providing the exit route for the substrate efflux into the external medium. The opening indicated by the white colour in the AcrB-trimer leads to the periplasmic space.

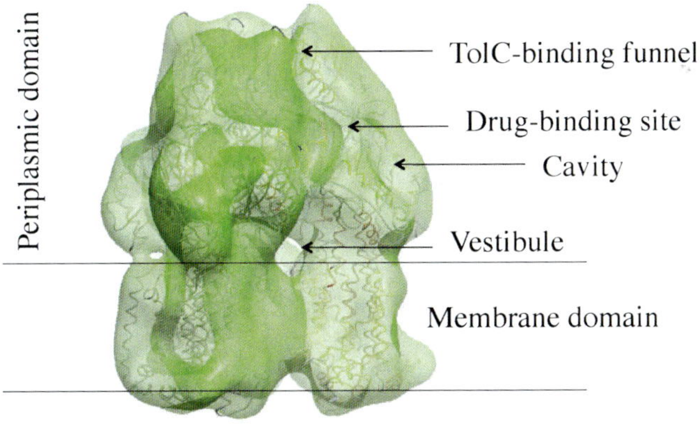

Figure 4.3 Structure of AcrB based on the asymmetric crystal. AcrB trimer contains a membrane domain with total of 36 transmembrane α-helices and a periplasmic domain with three vestibules and three cavities. In the asymmetric AcrB trimer, the conformations and orientations of drug binding sites differ for each AcrB protomer (see also Figure 4.2).

is also called the pore domain. In contrast, inside the membrane, the three protomers have very limited contact with each other.

Mutagenesis and structural studies have established that the periplasmic domains of RND transporters play a dual role. On the one hand, these domains are implicated in substrate recognition and binding (Mao *et al.*, 2002; Tikhonova *et al.*, 2002; Eda *et al.*, 2003; Elkins and Nikaido, 2003b). Several substrate binding pockets have been mapped in the pore domain of the transporter by structural and genetic studies. On the other hand, the periplasmic domains provide RND transporters with structural extensions that allow them to traverse halfway through the periplasm to gain contact with the outer membrane channel. The pore domain contains deep cavities, sites of possible interaction with a periplasmic protein, MFP. The contact between these proteins is believed to be extensive, involving the entire periplasmic length of the RND protein. In contrast, contact with the OM channel is very limited, with the two proteins barely touching each other.

The ability to capture substrates directly from the periplasm differentiates RND-transporters from all other MDR pumps. In all other cases, the amino acid residues that are responsible for substrate specificity have been mapped to transmembrane domains, and in several instances membrane capture of substrates has been confirmed in rigorous biochemical experiments, justifying their characterization as 'membrane vacuum cleaners' (Higgins and Gottesman, 1992). By the same token, capture of substrates by RND-type MDRs directly from the periplasm, coupled with extrusion across the outer membrane renders them periplasm vacuum cleaners (Elkins and Nikaido, 2003a).

AcrB-drug co-crystals have revealed drugs interacting with the protein in several, almost too numerous, locations (Fig. 4.4). In the first report, several structurally unrelated drugs (including the efflux pump inhibitor, MC-207110, see below) were seen in the central cavity on the membrane–periplasm interface, prompting a model that drug substrates first intercalate into the phospholipid bilayer and then laterally diffuse into the central cavity of AcrB (Yu *et al.*, 2003a). However, in the central RND cavity, residues involved in substrate binding are highly conserved, and binding of substrates in this site does not explain the documented differences in substrate specificities for different RNDs. Mutational analyses of AcrB and the related EmhB from *P. fluorescens* nevertheless show that amino acid residues in the central cavity do have an impact on transporter-mediated antibiotic resistance (Yu *et al.*, 2005; Hearn *et al.*, 2006).

In the second co-crystal structure of the same protein, an additional drug binding pocket was detected in the prominent cleft on the surface of the periplasmic domain. This site might be fully exposed to the periplasm with easy access for drug binding. A problem with this site is that it is not clear where drug can go from there. Mutations altering several amino acids within this site have also been reported to impact RND-related antibiotic resistance (Mao *et al.*, 2002; Middlemiss and Poole, 2004; Yu *et al.*, 2005). One intriguing possibility is that this site might play a role in regulation of the activity of the transporter by its substrates, rather than mediating the actual transport process directly.

Finally, a third co-crystal structure of AcrB has revealed yet another, non-overlapping, multidrug binding pocket, located deep inside the

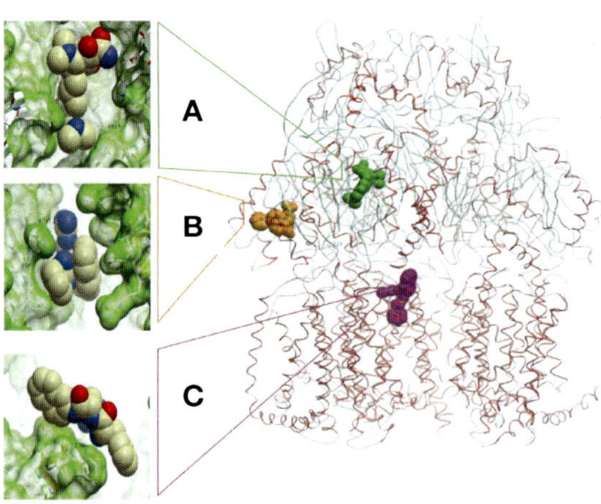

Figure 4.4 Three binding sites identified in X-ray crystallography studies of AcrB efflux pump from *E. coli*. Magenta and gold depict MC-207,110 (PBD qt8y); Minocycline in green (PBD 2drd).

periplasmic domain (Murakami *et al.*, 2006). This binding pocket is accessible through an uptake channel, a channel branching from a vestibule which runs parallel to the membrane plane and reaches the central cavity. There is also a clear path from this binding pocket to the top of the funnel, and therefore to TolC. This voluminous pocket is extremely rich in aromatic amino acid residues capable of hydrophobic and stacking interactions. There are also a few polar residues that can form hydrogen bonds. Interestingly, the co-crystallized substrates doxorubicin and minocycline were found to be interacting with different sets of amino acid residues (Fig. 4.5). Extensive mutagenesis studies position several amino acid residues responsible for substrate specificity in this site (Mao *et al.*, 2002; Aires and Nikaido, 2005; Bohnert *et al.*, 2007, 2008; Wehmeier *et al.*, 2009). Both structural and genetic studies with RND transporters fit very well with the rapidly emerging paradigm of a versatile multispecific recognition pocket, which was originally proposed based on the results obtained with soluble multidrug binding regulatory proteins such as BmrR (Zheleznova *et al.*, 1999; Heldwein and Brennan, 2001), QacR (Grkovic *et al.*, 2003; Schumacher *et al.*, 2004) and PXR (Watkins *et al.*, 2001). These early studies demonstrated (i) that different substrates can use different residues to bind in the same pocket; (ii) that the same substrates can assume multiple positions in the pocket; (iii) that two substrates can be bound simultaneously and (iv) that this binding can give rise to negative or positive co-operativity.

In a recent study, the interaction of about 30 compounds with AcrB was examined using computer docking (Takatsuka *et al.*, 2010). Substrates were predicted to bind to two distinct regions of the spacious binding pocket: a narrow groove at one end of the pocket or a wide cave at the other end. Competition assays using nitrocefin efflux and covalent labelling of Phe615Cys mutant AcrB with fluorescein-5-maleimide showed that presumed groove-binders competed against each other, but cave-binders did not compete against groove-binders, providing validation for the modelling studies and, importantly, further insights into multidrug recognition mechanism.

It is noteworthy that indifference or competition between substrates is not the only type of interactions between substrates. An alternative possibility is stimulation of the binding of one substrate by another and, as a result, a possible enhancement of efflux. Stimulation of drug efflux by small molecules was detected in the case of the ABC-transporter, P-glycoprotein (Kondratov *et al.*, 2001). Interestingly, it was recently shown that 1-(1-naphthylmethyl)-piperazine (NMP), a compound identified as an inhibitor of RND-mediated efflux of many antibiotics in *Enterobacteriaceae*, increased susceptibility of *Acinetobacter baumannii* to tetracycline and minocycline but decreased susceptibility to tigecycline

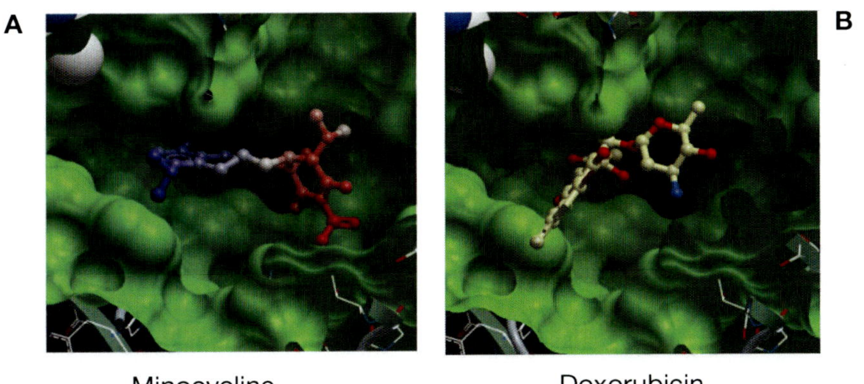

Figure 4.5 Minocycline (left) and doxorubicin (right) co-crystallized with AcrB. Two compounds interact with different sets of amino acids.

(Bean and Wareham, 2009). An intriguing possibility is that NMP increased tigecycline efflux from *Acinetobacter*.

While the possibility of periplasmic capture of substrates was confirmed in both genetic and biochemical experiments and was subsequently reinforced with the advent of high-resolution structures of protein-ligand co-crystals, there is also biochemical evidence for the cytoplasmic capture of substrates. Such evidence was obtained in reconstitution experiments with the AcrD transporter from *E. coli* that is capable of effluxing aminoglycosides (Aires and Nikaido, 2005). The preferential mode of substrate capture and location of the cytoplasmic substrate binding site remain to be determined. Mutations altering substrate specificity most frequently are found in the periplasmic domains consistent with periplasmic capture. Still, several such mutations were recently identified in transmembrane domain of the MexB by applying a random mutagenesis approach (Middlemiss and Poole, 2004). Further analysis of these mutations might help shed more light on the mechanism of cytoplasmic capture.

Transport mechanism

Besides transporter architecture and the location of potential substrate binding pockets, crystal structures provide further exciting and unexpected insights into possible transport mechanisms by RND transporters (Murakami *et al.*, 2006; Seeger *et al.*, 2006, 2008). The first X-ray structure of the RND pump AcrB was presented as a perfectly symmetrical trimer (Murakami *et al.*, 2002). It gave rise to the so-called 'elevator mechanism' of transport wherein it was proposed that substrates accumulating in the central cavity are actively transported into the upper portal space via a channel that opens along the central axis of the structure. However, for this model to work, a very significant conformational change associated with channel opening would have to be coupled with proton transport via the trans-membrane domain in order to accommodate the passage of substrates. In contrast, two new structures of AcrB trimers, while symmetrical overall, show each protomer in a distinct conformation (Murakami *et al.*, 2006; Seeger *et al.*, 2006). Although only one of these 'new generation' structures contains co-crystallized substrates (Murakami *et al.*, 2006) the conformations of protomers in both are strikingly similar. This argues that substrate is not needed to induce asymmetry.

In the periplasmic portion, the main differences between the old and new structures are in the substrate binding pocket. First, the substrate is present in only one of the protomers, dubbed the 'binding' protomer (B). The spacious drug binding pocket described earlier is open to the periplasm and expands to almost reach the TolC-docking funnel. The exit from the pocket into the funnel is blocked by an inclined α-helix of the central pore from the adjacent protomer. The binding pocket of this second protomer, called the 'extrusion' protomer (E), is closed to the periplasm, significantly reduced in size, and opened towards the funnel. The binding cavity of the third 'access' protomer (A) is largely inaccessible from either the periplasm or the exit funnel.

Based on this asymmetric structure, a new mechanism of drug transport has been proposed. This 'alternative occupancy' model implies that each protomer cycles through three consecutive conformations, named after the nomenclature for F_1F_0-ATPase, as loose (L), tight (T) and open (O), corresponding to three phases of efflux (Seeger *et al.*, 2006; Pos, 2009; Fig. 4.6). This cycling is sequential, rather than synchronous, such that at any given time each protomer exists in a different phase. Affinity of the substrate to the periplasm-accessible conformation of the pump subunit cavity is predicted to be higher than the affinity to the funnel-opened conformation. The transition of the ligand-bound protomer from a high-affinity to low-affinity state should require energy input, which is evidently provided by the coupled proton transport in the trans-membrane domains.

Until very recently this unexpected and beautiful mechanism remained just a hypothesis albeit a rather plausible one. Two approaches developed in the laboratory of Hiroshi Nikaido were used to confirm it unequivocally. In the first approach pairs of cysteine residues were introduced into AcrB such as to force the closure due to formation of disulfide bonds across the large periplasmic cleft that, according to the model,

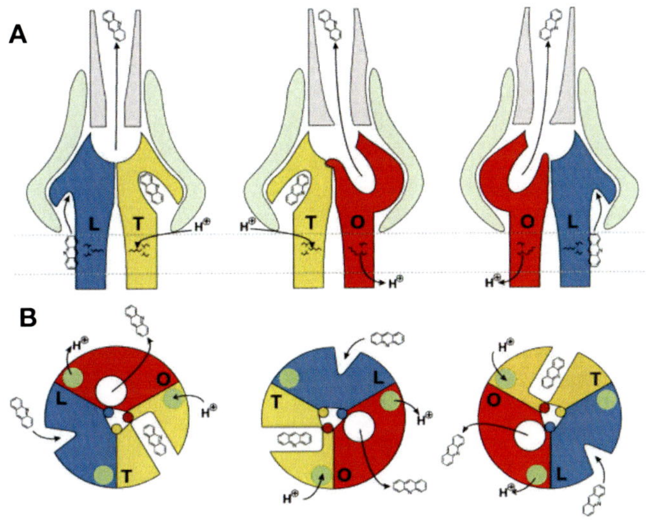

Figure 4.6 Schematic representation of the AcrB alternating site functional rotation transport mechanism. The conformational states loose (L), tight (T), and open (O) are coloured blue, yellow and red, respectively. (A) Side-view schematic representation of two of the three monomers of the AcrB trimer. AcrA and TolC are indicated in light green and light purple colours, respectively. (B) The lateral grooves in the L and T monomer indicate the substrate binding sites. The different geometric forms reflect low (triangle), high (rectangle), or no (circle) binding affinity for the transported substrates. Adapted from Pos (2009) with permission.

should undergo cyclical opening and closing. Addition of fast-acting disulfide cross-linking agents was shown to instantaneously stop efflux thus providing biochemical support for the rotating mechanism (Takatsuka and Nikaido, 2007). The second approach, elegant and intellectually superb, involved creating a giant gene in which three *acrB* sequences were fused together through short linkers. The linked AcrB trimer was capable of confirming drug resistance. Importantly, when an essential amino acid believed to be a part of the proton-relay pathway was altered in only one of the protomers, this resulted in an inactive efflux pump. Combining these two approaches, researchers created a pair of cysteine residues capable of being cross-linked, but only in one protomer; such cross-linking still resulted in AcrB inactivation (Takatsuka and Nikaido, 2009). These experiments provided solid proof of the functional rotating mechanism for RND-mediated efflux. Giant fused proteins with individual protomers trapped in a 'confirmation on-demand' might prove to be a powerful tool for further detailing the molecular mechanism of efflux. They also might be very useful for the discovery and elucidation of the mode of action of efflux pump inhibitors which can bind preferentially to a protein in a particular conformation.

Many details of the coupling between substrate efflux and proton transfer remain to be clarified, and existing studies provide just initial clues. Mutagenesis data indicate that AcrB has four electrostatically interactive residues that constitute the putative proton relay pathway, located in TM4, TM10 and TM11 (Goldberg *et al.*, 1999; Guan and Nakae, 2001). In the binding and access protomers, Lys 940 of TM10 and Asp407 and Asp408 of TM4 are coordinated by salt bridges. However, in the extrusion protomer, Lys940 is turned towards Thr978 and the salt bridges are absent. This in turn causes twisting of TM4 and TM10. Additional data are required to determine whether or not these subtle changes in TMDs are sufficient to produce the large conformational changes in the porter domain, ultimately resulting in efflux. However, what is absolutely clear is that the residues involved in proton and substrate translocation are, as expected, far apart.

Kinetics of efflux by RND transporters

Kinetic analysis of efflux has proven to be a very difficult task. The reason for this is that a transporter is functional only in the presence of other components (see below) as a complex spanning the whole cell envelope. This necessitates performing kinetic analyses in whole cells which imposes obvious difficulties in properly measuring substrate concentrations in the periplasm where substrate capturing occurs. This difficulty was recently overcome for a select group of substrates, beta-lactam antibiotics. For these experiments, researchers used cells simultaneously expressing the AcrB efflux pump and a beta-lactamase capable of degrading beta-lactams in the periplasm. This allowed determination of the periplasmic concentration of beta-lactams from the kinetic constants of the beta-lactamase (Lim and Nikaido, 2009; Nagano and Nikaido, 2009). It was shown that penicillins are generally better substrates as compared with cephalosporins: $K_{0.5}$ for penicillins was in the range of 1 μM while for cephalosporins it varied from 5 μM to 300 μM. These studies also showed that alterations of MICs upon overexpression or deletion of a pump were not always good predictors of efflux activity. For several antibiotics (cephaloridine, cephalothin), efflux was observed only at periplasmic concentrations (100 μM) that significantly exceeded periplasmic concentrations that kill bacteria (0.5 μM). The important implication is that in order for an agent to retain microbiological activity in the presence of an efflux pump, such an agent does not need to completely avoid the efflux pump: it just needs to kill bacteria at concentrations that are far below saturation of the pump.

Accessory proteins and assembly of RND drug efflux systems

As already mentioned above, in order to extrude drugs directly into the external medium, MDR pumps in Gram-negative bacteria engage the periplasmic and outer membrane proteins (OMP). AcrAB-TolC from *E. coli* is the best-characterized multidrug efflux complex capable of transenvelope transport (Nikaido and Zgurskaya 2001). In this complex, AcrA is the periplasmic membrane fusion protein (MFP) and TolC is the outer membrane channel.

Structure of outer membrane channels

The 3-D structures of three OMPs, TolC, OprM and VceC, from *E. coli*, *P. aeruginosa* and *Vibrio cholerae*, respectively, have been solved (Koronakis *et al.*, 2000; Akama *et al.*, 2004a; Federici *et al.*, 2005). Despite very little sequence similarity, they are structurally conserved. Like AcrB, TolC forms stable trimers organized into two barrel structures (Fig. 4.1). A 12-stranded 40-Å-long β-barrel inserts into the outer membrane to form an open pore 30 Å in diameter. The unusual α-helical barrel 100 Å in length protrudes deep into the periplasm, where it reaches the TolC docking domain of AcrB. The lower half of this barrel is bound by an equatorial domain of mixed α/β structure. The tip of the periplasmic end of the channel is closed in an iris-like manner by interacting loops of α-helices. This closed periplasmic entrance must be opened for substrates to gain access to the channel.

How TolC transitions from the closed to open conformation remains unclear. Structural and functional studies showed that disruption of salt bridges, which maintain the iris-like structure of the periplasmic end of TolC, leads to a dilatation of the channel (Andersen *et al.*, 2002b; Bavro *et al.*, 2008). *In vivo*, interactions with the other two components of the complex are suggested to produce the same result. Structural analysis of the possible extent of the interaction between the crown domain of AcrB and the periplasmic part of TolC indicates that such interaction is quite shallow and hence unlikely to be very stable (Bavro *et al.*, 2008). These interactions are probably restricted to the loops connecting helices H3/H4 and H7/H8 of TolC and the β-hairpins of the TolC-binding domain in AcrB. Thus, AcrA is believed to be the key to pump assembly, which stabilizes the weak AcrB-TolC contact and induces transition to the TolC open state (Fig. 4.2).

Structure of periplasmic membrane fusion proteins

Biochemical and genetic data demonstrate that MFPs interact with both the RND pump and the outer membrane channel (Tikhonova and Zgurskaya, 2004; Touze *et al.*, 2004). It is therefore proposed that the MFP stabilizes weak

RND–OMF interactions and promotes and maintains the tripartite complex. On the inner membrane, AcrA seems to actively contribute to transport by stimulating transport activities of AcrB and its close homologue AcrD (Zgurskaya and Nikaido, 1999a; Aires and Nikaido, 2005). Similarly, MacA, a close homologue of AcrA, is absolutely required for the ATPase activity of its cognate macrolide transporter MacB (Tikhonova *et al.*, 2007). Additional data emerged that some MFPs directly bind substrates and might initiate transport (Borges-Walmsley *et al.*, 2003). It is unclear, however, how common this feature is among structurally diverse MFPs. On the outer membrane, it is proposed that AcrA recruits TolC in a substrate-independent manner (Tikhonova and Zgurskaya, 2004). Once engaged into transport complexes, TolC is thought to undergo opening at the periplasmic end and that conformational transitions in AcrA are involved in this process (Andersen *et al.*, 2002a–c; Mikolosko *et al.*, 2006). Thus, AcrA appears not only to stabilize interactions between inner and outer membrane components but also actively contributes to transport events that occur at both membranes.

AcrA is a typical representative of MFPs (Zgurskaya *et al.*, 2009). The central domain of these proteins possesses two-fold symmetry of coiled-coil and lipoyl-binding motifs. In the recently solved structure of AcrA these domains fold back on themselves at the gap between helical regions, forming an α-helical hairpin (Fig. 4.1; Mikolosko *et al.*, 2006). Lipoyl domain connects the α–helical hairpin with the six stranded plus short α-helix β-barrel domain. The reported AcrA structure lacks the distal N- and C-terminal domains. However, the complete structure of MexA, the AcrA homologue from *P. aeruginosa*, was recently reported (Symmons *et al.*, 2009). In the MexA structure, the N- and C-termini together form a compact β-roll domain, which is the most inner membrane proximal domain of these proteins.

Biochemical and genetic studies implicated the α-helical hairpin of AcrA in complex formation with TolC, whereas all three (the lipoyl, the α–β barrel and the membrane proximal (MP) β-roll domain) form contacts with AcrB (Fig. 4.2; Tikhonova and Zgurskaya 2004; Touze *et al.*, 2004; Lobedanz *et al.*, 2007; Krishnamoorthy *et al.*, 2008; Symmons *et al.*, 2009). AcrA is conformationally flexible and can undergo large conformational changes (Ip *et al.*, 2003; Mikolosko *et al.*, 2006). This feature of AcrA was used to model the assembled structure of the three-component complex (Symmons *et al.*, 2009). To satisfy the chemical cross-linking data, AcrA was treated as a series of rigid bodies with interdomain flexibility. As a result the conformation of AcrA engaged into the modelled tripartite complex differs significantly from that in the crystal structure.

Assembly of three-component efflux pumps

There appears to be good agreement between functional studies of AcrAB-TolC and its modelled structure. Using a chemical mutagenesis approach, Krishnamoorthy and co-authors identified the S249N substitution in the β-barrel domain of AcrA, which enabled the functional fit between AcrA and the non-cognate *P. aeruginosa* MexB transporter (Krishnamoorthy *et al.*, 2008). The conclusion that this residue is located at the AcrA-transporter interface is in agreement with the finding that the cysteine residue introduced in position E250 (E196 in Symmons *et al.*, 2009) can be cross-linked to AcrB. The C-terminal domain of AcrA, which forms the membrane proximal β-roll, is the most conserved domain in very diverse MFPs (Zgurskaya *et al.*, 2009). In the purified AcrA and bi-partite AcrA–AcrB and AcrA–TolC complexes, this domain is unstructured and proteolytically labile (Ge *et al.*, 2009). However, when the functional tripartite complex is assembled this domain is protected from trypsin digest. In the same study, the importance of the C-terminal domain in AcrA function was further confirmed by site-directed mutagenesis. Mutation in G363 position of this domain significantly impaired the function of AcrAB–TolC complex. These results support the AcrAB-TolC model assembled by Symmons *et al.*, in which the β-roll domain interacts intimately with AcrB (Symmons *et al.*, 2009).

The inner membrane AcrAB complex is thought to pre-assemble without TolC. Furthermore, TolC interactions with both AcrB and AcrA are proposed to contribute to final assembly of the functional complex. A strong body of experimental evidence suggests that the α-helical hairpin of AcrA is the site of interaction with TolC (Misra and Bavro, 2009). In the most direct approach, Lobedanz *et al.*, carried out a comprehensive analysis of AcrA–TolC interactions through probing reciprocal cysteine-mediated cross-linking (Lobedanz *et al.*, 2007). Their data supported and confirmed a model in which the α-helical hairpin domain of AcrA docks at the intra-protomer grooves of TolC that are observed in the crystal structures of the partially open TolC.

The assembly of AcrA–TolC complexes appears to be a highly dynamic process (Tikhonova *et al.*, 2009). From the dissociation rate constants the life-times of TolC complexes with AcrA, EmrA and MacA, the three MFPs that function with RND, ABC and MFS multidrug transporters, respectively, were estimated to be within seconds to minutes. However, these short-lived complexes are formed with high affinity. The K_D values for MFP–TolC complexes were found to be in the low nanomolar and micromolar range.

One of the major steps in the assembly of the AcrAB–TolC complex is believed to be TolC transition into the open state. Based on the structural analysis of a TolC mutant, in which the stabilizing ion bridges are removed ('leaky' mutant), Bavro *et al.* proposed that in cells the interaction with the AcrB alone is likely to be sufficient for unlocking of the periplasmic gates, which, however, on its own is not sufficient to fully open the channel (Bavro *et al.*, 2008). The analysis of the TolC channel cavity reveals the presence of the second selectivity gate in it, which is formed by a double ring of aspartate residues, and which is not destroyed in the 'leaky' mutants, owing to its location further up the channel. Perhaps interaction with AcrA is necessary to open the second gate. However, the MFP-dependent transition of TolC into the open state has not been yet reconstituted and characterized *in vitro*.

The stoichiometry of the assembled tripartite complexes remains controversial. The major question is the oligomeric state of MFPs in the complex. The propensity for oligomerization *in vitro*, cross-linked dimers and trimers in living cells and the presence of paired MFPs in bacterial genomes, all support the notion that in transport complexes, MFPs function as homo- or hetero-oligomers (Mima *et al.*, 2007; Gristwood *et al.*, 2008; Tikhonova *et al.*, 2009; Zgurskaya *et al.*, 2009). Based on structures of AcrA and TolC, the most likely stoichiometry in the functional AcrA:AcrB:TolC complex is expected to be 6:3:3.

Other accessory proteins

As if three components of an efflux pump are not enough, recent structural studies raised a possibility of a fourth component. Endogenous AcrB that was expressed at 'normal' level was co-crystallized with an endogenously expressed membrane protein, YajC (Tornroth-Horsefield *et al.*, 2007) that was previously shown to interact with SecYEG and SecDF proteins of the *E. coli* protein translocation machinery. Moreover, yajC gene is the first gene of the yajF-secDE operon (reviewed in Sardis and Economou, 2010). It is not clear what the physiological significance of this finding is since deletion of *yajC* produced only a weak phenotype (hypersensitivity). It is possible that the role of YajC is more significant under conditions still to be identified.

Another open question is the role of the TonB protein in efflux. TonB is essential for energizing uptake of the siderophores, vitamin B_{12} and other substrates across the outer membrane of Gram-negative bacteria (reviewed in Schauer *et al.*, 2008)). It was also shown that the deletion of *tonB* significantly decreased efflux-mediated resistance in *P. aeruginosa* (Zhao *et al.*, 1998; Zhao and Poole, 2002) and *P. putida* (Godoy et *al.*, 2004).

Strategies to counter efflux-mediated resistance

The unquestionably significant impact of multidrug-resistance efflux pumps on the effectiveness of antibiotics in clinical settings and their emerging role in bacterial pathogenesis makes them attractive targets for inhibition. Efflux inhibitors, when used in combination with antibiotics, are expected to increase antibacterial potency, expand the spectrum of antibacterial activity, reverse resistance and dramatically reduce the rates of resistance development. An alternative approach is to identify and develop compounds which avoid efflux pumps altogether, or are not affected by efflux at concentrations which are sufficient for antibacterial activity (Lim and Nikaido, 2009). In the case of antibiotic-specific efflux pumps, this strategy proved to be rather successful. Tigecycline (Peterson, 2008) and PTK 0796 (Weir *et al.*, 2003) or telithromycin (Zuckerman *et al.*, 2009) are not recognized by tetracycline- or macrolide-specific transporters, respectively. Another example is florfenicol which is not effluxed by the chloramphenicol-specific transporter CmlA, although it is still a substrate of another specific-transporter Flo (Butaye *et al.*, 2003). Antibiotic-specific efflux pumps are single-component inner membrane proteins. To be as effective as they are in protection against naturally occurring antibiotics without any help from additional proteins or the outer membrane, they should be highly optimized to handle their specific substrates. Apparently, it is relatively easy to make antibiotic derivatives that cannot be expelled by these specialized efflux pumps.

Unfortunately, the same antibiotics are also effectively extruded by the MDR transporters. As discussed above, these pumps seem to have large and very flexible substrate binding pockets that contain multiple amino acid residues capable of interacting with substrates. While specific interactions can be prevented, a transporter will still find a way to capture its substrate using another set of amino acids. This conclusion is supported by mutagenesis data indicating that altering individual amino acids lining the substrate binding pockets of RND-transporters rarely has significant effect on the efflux of specific antibiotics.

Instead of making derivatives avoiding efflux, one may consider overwhelming an efflux pump. This strategy seems to be working in the case of Gram-positive bacteria. It is believed that the inability of MDR transporters from *Staphylococcus aureus* or *Streptococcus pneumoniae* to confer resistance to the newer fluoroquinolones, such as levofloxacin, moxifloxacin, gemifloxacin, or garenoxacin as opposed to norfloxacin and ciprofloxacin, can be attributed to the fact that they are sufficiently hydrophobic so that their rapid passive uptake overwhelms active efflux from the cell (Hooper, 2000). Unfortunately, the idea of making compounds more lipophilic to facilitate their uptake across the inner membrane will not work in Gram-negative bacteria, since uptake of the very same compounds will be strongly reduced by the outer membrane (see above). However, this approach might work for compounds with periplasmic rather than cytoplasmic targets. Another potential advantage of compounds working through periplasmic targets is that since they do not need to cross the inner membrane, they can be significantly less lipophilic as compared with compounds with cytoplasmic targets. Since lipophilicity seems to be an important requirement for the substrate recognition by the pump, more hydrophilic compounds can be less affected by efflux. In line with these considerations, the only known antibiotics that are not affected by MDR pumps are some beta-lactams that target penicillin-binding proteins located in the periplasm (Nikaido *et al.*, 1998).

By the same token, it is highly improbable that a compound could be identified that inhibits a cytoplasmic target while not being recognized by one or another of the MDR transporter in Gram-negative bacteria. This brings us back to inhibition of pumps by a separate chemical entity, the so-called efflux pump inhibitors (EPIs).

Inhibition of efflux – choice of partner antibiotic and targeting the right pump

While RND pumps present an attractive target for development of antibiotic potentiators, it is important to realize that efflux is often only one of several mechanisms of resistance in place for a given antibiotic. Thus, its inhibition will have a significant therapeutic effect only for those antibiotics and in those organisms where efflux is a major contributor to resistance. Tigecycline is a great example of an antibiotic for which efflux by RND-transporters is currently the only known clinically relevant resistance mechanism. The same can be true for several antibiotics in development with novel modes of action; they will not be affected by the known, widely spread resistance mechanisms and may therefore benefit significantly from efflux inhibition. Resistance to fluoroquinolones can arise from both target-site mutations and efflux via RND type pumps. However, in organisms harbouring both, loss of efflux alone dramatically reduces resistance (Lomovskaya *et al.*, 1999, 2001). Thus, inhibition of efflux should improve the effectiveness of fluoroquinolones against a majority of clinical isolates. Resistance to β-lactams in *P. aeruginosa* is mediated by the RND family MexAB-OprM efflux system and by β-lactamases, capable of degrading these antibiotics. It has been shown that loss of efflux in strains which express both mechanisms fails to overcome resistance (Nakae *et al.*, 1999). Efflux inhibition would not, therefore, be an effective means for countering β-lactam resistance in *P. aeruginosa*.

Pharmacodynamic considerations are also important for the choice of a partner antibiotic. In general, concentration-dependent antibiotics which are more effective at higher C_{max}/MIC or AUC/MIC ratios, are better partners for the EPIs than antibiotics whose action is time dependent (T > MIC, time above MIC). The effect of an EPI on efficacy of concentration-dependent antibiotics is directly proportional to the fold-decrease of MIC in the presence of an EPI. More EPI will require reaching the same effect for the time-dependent antibiotics: the increase in efficacy in the presence of an EPI is proportional to the natural logarithm of fold-decrease in MIC.

The pharmacokinetics of a potential partner antibiotic should also play a significant role. As an example, the fluoroquinolones levofloxacin and moxifloxacin at the highest approved human doses reach the free plasma C_{max} of ~8 μg/ml and ~2 μg/ml, respectively. Both antibiotics have the same MIC_{50} of 0.5 μg/ml in the presence of the broad-spectrum EPI, MC-207110 (Lomovskaya *et al.*, 2001), for a large panel of clinical strains of *P. aeruginosa*. According to the simplified criterion for clinical efficacy of fluoroquinolones (Drusano, 2004), free C_{max}/MIC > 10, it is expected that only levofloxacin/EPI but not moxifloxacin/EPI combination will be efficacious against *P. aeruginosa*.

Another consideration is that a single bacterium may contain multiple RND-systems capable of conferring resistance to a given antibiotic, as is the case for fluoroquinolones in *P. aeruginosa* and various *Enterobacteriaceae*. Consequently, in such a setting, only a broad-spectrum inhibitor able to interfere with multiple transporters would be effective. On the other hand, *Haemophilus influenzae* contains a single RND-transporter responsible for the efflux of macrolides; developing a selective inhibitor of this pump is therefore sufficient. Similarly, *Pseudomonas aeruginosa* contains a single tripartite efflux system, MexXY-OprM, which is responsible for the extrusion of aminoglycosides. Moreover, unlike the situation with *Enterobacteriaceae* where aminoglycoside modifying enzymes represent the major mechanism of resistance, in *P. aeruginosa*, MexXY-mediated efflux appears to play the most significant role (Poole, 2007). Therefore, disrupting this single pump with a selective inhibitor would be expected to be clinically effective in restoring aminoglycoside efficacy.

Different modes of inhibition of the activity of RND transporters

Several ways of interfering with the function of RND-transporters are apparent following elucidation of the structure and transport mechanism.

Inhibitors can target the substrate-binding pocket. This voluminous pocket can accommodate an extraordinary variety of structures; hence many distinct pharmacophore classes can be foreseen. Importantly, the binding site of the RND-transporter is located in the periplasm, so that an inhibitor might need to cross only one membrane.

As was evident in a recent structural study (Murakami *et al.*, 2006), various substrates can occupy different locations in the large substrate binding pocket. It is therefore expected that specific inhibitors will interact with, and potentiate only, a subset of substrate antibiotics. This is the case with MC-207110, the very first efflux pump inhibitor (EPI) to be identified, which is active against multiple RND pumps in a variety of Gram-negative pathogens. This compound (Fig. 4.7) was originally identified in a screen for potentiators of the fluoroquinolone levofloxacin in the RND pump-expressing organism *P. aeruginosa* (Lomovskaya *et al.*, 2001), and is now routinely used to evaluate the contribution of efflux in various Gram-negative bacteria using a variety of antibiotics (Kriengkauykiat *et al.*, 2005; Pages and Amaral 2009). Consistent with targeting the substrate-binding region of the pumps, the inhibitor potentiates the activity of only some antibiotic pump substrates (e.g. fluoroquinolones, macrolides, tetracyclines, etc., but not β-lactams)

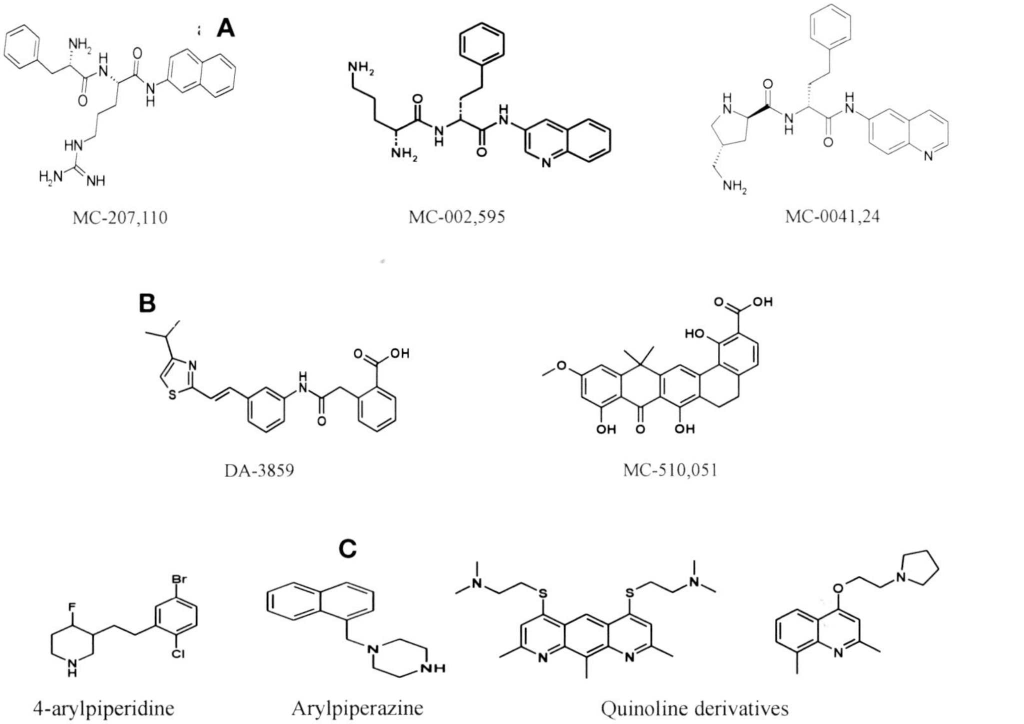

Figure 4.7 Various inhibitors of RND transporters (A) Broad-spectrum efflux pump inhibitors (EPIs) with activity against multiple RND pumps, including those in *P. aeruginosa*. These inhibitors are themselves substrates of efflux pumps and most probably interact with the transporter via a substrate-binding pocket. (B) Narrow-spectrum EPIs with selective activity against the MexAB–OprM complex from *P. aeruginosa*. These inhibitors might interact with the transporter at the allosteric 'modulator' site. (C) EPIs with activity against transporters from various species of Enterobacteriaceae. Their mode of action is mostly uncharacterized. Adapted from Lomovskaya *et al.* (2007) with permission.

and was itself shown to be exported by RND systems (Lomovskaya *et al.*, 2001). In addition to inhibition of efflux, it also permeabilized the outer membrane of Gram-negative bacteria without any effect on the integrity of the inner membrane (Lomovskaya *et al.*, 2001). This dual mode of action was considered strongly beneficial based on the synergy between efflux pumps and the outer membrane in conferring antibiotic resistance (see above). Many analogues of this compound were prepared in a collaboration between Microcide Pharmaceuticals and Daiichi Pharmaceuticals (Renau *et al.*, 2003; Watkins *et al.*, 2003), Fig. 4.7 shows MC-002595 and MC-004424. Compounds were able to potentiate the activity of levofloxacin and other fluoroquinolones, phenicols, macrolides and tigecycline against *P. aeruginosa* and many other Gram-negative bacteria. They showed good efficacy in animal models of infection. Importantly, prevention of levofloxacin resistance development in the presence of EPIs was demonstrated both *in vitro* and *in vivo* (Lomovskaya and Watkins, 2001; Lomovskaya *et al.*, 2001). An EPI programme based on more advanced analogues is actively being pursued by Mpex Pharmaceuticals in collaboration with GlaxoSmithKline.

A series of aryl-piperazines (Fig. 4.7) were identified as inhibitors of RND systems from Enterobacteriaceae (Bohnert and Kern, 2005; Schumacher *et al.*, 2006). As with MC-207110, they appear to potentiate the activity of some but not all of the antimicrobial substrates of these transporters. Interestingly, they differ from MC-207110 in the spectrum of the affected antibiotics (Schumacher *et al.*, 2006). The main advantage of inhibitors targeting the substrate binding site stems from the fact that it would be difficult to develop resistance to such compounds

– pump mutations leading to inhibitor resistance (i.e. reduced inhibitor affinity) would also probably compromise binding and, so, efflux of and resistance to antibiotic substrates. Indeed, despite attempts at recovering mutants 'resistant' to EPI-antibiotic (e.g. fluoroquinolone) combinations from amongst RND pump-expression Gram-negative bacteria, no mutations in RND pump components have yet been found (O. Lomovskaya, unpublished).

Efflux inhibitors might also act at sites distinct from those involved in substrate binding but whose disruption impacts overall pump activity. Such allosteric inhibitors would be expected to inhibit efflux of all substrates and therefore potentiate the activity of multiple antibiotics. A series of structurally very different inhibitors (lipophilic acids rather than lipophilic bases) with high selectivity towards the MexAB-OprM efflux pump from *P. aeruginosa* have been identified (Nakayama *et al.*, 2003a,b, 2004a,b; Yoshida *et al.*, 2006a,b; Fig. 4.7) and shown to negatively impact the export of all MexAB-OprM antimicrobial substrates equally (Nakayama *et al.*, 2003a,b). MexAB-OprM-selective EPIs (also lipophilic acids) were also isolated from the natural product extracts (MC-510, 051, Fig. 4.7) and were shown to have a similar mode of action (Lee *et al.*, 2001). Mutagen-induced mutants insensitive to these EPIs have been recovered; all mutations have been mapped to the *mexB* gene. Importantly, none of the mutations had any effect on broad-spectrum EPIs such as MC-207110 described above (O. Lomovskaya, unpublished). This indicated that the two types of inhibitors have different binding sites. It was hypothesized that pump-specific EPIs bind not to substrate-binding sites on the pump but rather to site(s) that modulate pump activity (i.e. modulation sites). An alternative possibility is that these compounds still interact with a transporter at a substrate binding pocket but these interactions are much more specific (hence, pump specificity) and the binding is much tighter. The transport mechanism of RND-transporters implies coordinated binding and release of substrates by different protomers: if one antibiotic substrate is being bound at the binding protomer while another substrate ('an inhibitor') is not being released from the extrusion protomer, this will bring whole process to a halt.

Several other classes of EPIs have been reported: arylpiperidines as inhibitors of AcrAB-TolC from *E. coli* (Thorarensen *et al.*, 2001), alkoxy- and alkylaminoquinolines (Fig. 4.6) showing activity against clinical strains of Enterobacteriaceae (Mallea *et al.*, 2002, 2003; Chevalier *et al.*, 2001, 2004; Hasdemir *et al.*, 2004), quinazoline derivatives that were able to potentiate multiple antibiotics against both Enterobacteriaceae and *Pseudomonas aeruginosa* (Chevalier *et al.*, 2010); phenothiazines, such as chlorpromazine and thioridazine, that in addition to Enterobacteriaceae (Martins *et al.*, 2008) showed antibiotic potentiation activity against *Burkholderia pseudomallei* as well (Chan *et al.*, 2007b). Very recently, plant phenolics, such as ellagic and tannic acids, were shown to possess EPI activity against efflux pumps from *Acinetobacter* (Chusri *et al.*, 2009). While all these compounds were shown to inhibit efflux, the specific mechanisms of inhibition remain to be clarified. Interestingly, chlorpromazine, in addition to directly inhibiting efflux, was recently shown to reduce expression of *acrB* from *S. typhimurium* (Bailey *et al.*, 2008). The potential dual mode of action could further improve the effectiveness of EPIs.

Several other possibilities for interfering with efflux can be envisaged. Examples of such approaches include blocking the TolC-like tunnel and targeting the assembly of the pump components. TolC 'plugs' might have a special advantage: no need to cross any membrane at all. The recent study demonstrating that TolC channel-forming activity can be inhibited by divalent and trivalent cations that target periplasmic entrance of TolC (Higgins *et al.*, 2004a,b), provides proof-of-principle for inhibition of efflux by blocking the TolC (i.e. OMF) channel. Despite the many possible approaches to efflux inhibition it is interesting to note that the sorts of screens used to date, based on antibiotic potentiation and therefore capable of identifying compounds interfering with efflux at any level, only returned binding- and, potentially, modulator-site inhibitors. Whether this reflects an inability to effectively target efflux via these other mechanisms or simply the lack of proper compounds in screening collections, remains to be determined. In any case, unless and until such inhibitors are found, these latter approaches will remain purely hypothetical.

Conclusions

Given the contribution of efflux pumps to resistance to numerous clinically relevant and experimental antimicrobials, there is a clear need to target these pumps in order to enhance and preserve the effectiveness of antibiotics. While the feasibly of inhibiting efflux pump activity both *in vitro* and *in vivo* was demonstrated, development of efflux pump inhibitors remains a challenging task. First, EPIs are new chemical entities (NCE) and there are obvious safety and efficacy risks associated with any NCE development. While humans do not have close homologues of RND transporters and therefore mechanism-based toxicity is not expected, off-target effects are still possible. This may potentially lead to a narrow therapeutic index, which is the main problem for the majority of antibacterial NCEs. Second, EPIs would be part of a combination therapy approach and the actual pharmacodynamics of this combination is driven by the antibiotic component. Pharmacokinetics of an EPI must be optimized in such a way so that it maximally enhances antibiotic performance. In addition, an antibiotic and an EPI should not engage in drug–drug interactions.

When a safe and efficacious antibiotic/EPI combination is finally ready for clinical trials, there are regulatory challenges. For example, the most important benefit provided by EPIs is the reversal of antibiotic resistance. To demonstrate this benefit, large clinical trials would need to be performed in order to enrol sufficient numbers of patients carrying resistant strains. This problem might be mitigated if the results of preclinical *in vitro* and *in vivo* PK/PD studies could be used to demonstrate that a combination drug has similar activity against strains that are susceptible or resistant to the antibiotic component of the combination. Based on the availability of these results, clinical data against susceptible strains may be supportive of efficacy against resistant strains (although some clinical data against resistant strains still will be necessary).

While the list of challenges can continue, they are still clearly overweighed by multiple benefits, so the efforts in both academic and applied research to stop these smart molecular machines continue. Demonstrating these multifactorial benefits of efflux pump inhibitors in clinical settings will provide the ultimate validation of an EPI-based combination approach.

Acknowledgements

We are grateful to David Griffith and Michael Dudley for helpful discussions and for critical reading of the manuscript. The research in H.I.Z's lab is supported by the National Institutes of Health grant AI052293-01A1.

References

Aires, J. R., and Nikaido, H. (2005). Aminoglycosides are captured from both periplasm and cytoplasm by the AcrD multidrug efflux transporter of *Escherichia coli*. J. Bacteriol. *187*, 1923–1929.

Akama, H., Kanemaki, M., Yoshimura, M., Tsukihara, T., Kashiwagi, T., Yoneyama, H., Narita, S., Nakagawa, A., and Nakae, T. (2004a). Crystal structure of the drug discharge outer membrane protein, OprM, of *Pseudomonas aeruginosa*: dual modes of membrane anchoring and occluded cavity end. J. Biol. Chem. *279*, 52816–52819.

Akama, H., Matsuura, T., Kashiwagi, S., Yoneyama, H., Narita, S., Tsukihara, T., Nakagawa, A., and Nakae, T. (2004b). Crystal structure of the membrane fusion protein, MexA, of the multidrug transporter in *Pseudomonas aeruginosa*. J. Biol. Chem. *279*, 25939–25942.

Alekshun, M. N., and Levy, S. B. (2006). Commensals upon us. Biochem. Pharmacol. *71*, 893–900.

Alonso, A., Morales, G., Escalante, R., Campanario, E., Sastre, L., and Martinez, J. L. (2004). Overexpression of the multidrug efflux pump SmeDEF impairs Stenotrophomonas maltophilia physiology. J. Antimicrob. Chemother. 53, 432–434.

Andersen, C., Hughes, C., and Koronakis, V. (2002a). Electrophysiological behavior of the TolC channel-tunnel in planar lipid bilayers. J. Membr. Biol. *185*, 83–92.

Andersen, C., Koronakis, E., Bokma, E., Eswaran, J., Humphreys, D., Hughes, C., and Koronakis, V. (2002b). Transition to the open state of the TolC periplasmic tunnel entrance. Proc. Natl. Acad. Sci. U.S.A. *99*, 11103–11108.

Andersen, C., Koronakis, E., Hughes, C., and Koronakis, V. (2002c). An aspartate ring at the TolC tunnel entrance determines ion selectivity and presents a target for blocking by large cations. Mol. Microbiol. *44*, 1131–1139.

Avrain, L., Garvey, M., Mesaros, N., Glupczynski, Y., Mingeot-Leclercq, M. P., Piddock, L. J., Tulkens, P. M.,

a robust survival strategy. Proc. Natl. Acad. Sci. U.S.A. *106*, 9051–9056.

Talbot, G. H. (2008). What is in the pipeline for Gram-negative pathogens? Expert Rev. Anti Infect. Ther. *6*, 39–49.

Talbot, G. H., Bradley, J., Edwards, J. E., Jr., Gilbert, D., Scheld, M., and Bartlett, J. G. (2006). Bad bugs need drugs: an update on the development pipeline from the Antimicrobial Availability Task Force of the Infectious Diseases Society of America. Clin. Infect. Dis. *42*, 657–668.

Tendolkar, P. M., Baghdayan, A. S., and Shankar, N. (2003). Pathogenic enterococci: new developments in the 21st century. Cell Mol. Life Sci. *60*, 2622–2636.

Tenover, F. C. (2006). Mechanisms of antimicrobial resistance in bacteria. Am. J. Infect. Control *34*, S3–10; discussion S64–73.

Thorarensen, A., Presley-Bodnar, A. L., Marotti, K. R., Boyle, T. P., Heckaman, C. L., Bohanon, M. J., Tomich, P. K., Zurenko, G. E., Sweeney, M. T., and Yagi, B. H. (2001). 3-Arylpiperidines as potentiators of existing antibacterial agents. Bioorg. Med. Chem. Lett. *11*, 1903–1906.

Tikhonova, E. B., and Zgurskaya, H. I. (2004). AcrA, AcrB, and TolC of *Escherichia coli* form a stable intermembrane multidrug efflux complex. J. Biol. Chem. *279*, 32116–32124.

Tikhonova, E. B., Wang, Q., and Zgurskaya, H. I. (2002). Chimeric analysis of the multicomponent multidrug efflux transporters from Gram-negative bacteria. J. Bacteriol. *184*, 6499–6507.

Tikhonova, E. B., Devroy, V. K., Lau, S. Y., and Zgurskaya, H. I. (2007). Reconstitution of the *Escherichia coli* macrolide transporter: the periplasmic membrane fusion protein MacA stimulates the ATPase activity of MacB. Mol. Microbiol. *63*, 895–910.

Tikhonova, E. B., Dastidar, V., Rybenkov, V. V., and Zgurskaya, H. I. (2009). Kinetic control of TolC recruitment by multidrug efflux complexes. Proc. Natl. Acad. Sci. U.S.A. *106*, 16416–16421.

Tornroth-Horsefield, S., Gourdon, P., Horsefield, R., Brive, L., Yamamoto, N., Mori, H., Snijder, A., and Neutze, R. (2007). Crystal structure of AcrB in complex with a single transmembrane subunit reveals another twist. Structure *15*, 1663–1673.

Touze, T., Eswaran, J., Bokma, E., Koronakis, E., Hughes, C., and Koronakis, V. (2004). Interactions underlying assembly of the *Escherichia coli* AcrAB-TolC multidrug efflux system. Mol. Microbiol. *53*, 697–706.

Truong-Bolduc, Q. C., Dunman, P. M., Strahilevitz, J., Projan, S. J., and Hooper, D. C. (2005). MgrA is a multiple regulator of two new efflux pumps in *Staphylococcus aureus*. J. Bacteriol. *187*, 2395–2405.

Truong-Bolduc, Q. C., Strahilevitz, J., and Hooper, D. C. (2006). NorC, a new efflux pump regulated by MgrA of *Staphylococcus aureus*. Antimicrob. Agents Chemother. *50*, 1104–1107.

Tseng, T. T., Gratwick, K. S., Kollman, J., Park, D., Nies, D. H., Goffeau, A., and Saier, M. H., Jr. (1999). The RND permease superfamily: an ancient, ubiquitous and diverse family that includes human disease and development proteins. J. Mol. Microbiol. Biotechnol. *1*, 107–125.

Vazquez-Laslop, N., Zheleznova, E. E., Markham, P. N., Brennan, R. G., and Neyfakh, A. A. (2000). Recognition of multiple drugs by a single protein: a trivial solution of an old paradox. Biochem. Soc. Trans. *28*, 517–520.

Watkins, R. E., Wisely, G. B., Moore, L. B., Collins, J. L., Lambert, M. H., Williams, S. P., Willson, T. M., Kliewer, S. A., and Redinbo, M. R. (2001). The human nuclear xenobiotic receptor PXR: structural determinants of directed promiscuity. Science *292*, 2329–2333.

Watkins, W. J., Landaverry, Y., Leger, R., Litman, R., Renau, T. E., Williams, N., Yen, R., Zhang, J. Z., Chamberland, S., Madsen, D., *et al.* (2003). The relationship between physicochemical properties, *in vitro* activity and pharmacokinetic profiles of analogues of diamine-containing efflux pump inhibitors. Bioorg. Med. Chem. Lett. *13*, 4241–4244.

Webber, M. A., Bailey, A. M., Blair, J. M., Morgan, E., Stevens, M. P., Hinton, J. C., Ivens, A., Wain, J., and Piddock, L. J. (2009). The global consequence of disruption of the AcrAB-TolC efflux pump in *Salmonella enterica* includes reduced expression of SPI-1 and other attributes required to infect the host. J. Bacteriol. *191*, 4276–4285.

Wehmeier, C., Schuster, S., Fahnrich, E., Kern, W. V., and Bohnert, J. A. (2009). Site-directed mutagenesis reveals amino acid residues in the *Escherichia coli* RND efflux pump AcrB that confer macrolide resistance. Antimicrob. Agents Chemother. *53*, 329–330.

Weir, S., Macone, A., Donatelli, J., Trieber, C., Taylor, D., Tanaka, S., and Levy, S. (2003). The Activity of PTK0796 Against Tetracycline Resistance, Paper presented at: 43rd Interscience Conference on Antimicrobial Agents and Chemotherapy (Chicago, IL).

Woodford, N. (2005). Biological counterstrike: antibiotic resistance mechanisms of Gram-positive cocci. Clin. Microbiol. Infect. *11* (*Suppl 3*), 2–21.

Wright, G. D. (2007). The antibiotic resistome: the nexus of chemical and genetic diversity. Nat. Rev. Microbiol. *5*, 175–186.

Yen, M. R., Chen, J. S., Marquez, J. L., Sun, E. I., and Saier, M. H. (2010). Multidrug resistance: phylogenetic characterization of superfamilies of secondary carriers that include drug exporters. Methods Mol. Biol. *637*, 47–64.

Yoshida, K., Nakayama, K., Kuru, N., Kobayashi, S., Ohtsuka, M., Takemura, M., Hoshino, K., Kanda, H., Zhang, J. Z., Lee, V. J., and Watkins, W. J. (2006a). MexAB-OprM specific efflux pump inhibitors in *Pseudomonas aeruginosa*. Part 5, Carbon-substituted analogues at the C-2 position. Bioorg. Med. Chem. *14*, 1993–2004.

Yoshida, K. I., Nakayama, K., Yokomizo, Y., Ohtsuka, M., Takemura, M., Hoshino, K., Kanda, H., Namba, K., Nitanai, H., Zhang, J. Z., *et al.* (2006b). MexAB-OprM specific efflux pump inhibitors in *Pseudomonas aeruginosa*. Part 6, Exploration of aromatic substituents. Bioorg. Med. Chem. *14*, 8506–8518.

Yu, E. W., Aires, J. R., and Nikaido, H. (2003a). AcrB multidrug efflux pump of *Escherichia coli*: composite substrate-binding cavity of exceptional flexibility generates its extremely wide substrate specificity. J. Bacteriol. *185*, 5657–5664.

Yu, E. W., McDermott, G., Zgurskaya, H. I., Nikaido, H., and Koshland, D. E., Jr. (2003b). Structural basis of multiple drug-binding capacity of the AcrB multidrug efflux pump. Science *300*, 976–980.

Yu, E. W., Aires, J. R., McDermott, G., and Nikaido, H. (2005). A periplasmic drug-binding site of the AcrB multidrug efflux pump: a crystallographic and site-directed mutagenesis study. J. Bacteriol. *187*, 6804–6815.

Zgurskaya, H. I. (2002). Molecular analysis of efflux pump-based antibiotic resistance. Int. J. Med. Microbiol. *292*, 95–105.

Zgurskaya, H. I., and Nikaido, H. (1999a). AcrA is a highly asymmetric protein capable of spanning the periplasm. J. Mol. Biol. *285*, 409–420.

Zgurskaya, H. I., and Nikaido, H. (1999b). Bypassing the periplasm: reconstitution of the AcrAB multidrug efflux pump of *Escherichia coli*. Proc. Natl. Acad. Sci. U.S.A. *96*, 7190–7195.

Zgurskaya, H. I., and Nikaido, H. (2000). Multidrug resistance mechanisms: drug efflux across two membranes. Mol. Microbiol. *37*, 219–225.

Zgurskaya, H. I., and Nikaido, H. (2002). Mechanistic parallels in bacterial and human multidrug efflux transporters. Curr. Protein Pept. Sci. *3*, 531–540.

Zgurskaya, H. I., Yamada, Y., Tikhonova, E. B., Ge, Q., and Krishnamoorthy, G. (2009). Structural and functional diversity of bacterial membrane fusion proteins. Biochim. Biophys. Acta *1794*, 794–807.

Zhao, Q., Li, X. Z., Mistry, A., Srikumar, R., Zhang, L., Lomovskaya, O., and Poole, K. (1998). Influence of the TonB energy-coupling protein on efflux-mediated multidrug resistance in *Pseudomonas aeruginosa*. Antimicrob. Agents Chemother. *42*, 2225–2231.

Zhao, Q., and Poole, K. (2002). Differential effects of mutations in tonB1 on intrinsic multidrug resistance and iron acquisition in *Pseudomonas aeruginosa*. J. Bacteriol. *184*, 2045–2049.

Zheleznova, E. E., Markham, P. N., Neyfakh, A. A., and Brennan, R. G. (1999). Structural basis of multidrug recognition by BmrR, a transcription activator of a multidrug transporter. Cell *96*, 353–362.

Zuckerman, J. M., Qamar, F., and Bono, B. R. (2009). Macrolides, ketolides, and glycylcyclines: azithromycin, clarithromycin, telithromycin, tigecycline. Infect. Dis. Clin. North Am. *23*, 997–1026, ix-x.

Metabolite-sensing Riboswitches as Antibacterial Drug Targets

5

Elaine R. Lee, Kenneth F. Blount and Ronald R. Breaker*

Abstract

The need for new antibacterial drug targets increases as antibiotic resistant pathogens continue to arise. Researchers have recently begun to investigate whether structured noncoding RNAs such as riboswitches can be exploited as targets for new classes of antimicrobial compounds. Riboswitches are gene control elements made entirely of RNA, and in bacteria they are usually located in the 5′ untranslated regions (UTRs) of messenger RNAs. These elements are capable of forming complex structures that selectively bind to specific fundamental metabolites and often control the expression of proteins critical for bacterial metabolism and survival. In principle, novel ligands could be designed that target specific riboswitches and alter the expression of the critical genes they regulate. Several riboswitch classes have begun to be examined as potential targets for new classes of antibacterial compounds. Herein we present some of the data generated by efforts to validate riboswitches as drug targets and discuss some of the key unanswered questions that will determine the ultimate success of antibacterial compounds that interact with these RNAs.

Introduction

As bacteria continue to evolve resistance to antibiotics, the search for new antibacterial drugs must evolve as well. Although there is a vast array of antibacterial drugs, many are becoming less clinically useful owing to increasing resistance. Moreover, commonly prescribed antibiotics inhibit only a few cellular processes including cell wall biosynthesis, protein synthesis, RNA polymerase and DNA gyrase (Coates *et al.*, 2002). Additional mechanisms for antibacterial chemotherapy involving new targets and new compounds will need to be exploited just to keep pace with the trend of increasing bacterial resistance. To maximize the chances for research and development success, it is appropriate to keep in mind that nucleic acids as well as proteins can be sources of novel targets for the development of future classes of antibacterial compounds. Indeed, almost half of all known natural antibiotics inhibit protein synthesis by targeting ribosomal RNA (Mankin, 2006), which highlights the possibility that other RNAs may make good target candidates for future drug development efforts.

Recently, we and others have begun to explore a class of non-coding RNAs called riboswitches as potential antibacterial drug targets (e.g. see Blount and Breaker, 2006; Breaker, 2009; Ott *et al.*, 2009; Serganov *et al.*, 2009). Metabolite-binding riboswitches are structured non-coding domains within messenger RNAs (mRNAs) that regulate gene expression in response to binding of a specific small-molecule ligand (Dambach and Winkler, 2009; Roth and Breaker, 2009). Many classes of riboswitches regulate the expression of genes responsible for synthesizing or transporting essential coenzyme intermediates or coenzymes such as thiamin pyrophosphate (TPP), coenzyme B_{12} (AdoCbl), flavin mononucleotide (FMN),

*Corresponding author

and S-adenosylmethionine (SAM). Many other classes regulate the synthesis or transport of key metabolites including lysine, guanine, adenine, glucosamine-6-phosphate, and prequeuosine-1. Still other classes sense metabolite second messengers such as c-di-GMP to regulate key adaptive pathways such as motility and biofilm formation.

In most instances, riboswitches repress expression of these metabolic genes (Barrick and Breaker, 2007), and therefore applying compounds that repress these genes at all times should disrupt normal bacterial growth. However, several riboswitch classes commonly turn on expression of their associated genes. For example, a riboswitch class has been found to sense glycine and activate the expression of genes needed to utilize this amino acid as an energy source (Mandal *et al.*, 2004). Also, a riboswitch class for S-adenosylhomocysteine (SAH) has been identified that activates the expression of proteins needed to break down this toxic by-product of SAM coenzyme use (Wang *et al.*, 2008). Even dysregulation of these riboswitches may cause cellular distress to an extent that may be therapeutically relevant. Given the growing diversity of riboswitch structures and mechanisms being discovered in bacteria, there appears to be multiple opportunities to design compounds that bind riboswitches and dysregulate key metabolic processes. This chapter discusses some of the key aspects regarding the prospects for developing new classes of antibacterial compounds that target riboswitches.

Riboswitch architectures and mechanisms

Riboswitches usually are composed of two functional domains called an 'aptamer' and an 'expression platform' (Fig. 5.1). The aptamer domain forms a highly structured pocket that binds selectively to a specific metabolite without the assistance of proteins. The expression platform domain alters the expression of the adjoining mRNA in response to metabolite binding by the aptamer (Barrick and Breaker, 2007; Breaker, 2008, 2009). Riboswitches are often classified by distinctive and highly conserved sequence and structural features of the aptamer domain and by the identity of the ligand that is recognized by the aptamer. For example, members of a class of riboswitches that bind FMN share conserved stretches of nucleotide sequence interspersed within a common six-stem junction (Gelfand *et al.*, 1999; Mironov *et al.*, 2002; Winkler *et al.*, 2002b). These conserved features collaborate to form a complex 3-D structure that selectively

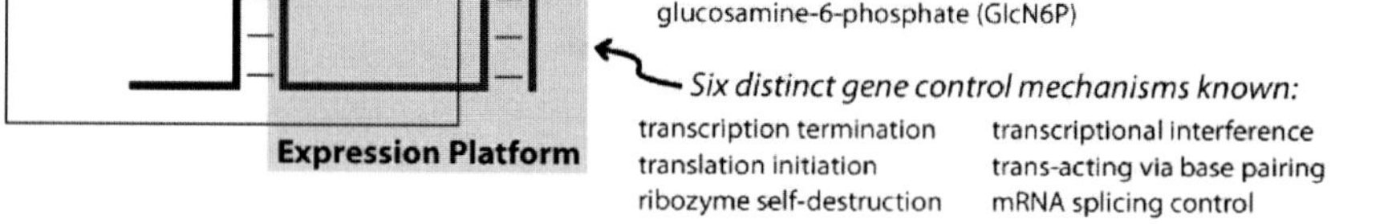

Figure 5.1 Riboswitch architectures, classes and mechanisms. Most riboswitches are composed of a ligand-binding aptamer and an expression platform that overlaps the downstream portion of the aptamer. Overlap of mutually exclusive structures of the aptamer and expression platform domains results in modulation of gene expression most commonly by transcription termination or by inhibition of translation initiation. Some gene control mechanisms do not have typical expression platform domains (e.g. GlcN6P-dependent ribozyme self-cleavage, mRNA splicing control). Since aptamer domains are the most highly conserved portion of riboswitches and expression platforms vary widely in sequence, structure and mechanism, aptamer structure and ligand binding characteristics are used to define classes listed here. Numbers in brackets designate the number of distinct aptamer classes known that sense the same ligand.

binds FMN, wherein some of the conserved nucleotides directly contact the ligand (Serganov *et al.*, 2009). Therefore, other RNAs that carry these critical nucleotides and substructures probably bind FMN, and would also be likely to bind drug compounds that mimic FMN.

In some cases, several distinct classes of aptamers have been identified that bind to the same ligand. For example, there are at least five different classes of riboswitches that all bind to *S*-adenosylmethionine (SAM) or *S*-adenosylhomocysteine (SAH; Wang and Breaker, 2008; Poiata *et al.*, 2009; Weinberg *et al.*, 2010). Global structure models of several of these riboswitch classes have been solved using x-ray crystallography or NMR and these models show that a single metabolite like SAM can be recognized in several different conformations by different RNA aptamers with distinct binding pockets (Wang and Breaker, 2008). A single riboswitch class is known to strongly discriminate against SAM and selectively bind SAH (Wang *et al.*, 2008), which is a metabolic by-product of SAM metabolism that differs from this coenzyme by a single methyl group. This high degree of specificity is important because intracellular concentrations of SAH can be toxic. This natural diversity of riboswitch aptamer structures and ligand binding characteristics observed for SAM and SAH may also occur for other metabolites, creating both challenges and opportunities for drug development as discussed in further detail below.

Riboswitch aptamers generally bind to their cognate metabolite with high selectivity and high affinity, with dissociation constant (K_D) values ranging from low picomolar to mid micromolar. For example, a guanine riboswitch from the *xpt-pbuX* operon from *Bacillus subtilis* binds guanine with a K_D no poorer than 5 nM (Mandal *et al.*, 2003) and discriminates against adenine by five orders of magnitude. Selectivity is often achieved through direct hydrogen bonding interactions between the riboswitch aptamer and the cognate ligand. Guanine riboswitches achieve high discrimination against adenine by forming a base pair interaction between the guanine ligand and a highly conserved cytosine nucleotide in the core of the aptamer (Batey *et al.*, 2004; Serganov *et al.*, 2004). Although additional contacts are made between the aptamer and ligand, mutation of this cytosine to a uridine alters the selectivity so that the aptamer preferentially binds to adenine (Mandal and Breaker, 2004). This selectivity is also exemplified in nature by adenine binding riboswitches, which share the same architecture as guanine riboswitches except for the key C-to-U variation noted above (Mandal and Breaker, 2004; Kim and Breaker, 2008). Such characteristics hint at the difficulty riboswitches will have in preventing drug binding, as mutations that disrupt analogue interactions are also likely to adversely affect binding of the natural ligand. In other words, riboswitch mutations that avoid analogue binding are also likely to disrupt normal regulation of the metabolic pathway, which may be deleterious to bacterial survival.

In contrast to riboswitch aptamers, expression platform domains and the manner in which they regulate gene expression are very diverse (Fig. 5.1) and can differ greatly in sequence and structure even within a specific class of riboswitches. The most common expression platform domains in bacteria regulate either transcription or translation of the adjoining mRNA (Barrick and Breaker, 2007). For example, a riboswitch that recognizes FMN is found upstream of the *ribD* operon in many Gram-positive bacteria, including *B. subtilis*. The operon in this bacterium codes for six proteins necessary for *de novo* riboflavin synthesis. When FMN binds to the *ribD* riboswitch, a transcription terminator hairpin (Yarnell and Roberts, 1999) can form immediately downstream of the FMN-binding aptamer, causing termination of transcription prior to synthesis of the first open reading frame (Fig. 5.2A; Winkler *et al.*, 2002b; Wickiser *et al.*, 2005). When FMN is not bound, the aptamer domain forms an alternative secondary structure called an antiterminator that prevents formation of the terminator stem. This latter configuration allows transcription of the entire *ribD* operon to proceed. However, the speeds of riboswitch folding, ligand binding, and RNA transcription can work together to create complicated pathways to genetic switch outcomes (Fig. 5.2B).

Riboswitches control gene expression at the level of translation using similar ligand-mediated structural changes. For example, a second FMN

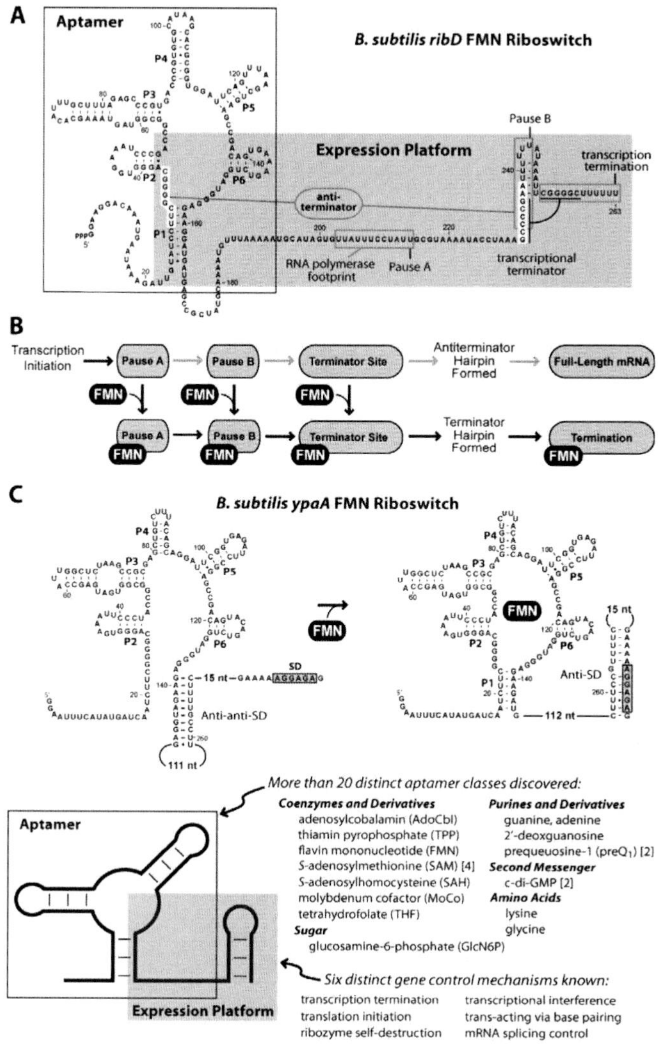

Figure 5.2 Transcription and translation control by flavin mononucleotide (FMN) riboswitches from *Bacillus subtilis*. (A) Sequence and secondary structure model for the *ribD* FMN riboswitch from *B. subtilis*. The features important for kinetic control of transcription termination are annotated. (B) Simplified scheme for kinetic control of the *B. subtilis ribD* riboswitch. The transcriptional pause sites provide time for RNA folding and ligand binding (Wickiser *et al.*, 2005). (C) Sequence and secondary structure model for the *ypaA* FMN riboswitch from *B. subtilis* (Winkler *et al.*, 2002b).

aptamer in *B. subtilis* is located in front of the *ypaA* gene, which encodes a transporter protein. In this case, formation of the ligand-bound *ypaA* aptamer structure sequesters the ribosome binding site or Shine–Dalgarno (SD) sequence, thereby preventing the initiation of translation (Fig. 5.2C; Winkler *et al.*, 2002b).

Despite the structural and functional diversity of expression platforms and their mechanisms for gene control, the aptamer domain establishes ligand binding specificity and subsequently guides the folding of the adjoining expression platform to regulate gene control. Therefore, analogues that induce aptamer shape changes similar to those triggered by the natural ligand should also trigger expression platform shape changes and functions regardless of the type of expression platform present. If the functions of riboswitch-controlled genes are critical for pathogen survival or disease-causing activities, and if riboswitch-binding compounds can cause sufficient dysregulation of gene expression, then the riboswitch class may be an excellent focus for new antibacterial compound development.

Riboswitches as antibacterial drug targets

Riboswitches have several characteristics that suggest they will be useful antibacterial drug targets. Already, non-coding RNA is a well-established drug target given the historical success of drugs that interfere with ribosome functions (Yonath, 2005; Mankin, 2006). Other non-coding RNAs have also been suggested as potential targets (e.g. see Ecker and Griffey, 1999; Wahlestedt, 2006; Vicens, 2009), but no compounds that target RNAs other than ribosomes have reached the clinic. In most of these examples, as in the case of ribosomes, the interaction of the ligand with the RNA is largely fortuitous and is based on inherently less-selective molecular interactions such as electrostatic shape complementation (Hermann and Westhof, 1998; Hermann and Tor, 2005). In contrast, riboswitches have evolved to form structured receptors that naturally bind to a specific small molecule ligand with very high selectivity. This suggests that novel small-molecule ligands could be designed to bind selectively to riboswitches using traditional medicinal chemistry approaches in much the same way that small-molecule drugs are designed for protein receptors.

Importantly, some riboswitch classes are very widespread among diverse bacterial species, including many that present the greatest clinical threats such as *Pseudomonas aeruginosa*, *Klebsiella pneumoniae*, and *Staphylococcus aureus* (Fig. 5.3; Blount and Breaker, 2006). Indeed over 2% of the genomes of some bacteria are regulated by riboswitches (Mandal *et al.*, 2003; Winkler, 2005). In many cases, the genes regulated by riboswitches are known to be essential for survival or virulence. Finally, no representatives of bacterial riboswitch classes have been identified in humans, which emphasizes the potential for pathogen-selective chemotherapeutic agents.

A number of lines of evidence are needed to build a case for pursuing a riboswitch class as a drug target. Most importantly, does the riboswitch control a gene or collection of genes whose proper regulation is essential for bacterial growth and survival? This question can be assessed by creating genetic knock-outs of the genes that are repressed by the riboswitch class of interest, and then determining the effects of these mutations on cell viability. However, an even more challenging question to address is whether binding of a ligand analogue to the riboswitch can repress the expression of these genes to a sufficiently low level to be detrimental to the bacterium. For some individual riboswitches, even a great abundance of the original metabolite under normal conditions may not completely repress expression because the gene control element may be naturally leaky and allow a small amount of mRNA transcription or translation. Therefore, to be a useful drug target, the riboswitch must be capable of reducing the expression of the adjoining gene to a level that adversely affects pathogen viability.

Another particularly challenging aspect of riboswitch drug development is to determine whether the riboswitch of interest is the exclusive target for the antibacterial compound. Confirming that targeting a given riboswitch is sufficient to bring about the desired antibacterial effect would be particularly valuable because every aspect of drug development from designing novel analogues to conducting biochemical efficacy assays could proceed with confidence if the true drug target is known. However, it is very difficult to rule out the possibility that important targets other than riboswitches are affected by an analogue, and that other interactions between the compound and RNA or protein targets may be required to bring about the observed antibacterial effect. The analysis of resistant mutants can provide some evidence to support the importance of the riboswitch target if mutations occur in the aptamer that disrupt analogue binding. Support for riboswitch importance also can be derived by testing a series of ligand analogues and establishing a correlation between antibacterial activity and riboswitch binding or functional modulation (Blount *et al.*, 2006; Kim *et al.*, 2009). Although these findings will not prove that the riboswitch

	TPP	AdoCbl	SAM I	SAM II	SAM III	SAM IV	FMN	Lysine	Glycine	Purine	c-di-GMP	Moco, Wco	$PreQ_1$-I	GlcN6P	SAH
Acinetobacter baumannii	1(1)	1(2)	–	–	–	–	1(1)	–	1(1)	–	–	–	–	–	–
Bacillus anthracis	7(19)	1(1)	17(36)	–	–	–	2(5)	4(4)	1(1)	6(9)	1(1)	–	2(5)	1(1)	–
Brucella melitensis	2(11)	4(8)	–	1(1)	–	–	1(1)	–	1(3)	–	–	–	–	–	–
Clostridium difficile	2(7)	3(25)	5(5)	–	–	–	2(5)	1(1)	2(2)	3(3)	12(21)	–	–	1(1)	–
Enterococcus faecalis	2(5)	2(5)	–	–	1(1)	–	1(1)	1(1)	–	1(2)	–	–	2(3)	1(1)	–
Escherichia coli	3(11)	1(2)	–	–	–	–	1(1)	1(1)	–	–	–	1(5)	–	–	–
Francisella tularensis	1(1)	–	–	1(1)	–	–	1(5)	–	–	–	–	–	1(1)	–	–
Haemophilus influenzae	3(11)	–	–	–	–	–	1(1)	1(1)	1(1)	–	–	1(1)	–	–	–
Helicobacter pylori	1(2)	–	–	–	–	–	–	–	–	–	–	–	–	–	–
Klebsiella pneumoniae	4(14)	2(21)	–	–	–	–	1(1)	1(1)	–	–	–	1(6)	–	–	–
Listeria monocytogenes	2(5)	2(20)	7(14)	–	–	–	1(1)	1(1)	1(3)	2(3)	–	–	1(6)	1(3)	–
Mycobacterium tuberculosis	2(6)	2(4)	–	–	–	1(1)	–	–	2(3)	–	–	–	–	–	–
Pseudomonas aeruginosa	1(1)	5(24)	–	–	–	–	1(2)	–	–	–	–	–	–	–	1(1)
Salmonella enterica	3(11)	2(22)	–	–	–	–	1(1)	–	–	–	–	1(1)	–	–	–
Shigella flexneri	3(12)	1(1)	–	–	–	–	1(1)	1(1)	–	–	–	1(5)	–	–	–
Staphylococcus aureus	2(7)	–	4(6)	–	–	–	2(5)	2(2)	1(3)	1(4)	–	–	1(3)	1(1)	–
Streptococcus pneumoniae	4(11)	–	–	–	1(1)	–	2(5)	–	1(1)	1(2)	–	–	–	–	–
Vibrio cholerae	2(7)	1(1)	–	–	–	–	1(1)	3(3)	1(1)	2(2)	–	1(5)	–	–	–
Yersinia pestis	3(8)	1(2)	–	–	–	–	1(1)	–	–	–	–	1(5)	–	–	–

Figure 5.3 Pathogenic bacteria with riboswitches. The distribution of representatives of 15 classes of riboswitches in various pathogenic bacteria. Each column reports the number of predicted riboswitch representatives and the number of genes (in parentheses) predicted to be directly regulated by riboswitch activity. This is an updated version of a chart from a previous publication (Blount and Breaker, 2006) created largely using riboswitch predictions from the Rfam database (Gardner *et al.*, 2009). Genes predicted to be in operons under riboswitch control are based on the presence of intergenic regions of fewer than 100 nucleotides.

is the only important target for antibacterial function, each data point will add further weight to the hypothesis.

Several different riboswitch classes have been recently studied as potential antibacterial drug targets (Blount and Breaker, 2006; Breaker, 2009). These studies address various aspects of riboswitch drug targeting, including an assessment of the potential for using traditional medicinal chemistry, structure-guided design, and high-throughput screening methods for discovering novel riboswitch-targeting lead compounds. Additionally, these reports demonstrated that some riboswitches are probable targets for several previously identified antibacterial compounds, including a naturally occurring compound (Lee *et al.*, 2009; Ott *et al.*, 2009). Perhaps most importantly, these studies have begun to examine how the functional principles of riboswitches will impact the success of targeting riboswitches with antibiotics. Below, we summarize the findings of these early studies and examine the key unanswered questions regarding riboswitch function in bacteria that will impact the success of targeting specific riboswitches.

TPP riboswitches as antibacterial drug targets

Thiamin pyrophosphate (TPP) is an essential coenzyme in all living systems. In bacteria, TPP is synthesized by the enzymatic condensation of two precursors: hydroxymethylpyrimidine pyrophosphate (HMP-PP) and 4-(β-hydroxyethyl)-5-methylthiazole phosphate (Thz-P; Taylor *et al.*, 1998). The non-phosphorylated form of each precursor can be transported into the cell, salvaged from other metabolic pathways or, most frequently, synthesized *de novo* (Begley *et al.*, 1999; Settembre *et al.*, 2003). In many bacteria, the expression of the genes involved in synthesis or salvage of TPP or its precursors is directly regulated by one or more riboswitches that recognize TPP (Mironov *et al.*, 2002; Rodionov *et al.*, 2002; Winkler *et al.*, 2002a; Barrick and Breaker, 2007). For example, in *E. coli* and many other bacteria, the expression of almost all genes responsible for HMP-PP and Thz-P synthesis (Park *et al.*, 2003; Settembre *et al.*, 2003; Leonardi and Roach, 2004), as well as genes responsible for the phosphorylation and transport of thiamin, are controlled by TPP riboswitches (Miranda-Ríos *et al.*, 2001; Rodionov *et al.*, 2002 Winkler *et al.*, 2002a; Sudarsan *et al.*, 2005). Similarly, two operons and three single genes in *B. subtilis* are regulated by TPP riboswitches.

Thus, in most bacterial species, TPP riboswitches participate in a feedback mechanism of regulation that attenuates the production of genes for TPP production as its concentration increases. Most commonly, TPP riboswitches accomplish this regulation by controlling formation of a typical bacterial transcription terminator, or by precluding translation initiation by sequestration of the SD sequence (Fig. 5.4A; Mironov *et al.*, 2002; Winkler *et al.*, 2002a; Barrick and Breaker, 2007).

The first experimentally validated examples of TPP riboswitches were those located in the 5′ UTRs of *E. coli thiM* and *thiC* genes (Mironov *et al.*, 2002; Winkler *et al.*, 2002a), which encode thiamin synthesis proteins. When these UTRs were fused to a *lacZ* reporter gene, β-galactosidase assays revealed that reporter gene expression levels in cells grown in minimal medium decreased when thiamin was supplemented (Winkler *et al.*, 2002a). Additionally, an RNA structure analysis method called in-line probing (Regulski and Breaker, 2008; Soukup and Breaker, 1999b) was used to show that TPP is bound by *thiM* and *thiC* riboswitches *in vitro* with high affinity and specificity. In-line probing is a technique that utilizes the spontaneous degradation of RNAs that occurs when the 2′ oxygen on a ribose attacks the adjacent phosphorus. Regions of RNA that are unstructured can adopt the conformation necessary for in-line attack more frequently than nucleotides that are involved in secondary or tertiary structures. Increased rates of RNA cleavage at specific internucleotide linkages upon ligand addition are usually due to increased floppiness of the flanking nucleotides, and this can be assessed by quantifying the cleavage product bands after separation by gel electrophoresis. Decreased rates of RNA cleavage indicate that the nucleotides are becoming more structured when ligand is present. Consequently, in-line probing can be used to identify regions of RNA that change in structure when ligand binds. Using in-line probing at concentrations ranging from 0 to 100 μM, it was determined that *thiC* and *thiM* riboswitches bind to TPP with an apparent K_D of approximately 100 nM and 600 nM, respectively. By comparison, the apparent K_D of the riboswitches with TP or thiamin is 1000-fold poorer (Winkler *et al.*, 2002a).

Prior to the discovery of riboswitches, researchers had synthesized numerous analogues of thiamin to enable the identification and characterization of thiamin biosynthesis enzymes. One such analogue, pyrithiamine (Tracy and Elderfield, 1941; Fig. 5.5A), was determined to be toxic to bacteria and fungi (Robbins, 1941; Woolley and White, 1943), but the precise mechanism of pyrithiamine toxicity remained a mystery for decades. More recently, the discovery of riboswitches led to the hypothesis that pyrithiamine could inhibit bacterial growth by targeting TPP riboswitches and thereby repressing thiamin biosynthesis (Sudarsan *et al.*, 2005). To investigate this possibility, pyrithiamine-resistant strains of *E. coli* that had been identified decades ago (Kawasaki and Nose, 1969) were examined for evidence of mutations in TPP riboswitch regions (Kawasaki *et al.*, 1976; Sudarsan *et al.*, 2005). Indeed, many of these resistant strains, as well as newly selected pyrithiamine-resistant strains, have point mutations in TPP riboswitches, and in each case these mutations decrease the binding affinity of the riboswitch for TPP and other ligands when measured by in-line probing. Moreover, when cloned upstream of a reporter gene, the mutated riboswitches are less sensitive to exogenous thiamin supplementation than wild-type TPP riboswitches (Sudarsan *et al.*, 2005).

Ligand binding assays also revealed that pyrithiamine is bound far less tightly by TPP riboswitches compared with its pyrophosphate derivative. This is similar to the observed binding specificity of TPP riboswitches, which favour TPP over the unphosphorylated thiamin form of the coenzyme (Winkler *et al.*, 2002a). These combined observations support a mechanism in which pyrithiamine is taken up by cells, phosphorylated to generate the diphosphate form, and inhibits

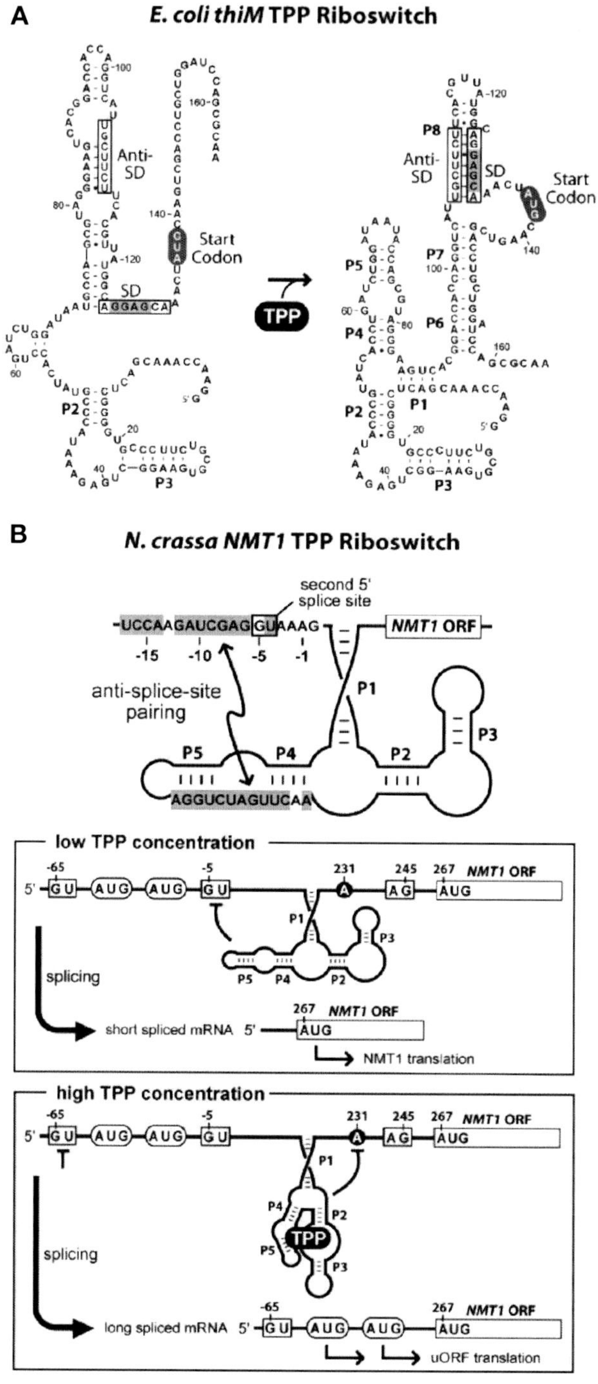

Figure 5.4 Examples of mechanisms for thiamin pyrophosphate (TPP) riboswitches from bacteria and fungi. (A) Structural models for translation control by the *Escherichia coli thiM* riboswitch (Winkler *et al.*, 2002a; Rentmeister *et al.*, 2007). (B) Schematic representation of TPP-mediated modulation of alternative splicing by the *Neurospora crassa thiM* riboswitch (Cheah *et al.*, 2008).

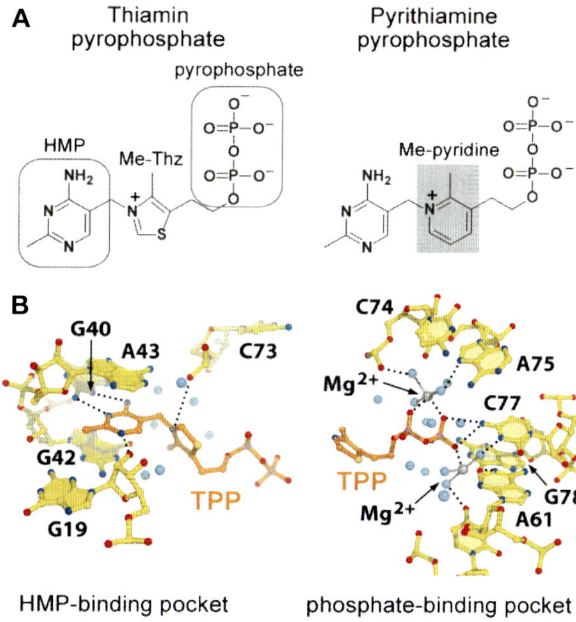

Figure 5.5 Chemical structures of thiamin pyrophosphate (TPP), pyrithiamine pyrophosphate, and the docking of TPP to a riboswitch aptamer. (A) Chemical structures of thiamin pyrophosphate and its antimicrobial analogue, pyrithiamine pyrophosphate. TPP is largely composed of three chemical moieties, HMP (hydroxymethylpyrimidine), Me-Thz (methylthiazole), and pyrophosphate. Pyrithiamine pyrophosphate differs from TPP by replacement of the Me-Thz group with Me-pyridine (methylpyridine). (B) Atomic-resolution structural model (Serganov *et al.*, 2006) of the binding pockets for the HMP (left) and pyrophosphate (right) moieties of TPP.

bacterial growth at least in part through targeting one or more TPP riboswitches. Moreover, these findings provided the first body of evidence that bacterial riboswitches could be targets for analogues of their natural ligands. However, conclusions regarding the action of possible riboswitch targeting drugs must be made with caution. Mutations in the putative riboswitch target to overcome drug binding frequently derepress gene expression. This mechanism of resistance could also overcome the effects of a drug that acts by binding some other target. Specifically, riboswitch derepression may cause overproduction of the natural metabolite, and a large increase in the concentration of the metabolite may simply allow it to outcompete the drug for its binding site on a completely different target.

TPP riboswitches are the only bacterial riboswitch class known to be carried by organisms from all three domains of life (Sudarsan *et al.*, 2003a). In eukaryotes, TPP riboswitches are common in many fungal and plant species, and their presence in fungi may explain why pyrithiamine is also toxic to these organisms. Indeed, pyrithiamine resistance can be used as a selectable genetic marker when cloning genes into fungi, and the resistance mutation has been mapped to a fungal TPP riboswitch (Kubodera *et al.*, 2000). As with bacterial TPP riboswitches, fungal representatives have been shown to bind pyrithiamine pyrophosphate (Sudarsan *et al.*, 2005).

In contrast to bacterial riboswitch gene control mechanisms, fungal TPP riboswitches have been shown to control gene expression by modulating mRNA splicing (Kubodera *et al.*, 2003; Cheah *et al.*, 2007). For example, each *NMT1* pre-mRNA of *Neurospora crassa* carries a TPP riboswitch aptamer embedded in an intron residing in its 5′ UTR (Fig. 5.4B). Under low TPP concentrations, the complete intron is removed via splicing to generate a mature mRNA that expresses protein from the main open reading frame (ORF). When

(e.g. Wickiser *et al.*, 2005). More importantly, this characteristic highlights the fact that the development of novel riboswitch ligands that trigger the desired gene regulation events may necessarily involve analysis of ligand association kinetics and folding pathways for riboswitches.

FMN riboswitches as drug targets

FMN riboswitches are not as common as are members of the TPP, AdoCbl, and SAM-I riboswitch classes (Fig. 5.7; Breaker, 2010; Ames and Breaker, 2010). However, FMN riboswitches almost always appear once or twice in numerous bacterial pathogens (Fig. 5.3; Blount and Breaker, 2006; Barrick and Breaker, 2007). The genomes of Gram-negative bacteria such as *E. coli* usually have only one FMN riboswitch representative located in the 5′ UTR of the *ribB* gene that codes for 3,4-dihydroxy-2-butanone-4-phosphate synthase (Mironov *et al.*, 2002; Winkler *et al.*, 2002b). The genomes of Gram-positive bacteria such as *B. subtilis* usually have two FMN riboswitch representatives. One is located in the 5′ UTR of an operon that encodes several proteins required for synthesizing riboflavin, which is subsequently phosphorylated to form FMN. The second FMN riboswitch is commonly found in the 5′ UTR of a riboflavin transporter protein called *ypaA* (also named *ribU*; Mack *et al.*, 1998; Gelfand *et al.*, 1999; Kreneva *et al.*, 2000; Lee *et al.*, 2001; Mironov *et al.*, 2002; Winkler *et al.*, 2002b; Vogl *et al.*, 2007).

An atomic-resolution structural model has been reported for an FMN riboswitch from *Fusobacterium nucleatum* bound to various ligands (Serganov *et al.*, 2009). The structure reveals a complex tertiary fold that creates a binding pocket using numerous nucleotides distributed among all of the regions joining the six base-paired stems (Fig. 5.2A). FMN binding specificity is established by intercalation of the isoalloxazine ring of FMN between two nucleobases, the formation of two hydrogen bonds between the ring and a highly conserved adenosine residue, and extensive interactions between the ligand's phosphate moiety and both divalent metal ion and RNA moieties. Surprisingly, there is only a single hydrogen bond formed between the aptamer and the ribotyl moiety of FMN. Despite the intricate fold of the RNA and its many contacts to FMN, it appears that there may be substantial opportunities for chemical manipulation of the ligand.

To be good drug targets, FMN riboswitches must control a gene or collection of genes that is essential for growth or survival of a bacterial

Figure 5.7 Riboswitch classes and the number of representatives in each class. The graphic is adapted from a previous publication (Breaker, 2010) and reflects the number of bioinformatics hits for each class found in 764 sequenced bacterial genomes.

pathogen. To assess the importance of riboswitch-controlled genes, the riboswitch in the 5′ UTR of the *ribD* operon from *B. subtilis* was replaced with an inducible promoter and the growth patterns of the modified strain were assessed under various levels of induction (Lee *et al.*, 2009). Compared with the growth of wild-type bacteria, the strain carrying the inducible *ribD* operon grew poorly without induction, but recovered near wild-type growth when induced. This suggests that FMN riboswitches control essential genes in *B. subtilis* and supports the finding that homologues of the regulated genes are essential in *E. coli* (Gerdes *et al.*, 2002).

Given that FMN riboswitches commonly control genes that are necessary to synthesize or import this essential metabolite, and given the apparent opportunities to manipulate the chemical structure of FMN without disrupting binding, this class appears to be a promising target for riboswitch drug development. Indeed, it appears that nature has already validated this riboswitch class as an antibacterial drug target. Roseoflavin (Fig. 5.8A) is a pigment synthesized by *Streptomyces davawensis* that has long been known to exhibit antimicrobial properties (Otani *et al.*, 1974, 1980). Bacteria that are resistant to roseoflavin have been shown to overproduce riboflavin and consequently, this phenomenon has been studied in detail by researchers who hope to create riboflavin-fortified foods (Burgess *et al.*, 2004; Sybesma *et al.*, 2004; LeBlanc *et al.*, 2005). Such observations are similar to those described above for pyrithiamine resistance and thiamine overproduction.

Researchers have also shown that many bacteria which are resistant to roseoflavin have mutations in the 5′ UTRs of riboflavin biosynthesis genes (Kil *et al.*, 1992; Ott *et al.*, 2009). These mutations have recently been found to map to FMN riboswitches (Lee *et al.*, 2009; Ott *et al.*, 2009; Serganov *et al.*, 2009). Since ligand binding to FMN riboswitches turns off the transcription of riboflavin synthesis genes, it was hypothesized that the increase in riboflavin production was caused by the inability of mutant riboswitches to bind to FMN (Lee *et al.*, 2009). This hypothesis was confirmed by conducting a series of experiments wherein wild-type and roseoflavin-resistant riboswitch mutants were fused to *lacZ* reporter genes and transformed into *B. subtilis*. When riboflavin or roseoflavin was added to the medium, the level of reporter gene expression controlled by a wild-type riboswitch dropped significantly (Lee *et al.*, 2009). Three

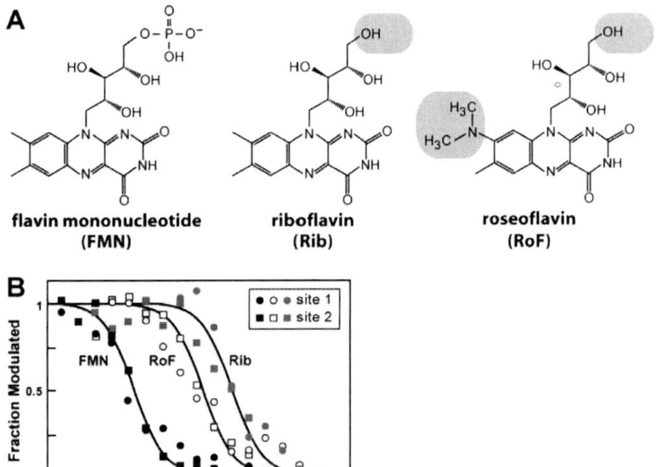

Figure 5.8 Roseoflavin structure and action. (A) The chemical structures of FMN, riboflavin and roseoflavin. (B) Binding curves generated by in-line probing for the FMN riboswitch aptamer from *Streptomyces davawensis* (Lee *et al.*, 2009).

point mutations to FMN riboswitches identified from roseoflavin-resistant strains of *B. subtilis* (Kil *et al.*, 1992) partially or completely resist reporter gene repression. Additionally, in-line probing data reveal ligand-binding characteristics for the wild-type and mutant RNAs that parallel the reporter gene results (Lee *et al.*, 2009). Therefore, roseoflavin resistance observed in some bacteria is caused by the inability of mutant FMN riboswitches to bind to the antimicrobial compound.

The chemical structure of roseoflavin differs from riboflavin by the addition of a single dimethylamino group (Otani *et al.*, 1974, 1980). In-line probing assays have shown that FMN riboswitch aptamers from *B. subtilis* bind to FMN very tightly, with a K_D of approximately 5 nM. In contrast, FMN riboswitches bind to riboflavin, the unphosphorylated precursor of FMN, with a K_D of approximately 2 μM (Winkler *et al.*, 2002b). Although FMN riboswitches do not bind roseoflavin as tightly as FMN, they bind to roseoflavin with a K_D of approximately 100 nM, which means that they have a stronger affinity for roseoflavin than riboflavin (Lee *et al.*, 2009) despite the fact that both compounds lack a phosphate group. Interestingly, the organism that produces this antibacterial compound, *S. davawensis,* also uses an FMN riboswitch to control the expression of its FMN biosynthesis operon. However, its riboswitch appears to be similarly receptive to roseoflavin binding (Fig. 5.8B), suggesting that this organism may have some strategy other than riboswitch discrimination to avoid being poisoned by its own antibacterial compound.

In the atomic-resolution structure models noted above, the riboswitch bound to roseoflavin does not appear to form distinct contacts with the dimethylamino group that would explain the tighter binding of roseoflavin (Serganov *et al.*, 2009). Rather, it may be that more subtle differences in the positions taken by binding pocket nucleotides explain the tighter binding of roseoflavin (Lee *et al.*, 2009). One such subtlety may be the observation that the ribityl moiety of roseoflavin appears to form five hydrogen bonding interactions with the aptamer whereas the riboflavin ribityl only forms four. Importantly, the proteins that phosphorylate riboflavin to form FMN have been shown to phosphorylate roseoflavin efficiently (Grill *et al.*, 2008). Therefore, it seems likely that the addition of a phosphate group to roseoflavin will increase binding affinity just as the phosphate moiety of FMN improves its affinity over riboflavin. If true, then the active form of the antibacterial compound may be roseoflavin phosphate (Lee *et al.*, 2009).

These results indicate that the antimicrobial properties of roseoflavin can be explained at least in part by binding to FMN riboswitches, which suggests that FMN riboswitches may be exploited as drug targets. Again, it may be prudent to identify compounds dissimilar to roseoflavin that are less likely to be bound by the many host flavoproteins. Given the propensity of riboswitch aptamers to make even subtle adaptive changes to their binding pockets, as is observed when roseoflavin binds (Serganov *et al.*, 2009), there should be substantial opportunities to create novel compounds that are accepted by this riboswitch class. Perhaps judicious synthesis and screening of FMN analogues will yield additional compounds that selectively target FMN riboswitches and that exhibit other properties suitable for antibiotic applications.

Guanine riboswitches as drug targets

Although guanine riboswitches (Mandal *et al.*, 2003) are even rarer than those discussed above (Fig. 5.7), these riboswitches are prevalent in several important pathogenic bacteria including *S. aureus* and *B. anthracis*. Furthermore, there is excellent functional and structural data for members of this class, which may be useful for those who seek to validate methods for riboswitch drug development. Guanine aptamers have a relatively simple secondary structure, consisting of a three-stem (P1, P2 and P3) junction with a short pseudoknot connecting stems P2 and P3 (Fig. 5.9A). The aptamers carry numerous well-conserved nucleotides, including one residue whose identity determines whether the aptamer selectively binds guanine (most commonly) or adenine (Mandal and Breaker, 2004; Fig. 5.9B).

Guanine riboswitches are usually located in the 5′ UTRs of genes that encode proteins involved in the synthesis, salvage or transport of purines. In

B. subtilis, the *pur* operon controls twelve genes that are responsible for the de novo synthesis of purines (Ebbole and Zalkin, 1987). A second operon called *xpt-pbuX* encodes a xanthine phosphoribosyltransferase and a xanthine-specific purine permease. This operon also had been known to be repressed in the presence of guanine, hypoxanthine, and xanthine (Christiansen *et al.*, 1997). By comparing the relatively long 5′ UTRs of these genes, researchers discovered the presence of a structured RNA that was also located in front of other purine-related genes (*pbuG* and *nupG*) in *B. subtilis* and other bacteria (Mandal *et al.*, 2003; Kim *et al.*, 2009). Using in-line probing, a representative aptamer located in front of the *xpt-pbuX* operon in *B. subtilis* was shown to bind tightly to guanine (K_D ~5 nM) and discriminate strongly against closely related purine compounds. When the guanine aptamer as well as the predicted downstream transcription terminator stem was fused to a *lacZ* reporter gene, gene repression was seen in the presence of guanine, hypoxanthine and xanthine. When mutations were made to the guanine aptamer, this gene repression disappeared. These results indicated that the riboswitch is capable of controlling gene expression in a guanine-dependent manner.

Although none of the genes in *B. subtilis* controlled by guanine riboswitches are individually essential (Kobayashi *et al.*, 2003), it seemed possible that the simultaneous repression of all genes controlled by guanine riboswitches could be lethal to cells. If true, this would provide a rationale for the possible antibacterial action of compounds that repress guanine riboswitch-mediated gene expression. To test this hypothesis, all guanine riboswitch-controlled genes were deleted from the *B. subtilis* genome with the exception of the *pur* operon, whose 5′ UTR was replaced with an inducible promoter. Interestingly, a viable strain in which all guanine riboswitch-mediated genes were knocked out could not be created, suggesting that comprehensive suppression of the entire regulon may be lethal to this organism. Furthermore, although the inducible strain can grow slowly in rich medium, this is probably due to the leaky production of pur operon mRNAs as detected by quantitative reverse transcription and polymerase chain reaction (qRT-PCR; Kim *et al.*, 2009). Likewise, compounds designed to target guanine riboswitches might not completely turn off transcription of guanine-regulated genes, but they may still be able to lower the expression to the point where bacterial growth is severely restricted owing to purine starvation.

Atomic-resolution structural analyses have revealed that guanine riboswitches completely envelop their cognate ligand (Fig. 5.9B), which could create challenges for those who wish to pursue the development of analogues that target this riboswitch. However, several studies reveal that analogues are reasonably tolerated by guanine aptamers (Mandal *et al.*, 2003; Gilbert *et al.*, 2006, 2009; Mulhbacher and Lafontaine 2007; Kim *et al.*, 2009). In some instances, substantial movement of binding site residues must occur, while other compounds carry modifications that should be largely accommodated by the original binding pocket configuration. By examining the atomic-resolution structures of the *B. subtilis xpt* guanine riboswitch (Batey *et al.*, 2004; Serganov *et al.*, 2004), it is evident that modifications to positions C2 and C6 of the purine ring should be tolerated without substantial adaptive changes. Indeed, of six guanine analogues with modifications at the C2 position, three are bound fairly tightly with K_D values ranging from 500 pM to 200 nM. These affinities compare favourably with the 5 nM K_D measured for guanine. Ten compounds were designed with modifications to the C6 position of guanine and all of these compounds bound to the riboswitch with K_D values ranging from 8 to 500 nM. These findings support the hypothesis that compounds can be rationally designed to target riboswitches using information obtained from structural models (Kim *et al.*, 2009).

All guanine analogues noted above were tested to determine whether they could inhibit *B. subtilis* growth (Kim *et al.*, 2009). Although five of the compounds were shown to negatively affect growth, only three of these compounds were able to completely inhibit growth in minimal medium. Only one of these compounds had an MIC (3.9 μM) within the range typically seen for commercial antimicrobial compounds. Furthermore, MIC values for all of the compounds became poorer when cells were grown in rich medium, which probably contains numerous

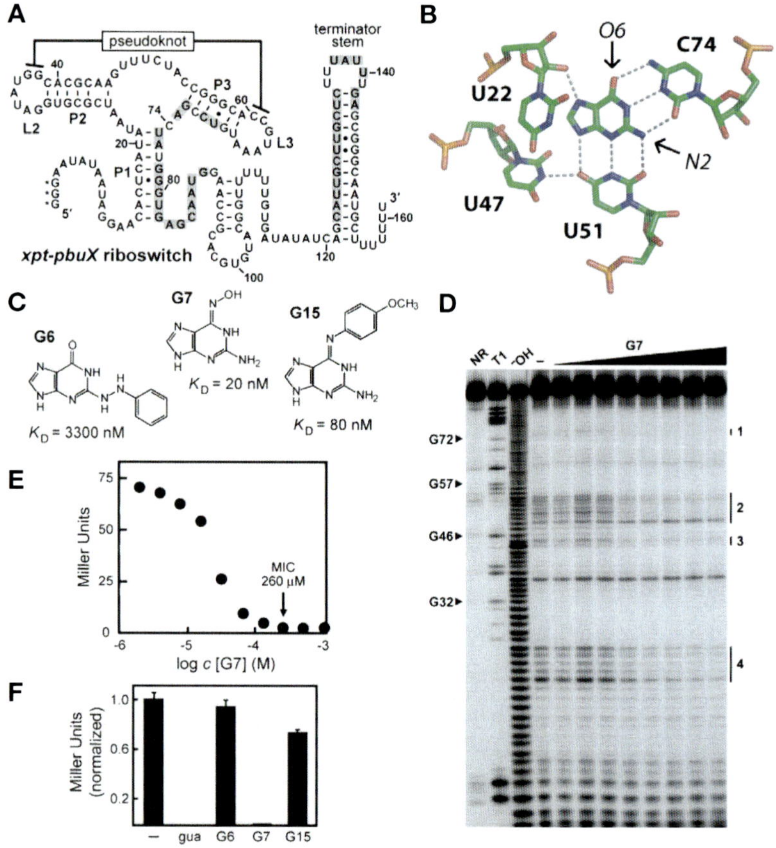

Figure 5.9 Targeting guanine riboswitches with guanine analogues. (A) Sequence and secondary structure of the guanine riboswitch associated with the *xpt-pbuX* operon from *Bacillus subtilis*. Shaded nucleotides identify the putative anti-terminator stem that forms in the absence of guanine. (B) Ligand binding site for the *B. subtilis xpt-pbuX* aptamer (Serganov *et al.*, 2004). (C) Representative guanine analogues that are bound by guanine riboswitch aptamers. (D) In-line probing assay for binding of compound G7 (ranging from 0 to 1 μM) by the *B. subtilis xpt-pbuX* aptamer. (E) Plot of β-galactosidase expression versus G7 concentration in the medium. (F) Activity of exogenous guanine (gua) or analogues depicted in C on expression of a reporter gene fused to the *B. subtilis xpt-pbuX* riboswitch. Figures are adapted from Kim *et al.* (2009).

purine metabolites. Despite the fact that some of the analogues bound to the riboswitches tighter than guanine, they did not significantly suppress bacterial growth. This could be because the compounds did not accumulate in cells, or the compounds may have been chemically modified in a manner that decreased their riboswitch affinity or converted them into precursors for purine synthesis.

The three analogues (Fig. 5.9C) that best inhibited bacterial growth were examined to determine whether they also inhibit guanine riboswitch-controlled gene expression (Kim *et al.*, 2009). The 5′ UTR of the *B. subtilis xpt-pbuX* operon containing wild-type and mutant riboswitches was fused to a *lacZ* reporter gene. As expected, reporter genes controlled by wild-type riboswitches were repressed in the presence of exogenous guanine (Fig. 5.9D), but no repression was seen for the mutant riboswitches that have lost the ability to bind guanine (data not shown). Two of the compounds, including the one with the lowest MIC, did not repress gene expression in a riboswitch-dependent manner, which reveals that mechanism of their toxicity probably does not involve binding to riboswitches. However, one of the compounds (G7) was shown to repress the expression of the reporter gene controlled by the wild-type riboswitch. Additional data revealed that this compound gives near complete repression near its MIC value. Although these results to not provide conclusive evidence that guanine riboswitches will make good drug targets, the data for G7 is consistent with a riboswitch-based mechanism for cell toxicity.

Prospects for future riboswitch drug development efforts

The riboswitch classes described above are not the only candidate targets for novel antibacterial discovery efforts. Several additional riboswitches, including SAM-I (Wang and Breaker, 2008) and GlcN6P (Winkler *et al.*, 2004) control genes for essential metabolic processes and are widespread in pathogens. Already, SAM-I riboswitches are known to bind the SAM analogue sinefungin (Montange *et al.*, 2010), although it is unknown whether the antibacterial properties of this compound are related to riboswitch binding. Furthermore, a riboswitch class that selectively binds the bacterial second messenger c-di-GMP (Sudarsan *et al.*, 2009) may offer drug developers a target for the manipulation of broader cell physiology changes (Breaker, 2010). Atomic-resolution structural models are available for each of these riboswitch classes (Klein and Ferré-D'Amaré, 2006; Kulshina *et al.*, 2009; Montange and Batey, 2006; Smith *et al.*, 2007, 2009; Montange *et al.*, 2010), which should assist drug developers as they consider analogue designs that trigger riboswitch function.

Some of the known riboswitch classes are very common in bacteria (Fig. 5.7). However, few classes are nearly universal in pathogens, and therefore compounds that target riboswitches are unlikely to serve as broad-spectrum antibiotics. Expanding genomic DNA database sizes and improving riboswitch discovery methods will undoubtedly produce many new riboswitches classes (Barrick and Breaker, 2007; Weinberg *et al.*, 2007), but the vast majority of these are likely to be far more rare that the classes currently known (Ames and Breaker, 2010). However, even riboswitches that are only in a few bacterial species may prove useful for the development of valuable narrow-spectrum antibacterial compounds. For example, multiple riboswitch classes are found in *S. aureus*. Because methicillin-resistant *S. aureus* is becoming a severe problem clinically, any drugs that can kill these bacteria would make an enormous impact. With ongoing improvements in technologies for rapid diagnostics, pathogen strain identification coupled with narrow-spectrum antibiotics treatment may obviate the need for broad-spectrum drugs for most patients. Even riboswitches with limited pathogen distributions may merit attention and be preferred by drug developers.

As with all antimicrobial agents, resistance to riboswitch-targeting drugs is of great concern. To date, very little data has been collected on how bacteria can evolve resistance to antimicrobials that target riboswitches. The most obvious mechanism of resistance is to acquire mutations that prevent analogue binding. This mechanism of resistance has been seen in some bacteria that have developed resistance to pyrithiamine (Sudarsan *et al.*, 2005), lysine analogues (Vold *et al.*, 1975; Lu *et al.*, 1992; Blount *et al.*, 2006), and roseoflavin (Kil *et al.*, 1992; Burgess *et al.*, 2004; Ott *et al.*, 2009; Sybesma *et al.*, 2004). Interestingly, such mutations also cause the riboswitch to fail to bind its natural ligand, resulting in excess expression of the genes under riboswitch regulation. Given the potential metabolic cost of excess expression of certain metabolic pathways, there may be selective pressure against the evolution of deactivating mutations. Thus, although these riboswitch mutations do appear in laboratory settings, it is unclear how quickly bacteria could evolve resistance in a human host. Moreover, many bacteria contain multiple representatives for some riboswitch classes, and it may take several spontaneous mutations in different riboswitch representatives to completely overcome the drug.

Riboswitch-targeting compounds that resemble the natural metabolite may also be susceptible to other mechanisms for antibiotic resistance (Blount and Breaker, 2006), including chemically modifying the drug or transporting the compound out of the cell. For example, the overexpression of the TenA protein may cause resistance to pyrithiamine, perhaps because it allows bacteria to hydrolyse the drug (Haas *et al.*, 2005). Also, it had previously been shown that *B. subtilis* can become resistant to 2-fluoroadenine by overexpressing a purine transporter capable of exporting the compound (Saxild and Nygaard, 1987). This same mutation, surprisingly occurring in an adenine-sensing riboswitch that controls expression of the transporter, was observed in a strain of *B. subtilis* that became resistant to a purine analogue designed to bind to guanine riboswitches (Kim *et al.*, 2009). Even so, the proteins that enable enzymatic or transport-related resistance to riboswitch-targeting agents are unlikely to be rapidly spread among bacterial species initially as is often the case for other antibiotics (e.g. β-lactamases on mobile genetic elements).

The availability of numerous atomic-resolution structural models for the most common riboswitch classes, including TPP, SAM-I, FMN, lysine, and guanine provide drug developers with much insight into how these RNAs bind their corresponding ligands and how the ligands may be modified but still retain their function. As noted above, these models already have been used to understand how some analogues are bound by riboswitches. In addition, these models can guide future analogue design to create novel compounds that retain binding and gene control activities, as has been done with guanine riboswitches (Kim *et al.*, 2009). Some guanine compounds were found to bind even tighter to the aptamer than the natural ligand, highlighting the possibility of creating 'superligands' that trigger riboswitch-mediated gene control at lower concentrations than their natural ligands.

Even without atomic-resolution structural models, riboswitches can be configured to serve as reporters for high-throughput screening assays. One straightforward application of riboswitches in high-throughput screening was demonstrated using the riboswitch for glucosamine-6-phosphate (Blount *et al.*, 2006; Mayer and Famulok, 2006). This riboswitch functions as a self-cleaving ribozyme whose activity is triggered by ligand binding (Winkler *et al.*, 2004). Fluorescent labels attached to key parts of the RNA can report ribozyme activity in real time, allowing robotic analysis of thousands of test samples containing different compounds. Such assays can be conducted in microplates using methods similar to those for many screens conducted with protein drug targets. Because riboswitch aptamers could be fused to self-cleaving ribozymes to create additional reporter constructs (Seetharaman *et al.*, 2001), similar fluorescence-based assays could be used for any riboswitch class.

Of course, the identification of new riboswitch-targeting compounds addresses only part of the challenge of creating new classes of antibacterial compounds. Great hurdles still remain regarding compound bioavailability and toxicity. Fortunately, there are no known human homologues of the common bacterial riboswitches. Therefore, unintentional misregulation of human genes by riboswitch-targeting compounds could be avoided. However, many of the compounds bound by riboswitches are essential vitamin derivatives that utilized by human proteins. Although the compound must mimic the natural metabolite in order to bind to riboswitches, it cannot be so similar that it binds and disrupts the function of human proteins to produce unwanted side-effects. Efficient binding and function assays for screening riboswitch-targeting compounds will allow researchers to accrue collections of candidate compounds that then can be further tested for necessary properties beyond antibacterial activity.

The findings described above suggest that noncoding RNAs such as riboswitches may be good targets for the development of new classes of antibacterial compounds. Hopefully, some of these efforts will produce antibacterial agents that can be used in the battle against antibiotic-resistant pathogens. Riboswitch discovery efforts are sure to produce an ever increasing diversity of possible drug targets in many pathogens. These efforts are also revealing a much larger collection of structured noncoding RNAs in bacteria that are not riboswitches, but have other unknown functions (e.g. Puerta-Fernandez *et al.*, 2006; Weinberg *et al.*, 2007; Meyer *et al.*, 2009; Tseng *et al.*, 2009; Weinberg *et al.*, 2010). The functions of these RNAs are often difficult to determine and some of these mysteries may take decades to resolve, but eventually these RNAs may themselves be recognized as potential antimicrobial drug targets. Perhaps, the technologies developed to design antibacterial compounds that target riboswitches will find broader use in efforts to target these new and mysterious noncoding RNAs that appear to be so prevalent in bacteria.

Acknowledgements

We thank Philip McCown for information included in the pathogen riboswitch chart, and we thank members of the Breaker laboratory for helpful discussion. This work is supported by a grant from the NIH and by support to RRB from the Howard Hughes Medical Institute.

References

Ames, T.D., and Breaker, R.R. (2010). Bacterial riboswitch discovery and analysis. In: The Chemical Biology of Nucleic Acids, Meyer, G. ed., Wiley-VCH.

Ataide, S.F., Wilson S.N., Dang S., Rogers T.E., Roy B., Banerjee R., Henkin T.M., Ibba M. (2007). Mechanism of resistance to an amino acid antibiotic that targets translation. ACS Chem. Biol. 2, 819–827.

Barrick, J.E., and Breaker, R.R. (2007). The distributions, mechanisms, and structures of metabolite-binding riboswitches. Genome Biol. *8*, R239.

Barrick, J.E., Corbino, K.A., Winkler, W.C., Nahvi, A., Mandal, M., Collins, J., Lee, M., Roth, A., Sundarson, N., Jona, I., *et al.* (2004). New RNA motifs suggest an expanded scope for riboswitches in bacterial genetic control. Proc. Natl. Acad. Sci. U.S.A *101*, 6421–6426.

Batey, R.T., Gilbert, S.D., and Montange, R.K. (2004). Structure of a natural guanine-responsive riboswitch complexed with the metabolite hypoxanthine. Nature *432*, 411–415.

Begley, T.P., Dams, D.M., Ealick, S.E., McLafferty, F.W., Van Loon, A.P., Taylor, S., Campobasso, N., Chiu, H.J., Kinsland, C., Reddick, J.J. (1999). Thiamine biosynthesis in prokaryotes. Arch. Microbiol. *171*, 293–300.

Blount, K.F., and Breaker, R.R. (2006). Riboswitches as antibacterial drug targets. Nat. Biotechnol. *24*, 1558–1564.

Blount, K. F., and Breaker, R.R. (2008). Riboswitches as new antibiotics targets. FASEB J. *22*, 264.3.

Blount, K., Puskarz, I., Penchovsky, R., and Breaker, R. (2006). Development and application of a high-throughput assay for *glmS* riboswitch activators. RNA Biol. 3, 77–81.

Blount, K.F., Wang, J.X., Lim, J., Sudarsan, N., and Breaker, R. R. (2007). Antibacterial lysine analogs that target lysine riboswitches. Nat. Chem. Biol. 3, 44–49.

Breaker, R.R. (2008). Complex riboswitches. Science *319*, 1795–1797.

Breaker, R.R. (2009). Riboswitches: from ancient gene-control systems to modern drug targets. Future Microbiol. 4, 771–773.

Breaker, R.R. (2010) RNA second messengers and riboswitches: relics from the RNA World? Microbe 5, 13–20.

Burgess, C., O'Connell-Motherway, M., Sybesma, W., Hugenholtz, J., and van Sinderen, D. (2004). Riboflavin production in *Lactococcus lactis*: potential for in situ production of vitamin-enriched foods. Appl. Environ. Microbiol. *70*, 5769–5777.

Cheah, M.T., Wachter, A., Sudarsan, N., and Breaker, R.R. (2007) Control of alternative splicing and gene expression by eukaryotic riboswitches. Nature *447*, 497–500.

Christiansen, L.C., Schou, S., Nygaard, P., and Saxild, H.H. (1997). Xanthine metabolism in *Bacillus subtilis*: characterization of the xpt-pbuX operon and evidence for purine- and nitrogen-controlled expression of genes involved in xanthine salvage and catabolism. J. Bacteriol. *179*, 2540–2550.

Coates, A., Hu, Y., Bax, R., and Page, C. (2002). The future challenges facing the development of new antimicrobial drugs. Nat. Rev. Drug Discov. *1*, 895–910.

Cochrane, J.C., Lipchock, S.V., and Strobel, S.A. (2007). Structural investigation of the GlmS ribozyme bound to its catalytic cofactor. Chem. Biol. *14*, 97–105.

Dambach, M.D., and Winkler, W.C. (2009) Expanding roles for metabolite-sensing regulatory RNAs. Curr. Opin. Microbiol. *12*, 161–169.

Di Girolamo, M., Busiello, V., Coccia, R., and Foppoli, C. (1990). Aspartokinase III repression and lysine analogs utilization for protein synthesis. Physiol. Chem. Phys. Med. NMR *22*, 241–245.

Ebbole, D.J., and Zalkin, H. (1987). Cloning and characterization of a 12-gene cluster from *Bacillus subtilis* encoding nine enzymes for de novo purine nucleotide synthesis. J. Biol. Chem. *262*, 8274–8287.

Ecker, D.J., and Griffey, R.H. (1999) RNA as a small-molecule drug target: doubling the value of genomics. Drug Disc. Today *4*, 420–429.

Edwards, T.E., and Ferré-D'Amaré, A.R. (2006). Crystal structures of the thi-box riboswitch bound to thiamine pyrophosphate analogs reveal adaptive RNA-small molecule recognition. Structure *14*, 1459–1468.

Gardner, P.P., Daub, J., Tate, J.G., Nawrocki, E.P., Kolbe, D.L., Lindgreen, S., Wilkinson, A.C., Finn, R.D., Griffiths,-Jones, S., Eddy, S.R., *et al.* (2009), Rfam: updates to the RNA families database. Nucleic Acids Res. 37, D136–140.

Garst, A.D., Héroux, A., Rambo, R.P., and Batey, R.T. (2008). Crystal structure of the lysine riboswitch regulatory mRNA element. J. Biol. Chem. *283*, 22347–22351.

Gelfand, M.S., Mironov, A.A., Jomantas, J., Kozlov, Y.I., and Perumov, D.A. (1999). A conserved RNA structure element involved in the regulation of bacterial riboflavin synthesis genes. Trends Genet. *15*, 439–442.

Gerdes, S.Y., Scholle, M.D., D'Souza, M., Bernal, A., Baev, M.V., Farrell, M., Kurnasov, O.V., Daugherty, M.D., Mseeh, F., Polanuyer, B.M., (2002). From genetic footprinting to antimicrobial drug targets: examples in cofactor biosynthetic pathways. J. Bacteriol. *184*, 4555–4572.

Gilbert, S.D., Mediatore, S.J., and Batey, R.T. (2006). Modified pyrimidines specifically bind the purine riboswitch. J. Am. Chem. Soc. *128*, 14214–14215.

Gilbert, S.D., Reyes, F.E., Edwards, A.L., and Batey, R.T. (2009). Adaptive ligand binding by the purine riboswitch in the recognition of guanine and adenine analogs. Structure *17*, 857–868.

Goodwin, T. (1963). The Biosynthesis of Vitamins and Related Compounds. (Academic Press, New York).

Grill, S., Busenbender, S., Pfeiffer, M., Köhler, U., and Mack, M. (2008). The bifunctional flavokinase/flavin adenine dinucleotide synthetase from Streptomyces davawensis produces inactive flavin cofactors and is not involved in resistance to the antibiotic roseoflavin. J. Bacteriol. *190*, 1546–1553.

Grundy, F.J., Lehman, S.C., and Henkin, T.M. (2003). The L box regulon: lysine sensing by leader RNAs of bacterial lysine biosynthesis genes. Proc. Natl. Acad. Sci. U.S.A. *100*, 12057–12062.

Haas, A.L., Laun, N.P., and Begley, T.P. (2005). Thi20, a remarkable enzyme from *Saccharomyces cerevisiae* with dual thiamine biosynthetic and degradation activities. Bioorg. Chem. *33*, 338–344.

Hermann, T., and Westhof, E. (1998) RNA as a drug target: chemical, modelling, and evolutionary tools. Curr. Opin. Biotechnol. *9*, 66–73.

Hermann, T., and Tor, Y. (2005) RNA as a target for small molecule therapeutics. Expert Opin. Ther. Pat. *15*, 49–62.

Hirshfield, I.N., Tomford, J.W., and Zamecnik, P.C. (1972). Thiosine-resistant mutants of *Escherichia coli* K-12 with growth-medium-dependent lysyl-tRNA synthetase activity.II. Evidence for an altered lysyl-tRNA synthetase. Biochim. Biophys. Acta *259*, 344–356.

Jenkins, A.L., Zhang, Y., Ealick, S.E., and Begley, T.P. (2008) Mutagenesis studies on TenA: a thiamine salvage enzyme from *Bacillus subtilis*. Bioorg. Chem. *36*, 29–32.

Joyce, G.F. (2002). The antiquity of RNA-based evolution. Nature *418*, 214–221.

Kawasaki, T. and Nose, Y. (1969). Thiamine regulatory mutants in *Escherichia coli*. J. Biochem. (Tokyo) *65*, 417–425.

Kawasaki, T., Sanemori, H., Egi, Y., Yoshida, S., and Yamada, K. (1976). Biochemical studies on pyrithiamine-resistant mutants of *Escherichia coli* K12. J. Biochem. (Tokyo) *79*, 1035–1042.

Kil, Y.V., Mironov, V.N., Gorishin, I.Y., Kreneva, R.A., and Perumov, D.A. (1992). Riboflavin operon of *Bacillus subtilis*: unusual symmetric arrangement of the regulatory region. Mol. Gen. Genet. *233*, 483–486.

Kim, J. N., and Breaker, R. R. (2008). Purine sensing by riboswitches. Biol. Cell *100*, 1–11.

Kim, J.N., Blount, K.F., Puskarz, I., Lim, J., Link, K.H., and Breaker, R.R. (2009). Design and antimicrobial action of purine analogues that bind guanine riboswitches. ACS Chem. Biol. *4*, 915–927.

Klein, D.J., and Ferré-D'Amaré, A.R. (2006). Structural basis of *glmS* ribozyme activation by glucosamine-6-phosphate. Science *313*, 1752–1756.

Kobayashi, K. *et al.* (2003). Essential *Bacillus subtilis* genes. Proc. Natl. Acad. Sci. U.S.A. *100*, 4678–4683.

Kochhar, S., and Paulus, H. (1996). Lysine-induced premature transcription termination in the *lysC* operon of *Bacillus subtilis*. Microbiology *142*, 1635–1639.

Kreneva, R.A., Gel'fand, M.S., Mironov, A.A., Iomantas, I.A., Kozlov, I.I., Mironov, A.S., and Perumov, D.A. (2000). Study of the phenotypic occurrence of ura gene inactivation in *Bacillus subtilis*. Genetika *36*, 1166–1168.

Kubodera, T., Yamashita, N., and Nishimura, A. (2000). Pyrithiamine resistance gene (*ptrA*) of *Aspergillus orizae*: cloning, characterization and application as a dominant selectable marker for transformation. Biosci. Biotechnol. Biochem. *64*, 1416–1421.

Kubodera, T., Watanabe, M., Yoshiuchi, K., Yamashita, N., Nishimura, A., Nakai, S., Gomi, K., and Hanamoto, H. (2003). Thiamine-regulated gene expression of *Aspergillus oryzae thiA* requires splicing of the intron containing a riboswitch-like domain in the 5'-UTR. FEBS Lett. *555*, 516–520.

Kulshina, N., Baird, N.J., and Ferré-D'Amaré, A.R. (2009) Recognition of the bacterial second messenger cyclic diguanylate by its cognate riboswitch. Nat. Struct. Mol. Biol. *16*, 1212–1217.

LeBlanc, J.G., Burgess, C., Sesma, F., de Giori, G.S., and van Sinderen, D. (2005). *Lactococcus lactis* is capable of improving the riboflavin status in deficient rats. Br. J. Nutr. *94*, 262–267.

Lee, J.M., Zhang, S., Saha, S., Santa Anna, S., Jiang, C., and Perkins, J. (2001). RNA expression analysis using an antisense *Bacillus subtilis* genome array. J. Bacteriol. *183*, 7371–7380.

Lee, E.R., Blount, K.F., and Breaker, R.R. (2009). Roseoflavin is a natural antibacterial compound that binds to FMN riboswitches and regulates gene expression. RNA Biol. *6*, 187–194.

Leonardi, R., and Roach, P.L. (2004). Thiamine biosynthesis in *Escherichia coli*: *in vitro* reconstitution of the thiazole synthase activity. J. Biol. Chem. *279*, 17054–17062.

Lu, Y., Shevtchenko, T.N., and Paulus, H. (1992). Fine-structure mapping of cis-acting control sites in the lysC operon of *Bacillus subtilis*. FEMS Microbiol. Lett. *71*, 23–27.

Mack, M., van Loon, A.P., and Hohmann, H.P. (1998). Regulation of riboflavin biosynthesis in *Bacillus subtilis* is affected by the activity of the flavokinase/flavin adenine dinucleotide synthetase encoded by *ribC*. J. Bacteriol. *180*, 950–955.

Mandal, M., and Breaker, R.R. (2004). Adenine riboswitches and gene activation by disruption of a transcription terminator. Nat. Struct. Mol. Biol. *11*, 29–35.

Mandal, M., Boese, B., Barrick, J.E., Winkler, W.C., and Breaker, R.R. (2003). Riboswitches control fundamental biochemical pathways in *Bacillus subtilis* and other bacteria. Cell *113*, 577–586.

Mandal, M., Lee, M., Barrick, J.E., Weinberg, Z., Emilsson, G.M., Ruzzo, W.L., and Breaker, R.R. (2004). A glycine-dependent riboswitch that uses cooperative binding to control gene expression. Science *306*, 275–279.

Mankin, A. (2006). Antibiotic blocks mRNA path on the ribosome. Nat. Struct. Mol. Biol. *13*, 858–860.

Mayer, G., and Famulok, M. (2006). High-throughput-compatible assay for glmS riboswitch metabolite dependence. Chem. BioChem. 7, 602–604.

Meyer, M.M., Ames, T.D., Smith, D.P., Weinberg, Z., Schwalbach, M.S., Giovannoni, S.J., and Breaker, R.R. (2009). Identification of candidate structured RNAs in the marine organism 'Candidatus Pelagibacter ubique'. BMC Genomics *10*, 268.

Miranda-Ríos, J., Navarro, M., and Soberón, M. (2001). A conserved RNA structure (thi box) is involved in regulation of thiamine biosynthetic gene expression in bacteria. Proc. Natl. Acad. Sci. U.S.A. *98*, 9736–9741.

Mironov, A.S., Gusarov, I., Rafikov, R., Lopez, L.E., Shatalin, K., Kreneva, R.A., Perumov, D.A., and Nudler, E. (2002). Sensing small molecules by nascent RNA: a mechanism to control transcription in bacteria. Cell *111*, 747–756.

Montange, R.K., and Batey, R.T. (2006) Structure of the S-adenosylmethionine riboswitch regulatory mRNA element. Nature *441*, 1172–1175.

Mulhbacher, J., and Lafontaine, D.A. (2007). Ligand recognition determinants of guanine riboswitches. Nucleic Acids Res. 35, 5568–5580.

Otani, S., Takatsu, M., Nakano, M., Kasai, S., and Miura, R. (1974). Letter: Roseoflavin, a new antimicrobial pigment from *Streptomyces*. J. Antibiot. *27*, 86–87.

Otani, S., Kasai, S., and Matsui, K. (1980). Isolation, chemical synthesis, and properties of roseoflavin. Meth. Enzymol. *66*, 235–241.

Ott, E., Stolz, J., Lehmann, M., and Mack, M. (2009). The RFN riboswitch of *Bacillus subtilis* is a target for the antibiotic roseoflavin produced by *Streptomyces davawensis*. *RNA Biol. 6*, 276–280.

Park, J., Dorrestein, P. C., Zhai, H., Kinsland, C., McLafferty, F.W., and Begley, T.P. (2003). Biosynthesis of the thiazole moiety of thiamine pyrophosphate (vitamin B1). Biochemistry *42*, 12430–12438.

Patte, J.C. (1996). Biosynthesis of lysine and threonine. In: *Escherichia coli* and *Salmonella*: Cellular and Molecular Biology, Vol. 1, Neidhardt, F. C., Curtis, R., eds. (ASM Press, Washington, DC) pp. 528–541.

Patte, J.C., Akrim, M., Méjean, V. (1998). The leader sequence of the *Escherichia coli lysC* gene is involved in the regulation of LysC synthesis. FEMS Microbiol. Lett. *169*, 165–170.

Poiata, E., Meyer, M.M., Ames, T.D., and Breaker, R.R. (2009). A variant riboswitch aptamer class for S-adenosylmethionine common in marine bacteria. RNA *15*, 2046–2056.

Puerta-Fernandez, E., Barrick, J.E., Roth, A., and Breaker, R.R. (2006). Identification of a large noncoding RNA in extremophilic eubacteria. Proc. Natl. Acad. Sci. U.S.A. *103*, 19490–19495.

Regulski, E.E., and Breaker, R.R. (2008). In-line probing analysis of riboswitches. Methods Mol. Biol. *419*, 53–67.

Rentmeister, A., Mayer, G., Kuhn, N., and Famulok, M. (2007). Conformational changes in the expression domain of the *Escherichia coli thiM* riboswitch. Nucleic Acids Res. *35*, 3713–3722.

Robbins, W.J. (1941). The pyridine analog of thiamine and the growth of fungi. Proc. Natl. Acad. Sci. U.S.A. *27*, 419–422.

Rodionov, D.A., Vitreschak, A.G., Mironov, A.A., and Gelfand, M.S. (2002). Comparative genomics of thiamine biosynthesis in procaryotes. New genes and regulatory mechanisms. J. Biol. Chem. *277*, 48949–48959.

Roth, A., and Breaker, R.R. (2009). The structural and functional diversity of metabolite-binding riboswitches. Annu. Rev. Biochem. *78*, 305–334.

Saxild, H.H., and Nygaard, P. (1987). Genetic and physiological characterization of *Bacillus subtilis* mutants resistant to purine analogs. J. Bacteriol. *169*, 2977–2983.

Seetharaman, S., Zivarts, M., Sudarsan, N., and Breaker, R.R. (2001). Immobilized RNA switches for the analysis of complex chemical and biological mixtures. Nat. Biotechnol. *19*, 336–341.

Serganov, A. Yuan Y.R., Pikovskaya O., Polonskaia A., Malinina L., Phan A.T., Hobartner C., Micura R., Breaker R.R., Patel D.J. (2004). Structural basis for discriminative regulation of gene expression by adenine- and guanine-sensing mRNAs. Chem. Biol. *11*, 1729–1741.

Serganov, A., Polonskaia, A., Phan, A.T., Breaker, R.R., and Patel, D.J. (2006). Structural basis for gene regulation by a thiamine pyrophosphate-sensing riboswitch. Nature *441*, 1167–1171.

Serganov, A., Huang, L., and Patel. D.J. (2008). Structural insights into amino acid binding and gene control by a lysine riboswitch. Nature *455*, 1263–1267.

Serganov, A., Huang, L., and Patel, D.J. (2009). Coenzyme recognition and gene regulation by a flavin mononucleotide riboswitch. Nature *458*, 233–237.

Settembre, E., Begley, T.P., and Ealick, S.E. (2003). Structural biology of enzymes of the thiamine biosynthesis pathway. Curr. Opin. Struct. Biol. *13*, 739–747.

Smith, K.D., Lipchock, S.V., Ames, T.D., Wang, J., Breaker, R.R., and Strobel, S.A. (2009) Structural basis of ligand binding by a c-di-GMP riboswitch. Nat. Struct. Mol. Biol. *16*, 1218–1223.

Soukup, G.A., and Breaker, R.R. (1999a). Nucleic acid molecular switches. Trends Biotechnol. *17*, 469–476.

Soukup, G.A., and Breaker, R.R. (1999b). Relationship between internucleotide linkage geometry and the stability of RNA. RNA *5*, 1308–1325.

Soukup, G.A., and Breaker, R.R. (2000). Allosteric nucleic acid catalysts. Curr. Opin. Struct. Biol. *10*, 318–325.

Sudarsan, N., Barrick, J.E., and Breaker, R.R. (2003a) Metabolite-binding RNA domains are present in the genes of eukaryotes. RNA *9*, 644–647.

Sudarsan, N., Wickiser, J.K., Nakamura, S., Ebert, M.S., and Breaker, R.R. (2003b). An mRNA structure in bacteria that controls gene expression by binding lysine. Genes Dev. *17*, 2688–2697.

Sudarsan, N., Cohen-Chalamish, S., Nakamura, S., Emilsson, G.M., and Breaker, R.R. (2005). Thiamine

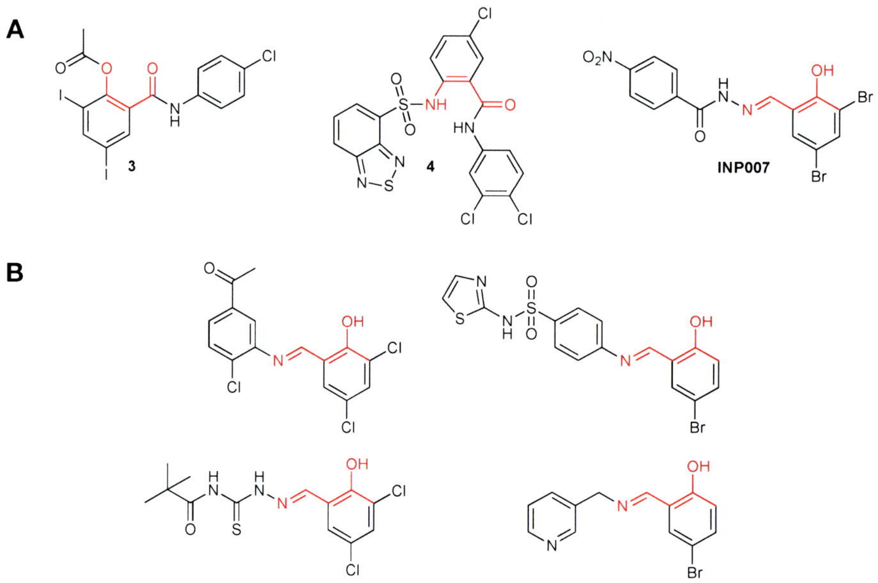

Figure 6.2 (A) Structures of the three positive hits from a HTS performed by Kauppi *et al.* (2003) for T3S inhibitors in *Yersinia pseudotuberculosis*. The acylated salicylaldehyde hydrazone INP007 is indicated. Although the structures are different, they have in common an X-C=C-C=X where X is an N or O. (B) Structures of positive hits independently identified in a HTS performed by Gauthier *et al.* (2005) for T3S inhibitors in enteropathogenic *Escherichia coli*. The conserved X-C=C-C=X motif for both sets of structures are shown in red.

shown to inhibit secretion by the evolutionarily related flagellar-specific T3 secretion apparatus, which is required for assembly of this organelle, suggesting that this compound may represent a potential broad spectrum inhibitor of T3S systems. Compounds that inhibit flagellar-specific T3S may have additional utility as therapeutic agents, as flagellin protein is a potent stimulator of the innate immune system. Subsequent studies demonstrated that several structures in the acyl salicylhydrazone class inhibit T3S systems and T3-dependent processes in a wide array of pathogens including *Yersinia* spp., *Chlamydia* spp., *S. typhimurium* and *Shigella flexneri* (Kauppi *et al.*, 2003; Bailey *et al.*, 2006; Muschiol *et al.*, 2006, 2009; Hudson *et al.*, 2007; Negrea *et al.*, 2007; Veenendaal *et al.*, 2009; Fig. 6.3).

Interestingly, the X-C=C-C=X motif along with halogenated aryl rings reappeared in four of six positive hits found in a HTS performed by Gauthier *et al.* (2005; Fig. 6.2B). For this study, enteropathogenic *E. coli* (EPEC) were sub-cultured in the presence of 20,000 commercial compounds at a concentration of 40 μM arrayed in a 96-well format and grown under T3S-inducing conditions. EspB secretion, an EPEC T3S substrate, was monitored by enzyme-linked immunosorbent assay (ELISA). Secondary screens were performed to discard compounds that inhibited growth of EPEC and/or degraded secreted proteins. Positive hits were confirmed by SDS-PAGE and Western blotting techniques to determine the compound(s) effect on the secretion of other T3 substrates as well as non-T3 secreted substrates. Compounds that resulted in decreased secretion and/or a specific decrease in the expression of T3S-associated or virulence-specific proteins were further characterized. Using transcriptional fusions Gauthier *et al.* (2005) examined the affect of their acyl salicylhydrazones on the expression of the T3S-associated and non-related virulence genes and found that increasing concentrations of these compounds reduced the expression of these genes. The authors did not find an effect on flagellin expression or motility. These data suggest that these compounds are transcriptionally regulating virulence-associated gene products. Further studies with these acyl

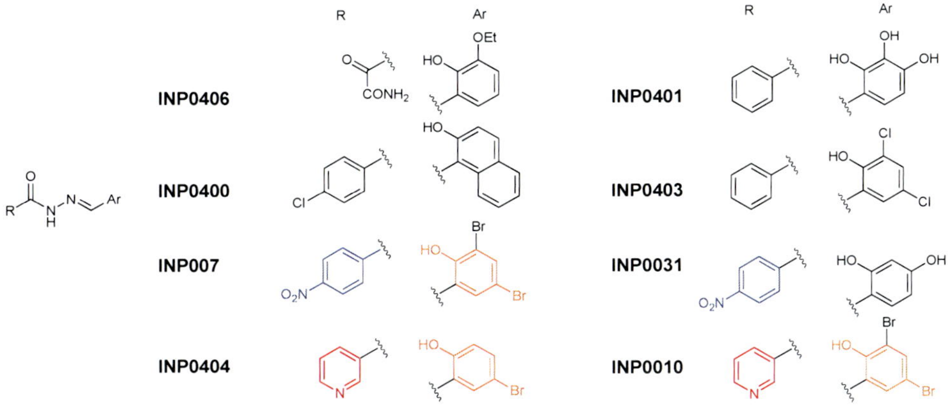

Figure 6.3 Representative structural analogues within the acyl salicylhydrazone class, which have been analysed in various studies for their effect on T3S in a number of different organisms. The compounds INP0400, lacking any aryl ring at R, and INP0406, in which Ar is naphthyl, are included to illustrate the structural diversity explored during the course of this programme. The colours highlight common structural features shared by compounds that nonetheless have been shown to have different phenotypic profiles. For example, compounds on the right hand side have been shown to inhibit bacterial growth at concentrations within the range of virulence inhibition. The 2-hydroxy-5-bromophenyl Ar substituent (gold) is found in both left- and right-hand column compounds. Compounds INP007 (left hand column) and INP0031 (right hand column) both have as R a 4-nitrophenyl ring (blue); compounds INP0404 and INP0010 both have as R a 3-pyridiyl group (red). From these observations it becomes clear that overall compound activity is more complex than a simple function of the R and Ar groups.

salicylhydrazones have demonstrated that in the presence of inhibitor the formation of A/E lesions on EBL (embryonic bovine lung) epithelial cells by enterohaemorrhagic *E. coli* is decreased (Tree *et al.*, 2009). Whereas significant progress has been made on the spectrum of activity for acyl salicylhydrazones on T3S systems, their target and mechanism of action has not been unambiguously identified. Whether these compounds inhibit virulence gene expression (Gauthier *et al.*, 2005; Tree *et al.*, 2009) or secretion assembly (Veenendaal *et al.*, 2009), they were the first demonstrations of chemical attenuation of T3S and associated virulence in tissue culture and animal models of infection.

Several distinct compounds were identified as inhibitors of T3 secretion in *Yersinia pestis*, the causative agent of bubonic plague (Pan *et al.*, 2009; Fig. 6.4). Similar to the Kauppi study (2003), T3S was monitored by luciferase expression, but in the case of *Y. pestis* growth is known to arrest upon induction of T3S by a shift in growth temperature to 37°C and depletion of Ca^{2+} from the media. Therefore the authors measured luciferase expression from plasmid encoded *luxABC and D* genes in a *Y. pestis* reporter strain, when grown in the presence of approximately 70,000 small molecules and plant extracts under T3S-inducing growth conditions. Several lead compounds appear to demonstrate extended specificity as they inhibited T3-dependent secretion of the Tir protein in enteropathogenic *E. coli*, but again the molecular mechanism of action of these compounds remains to be determined. These screening hits bear no structural similarity to each other or the Kauppi hits, lacking both halogenated aryl rings and the X-C=C-C=X motif. A major challenge in the future will be to reconcile, probably through target identification and molecular mechanism of action characterization, the wide diversity of structures for positive hits found both within and across independent HTS for T3S inhibitors.

Aiello *et al.* (2010) used an analogous cell-based bioluminescent reporter approach as described above to identify three distinct classes of small molecule inhibitors of *P. aeruginosa* T3S. In *P. aeruginosa*, T3-dependent gene expression is tightly controlled by the negative transcriptional regulator ExsE, which is secreted in a T3-dependent manner. Therefore a reporter strain was

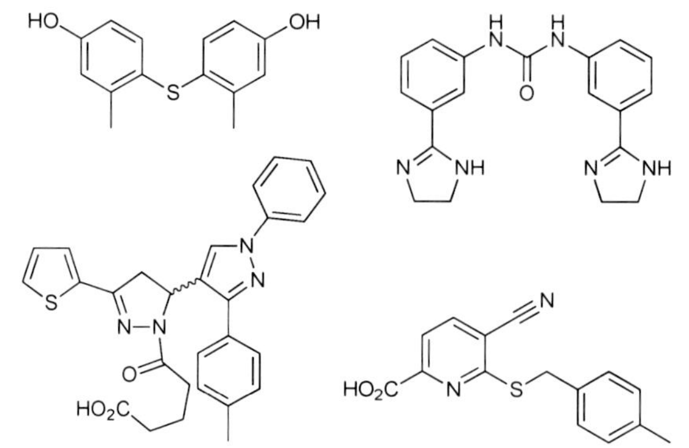

Figure 6.4 Hits reported by Pan *et al.* (2009) that were identified in an HTS for inhibitors of T3S in *Yersinia pestis*.

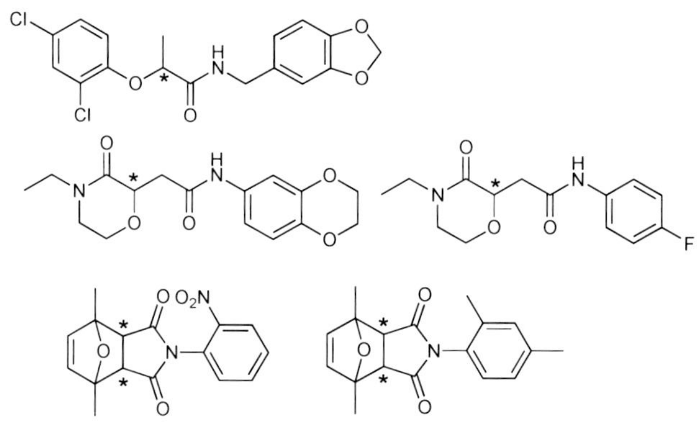

Figure 6.5 Structures of positive hits identified from a HTS for inhibitors T3S in *P. aeruginosa* performed by Aiello *et al.* (2010). The asterisk denotes chiral centres, and several of the hits identified in this work can be distinguished from the majority of identified T3SS inhibitory structures which are achiral. Synthetic expansion of the lactate benzylamide (top compound) showed that the R enantiomer was 20-fold more potent than the S enantiomer.

constructed that had a transcriptional fusion between the luminescent genes of *Photorhabdus luminescens* and the T3 effector gene *exoT*. Luminescence production was then evaluated when bacteria were grown under T3-inducing and repressing conditions in the presence of 80,000 arrayed small molecules. After evaluating directly for the inhibition of T3S using an effector-reporter protein fusion to TEM1 β-lactamase lacking it's secretion signal, three distinct classes of compounds were found to inhibit T3S in *P. aeruginosa* (Fig. 6.5). In contrast, to the generally achiral hits of previous screening programs, these structures all contained stereocentres. In subsequent assays, MBX 1641, a lactate benzylamide, was also found to block T3S-dependent growth of *C. trachomatis* L2 in Hep-2 cells and inhibit secretion of a *Y. pestis* YopE-TEM1 β-lactamase reporter fusion protein. These data suggest a broad spectrum of activity similar to the acylated salicylaldehyde hydrazones, likely reflecting the high degree of conservation of T3S structural components among Gram-negative pathogens. The molecular mechanism of action of the lactate benzylamides is yet to be determined, but a preliminary SAR study with three nodes of diversity demonstrated that the stereocentre of these compounds was important for inhibitory activity (Aiello *et al.*, 2010). This suggests that a specific molecular target is being inhibited by these compounds.

V. cholerae is a Gram-negative pathogen that causes the diarrhoeal disease cholera and whose

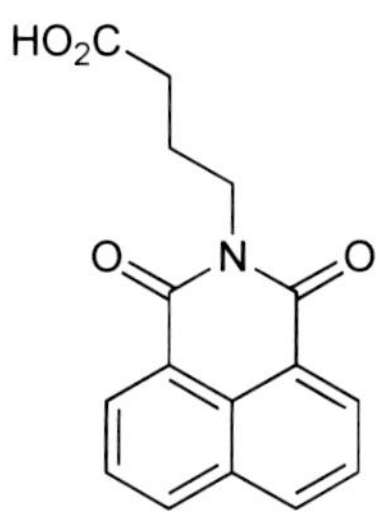

Figure 6.6 The structure of virstatin, an inhibitor of virulence gene regulation and toxin secretion in *Vibrio cholerae*, is shown.

pathology is dependent upon the expression of cholera toxin (CT) and the toxin coregulated pilus (TCP). Using a phenotypic HTS screen, the small molecule 4-[*N*-(1,8)-naphthalimide-*n*-butyric acid, or virstatin, was identified as an inhibitor of virulence gene regulation and toxin secretion in *V. cholerae* (Fig. 6.6; Hung *et al.*, 2005). The authors of this study constructed a screening strain of the classical biotype strain of *V. cholerae* 0395 that carries a chromosomally integrated tetracycline resistance gene (*tetA*) controlled by the cholera toxin (*ctx*) promoter. In a multiwell format 50,000 compounds were screened for their ability to confer tetracycline sensitivity when the reporter strain was grown under *in vitro* conditions that induce CT expression.

In the canonical model for *V. cholerae* virulence regulation, environmental signals stimulate virulence by activating a cascading series of transcriptional regulators that ultimately induce transcription of the CT and TCP genes. Using transcriptional reporter fusions, the effect of virstatin on these known virulence regulators was investigated to determine that in the presence of virstatin, the transcript levels of these genes including the most downstream gene, *toxT*, were all relatively unaffected (Hung *et al.*, 2005). Subsequent studies, using the bacterial two-hybrid system in *E. coli* demonstrated that virstatin disrupts the homodimerization of the ToxT transcriptional regulator thus preventing expression of CT and TCP (Shakhnovich *et al.*, 2006). In animal studies, inoculation into an infant mouse with *V. cholerae* in the presence of virstatin demonstrated a marked reduction in intestinal colonization (Hung *et al.*, 2005).

Recently, a separate group of small molecules were identified as bacterial secretion inhibitors using another whole-cell based HTS (Felise *et al.*, 2008). For this study, a strain of *S. typhimurium* was engineered that secreted a phospholipase A2 reporter construct in a T3S-dependent manner. The strain contained a secreted reporter protein fusion between the *S. typhimurium* T3S substrate SipA and the *Y. enterocolitica* T3 secreted phospholipase YplA. T3-dependent secretion of the reporter resulted in the cleavage of a self-quenched fluorescent substrate followed by a readily detectable increase in fluorescence. For the HTS, bacteria were grown overnight in an arrayed format in the presence of over 90,000 natural and synthetic compounds. Fluorescent substrate was added and fluorescence measured the following day. Using operon and gene fusions compounds researchers were able to identify those compounds that affected general bacterial processes, such as transcription and translation were eliminated. Six compounds were found to specifically inhibit T3S-dependent gene expression and one compound, a 2-imino-5-arylidene thiazolidinone, appeared to inhibit the process of secretion (Fig. 6.7A). Similar to the above compounds, this molecule had broad-spectrum activity against T3S systems in *Salmonella* and *Yersinia*, but in addition it also inhibited T2S and T4P mediated twitching motility in *P. aeruginosa*. This compound protected bone marrow-derived macrophages from lysis after infection with *S. typhimurium* and reduced the virulence of *P. syringae* in a plant infection model (Felise *et al.*, 2008). Synthesis and screening of over 100 analogues demonstrated that key structural features were essential for activity while other functional groups displayed wide permissiveness for modification (see Fig. 6.7B and C; Felise *et al.*, 2008; Kline *et al.*, 2008, 2009). These data suggest a specific interaction(s) with a molecular target. Although the target and molecular mechanism of action of these compounds also remains to be elucidated, this study demonstrates the potential for the development of anti-virulence agents with the ability to inhibit virulence protein and toxin secretion from multiple classes of secretion systems found in Gram-negative bacterial pathogens.

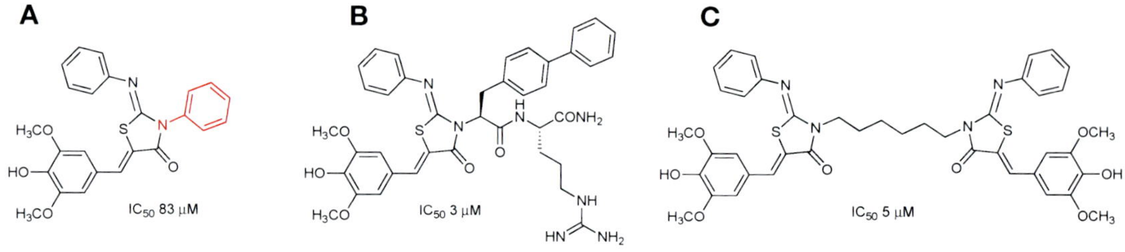

Figure 6.7 (A) Structure of a positive hit 2-imino-5-arylidine thiazolidinone identified in a HTS performed by Felise *et al.* (2008) for inhibitors of T3S in *Salmonella typhimurium*. This scaffold was amenable to structural modification at N-3 (highlighted in red). Extensive changes could be introduced at this site, whereas modifications at the 2- or 5-aryl rings were deleterious (Felise *et al.*, 2008). (B) and (C) Two N-3 modified analogues with greater activity than the original hit are shown, illustrating the breadth of diversity tolerated at this site (Felise *et al.*, 2008; Kline *et al.*, 2008, 2009). Compounds were assayed for inhibition of the T3S of the virulence effector SipA from *S. typhimurium*, as determined by Western blot. The IC_{50} values for these compounds in this assay are shown.

Target based high-throughput screens to identify secretion inhibitors

As indicated above, a drawback of whole-cell based HTS is that the target of inhibition is not immediately apparent and it can be difficult to determine the molecular mechanism of action. Therefore target based screens are attractive assays. But it is important to consider that positive hits from target-based screens may be less amenable to further development for anti-microbial treatment, as lead compounds need to have an optimum hydrophilicity and polar surface area in order to penetrate the outer membrane and reach their cytoplasmic or periplasmic protein target(s) (O'Shea and Moser, 2008). In addition, typical compound libraries available for HTS are often dominated by hydrophobic molecules originally designed to inhibit mammalian targets and therefore it can be challenging, but not impossible to transition from *in vitro* potency to whole cell activity.

Transcriptional inhibitors

Multiple adaptional response (MAR) transcriptional factors are present in both Gram-positive and Gram-negative bacteria and are known to regulate diverse phenotypes including virulence gene expression, antibiotic resistance and stress-responses. These transcriptional regulators are characterized by two highly conserved helix–turn–helix DNA-binding domains, are required for virulence and are not present in eukaryotic cells; therefore these proteins are potentially attractive drug targets.

LcrF is a MAR that regulates expression of the T3S system in *Y. pestis* and *Y. pseudotuberculosis*. Using small molecule docking experiments with the co-crystal structures of MarA-DNA and Rob-DNA (MAR proteins from *E. coli*), *N*-hydroxybenzimidazole compounds were predicted and subsequently shown to be inhibitors of these transcriptional regulators (Bowser *et al.*, 2007). Therefore, Kim *et al.* (2009) synthesized and screened an acrylic amide series of *N*-hydroxybenzimidazole for activity in a cell-free LcrF-DNA binding assay as well as inhibition of T3S-dependent *Y. pseudotuberculosis* cytotoxicity of J774A.1 macrophages at 50 μg/ml (Fig. 6.8). A number of these compounds inhibited T3S-dependent cytotoxicity by *Y. pseudotuberculosis* as well as DNA binding of the closely related ExsA MAR transcriptional regulator of *P. aeruginosa*, suggesting that targeting global transcriptional regulators of virulence functions may be a viable approach for the development of novel broad spectrum anti-virulence agents.

ATPase inhibitors

Secretion system assembly as well as substrate translocation requires energy and associated ATPases. Although the sequence, number and location of ATPases vary between secretion systems, they all contain conserved nucleotide

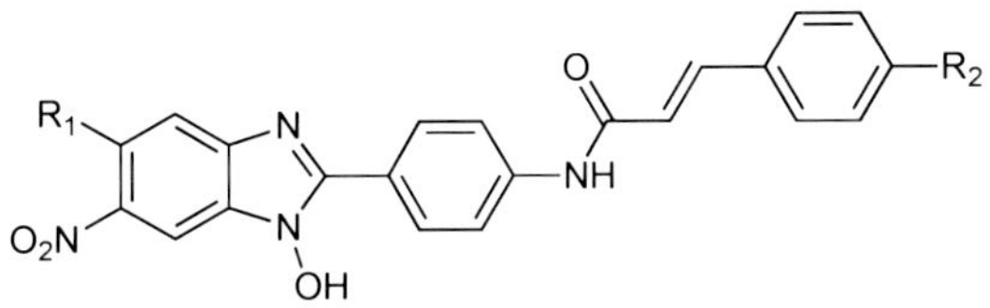

	R_1	R_2	IC_{50} (μM)
14	H	F	3.9
22	H	OCH_3	8.0
36	CH_3	F	7.6

Figure 6.8 Structures and IC_{50} values for three representative benzimidazole inhibitors of DNA binding by the T3S transcriptional regulator LcrF in *Yersinia pseudotuberculosis* are shown. The cell-free LcrF-DNA binding assay was conducted with biotinylated double-stranded DNA containing the LcrF binding sequence and hexahistidinyl LcrF. After washing unbound DNA and protein, the bound DNA–protein complexes were detected by immunoconjugated horseradish peroxidase acting on chemoluminescent substrate.

binding motifs. Recent crystal structures and cryo-electron microscope analyses of ATPases for T4 secretion systems suggest that they function as dynamic hexameric assemblies (Yeo *et al.*, 2000; Savvides *et al.*, 2003). Therefore it is possible that conserved enzymatic and/or structural features of bacterial secretion NTPases may be viable targets for the development of anti-virulence agents (Baron and Coombes, 2007). As measurement of nucleosidase activity is straightforward, high-throughput assays to screen for inhibition of ATPase activity can easily be designed.

Hilleringmann *et al.* (2006) performed an *in vitro* enzymatic assay using purified Caga ATPase from the T4S system in *H. pylori* to screen more than 500,000 compounds for inhibitors of activity. Three compounds were identified (CHIR1-3; Fig. 6.9A). CHIR-1 also inhibited transfer of the substrate protein CagA from *H. pylori* to eukaryotic host cells in an *in vitro* model of infection. In addition, *H. pylori* bacteria pretreated with this compound prior to infection showed an impaired ability to colonize the gastric mucosa in mouse models serving as the first example of chemical attenuation of virulence with a T4S inhibitor. The chemotype of CHIR-1 and -2 is most closely associated with kinase inhibitors, such that target versus off-target effects in future studies and an analogue programme with this class of compounds will have to be carefully monitored.

Using the published *E. coli* crystal structure of the GSP ATPase SecA, Li *et al.* (2008) identified 31 potential inhibitors using a virtual HTS. The original compounds were weak inhibitors, with *in vitro* IC_{50} values around 100 μM, but an extensive synthetic analogue campaign resulted in two derivatives of an initial hit that had activity in the low micromolar range *in vitro* (Chen *et al.*, 2010; Fig. 6.9B). These compounds are still quite hydrophobic in nature and show whole cell activity only against membrane compromised bacteria; thus their utility *in vivo* remains to be determined. Although the transition from *in vitro* promise to drug candidate has been difficult, these studies demonstrate that inhibitors of ATPase activity can be identified and activity improved in hit to lead programs. Despite the caveats regarding ATPases as drug targets, specificity has been successfully engineered into many drugs currently in clinical trials or practice, for example those that bind non-adenosine binding sites of specific ATPases (Chene, 2002). Thus, ATPases may be valid bacterial targets, and suitable molecules may emerge from these screens and follow-up studies.

Outer membrane secretin protein inhibitors

Secretins are large protein complexes, which function to form a pore in the bacterial outer membrane for several export pathways, including T2S, T3S, T4P assembly systems and filamentous phage release (Bayan *et al.*, 2006). Despite the diversity of systems utilizing these proteins they have been shown to share similar structure and organization in the C-terminal portion of the protein (Thanassi, 2002). Therefore, it is possible that there are conserved features of these proteins that can be targeted for the development of broad spectrum inhibitors of Gram-negative bacterial secretion systems.

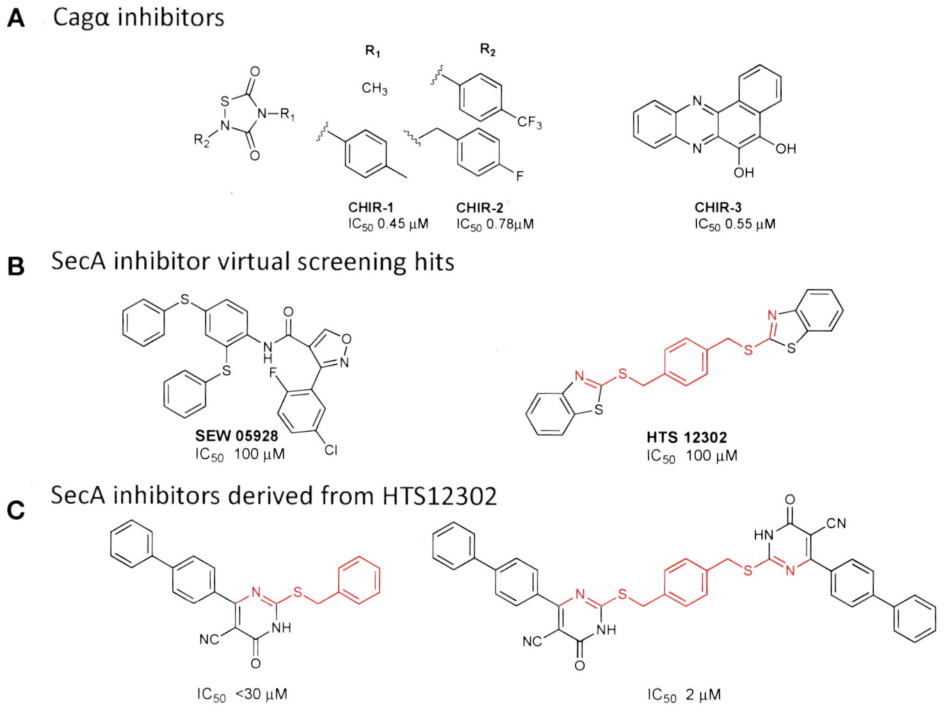

Figure 6.9 (A) Compounds identified by Hilleringman *et al.* (2006) that are active against the T4S ATPase Cagα in *Helicobacter pylori*. (B) Initial hits from a virtual HTS performed by Li *et al.* (2008) against the GSP ATPase SecA had IC_{50} values in the 100 μM range. The S-benzyl heterocyclic motif (highlighted in red) was expanded (Chen *et al.*, 2010). (C) With the advent of the S-benzyl 2-thiouracils group (highlighted in red) analogues of HTS-12302 that were significantly more potent were obtained. Note that both HTS12302 and the 2 μM analogue are symmetrical dimers about a common *bis*-thiobenzyl core.

Mouratou *et al.* (2007) recently engineered a class of proteins that bind specifically to PulD, a T2 secretin protein of *Klebsiella oxytoca* and showed that it inhibited PulD multimerization as well as secretion of the T2 substrate pullunase when co-expressed in *E. coli*. An argument against chemical attenuation of secretin oligomerization is that interruption of protein–protein interactions would probably be required, rather than inhibition of enzymatic functions similar to classical antibiotics. This study, however, demonstrates the proof-of-principle that secretin oligomerization in the outer membrane can be inhibited and that this inhibition results in a corresponding loss of substrate secretion. The continued advancement of our understanding of secretin biology will probably facilitate the targeted design of broad-spectrum functional inhibitors of these macromolecular secretion systems.

Prospects for the use of Gram-negative secretion inhibitors for prophylaxis and treatment of bacterial infections

There are a number of situations where one could envisage that the use of virulence protein secretion inhibitors may be a viable, if not also advantageous approach to the treatment of infectious diseases. Specific disease syndromes that are candidates for this type of therapeutic intervention (either with or without classical antibiotics) include nosocomial infections, gastroenteritis, and venereal disease.

Pseudomonas aeruginosa and nosocomial infections

P. aeruginosa is a ubiquitous Gram-negative bacterium that is the leading cause of opportunistic infections in immunocompromised individuals and debilitated patients (Engel, 2003). While it rarely infects healthy individuals, when defences are compromised it can infect virtually all tissue types posing a serious threat to infected patients. Many of these opportunistic infections are acute, including sepsis and infections in post-operative wound and burn patients. Despite treatment with antibiotics, mortality remains as high as 40% in acute infections (Engel and Balachandran, 2009). *P. aeruginosa* is also particularly well adapted to the respiratory tract environment. It is a causative agent of ventilator-associated pneumonia (Hunter, 2006; Wright and Romano, 2006) and chronically colonize the lung of cystic fibrosis and bronchiectasis patients (Gibson *et al.*, 2003) – leading to severe pulmonary damage and potentially death. Unfortunately, it demonstrates a remarkable ability to resist antibiotics, partly due to the constitutive expression of β-lactamases, presence of efflux pumps, low outer-membrane permeability and acquisition of resistance genes (such as enzymes inactivating aminoglycosides; Mesaros *et al.*, 2007). The combination of these mechanisms results in multiresistant phenotypes, making *P. aeruginosa* a difficult target for antibiotic therapy.

P. aeruginosa strains found associated with opportunistic infections are often invasive and toxigenic, as a result of the coordinated expression of surface proteins and structural features that facilitate bacterial attachment, colonization, and invasion (Engel 2003). In addition, secreted virulence factors, including toxins, proteases and phospholipases, damage tissues and/or trigger the production of host cytokines that are essential for pathogenesis as demonstrated in animal models of infection. For example, the *P. aeruginosa* T3S substrates (such as ExoU, ExoS and ExoT) have been shown to inhibit wound repair as well as the host innate immune response perpetuating tissue injury and rendering the host susceptible to secondary infections (Engel and Balachandran, 2009). As this is an important issue with nosocomial infections of the airway, the use of anti-secretion therapies may serve as an important adjuvant to antimicrobial therapy to limit associated inflammation and host cell death in these patients. In support of this idea, Sawa *et al.* (1999) demonstrated that vaccination against PcrV, a T3S protein involved in the translocation of T3 toxins, ensured survival of challenged mice and decreased lung inflammation and injury. They also found that monoclonal antibodies to PcrV administered at the time of infection prevented sepsis and death (Frank *et al.*, 2002). Finally, anti-secretion agents could hypothetically be used alone to treat infections arising from multidrug resistant organisms for which there is no current therapy available.

Enteric pathogens and gastroenteritis

Diarrhoeal diseases are one of the leading causes of death and disability worldwide, with an estimated 2.5 million deaths occurring annually and the second leading cause of death in children under five years of age (Johansson and Wardlaw, 2009). Although a significant global problem, they are especially prevalent in developing countries where poor environmental sanitation conditions, inadequate water supplies and limited education prevail. Diarrhoeal infections have significant impact on malnutrition in these regions, as these illnesses may result in a reduction in appetite and decreased absorption of nutrients. In addition, they are often associated with other infections, such as malaria and HIV, and contribute to the morbidity associated with these pathogens.

A challenge in treating diarrhoeal disease is the wide breadth of causative agents; including bacteria, viruses and parasites, with approximately 50% of gastroenteritis being of bacterial aetiology. *Shigella* species, non-typhoid *Salmonella* species, enterotoxigenic *E. coli* (ETEC), *V. cholerae* and *Campylobacter* species are Gram-negative bacteria commonly associated with diarrhoeal diseases, including dysentery and traveller's diarrhoea. Current treatment is typically a regime of rehydration and possible antibiotics in severe cases, regardless of the causative agent. However, antibiotics can be detrimental in some cases of haemorrhagic colitis caused by enterohaemorrhagic *E. coli* O157:H7 and indiscriminate use for non-typhoidal salmonellosis can prolong intestinal colonization (Aserkoff and Bennett, 1969; Wong *et al.*, 2000). Even in cases of bacillary dysentery in which treatment with antibiotics is the standard of care a progressive increase in

Recent Developments in Natural Products: Potential Impact on Antibacterial Drug Discovery

7

Ronald J. Quinn and Jeffrey E. Janso*

Abstract

Natural products and derivatized natural products, produced mainly by actinomycetes, have been one of the most successful sources of drugs used to treat and cure infectious diseases. However, many bacteria have quickly become resistant to the majority of antibiotics in use today prompting an urgent need to discover new classes of antibacterial compounds. The goal of this chapter is to summarize some of the recent advances that favourably position natural products drug discovery in the quest to discover new antibacterial agents. This includes new sources of microbial biodiversity such as plants and the oceans as well as overlooked actinomycete biodiversity present in soil. Other encouraging advancements include (1) the development of new culturing techniques, which have enabled the isolation of microbes that were once thought to be uncultivable, (2) the impact of sequencing technology and bioinformatics that have made strain dereplication more reliable and revealed that actinomycete genomes encode far more secondary metabolite gene clusters than originally thought and (3) the use of innovative methods to express and exploit these orphan biosynthetic pathways. Finally, the ability to dereplicate, isolate and elucidate the structure of natural products from minimal sample quantities will also be discussed.

Introduction

Natural products are organic compounds produced by living organisms at some point during their life cycle. More specifically, in drug discovery research, natural products are considered to be secondary metabolites produced by and isolated from plants, marine macroorganisms and the fermentation cultures of fungi and bacteria. Culture collections containing diverse microorganisms collected from various habitats throughout the world are the foundation of any natural products discovery effort. These microbes are cultured in various liquid media and solid substrates under many different conditions (temperature, aeration, agitation, etc.) to generate 'libraries' of crude extracts, semi-pure mixtures (fractions), or purified compounds. For plants, these libraries are created by extracting various carefully separated tissues with different organic solvents. For marine macroorganisms such as sponges and tunicates, pieces of organisms or sometimes entire organisms are extracted. The natural product libraries are tested for bioactivity as either activators or inhibitors of specific biological targets. 'Actives' from extracts and semi-pure mixtures are dereplicated by a series of skilful and sometimes tedious chemical separations and purifications with the aid of high-performance liquid chromatography coupled with mass spectrometry (HPLC-MS) until the single component is realized and the structure elucidated by nuclear magnetic resonance (NMR) and/or high resolution mass spectrometry (HRMS).

Historically, natural products and their derivatives have been very successfully developed as antimicrobial agents. The 1940s through the 1950s, often referred to as the 'golden age' of

*Corresponding author

antibiotic discovery, achieved a spectacular array of antibiotics including the sulphonamides, penicillin, streptomycin and other aminoglycosides, tetracyclines, erythromycin, vancomycin, cephalosporins and nalidixic acid (Baltz, 2005; Overbye and Barrett, 2005). In the subsequent 50 years, approximately 200 antibacterial agents, most of them derivatives of natural products, were introduced into clinical practice as a result of medicinal chemistry efforts that focused on numerous modifications to only about 12 core chemical scaffolds (Overbye and Barrett 2005). Most recently, Newman and Cragg (2007) reported that 68% of the 109 new chemical entities launched as antibacterials from 1981 to 2006 were natural products or derived from natural products typically as semisynthetic derivatives.

By the end of the 1960s, infectious diseases were no longer considered a major health threat to the general population. This was followed by a departure by many pharmaceutical companies from infectious disease research during the 1980s, 1990s, and early 2000s to pursue more lucrative markets for chronic diseases. Also during this time, the pharmaceutical industry deemphasized or eliminated natural products programs and focused on screening synthetic or combinatorial chemistry libraries which were more amenable to enzyme-based high-throughput screens. As a result, the only new classes of antibacterial agents that have been marketed within the past 30 years for human use are the synthetically derived oxazolidinones such as linezolid, the natural product daptomycin, a lipopeptide originally described in 1988 (Debono *et al.*, 1988), and the natural product derivative retapamulin, a pleuromutilin used as a topical agent (Novak and Shlaes, 2010). This is unfortunate, because it is now evident that emerging antibacterial-resistant pathogens are causing increasingly difficult-to-treat bacterial infections (Overbye and Barrett 2005), and we have fewer new classes of antibiotics in our arsenal with which to combat them.

Fortunately, many believe that natural products will continue to be an important source of new antibiotics (see Clardy *et al.*, 2006; Harvey, 2008; Li and Vederas, 2009; Donadio *et al.*, 2010), that could still provide new chemical scaffolds active against multidrug-resistant microorganisms. Clearly, microbes are especially adept at producing compounds with antibiotic activity; more than 16,500 microbial natural product antibiotics have been reported (Bérdy, 2005; Fig. 7.1). Over half (8700) of these have been isolated from various

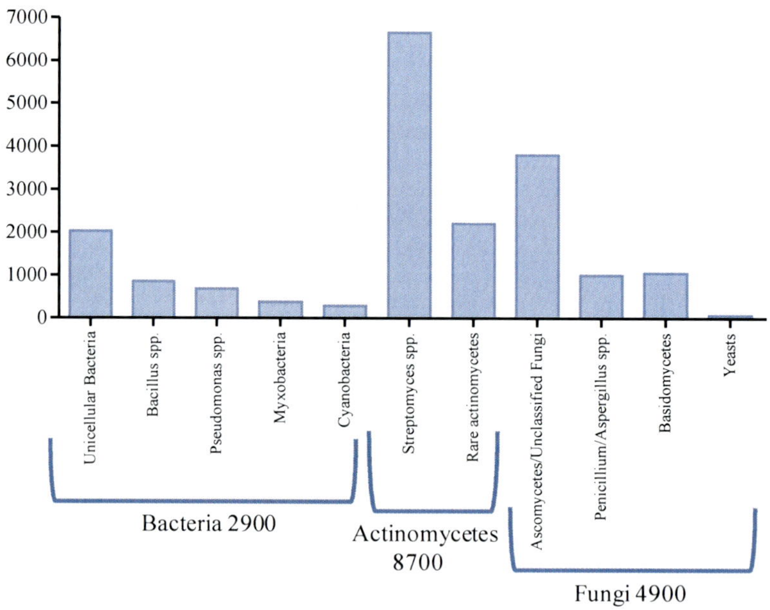

Figure 7.1 Most prolific microbial producers of antibiotics compiled through the end of 2003. Adapted from Table 4 in Bérdy (2005).

filamentous species of the order Actinomycetales, particularly *Streptomyces* (6550) and non-streptomycetes (2250), the so-called rare actinomycetes. Other prolific bacterial producers are *Bacillus* spp. (795 antibiotics), *Pseudomonas* spp. (610 antibiotics), myxobacteria, (400 antibiotics) and cyanobacteria (300 antibiotics). Interestingly, all of these bacteria have relatively large genomes, generally > 5 Mb, with large percentages of genes dedicated to regulating, producing and transport of secondary metabolites. Filamentous fungi, namely the ascomycetes and basidiomycetes, are also quite talented antibiotic producers The most commonly isolated and ubiquitous fungi, *Aspergillus* spp. and *Penicillum* spp., have produced a large percentage, about 20%, of the 4900 antibiotics reported from fungi.

Nearly all of the antibiotics that have been isolated over the last 50–60 years come from microbes that are easy to culture and are amenable to growth in the laboratory. However, the number of species of microorganisms that are thought to exist (overall microbial biodiversity) is tremendously greater than what is currently known (Keller and Zengler, 2004). There may be hundreds of thousands if not millions of microbes that are still waiting to be cultured in the laboratory under the appropriate conditions. This untapped biodiversity is an important source of new natural product chemical diversity especially since the re-isolation of known compounds from the more easily cultivated microbes is quite frequent. Focusing only on the soil-derived actinomycetes, Baltz (2005) has shown that the frequency of isolation of the most commonly produced antibiotics ranges from 0.4% for tetracycline to 20% for streptothricin. To find one new antibiotic using traditional fermentation and extract library screening approaches, more than one to ten million actinomycetes would have to be examined.

Luckily, new approaches and developments in natural products discovery are providing researchers with the tools needed to correlate taxonomic diversity with chemical diversity, thereby finding new antibiotics from old sources as well as new sources of microbes. Sequencing, genomics and bioinformatics have revealed that the secondary metabolic biosynthetic capabilities of even the most commonly isolated microbes are far greater than originally thought. The sequencing of 16S rRNA genes for bacteria has become quite routine and large public databases of these sequences have made it easier to compare strains or identify new strains. The marine environment and terrestrial plants are proving to be largely untapped sources of new cultivable biodiversity. In addition, new culturing methods have been developed that enable the cultivation of microorganisms that were previously thought to be unculturable or only known to exist through detection of their DNA in metagenomic analyses of environmental samples. Finally, new chemical technologies and increased sensitivity of HPLC, MS, and NMR have made the dereplication of complex extracts easier allowing natural products chemists to solve chemical structures with only micrograms of purified material (Koehn and Carter, 2005). This chapter will explore each of these topics, using actinomycetes where possible as examples since they have historically been the most productive group of microorganisms in antibacterial drug discovery.

New biodiversity and biosynthetic potential from soil-derived actinomycetes

Actinomycetes are saprophytic microbes that achieve the highest population density and biodiversity in soil. Since the first antibiotic was isolated from a soil dwelling actinomycete more than 60 years ago, the pharmaceutical industry has invested a lot of effort in the isolation and screening of actinomycetes from soils collected around the world. From the late 1940s through the mid 1960s, there was an exponential increase in the discovery of new agents with antimicrobial activity from *Streptomyces* spp. alone (Watve *et al.*, 2001). This phase of rapid discovery then slowed to 70–100 new compounds per year until the early 1980s. Unfortunately, the discovery of new compounds with antimicrobial activity continued to decline through the 1990s as known compounds were isolated more frequently and pharmaceutical companies invested in high throughput screening of genomics based targets to find the next blockbuster drugs (Baltz, 2005; Li and Vederas, 2009). However, the traditional soil derived actinomycetes continue to be a robust source of new antimicrobials.

Watve *et al.* (2001) estimated that actinomycetes in the genus *Streptomyces* are capable of producing 150,000 compounds with antimicrobial activity. So far, only a fraction of these have been discovered. Within the last few years, the platensimycins (Singh *et al.*, 2006) and platencins (Jayasuriya *et al.*, 2007), two series of compounds with novel chemical scaffolds and unprecedented ketolide units, were discovered from the fermentation extracts of strains of *Streptomyces platensis*. A novel assay design involving antisense-based differential sensitivity screening with whole cells afforded the discovery of both of these potent inhibitors of bacterial fatty acid biosynthesis. The lucensimycins (Singh *et al.*, 2009) isolated from the fermentation extracts of a strain of *Streptomyces lucensis* were also detected by an antisense-based assay but in this version, the *rps*D gene encoding the small ribosomal protein S4 was the target. Although the lucensimycins are not extremely potent, they are new chemically and biosynthetically interesting structures isolated from a common soil-derived actinomycete.

Actinomycetes have not been ignored in the recent explosion of genome sequencing. In the last decade 232 Actinobacteria genomes have been sequenced, including 41strains from the four suborders, Micromonosporineae, Pseudonocardineae, Streptomycineae, and Streptosporangineae, that have historically been the most prolific and productive antibiotic producers (Fig. 7.2). Nearly all of the natural product-derived antibacterial agents on the market are produced by large multimodular enzymatic assembly lines referred to as type I modular and iterative polyketide synthases (PKS I), type II iterative polyketide synthases (PKSII), non-ribosomal peptide synthetases (NRPS) and hybrid PKS-NRPSs (Fischbach and Walsh, 2006). For type I and II PKS systems, monomers of acyl-CoA thioesters are incorporated into the growing polyketide chain whereas for NRPS systems, proteinogenic and non-proteinogenic amino acids and other carboxylic acids are the building blocks (Fischbach and Walsh, 2006). Genome sequencing analyses have revealed that the secondary metabolite biosynthetic potential in terms of the numbers of gene clusters that encode these multimodular enzymes is greater than originally expected even for commonly isolated soil inhabiting strains. The ~8.7 Mb genome of *Streptomyces coelicolor* strain A3(2) encodes 22 different secondary metabolite gene clusters but only 13 compounds have been isolated from the strain (Bentley *et al.*, 2002). Nearly 7% of the 9 Mb genome of *Streptomyces avermitilis*, best

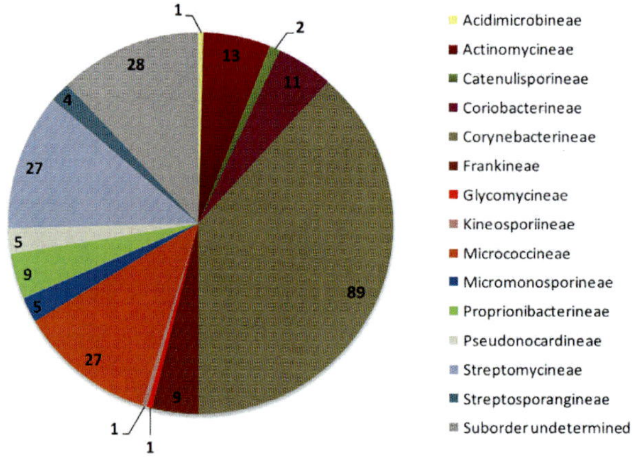

Figure 7.2 Distribution of the 232 sequenced actinomycete genomes by suborder classification. Suborders listed start at the top of the chart and continue clockwise. Data compiled from NCBI and DOE JGI websites (1 October 2010).

known as the producer of the antiparasitic agent avermectin, is composed of genes involved in secondary metabolite biosynthesis (30 gene clusters; Ikeda *et al.*, 2003). Perhaps most intriguing is that the genome of *Streptomyces griseus*, which produces streptomycin, one of the most commonly isolated antibiotics, contains 33 other gene clusters that encode for secondary metabolites; 12 encode for type I PKS, type II PKS, NRPS or PKS–NRPS hybrid molecules that have not yet been elucidated (Ohnishi *et al.*, 2008). How these orphan pathways can be exploited such that the full biosynthetic potential of these strains is realized will be discussed later.

There are 70 validly described culturable genera of actinomycetes within the suborders Micromonosporineae, Pseudonocardineae, Streptomycineae, and Streptosporangineae; 29 of those have been published since 1998 and four are reclassifications of previously described genera. Nearly 76% of recently described new genera were isolated from soil, underscoring that soil continues to be a source of new culturable actinomycete biodiversity. By plating a series of Australian soil core samples onto several different media and incubating for up to three months, Joseph *et al.* (2003) showed that it is possible to isolate previously uncultured bacteria. Of the 350 bacteria isolated, 27% grouped by 16S rRNA sequence analysis to 20 different potentially new, undescribed families; four of which belong to the subclass Actinobacteridae (Joseph *et al.*, 2003). Several strains from one of these actinomycete groups, 'Ellin5034 group' (Joseph *et al.*, 2003), have since also been isolated by researchers at Vicuron Pharmaceuticals (now KtedoGen), who declared that they are a new genus, *Catenulispora*, belonging to the new family, *Catenulisporaceae* (Busti *et al.*, 2006a). From the same soil sample collected from a wooded area in Italy, the researchers at Vicuron discovered additional strains that they described as a second new actinomycete genus, *Actinospica*, belonging to the new family *Actinospicaceae* (Cavaletti *et al.*, 2006). Both the *Catenulisporaceae* and *Actinospicaceae* formed the new suborder Catenulisporinae (Cavaletti *et al.*, 2006). These strains, which were isolated from acidic soils by simply decreasing the pH of the isolation media below 6.0, were found to have large genomes (8–10 Mb) that encode for several different type I PKS, type II PKS and NRPS biosynthetic gene clusters (Busti *et al.*, 2006b). The fermentation extracts of several *Catenulispora* spp. and *Actinospica* spp. displayed anti-Gram-positive activity in whole cell assays indicating that they may be a promising new source of antimicrobials.

New sources of actinomycetes

Marine habitats

Marine-derived microorganisms, including actinomycetes, have been isolated from the marine environment for at least the last 50 years (Goodfellow, 1984; Fenical and Jensen, 1993). However, it was not until the 1990s that the isolation of marine microbes intensified and marine-derived microorganisms were truly investigated for the production of natural products in drug discovery research. Today, the impact of marine-derived microorganisms in the discovery of new natural products is evident from the recent increase in occurrence of compounds produced by marine microbes in annual reviews of marine natural products literature. A summary of the new natural products from marine microbes published over a 40-year span (1965–2005) revealed that 45% were reported in just the last five years of that period (2001–2005; Blunt *et al.*, 2007). More recently, Blunt *et al.* (2009) reported that an average of fewer than 50 compounds per year were reported from marine microorganisms during the period 1965–2005 compared with ~150 reported for the year 2006 and ~240 for the year 2007; an increase of nearly 600% for 2007 over the period 1965–2005. Amongst marine-derived prokaryotes, cyanobacteria and various proteobacteria have been excellent sources of natural products, but an increasing number of new natural products are being reported from marine-derived actinomycetes.

Marine microorganisms are typically isolated and cultured in media composed with 50–100% natural or artificial (e.g. Instant Ocean®, United Pet Group, Inc., Cincinnati, OH) seawater instead of deionized water. For the isolation of marine-derived actinomycetes, the isolation media are usually nutritionally lean and contain low levels of antibacterial agents such as nalidixic acid to reduce the growth of the more abundant and faster growing unicellular bacteria. Heating of samples such as marine sediments to 50°C for 30–60 minutes is another selective technique for the isolation of actinomycetes, whose spores are fairly heat resistant (Jensen *et al.*, 1991).

Marine-derived actinomycetes have been isolated from various marine substrates including sediments from near shore samples (Jensen *et al.*, 1991) as well as deep-sea trenches (Goodfellow, 1984), sponges (Kim *et al.*, 2005; Montalvo *et al.*, 2005; Zhang *et al.*, 2006) and ascidians (Charan *et al.*, 2004). Although actinomycete spores are undoubtedly washed into the oceans from terrestrial habitats, it is becoming increasingly evident that certain actinomycetes are indigenous to marine environments and that others can adapt to them (Jensen *et al.*, 1991). Since the description of the first actinomycete, *Rhodococcus marinonascens* (Helmke and Weyland, 1984), that requires seawater for growth, several other genera of marine-derived actinomycetes such as *Salinibacterium* (Han *et al.*, 2003), *Sciscionella* (Tian *et al.*, 2009) and *Serinicoccus* (Yi *et al.*, 2004) have been discovered that grow optimally in media containing 2–5% sodium chloride but do not require seawater for growth. But perhaps most significant to natural products research has been the discovery of the marine obligate genus *Salinispora* (Mincer *et al.*, 2002; Maldonado *et al.*, 2005), which requires seawater or a balance of salts (Tsueng and Lam, 2008) for growth. All three *Salinispora* spp., *S. arenicola*, *S. tropica*, and '*S. pacifica*' (*S. pacifica* has not yet been formally described) are all prolific producers of secondary metabolites (Fenical and Jensen, 2006; Jensen *et al.*, 2007).

In 1991, researchers at Scripps Institution of Oceanography reported on the isolation and cultivation of several *Micromonospora*-like strains, formally described as *Salinispora* (Maldonado *et al.*, 2005), that occurred most commonly in sediments further from shore and that did not grow in the absence of seawater (Jensen *et al.*, 1991). Since then, strains of *Salinispora* spp. have been found to be ubiquitous in the marine environment occurring not only in sediments but also in sponges (Kim *et al.*, 2005) and ascidians (Asolkar *et al.*, 2010). All three species of *Salinispora* share 99% 16S rRNA sequence identity and differ by as few as five nucleotides through pair-wise species comparisons (Jensen, 2010). Because *Salinispora* have a widespread biogeographical distribution and are phylogenetically very closely related, it has been possible to thoroughly investigate both interspecific and intraspecific secondary metabolite production (Jensen *et al.*, 2007). For example, different secondary metabolite profiles were found in the fermentation extracts of several strains of *S. pacifica* that differ by only two or three nucleotides in their 16S rRNA genes. Also, the recently sequenced genomes of *S. tropica* (Udwary *et al.*, 2007) and *S. arenicola* have revealed that, like the genomes of many soil-derived *Streptomyces*, they harbour many secondary metabolite gene clusters that encode natural products not yet discovered (Penn *et al.*, 2009; e.g. salinilactam A described later in this chapter). Interestingly, a comparison of the two genomes revealed that most of the species-specific secondary metabolite pathways were concentrated in genomic islands, which are generally acquired by horizontal gene transfer (HGT).

A phylogenetically unique taxon of marine-derived actinomycetes within *Streptomycetaceae*, for which the genus '*Marinispora*' has been proposed (Fenical and Jensen, 2006), also appear to be an excellent source of new natural products. Strain CNQ-140, isolated from a sediment sample taken 56 m offshore of La Jolla, California, produced unusual macrodiolides, marinomycins A–D (Kwon *et al.*, 2006). Marinomycin A was the most potent antibacterial agent of the series exhibiting *in vitro* minimum inhibitory concentrations (MIC_{90}) of 0.13 μM against both methicillin-resistant *Staphylococcus aureus* (MRSA) and vancomycin-resistant *Enterococcus faecium* (VREF). The marinomycins also displayed potent anti-tumour activity against several melanoma cell lines. A second '*Marinispora*' sp., strain NPS00892, which was isolated from a sediment sample collected in Cocos Lagoon, Guam, produced the novel antibacterial agents, lipoxazolidinones A–C (Macherla *et al.*, 2007). Although strain NPS00892 does not require seawater for growth, the lipoxazolidinones were only produced in culture media containing natural or artificial seawater (Sunga *et al.*, 2008). Lipoxazolidine A displayed broad spectrum activity with MIC values ranging from 0.5 to 5 μM/ml against two strains of *Haemophilus influenzae*. Interestingly, the lipoxazolidinones, which are 4-oxazolidinones, exhibited an antibacterial spectrum of activity and potency that was similar to that of linezolid, which has a 2-oxazolidinone skeleton (Macherla *et al.*, 2007).

A large part of the earth's microbial biodiversity is harboured in the world's oceans, which cover more than 70% of the earth's surface (Fenical and Jensen, 2006). In addition, they are an excellent source of new biodiversity as evidenced by the recent identification of 148 previously unknown bacterial phylotypes by whole-genome shotgun sequencing of microbial populations in seawater from the Sargasso Sea near Bermuda (Venter *et al.*, 2004). Considering that the exploration of deep ocean environments and the search for symbiotic marine bacteria in marine flora and fauna is just beginning to gain attention, the potential for new natural product discovery from marine microorganisms, including those with antibacterial activity, looks promising.

Plants

There is no one specific definition for what classifies a bacterium as an endophyte, but it is generally accepted that a bacterial endophyte is any bacterium that lives within plant tissues without harming or gaining any benefit other than residency from its host and can be isolated from surface disinfected plant tissues and/or can be shown microscopically to exist inside plant tissues (Kobayashi and Palumbo, 2000; Hasegawa 2006). The procedure to isolate bacterial endophytes including actinomycetes is rather straightforward and commonly involves immersing small pieces of plant tissues in a series of disinfecting solutions such as 70–90% ethanol, followed by 3–5% sodium hypochlorite to kill any epiphytic microbes attached to the surfaces. The plant tissue pieces are then plated onto the surface of nutritionally lean agar or gellan gum based media and incubated for several weeks to months to allow the endophytes to emerge and grow onto the solid substrate. Endophytic actinomycetes can be isolated from virtually any plant tissue (leaves, stems, petioles, roots, etc.) but seem to occur more commonly in roots (Hasegawa 2006; Verma *et al.*, 2009; Janso and Carter, 2010). Root exudates such as ions, free oxygen and water, enzymes, mucilage, and primary and secondary metabolites are secreted by plants into the surrounding soil or rhizosphere (Bais *et al.*, 2006). Some bacteria, which are attracted by root exudates, are stimulated to form protective biofilms or secrete antimicrobial agents into the surrounding rhizosphere to protect the plant against potential pathogens (Bais *et al.*, 2006). Other bacteria become endophytic by penetrating the roots through the damaged epidermal layer created when lateral roots grow out from the existing root structure. Microscopic evidence has shown that endophytic actinomycetes can colonize plants through the stomas found in leaves and stems (Igarashi, 2002).

Plant biodiversity is well documented; there are seventeen megadiverse countries including Australia, Brazil, China, Colombia, Democratic Republic of the Congo, Ecuador, India, Indonesia, Madagascar, Malaysia, Mexico, Papua New Guinea, Peru, Philippines, South Africa, USA and Venezuela that not only account for >70% of terrestrial biodiversity but are also rich in endemic plant species (Williams *et al.*, 2001; Fig. 7.3). Considering that every species of plant has the potential to harbour endophytic microbes, plants are an attractive source of new actinomycete diversity. Intensive investigations for endophytic actinomycetes has really only occurred since the 1990s but has resulted in the isolation of actinomycetes from a range of plants including crops (Coombs and Franco, 2003; Cao *et al.*, 2005; Lee *et al.*, 2008), native plants (Castillo *et al.*, 2007; Janso and Carter, 2010) and medicinal plants (Taechowisan *et al.*, 2003; Qin *et al.*, 2009a). This has led not only to the identification of several new species of endophytic actinomycetes (Qin *et al.*, 2008a,b; Thamchaipenet *et al.*, 2010), but also to the recent discovery of four new actinomycete genera, *Plantactinospora* (Qin *et al.*, 2009b), *Actinophytocola* (Indananda *et al.*, 2010) *Jishengella* (Xie *et al.*, 2010), and *Flindersiella* (Kaewkla and Franco, 2010) initially all isolated from roots. Metagenomic analyses also support that plants are a rich source of new bacterial biodiversity.

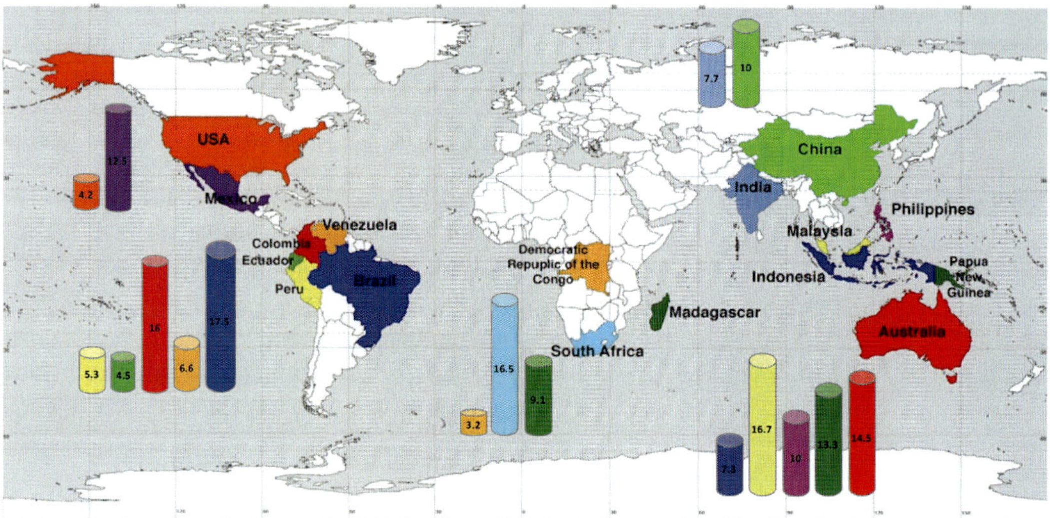

Figure 7.3 Number of endemic vascular plant species (in the thousands) in 17 megadiverse countries. The data represented in this figure was provided by the Department of Sustainability, Environment, Water, Population and Communities. This information appears in the 'Australia State of the Environment Report 2001', first published in 2001. No part may be reproduced without the prior written permission from the Commonwealth.

Wang *et al.* (2008) found that a metagenomic library enriched for plant microbiota from the tropical tree *Mallotus nudiflorus,* which is one source of the anti-tumour natural product maytansinoids, was dominated by Actinobacteria and α-Proteobacteria. In addition, analyses of PCR-amplified 16S rRNA libraries and 16S rRNA clones suggested that most of the microbes might belong to previously unknown phylotypes.

Endophytic actinomycetes are a promising new source of antimicrobial compounds. Because of their association with plants, most endophytes are screened against plant pathogenic fungi for potential applications as biocontrol agents. However, the fermentation extracts of some endophytic actinomycetes have been shown to inhibit the growth of pathogenic bacteria such as *Staphylococcus aureus, Pseudomonas aeruginosa, P. fluorescens,* and *E. coli* in whole cell assays (Li *et al.*, 2008; Qin *et al.*, 2009a; Verma *et al.*, 2009). In addition, a large percentage of many non-streptomycetes that commonly occur in plants such as *Microbispora, Streptosporangium, Actinoallomurus, Sphaerisporangium,* and *Planotetraspora* encode for type I PKS, type II PKS and NRPS biosynthetic gene clusters (Janso and Carter, 2010). Many of the antibacterial compounds from these endophytic actinomycetes remain unknown, but the hope is that structures will soon be solved and the compounds revealed within the next few years.

Novel cultivation techniques

The number of bacterial species estimated to exist in the oceans and in soil is on the order of millions (Keller and Zengler, 2004). These diversity estimates are largely based on cultivation independent investigations or molecular ecological surveys based on the detection of PCR-amplified 16S rRNA genes from environmental samples (Janssen, 2006). The fact that so many more bacterial groups are detected by culture-free methods than are actually isolated and cultured in the laboratory suggests that the vast majority (>99%) of all bacterial species remain uncultivated (Gillespie *et al.*, 2002). This is also true for actinomycetes, despite that the actinomycetes have been intensively isolated and cultured for decades and that there are more described species of *Streptomyces* than any other prokaryotic taxon (Jensen, 2010). For example, the analysis of ~3400 soil-derived 16S rRNA gene clones from various soils revealed that less than 2% of Actinobacteridae clones could be confidently assigned to known genera (Janssen, 2006). As mentioned earlier, simple changes to culturing techniques such as reducing the pH of isolation media can effectively result in the isolation of novel genera initially detected through metagenomic analyses. This section will briefly discuss the few but significant advancements that have been made in culturing previously uncultivated bacteria using innovative approaches.

High concentrations of organic carbon and nitrogen often found in rich, complex media is not appropriate for the cultivation of the most dominant environmental microbes, which are adapted to oligotrophic growth conditions (Connon and Giovannoni, 2002). Using a high-throughput extinction culturing method, where seawater samples were diluted to 1.1–5 cells per millilitre in very low nutrient media in 48-well plates, Connon and Giovannoni (2002) were able to culture four unique lineages from previously uncultured Proteobacteria. Most of the cultures achieved cell densities of less than $1.0–9.9 \times 10^4$ after 3 weeks of incubation, but the culturability by this method was 1.4–120 times higher than plating the same inoculum on nutrient rich agar-based media. Another successful method to culture previously uncultured bacteria employed a diffusion growth chamber based on the probability that most microorganisms need various signalling molecules and nutrients from their natural environment to grow (Kaeberlein *et al.*, 2002). The chamber is constructed by sandwiching a mixture of warm agar or very dilute media with diluted seawater samples, water samples or sediment derived samples between two polycarbonate membranes affixed to a washer (Bollmann *et al.*, 2007; Kaeberlein *et al.*, 2002). The diffusion chamber is incubated with one side in contact with the natural substrate to allow signalling molecules and chemicals to permeate the growth medium. In this way, many bacteria grow as microcolonies that can be passaged to new chambers but do not grow well on petri dishes without coculture. The diversity and uniqueness of cultures obtained by the diffusion chamber is far superior to plating an inoculum on traditional Petri plates (Bollmann *et al.*, 2007).

Most of the organisms obtained by using oligotrophic media are too fastidious and slow growing to culture in traditional flask or bioreactor fermentation conditions. However, two strains of actinomycetes obtained from soil samples using the diffusion chamber technique have been reported to produce new natural products under standard fermentation conditions. *Nocardia* sp. strain G0655 produces the neocitreamicins, which belong to the polycyclic xanthones (Peoples *et al.*, 2008). These compounds displayed very good activity against multidrug resistant Gram-positives with MICs of 0.06–0.50 μg/ml but were inactive against Gram-negatives. *Actinoplanes* sp. strain ISO06811 produced a new isofuranonaphthoquinone, a class of compounds rarely isolated from bacteria (Zhang *et al.*, 2009). These compounds did not inhibit the growth of *B. subtilis* or *E. coli* but may complex iron and act as siderophores.

The final technique discussed in this section involves the high throughput cultivation of bacteria from single cells encapsulated within gel micro- or macrodroplets. At Cubist Pharmaceuticals, individual actinomycete spores from soil samples are encapsulated within calcium alginate gel macrodroplet beads about 2 mm in diameter (Baltz, 2006). The macrodroplets contain growth media for actinomycete cultivation and antibiotics to suppress the growth of non-actinomycetes. Millions of these microfermentations can be performed per year and screened against a multidrug-resistant *E. coli* strain for the production of novel antimicrobials. Similarly, at Diversa Corporation (now Verenium), scientists used agarose gel microdroplets to encapsulate single bacterial cells from seawater and soil sample suspensions (Keller and Zengler, 2004). The individual cells are transferred to a chromatography column where dilute nutrients for growth are continually pumped through. In this way, the system mimics a natural environment, allowing for the exchange of signalling molecules, growth regulators and other metabolites favoured by slower growing and previously uncultured bacteria (Zengler *et al.*, 2002). Microdroplets with growth, some containing as few as 20–100 cells, can be separated by flow cytometry into microtitre plates for fermentation with richer media and subsequent screening for novel secondary metabolites.

Certainly the gap between the number of cultured versus uncultured microbes will become smaller as scientists continue to develop novel cultivation techniques. And given the advances within the last two decades in culturing bacteria, it is probably incorrect to speak of the majority of bacterial species as being unculturable. However, many of the strains of rarely isolated groups are fastidious, making them difficult to maintain in the laboratory, and very slow growing, requiring weeks or months to form visible colonies rather than hours or days (Janssen, 2006). As can be seen in the examples above, a little bit of ingenuity and patience will undoubtedly result in the successful isolation of novel cultures.

Advances in strain dereplication

The development of PCR and rapid improvements in DNA sequencing provided a revolutionary way to classify and compare microorganisms taxonomically. As DNA sequencing became more accessible and routine, molecular-based systematics became more prevalent and was found to complement phenotypic taxonomic methods such as morphology and fatty acid, peptidoglycan, whole cell sugar and menaquinone profile analyses. The small subunit rRNA genes, in particular 16S rRNA operons, were established as the ideal genes for determining phylogenetic relationships between bacteria. This was owing to their universal distribution, structural conservation, lack of evidence for horizontal gene transfer and the presence of conserved regions, which were helpful in establishing higher level relationships as well as variable regions useful in affiliating bacteria to genera and species (Neufeld and Mohn, 2006). The creation of large public sequence repository databases (e.g. taxonomy database – National Centre for Biotechnology Information (NCBI) http://www.ncbi.nlm.nih.gov/taxonomy) and the deposition of 16S rRNA sequences of validly described and published type species enabled laboratories from around the world to share, compare and determine the taxonomic position of isolates.

Since chemical diversity is often equated with microbial diversity, molecular-based strain dereplication methods have become essential in choosing which microorganisms or groups of microorganisms to focus on for the potential to produce new bioactive natural products. In addition, the need for new strains of antibiotic producers has never been more urgent because the same antibiotics and antibiotic classes are reisolated and rediscovered frequently by the isolation and fermentation of similar strains of microorganisms (Baltz, 2006). Morphological characteristics of different species and even some genera of actinomycetes can be very similar lending to the difficulty and tediousness involved in dereplicating strains by phenotypic observations. By contrast, the integrity of genotypic methods which involve the DNA of whole genes or gene fragments is greater because they are not subject to the variability found in media and culturing conditions.

After cultures are isolated from environmental samples, 16S rRNA sequencing can be used to efficiently classify and group strains taxonomically. Often, the phylogenetic relationships between strains are established by 16S rRNA sequence comparisons through BLAST analyses (Altschul *et al.*, 1990), sequence alignments, and the creation of phylogenetic trees. In this way, common antibiotic –producing grey spore *Streptomyces* spp. such as *S. griseus* can be excluded from natural products pure compound or extract screening libraries while increasing the diversity of potentially new species. Alternatively, fragments of several other housekeeping genes such as *gyrB*, *rpoB*, *recA*, etc. can also be sequenced and used to achieve a finer level of strain discrimination by forming concatenated phylogenetic trees of more than one gene. This process termed multilocus sequence analysis (MLSA) has been used successfully to dereplicate very closely related *Streptomyces* spp. (Guo *et al.*, 2008).

Many non-sequencing-based dereplication methods also exist to discriminate between strains of species; two commonly applied methods will be briefly discussed here. Restriction fragment length polymorphism (RFLP) is based on the principle that recognition sites for restriction endonucleases within a genome or PCR-amplified gene of related strains may have been altered over time by single base substitutions, insertions, deletions or transpositions (Pukall 2006). Fragments resulting from restriction enzyme digestion with two or more enzymes are separated out by size by gel electrophoresis. Strains sharing similar electrophoresis profiles are closely related while distantly related strains may not share any fragments of the same size. When this process is applied with several restriction enzymes to PCR-amplified ribosomal genes (16S, 16S-23S intergenic spacer and/or 23S) it is referred to as amplified ribosomal DNA restriction analysis (ARDRA). Lanoot *et al.* (2005) applied this method to 463 type strains of the most common antibiotic producing family, *Streptomycetaceae*, and demonstrated that it was just as effective as 16S rDNA sequencing in delineating species groups. Interestingly, Ayuso *et al.* (2005) used RFLP profiles of PCR-amplified fragments of type I PKS, type II PKS and NRPS biosynthetic gene clusters to dereplicate actinomycetes isolated from three different soils. Correlated with antimicrobial activity, they found a rich diversity of amplification profiles on a group of actinomycetes with enriched biosynthetic potential. Also, some strains shared RFLP patterns for certain biosynthetic gene clusters suggesting that they make the same or similar secondary metabolites.

Similar to RFLP and ARDRA, ribotyping involves endonuclease restriction digest profiles and ribosomal genes. However, in ribotyping, total genomic DNA is digested and chemically labelled or chemiluminescent rDNA gene probes based on one of the ribosomal operons from *E. coli* are allowed to hybridize to restricted, denatured DNA fixed to a Nylon membrane (Southern hydridization; Pukall 2006). Because bacteria contain more than one rRNA operon spread throughout the genome, a unique ribotype or genetic fingerprint is formed for different strains of the same species of bacteria. An automated instrument, the RiboPrinter® microbial characterization system, has been developed to standardize and facilitate ribotyping (Bruce 1996). It provides a rapid and reliable means to dereplicate bacterial isolates by comparing ribotype patterns in a computerized database. The RiboPrinter® has been shown to be effective at discriminating between strains of actinomycetes (Ritacco *et al.*, 2003; Suihko *et al.*, 2006) and has been used in the Wyeth/Pfizer natural products programme to effectively dereplicate soil-derived (Ritacco *et al.*, 2003), marine and endophytic (Janso and Carter, 2010) actinomycetes for over a decade.

Advances in chemical dereplication

Natural products are chemically complex compounds containing many chiral centres, a high degree of oxygen atoms and hydrogen bond donors and acceptors, and diverse ring systems compared with synthetic molecules (Koehn, 2008). Often, they are produced in very small quantities (submilligram) as mixtures of compounds that then need to be isolated and purified from an extract. Therefore, the identification of known and chemically unattractive compounds in a complex extract, a process known as dereplication, is extremely important so that the chemist does not duplicate previous efforts by reisolating the same compound(s) but instead can focus on the elucidation of novel bioactive components.

Most bioactive natural products are discovered by screening extracts against whole cells such as bacteria or a defined molecular target, usually an enzyme. A sample that displays desirable activity such as inhibition or activation is deemed a 'hit'. In the creation of a crude extract screening library, virtually no upfront chemical dereplication has been performed, and a hit can be one compound in a sample containing tens to hundreds of components (Koehn and Carter, 2005). In a pre-fractionated library, crude extracts are fractionated, frequently by reversed-phase HPLC, and very hydrophilic (early eluting) and hydrophobic (late eluting) compounds are excluded so that each sample (fraction) contains semi-pure mixtures of <10 compounds (Wagenaar, 2008). Finally, a pure compound library of natural products is formed through extensive upfront chemical dereplication of crudes and fractions by highly sensitive instrumentation such as LC-MS so that only compounds unique to the library are included and each sample contains a single purified compound. Whether fractionation and dereplication are performed at the stage of library creation or in the isolation, purification and structural elucidation of natural products 'hits', significant improvements particularly in separation (chromatography), MS and NMR have been made within the last decade that can facilitate the discovery of new antibacterials from natural product crude extracts, fractions or pure compounds.

Traditionally, the separation of natural product extracts into fractions has been performed by reversed-phase HPLC where the sample is dissolved in a solvent and eluted from a non-polar stationary phase by a gradient of solvents designated as the mobile phase. More recently, supercritical fluid chromatography (SFC) has emerged as an alternative to HPLC (Farrell *et al.*, 2009). In SFC, the mobile phase is liquid carbon dioxide (CO_2) but there is more flexibility in the choice of stationary phase between polar and non-polar phases or coupling of both types of columns together (Farrell *et al.*, 2009). The use of liquid CO_2, sometimes with modifiers added to increase polarity range, allows higher flow rates to be used which result in faster analyses and higher throughput than HPLC (Ramirez *et al.*, 2007). In natural products research, SFC has mainly been used to fractionate plant extracts. For example, Ramirez *et al.* (2007) demonstrated the use of preparative SFC to separate a complex extract of rosemary (*Rosmarinus officinalis*) into fractions containing improved antimicrobial activity.

Both HPLC and SFC use a solid matrix stationary phase that binds the compounds which are eluted by a liquid mobile phase. However, sometimes a sample can strongly or even irreversibly adsorb to the solid support matrix resulting in poor elution or sample loss (Case *et al.*, 2007). Counter-current chromatography (CCC) is a separation method that uses liquid–liquid countercurrent distribution with two immiscible liquid phases thereby eliminating the need of a solid stationary phase (Wu *et al.*, 2008). By mixing different solvents to tailor a solvent system for a particular sample, the resolution of separation can be maximized facilitating downstream dereplication (Case *et al.*, 2007). Like SFC, the potential of CCC in natural products research is being realized mainly in plant extract separations (Si *et al.*, 2006). For example, the high speed CCC fractionation of a crude dichloromethane extract of *Dracaena angustifolia* leaves resulted in the isolation and identification of three anti-TB active compounds, ergosterol-5, 8-endoperoxide, linoleic acid and E-phytol, with MICs less than or equal to 2 μg/ml (Case *et al.*, 2007). While most separation systems fractionate one sample at a time, Wu *et al.* (2008) developed a multichannel CCC system for natural products drug discovery that can fractionate up to three samples at a time increasing throughput and shortening the time needed to build a fractionated library or prepare samples for dereplication.

After separation of extracts into fractions, the structures of the components need to be elucidated. LC-MS in combination with searches of the available natural products databases for matches or similarities in spectra have had a major impact in dereplicating the known compounds so that more attention can be focused on the structural elucidation of new compounds (Koehn and Carter, 2005). Sometimes the information that can be obtained by LC-MS such as UV-VIS absorbance spectra, mass abundance and molecular mass may be enough to reveal if a compound is new or known. However, high-resolution MS (HRMS) and NMR are critical to the structure determination of new natural products. Accordingly, the most significant advances in natural products structure elucidation can be attributed to the increased detection sensitivity in MS and NMR instrumentation.

The determination of the molecular formula or composition of a natural product is essential to solving the structure. The sensitivity of HRMS, approaching attomole (10^{-18} mole) levels, has enabled this to be accomplished on micrograms or less of material (Koehn and Carter, 2005; Molinski, 2009). The most accurate MS instrument with the highest resolution that is now more widely used is Fourier transform ion-cyclotron resonance mass spectrometry (FTICR-MS). This type of analysis has been particularly valuable in determining the molecular formulas of natural products with MW more than 1000. The complex structures of the major and minor analogues of two different classes of cell wall-active antibacterial natural products, the muraymicins (17 different analogues, ~900–1200 MW) and mannopeptimicins (five different analogues, ~970–1380

MW), were, in part, assembled by the diagnostic fragment ions obtained by FTICR-MS (He *et al.*, 2002; McDonald *et al.*, 2002). In addition, the biosynthesis of natural products can be studied using FTICR-MS to detect the incorporation of isotopically labelled biosynthetic precursors 'fed' to the producing organism during fermentation (Feng *et al.*, 2009). The information gained from these types of experiments could then be used to generate new or more potent analogues.

Recently, the Dorrestein lab reported on two different MS analyses that have been used to study the spatial and temporal production (metabolic exchange) of metabolites produced by the prolific antibiotic producer *Bacillus subtilis* grown alone and between two different species, *Bacillus subtilis* and *Streptomyces coelicolor*, grown in proximity to each other (Watrous *et al.*, 2010; Yang *et al.*, 2009). In one study, matrix-assisted laser desorption/ionization (MALDI) mass spectrometry was used to detect the range of metabolites produced by cultures grown on a thin film of agar layered on top of a MALDI plate (Yang *et al.*, 2009). The accurate mass of secondary metabolites secreted into the agar or associated with the cells or mycelium could be monitored and recorded. In the other study, an imprint of the bacterial coculture grown on agar was taken with a filter membrane, which was analysed by desorption electrospray ionization (Watrous *et al.*, 2010). With this technique, only the compounds that absorb onto the membrane could be detected. Both of these techniques detected the production of growth inhibitory antibiotics and signalling molecules that affected antibiotic production between the two different species. They represent an intriguing possibility to study and discover new antibacterials that are only produced by the interaction of two or more microorganisms in coculture, similar to the native habitat. Furthermore, these types of analyses are only possible with the extreme sensitivity and resolution of MS.

Although not as sensitive as MS, NMR is the key spectroscopic method used to assign functional groups and assemble the structures of natural products. In the past, the relatively low sensitivity of NMR restricted analyses to only the most abundant extract components. However, two recent technologies, cryoprobe and capillary or microcoil probe NMR, have been developed that reduce the signal to noise ratio and increase sensitivity while using very low sample volumes (7 μl; Molinski, 2009). These developments have improved the chances of finding new compounds that are produced in very small quantities and have enabled the use of NMR in the dereplication of complex mixtures. Recently, Lang *et al.* (2008) combined the power of capillary probe NMR with the searchable AntiMarin database (www.chem.canterbury.ac.nz/marinlit/marinlit.shtml) containing ~50,000 marine and terrestrial microbial and marine macroorganism derived natural product compounds to effectively dereplicate crude extracts from microbes. After HPLC separation of the crude extracts, the NMR spectra were acquired on samples that contained as little as 2–40 μg of bioactive compound.

Genome mining, orphan biosynthetic pathways, and metagenomics

The wealth of genomic data that is currently available in public databases will continue to grow as more and more microbial genomes are sequenced. This ever increasing pool of sequence data provides the basis of a largely bioinformatics-dependent effort to find new natural products through 'genome mining'(Zerikly and Challis, 2009). Bioinformatics tools have made it easier to identify and analyse the genes and gene clusters involved in secondary metabolite biosynthesis. The analysis of the data has contributed to the understanding of how natural products are generated at the genetic level by the producing organisms. For example, the basic modular architecture of NRPSs consists of an adenylation (A) domain to select and activate the amino acid to be incorporated into the growing peptide chain, a peptidyl carrier protein (PCP) or thiolation (T) domain, which covalently attaches and transports the amino acid, and the condensation (C) domain that catalyses peptide bond formation (Finking and Marahiel 2004). Similarly, PKSs consist minimally of three domains: an acyltransferase (AT) that selects an acyl-CoA building block and transfers it to an acyl carrier protein (ACP), which loads the extender unit, and a ketosynthase

(KS) to extend the polyketide chain through decarboxylative condensation (Fischbach and Walsh, 2006). Very often, the genes of secondary metabolite pathways are organized in clusters, and the structures of the compounds encoded by the genes can be reasonably predicted from the collinear or assembly line domain architecture, although exceptions to the co-linearity rule such as iterative domain and module use, module skipping and variations in domain assembly have been more recently reported (Nguyen *et al.*, 2008; Wenzel and Mueller, 2005).

As mentioned earlier in this chapter, the analysis of microbial and even plant genomes has resulted in the discovery that there are many more secondary metabolic pathway encoding gene clusters per organism than there are isolated and characterized natural products for that organism. In addition, the PCR-based screening for the presence of PKS and NRPS gene clusters in bacteria, particularly actinomycetes, has revealed that even genera that have never been reported to produce natural products have the potential to do so (Gontang *et al.*, 2010; Janso and Carter, 2010). The biosynthetic pathways that putatively encode metabolites that are unknown or have not yet been identified have been classified as cryptic or orphan biosynthetic pathways (Gross, 2007; Zerikly and Challis, 2009). Given the huge biodiversity of microbes, these orphan pathways could be an incredible source of new or novel natural products with antibacterial or other bioactivity. Guided by genome mining, several strategies have arisen to determine the natural products that are encoded by the many biosynthetic pathways of cultured microorganisms for which no metabolite has been reported.

The regulation of secondary metabolite production in actinomycetes is affected by the expression of pleiotropic regulators and/or pathway specific regulators. It is quite common to find one or more pathway specific regulator(s), which can act as transcriptional activators or repressors, encoded in secondary metabolite biosynthetic gene clusters (Chen *et al.*, 2010). Overexpression of a pathway activator or deletion of a pathway repressor could result in the increased production of a natural product. These techniques have been used successfully to dramatically increase the yields (~100-fold) of previously discovered antibacterials such as platencin and platensimycin (Chen *et al.*, 2010). However, this method has also been used to discover new antibiotics from orphan gene clusters detected by genome mining. Gottelt *et al.* (2010) inactivated the transcriptional repressor *scbR2* in *S. coelicolor* that is associated with the orphan type I PKS gene cluster *cpk*. The deletion of the repressor resulted in expression of the *cpk* cluster and detection of antibacterial activity. Although the structure of the antibacterial compound encoded by cpk has not yet been elucidated, this method shows that some orphan pathways do encode undiscovered antibacterial compounds even from well-studied strains.

The inactivation of a key gene in the biosynthesis of an identified natural product is one technique used to prove that the putative biosynthetic pathway definitively encodes that particular compound. Gene inactivation can also be used to reveal which metabolite is the product of an orphan pathway (Gross, 2007). In a comparison of the fermentation extract profiles by HPLC or LC-MS of a wild type strain and a mutant with a disruption in the orphan biosynthetic gene cluster, the disappearance of a peak or multiple peaks from related structures is suggestive that the compound is encoded by the cluster in question. This approach was effectively applied to the *cch* gene cluster in *S. coelicolor* resulting in the discovery of the new siderophore, coelichelin (Lautru *et al.*, 2005). By inactivating the *cchH* gene which encodes the trimodular NRPS of coelichelin, and culturing the strain in an iron-deficient medium, Lautru *et al.* (2005) identified the siderophore as the product of the *cch* orphan pathway.

Bioinformatics tools can be used to predict the structure of a compound or its physicochemical properties by examination of the domains and modules encoded by an orphan biosynthetic gene cluster. It may then be possible to tailor the fermentation conditions and chemical analyses to identify the compound in a fermentation extract (Gross, 2007; Zerikly and Challis, 2009). This method has been successfully applied to identify new compounds encoded by orphan pathways detected in whole genome sequences (Udwary *et al.*, 2007; Nguyen *et al.*, 2008), on cosmids or in bacterial artificial chromosome (BAC) vectors

(Zazopoulos *et al.*, 2003). The colinearity rule was used to predict that the *slm* gene cluster, containing ten modules and 49 domains, detected in the genome sequence of *Salinispora tropica* strain CNB-440 encoded a novel lysine-primed polyene macrolactam (Udwary *et al.*, 2007). After close inspection of a fermentation broth for compounds that absorb UV light at the characteristic wavelengths of a polyene, the new polyene macrolactam, salinilactam A, was isolated and characterized by NMR and MS.

It is not necessary to have the entire genome of an organism sequenced to detect secondary metabolite gene clusters and predict the encoded structure. A method called genome scanning was developed as a high-throughput approach to detect and then sequence entire secondary metabolic gene clusters from libraries of large gene fragments cloned into cosmids or BAC vectors (Zazopoulos *et al.*, 2003). Briefly, short (700 bp) genomic DNA fragments are cloned and sequenced to identify those fragments encoding natural product biosynthetic genes, which are then used as probes to identify the larger subcloned fragments that contain the genes of interest as well as neighbouring genes. This technique has been successful because the genes required for secondary metabolite biosynthesis are typically clustered together in the genome and the small fragments provide reasonable coverage of the entire genome. Genome scanning coupled with proprietary bioinformatics software, DECIPHER®, analyses led to the discovery of a novel polyketide, ECO-0501, from a vancomycin producer, *Amycolatopsis orientalis* (Banskota *et al.*, 2006). This new glycosidic polyketide, containing guanidine, glucuronic acid and aminohydroxycyclopentenone groups, has very good growth inhibition of several Gram-positives but is unfortunately not any more active against vancomycin-resistant *E. faecalis* than vancomycin (Banskota *et al.*, 2006).

Finally, metagenomic DNA isolated from various environmental substrates including soil (Courtois *et al.*, 2003; Gillespie *et al.*, 2002), marine and terrestrial invertebrates (Fisch *et al.*, 2009; Piel 2006) and plants (Wang *et al.*, 2008) is a promising source of new natural products. One advantage of analysing metagenomic DNA is that unculturable as well as culturable microbes can be exploited for the discovery of new antibacterials. After DNA is isolated from environmental samples, it is cloned into BAC, cosmid, or fosmid vectors that can accommodate large insert fragments (usually up to 50 kb) and can stably maintain the DNA in a heterologous host such as *E. coli*. It is worth noting that some BAC vectors have been developed that can maintain the stability of DNA inserts greater than 100 kb (Donadio *et al.*, 2002; Liu *et al.*, 2009). Clone libraries containing thousands or even millions (Banik and Brady, 2008; Wang *et al.*, 2008) of clones can then be screened by PCR for virtually any type of secondary metabolic biosynthetic pathway gene. By analysing the sequences of the amplified PCR products, dereplication can be performed to focus on the clones that potentially encode novel natural product biosynthetic genes instead of those that share very high sequence identity to genes of known natural products. At the same time this method can be effectively used to discover new members of a particular class of natural products. In one noteworthy study, Banik and Brady (2008) screened a 10,000,000-membered soil DNA 'megalibrary' for the highly conserved *oxyC* genes involved in the biosynthesis of glycopeptides such as vancomycin. Through this amazing effort, two clones were found to contain OxyC like sequences. The gene clusters, one entire vancomycin-like and one partial teicoplanin-like, were sequenced and predicted to encode novel glycopeptides. The teicoplanin-like cluster encoded three sulfotransferases suggesting that it would produce the first known polysulfated glycopeptides. Owing to their altered physicochemical properties from known glycopeptides, polysulfated glycopeptides could potentially display more potent antibacterial activity (Banik and Brady, 2008).

The next step in the discovery of new natural products from metagenomic DNA libraries is to then heterologously express the biosynthetic pathway in a surrogate host that is easily cultured. This has been performed in *E. coli* where the expression of several clones from a ~25,000 soil metagenomic library resulted in the isolation of the turbomycins, which displayed good antibacterial activity against Gram-positives but poor or no activity against Gram-negatives (Gillespie *et al.*, 2002). Because actinomycetes have been such prolific antibiotic producers, the manipulation of DNA and the heterologous expression of genes in various actinomycetes, *Streptomyces* in particular, has been well developed. Several *Streptomyces* spp. such as *S. coelicolor*, *S. lividans*, *S. albus*, and *S. avermitilis* have been used as surrogate hosts because they are genetically well characterized, have the metabolic capability to express natural product gene clusters, and can efficiently transcribe heterologous promoters (Chen *et al.*, 2010; Donadio *et al.*, 2002). DNA is introduced into *Streptomyces* by intergenic conjugation with *E. coli* that contains the clone library in *E. coli–Streptomyces* shuttle BAC (Donadio *et al.*, 2002; Liu *et al.*, 2009) or cosmid (Courtois *et al.*, 2003) vectors. These shuttle vectors encode an integrase (*int*) gene for site-specific integration of the cloned DNA into the *Streptomyces* genome via the *attB* site (Donadio *et al.*, 2002). Courtois *et al.* (2003) successfully used this method to introduce soil DNA inserts containing PKS genes into *S. lividans*. This method has reliably been used to heterologously express entire biosynthetic gene clusters in *Streptomyces* (Chen *et al.*, 2010) from other cultured actinomycetes and clearly shows potential to express novel natural product biosynthetic gene clusters from uncultured microbes.

Concluding remarks

It is unlikely that future natural products research will have the same degree of impact on antibacterial drug discovery that it had during the 'golden age' of antibiotics. However, two of the main bottlenecks in natural products drug discovery, namely the reisolation of known compounds and the inability to isolate and characterize minor components from complex mixtures, are becoming less of an impediment. Newer genetic-based strain dereplication methods are more discriminatory than phenotypic techniques. Thus, it is possible to exclude common antibiotic producing organisms from natural product extract libraries and focus on new taxa. Fortunately, new biodiversity awaits discovery from largely untapped sources such as the world's immense oceans and spectacular plant diversity. Even soil is by no means an exhausted source. At the same time, the genomics era has revealed the true microbial biosynthetic potential in both cultured organisms and those that are yet to be cultured. By analysing fragments of secondary metabolite gene clusters amplified by PCR or analysing sequenced genomes for unusual or orphan pathways, it is now possible to invest resources in the characterization and expression of pathways likely to produce new chemical diversity. Although the days of 'low hanging fruit' are long gone, the toolbox available to natural products chemists for chemical chromatography and dereplication are more powerful today than they were 20 years ago, allowing access to components that are present in submilligram amounts. These recent developments and future ingenuity ensure that natural products will continue to impact drug discovery through the production of novel antibacterial agents that are desperately needed to combat multiresistant bacteria and other infectious diseases.

References

Altschul, S.F., Gish, W., Miller, W., Myers, E.W., and Lipman, D.J. (1990). Basic local alignment search tool. J. Mol. Biol. *215*, 403–410.

Asolkar, R.N., Kirkland, T.N., Jensen, P.R., and Fenical, W. (2010). Arenimycin, an antibiotic effective against rifampin- and methicillin-resistant *Staphylococcus aureus* from the marine actinomycete *Salinispora arenicola*. J. Antibiot. *63*, 37–39.

Ayuso, A., Clark, D., Gonzalez, I., Salazar, O., Anderson, A., and Genilloud, O. (2005). A novel actinomycete strain de-replication approach based on the diversity of polyketide synthase and nonribosomal peptide synthetase biosynthetic pathways. Appl. Microbiol. Biotechnol. *67*, 795–806.

Bais, H.P., Weir, T.L., Perry, L.G., Gilroy, S., and Vivanco, J.M. (2006). The role of root exudates in rhizosphere interations with plants and other organisms. Ann. Rev. Plant Biol. *57*, 233–266.

Baltz, R.H. (2005). Antibiotic discovery from actinomycetes: Will a renaissance follow the decline and fall? SIM News *55*, 186–196.

Baltz, R.H. (2006). Marcel Faber Roundtable: is our antibiotic pipeline unproductive because of starvation, constipation or lack of inspiration? J. Ind. Microbiol. Biotechnol. *33*, 507–513.

Banik, J.J., and Brady, S.F. (2008). Cloning and characterization of new glycopeptide gene clusters found in an environmental DNA megalibrary. Proc. Natl. Acad. Sci. U.S.A. *105*, 17273–17277.

Banskota, A.H., McAlpine, J.B., Sorensen, D., Ibrahim, A., Aouidate, M., Piraee, M., Alarco, A.M., Farnet, C.M., and Zazopoulos, E. (2006). Genomic analyses lead to novel secondary metabolites. Part 3. ECO-0501, a novel antibacterial of a new class. J. Antibiotics *59*, 533–542.

Bentley, S.D., Chater, K.F., Cerdeno, T.A.M., Challis, G.L., Thomson, N.R., James, K.D., Harris, D.E., Quail, M.A., Kieser, H., Harper, D., *et al.* (2002). Complete genome sequence of the model actinomycete *Streptomyces coelicolor* A3(2). Nature *417*, 141–147.

Bérdy, J. (2005). Bioactive microbial metabolites. J. Antibiot. *58*, 1–26.

Blunt, J.W., Copp, B.R., Hu, W.-P., Munro, M.H.G., Northcote, P.T., and Prinsep, M.R. (2007). Marine natural products. Nat. Prod. Rep. *24*, 31–86.

Blunt, J.W., Copp, B.R., Hu, W.-P., Munro, M.H.G., Northcote, P.T., and Prinsep, M.R. (2009). Marine natural products. Nat. Prod. Rep. *26*, 170–244.

Bollmann, A., Lewis, K., and Epstein, S.S. (2007). Incubation of environmental samples in a diffusion chamber increases the diversity of recovered isolates. Appl. Environ. Microbiol. *73*, 6386–6390.

Bruce, J. (1996). Automated system rapidly identifies and characterizes microorganisms in food. Food Technol. *50*, 77–81.

Busti, E., Cavaletti, L., Monciardini, P., Schumann, P., Rohde, M., Sosio, M., and Donadio, S. (2006a). *Catenulispora acidiphila* gen. nov., sp. nov., a novel, mycelium- forming actinomycete, and proposal of *Catenulisporaceae* fam. nov. Int. J. Syst. Evol. Microbiol.*56*, 1741–1746.

Busti, E., Monciardini, P., Cavaletti, L., Bamonte, R., Lazzarini, A., Sosio, M., and Donadio, S. (2006b). Antibiotic-producing ability by representatives of a newly discovered lineage of actinomycetes. Microbiology *152*, 675–683.

Cao, L., Qiu, Z., You, J., Tan, H., and Zhou, S. (2005). Isolation and characterization of endophytic streptomycete antagonists of Fusarium wilt pathogen from surface-sterilized banana roots. FEMS Microbiol. Lett. *247*, 147–152.

Case, R.J., Wang, Y., Franzblau, S.G., Soejarto, D., Matainaho, L., Piskaut, P., and Pauli, G.F. (2007). Advanced applications of counter-current chromatography in the isolation of anti-tuberculosis constituents from *Dracaena angustifolia*. J. Chromatogr. A *1151*, 169–174.

Castillo, U.F., Browne, L., Strobel, G., Hess, W.M., Ezra, S., Pacheco, G., and Ezra, D. (2007). Biologically active endophytic streptomycetes from *Nothofagus* spp. and other plants in Patagonia. Microb. Ecol. *53*, 12–19.

Cavaletti, L., Monciardini, P., Schumann, P., Rohde, M., Bamonte, R., Busti, E., Sosio, M., and Donadio, S. (2006). *Actinospica robiniae* gen. nov., sp. nov. and *Actinospica acidiphila* sp. nov.: Proposal for *Actinospicaceae* fam. nov. and Catenulisporinae subord. nov. in the order Actinomycetales. Int. J. Syst. Evol. Microbiol.*56*, 1747–1753.

Charan, R.D., Schlingmann, G., Janso, J., Bernan, V., Feng, X., and Carter, G.T. (2004). Diazepinomicin, a new antimicrobial alkaloid from a marine *Micromonospora* sp. J. Nat. Prod. *67*, 1431–1433.

Chen, Y., Smanski, M.J., and Shen, B. (2010). Improvement of secondary metabolite production in *Streptomyces* by manipulating pathway regulation. Appl. Microbiol. Biotechnol. *86*, 19–25.

Clardy, J., Fischbach, M.A., and Walsh, C.T. (2006). New antibiotics from bacterial natural products. Nature Biotechnol. *24*, 1541–1550.

Connon, S.A., and Giovannoni, S.J. (2002). High-throughput methods for culturing microorganisms in very-low- nutrient media yield diverse new marine isolates. Appl. Environ. Microbiol. *68*, 3878–3885.

Coombs, J.T., and Franco, C.M.M. (2003). Isolation and identification of actinobacteria from surface-sterilized wheat roots. Appl. Environ. Microbiol. *69*, 5603–5608.

Courtois, S., Cappellano, C.M., Ball, M., Francou, F.X., Normand, P., Helynck, G., Martinez, A., Kolvek, S.J., Hopke, J., Osburne, M.S., *et al.* (2003). Recombinant environmental libraries provide access to microbial diversity for drug discovery from natural products. Appl. Environ. Microbiol. *69*, 49–55.

Debono, M., Abbott, B.J., Molloy, R.M., Fukuda, D.S., Hunt, A.H., Daupert, V.M., Counter, F.T., Ott, J.L., Carrell, C.B., Howard, L.C., *et al.* (1988). Enzymatic and chemical modifications of lipopeptide antibiotic A21978C: the synthesis and evaluation of daptomycin LY14603. J. Antibiot. *41*, 1093–1105.

Donadio, S., Maffioli, S., Monciardini, P., Sosio, M., and Jabes, D. (2010). Antibiotic discovery in the twenty-first century: current trends and future perspectives. J. Antibiot. *63*, 423–430.

Donadio, S., Monciardini, P., Alduina, R., Mazza, P., Chiocchini, C., Cavaletti, L., Sosio, M., and Puglia, A.M. (2002). Microbial technologies for the discovery of novel bioactive metabolites. J. Biotechnol. *99*, 187–198.

Farrell, W.P., Aurigemma, C.M., and Masters-Moore, D.F. (2009). Advances in high throughput supercritical fluid chromatography. J. Liquid Chromatogr. Rel. Technol. 32, 1689–1710.

Feng, X., Ratnayake, A.S., Charan, R.D., Janso, J.E., Bernan, V.S., Schlingmann, G., He, H., Tischler, M., Koehn, F.E., and Carter, G.T. (2009). Probing natural product biosynthetic pathways using Fourier transform ion cyclotron resonance mass spectrometry. Bio. Med. Chem. *17*, 2154–2161.

Fenical, W., and Jensen, P.R. (1993). Marine microorganisms: a new biomedical resource. In Marine Biotechnology, Volume *1*, Pharmaceutical and Bioactive Natural Products, Attaway, D.A., and Zaborsky, O.R., eds. (Plenum Press, New York), pp. 419–457.

Fenical, W., and Jensen, P.R. (2006). Developing a new resource for drug discovery: marine actinomycete bacteria. Nat. Chem. Biol. *2*, 666–673.

Finking, R., and Marahiel, M.A. (2004). Biosynthesis of nonribosomal peptides. Ann. Rev. Microbiol. *58*, 453–488.

Fisch, K.M., Gurgui, C., Heycke, N., van der Sar, S.A., Anderson, S.A., Webb, V.L., Taudien, S., Platzer, M., Rubio, B.K., Robinson, S.J., *et al.* (2009). Polyketide assembly lines of uncultivated sponge symbionts from structure-based gene targeting. Nat. Chem. Biol. *5*, 494–501.

Biosynthetic Engineering of Antibacterial Natural Products

8

Jay Fitzgerald, Younjoo Lee and Chaitan Khosla*

Abstract

Since the discovery of penicillin, the development of anti-infective drugs has been a central theme in the pharmaceutical industry through much of the twentieth century. However, the pace of developing new anti-infective agents has precipitously declined in the past two decades. The main reason for this change is an economic one – whereas the technical and regulatory risks associated with the development of a new broad-spectrum antibiotic are deemed unacceptably high, the financial returns derived from a targeted (narrow-spectrum) antibiotic are unattractive to the pharmaceutical industry. Meanwhile, the need for new anti-infective agents continues to be as urgent as ever. New business models are called for, ones that are grounded in the possibilities and realities of twenty-first century technologies for antibiotic discovery and development. This chapter discusses, using four selected examples, the opportunities for harnessing modern biosynthetic insights and engineering methods to discover new antibiotics.

Introduction

Polyketide and peptide natural products have been an extremely rich source of new anti-infective agents. Antibiotics such as beta-lactams, erythromycin, tetracyclines, amphotericin and avermectin have been the mainstay of antibacterial, antifungal and antiparasitic therapy for decades. Although new polyketide and peptide antibiotics continue to be discovered at a steady pace, their translation into clinical practice has almost entirely tapered off over the past two decades. Recent insights into polyketide and non-ribosomal peptide biosynthesis have spawned new engineering technologies that have the potential to improve access to scarce natural products, and to facilitate the introduction of chemo- and regio-selective functional handles into these complex molecules. At the same time, new synthetic methods, especially cross-coupling chemistry, have reached a level of maturity where they can be applied to complex natural products such as polyketides and peptides.

Our choice of examples in this chapter is based on a common theme. In each case, a recent biological discovery has set the stage for efficiently leveraging a large body of existing chemical and pharmacological information to 're-tread' a known polyketide or peptide for a targeted anti-infective application.

Erythromycin

First isolated from the fermentation extracts of the bacterium *Saccharopolyspora erythraea* in the early 1950s (Mcguire *et al.*, 1952), erythromycin A is one of the most clinically important macrolide antibiotics. It has broad-spectrum activity against Gram-positive bacteria, Gram-negative cocci, as well as atypical and intracellular pathogens. The agent binds to the exit tunnel in the bacterial 50S ribosomal subunit making specific interactions with the 23S RNA and interfering with protein

*Corresponding author

synthesis. Erythromycin A is commonly used in the treatment of upper and lower respiratory tract infections, skin and soft tissue infections, and genital tract infections. Erythromycin A displays low toxicity and is fairly well tolerated, making it useful for treatment of bacterial infections in infants and pregnant women. However, adverse effects such as poor bioavailability, a limited spectrum of activity, and gastrointestinal contractions limit its use (Welling *et al.*, 1978; Itoh *et al.*, 1984). Erythromycin A's adverse physiological effects have been attributed to degradation of the molecule in gastric acid to form the inactive 8,9-anhydro-6,9-hemiketal.

To improve acid instability, semi-synthetic erythromycin A derivatives such as clarithromycin and azithromycin were introduced. More recently, telithromycin has shown efficacy against certain macrolide-resistant bacteria, although more general solutions to the problem of drug resistance are still lacking (Denis *et al.*, 1999). With the recent solution of X-ray structures of the bacterial ribosome, including ribosomes bound to a variety of macrolides, an entirely new foundation has emerged for the systematic enhancement of erythromycin activity against challenging resistance mechanisms (Schlunzen *et al.*, 2001; Hansen *et al.*, 2002; Schlunzen *et al.*, 2003). By introducing new functional handles into the complex natural product, biosynthetic engineering can facilitate these structure-based efforts.

Much is known about the biosynthesis of erythromycin A (Fig. 8.1) which consists of a 14-membered lactone ring and two unusual sugar moieties: D-desosamine and L-cladinose. One propionyl-CoA and six methylmalonyl-CoA units

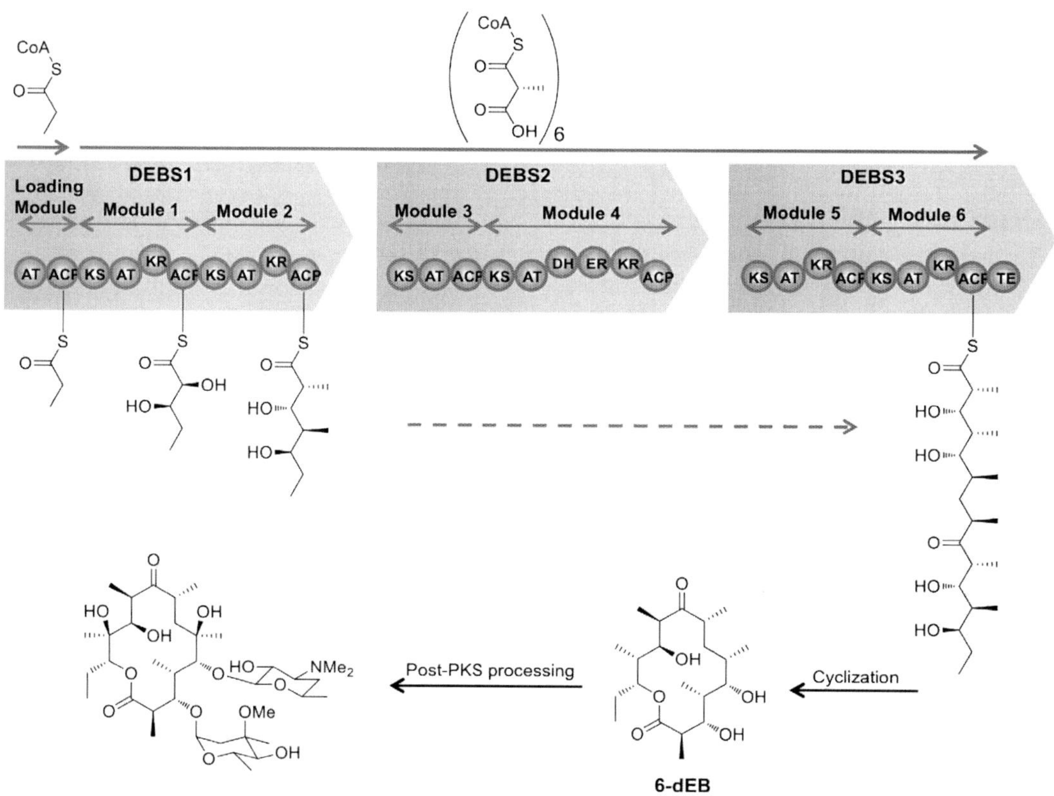

Figure 8.1 Biosynthesis of erythromycin A is achieved through modular polyketide synthases (PKSs) collectively known as 6-dexoyerhythronolide B synthase (DEBS) 1, 2, and 3. The heptaketide chain produced from one starter unit of propionyl-CoA and six extender units of methylmalonyl-CoA is cyclized and released from the enzyme complex to generate the first free intermediate 6-deoxyerythronolide B (6-dEB). The aglycone 6-dEB undergoes further post-PKS modifications including two hydroxylations, two glycosylations, and methylation to form erythromycin A. Abbreviations: AT, acyl transferase; ACP, acyl carrier protein; KS, ketosynthase; KR, ketoreductase; DH, dehydratase; ER, enoyl reductase; TE, thioesterase.

are incorporated into a growing polyketide chain, and the extended chain is cyclized and released from the enzyme complex to generate a macrolide intermediate 6-deoxyerythronolide B (6-dEB). 6-dEB is further acted upon by two macrocycle-modifying hydroxylases, two glycosyltransferases, and a methyltransferase to yield erythromycin A.

The basic mechanism of the 6-deoxyerythronolide B synthase (DEBS), a modular polyketide synthase (PKS), is shown in Fig. 8.1. Each of three DEBS proteins consists of two elongation modules containing ketosynthase (KS), acyl transferase (AT), and acyl carrier protein (ACP) domains, as well as additional reductase domains, such as ketoreductase (KR) or enoylreductase (ER) domains, as required. The propionyl-CoA derived primer unit is transferred to a loading domain and subsequently onto the KS domain of module 1. Methylmalonyl-CoA derived extender units are subsequently incorporated into the growing chain at each extender module. Reductive modifications occur after each round of chain elongation, except for the elongation catalysed by module 3. The full-length chain is then cyclized and released by a thioesterase (TE) domain at the C-terminal end of module 3, and the resulting 6-dEB is hydroxylated to produce erythronolide B. Convergently, mycarosyl-TDP and desosaminyl-TDP are synthesized from glucose 6 phosphate by dedicated enzymes, and the deoxysugars are transferred to the C-3 and C-5 hydroxyl groups, respectively, by distinct glycosyltransferases. The last steps in erythromycin A biosynthesis involves C-12 hydroxylation and methylation of the mycarose sugar to result in the cladinose, the non-nitrogenous deoxysugar, mentioned above. Several of the biosynthetic intermediates have antibacterial activity, though none is as potent as erythromycin A.

To overcome the poor bioavailability and gastrointestinal side-effects of erythromycin A, several semi-synthetic derivatives of erythromycin A have been developed including roxithromycin, dirithromycin, clarithromycin, and azithromycin (Fig. 8.2). However, the second-generation derivatives are no more efficacious against macrolide-resistant bacteria than the natural product. The emergence of macrolide resistance in a number of Gram-positive pathogens such as streptococci, pneumococci and enterococci arose soon after the launch of erythromycin A (Jones *et al.*, 1956). Three mechanisms account for much of the clinically relevant macrolide resistance: modification of the drug binding site on the ribosome, efflux of the drug from the bacterial cell, and (to a lesser extent) cleavage of the macrocyclic lactone ring *via* an esterase. Methylation of the 23S ribosomal RNA at adenine residue 2058 by erythromycin resistance methylase genes (*erms*) prevents access of antibiotics such as macrolides, lincosamides, and streptogramin B to their ribosomal binding sites causing MLS_B cross-resistance (Roberts *et al.*, 1999). As an alternative, macrolide efflux pumps such as *msr*A from staphylococci or *mef*A from streptococci mediate a decrease in the intracellular drug concentration.

Efforts to overcome these macrolide resistance mechanisms led to the discovery of a new class of macrolides, the ketolides. In these semi-synthetic antibiotics, the L-cladinose at the C-3 position of clarithromycin is removed and substituted with a carbonyl group. The 3-keto functional group promotes greater acid stability, and prevents the MLS_B resistance phenotype. In addition to 3-keto functionality, Telithromycin also harbours a carbamate ring between the C-11 and C-12 positions that is further modified by a butyl imidazolyl pyridinyl side chain. The cyclic carbamate stabilizes the active conformation of telithromycin, whereas the aryl group enhances its binding affinity to the methylated ribosome.

A variety of rational engineering strategies have been developed for the rapid genetic and chemo-biosynthetic manipulation of erythromycin biosynthesis. For example, by genetically inactivating the KS domain of module 1 with the active site $Cys^{729} \rightarrow Ala$ mutation, synthetic diketides can be transformed into erythromycin analogues (Jacobsen *et al.*, 1997). This methodology was used to prepare 15-fluoro- and 15-azido erythromycin A, novel analogues with unique physicochemical properties containing functional groups that allow for selective derivatization (Ashley *et al.*, 2006). Alternatively, by replacing the loading module of an erythromycin PKS with a heterologous loading module from an avermectin PKS, unnatural starter acyl groups can

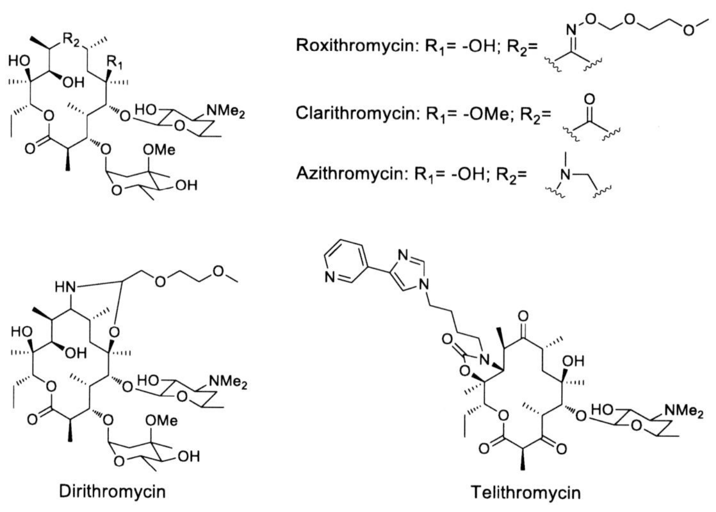

Organism	Erythromycin	Azithromycin	Clarithromycin	Roxithromycin	Dirithromycin	Telithromycin
Gram-positive aerobes						
Streptococcus pyogenes						
(Ery. Susceptible)	0.06–0.12	0.11–0.25	0.05–0.12	0.11	0.11	0.03
(ermA resistance)	1-32	16-32	2-16			0.015-0.25
(ermB resistance)	>32	>32	>32	>32		>8
(mefA resistance)	8-16	8	8-16	16		>0.25-1
Streptococcus pneumoniae						
(Ery. Sensitive)	0.03–0.12	0.06–0.25	0.03-0.12			0.008–0.03
Staphylococcus aureus	0.25-0.45	0.95	0.24	1.7	0.9	0.03-0.06
Gram-negative aerobes						
Haemophilus influenza	8	2-4	4-16			2-4
Moraxella catarrhalis	0.125–0.25	0.06–0.12	0.12–0.25	0.25	4	0.12
Legionella pneumophila	0.12-0.5	0.25-2	0.06–0.25			0.015–0.006
Neisseria gonorrhoeae	0.5	0.25				0.12
Other pathogens						
Chlamydophila pneumonia	0.06–0.25	0.125–0.25	0.03– 0.06			0.06-0.25
Mycoplasma pneumoniae	≤0.015–0.06	≤0.015	≤0.015–0.06	≤0.015	≤0.015	≤0.015

Figure 8.2 Erythromycin A analogues with antibacterial activity expressed as MIC_{90} (mg/l); ranges are different values reported in references (Hamilton-Miller 1992; Johnson *et al.*, 1992; Kataja *et al.*, 1999; Bebear *et al.*, 2000; Bemer-Melchior *et al.*, 2000; Jacobs *et al.*, 2003; Zuckerman 2004; Shimizu *et al.*, 2007).

be incorporated into the macrolide lactone (Marsden *et al.*, 1998). The methyl substituents of the 14-membered ring can be modified by engineering targeted AT domains, which are responsible for extender unit incorporation (Oliynyk *et al.*, 1996; Liu *et al.*, 1997; Ruan *et al.*, 1997; Stassi *et al.*, 1998). Structural changes have also been achieved by manipulating the reductive loop domains. Either loss of function (deletion) or gain of function (insertion) mutations in the KR, DH and ER domains lead to different oxidation levels of β-carbonyl (Donadio *et al.*, 1991, 1993; Kao *et al.*, 1997; McDaniel *et al.*, 1997). Polyketide chain length can be also controlled by altering the number of elongation modules (Kao *et al.*, 1995). In addition, some enzymes involved in the post-PKS tailoring of 6-dEB are highly versatile enabling the efficient production of novel compounds (Weissman *et al.*, 2005). For example, multiple genetic modifications of DEBS were achieved by introducing multiple foreign domains such as the AT domains and β-carbon processing domains (KR, DH and ER domains) from the rapamycin PKS (RAPS) in place of DEBS. Leading to a diverse library of novel erythromycin compounds (McDaniel *et al.*, 1999). Many types of chemical modifications, however, still remain beyond the realm of genetic manipulation. For example, general strategies for regiospecific incorporation of alkene, alkyne, azide or amine groups into the macrolide would significantly increase the range of available erythromycin derivatives for medicinal chemistry.

The structure of ribosomal subunits and their complexes with macrolides have been determined by X-ray crystallography for *Haloarcula*, *Deinococcus* and *Escherichia* species. Macrolides interact with specific nucleotides on domain V of the 23S rRNA through hydrogen bonds as well as hydrophobic interactions (Schlunzen *et al.*, 2001). The 4, 5, and 6 positions of the macrolactone ring bind to the ribosomal exit tunnel wall via hydrophobic interactions, and the reactive groups on the lactone ring and the desosamine sugar hydrogen bond with ribosomal nucleotides. The 6-OH, 11-OH and 12-OH groups are positioned within hydrogen bonding distance of the 23S rRNA. The 11-OH and 12-OH groups form hydrogen bond interactions with O4 of U2609 in *Escherichia coli* (Ec). In the case of clarithromycin, the 6-OH group is replaced with a methoxy group but bioactivity comparable to native erythromycin is maintained as the oxygen of the 6-methoxy group still interacts with N6 of A2062(Ec). The cladinose sugar does not appear to form hydrogen bonds with the 23S rRNA, confirming SAR data showing that the 4-OH is dispensable. The ketolides such as telithromycin and cethromycin, which lack a cladinose sugar, have been shown to bind more strongly to ribosomes than macrolides. The 2 -OH group on the desosamine sugar forms hydrogen bonds with N6 of A2059(Ec), and N6 and N1 of A2058(Ec), nucleotides which are implicated in macrolide resistance. The N6 methylation of A2058(Ec) by the *Erm* family of methyltransferases hinders formation of hydrogen bonds to the 2′-OH group. The ribosomal proteins, L4 and L22 are also involved in erythromycin resistance (Wittmann *et al.*, 1973). Although these proteins do not directly interact with macrolides, mutations in them cause a perturbation of the 23S rRNA, resulting in macrolide resistance. Telithromycin displays improved bioactivity against macrolide-resistant pathogens. Its tight binding to ribosomes is a result of the lack of a cladinose sugar as well as the unique stacking arrangement of its extended aryl-alkyl arm, which interacts with domain II even when the macrolide binding site in

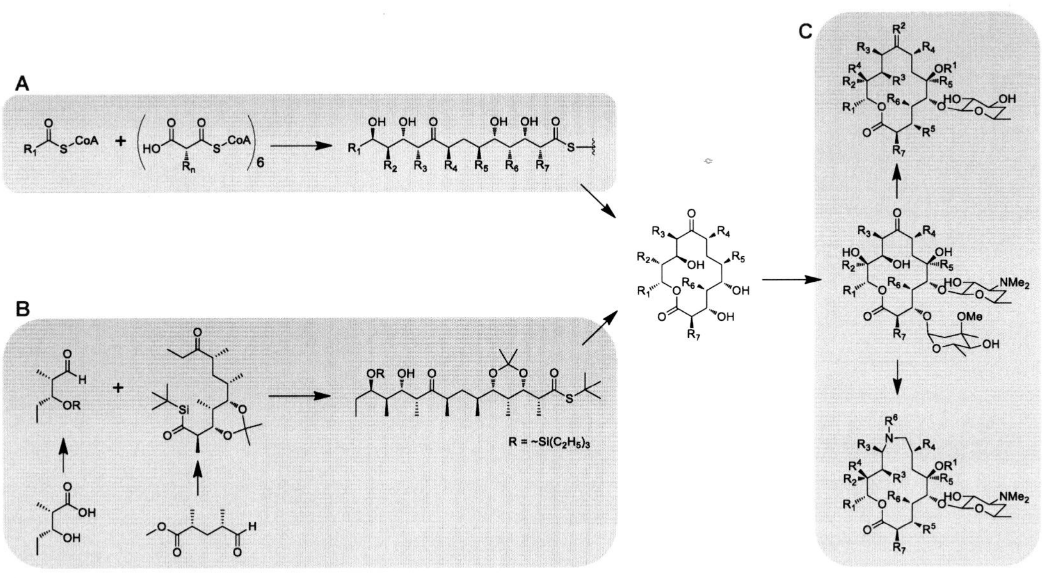

Figure 8.3 The complementary nature of synthetic and biosynthetic approaches toward erythromycin modification. (A) One starter unit of propionyl-CoA and six extender units of methylmalonyl-CoA are required for the biosynthesis of 6-deoxyerythronolide B (6-dEB). (B) Total synthesis of 6-dEB. The left-hand fragment (C11–C13) and the right-hand fragment (C1–C10) are cyclized to afford 6-dEB (Myles *et al.*, 1981). (C) Semisynthetic derivatives from erythromycin A.

domain V is perturbed (Douthwaite *et al.*, 2000). Structural insights gained from first generation macrolides which overcome antibiotic resistance will undoubtedly pave the way for further modification of the erythromycin scaffold to overcome bacterial resistance in the future (Fig. 8.3).

Tetracyclines

Since the introduction of the type II polyketide chlortetracycline to the clinic in 1948, the tetracycline scaffold has provided an enormously useful starting point for the treatment of bacterial infection. Following quickly on the heels of chlortetracycline's isolation from *Streptomyces aureofaciens* was the isolation of oxytetracycline from *Streptomyces rimosus* and the discovery of tetracycline, the namesake of the class, in 1953. The tetracyclines were rapidly accepted by the medical community as complements to the other dominant antibacterial class, the penicillins, owing to their vastly improved spectrum of activity. Clinically successful semisynthetic analogues with improved pharmacokinetics soon appeared beginning with rondomycin and doxycycline in 1963, followed by minocycline in 1967. These compounds resulted from extensive efforts to semi-synthetically modify the tetracycline scaffold, and in turn led to a wealth of structure–activity relationship data.

Tetracycline derived antibacterials appear to function through inhibition of the 30S ribosomal subunit. A high-resolution crystal structure of tetracycline bound to two separate locations on this subunit has been solved (Brodersen *et al.*, 2000). Tetracycline primarily sits directly above the A site tRNA binding pocket, stabilized by hydrogen bonding interactions between the intricately oxidized 'southern' and 'eastern' faces of the polyketide and the sugar–phosphate backbone of H34, in addition to a coordinated magnesium atom which makes salt bridges to the phosphate oxygens of G1197. This area of the ribosome is critical for the binding of the elongation factor Tu (EF Tu), aa-tRNA, GTP complex. Tetracyclines are thought not to interfere with initial binding of the EF Tu complex to the ribosome, but instead to prevent rotation of the tRNA into the A site after binding has occurred. This mode of action explains tetracyclines effectiveness as they not only prevent protein manufacture, but allow for GTP hydrolysis by EF Tu, draining cellular resources.

As tetracyclines became front-line antibiotics, resistance began to develop. By 2005, almost 40 tetracycline resistance genes were characterized with the majority (~60%) providing more efficient efflux of tetracyclines from the cell and another large fraction (~30%) related to ribosomal protection modifications (Roberts 2005). This emerging resistance spurred extensive semisynthetic efforts with the tetracycline scaffold, most recently leading to the introduction of tigecycline to the clinic in 2005.

The tetracyclines, like all type II polyketides, are produced by a series of discrete enzymes. Biosynthesis proceeds through iterative decarboxylative condensations of malonyl-CoA derived building blocks onto a growing polyketide chain. This process is catalysed by a ketosynthase-chain length factor heterodimer (KS/CLF), which in addition to extending the nascent chain determines the chain length of the resulting polyketide (Tang *et al.*, 2003). In combination with a dedicated acyl carrier protein (ACP), and a malonyl-CoA:ACP transacylase (MAT), borrowed from the fatty acid synthesis machinery, these enzymes constitute the minimal polyketide synthase necessary for formation of the core, highly unstable poly β-keto chain (Fig. 8.4). A variety of tailoring enzymes including ketoreductases, aromatases, cyclases, and oxygenases can then transform this skeleton into an active antibiotic molecule.

Early investigations of tetracycline biosynthesis by McCormick *et al.* in the 1960s identified pretetramid, 6-methylpretetramid, and anhydrotetracycline (ATC) as key intermediates (Fig. 8.4) in a series of papers detailing blocked mutant studies (Blackwood *et al.*, 1985). The investigators also established that tetracycline production could be recovered in a series of mutants blocked in early tetracycline biosynthesis through feeding of these later stage intermediates. Molecular biology techniques developed in the 1980s led to the identification of a cosmid capable of heterologous production of oxytetracycline in *Streptomyces lividans* (Binnie *et al.*, 1989). This confirmed that all enzymes responsible for

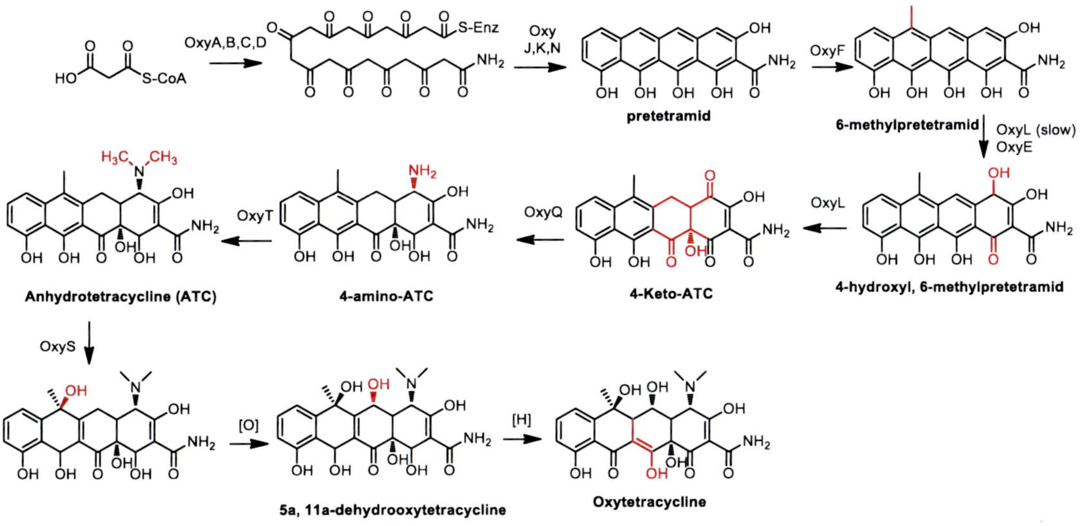

Figure 8.4 Biosynthesis of oxytetracycline. Heterologous reconstitution of the oxytetracycline pathway to the intermediate ATC has been achieved in *S. coelicolor* CH999 using a pRM5-based vector system. The enzymes responsible for the last two steps involving the oxygenation of the C5 position as well as the final reduction of 5a 11a-dehydrooxytetracycline to oxytetracycline remain elusive.

steps unique to oxytetracycline biosynthesis are encoded in a compact region of the genome, as is common for many polyketides. More recently, in a thorough investigation of oxytetracycline biosynthesis in *S. rimosus*, the minimal set of genes necessary to produce the late stage intermediate anhydrotetracycline (ATC) has been identified using heterologous reconstitution in a CH999 based host-vector system (Zhang *et al.*, 2008).

The oxytetracycline minimal PKS, oxyABC, when expressed on a pRM5 based plasmid (McDaniel *et al.*, 1993) in *S. coelicolor* CH999 gives SEK15, a previously reported decaketide (Hopwood *et al.*, 1994), implying that the minimal PKS alone can maintain decaketide chain length specificity. The addition of OxyJ, a C-9 ketoreductase, and OxyD, an asparagines synthase homologue, was shown to be sufficient for the incorporation of a malonamate starter unit giving an isolable malonamyl-primed decaketide shunt product (Zhang *et al.*, 2006). Despite the presence of four rings, only two cyclases, OxyK and OxyN, were sufficient to produce the fully cyclized pretetramid product. Surprisingly, the incorporation of the malonamate starter unit was essential for formation of the A ring in pretetramid (Zhang *et al.*, 2007). After formation of the tetracyclic core, a variety of enzymes decorate the molecule, transforming it into a fully active natural product. First, installation of a C-6 methyl group is performed by OxyF. Then the combined action of OxyE, a fast-acting C-12a oxygenase and OxyL a slow-acting C-12a oxygenase and a C-4 oxygenase form the hypothetical intermediate 4-keto-ATC (Wang *et al.*, 2009). This in turn is acted upon by OxyQ, a novel reductive transaminase, to give the short-lived 4-amino-ATC intermediate, which is then rapidly converted to ATC by the SAM-dependent *N*,*N*-dimethyltransferase OxyT (Zhang *et al.*, 2008). From ATC two oxygenations and a reduction remain to access the fully active oxytetracycline molecule. OxyS catalyses the first C-6 oxygenation (Binnie *et al.*, 1989). The enzymes responsible for the later C-5 hydroxylation and the final reduction steps remain to be definitively identified. Two oxygenases OxyG and OxyR in the OTC gene cluster have not been assigned a definitive function and are likely candidates to perform the oxygenation, though homologues are also found in the non-C-5 oxygenated chlorotetracycline gene cluster (Zhang *et al.*, 2006). The final reduction step is probably performed by an enzyme outside the tetracycline gene cluster

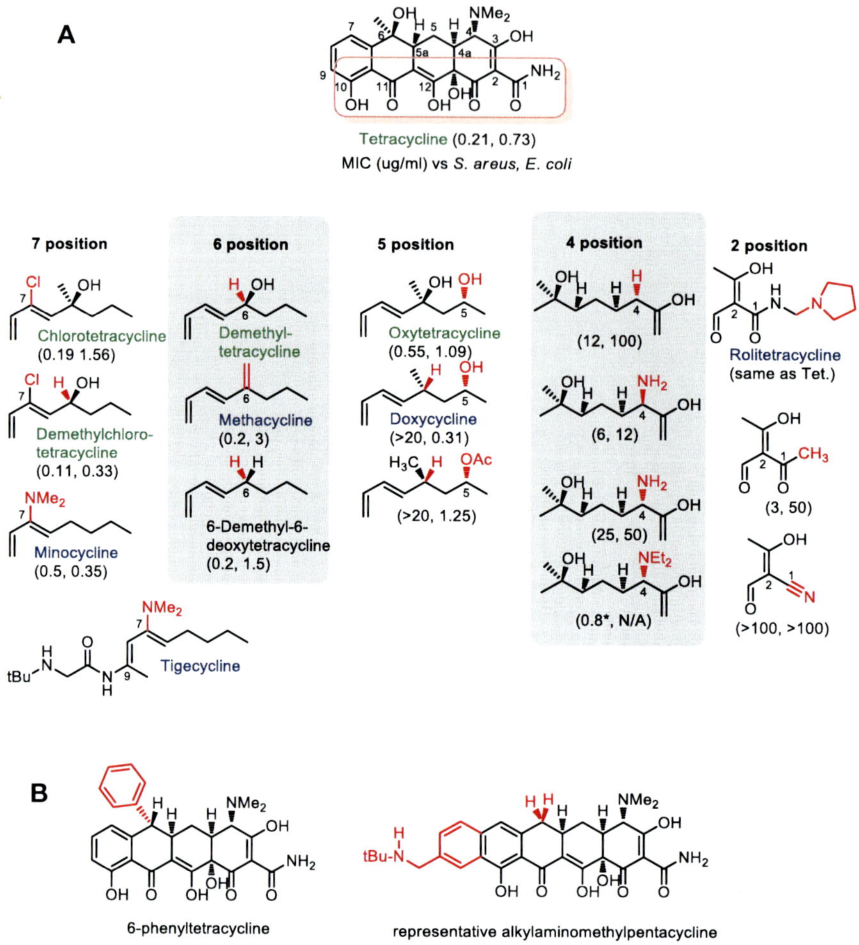

Figure 8.5 Semisynthetic and totally synthetic tetracycline derivatives. (A) Tetracycline is shown with a pink box around the southern and eastern sections indicating regions important in binding the ribosome. Semisynthetic tetracycline derivatives are also shown with the relevant portion of the parent tetracycline molecule in black and modifications in red. Green indicates compounds isolated as natural products. Blue indicates medicinally useful semisynthetic derivatives. The historical numbering scheme for tetracyclines begins with the carbon of the priming amid as carbon 1 and continues around the scaffold with carbons internal to the tetracyclic structure being labelled 4a 5a, etc. MIC values versus *S. aureus* and *E. coli* are shown in parenthesis (Blackwood *et al.*, 1985). (B) Bioactive totally synthetic tetracycline derivatives produced by the Myer group with modifications shown in red.

similar to the reductase responsible for chlorotetracycline production in *S. aurofaciens*, TchA, which resides outside of the mapped cluster (Nakano *et al.*, 2004).

Semi-synthetically derived structure–activity relationships (Fig. 8.5) provide a wealth of information about useful substitution positions on the tetracycline scaffold and should be carefully considered in the design of new biosynthetically engineered tetracyclines. For example, the carboxamido moiety at the C-2 position arises from the malonamyl starter unit in tetracycline biosynthesis. The amide at this position has been found to be important for tetracycline activity. Substitution of the carboxamido group for an acetyl group leads to a 15- to 50-fold drop in activity, whereas substitution with a nitrile group abolishes activity (Blackwood *et al.*, 1985). An apparent exception is the semisynthetic analogue, rolitetracycline, which contains a pyrrolidine group at the C-2 position that acts as a more soluble pro-drug. The α-dimethylamino group at the C-4 position is also crucial for tetracycline activity. Similarly, the substituents at C-3, -10, -11 and -12 directly interact with the ribosome binding site, greatly decreasing activity when modified (Nelson 2002). In contrast, substitutions at the C-5 and C-6 positions of tetracycline are well tolerated. The therapeutically useful natural product oxytetracycline contains an alcohol group at C-5 which prevents epimerization at C-6, conferring acid stability. In addition, doxycycline, which contains the same C-5 alcohol, proved extremely clinically useful against Gram-negative bacteria. Increased activity with the alcohol substitution at C-5 is probably due to hydrogen bonding contacts as acetylation or methylation lowers activity (Blackwood *et al.*, 1985). Demethyltetracycline, produced by *S. aureofaciens*, provided a new starting scaffold for analogue exploration granting access to 6-demethyl-6-deoxytetracycline, which has comparable activity to tetracycline as well as the clinically useful methacycline. The naturally occurring C-7 substituted chlorotetracyclines show similar or slightly improved potency compared with tetracycline. Classic semisynthetic analogues at the C-7 and C-9 positions are available via nucleophilic aromatic substitution of the acid stable 6-deoxytetracyclines. The widely used semisynthetic derivative minocycline, harbouring a dimethylamino group at the C-7 position, is prepared in this way. In addition tigecycline, the newest tetracycline derivative approved for clinical use contains a large bulky amide appended to the C9 position allowing it to evade most common tetracycline resistance mechanisms. Additionally, the 2000 co-crystal structure of tetracycline with the ribosome provides an alternative structural guide to the biosynthetic engineer. Models developed around an electrostatic representation of the tetracycline binding pocket have been used scan large databases of compounds and identify some potential features of compounds with good ribosomal binding characteristics (Foloppe *et al.*, 2004).

In a recent series of publications, the Myers group pioneered a fundamentally new approach to the synthesis of diverse tetracycline derivatives (Charest *et al.*, 2005). As opposed to earlier syntheses by Woodward, Shemyakin and Muxfeldt (Gurevich *et al.*, 1967; Korst *et al.*, 1968; Muxfeldt *et al.*, 1979), which began with a bicyclic fused C–D ring precursor to which they appended the A and B rings, the Myers group set about to end the synthesis with the addition of the D-ring. In this way an easily modifiable D-ring precursor is attached enantioselectively to an AB ring precursor through a Michael–Dieckmann condensation cascade yielding a very late stage synthetic intermediate (Fig. 8.6). Though the first series of compounds with heteroatom containing D-ring derivatives appear relatively inactive (Charest *et al.*, 2005), two interesting classes of compounds emerged from this work: the 6-aryltetracyclines and the alkylaminomethylpentacyclines. These scaffolds show greatly improved activity against tetracycline resistant strains of a broad range of bacteria including *S. aureus* and *S. pneumoniae* (Sun *et al.*, 2008). For example 6-phenyltetracycline has an MIC of <0.06 μg/ml against penicillin and tetracycline resistant *S. pneumoniae* whereas tetracycline has only an MIC of only 64 μg/ml.

Biosynthetic engineering of type II polyketide synthases has gained momentum in recent years. A biosynthetically modified polyketide can be created in one of two complementary approaches: gene deletion or heterologous expression and gene supplementation. The former is generally easier to achieve and has been extensively employed to investigate tetracycline biosynthesis. The latter, though more labour intensive, has the potential to create a wider range of analogues as incorporation of enzymes from a diverse bacterial pool is possible. Deletion of enzymes from the tetracycline biosynthetic pathway, such as the SAM-dependent methyl transferase (OxyF) will probably allow access to 6-demethyltetracyclines. Tang and co-workers have demonstrated that heterologous expression of OxyABCDJKNLTQ, the enzymes required to produce ATC minus OxyF, can make 6-desmethyl-ATC (Zhang *et al.*, 2008). Similarly, deletion of the oxygenase (OxyS) may be able to produce 6-deoxytetracyclines. Though deletion of OxyS from the *S. rimosus* reportedly produces only truncated polyketides (Peric-Concha *et al.*, 2005), when the minimal cluster is expressed in pRM5 as discussed in the biosynthesis section,

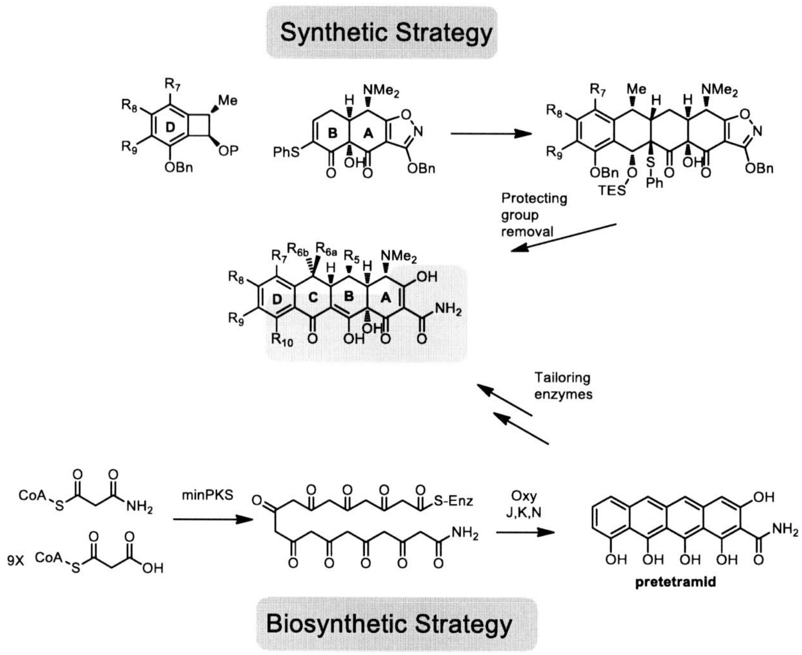

Figure 8.6 Combination of synthetic and biosynthetic approaches to tetracycline highlighting the readily modifiable positions at R5, 6a 6b, 7, 8, 9, and 10. Enzymes discovered by Tang *et al.* (2003) in the SF2575 pathway can potentially modify positions 4, 6a, and 9. The Myers group's synthetic methodology can easily access analogues at R6, 7, 8, 9, and 10.

full length polyketides are the major product. It has also been suggested that when OxyS is deleted that other oxygenases may oxidize the molecule at this position, but this is probably host dependent.

Perhaps the most compelling example of generating biosynthetically modified tetracyclines through heterologous expression involves reconstitution of the biosynthetic pathway for the anticancer tetracycline SF2575 by the Tang group (Pickens *et al.*, 2009). A few of the strikingly useful tetracycline modifying enzymes in this pathway include an enzyme able to perform C-C glycosylation with D-olivose at the C9 position (SsfS6), an enzyme capable of modifying the C-4 position with a salicylic acid moiety (SsfX3), and a C-6 *O*-methyltransferase (likely SsfOI). The effect of sugars at the C-9 position of tetracyclines is unexplored, but as the Myers group demonstrated, strategic modification of the D-ring can help to overcome resistance mechanisms. At first glance salicylation of the C-4 group may appear to be without merit as it leads to greatly reduced antibiotic activity (Hatsu *et al.*, 1992). But, SsfX3 has shown to have reasonable substrate flexibility (Pickens *et al.*, 2009) and an involved structure activity relationship has not been conducted at this position with aryl groups to date. Owing to the inherent instability conferred by the C6 alcohol moiety, the C-6 *O*-methyltransferase may lead to more acid stable analogues as suggested by Pickens *et al.*, Other potentially combinatorially useful members of the tetracycline family include the dactylocyclines, which are glycosylated with a C-6 (R) configuration and have an alcohol or a methoxy group at the semi-synthetically inaccessible C-8 position. These compounds, though generally less potent than their tetracycline counterparts, maintain activity against tetracycline-resistant bacteria (Tymiak *et al.*, 1993).

The semisynthetic tools developed in the exploration of the first tetracyclines could also be harnessed to modify new scaffolds derived through genetic engineering. For example, the amide group installed at C-9 in the semisynthetic derivative tigecycline, which helps to overcome tetracycline resistance, might be installed in conjunction with a biosynthetic C-6 modification. Additionally, as the purely synthetic compounds created by the Myers group highlight, it would be quite useful to have semisynthetic techniques capable of appending an aromatic moiety to the C6b position or fused to the C8 and 9 positions of a metabolically engineered 6-deoxy-6-deoxytetracycline (Fig. 8.6).

Tyrocidine

The tyrocidines are cyclic peptide antibiotics produced by the soil bacteria *Bacillus brevis* ATCC8185. They consist of a mixture of four cyclic decapeptides with slightly different amino acid sequences, providing tyrocidine A, B, C, and D. The function of the tyrocidines in *Bacillus brevis* is not firmly established, (Katz *et al.*, 1977) but they probably act as regulators in the sporulation process (Lee *et al.*, 1975; Ristow *et al.*, 1975). In addition, tyrocidine has antibacterial activity against Gram-positive and Gram-negative bacteria. These amphipathic molecules harbour a hydrophobic and a hydrophilic face forming a β-pleated sheet-like structure (Kuo *et al.*, 1980). A single basic ornithine (Orn) residue on hydrophilic face is located in the middle of one strand of the β-sheet, while several hydrophobic residues comprise the other face. The β-sheet conformation of tyrocidine is stabilized by intramolecular hydrogen bonding between carbonyl oxygens and backbone amides on the opposite strand. Owing to their amphipathic nature, the tyrocidines physically associate with cell membranes, resulting in loss of structural integrity and ultimately membrane fragmentation and cell death (Salgado *et al.*, 2001; Shai 2002; Lewis *et al.*, 2005). Unfortunately, these peptide natural products also disrupt mammalian cell membranes, as demonstrated by their high haemolytic properties (Lambert *et al.*, 1992), limiting their clinical impact over the decades. Notwithstanding this toxicity, they possess a greater affinity for bacterial cells than mammalian cells, owing to their preferential interactions with negatively charged phosphate groups within bacterial walls as opposed to the zwitterionic phospholipids in eukaryotic membranes (Zasloff, 2002). Moreover, cholesterol dispersed through eukaryotic cell membranes further attenuates binding of these cyclic peptides (e.g. gramicidin S) with phospholipid bilayers (Prenner *et al.*, 2001). Importantly, bacteria have not yet developed significant resistance to tyrocidine, since it is difficult for pathogens to alter their membrane composition and organization (Zasloff 2002). This makes these natural products appealing as prospective leads for next-generation antibiotics. In particular, structural optimization of tyrocidine may further diminish its haemolytic activity while enhancing its bactericidal activity.

Tyrocidine is synthesized by a non-ribosomal peptide synthetase (NRPS) assembly line (Fig. 8.7). This megasynthetase consists of three large proteins: TycA, TycB and TycC. Starting with D-phenylalanine, individual amino acids are incorporated into a growing peptide chain, and the fully elongated linear decapeptide is ultimately cyclized. Tyrocidine A contains D-Phe-Pro-Phe-D-Phe-Asn-Gln-Tyr-Val-Orn-Leu whereas in other tyrocidines, the Phe and Tyr residues at positions 3, 4 and 7 are substituted by Trp. Each compound has two D-amino acids and one non-proteinogenic amino acid, L-Orn.

In analogy to polyketide synthases such as DEBS, NRPSs have a modular architecture. The number of modules corresponds to the number of amino acids in the peptide backbone (Mootz *et al.*, 2002). TycA, TycB and TycC contain one, three and six modules, respectively. Each module executes all the catalytic reactions necessary for incorporation and epimerization (if required) of a defined amino acid into the growing peptide. The simplest elongation modules possess adenylation (A), thiolation (T) and condensation (C) domains. The first module does not carry a C-domain, while the last module has an appended TE domain for release via cyclization.

Adenylation domains are responsible for substrate recognition and activation. The carboxyl group of the target amino acid is activated with ATP to form the corresponding aminoacyl-adenylate. This mixed anhydride reacts with the free thiol group of the adjacent thiolation domain (also referred to as a peptidyl carrier protein). The condensation domain catalyses amide bond formation, and is invariably positioned between consecutive pairs of A-T domains. The epimerization (E) domains of modules 1 and 4 racemize the

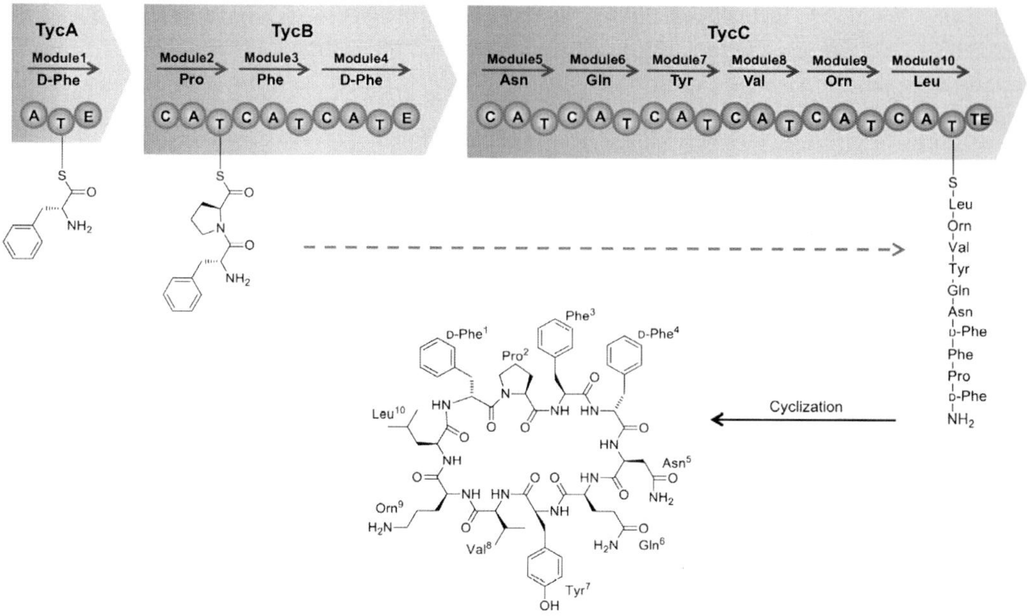

Figure 8.7 Biosynthesis of the cyclic decapeptide, tyrocidine A. The tyrocidine synthetase consists of three large synthetase proteins, TycA, TycB and TycC, which comprise one, three and six modules, respectively. A total of ten modules generate the growing peptide chain as it passes along the enzyme template. The linear decapeptide is cyclized and released through TE (thioesterase)-catalysed macrocyclization. Abbreviations: A, adenylation; T, peptidyl carrier protein (PCP); E, epimerization; C, condensation.

C_α-carbons of the corresponding L-aminoacyl-thioesters prior to condensation.

In order to improve antibacterial potency and selectivity, libraries of synthetic tyrocidine derivatives have been prepared and tested. Using solid-phase peptide synthesis (SPPS), a series of cyclic decapeptides were prepared, whose 1 and 4 positions were substituted with non-native D-amino acids (Kohli *et al.*, 2002). The recombinant TE domain from TycC was used to catalyse cyclization of this library of linear peptides (Trauger *et al.*, 2000). It was found that D-amino acids at position 4 are needed for the formation of the cyclic β-sheet. In addition, a decapeptide with a negatively charged D-Glu or D-Asp at position 4 does not display significant antibiotic activity, whereas analogues with cationic D-amino acids such as D-Arg, D-Lys, or D-Orn are more potent than tyrocidine A. Only aromatic D-amino acids at position 1 are suitable for the TE-catalysed cyclization. The dibasic peptides 1 (D-furylalanine at position 1 and D-arginine at position 4) and 2 (D-*p*-F-phenylalanine at position 1 and D-lysine at position 4) not only exhibit broad-spectrum activity against Gram-positive and Gram-negative bacteria but have improved ability to discriminate against human erythrocytes (Fig. 8.8).

Modification at position 6 also improves the therapeutic profile (MHC/MIC) of tyrocidine analogues (Qin *et al.*, 2003). In particular, introduction of cationic residues such as Orn, Lys, or Arg at this position increases the MHC/MIC ratio by 90–140-fold (Fig. 8.8), highlighting the importance of the basic face of tyrocidine A. However, further increasing the polarity by substitution of cationic amino acids on the hydrophilic face does not provide additional increases in antibacterial potency (Marques *et al.*, 2007). Increasing hydrophilicity with the insertion of two lysine residues at positions 4 (D-configuration) and 6 on the polar face of the amphipathic structure does not increase antibacterial potency against Gram-positive strains but has improved activity against the Gram-negative bacteria *E. coli*. These results suggest a path towards structural modification of the tyrocidines to increase

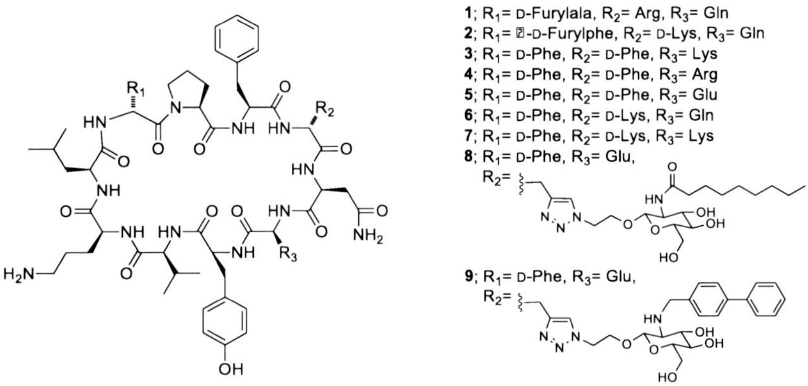

	Eukaryotic (MHC)	Gram-positive (MIC)			Gram-negative (MIC)		MHC/MIC
	Human erythrocytes	Bacilus subtilis	Staphylococcus epidermidis	MRSA	Escherichia coli	Pseudomonas aeruginosa	
Tyrocidine	5-8.2	1.3-8.3	2.5	3.8	>368	368	1-4
1	77	2.6	9	5	155	39	
2	203	6.8	24	12	392	50	
3	99	1.1-2		4	128		50-90
4	42	0.3					140
5	30	7.5					4
6		16		64	128		
7		2		16	32		
8	>482	4.8					>100
9	>498	5					>100

Figure 8.8 Antimicrobial activities and therapeutic indices of tyrocidine A analogues. MHC, minimum haemolytic concentration (μg/ml); MIC, minimum inhibitory concentration (μg/ml); MHC/MIC, MHC (human erythrocytes)/MIC (*B. subtilis*); MRSA, methicillin-resistant *S. aureus* (Kohli *et al.*, 2002; Qin *et al.*, 2003; Lin *et al.*, 2004; Marques *et al.*, 2007).

their activity against both Gram-positive and Gram-negative bacteria.

Another series of promising antibiotic candidates, glycosylated derivatives of tyrocidine A, were reported in 2004 (Lin *et al.*, 2004; Lin *et al.*, 2004). Glycosylation is common in late stage modification of natural products, and enhances biological activity and/or bioavailability (Ge *et al.*, 1999). To improve the therapeutic index of tyrocidine, alkyne-containing peptides were synthesized by SPPS, and cyclized with a thioesterase domain. Then, azido sugars were tethered to the cyclic peptides by Cu(I) catalysed (3 + 2) cycloaddition, and acyl chains or biphenyl groups were attached to these azido sugars to increase membrane affinity (Kerns *et al.*, 2000). Gratifyingly, lipophilic glycosylation at position 4 of tyrocidine retains the antibacterial activity of the natural product, but diminishes its haemolytic activity dramatically.

In order to obtain novel peptide products, the redesign of NRPS modules has been undertaken at the genetic level. Module swapping of tyrocidine synthetase was first examined by fusing the second, proline-specific module with the ninth, ornithine- and the tenth, leucine-specific modules (Mootz *et al.*, 2000). The D-Phe activating module 1 was incubated with the hybrid enzymes, yielding unnatural tripeptides, D-Phe-Pro-Orn and D-Phe-Pro-Leu. The tripeptide products were released in the presence of a TE domain at the C-terminal end of the last module. A precursor of the high-intensity sweetener (aspartame), α-Asp-Phe, also was obtained by redesign of the NRPS

Tang, Y., Tsai, S., and Khosla, C. (2003). Polyketide chain length control by chain length factor. J. Am. Chem. Soc. *125*, 12708–9.

Thompson, J., Schmidt, F., and Cundliffe, E. (1982). Site of action of a ribosomal RNA methylase conferring resistance to thiostrepton. J. Biol. Chem. 257, 7915.

Trauger, J.W., Kohli, R.M., Mootz, H.D., Marahiel, M.A., and Walsh, C.T. (2000). Peptide cyclization catalysed by the thioesterase domain of tyrocidine synthetase. Nature *407*, 215–218.

Tymiak, A., Aklonis, C., Bolgar, M., Kahle, A., Kirsch, D., O'Sullivan, J., Porubcan, M., Principe, P., and Trejo, W. (1993). Dactylocyclines: novel tetracycline glycosides active against tetracycline-resistant bacteria. J. Org. Chem. *58*, 535–537.

Wang, P., Zhang, W., Zhan, J., and Tang, Y. (2009). Identification of OxyE as an ancillary oxygenase during tetracycline biosynthesis. ChemBioChem *10*, 1544–1550.

Weissman, K.J., and Leadlay, P.F. (2005). Combinatorial biosynthesis of reduced polyketides. Nat. Rev. Microbiol. 3, 925–936.

Welling, P.G., and Craig, W.A. (1978). Pharmacokinetics of intravenous erythromycin. J. Pharm. Sci. *67*, 1057–1059.

Wichtowski, J.A. *et al.* (2004). Nocathiacin I analogues: synthesis, *in vitro* and *in vivo* biological activity of novel semi-synthetic thiazolyl peptide antibiotics. Bioorg. Med. Chem. Lett. *14*, 5573–5577.

Wieland-Brown, L.C., Acker, M.G., Clardy, J., Walsh, C.T., and Fischbach, M.A. (2009). Thirteen posttranslational modifications convert a 14-residue peptide into the antibiotic thiocillin. Proc. Natl. Acad. Sci. U.S.A. *106*, 2549–2553.

Wittmann, H.G. *et al.* (1973). Biochemical and genetic studies of two different types of erythromycin resistant mutants of *Escherichia coli* with altered ribosomal proteins. Mol. Gen. Genet. *127*, 175–189.

Xu, L., Farthing, A.K., Dropinski, J.F., Meinke, P.T., McCallum, C., Leavitt, P.S., Hickey, E.J., Colwell, L. Barret, J., and Liu, K. (2009). Nocathiacin analogs: Synthesis and antibacterial activity of novel water-soluble amides. Bioorg. Med. Chem. Lett. *19*, 3531–3535.

Zak, M., Rahimipour, S., Estrada, A.A., Lee, S.H., Giannakakou, P., and Ghadiri, R. (2005). Discovery of a biologically active thiostrepton fragment. J. Am. Chem. Soc. *127*, 15042–15044.

Zasloff, M. (2002). Antimicrobial peptides of multicellular organisms. Nature *415*, 389–395.

Zhang, W., Ames, B., Tsai, S., and Tang, Y. (2006). Engineered biosynthesis of a novel amidated polyketide, using the malonamyl-specific initiation module from the oxytetracycline polyketide synthase. Appl. Environ. Microbiol. 72, 2573.

Zhang, W., Watanabe, K., Wang, C., and Tang, Y. (2007). Investigation of early tailoring reactions in the oxytetracycline biosynthetic pathway. J. Biol. Chem. *282*, 25717.

Zhang, W., Watanabe, K., Cai, X., Jung, M., Tang, Y., and Zhan, J. (2008). Identifying the minimal enzymes required for anhydrotetracycline biosynthesis. J. Am. Chem. Soc. *130*, 6068.

Zuckerman, J.M. (2004). Macrolides and ketolides: azithromycin, clarithromycin, telithromycin. Infect. Dis. Cllin. North Am. *18*, 621–49.

Part III

Microbial Communities and Interactions with the Host

Bacterial Phenotypes Refractory to Antibiotic-mediated Killing: Mechanisms and Mitigation

9

Alex J. O'Neill

Abstract

Antibiotic resistance is conferred by heritable genetic determinants that enable a bacterium to grow and cause disease in the presence of therapeutically achievable concentrations of the corresponding antibiotic. However, bacteria may also become refractory to the killing action of antibacterial agents in ways that do not fit this definition, and which are collectively referred to here as 'antibiotic survival'. These phenomena, which include drug indifference, tolerance, persistence, and the recalcitrance of biofilms to antibacterial agents, are believed to play a central role in antibacterial treatment failure. In addition, they can extend the duration of treatment required to resolve bacterial infections, and facilitate the emergence of acquired antibiotic resistance. This chapter will provide an overview of the different types of antibiotic survival, and will discuss chemotherapeutic approaches to minimizing or overcoming the problems that they present to effective antibacterial treatment.

Introduction

Antibiotic resistance involves the acquisition of genetic determinants that enable bacterial pathogens to grow and cause disease in the presence of therapeutically achievable concentrations of the corresponding antibiotic in the body. This phenomenon represents an ever-growing problem that is progressively restricting our ability to successfully treat and cure bacterial infections (Chopra *et al.*, 2008). Compounding this is the dire lack of novel antibacterial agents under advanced development to replace those whose activity has become compromised by resistance (Chopra *et al.*, 2008; O'Neill 2008), a situation arising out of a multiplicity of issues (technical, financial, regulatory) that have been extensively discussed (Projan and Shlaes, 2004; Payne *et al.*, 2007; Chopra *et al.*, 2008).

The evolving issue of acquired antibiotic resistance is often viewed as the most important, if not the sole, driver for antibacterial drug discovery and development, and it is certainly true that it is currently the most pressing. However, it is important to acknowledge that antibiotic resistance is only one of the reasons for antibiotic treatment failure, and may not even represent the primary reason (Norrby, 1991; Levin and Rozen, 2006; Smith and Romesberg, 2007). Illustrating this idea, examples of reported treatment problems and failure not attributable to acquired antibiotic resistance are shown in Box 9.1.

A variety of factors may participate in chemotherapeutic failure associated with antibacterial agents (Levin and Rozen, 2006). However, a central thesis of this chapter is that a major contributor to treatment failure, and an important factor underpinning other problems associated with our present approach to antibacterial chemotherapy (see below), is the inability of most of our existing antibiotics to achieve a sufficient kill of the infecting bacterial population within the patient. This concept is supported by findings from laboratory studies and clinical observation, examples of which are described below, and also

Box 9.1 Clinical reports demonstrating that factors other than antibiotic resistance negatively impact antibacterial treatment

- Thirty-seven of 81 (46%) patients who received vancomycin alone or in combination with other antibiotics for osteomyelitis caused by vancomycin-susceptible *Staphylococcus aureus* experienced treatment failure (Dombrowski and Winston, 2008).
- The mortality rate for pneumococcal pneumonia is 12–17%, a figure that stands unchanged since the introduction of penicillin, despite a substantial increase in the prevalence of penicillin resistance in strains of the causative organism (Levin and Rozen, 2006).
- Seventeen of 49 (35%) β-lactam-susceptible Gram-negative strains isolated from patients being treated with cephalosporins persisted for the duration of therapy without change in their antibiotic susceptibility (Dworzack *et al.*, 1987).
- In 77 cystic fibrosis patients infected with *Pseudomonas aeruginosa*, no correlation was observed between antibiotic susceptibility of the infecting strain and the clinical response following antibacterial chemotherapy (Smith *et al.*, 2003).
- Antibacterial treatment failure was observed in 8 of 8 patients with foreign body-associated infections in whom therapy was instituted without surgical removal of the infected foreign body (Falagas *et al.*, 2009).

by mathematical models (Wiuff *et al.*, 2005; Levin and Rozen, 2006; Wiuff and Andersson, 2007). Current antibacterial treatment regimens do not themselves eradicate bacterial infection, and their role must therefore be considered supportive (Levin and Rozen, 2006), providing the host's immune system with the opportunity to gain control of the infection and ultimately clear the invading bacteria from the body. Accordingly, in a patient where these host processes are working suboptimally (for example, owing to underlying medical conditions or immunodeficiency), or where the infection is poorly accessible to components of the immune system, even extensive and intensive antibacterial treatment may be unable to effect a cure.

The inability of even bactericidal antibiotics to achieve sterilization of infections is the consequence of several bacterial traits, most of which are intrinsic and probably universal to prokaryotes. In their excellent review of this topic, Levin and Rozen (2006) employ the expression 'non-inherited antibiotic resistance' to collectively describe these phenomena. However, this term is potentially confusing, especially since it does not accurately describe processes which are indeed genetically determined, and can therefore be inherited (e.g. tolerance). Consequently, the collective term 'antibiotic survival' (AS) has been used here, defined as the ability of bacteria to avoid being killed in the presence of bactericidal antibiotics. AS mechanisms may be distinguished from antibiotic resistance in that they attenuate only the killing activity, but not the inhibitory action, of an antibiotic, i.e. they do not enable bacteria to grow in the presence of antibiotics. Furthermore, unlike antibiotic resistance, AS is in most cases not an acquired genetic trait, but usually results from bacteria temporarily entering a protected phenotypic state in which they become refractory to antibiotic-mediated killing.

Besides the role it plays in treatment failure, AS appears to be a major reason that many bacterial diseases require an extended course of treatment to achieve resolution of infection (Table 9.1). A prominent example of this phenomenon is tuberculosis, for which patients require chemotherapy involving multiple antibacterial agents for a minimum of six months. Nonetheless, even acute bacterial infections usually require treatment with courses of antibiotics lasting several days or weeks (Table 9.1).

In addition, AS probably contributes to the development of acquired antibiotic resistance (Coates *et al.*, 2002). Bacteria that are not killed

Table 9.1 Representative duration of antibacterial treatment required to resolve different types of bacterial infection (Johnson and Stamm, 1989; Phillips *et al.*, 2000; Hayes and Williamson, 2001; Elliott *et al.*, 2004; James *et al.*, 2008; Bignell, 2009; Hall *et al.*, 2009)

Infection	Treatment interval
Chronic wound (e.g. venous leg ulcer)	1 to >12 months
Tuberculosis	6–12 months
Staphylococcal endocarditis	4–6 weeks
Streptococcal pharyngitis	10 days
Complicated urinary tract infection	7 days
Uncomplicated urinary tract infection	3 days
Gonorrhoea	Single dose

but survive for extended periods in the presence of antibiotics are subject to an evolutionary bottle-neck that will strongly favour the selection of resistance. This process is exacerbated by the ability of some antibiotics to facilitate the emergence of resistance by inducing increases in mutation frequency (Alonso *et al.*, 1999), triggering horizontal DNA transfer (Beaber *et al.*, 2004; Maiques *et al.*, 2006; Guerin *et al.*, 2009), and stimulating recombination between divergent DNA sequences (Lopez *et al.*, 2007). Furthermore, longer treatment intervals are associated with lower patient compliance, a factor that has been well documented to predispose towards the development of resistance (Coates *et al.*, 2002).

The recognition of AS as a complicating factor in antibacterial chemotherapy raises the possibility of developing strategies to address it. Establishment of chemotherapeutic interventions to overcome or mitigate AS could offer dramatic improvements in antibacterial treatments in terms of improved efficacy, reduced length of treatment interval, and diminished resistance development. These outcomes are consistent with existing goals in antibacterial drug development, and approaches to overcome AS should therefore be encompassed by current and future antibacterial drug discovery programmes.

This chapter will review the various AS mechanisms that bacteria display which negatively impact our ability to achieve rapid treatment and cure of the infections they cause, and will discuss potential chemotherapeutic approaches to overcoming them.

Types and mechanisms of antibiotic survival

Prior to discussing the different types of AS, it should be emphasized that the scientific literature relating to this topic features inconsistent and conflicting nomenclature. Here I have used established naming conventions for the different AS phenomena, following the designations reiterated by Levin and Rozen (2006). However, for the sake of coherence with previously published work, alternative names that have been used to describe these phenomena are also given here.

Drug indifference

Bacteria that are non-growing or metabolically inactive become refractory to the killing action of most antibiotics, a phenomenon known as drug indifference, or phenotypic tolerance (Tuomanen *et al.*, 1986; Fig. 9.1). First described in studies conducted in the 1940s to investigate the killing of *Staphylococcus aureus* by penicillin (Lee *et al.*, 1944), this phenomenon has subsequently been observed with a variety of bacterial species and antibacterial classes (Eng *et al.*, 1991; Herbert *et al.*, 1996; Schierholz *et al.*, 1998; Levin and Rozen, 2006). For example, Eng *et al.* (1991) reported that of a range of antibacterial agents capable of killing actively growing bacteria, only fluoroquinolones demonstrated cidality against non-growing cultures of several Gram-negative species, and no antibiotic class tested retained a cidal effect against non-growing cultures of *S. aureus*. In the case of some antibiotics, notably the β-lactams, lack of growth can result in essentially complete protection of bacteria from antibiotic-mediated

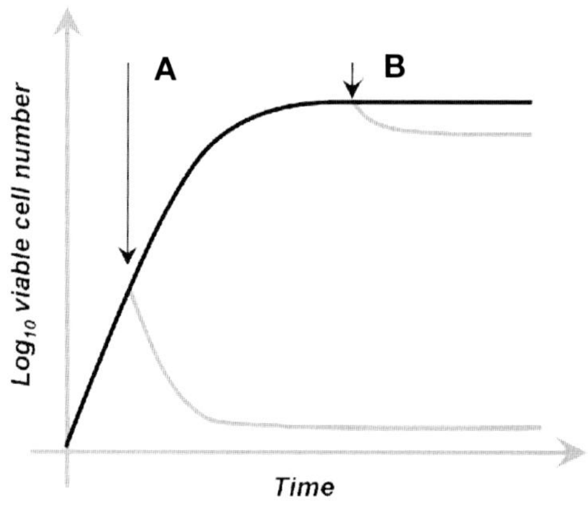

Figure 9.1 Examples of antibiotic survival *in vitro*. Shown in black is part of a bacterial growth curve in the absence of antibiotic challenge: a similar curve would be observed for an antibiotic-resistant strain grown in the presence of antibiotics. Persistence: addition of a bactericidal antibiotic to an actively growing culture (A) will kill the majority of bacteria, but will not reduce the viable cell number to zero – the surviving cells are persisters. Drug indifference: addition of a bactericidal antibiotic to a stationary-phase culture (B) will result in little or no bacterial killing. Tolerance: antibiotic challenge of a tolerant strain will result in a similar response to that for drug indifference, but is not dependent on the culture being in stationary-phas, i.e. it will be observed even if the antibiotic challenge occurs at (A).

killing, whilst other classes may mediate some killing, though severely attenuated compared with that seen against rapidly growing bacteria (Eng *et al.*, 1991; Levin and Rozen, 2006). This indifference to lethal challenge in the non-growing state is not restricted to antibacterial agents: starved *Escherichia coli* are more resilient to a range of stresses, including osmotic challenge, oxidative stress and heat shock (Kolter *et al.*, 1993).

In the natural environment, the non-growing state is the default for bacteria, punctuated by brief periods of growth when nutrients become available – the 'feast or famine' lifestyle (Kolter *et al.*, 1993). A similar situation occurs during the infection process. Upon infection, a 'feast' usually ensues during which bacteria divide rapidly, following which growth slows and may stop owing to lack of available nutrients (Levin and Rozen, 2006). The period of rapid growth may only be short-lived: for example, in an *E. coli* K1 mouse thigh infection model, bacterial cell division ceases eight hours post-inoculation (Levin and Rozen, 2006). At this stage, the non-growing state of the bacterial population renders it indifferent to the killing action of antibacterial agents. This effect can be demonstrated in animal infection models, wherein the duration of infection prior to initiation of antibiotic treatment has a dramatic influence upon outcome, with more well-established infections failing to respond to antibacterial chemotherapy (Levin and Rozen, 2006).

Since most bacterial infections will contain drug-indifferent bacteria by the time the patient receives antibacterial chemotherapy, this phenomenon is anticipated to make a significant contribution to the requirement for an extended treatment interval to enable resolution of a bacterial infection. This is best exemplified by tuberculosis (Coates *et al.*, 2002; Levin and Rozen, 2006). Within the tubercular lung, the mycobacterial population comprises at least two subpopulations, one of which is actively multiplying, and the other which is dormant, and considered clinically latent (Coates *et al.*, 2002). Upon initiation of antibacterial chemotherapy, the growing mycobacteria will undergo rapid killing, whilst the dormant fraction will exhibit a degree of drug indifference. Achieving complete clearance of this residual population requires the consistent administration of a multidrug regimen, comprising agents with different modes of action, over a period of many months.

How do non-growing bacteria become refractory to antibiotic mediated killing? In considering this question, it is useful to recall that inhibition of bacterial growth does not necessarily lead to bacterial death, as illustrated by the many antibacterial agents whose mode of action is predominantly bacteriostatic (e.g. inhibitors of folate biosynthesis and the majority of antibiotics targeting protein translation; O'Neill and Chopra 2004). Accordingly, it follows that inhibiting an enzyme target and/or arresting growth do not automatically lead to death, but instead, that death is contingent upon processes occurring subsequent or in addition to inhibition. These processes have been referred to as 'corruption' of antibacterial targets (Lewis, 2007). For example, β-lactams and glycopeptide antibiotics not only inhibit peptidoglycan synthesis, but also trigger the action of autolysins (peptidoglycan hydrolases), leading to enzymatic breakdown of the cell-wall and ultimately death. Fluoroquinolones, whilst not affecting the DNA-nicking activity of topoisomerase enzymes, inhibit their ability to catalyse the subsequent religation of the DNA backbone. As a consequence, topoisomerases are effectively converted to endonucleases which act to destroy the structural integrity of the bacterial chromosome.

In the non-growing state, bacteria can evade killing by avoiding lethal corruption of essential functions. Several mechanisms might prevent an antibacterial drug target from becoming corrupted, thereby attenuating bacterial killing. For example, the target may be absent, present at low concentration, or present but temporarily rendered non-functional, such that antibiotic-mediated target corruption will not occur on a scale sufficient to cause cell death. Alternatively, a target may form part of a biosynthetic pathway that is inactive. Non-growing cells may also restrict antibiotic uptake: changes in membrane fatty acid composition which occur in non-growing *E. coli* cells cause reductions in membrane fluidity and permeability (Kolter *et al.*, 1993).

Tolerance

Antibiotic tolerance describes a situation that is superficially similar to indifference, in that the anticipated bactericidal response of a population following antibiotic challenge becomes partially or completely attenuated (Fig. 9.1). However, several distinctions can be drawn. In contrast to indifference, tolerance is not linked to the growth or metabolic status of the organism, but results from genetic changes that lead to uncoupling of antibiotic-mediated inhibition from killing (Handwerger and Tomasz, 1985). Since the ability of an antibacterial agent to inhibit the growth of a tolerant strain is not affected, tolerance is not detected by standard susceptibility determinations of the minimum inhibitory concentration (MIC), but only becomes apparent upon examination of the ratio between minimum bactericidal concentration (MBC) and MIC (Handwerger and Tomasz, 1985). Of the AS mechanisms discussed here, tolerance is the sole example that does not represent an intrinsic property of bacteria, but develops only in select strains, potentially as a consequence of antibiotic exposure (Handwerger and Tomasz, 1985). The term 'tolerance' has been indiscriminately employed in the scientific literature to describe any and all manifestations of AS: in consequence, the expression 'genotypic tolerance' has sometimes been used to mark the contrast between this and the other AS phenomena (Tuomanen *et al.*, 1986).

Tolerance was first described in laboratory mutants of *Streptococcus pneumoniae* (Tomasz *et al.*, 1970). However, it was not until the landmark paper of Sabath *et al.* (1977), which described the isolation of tolerant strains of *S. aureus* from patients with a variety of infection types (endocarditis, osteomyelitis or pneumonia) who had responded poorly to penicillin treatment, that the full clinical implications of tolerance were appreciated. Subsequent studies have reinforced the connection between tolerance and adverse treatment outcomes when β-lactams or vancomycin are used for chemotherapy of particular types of staphylococcal and streptococcal infection (Gopal *et al.*, 1976; Rajashekaraiah *et al.*, 1980; Musher and Fletcher, 1982; Liu and Tomasz, 1985; Thometz *et al.*, 1996; May *et al.*, 1998; McCullers *et al.*, 2000). For reasons that are not clear, tolerance remains a relatively restricted phenomenon in the clinical setting, primarily associated with the attenuated killing activity of cell wall-active antibiotics against Gram-positive bacterial species (Handwerger and Tomasz, 1985).

Although there is broad appreciation of the processes underlying tolerance (Handwerger and Tomasz, 1985), mechanistic details are lacking. As alluded to above, antibiotics that inhibit peptidoglycan biosynthesis do not immediately cause bacterial cell death, but trigger a self-induced killing process (Henriques Normark and Normark, 2002). Tolerance arises when this bactericidal process becomes defective or blocked as a consequence of genetic mutation, and multiple routes to tolerance are apparent from the varied properties of natural and laboratory-selected tolerant strains (Handwerger and Tomasz, 1985). One class of tolerant strain exhibits defects in autolysis, thereby preventing the enzymatic dismantling of the peptidoglycan sacculus which is usually triggered upon inhibition of cell-wall biosynthesis. This includes strains in which autolysins are absent or defective, subject to accelerated turnover by the action of extracellular proteases, or inhibited as a consequence of alterations in the peptidoglycan substrate/through increased production of autolysin inhibitors (Handwerger and Tomasz, 1985).

Another class of tolerant strain promptly enters a state of metabolic shut-down upon inhibition of cell-wall biosynthesis, thereby rendering it indifferent to antibiotic-mediated killing. Tolerant variants of both streptococci and staphylococci have been reported which undergo rapid inhibition of RNA and protein synthesis following challenge with β-lactams (Mychajlonka *et al.*, 1980; Goessens, 1993). How this might be mediated remains to be established, although involvement of the stringent response (see below) has been mooted (Goessens, 1993).

Persistence

The phenomenon of persistence [which has also been referred to as phenotypic tolerance (Wiuff *et al.*, 2005) or adaptive resistance (Barclay *et al.*, 1992)] was first observed in the early days of the antibiotic era (Bigger, 1944). Persistence describes the failure of bactericidal antibiotics to achieve sterilization of an actively growing bacterial culture: a small subpopulation of persisters (often expressed as $\sim 10^{-6}$ cells for *E. coli*) is able to survive a challenge that is lethal for the majority (Bigger, 1944; Vazquez-Laslop *et al.*, 2006; Fig. 9.1). Persisters are not genetic variants, since their progeny are no more resistant to antibiotics than the ancestral population from which the persisters derived, and they do not yield a larger subpopulation of persisters than the ancestral population upon cultivation and re-exposure to antibiotics (Bigger, 1944). Instead, persistence appears to represent a transient phenotypic property of bacterial cells.

Early work on persisters suggested that they survive antibiotic exposure because they are non-growing cells (Bigger, 1944), a view confirmed by more recent studies. For example, microscopic examination of single bacterial cells trapped within the channels of a microfluidic device established that those persisting following antibiotic challenge were non- or slow-growing prior to drug exposure (Balaban *et al.*, 2004). Transcriptional profiling of persister cells has revealed down-regulation of a substantial number of metabolic genes, including operons involved in oxidative phosphorylation, further reinforcing the idea that persisters are quiescent (Shah *et al.*, 2006). Therefore, the reasons outlined above to explain the indifference of non-growing bacteria to antibiotic-mediated killing may also be applied to understanding the antibiotic-refractory nature of non- or slow-growing persister cells (Lewis, 2007). It is plausible, based on their phenotypic properties, that persister cells exist in a growth-arrested state distinct from that predominating in the drug-indifferent bacterial population in stationary-phase. This is suggested, for example, by the observation that whilst fluoroquinolones retain strong killing activity against a non-growing population of *E. coli* (Eng *et al.*, 1991), they are nonetheless unable to kill *E. coli* persister cells (Wiuff *et al.*, 2005). Nevertheless, it remains to be established whether persisters are experiencing a truly distinct state from that of the stationary-phase cell, or simply represent those cells which are most refractory to antibiotic-mediated killing within a population distribution (Hall-Stoodley *et al.*, 2004).

A growing number of processes and genetic loci have been implicated in persister formation, including stress-response pathways known to affect growth. For example, *E. coli* strains defective for induction of the SOS DNA repair response, a process during which cells transiently experience growth arrest, were found to yield 10- to 100-fold less persisters upon exposure to several different antibiotic classes (Debbia *et al.*, 2001). Several observations link persistence with the stringent response, a pathway induced in response to amino acid starvation, and which is registered as an increase in the ratio between uncharged and charged tRNA species at the ribosome (Potrykus and Cashel, 2008). Stringency is signalled by accumulation of the alarmone 3′,5′-bispyrophosphate [(*p*)ppGpp], synthesized by the ribosome-bound RelA protein, and mediates reductions in RNA synthesis, almost complete shut-down of protein synthesis, and inhibition of growth (Potrykus and Cashel, 2008). A strain of *Pseudomonas aeruginosa* producing elevated basal levels of (*p*)ppGpp owing to deletion of the (*p*)ppGpp hydrolase, SpoT, displayed a 40-fold increase in persister production (Viducic *et al.*, 2006). Furthermore, the ability of a high-persistence mutant of *E. coli* (*hipA7*) to generate a large persister subpopulation was eliminated in the absence of a functional stringent response (Korch *et al.*, 2003).

Elevated expression of proteins which slow or prevent growth results in an increase in the size of the persister subpopulation (Keren *et al.*, 2004; Shah *et al.*, 2006; Vazquez-Laslop *et al.*, 2006). A particular focus of interest in this regard has been the postulated involvement in persistence of toxin-antitoxin (TA) modules (reviewed by Lewis, 2007, 2008): two-gene cassettes in which one gene encodes a stable toxin, and the other an unstable cognate antitoxin capable of complexation with, and neutralization of, the toxin (Gerdes *et al.*, 2005). There are >10 TA modules in the *E. coli* chromosome, the encoded toxins of which are capable of mediating reversible bacterial stasis through inhibition of essential cellular functions, such as protein translation (Gerdes *et al.*, 2005). Artificial overexpression of several of these toxins, including HipA (Falla and Chopra, 1998), RelE (Keren *et al.*, 2004) and YgiU (Shah *et al.*, 2006), dramatically increases persister formation. Furthermore, up-regulation of genes encoding TA toxins has been observed in the transcriptional profiles of persister cells (Keren *et al.*, 2004; Shah *et al.*, 2006).

However, the role played by TA modules in persistence remains unclear. Enhanced persister formation is not unique to cells overexpressing TA toxins, but can also be induced in *E. coli* through ectopic overexpression of other unrelated proteins which retard growth (Vazquez-Laslop *et al.*, 2006), suggesting that any factor which acts to slow the growth rate can potentially contribute to persister formation. In addition, disruption of individual TA modules in *E. coli* has no impact upon persister formation in actively growing cultures (Keren *et al.*, 2004; Shah *et al.*, 2006), although it is possible that the redundancy of these elements could serve to obscure such an effect. Furthermore, whilst persistence is thought to be a universal phenomenon amongst bacteria, TA modules are not present in all bacterial chromosomes (Pandey and Gerdes, 2005).

Numerous other genes, including those involved in global regulation and metabolism, impact the size of the persister subpopulation in *E. coli* when either disrupted or artificially overexpressed (Spoering *et al.*, 2006; Li and Zhang, 2007; Hansen *et al.*, 2008; Ma *et al.*, 2009), although in many cases, the impact on persister formation is modest. Several genes exerting a more pronounced effect upon the size of the persister pool are associated with pleiotropic effects, including growth defects and alterations in antibacterial susceptibility (Li and Zhang, 2007; Ma *et al.*, 2009), potentially complicating an interpretation of their role in persistence. Overall, multiple, redundant processes appear to participate in persister formation.

Like other forms of phenotypic variation exhibited by clonal populations (Elowitz *et al.*, 2002), persister formation appears to represent a stochastic process (Lewis, 2007). Noise in expression of a broad variety of proteins or pathways which impact bacterial growth may be responsible for triggering the development of persisters (Vazquez-Laslop *et al.*, 2006). Supporting this concept, stochastic expression of the SOS response in *E. coli* (McCool *et al.*, 2004), and the stringent response in *Mycobacterium smegmatis* (Sureka *et al.*, 2008), has been documented. Although it remains to be established why a proportion of an actively growing bacterial population should adopt a non-growing state, theoretical considerations argue for a 'bet-hedging' strategy that balances the benefits of rapid growth and a state protected from environmental challenge (Kussell *et al.*, 2005).

Despite the multiple routes leading to persister formation, it has been suggested that persisters may represent a single, discrete subpopulation (LaFleur *et al.*, 2006; Hansen *et al.*, 2008). However, there is also evidence which points to heterogeneity in the persister state. In *E. coli*, both the concentration and type of antimicrobial agent used to challenge the culture affects the size of the resulting persister population, and persisters are not fully protected from a second challenge with a higher concentration of the same antibiotic (Wiuff *et al.*, 2005). Furthermore, although a degree of cross-persistence is observed when persister cells are subsequently challenged with a different antibacterial class, the level of persistence is in general not consistent (Wiuff *et al.*, 2005). A similar observation has been reported for *M. tuberculosis*, whereby fluoroquinolones were able to substantially reduce the size of a persister population formed following a prior challenge with rifampin (Hu *et al.*, 2003). The apparent implication of these observations is that different antibiotics at different concentrations impact different portions of the persister subpopulation. As mentioned in respect of the relationship between stationary-phase cells and persisters, it is unclear whether this apparent heterogeneity reflects the existence of persisters in distinct states, or rather, represents a continuum which describes a single persister state (Ma *et al.*, 2009).

Biofilms

Biofilms are structured communities of bacteria adhered to a biological or inert surface and encased in an extracellular matrix which is predominantly composed of exopolysaccharide of bacterial origin (del Pozo and Patel, 2007). They represent a primary mode of bacterial growth both in the natural environment, and during infection: it has been estimated that approximately 80% of bacterial infections of humans involve a biofilm component (NIH, 2002). Biofilms have gained particular notoriety for the role they play in infections associated with indwelling medical devices, such as intravenous catheters and joint prostheses, often caused by staphylococci (*S. aureus*, *S. epidermidis*) or *P. aeruginosa* (Hall-Stoodley *et al.*, 2004; del Pozo and Patel, 2007).

A defining property of biofilms is their recalcitrance to killing by antibacterial agents (Donlan and Costerton, 2002), even when the biofilm comprises cells that demonstrate full antibiotic susceptibility upon growth in planktonic (free-living) culture. This effect appears to have been first noted by Anthony van Leeuwenhoek in the seventeenth century whilst investigating the effect of vinegar on bacteria in dental plaque, observing that it '... kill'd only those animals [bacteria] which were on the outside of the scurf [plaque] ...' (van Leeuwenhoek, 1684). Since currently available antibiotic chemotherapy usually fails to treat biofilm infections (Falagas *et al.*, 2009), successful treatment of infections associated with indwelling medical devices or foreign bodies frequently involves surgical intervention to remove the biofilm-associated surface (Marr *et al.*, 1997; Falagas *et al.*, 2009). AS in the biofilm also presents a significant problem in the case of chronic wounds, where biofilms appear to be a major factor confounding healing (James *et al.*, 2008): the deployment of treatments lacking potent activity against wound biofilms leads to the requirement for extended treatment (Table 9.1), which often ultimately fails, necessitating limb amputation (James *et al.*, 2008). In view of their ubiquity and their intransigence to antibacterial chemotherapy, biofilms may be considered the most extreme and clinically problematic presentation of AS.

The antibiotic-refractory nature of the biofilm appears to result primarily from drug indifference and persistence (Hall-Stoodley *et al.*, 2004), and indeed the biofilm response to antibiotics *in vitro* is similar to that of a non-growing planktonic culture (Spoering and Lewis, 2001). This correlates with the observation that the majority of cells within a mature biofilm are non-dividing (Walters *et al.*, 2003). In most instances this is anticipated to reflect nutrient limitation, but other factors, for example oxygen limitation below the surface of the biofilm, can also restrict growth (Borriello *et al.*, 2004). Augmenting the AS of free-living, non-growing cells, the physical structure of the biofilm acts to assist bacterial survival following antibiotic challenge. Where indifference or persistence occur in the absence of a biofilm, the host's immune system may be able to clear the non-growing bacteria that remain following antibacterial treatment. However, the extracellular matrix of the biofilm shields its inhabitants from components of the immune system (del Pozo and Patel, 2007; Spoering and Lewis, 2001), thereby protecting the survivors from immunological challenge. The biofilm structure, comprising between 65% and 95% extracellular matrix (del Pozo and Patel, 2007), is essentially unaffected by antibiotic exposure, and provides a scaffold that surviving bacteria can rapidly repopulate upon discontinuation of antibacterial chemotherapy.

Additional factors may also contribute to AS in the biofilm in specific instances. The negatively charged extracellular matrix may impede diffusion of positively charged antibiotics such as the aminoglycosides by sequestration (Shigeta *et al.*, 1997), and is also capable of electrostatic repulsion of negatively charged antibacterial agents, including antimicrobial peptides (Otto, 2008).

Strategies to address antibiotic survival

It seems probable that all bacterial infections treated with cidal antibiotics will display some degree of AS, since it appears to be part of a fundamental protective strategy intrinsic to prokaryotes. Nevertheless, even a course of antibacterial chemotherapy that fails to kill bacteria directly, in conjunction with a robust immune response, often ultimately leads to resolution of the infection. Therefore, it is important to emphasize that drugs whose activity is solely bacteriostatic are effective antibacterial agents and continue to be successfully employed for the treatment of a wide variety of bacterial infections.

However, especially when there is suboptimal or defective host response, or the infection is taking place either in a physical location or in a manner that renders it poorly accessible to components of the immune system, AS may significantly influence its course, extending the necessary treatment interval or causing therapeutic failure. The successful development of approaches to overcome AS would be anticipated to result in higher rates of successful treatment, even in instances where the physiological state of the infecting bacterium or host factors, such as immunodeficiency, would negatively impact current antibiotic treatment regimens. Furthermore, it would in principle enable the use of substantially shorter courses of antibacterial chemotherapy in many bacterial infections, potentially reducing drug side-effects, lowering treatment costs and prompting improved patient compliance, thereby slowing the emergence of acquired antibiotic resistance (Coates and Hu, 2008).

In the discussion of strategies to address AS that follows, I have concentrated solely on chemotherapeutic approaches that aim to target bacterial physiology; however, there may also be merit in approaches that fall outside this remit, e.g. vaccination, immunotherapy, etc. Two main routes to addressing AS will be considered: first, strategies to augment existing antibacterial chemotherapy to include coverage for AS, and second, development of novel antibacterial agents uncompromised (or less compromised) by AS. The question of where novel chemical entities might be sourced with the desired properties to achieve either of these aims is ultimately the same as that raised in regard of antibacterial drug discovery in general, and the reader is referred to recent perspectives on the latter (Payne *et al.*, 2007; Baltz, 2008; Fischbach and Walsh, 2009; Galloway *et al.*, 2009). Tolerance will not be further considered here, since it is a phenomenon restricted to specific antibacterial classes in the treatment of certain infections, and can be circumvented through deployment of alternative established antibacterial agents possessed of a distinct mode of action. Our current understanding of the other three AS mechanisms – indifference, persistence and biofilms – can be summarized as a lack of killing activity of antibiotics against various types of non-growing cells (compounded with, in the case of biofilms, physical compartmentalization away from components of the immune system), and may in fact represent different manifestations of a single phenomenon.

Whilst augmenting antibiotic treatment with adjunctive therapy targeted at the AS mechanism has intuitive appeal, with a few potentially important exceptions (see below), this approach is unlikely to be widely practicable. AS apparently differs between bacterial pathogens, seems to entail multiple routes to growth arrest, and generally involves a level of complexity and redundancy that would make identifying an appropriate broad-spectrum anti-AS target extremely difficult, if not impossible. Aside from situations in which antibacterial agents are administered as prophylaxis, bacteria exhibiting AS will pre-exist initiation of antibacterial chemotherapy in the patient. This may further complicate matters, since suitable anti-AS targets, even if there are such, may be down-regulated or not expressed in the absence of growth (Coates and Hu, 2007). Ensuring delivery of anti-AS agents into persisters and other non-growing cells might also prove challenging owing to the aforementioned reductions in bacterial membrane permeability (Kolter *et al.*, 1993). Finally, since anti-AS agents would alone not be expected to demonstrate antibacterial activity or efficacy against bacterial infections *in vivo*, they would probably be subject to substantially greater development and regulatory hurdles than those already facing novel antibacterial candidates (Lewis, 2008).

Notwithstanding the above, there are particular types of infection where the AS problem at issue has perhaps been adequately defined, and is of sufficient clinical importance, to justify considering the development of specific anti-AS agents. Biofilms, probably the most problematic example of AS, present one such case.

Providing proof of principle for such an approach, a number of agents capable of inhibiting or disrupting the biofilms of some bacterial species have been identified. These include the enzymes dispersin B, deoxyribonuclease I and lysostaphin (Wu *et al.*, 2003; Kaplan, 2009), and the chemical agents farnesol, ursolic acid and gallium (Ren *et al.*, 2005; Jabra-Rizk *et al.*, 2006; Kaneko *et al.*, 2007). It has also been demonstrated that compounds which inhibit bacterial efflux pumps are capable of abolishing biofilm formation in *E. coli* (Kvist *et al.*, 2008). Development of inhibitors of bacterial quorum sensing, a process coordinating the transition of some bacterial species from the free-living bacterial lifestyle to that of a sessile biofilm community, represents an approach with potential, and this is addressed in another chapter in this volume. All approaches aimed specifically at interfering with the formation or structure of biofilms would need to take account of the fact that these properties are often multifactorial, even within a single bacterial species (Goller and Romeo, 2008; Otto, 2008). Furthermore, it must be recognized that some medical biofilms (e.g. those found in wounds) represent polymicrobial consortia (Dowd *et al.*, 2008), and effective anti-biofilm strategies would in such cases probably need to demonstrate coverage for multiple species.

Development of anti-AS agents to counter the drug indifference exhibited by *M. tuberculosis* might also prove useful and feasible. Co-administration with antimycobacterial drugs of anti-AS agents capable of either blocking mycobacterial dormancy, or of triggering the growth of non-growing subpopulations, might dramatically expedite clearance of the infection, potentially shortening treatment intervals for tuberculosis to a matter of days or weeks instead of the current time-scale of months to years. Attempts to realize such an approach are already under way: Gupta *et al.* (2009) recently reported the discovery of a phenylcoumarin derivative capable of blocking the dormancy regulator, DevR, thereby reducing survival of hypoxic mycobacteria. However, even in the case of *M. tuberculosis* our understanding of AS may not be sufficiently complete for such an approach to be progressed with confidence. Studies have shown that mycobacterial mutants defective in dormancy are not necessarily defective in long-term survival *in vivo*, and it has been proposed that the recalcitrance of tuberculosis to antibacterial chemotherapy may in part be attributable to growth of the organism as a biofilm (Ojha *et al.*, 2008).

Developing novel antibacterial agents whose killing activity is uncompromised (or less compromised compared with existing drugs) against bacterial cells demonstrating AS represents, at least in principle, a more straightforward approach to addressing the problem. The majority of existing antibacterial agents were discovered by their ability to inhibit the growth of rapidly dividing bacteria (Coates and Hu, 2008). Accordingly, the fact that most of them lack useful killing activity against non-growing cells may in part reflect a bias in the approach by which they were discovered, rather than suggesting scarcity of such compounds in nature. Furthermore, a growing number of examples indicate that antibacterial agents with activity against non-growing bacteria exist. For example, and as mentioned above, the fluoroquinolones exhibit killing against the majority of non-growing Gram-negative bacterial populations *in vitro* (Eng *et al.*, 1991). Oritavancin, rifalazil (a rifampin derivative) in combination with vancomycin and the investigational agents XF-70 and XF-73 have all been shown to possess killing activity against non-growing and/or biofilm cultures of *S. aureus* (Osburne *et al.*, 2006; Belley *et al.*, 2009; Ooi *et al.*, 2010). TG44, an investigational agent under preclinical evaluation, displays potent bactericidal action against non-growing *Helicobacter pylori* (Kamoda *et al.*, 2006).

Although counter-intuitive from the perspective of our current paradigm of antibiotic-mediated killing, it is nonetheless possible that antibacterial agents could be identified that exhibit dramatically improved, or even exclusive, killing of non-dividing cells/persisters, compared with actively growing cells (Coates and Hu, 2008). In support of this concept, it appears that pyrazinamide, an established anti-tuberculosis agent, exhibits greater bactericidal effects the lower the metabolic activity of the mycobacteria under test, at least *in vitro* (Hu *et al.*, 2006).

These examples provide confidence that antibacterial agents can in principle be identified which exhibit substantially improved killing activity against AS cells over the vast majority of current clinically deployed antibiotics. To identify such agents, our existing paradigm of drug discovery and development will require some refinement. This is already under way as a consequence of the growing recognition of the extent to which AS impacts antibacterial chemotherapy, and the importance of developing agents that can, for example, kill non-growing staphylococci residing within a biofilm, or non-growing mycobacteria. Thus, it is becoming more commonplace to see scientific articles which describe the microbiological properties of new or candidate antibacterial agents including evaluation of the activity of these agents against growth-arrested or biofilm cultures (Kamoda *et al.*, 2006; Osburne *et al.*, 2006; Mascio *et al.*, 2007; Belley *et al.*, 2009; Ooi *et al.*, 2010).

A variety of straight-forward experimental approaches, many of which should be amenable to miniaturization and therefore high-throughput screening, have been described that enable an evaluation of antibacterial effects upon cells exhibiting AS. As a consequence, empirical screening for antibacterial agents with activity against bacterial cells in these states is feasible (Byrne *et al.*, 2007). Assays to identify or evaluate killing of non-growing bacteria include the use of stationary-phase cultures (Byrne *et al.*, 2007), bacteria resuspended in saline (Levin and Rozen, 2006) or nutrient-depleted growth medium (Eng *et al.*, 1991; Belley *et al.*, 2009), bacterial cultures induced to exhibit the stringent response by the addition of mupirocin (Ooi *et al.*, 2010), and cultures arrested by the addition of inhibitors or growing at reduced temperature (Mascio *et al.*, 2007). Standardization in approaches to conduct, and report the results of, these assays are currently lacking and would be beneficial. Indeed, Coates and colleagues (2002) have proposed adoption of such concepts as the minimum stationarycidal concentration (MSC) to report the degree of kill achieved against non-growing cells, e.g. MSC_{50} or MSC_{99}, the antibacterial concentration required to bring about 50% or 99% kill of a stationary-phase culture.

To what extent such assays accurately reflect the 'natural' non-growing state, particularly as it manifests *in vivo*, is not known (Coates and Hu, 2008). This is particularly the case given the current lack of clarity over whether cells exhibiting AS exist in distinct states of growth-arrest, or whether they represent different manifestations of a single non-growing state. Furthermore, it is already apparent that the results achieved with these different assays will not necessarily concur: despite retaining cidal activity against *S. aureus* cultures in which growth had been arrested by the addition of CCCP or erythromycin, daptomycin exhibited attenuated killing of stationary-phase cultures (Mascio *et al.*, 2007). Whilst it will therefore need to be established empirically which of these assays are most predictive of *in vivo* activity, it seems reasonable to suppose that the ability of a candidate antibacterial agent to produce substantial killing of non-growing bacteria in several of these assays may translate into enhanced killing of non-growing bacteria *in vivo*.

An assay for identifying or evaluating killing activity specifically against persister cells has been detailed for *M. tuberculosis*, and involves an initial challenge with rifampin to isolate the persister subpopulation, followed by exposure to the test agent(s) (Hu *et al.*, 2003). This approach could readily be adapted for other bacterial species. For evaluating the activity of candidate antibacterial agents against biofilm cultures, methods amenable to medium- to high-throughput screening have also been described, including the MBEC system (formerly the Calgary Biofilm device; Ceri

Shah, D., Zhang, Z., Khodursky, A., Kaldalu, N., Kurg, K., and Lewis, K. (2006). Persisters: a distinct physiological state of *E. coli*. BMC Microbiol. *6*, 53.

Shigeta, M., Tanaka, G., Komatsuzawa, H., Sugai, M., Suginaka, H., and Usui, T. (1997). Permeation of antimicrobial agents through *Pseudomonas aeruginosa* biofilms: a simple method. Chemotherapy *43*, 340–345.

Smith, A.L., Fiel, S.B., Mayer-Hamblett, N., Ramsey, B., and Burns, J.L. (2003). Susceptibility testing of *Pseudomonas aeruginosa* isolates and clinical response to parenteral antibiotic administration: lack of association in cystic fibrosis. Chest *123*, 1495–1502.

Smith, P.A., and Romesberg, F.E. (2007). Combating bacteria and drug resistance by inhibiting mechanisms of persistence and adaptation. Nat. Chem. Biol. *3*, 549–556.

Spoering, A.L., and Lewis, K. (2001). Biofilms and planktonic cells of *Pseudomonas aeruginosa* have similar resistance to killing by antimicrobials. J. Bacteriol. *183*, 6746–6751.

Spoering, A.L., Vulic, M., and Lewis, K. (2006). GlpD and PlsB participate in persister cell formation in *Escherichia coli*. J. Bacteriol. *188*, 5136–5144.

Sureka, K., Ghosh, B., Dasgupta, A., Basu, J., Kundu, M., and Bose, I. (2008). Positive feedback and noise activate the stringent response regulator *rel* in mycobacteria. PLoS One 3, e1771.

Thometz, J.G., Lamdan, R., Kehl, K.S., and Chusid, M.J. (1996). Microbiological tolerance in orthopaedic infections: delayed response of septic arthritis and osteomyelitis of the hip due to infection with tolerant *Staphylococcus aureus*. J. Pediatr. Orthop. *16*, 518–521.

Tomasz, A., Albino, A., and Zanati, E. (1970). Multiple antibiotic resistance in a bacterium with suppressed autolytic system. Nature *227*, 138–140.

Tuomanen, E., Durack, D.T., and Tomasz, A. (1986). Antibiotic tolerance among clinical isolates of bacteria. Antimicrob. Agents Chemother. *30*, 521–527.

Vazquez-Laslop, N., Lee, H., and Neyfakh, A.A. (2006). Increased persistence in *Escherichia coli* caused by controlled expression of toxins or other unrelated proteins. J. Bacteriol. *188*, 3494–3497.

Viducic, D., Ono, T., Murakami, K., Susilowati, H., Kayama, S., Hirota, K., and Miyake, Y. (2006). Functional analysis of *spoT*, *relA* and *dksA* genes on quinolone tolerance in *Pseudomonas aeruginosa* under nongrowing condition. Microbiol. Immunol. *50*, 349–357.

Walters, M.C. 3rd, Roe, F., Bugnicourt, A., Franklin, M.J., and Stewart, P.S. (2003). Contributions of antibiotic penetration, oxygen limitation, and low metabolic activity to tolerance of *Pseudomonas aeruginosa* biofilms to ciprofloxacin and tobramycin. Antimicrob. Agents Chemother. *47*, 317–323.

Wiuff, C., and Andersson, D.I. (2007). Antibiotic treatment *in vitro* of phenotypically tolerant bacterial populations. J. Antimicrob. Chemother. *59*, 254–263.

Wiuff, C., Zappala, R.M., Regoes, R.R., Garner, K.N., Baquero, F., and Levin, B.R. (2005). Phenotypic tolerance: antibiotic enrichment of noninherited resistance in bacterial populations. Antimicrob. Agents Chemother. *49*, 1483–1494.

Wu, J.A., Kusuma, C., Mond, J.J., and Kokai-Kun, J.F. (2003). Lysostaphin disrupts *Staphylococcus aureus* and *Staphylococcus epidermidis* biofilms on artificial surfaces. Antimicrob. Agents Chemother. *47*, 3407–3414.

Quorum-sensing Inhibitors Disable Bacterial Biofilms

10

Thomas Bjarnsholt, Tim Tolker-Nielsen and Michael Givskov*

Abstract

It is now evident that bacteria assume the biofilm mode of growth during chronic infections. The important hallmarks of biofilm infections are development of local inflammations, extreme tolerance to the action of conventional antimicrobial agents and an almost infinite capacity to evade the host defence systems in particular innate immunity. In the biofilm mode, bacteria use cell to cell communication termed quorum-sensing (QS) to coordinate expression of virulence, tolerance towards a number of antimicrobial agents and shielding against the host defence system. Chemical biology approaches may allow for the development of new treatment strategies focusing on interference with cell to cell communication with the aim of primarily disabling expression of virulence, immune shielding and antibiotic tolerance. Here we present our experience with screening and testing small molecule chemistry for *N*-acyl homoserine lactone-dependent QS inhibition. In addition we present our thoughts with respect to advantages and potential limitations of the intervention strategies described.

Quorum sensing – an introduction

The first description of cooperative behaviour among bacteria dates back to the late 1960s and early 1970s, when Alexander Tomasz (Tomasz, 1965) first described the development of genetic competence in *Streptococcus pneumoniae*, transmitted from cell to cell by a specific extracellular factor. Competence is the ability of bacteria to take up biologically active DNA molecules, and integrate them into their genomes. Around the same time, Nealson and co-workers (1970) were studying the biology of light-producing organelles in deep-sea fish which turned out to be colonized by bacteria of the species *Vibrio fischeri*, which, when present at high densities, produced bioluminescence. This bioluminescence originated from two luciferase genes, expression of which (under *in vitro* conditions) rapidly increased when bacteria entered the late exponential phase and early stationary phase of growth. Furthermore, they found that the luminescence of bacteria in early log phase was inducible by the addition of cell-free medium derived from stationary phase bacteria. The bacteria produce diffusible signal compounds (previously termed autoinducers) which bind to a receptor forming a transcriptional regulatory complex. These systems function to measure bacterial density. To reflect the number of cells, the concentration of signal molecules increases proportionally with the number of bacteria during log phase. At high density, the concentration of signal molecules in complex with the cognate receptor protein triggers a positive feed back loop operating on signal synthesis (hence the term autoinducer). This relatively simple system enables bacteria to monitor their own population size and is termed quorum sensing (QS)(Fuqua *et al.*, 1994). Transcription of the luminescence genes is activated when a threshold concentration of the autoinducers has been reached, i.e. the bacteria have reached

*Corresponding author

the sufficient density for this behaviour to take place. The autoinducer of *V. fischeri* was the first to be identified as *N*-3-oxo-hexanoyl-homoserine lactone (Eberhard *et al.*, 1981). In 1983, the *lux* operon was isolated and the mechanism of the genetic control of the light production was described by Engebrecht *et al.* (1983). *N*-acyl-l-homoserine lactone (AHL) QS signalling systems are the most well studied examples of QS. More than 70 Gram-negative bacteria have been found to employ such *lux*–homologue systems (Fuqua *et al.*, 2001; Taga and Bassler, 2003; see Fig. 10.1). A few *Vibrio* spp. bacteria use AHLs as signal molecules but these signalling pathways do not function like the *lux* system (Bassler *et al.*, 1993; Hanzelka *et al.*, 1999). QS controls a wide range of functions, except for housekeeping processes, such as Ti-plasmid conjugation in *Agrobacterium tumefaciens*, antibiotic production in *Erwinia carotovora* (Taga and Bassler, 2003), virulence gene expression in *Vibrio cholerae* (Zhu *et al.*, 2002) *Burkholderia cepacia* (Gilligan, 1991), and *Pseudomonas aeruginosa* (Whiteley *et al.*, 1999), and surface motility by means of swarming in *Serratia liquefaciens* (Eberl *et al.*, 1996, 1999), *P. aeruginosa* (Kohler *et al.*, 2000) and *B. cepacia* (Huber *et al.*, 2001). In particular, QS in *P. aeruginosa* features a third intertwined system, the PQS system, based on quinolone-derived signal molecules (see Fig. 10.1). PQS is produced during logarithmic growth (Dubern and Diggle, 2008) through early stationary phase peaking in late stationary phase of growth (McKnight *et al.*, 2000). The PQS system has been shown to be a part of the *P. aeruginosa* QS hierarchy (Diggle *et al.*, 2003). Instead of AHL molecules, Gram-positive bacteria employ small peptides as signalling molecules, usually 5–17 amino acids in length (Federle and Bassler, 2003; see Fig. 10.1). Examples of Gram-positive QS controlled behaviour are development of genetic competence and sporulation in *Bacillus subtilis* (Magnuson *et al.*,

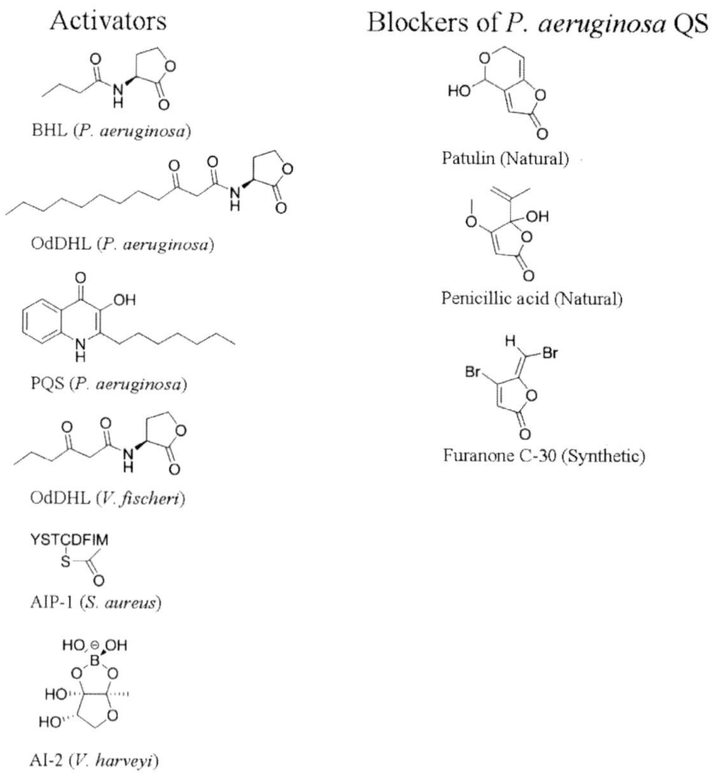

Figure 10.1 Examples of compounds that control quorum sensing (QS) in different bacteria and virulence of *P. aeruginosa* by blocking its QS.

1994) virulence expression in *Enterococcus faecalis* (Qin *et al.*, 2000) and *Staphylococcus aureus* (Kong *et al., 2006)* and biofilm development in *S. aureus* (Boles and Horswill, 2008).

The existence of a fourth type of QS system has been proposed (Winans and Bassler, 2002). This system functions by means of a two component signalling pathway, the first using 3-hydroxy-C6 HSL referred to as AI-1, the second using a furanosyl borate diester referred to as AI-2. It has been proposed, that this AI-2 system, in addition to controlling the light production of *V. harveyi*, is a global QS regulatory system. By DNA sequence homology, *V. harveyi luxS* analogues (which are believed to encode signal synthetases) have been identified in more than 80 bacterial species (Miller and Bassler, 2001; Roy *et al.*, 2010). Its universal appearance suggests a role of AI-2 in interspecies communication. However, it has also been proposed that AI-2 is a toxic waste product of the central metabolism, and therefore actively secreted (Winzer *et al.*, 2002a,b; see Fig. 10.1). Sun *et al.* (2004) performed comparative genomic studies and phylogenetic analysis on 138 completed genomes of Bacteria, Archaea and Eukaryotes, searching for genes involved in the synthesis and detection of the AI-2 signal. Their conclusion was not in support of a multispecies signalling system present in non *Vibrio* organisms; however the role of AI-2 is still under debate.

Quorum sensing and AHL signals

AHL signal molecules vary between bacteria and some bacteria produce more than one type. They all exhibit the same basic structure of an acyl chain of variable length usually from 4 to 16 carbons and in most cases even numbered (C4, C6 and C8, etc.; Fuqua *et al.*, 2001). Certain positions of the hydrocarbon chain may be substituted; a hydroxyl- or an oxo-group at the third carbon (Fuqua *et al.*, 1996) is the most common substitution. The acyl chain is conjugated to a lactonized homoserine through an amide bond (see Fig. 10.1).

The *lux* system comprises the *luxI* gene encoding the AHL synthetase and the *luxR* gene encoding the receptor. During the growth of *V. fischeri* in the light organ of certain deep sea fish, the number of the bacteria rises to 10^{10}. During the growth of the bacteria the signal molecule 3-oxo-C6-HSL (*N*-(3-oxo-hexanoyl-)-L-homoserine lactone) is produced by the synthetase LuxI. The acyl portion of the AHL is derived from fatty-acid precursors conjugated to the acyl carrier protein (acyl-ACP), and the homoserine lactone moiety is derived from *S*-adenosylmethionine (Fuqua *et al.*, 2001).

The short chained AHL signal molecules diffuse passively across the cell membrane (Kaplan and Greenberg, 1985). Longer chained AHL signal molecules, and to some extent molecules with different substitutions of the *N*-acyl side chain, seem to require active transport by efflux pumps to be secreted (Pearson *et al.*, 1999). The signal molecules are secreted into the surroundings and are retrieved by the neighbouring bacteria, thus in theory the internal concentration is equivalent to the external concentration of signal molecules. Active influx has not adequately been described in the literature, but a *P. aeruginosa* strain deficient in producing signal molecules responds to externally added signal molecules (Hentzer *et al.*, 2003). The signal molecules produced by the bacteria bind to the receptor, the LuxR protein, an AHL-responsive transcriptional activator. The LuxR type proteins contain two functional domains, the AHL binding region located in the amino-terminal portion of the protein, and a helix–turn–helix motif (HTH) located in the carboxy-terminus responsible for binding to the target promoters (Choi and Greenberg, 1991; Choi and Greenberg, 1992; Hanzelka and Greenberg, 1995). It has been demonstrated that the LuxR protein and the AHL signal molecule interact on a 1:1 molar ratio *in vitro* (Zhu and Winans, 1999). The interaction of the AHL signal molecule with the LuxR protein promotes binding to promoters of the target genes. Binding of AHL to LuxR introduces a conformational change in the protein and the HTH motif is exposed to binding to DNA recognition sites denoted the *lux* box. It has been shown for TraR, a homologue of LuxR, that the AHL functions as a scaffold for the correct folding of the R protein, 'hiding' the AHL molecule (Zhu and Winans,

2001). If the R protein is folded in the absence of the signal molecule, it is subject to rapid proteolysis (Zhu and Winans, 2001). The DNA sequence of the QS recognition signature (the *lux* box) is located upstream in the –35 region overlapping with the positions of most QS regulated promoters (Fuqua *et al.*, 1994). As a rule of thumb, LuxR-type proteins act as positive transcriptional regulators although a few act as repressors. Such transcriptional LuxR-type repressors have been identified in *Pantoea stewartii* (the EsaR protein), and in *Yersinia pseudotuberculosis* (the YpsR protein; von Bodman *et al.*, 1998; Atkinson *et al.*, 1999).

Biofilms

The biofilm life style has probably dominated since the first bacteria evolved. With respect to medical relevance, biofilms are typically involved in chronic infections, in contrast to acute infections which are dominated by planktonic bacteria (Høiby *et al.*, 2010). Chronic infections, such as tuberculosis and leprosy, slowly degrade a patient's health and eventually result in death. Prior to the advent of the antibiotic era, bacterial infections caused high rates of mortality (up to one third of the population). With the development of antibiotics, an increase in 'low-grade' infections, such as middle ear infections, urinary tract infections, prostatitis etc., were observed. The patients experience discomfort, fever and other clinical signs of an infection. This is in contrast to acute infections which are relatively easier to treat with antibiotics (if diagnosed in time). The origins of these persistent infections were investigated through a series of both *in vitro* and *in vivo* observations 20–30 years ago, and were identified as being caused by bacterial biofilms (Costerton *et al.*, 1978, 1987; Lam *et al.*, 1980; Hall-Stoodley *et al.*, 2006; Christensen *et al.*, 2007b; Brady *et al.*, 2008; Bjarnsholt *et al.*, 2008, 2009a,b; Homoe *et al.*, 2009). Noticeably, biofilms and their extreme tolerance towards antimicrobial agents were discovered 300 years earlier by Leeuwenhoek (1684). He observed that the 'animals' (bacteria) inside the 'scurfs' (plaques) were protected against the disinfectant vinegar in contrast to the 'animals 'outside the 'scurfs' which were killed. This ancient discovery underscores one of the remarkable features of bacterial biofilms – the extreme tolerance against antimicrobial agents and disinfectants.

It should be noted that biofilm tolerance to antibiotic is not to be mistaken with antibiotic resistance. Tolerance is a reversible phenomenon – the bacteria in biofilm survive antibiotic treatment but if the biofilm is disrupted the bacteria become susceptible to the treatment (Bayles, 2007).

Today we know from numerous *in vitro* and *in vivo* observations that the origin of most persistent infections involves aggregated bacteria. Bacteria in these aggregates are physically joined and produce an extracellular matrix consisting of many kinds of extracellular polymeric substances (EPS) such as proteins, DNA and polysaccharides. It is evident that these aggregates withstand the highest achievable exposures of antibiotics which would normally readily kill planktonic cells (as long as they have not acquired resistance). The exopolymeric substances (EPS) of the biofilm matrix seem to constitute a scaffolding component of the aggregating bacteria (Nivens *et al.*, 2001; Bjarnsholt *et al.*, 2005a; Allesen-Holm *et al.*, 2006) acting as a scavenger of free oxygen radicals (Simpson *et al.*, 1989), and annihilating the effect of antibiotics including aminoglycosides (Allison and Matthews, 1992; Hentzer, 2001). In addition, their tolerance towards the host immune defence is dramatically increased. Many such infections develop on abiotic surfaces such as on implants, catheters, artificial heart valves, teeth and contact lenses. However, the biofilm mode of life does not necessarily involve association to either a biotic or abiotic surface. Accordingly, our definition of a biofilm is *a coherent cluster of matrix-embedded bacterial cells which are more tolerant to most antimicrobials and host defences, than planktonic bacterial cells* (see Fig. 10.2). This is in line with the many observations from non-surface related infections such as CF, otitis media, chronic wounds, and osteomyelitis. Based on this, chronic infections are defined as infections that (A) persist in spite of antibiotic therapy and the host's innate and adaptive immune responses and (B) are characterized by persisting pathology (Høiby, 2008).

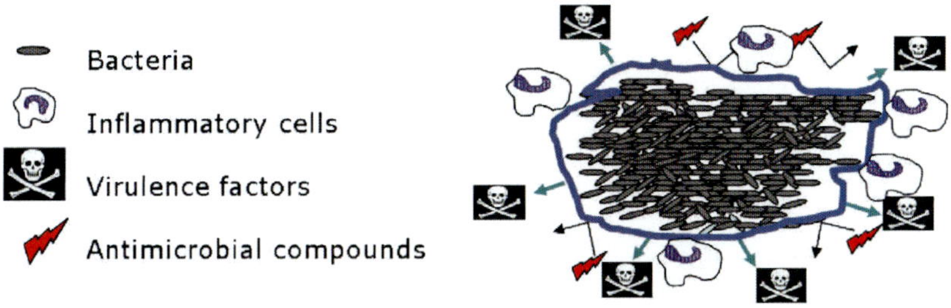

Figure 10.2 Schematic drawing of a bacterial biofilm. Biofilms are bacterial aggregates. The high density of bacteria within a small space enable density-dependent gene regulation like quorum sensing (QS). Subsequently, the QS-regulated virulence factors, such as rhamnolipids, a leucocidal toxin, are produced and function as a protective shield (indicated as a blue line for the purpose of visualization). Within these biofilms the bacteria are also well protected from antimicrobial agents.

Quorum sensing and the biofilm mode of life

Many believe that biofilm formation is a developmental progression that involves a number of complex and highly regulated events (O'Toole *et al.*, 2000) and is, at least in part, controlled by QS and other signalling systems. Sauer *et al.* (2003) have generated proteomic data suggesting that *in vitro*, *P. aeruginosa* biofilm formation proceeds through an ordered, temporal series of events which inevitably relies on the ordered expression of a collection of biofilm-specific genes (the biofilm programme; O'Toole *et al.*, 2000; Sauer, 2003). What presently speaks against a particular developmental biofilm programme is that specific genes, expressed independently of the conditions under which biofilms have been formed, still have not been identified in *P. aeruginosa* (Hentzer *et al.*, 2003). In our view, the available data support a model by which *P. aeruginosa* biofilm formation, and the successive growth and development, may proceed through different but converging paths, each of which may contain chaotic elements. So instead of one particular, predetermined biofilm programme, much of the available data suggest that the *P. aeruginosa* biofilm adapt to the available nutritional and environmental conditions (Purevdorj *et al.*, 2002; Reimmann *et al.*, 2002; Hentzer *et al.*, 2005a,b). However, what is needed to get the full picture is increased sensitivity of the present transcriptomic and other 'omics' analyses performed under relevant *in vivo* conditions.

Throughout the years, great emphasis has been put on mushroom-shaped, multicellular structures as a phenomenon that underscores the presence of confined cell differentiation and divergent activities of subpopulations in development of biofilm architecture (see Fig. 10.3A). Davies *et al.* (1998) concluded that *P. aeruginosa* required QS signalling to enable development of differentiated biofilms with these mushroom-shaped, multicellular structures. There has been some controversy to this claim. Heydorn *et al.* (2002) showed that with citrate as the carbon source, the wild-type and QS mutant form similar, flat biofilms. Apparently the involvement of signalling in the biofilm forming process depends on the specific conditions of each *in vitro* laboratory set up and is therefore conditional (Kjelleberg and Molin, 2002). Another contribution to the dispute of the involvement of QS in biofilm development is the application of QS mutants with uncertain genetic backgrounds. It was reported by Beatson *et al.* (2002) that QS mutants accumulate secondary mutations, which lead to misinterpretation of the QS effect on biofilm formation. We would like to emphasize, therefore, that the available evidence suggests that formation of the mushroom-shaped structures in *P. aeruginosa* biofilms is first and foremost subject to the physical and chemical conditions present in the laboratory flow-cells, which argues against the hypothesis that biofilm formation in essence follows a predestined developmental programme which, as indispensable components, include surface sensing and concomitant QS signalling.

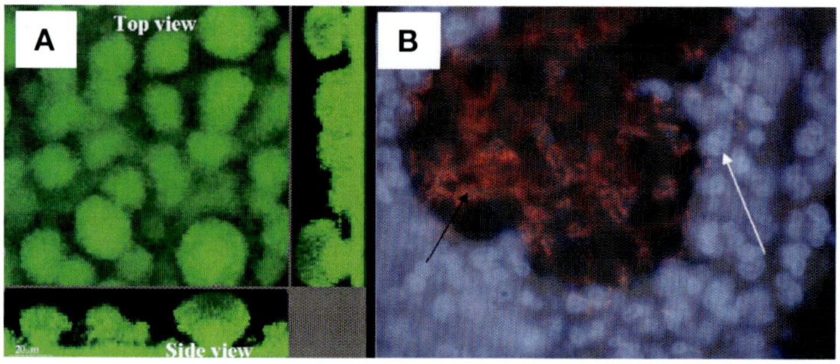

Figure 10.3 Examples of (A) an *in vitro P. aeruginosa* biofilm grown in a flow cell where the bacteria are visualized by confocal laser scanning microscopy owing to constitutive production of GFP. (B) *In vivo P. aeruginosa* biofilm (visualized by specific PNA FISH in red, black arrow) during a chronic lung infection of a cystic fibrosis patient. Here the cellular components of the host defence (visualized in blue by DAPI, white arrow) are unable to phagocytose the biofilm.

Furthermore, the presence of mushroom structures has to our knowledge never been reported in *ex vivo* samples (for a typical *ex vivo* biofilm, see Fig. 10.3B). However, although they may be limited to flow-cells, the studies involving these structures have provided important insights regarding the fundamental processes of biofilm formation and tolerance development that may be of relevance outside the laboratory model systems.

Characterization of the role of intracellular signalling in the biofilm lifecycle has gained momentum. Biofilms are thought to maintain 'steady state' by means of growth and dispersal. The dispersal occurs either as single cells or as small microcolonies ripped off of the biofilm (Costerton *et al.*, 1999; Stoodley *et al.*, 2001; Webb *et al.*, 2003; Stapper *et al.*, 2004). Evidence is accruing that the shift between the biofilm lifestyle (with production of matrix components such as Pel/Psl polysaccharides and Cup fimbriae) and the planktonic lifestyle (without production of these matrix components) is regulated via proteins that contain diguanylate cyclase or phosphodiesterase activities. Such proteins influence the intracellular level of the second messenger molecule cyclic diguanosine-5′-monophosphate (c-di-GMP; Hickman *et al.*, 2005; Meissner *et al.*, 2007). It appears that high intracellular c-di-GMP levels up-regulate matrix production and biofilm formation, whereas low intracellular c-di-GMP levels down-regulate matrix production and induce a planktonic lifestyle. Several factors may contribute to dispersal of biofilms including phage induction (Webb *et al.*, 2003), carbon starvation (Gjermansen *et al.*, 2005, 2010; Schleheck *et al.*, 2009) and nitric oxide signalling (Barraud *et al.*, 2006, 2010). Carbon starvation and nitric oxide signalling were shown to mediate phosphodiesterase activity in *P. aeruginosa* causing decreased intracellular c-di-GMP levels. In addition, Davies and Marques (2009) have recently shown that *P. aeruginosa* produces a small organic compound, *cis*-2-decanoic acid which, when added exogenously at native concentrations (2.5 nM), was able to induce dispersion in an *in vitro P. aeruginosa* biofilm (Davies and Marques, 2009). Taken together, this suggests that dispersal is both a programmed process and subject to physical conditions such as shear forces and sloughing.

The above listed conclusions are based solely on *in vitro* observations, but if massive dispersal can be chemically promoted and occurs in medical biofilms, it may have severe implications since it is a mechanism by which biofilm bacteria might disseminate through the infected organ to the entire body and thereby cause rigorous systemic infections. For this 'worst case scenario' to develop, it would require flow conditions as those present in blood vessels or in the synovial fluid surrounding a moving artificial limp. If the biofilm, however, is located within tissue or organs like in the CF lung, a chronic wound or permanent tissue fillers, the absence of direct flow limits the dissemination of bacteria, but localized increase in inflammation may occur as a consequence of biofilm spreading to vacant areas. As we will discuss later, the role of QS is to offer protection and enable the biofilm bacteria to evade host immunity and the detrimental action of antimicrobial compounds.

QS scrambling disables expression of virulence and biofilm tolerance

The rationale of QS scrambling

Recently, several studies have supported the concept of inhibiting QS as a functional and relevant antimicrobial strategy (Hentzer *et al.*, 2003; Rasch *et al.*, 2004; Wu *et al.*, 2004; Bjarnsholt *et al.*, 2005b). Despite the fact that QS systems do not control expression of essential genes *per se*, scrambling of the QS systems leads to attenuation in animal models (Hentzer and Givskov, 2003; Wu *et al.*, 2004; Rasch *et al.*, 2004; Bjarnsholt *et al.*, 2005a,b). Interestingly, we also recently demonstrated that knocking out the *rhlA* gene (which results in rhamnolipid deficiency but leaves an otherwise functional QS system behind) leads to significant attenuation in two animal infection models (van *et al.*, 2009). We therefore propose that QS inhibitors (QSI) conduct functional antimicrobial effects *in vivo* mainly by preventing the establishment of a rhamnolipid PMN shield, and consequently reinstating proper antimicrobial functions of the PMNs as follows. The role of rhamnolipids in shielding *P. aeruginosa* biofilms was elucidated by Jensen *et al.* (2007) and later Alhede *et al.* (2009). They showed that rhamnolipids induce fast necrosis and lysis of human PMNs and that biofilms respond to the presence of PMN by further up-regulating the production of rhamnolipids. For this to happen, a functional QS system is required (Alhede *et al.*, 2009). Accordingly, since PMN lysis is likely to be reduced when QS is blocked by a QS scrambling drug, local tissue destruction and inflammation will be kept to a minimum. This in turn reduces release of PMN DNA which will be expected to further contribute to reduction of the viscosity of the bronchial secretion in CF. This may help maintaining or even improving lung function (Hodson *et al.*, 2003). Both effects are expected to tip the balance in favour of bacterial eradication and subsequent healing. QS inhibitors have proven equally active on planktonic and biofilm cells; they do not seem to be as subject to matrix shielding as many of today's antibiotics (Hentzer *et al.*, 2003). Disruption of the QS signalling cascade may therefore be a particularly valuable strategy in the treatment of chronic, biofilm infections. Likewise, it must be emphasized that the rationale behind this approach is to specifically interfere with disease-causing traits and development of inflammation rather than restricting the growth of the bacteria. As production of virulence factors is inhibited, the pathogen can no longer maintain its position in an increasingly more hostile host environment and consequently becomes cleared by the innate host defences. Given that the mere attenuation of the pathogen does not impose a harsh and direct selective pressure on the cells, it can be hypothesized that resistant mutants will not arise as readily as in the case of conventional antibiotic treatment. We have denoted this the 'anti-pathogenic drug principle' (Hentzer and Givskov, 2003). In addition to this, we strongly recommend a full transcriptional validation of potential QSIs, rather than just the use of single gene fusions (Bjarnsholt *et al.*, 2010).

The first QS inhibitors

The first example that nature already provided for utilization of QSIs as an antibacterial strategy is the halogenated furanones produced by the Australian macroalga *Delisea pulchra* (de Nys *et al.*, 1993; Givskov *et al.*, 1996; Kjelleberg *et al.*, 1997). The authors showed that the halo-furanones, administered within a certain concentration window, specifically inhibit bacterial QS thereby blocking surface colonization and virulence factor expression in *Serratia liquefaciens*. Western blot analysis indicated that the halo-furanones are able to promote rapid turnover of LuxR homologue proteins, thereby reducing the amount of protein available to interact with AHL and act as transcriptional regulators (Manefield *et al.*, 2002). These observations, taken together with a mutational and *in silico* docking analysis of compounds into the LuxR crystal structure, indicate a direct interaction with the receptor protein (Koch *et al.*, 2005). The efficacy against *P. aeruginosa* QS and virulence was greatly enhanced after subsequent development of a series of synthetic furanone analogues. Among these, compound C-30 (Hentzer *et al.*, 2003) and C-56 (Hentzer *et al.*, 2002) showed QSI drug potential in *P. aeruginosa* (see Fig. 10.1). Hentzer *et al.* (2003) showed that biofilms grown with and without furanone C-30 were not significantly different with respect to formation and biomass. However, when treating three day old biofilms with 100 μg/ml tobramycin (grown either in the presence or in the absence of furanone C-30), a significantly high fraction of cells was eradicated in the biofilm that had been grown in the presence of furanone C-30. The antimicrobial properties of both furanone C-30 and furanone C-56 were also demonstrated in a chronic pulmonary mouse model of infection (Hentzer *et al.*, 2003; Wu *et al.*, 2004) an implant model (Christensen *et al.*, 2007a) and a fish model of vibriosis infection (Rasch *et al.*, 2004). Other furanone derivatives have been synthesized by Kim *et al.* (2008) that, in contrast, also blocked *in vitro* biofilm formation of *P. aeruginosa*. This indicates that these furanone derivatives are highly toxic to the bacteria since genetic deletion analysis shows that a functional QS system is not required for biofilm formation *in vitro*.

Natural production of QSIs

Several years ago, we developed a new tool for high-throughput screening of fresh samples as well as crude extracts and pure compounds of QSI (Rasmussen *et al.*, 2005a; Bjarnsholt *et al.*, 2009c; see Fig. 10.1). Using this tool, we have found that QSI compounds are present in a number of food, herbal and fungal sources (to be published). The presence of QSI compounds in natural foods is extremely interesting because, in most cases, vegetables and herbs are non-toxic to humans and are easily available. One of the most potent QSIs we have detected is contained in extracts of crushed garlic (Bjarnsholt *et al.*, 2005b; Rasmussen *et al.*, 2005a). Fungi (Rasmussen *et al.*, 2005b) and marine organisms (Skindersoe *et al.*, 2008b) also produce QSIs, supporting the view that organisms devoid of functional immune defences rely on chemical defence systems against competing or harmful bacteria.

Available QSIs

In 2001 Teteda *et al.* (2001) published results showing that the antibiotic azithromycin inhibits QS of *P. aeruginosa*. This led us to investigate whether QS inhibition is a common feature of antibiotics. We screened 12 antibiotics for their QS-inhibitory activities using the QSI selector system (Rasmussen *et al.*, 2005a). Three of the antibiotics tested (azithromycin, ceftazidime, and ciprofloxacin) were active in the assay and were further examined for their effects on QS-regulated virulence factor production in *P. aeruginosa*. Consistent results from the virulence factor assays, reverse transcription-PCR, and DNA microarray-based transcriptomics support the finding that azithromycin, ceftazidime, and ciprofloxacin (in sub-lethal and sub-growth inhibitory doses) decrease the expression of a range of QS-regulated virulence factors. The data suggest that the underlying mechanism may be mediated by changes in membrane permeability, thereby influencing the flux of 3-oxo-C12-HSL (Skindersoe *et al.*, 2008a). Azithromycin inhibits or reduce the production of several of the virulence factors of *P. aeruginosa*, such as elastase and rhamnolipids (Skindersoe *et al.*, 2008a), and interferes with alginate polymerization (Hoffmann *et al.*, 2007). Recently another group reported that azithromycin inhibits transcription of AHL synthesis by blocking unknown *lasI* and *rhlI* regulation (Kai *et al.*, 2009). It is proposed that the azithromycin interferes with one of the precursors of the AHL molecules, indicating that the QSI activity is a secondary effect (Kai *et al.*, 2009). Azithromycin treatment does improve the respiratory function in CF patients (Saiman, 2004; Southern and Barker, 2004). The principal role of QSIs is to attenuate the virulence of the bacteria, and in our opinion if the excreted rhamnolipids are reduced, *P. aeruginosa* biofilms are dramatically weakened. It is therefore of great interest that Park *et al.* (2008) showed that Solenopsin A extracted from the fire ant *Solenopsis invicta* specifically targets the *rhl* QS system (which controls rhamnolipid synthesis).

Synthesized QS inhibitors

Since the discovery of the halo-furanones, a number of researchers have reported on synthesis of QSI compounds by means of organic chemistry approaches (Olsen *et al.*, 2002; Persson *et al.*, 2005a,b; Muh *et al.*, 2006; Riedel *et al.*, 2006; Geske *et al.*, 2007, 2008a–c; Mattmann *et al.*, 2008; Borlee *et al.*, 2008). Many of these compounds have been developed using agonist-based designs and probed for their action by means of specific QS screens (see below). The Greenberg group has employed *in silico* analysis for drug development (Muh *et al.*, 2006), where the 12 alkyl chain substituted-tetrazol PD12 showed an IC_{50} of a staggering 30 nM (based on *in vitro* analysis). The target specificity and functionality of C-30 have been validated by transcriptomic analysis and animal experiments, and C-30 is so far the most active, functional QSI drug ever designed with an IC_{50} of 2 µM and full QS inhibition (without any growth inhibition) exerted even at 10 µM (Hentzer, *et al.*, 2003). Note that physiologically relevant concentrations of *P. aeruginosa* QS signals are likely to be in the range of 1–20 M (Whiteley *et al.*, 2001).

Recently, a crystal structure of a LasR-peptide with OdDHL bound became available (Bottomley *et al.*, 2007). Therefore, it has now become possible to apply *in silico* analysis of receptor–ligand interactions such as those performed by Amara *et al.* (2009). The authors demonstrated direct covalent interaction of a reactive isothiocyanate AHL derivative with an amino acid present in the binding site. Based on the three-dimensional structure of the LasR protein (without ligand), we have used computer-based virtual screening in search of putative quorum-sensing inhibitors from a database comprising approved drugs and natural compounds (Yang *et al.*, 2009). Three potential QSIs which showed effects on *P. aeruginosa* virulence factor expression were identified, but their activity was significantly lower than that of C-30.

In conclusion, in silico docking and subsequent synthesis of chemical libraries as source of QSIs suffer from the lack of proper crystal models and problems with membrane permeability. Small molecule chemistry which has been optimized for functionality in the environment circumvents many of these deficiencies. We believe strongly that the use of natural structures as scaffolds for drug development is so far the most promising strategy.

Advantages and limitation using QSI drugs

The central role of bacterial communication and biofilm formation in pathogenesis has been firmly established. QS in particular seems to be one obvious new antimicrobial drug target, as it plays an important role in the production of virulence factors, antibiotic tolerance, and modulation of the host immune system. We believe that QSI and also c-di-GMP modulating drugs could be beneficial in the fight against biofilm infections, though much effort and enthusiasm from the industry is needed to develop pharmaceutically relevant molecules. This endeavour is of global, societal, and economic significance. The goal of interfering with the bacterial communication is to control the pathogenicity of the bacteria without interfering with growth or inducing selection of specific survival relevant phenotypes. Recent work on *P. aeruginosa* has shown that QSIs disable the bacterial shield against the cellular immunity and lower their tolerance towards antibiotics. Because the major problem of traditional antibiotics is the selection for resistant subpopulations, we hypothesize that a combination of QSIs and conventional antibiotics would enable killing of most of the bacteria by the antibiotics, with the remaining being eliminated by the professional phagocytes of the host defence system. We hypothesize that owing to their nonlethal effects, target-specific QSI drugs are not expected to eliminate communities of helpful and protective microorganisms. Consequently, administration of such drugs would not liberate niches that would attract other pathogens. On the other hand, Gram-positive and Gram-negative do not share signal molecules. Therefore, owing to this rarity of cross-talk, a broad-spectrum QSI drug is not likely to ever be available. Finally, although the selective force imposed by signalling inhibitors may be less stringent as compared with conventional antibiotics, it would be unwise to assume that resistance at some point would not occur; this probably is why the battle against infections is a never-ending story.

References

Alhede, M., Bjarnsholt, T., Jensen, P.O., Phipps, R.K., Moser, C., Christophersen, L., Christensen, L.D., van, G.M., Parsek, M., Hoiby, N., Rasmussen, T.B., and Givskov, M. (2009). *Pseudomonas aeruginosa*

Taga, M.E. and Bassler, B.L. (2003). Chemical communication among bacteria. Proc. Natl. Acad. Sci. U.S.A. *100*, 14549–54.

Tomasz, A. (1965). Control of the competent state in Pneumococcus by a hormone-like cell product: an example for a new type of regulatory mechanism in bacteria. Nature *208*, 155–159.

van, G.M., Christensen, L.D., Alhede, M., Phipps, R., Jensen, P.O., Christophersen, L., Pamp, S.J., Moser, C., Mikkelsen, P.J., Koh, A.Y., *et al.* (2009). Inactivation of the rhlA gene in *Pseudomonas aeruginosa* prevents rhamnolipid production, disabling the protection against polymorphonuclear leukocytes. APMIS *117*, 537–546.

Webb, J.S., Thompson, L.S., James, S., Charlton, T., Tolker-Nielsen, T., Koch, B., Givskov, M., and Kjelleberg, S. (2003). Cell death in *Pseudomonas aeruginosa* biofilm development. J. Bacteriol. *185*, 4585–4592.

Whiteley, M., Bangera, M.G., Bumgarner, R.E., Parsek, M.R., Teitzel, G.M., Lory, S., and Greenberg, E.P. (2001). Gene expression in *Pseudomonas aeruginosa* biofilms. Nature *413*, 860–864.

Whiteley, M., Lee, K.M., and Greenberg, E.P. (1999). Identification of genes controlled by quorum sensing in *Pseudomonas aeruginosa*. Proc. Natl. Acad. Sci. U.S.A. *96*, 13904–13909.

Winans, S.C., and Bassler, B.L. (2002). Mob psychology. J. Bacteriol. *184*, 873–883.

Winzer, K., Hardie, K.R., Burgess, N., Doherty, N., Kirke, D., Holden, M.T., Linforth, R., Cornell, K.A., Taylor, A.J., Hill, P.J., and Williams, P. (2002a). LuxS: its role in central metabolism and the *in vitro* synthesis of 4-hydroxy-5-methyl-3(2H)-furanone. Microbiology *148*, 909–922.

Winzer, K., Hardie, K.R., and Williams, P. (2002b). Bacterial cell-to-cell communication: sorry, can't talk now – gone to lunch! Curr. Opin. Microbiol. *5*, 216–222.

Wu, H., Song, Z., Hentzer, M., Andersen, J.B., Molin, S., Givskov, M., and Høiby, N. (2004). Synthetic furanones inhibit quorum-sensing and enhance bacterial clearance in *Pseudomonas aeruginosa* lung infection in mice. J. Antimicrob. Chemother. *53*, 1054–1061.

Yang, L., Rybtke, M.T., Jakobsen, T.H., Hentzer, M., Bjarnsholt, T., Givskov, M., and Tolker-Nielsen, T. (2009). Computer-aided identification of recognized drugs as *Pseudomonas aeruginosa* quorum-sensing inhibitors. Antimicrob. Agents Chemother. *53*, 2432–2443.

Zhu, J., and Winans, S.C. (1999). Autoinducer binding by the quorum-sensing regulator TraR increases affinity for target promoters *in vitro* and decreases TraR turnover rates in whole cells. Proc. Natl. Acad. Sci. U.S.A. *96*, 4832–4837.

Zhu, J., and Winans, S.C. (2001). The quorum-sensing transcriptional regulator TraR requires its cognate signaling ligand for protein folding, protease resistance, and dimerization. Proc. Natl. Acad. Sci. U.S.A. *98*, 1507–1512.

Zhu, J., Miller, M.B., Vance, R.E., Dziejman, M., Bassler, B.L., and Mekalanos, J.J. (2002). Quorum-sensing regulators control virulence gene expression in *Vibrio cholerae*. Proc. Natl. Acad. Sci. U.S.A. *99*, 3129–3134.

The Indigenous Human Microbiota

11

Adam M. Nelson and Vincent B. Young*

Abstract

Recent technological advances have expanded the tools available for study of the indigenous human microbiota. One of the early limitations in this field was the difficulty in recovering most residents of the community via standard culture-based methods. Many residents of the flora are anaerobic or microoxic, require specific nutrients, or are dependent on microbe–microbe/microbe–host interactions that are difficult to replicate *in vitro*, thus making their cultivation difficult. Naturally, the easiest species to grow in the laboratory have been the best studied. However, these cultivatable species are only a fraction of the total population of the microbiota. This chapter will introduce both the culture and non-culture based techniques being used to look deeper into the population structure both on a temporal and spatial scale. It will also discuss how disruptions (including those mediated by the administration of antibiotics) of the microbiota can produce changes in human health, and outline ongoing efforts by the National Institutes of Health and international investigators to study the indigenous microbiota.

Introduction: human-associated microbial communities

The microbial communities that inhabit various body sites are complex, dynamic associations of bacteria with members that can vary in their interaction with the host from mutualistic to pathogenic. This microbial community is referred to as the microbiota, and it contains collections of species that can vary greatly from person to person, and also topographically and temporally within the same individual.

Regions of the body with known, stable bacterial communities include the gut (Eckburg *et al.*, 2005; Gill *et al.*, 2006; Kinross *et al.*, 2008) and oral cavity (Kroes *et al.*, 1999; Avila *et al.*, 2009; Lazarevic *et al.*, 2009; Zaura *et al.*, 2009). Recent work has expanded the sampling sites to include many additional regions such as the vagina (Oakley *et al.*, 2008), ear (Frank *et al.*, 2003), lung (Hilty *et al.*, 2010), nasal cavity (Hilty *et al.*, 2010), and numerous sites on the skin (Costello *et al.*, 2009; Grice *et al.*, 2009). When sequencing efforts began, it quickly became evident that these regions of the body contained large numbers of uncultivable organisms. New techniques in culture-independent analysis, which will be discussed in detail later, have allowed for many more members of a complex community to be identified, even when the appropriate culture conditions have not yet been discovered. Variability in community structure at sampling sites was seen from person-to-person, topographically, and temporally (Costello *et al.*, 2009).

The stigma of the presence of bacteria as harmful has been erased with a large body of evidence showing microbial communities, especially in the gut, to be beneficial for nutrient utilization, immune regulation, and protection from pathogens (Fava *et al.*, 2006; Thompson-Chagoyan *et*

*Corresponding author

al., 2007; Round and Mazmanian, 2009; Velagapudi *et al.*, 2010). The concept of exactly what type of interactions the microbiota have with the host is often unclear. How the host is able to recognize and tolerate the growth of a large community of bacteria within the gut and other areas of the body, while mounting active immune responses to pathogens, is another issue that is currently under investigation (Chen and Nunez, 2009). There are at least two classes of receptors involved in the host's ability to distinguish between pathogens and non-pathogens. These include the extracellular surface proteins called Toll-like receptors (TLRs), and cytoplasmic proteins called Nod-like receptors (NLRs; Akira *et al.*, 2006) that both have the ability to recognize patterns of microbial cell-surface proteins, and then activate downstream transcription of anti-microbial compounds (including lysozyme, α-defensins, and secretory phospholipase A_2; Wehkamp *et al.*, 2004; Chen and Nunez, 2009). The indigenous microbiota can provide other beneficial functions. Specifically, TLR2 has been shown to protect against mucosal injury and stress-induced damage by preserving the tight junction-associated barrier assembly. The preservation of barrier function is important because it prevents normal intestinal microbiota from inducing an ongoing inflammatory response by the immune system (Cario *et al.*, 2007).

The stability of the microbiota in the gut is important for maintaining the beneficial interactions of the host and the numerous species contained within the microbiota. Changes to the community structure of the microbiota, through antibiotics and other stresses, can allow for opportunities for pathogens to cause disease. *Clostridium difficile* is one opportunistic pathogen that is able to cause serious disease in the host following antibiotic treatment. *C. difficile* is able to establish itself in the gut, following antibiotic disruption of the stable communities residing there, leading to the creation of a new niche (Khoruts *et al.*, 2010).

The phenomenon by which the microbiota is thought to protect the host from pathogens is called colonization resistance (Freter, 1962; Hentges and Freter, 1962; Vollaard and Clasener, 1994). The mechanisms for colonization resistance include limiting sites of attachment for pathogens in the gut epithelium, utilization of resources by commensal or non-pathogenic species to reduce availability for pathogens, and also the production of substances that inhibit growth of pathogens, such as those activated by TLRs and NLRs (Brandl *et al.*, 2008; Momose *et al.*, 2008; Pultz *et al.*, 2005; Wagner *et al.*, 2008). In addition to recognizing patterns of microbial extracellular surface proteins that may indicate potential pathogens, TLRs and NLRs also help promote intestinal homeostasis by regulating the development of gut immunity (Chen and Nunez, 2009). Fragments of peptidoglycan from non-pathogenic species in the gut can activate a NLR called Nod1. This activation leads to both the production of antimicrobial compounds against pathogens, but also triggers recruitment of immune cells by chemokine production (Chen and Nunez, 2009). Mice deficient in Nod1 have a resident microbiota that is profoundly altered in composition (Bouskra *et al.*, 2008). Disruptions in this resistance to colonization can result both in increased risk of susceptibility to pathogens, and alter beneficial microbe–microbe interactions that help regulate metabolism or innate immune response.

The microbiome is a term used to describe the total genetic content of the host and the resident bacteria living on and within the host. It has been estimated that the microbiome contains 100 times as many genes as the human genome (Hooper and Gordon, 2001). This provides an extraordinary array of abilities when the collective genome of microbes and humans are viewed as one functional unit – the 'super-organism' concept. Many of these genes have remained uncharacterized and some may encode for a variety of abilities, including antibiotic resistance. Given the rise of antibiotic resistance amongst numerous pathogens like methicillin-resistant *Staphylococcus aureus* (Pantosti and Venditti, 2009), and multidrug resistant *Mycobacterium tuberculosis* (Lalloo *et al.*, 2006), there is concern amongst the medical community for this problem to get worse in the future. One area that has only recently been investigated is the role of the microbiota as a reservoir for novel antibiotic resistance mechanisms (Sommer *et al.*, 2009). Many of these antibiotic resistance genes are encoded on mobile genetic elements. Given that microbiome contains so many more genes than the human genome, and the frequency of lateral gene transfer, the opportunities for both indigenous bacteria and pathogens to acquire new resistance abilities may be increasing. If the concept of a super-organism is in fact true, then analysis of the community structure could grant insights into possible functions of the community.

Another area of work has been focused on probiotics, the use of bacteria (or bacterial products) to benefit human health by indirectly controlling pathogen growth (Britton and Versalovic, 2008). The effectiveness probiotics is still under active investigation, but probiotics have been shown to be beneficial to a variety of disorders of the GI tract. Using bacterial species control the growth of pathogens allows for therapeutic manipulation of the microbiota in ways beyond the use of traditional antibiotics (Parkes *et al.*, 2008, 2010).

Experimental methods for studying the indigenous microbiota

Early studies of the gut microbiota first began with analysis of bacteria isolated from human faecal samples. Although these initial surveys were based only on differences in colony and cellular morphology, large numbers of species were isolated in pure culture and subsequently characterized (Zubrzycki and Spaulding, 1962; Moore and Holdeman, 1974; Holdeman *et al.*, 1976; Bornside, 1978).

Microscopic evaluation of the faecal community showed many additional species that were present, but not retrieved by the available cultivation techniques. This observation was verified by recent findings using culture-independent techniques, including DNA sequencing, which illustrate that cultivatable species are only a small fraction of the total bacterial community in the gut. Current advances in both culture-based and culture-independent methods to examine the human microbiome will be discussed here.

Culture-dependent techniques have improved with greater understanding of the specific conditions needed for many of bacterial residents of the gut to grow. Often it requires manipulation of both gas levels (especially oxygen and nitrogen), but also the introduction of specific metabolites or nutrients that are often absent in conventional media. This has helped obtain a better understanding of the diversity in complex communities. Despite even the best techniques aimed at growing difficult-to-culture species, any manipulation of culture techniques will undoubtedly introduce selection, and therefore bias the analysis to a specific subpopulation suited for the particular growth conditions used (Torsvik *et al.*, 1998). Thus, in part because of concerns about missing the majority of bacterial species in a sample that are unable to be cultured, investigations into new techniques to classify and examine the microbiota using culture-independent techniques began, primarily using DNA sequencing technology.

It is important to remember some potential complications with the classification of species using sequencing data. When sequences are generated, they can be classified into genus and species based reference sequences in online databases. This classification is based only on the sequences of organisms that have been cultured, so precise classification to the genus and species level is only possible for organisms that have been cultured. Sequences from organisms that have yet to be cultured will often only provide identification to the order or family level, and with more specific genus or species classification impossible. However, despite some limitations, the use of high-throughput sequencing has greatly expanded the ability to analyse large and complex samples with high resolution.

The massive data sets generated by high-throughput sequencing necessitated the development of new tools to manage and analyse information. Because sequence-based identification of species or strains is not equivalent to conventional classifications using culture-dependent techniques, sequence fragments called operational taxonomical units (OTUs) are used as a proxy for diversity. Just as with species classification, the precise description of what defines an OTU is debatable, and is subject to manipulation depending on the type of analysis or level of resolution used (Dethlefsen *et al.*, 2008; Schloss and Handelsman, 2008). Regardless of the intricacies of what defines a species (or OTU), high-throughput techniques have lowered the significant costs, and improved the poor resolution and sensitivity of previous methods.

A major advantage of examining the community dynamics by culture-independent techniques is that it allows for a 'big picture' view of the community that is inclusive to nearly every species forming that community. Particularly in environmental samples, where the uncultured portion of the bacterial population can reach 99% of all species present, it is critical to use culture-independent tools to determine community structure. While a higher percentage of gut bacteria have been isolated by culture compared with environmental samples, the proportion of bacteria in the distal gut for which cultivars have not been isolated is still 60–80% (Riesenfeld *et al.*, 2004; Walk and Young, 2008).

One technique used frequently for analysis of the microbiota is DNA sequencing of the 16S rRNA-encoding gene (Pace 1997, 1985). This is highly conserved in Bacteria and Archaea and is an effective target for sequencing and subsequent phylogenetic analysis because it is a large gene with conserved segments under stabilizing selection, but also regions of high variability. Many early surveys of microbial diversity involved the construction of 16S rRNA-encoding gene libraries using the following procedure (Giovannoni *et al.*, 1990; Antonopoulos *et al.*, 2009; Chang *et al.*, 2008; Li *et al.*, 2009):

1. collection and extraction of total DNA from a bacterial community;
2. PCR using primers targeting specific regions (or the entire 16S region) to amplify large numbers of amplicons that represent the population as a whole;
3. cloning the 16S fragments into a plasmid, then transformation of the plasmid + 16S insert into competent bacterial cells (such as *E. coli*);
4. overnight growth using antibiotic markers for selection;
5. selection of individual colonies carrying the plasmid + 16S insert;
6. overnight growth;
7. PCR amplification of 16S inserts using plasmid specific primers; and
8. submission for sequencing.

The greater number of individual colonies that are selected in step 5 leads to greater resolution of the community structure.

Culture-independent techniques were initially developed using historically modern laboratory techniques, such as the use of metagenomic clone libraries followed by DNA sequencing, screening for particular sequences by PCR or hybridization, or screening for expression phenotypes (Riesenfeld *et al.*, 2004). Recent advances in massively parallel DNA sequencing has allowed for an exponential increase in the number of sequence reads per sample (Sogin *et al.*, 2006; Huse *et al.*, 2008; McLellan *et al.*, 2009). By using fibre-optic slides with 1.6 million wells, 25 million bases or more can be sequenced in one run with 454 pyrosequencing technology (Margulies *et al.*, 2005). This allows for hundreds of thousands of sequence reads to be performed per sample, lowering the threshold of detection and revealing nearly all OTUs in a given sample.

While 16S rRNA gene analysis has been a tremendously important tool for community analysis, it does have limitations and biases that have caused other researchers to look for additional tools for study. One technique using DNA directly from mixed samples is called metagenomics (Handelsman *et al.*, 1998), a new field aimed at analysis of these large-scale data sets, which can cover nearly all sequence information available from a given sample (Riesenfeld *et al.*, 2004). DNA is isolated, sheared to smaller, more manageable sizes, ligated into an expression vector, and then transformed into a recipient cell (usually *E. coli*). Transformants are then screened using functional assays to isolate the cells carrying the insert of interest. The inserts can also be sequenced to determine nucleotide diversity. Several studies have used metagenomics for analysis of complex community structures (Schloss and Handelsman, 2008; Tuohy *et al.*, 2010), including marine microbial communities (Bohannon, 2007; Kennedy *et al.*, 2008), viral interactions (Ng *et al.*, 2009; Schoenfeld *et al.*, 2010), and even comparisons of the woolly mammoth genome to modern relatives (Poinar *et al.*, 2006). Many of the tools used for analysis of the microbiome have been used for other communities as well.

Results from culture-independent studies

In a study examining 16S rRNA segments, isolated from twenty distinct skin sites in ten healthy adult humans, physiologically comparable sites were shown to have similar bacterial communities in all volunteers, when stratified by microenvironment (sebaceous, moist, dry; Grice *et al.*, 2009). Another study examining 27 sites from seven to nine healthy adult humans also showed that while variability of sequences was high within individuals, the community composition was determined primarily by body habitat (Costello *et al.*, 2009). The 27 sampling sites from this study are illustrated in Fig. 11.1, along with the distribution of the microbial community at each site as identified by 16S rRNA sequencing. The idea of a core microbiome is bolstered by evidence that while human microbiota at various sites appears to be personalized, there are some patterns and similarities across multiple individuals at the same body habitat or time (Costello *et al.*, 2009).

Several attempts have been made to identify a 'core' human microbiome, a set of shared bacterial species present in all (or nearly all) individuals at the same sampling site. The gut contains the densest collection of microbes known, with about 10–100 trillion organisms (Turnbaugh *et al.*, 2009). Despite this species richness, and the confounding issues of variability, a few recent studies

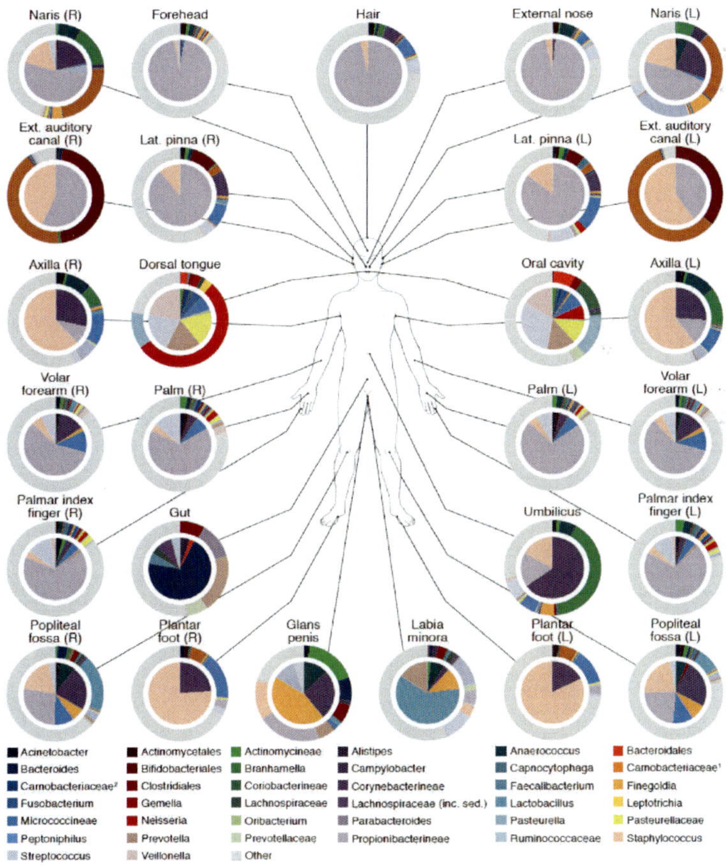

Figure 11.1 A comprehensive survey showing the variation in both core and peripheral taxa at 27 locations on the human body was performed by Costello and colleagues (2009). Here they show the 40 most abundant bacterial lineages detected in human volunteers at the sites indicated. The central pie shapes represent core taxa found in all subjects studied at that particular site, while the outer rings correspond to peripheral taxa that were not found in all subjects. Variability in community membership is seen from site to site, but also between corresponding sites on the right (R) and left (L) sides of the body. Reprinted with permission from Costello *et al.* (2009).

have attempted to identify a core microbiome. Turnbaugh and Gordon addressed this question by comparing the microbiome structure of adult female twins (Turnbaugh and Gordon, 2009). A total of 154 adult females were included in the study (31 monozygotic and 23 dizygotic young adult female twin pairs and the majority of their mothers). While it is probable the microbiota varies at different locations in the GI tract, the exact differences are not well understood. Stool was collected from these volunteers because it is considered to be a representative sample of the total microbial diversity of the gut (Eckburg *et al.*, 2005). While there was some co-variation amongst both monozygotic and dizygotic twin pairs, no single bacterial species was found in all 154 individuals. The co-variance in twin pairs may be evidence that early environmental exposures are important for determining the structure of the gut community as adults (Turnbaugh and Gordon, 2009).

There does appear to be some similarity in gut microbiota between family members versus unrelated individuals, but there is a high degree of variation between unrelated individuals. Several reasons for the variation between unrelated individuals may include differences in diet, lifestyle, and genetic background that may affect both immune function and metabolism. The high variability between individuals makes the identification of a core collection of species difficult. Rather than attempting to identify a core of species found in all individuals, Turnbaugh and Gordon suggested a core microbiome might only exist on a gene level, rather than a species level (Turnbaugh and Gordon, 2009); however, this topic is still open for debate.

Given the high degree of variability in human microbiota, and the dense collection of total bacterial numbers in the distal gut, it is surprising that the diversity of bacteria in this region is largely limited to only two phyla, the Bacteroidetes and the Firmicutes (Ley *et al.*, 2008; Mahowald *et al.*, 2009). Humans, as well as our other mammalian relatives, have gut microbiota dominated by these two phyla. Bacterial diversity increases from mammalian carnivores to omnivores to herbivores. The gut microbiota of humans is similar to other mammalian omnivores and is influenced by diet and phylogeny (Ley *et al.*, 2008).

While a recent study examining the microbiota of 60 different mammals, including humans, found members of 17 different phyla, the most predominant members were, again, the Bacteroidetes and the Firmicutes (Ley *et al.*, 2008). This is evidence of a great overall species diversity in the microbiota, but with an ecological environment that selects for specific phyla. The reasons for this selection are unclear, but may be related to an advantage in some species to better utilize available nutrients from the host's diet, including metabolic intermediates during digestion. The restriction of the majority of the species in the microbiota in the gut to only the Bacteroidetes and the Firmicutes poses to a questions regarding how the ecology of the gut shapes the populations living there (or perhaps how the microbiota shapes gut ecology). It also illustrates some problems faced when sampling is done to determine the makeup of a complex bacterial community.

Because the Bacteroidetes and the Firmicutes dominate the microbiota of mammals, and perhaps other non-mammalian species, detection of other, rare phylotypes requires looking at large numbers of sequences. Recent developments in sequencing technology have been leveraged to allow deeper sampling of the community to reveal these rare phylotypes. This allows for a fine-scale analysis of nearly all the members of a complex bacterial community by sampling thousands or millions of sequences. By increasing the sample size exponentially compared with previous sequencing methods, rare phylotypes can be seen amongst the background of a few predominant phyla. Larger scale sequencing techniques also allows finer measurements of the composition of bacterial communities (% detection), so tracking changes in community structure during perturbation is more accurate. Detection and inclusion of rare phylotypes into a bacterial 'census' is important not only for accurate phylogenetic analysis, but also because some of these rare species may play important roles in the microbial community structure and function.

Another recent example of the application of 16S rRNA gene sequencing to complex communities is seen in a metagenomic study of oral microbiota using an 82 bp segment of the V5 region of the 16S rRNA gene. This analysis resolved the major common phyla in the oral cavity (Firmicutes, Proteobacteria, Actinobacteria, Fusobacteria, Bacteriodetes, and TM7). While this technique generated vast amounts of sequence data (over 1.3 million sequences), this study also demonstrated the trade-off between sequence reads and OTU discovery. The detection of rare OTUs in complex samples requires an increasingly large amount of sequencing. Because of this, the authors estimate it would take another 30,000 sequence reads to find just one additional OTU that was not seen in the 1.3 million reads already performed (Lazarevic *et al.*, 2009).

Antibiotic-induced alterations of microbial communities

Much work has also been aimed at elucidating effect of the microbiota on human health. The largest assemblage of bacterial species of any area in the human body is in the colon (Ley *et al.*, 2006). A stable intestinal microbiota is important for metabolism, immune response regulation, and nutrition (Dethlefsen *et al.*, 2008; Li *et al.*, 2008). Disruptions that alter the microbial community structure can happen from a variety of stresses, including changes in diet and antibiotic use (Dethlefsen *et al.*, 2007).

The microbiota of the gut also plays important roles in the utilization of energy from ingested food. Comparisons of the microbiota between lean and obese mice suggest the differences observed in the two communities may be due to differential levels of calorie harvest and energy storage efficiency (Li *et al.*, 2008; Turnbaugh *et al.*, 2008). A single uncultured clade of Mollicutes (a class within the Firmicutes) was found at increased levels of mice with dietary-induced obesity. The microbiota from diet-induced obese mice was transplanted into lean, germ-free mice. These mice showed greater fat deposition when compared with lean, germ-free mice with microbiota transplanted from lean donors (Turnbaugh *et al.*, 2008).

Both short and long-term changes to the gut microbiota can result from antibiotic use, with acute symptoms ranging from a mild, self-limiting diarrhoea to a severe life-threatening pseudomembranous colitis (Wilcox, 2003). The long-term effects of antibiotic use are less clear. Some research suggests that alterations of the microbiota early in life can lead to the development asthma and other chronic diseases related to immune system dysfunction (Noverr and Huffnagle, 2005; Marra *et al.*, 2006). Antibiotic disruption of stable communities causes dramatic decreases in both the overall numbers of bacteria in the gut, and also the overall diversity of species. This disruption eliminates colonization resistance and allows an opportunity for pathogens to establish growth in the gut. Murine models of infection by *Salmonella enterica* serovar Typhimurium following streptomycin treatment showed substantial changes in the microbial composition of the small intestine, particularly amongst members of the phylum Firmicutes (Garner *et al.*, 2009).

Because the gut community between different individuals can vary greatly, studying the effect of antibiotics and the recovery of the microbiota after perturbation is difficult unless variability is controlled. One way this has been accomplished is through the use of inbred mouse models to provide a uniform genotype and diet to limit as much variability as possible. These models then help remove the inherent environmental and genetic variation that influences the population of the microbiota (Dethlefsen *et al.*, 2006). Measurements of stable baseline communities, and the effects of severe antibiotic treatment on the microbiota, can then be measured more accurately. This type of analysis is based on two assumptions that, until recently, had not been addressed in detail: (1) genetically identical mice have a similar and consistent microbiota, and (2) antibiotic use changes the microbiota in a consistent manner that is responsible for the changes seen in associated disease states (Dethlefsen *et al.*, 2006).

A clearer picture of the precise nature of community disruption and recovery following antibiotic treatment was seen in a recent study from our lab using antibiotic-treated C57BL/6 interleukin-10-deficient mice (Antonopoulos *et al.*, 2009). Fig. 11.2 shows the different groups of mice and the design of this experiment. Mice were treated with a cocktail of amoxicillin, metronidazole, and bismuth (AMB) for 10 days. The

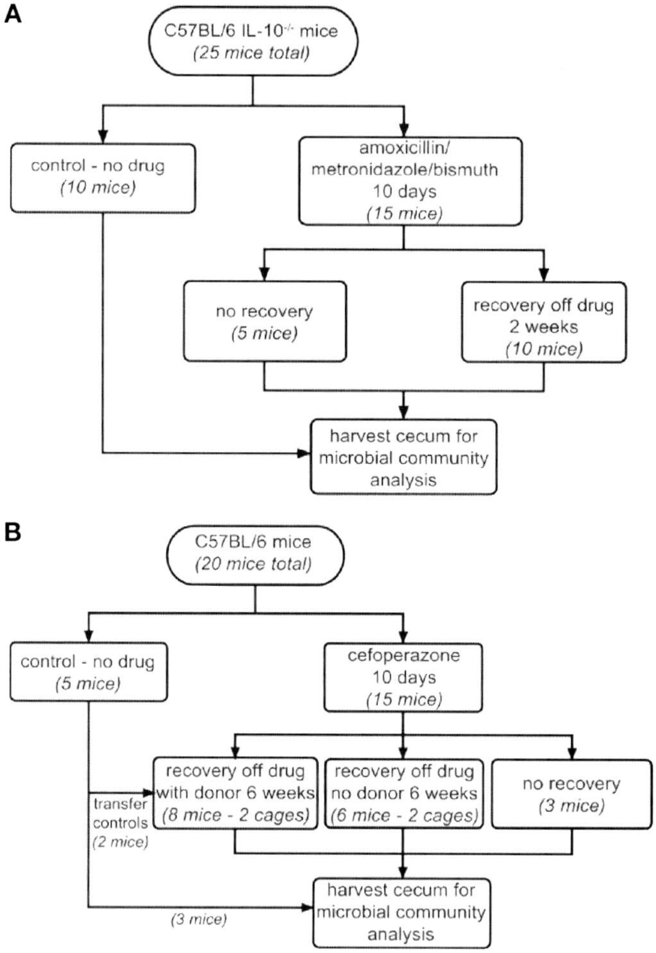

Figure 11.2 The schemata for mouse antibiotic administration used by Antonopoulos and colleagues. In (A), C57BL/6 IL IL-10$^{-/-}$ mice were given a cocktail of amoxicillin, metronidazole, and bismuth (AMB) in their chow for 10 days, and compared with mice on control chow. Mice given AMB were divided into two groups, one euthanized immediately following antibiotic administration, and another group allowed to recover for 14 days on non-medicated chow, before euthanasia. In (B), C57BL/6 mice were divided into two groups, a control that received sterile drinking water and a treatment group that received sterile drinking water with 0.5 mg/ml of cefoperazone for 10 days. Mice on antibiotic-treated water were further divided into three groups; the first group euthanized immediately, a second group allowed to recover for 6 weeks without antibiotics, and a third group also allowed to recover for 6 weeks without antibiotics, but also given an untreated control mouse to be co-housed during the recovery period. Community analysis was then performed on caeca-associated bacteria following euthanasia. Reprinted with permission from Antonopoulos *et al.* (2009).

AMB treated mice were then divided into two groups, five mice with no recovery period, and 10 mice given a 2-week recovery without antibiotics. Both groups were euthanized and caecal contents were harvested for community analysis. These were then compared with control mice given no antibiotics.

Deep sequencing analysis, using primers targeted to the V6 hypervariable region of the 16S rRNA gene, produced over one million sequence tags from 21 mice (nine controls, two antibiotic treated, and ten antibiotic-treated with the 2-week recovery period). The antibiotic treated mice showed major differences in the gut community structure. The microbiota of the control mice were dominated by the phyla Bacteroidetes and Firmicutes, with only 1% of sequences belonging to Proteobacteria. This community structure is in sharp contrast to antibiotic-treated mice, where 71% of sequence tags matched Proteobacteria, with two specific sequences within family Enterobacteriaceae accounting for 93% of all of the Proteobacteria tags (Antonopoulos *et al.*, 2009). This provides evidence that both the overall community structure and the diversity of sequences are dramatically altered with antibiotic perturbation.

Mice given a 2-week period of recovery following antibiotic treatment showed a return of higher levels of Firmicutes (70%) and Bacteroidetes (22%). The Proteobacteria, which had surged to 71% in the antibiotic-treated group, had fallen to about 5% in this population (Antonopoulos *et al.*, 2009). The recovery of the Firmicutes and Bacteroidetes groups was consistent in the antibiotic-treated mice. This finding may be limited to the particular antibiotic cocktail used for treatment.

In a concurrent experiment, when wild-type C57BL/6 mice were given the broad-spectrum antibiotic cefoperazone, significant and long-term effects were seen. Fifteen mice were given cefoperazone in their drinking water (0.5 mg/ml) for 10 days. The mice were then separated into three groups. The microbiota of three mice immediately euthanized showed a baseline community structure after cefoperazone treatment. The other 12 mice were allowed to recover for 6 weeks. Six of the recovery mice were isolated and housed three per cage during the recovery period. The other six mice were also housed three per cage, but included an additional mouse from the untreated control group (Antonopoulos *et al.*, 2009). This last group, co-housed with a control mouse, was included to examine the effects of microbiota recovery with the potential for corprophagy to reintroduce gut species from the control mouse to the co-housed recovery mice.

The cefoperazone-treated, wild type C56BL/6 mice without a recovery period had an average of 4300-fold decrease in bacterial load compared with the untreated controls. In both of the 6-week recovery groups (with and without the co-housed control mouse), the bacterial load returned to levels comparable with the control animals, but the composition of their microbiota was very different. The mice co-housed with donor control mice showed a recovery consistent with the untreated control animals (Antonopoulos *et al.*, 2009). The total bacterial load and species diversity were similar after a 6-week antibiotic free recovery period. However, as expected, the mice housed in isolated cages during the recovery period had a lower level of phylotype richness than both the untreated controls and the recovery group with a donor mouse (Antonopoulos *et al.*, 2009).

It should be noted however, that antibiotic disruption of the microbiota is not always necessary for a pathogen to establish infection. Disruption probably aids in the process, and possibly creates a niche for additional obligate or opportunistic pathogens to establish infection, but some pathogens do not require prior disruption to cause disease. Previous work in our lab using 16S rDNA analysis has also shown that the indigenous microbiota in wild-type mice can be disrupted by colonization of the caecal mucosa with *Helicobacter hepaticus* (Kuehl *et al.*, 2005). This is in mice that were not previously treated to disrupt the microbiota, and shows that disruptions are not always mediated by antibiotic use, but also can be caused by pathogens able to overcome colonization resistance and establish infection.

An emerging medical issue relating to disruptions of the microbiota is the rise of *Clostridium difficile*-associated disease (CDAD). Infection with *C. difficile*, the major cause of nosocomial antibiotic-associated diarrhoea and colitis, is increasing in frequency in hospital settings. *C. difficile* is a Gram-positive, spore-forming bacterial rod that causes disease in the gut. It encodes a variety of toxins, most predominantly exotoxins A and B, which produce acute inflammation, fluid secretion, and necrosis of epithelial cells in the gastrointestinal tract (Chen *et al.*, 2008). Shortly after *C. difficile* was identified as the etiologic agent of antibiotic-associated diarrhoea, a variety of hypotheses sought to explain the phenomenon. One regarded disruptions of the normal gut microbiota leading to a loss of colonization resistance and providing an opportunity for an overgrowth of *C. difficile* spores or vegetative cells dormant in the gut before antibiotic therapies (Vollaard and Clasener, 1994). This idea has been supported by more recent observations that specific antibiotics that adversely affect the gut microflora allow a new niche for growth of *C. difficile* (Wilcox, 2003). In immunocompetent individuals, *C. difficile* overgrowth generally does not occur without antibiotic treatment to reduce the resident microbiota. Also, individuals with HIV or other diseases that reduce immune function are more susceptible to *C. difficile* infection (Makarewicz, 2008).

One of the historical difficulties in studying CDAD was the lack of a functional animal model. Recently, a model of CDAD was developed in mice by using a strong cocktail of five antibiotics to obliterate the resident microbiota (Chen *et al.*, 2008). Kanamycin, gentamicin, colistin, metronidazole and vancomycin were given to the mice for 3 days. The mice were then injected with a single dose of clindamycin, followed by intragastric challenge of *C. difficile*. A common treatment of CDAD in humans is cessation of the antibiotic administration that caused the CDAD, followed by treatment using additional antibiotics like vancomycin or metronidazole. For the development of this mouse model of CDAD, after the *C. difficile* exposure, the mice were given oral vancomycin to examine the effects of this therapeutic regimen on recovery and survival rates. The crucial next step in the development of disease similar to CDAD in humans was the re-inoculation of *C. difficile* into these mice to challenge them a second time (Chen *et al.*, 2008).

Interestingly, the mice that survived an initial episode of CDAD do not develop the disease again upon subsequent rechallenge. The mice developed an immune system-mediated protection to *C. difficile*. The proportion of mice that died from severe CDAD was proportional to the dose of *C. difficile* given, so the dose can be adjusted to make the mortality rate in mice similar to what is seen in humans (1–7%; Chen *et al.*, 2008). These mice also had similar pathology to humans, including the submucosal oedema and marked bowel wall thickening. An accurate model of CDAD is critical for further study of this disease and an understanding of how *C. difficile* takes advantage of microbiota perturbation.

We have previously shown that decreased diversity of the faecal microbiome is associated with *C. difficile* infection in humans (Chang *et al.*, 2008). A total of ten patients (seven with CDAD and three controls) submitted stool samples for DNA extraction and 16S rRNA encoding gene sequence analysis. Four of the seven CDAD patients had only an initial occurrence of CDAD, while the other three had recurrent CDAD. In the control patients and those with only an initial case of CDAD, analysis of the bacterial communities at the phylum level showed the Bacteroidetes and Firmicutes to be the predominant phyla. Patients with recurrent CDAD had more variation and deviation from the controls, and also had lower species richness compared with initial CDAD patients and controls (Chang *et al.*, 2008). The lack of species diversity in recurrent CDAD may indicate a failure of the community to re-establish the baseline community structure that existed prior to disruption. The loss of colonization resistance has been proposed to be one of the reasons why CDAD can recur in some patients (Maroo and Lamont, 2006).

The microbiota as a reservoir for antibiotic resistance

Antibiotic resistance genes have been found with increasing frequency in the past few decades, and resistance to multiple antibiotics is a continuing problem for public health. Many resistance genes are found on mobile genetic elements (plasmids, pathogenicity islands, or bacteriophages), and can be horizontally transferred between species in close physical association. Given the number of genes in the collective microbiome of humans, it is not surprising that many species carry antibiotic resistance genes in their genomes. This may provide an opportunity for horizontal acquisition of antibiotic resistance. This phenomenon has not been extensively studied, but there is some evidence for exchange of antibiotic resistance genes (particularly the erythromycin resistance gene, *ermB* (Shoemaker *et al.*, 2001) in distantly related *Bacteriodes* species, and the transfer of the vancomycin resistance gene, *vanA*, from *Enterococcus faecalis* to *Staphylococcus aureus* (Noble *et al.*, 1992; Chang *et al.*, 2003).

A recent study by Sommer and colleagues (Sommer *et al.*, 2009), aimed to integrate metagenomics and functional characterization of the microbiota in humans and explore possible reservoirs for antibiotic resistance. DNA fragments from the microbiota of healthy patients who had not been on antibiotics for at least one year were cloned into an expression vector and transformed into a recipient *E. coli* strain. After screening using a panel of 13 antibiotics, inserts from resistant *E. coli* colonies were sequenced. A total of 95 unique inserts were found that confer functional resistance to antibiotics (Sommer *et al.*, 2009). Only 22% of these inserts had high homology (90% amino acid identity) to previously known genes. The others averaged only 65% amino acid identity to known antibiotic resistance genes. When comparing the amino acid sequences to those from known pathogens, the commensal species had less than 55% sequence homology. It is unclear exactly why there is not a higher sequence similarity between the functional isolated antibiotic genes from commensal species and those from known pathogens. This may indicate a barrier to horizontal transfer of these genes by an unknown mechanism, or be evidence for a lack of opportunity for horizontal transfer. Given the high frequency of antibiotic resistance gene diversity in bacteria isolated from healthy adults, it is surprising that more transfer is not occurring. Perhaps the commensal species harbouring these antibiotic resistance cassettes are actually pathogens that are dormant or in disguise (Sommer *et al.*, 2009). Until the mechanisms for transfer of antibiotic resistance abilities is more clearly understood, this phenomenon will continue to be of great interest as a potential source for further resistance in the future.

Indigenous microbiota and other disease states

While this chapter has focused largely on the interactions of the microbiota to specific pathogens, it is important to also consider the interactions of the indigenous bacteria to the host. This is a field of research that is less clear, but may be important for human health. When the population of indigenous species in the community has negative interactions with the host (called dysbiosis), conditions such as inflammatory bowel disease (IBD), bacterial vaginosis, and other medical issues may occur (Salzman and Bevins, 2008). The precise mechanisms for dysbiosis are unknown, but may include a variety of causes such as ineffective immune response relating to microbe recognition, barrier function, or intracellular communication between commensal species. There is evidence supporting both improper bacterial and improper host-mediated responses as the cause for dysbiosis triggering IBD (Salzman and Bevins, 2008).

Crohn's disease (CD) is a type of IBD that is characterized by severe chronic diarrhoea, often accompanied by blood or mucus loss in the stool, malaise, abdominal pain, weight loss, and other symptoms (Panes *et al.*, 2007). Two different bacterial species have been implicated in CD, although in both cases the evidence is sometimes conflicting. A specific intestinal pathogen called adherent-invasive *E. coli* (AIEC) has been found restricted to CD patients (Rolhion and Darfeuille-Michaud, 2007). In CD patients, AIEC is highly associated with the mucosa of the ileum (36.4% vs. 6% in non-CD control patients). The inflammation caused by mucosally associated AIEC may be a trigger for the development of CD (Rolhion and Darfeuille-Michaud, 2007).

The other pathogen implicated in CD is *Mycobacterium avium paratuberculosis* (MAP), which has been isolated from blood, intestines, and breast milk in some patients with CD, but was not found in elevated levels of other CD patients (Salzman and Bevins, 2008). MAP causes Johne's disease in cattle, leading to inflammation of the ileum similar to what is seen in CD. However, only a correlation between the presence of specific pathogens like MAP or AIEC to Crohn's disease has been seen, not a direct causative link, so this is an area of research that is still ongoing (Salzman and Bevins, 2008) and it should be stressed that the dynamics of whole communities relating to CD may be more important, rather than a specific pathogen.

Improper host-mediated responses triggering dysbiosis may also be responsible for disease states like CD (Packey and Sartor, 2008; Sartor and Muehlbauer, 2007). Instead of a pathogen triggering the inflammation associated with IBD, a shift in the normal healthy microbiota to one that triggers a greater inflammatory response from the host may be responsible for dysbiosis

(Salzman and Bevins, 2008). As we have outlined in this chapter, there are a variety of stresses that can disrupt the community of bacteria in the gut, without including pathogenic infection. Often disruptions are only transient and the microbiota will return to a normal composition following the removal of stresses. However, dysbiosis can occur when the healthy microbiota is not re-established. Studies in CD patients have shown lower diversity of species in the intestinal microbiota, including decreased levels of *Bacteriodes* and *Lactobacillus* species (Swidsinski *et al.*, 2005). The precise level of disruption necessary to trigger sufficient inflammation for the onset of IBD is unknown, and susceptibility may vary between individuals depending on a number of factors, including diet and genetics.

The NIH human microbiome project

The formation of a collaborative, multinational research team funded by the National Institutes of Health in 2007 established the NIH Human Microbiome Project. The goals of the initiative are to more fully characterize the human microbiome by studying the microbiota from various body sites in 250 'normal' individuals, examine potential associations in health by studying changes in the microbiome, and to standardize data collection and technological approaches (Peterson *et al.*, 2009).

This is a five-year project with funding of US$150 million. Initial aims include large-scale sequencing of 500 new bacterial genomes divided amongst the GI tract, oral cavity, skin, airways, urogenital tract, and blood. Also, the development of protocols to sample body sites of 'normal' volunteers, and performing 16S rRNA sequencing on those samples are key areas of interest. Volunteers were specifically selected to represent a reasonably diverse sample of the population of the USA. One feature used to benchmark a data set is the use of a 'mock' community as a reference point where other data sets can be compared with help verify the accuracy of sequence information (Hsiao and Fraser-Liggett, 2009; Peterson *et al.*, 2009).

This project includes a number of ethical implications, which have been taken into consideration during the study design. These include privacy issues relating the potential identification of unique individuals based on their microbiome, potential stigmatization due to particular characteristics of the microbiome, and challenges dealing with informed consent. Intellectual property rights and other legal issues are also of concern (McGuire *et al.*, 2008; Peterson *et al.*, 2009).

In addition to the Human Microbiome Project, a new group was formed in 2008 aimed at international cooperation with microbiome research. The International Human Microbiome Consortium contains ten member countries and is aimed a coordinating microbiome work around the globe in an effort to ensure rapid release of data and eliminate major duplication of research efforts from multiple projects (Qin *et al.*, 2010).

Summary

The microbiota of humans is an extraordinarily complex and dynamic community that is important to maintaining good health and protection from a variety of pathogens. Disruptions of this community by antibiotics or other stresses can allow an opportunity for pathogens to establish infection. In most cases, following the cessation of antibiotic therapy, the microbiota will recover to a stable community that existed prior to disruption. However, in some cases, particularly when a strong cocktail of antibiotics are used to cause major or prolonged disruptions, the microbiota may not fully re-establish the community that existed before disruption, leading to chronic gastrointestinal problems.

The molecular toolbox for researchers studying the microbiota has increased tremendously in recent years. The use of improved culture techniques and culture-independent methods have allowed for a much cheaper, faster, and more accurate generation of data. Nearly all members of a complex community can now be detected, allowing a big picture view of how community dynamics can change under a variety of conditions.

Metagenomics and other tools to study communities have been used in a variety of systems, including a collaborative effort to investigate the

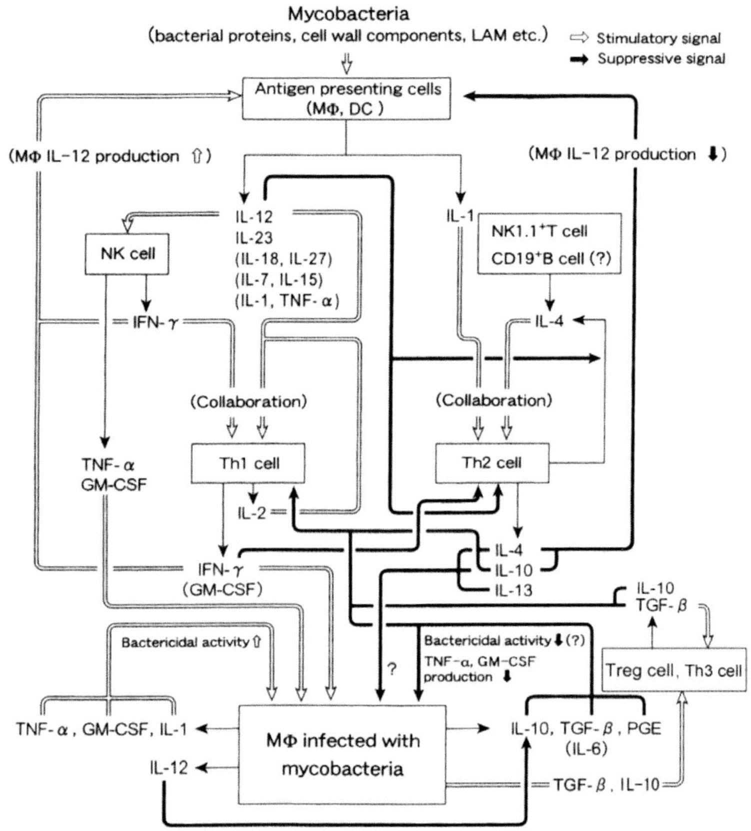

Figure 12.1 Cytokine network in hosts with mycobacterial infection. In this cytokine network, proinflammatory cytokines such as IL-12 and IL-18, which are required for induction/activation of Th1 cells, and Th1 and Th1-like cytokines, such as IFN-γ, IL-2, TNF-α and GM-CSF, play crucial roles in the expression of host resistance against mycobacterial infections. In addition, immunosuppressive/macrophage-deactivating cytokines or humoral factors, such as IL-4, IL-10, TGF-β and prostaglandin E (PGE), which are produced by Th2 cells, Treg cells, Th3 cells, and macrophages (MΦs), appear to play important roles in the establishment of immunodeficiency frequently encountered in persistent and advanced infection with mycobacterial pathogens including MTB.

pathogenic mycobacteria inhibits an increase in the Ca^{2+}/calmodulin concentration, preventing phagosome–lysosome fusion (P–L fusion). Man-LAM also down-regulates mitogen-activated protein kinase (MAPK) pathways by activating Src-homology 2 domain-containing tyrosine phosphatase 1, resulting in the suppression of macrophage functions, such as the production of proinflammatory cytokines, that are required for the expression of innate immunity against mycobacterial infection (Tse *et al.*, 2002). The induction of these evasive mechanisms in MTB most probably involves the ability of the organism to adapt its responses to external signals. In this context, the MTB genome encodes some proteins, especially protein kinases, that act on and intervene in host cell pathways, thereby yielding an intracellular milieu suitable for bacterial survival and growth in host cells, including macrophages. Our efforts in targeting mycobacterial signalling proteins are greatly encouraged by the recent findings of cross-talk between mycobacterial and host macrophage signalling systems enabling successful infection by mycobacterial pathogens. The identification of these proteins as targets in mycobacteria, especially MTB, may lead to the development of new antituberculous drugs. This section deals with potential targets in mycobacterial pathogens related to bacterial and host cellular signalling.

Effects of mycobacterial components on macrophage signalling

TLR-mediated signalling pathways

Toll-like receptors (TLRs) generally serve as pattern-recognition receptors (PRRs) to recognize a wide range of pathogen-associated molecular patterns (PAMPs) including lipids, lipoproteins, proteins, glycans and nucleic acids. The TLRs play a central role in initiating innate immune responses (Kawai and Akira 2007, 2009). The intracellular signalling pathways downstream of TLRs, especially TLR4 and TLR2, are indicated in Fig. 12.2. In particular, TLR2 is a PRR that is activated by microbial components including lipoproteins, lipoteichoic acids and peptidoglycan as shown in Fig. 12.3 (Sutterwala and Mosser, 1999). Lipoglycans, such as lipoarabinomannan (LAM) and trehalose 6,6′-dimycolate (TDM), are strong TLR2 agonists produced by *Mycobacterium* and related bacteria. Interestingly, lipoglycan activity is directly determined by mannan chain length (Nigou *et al.*, 2008).

TDM is the most abundant lipid of the MTB cell wall. TDM signals are known to be transmitted via TLRs and myeloid differentiation factor 88 (MyD88) in macrophages and their downstream MAPK kinase (MAPKK) pathways (Geisel *et al.*, 2005). Bowdish *et al.* identified a class A scavenger receptor MARCO (macrophage receptor with collagenous structure) as a receptor for TDM (Bowdish *et al.*, 2009). They indicated that MARCO, which is expressed by peritoneal macrophages but not by bone marrow-derived macrophages, is utilized to tether TDM molecules to macrophage surface and to activate the TLR2 signalling pathway resulting in the activation of NF-κB and presumably TLR1/6 signalling causing the phosphorylation and activation of MAPK such as in the extracellular-regulated kinase 1/2 (ERK1/2) pathway (Bowdish *et al.*, 2009). MARCO-expressing macrophages secreted pro-inflammatory cytokines including TNF-α more effectively in response to TDM owing to cooperation between MARCO and TLR2/CD14 (TLR4) than did MARCO-deficient macrophages (Fig. 12.2). These findings imply that the scavenger receptor plays an essential role as a receptor for TDM. It has also been indicated that TDM activated macrophages and dendritic cells (DCs) via Syk-Car9-Bcl10-Malt1 pathway to induce specific innate activation events distinct from TLR-mediated signalling pathways and induced production of Th1 and Th17 in response to a MTB subunit vaccine and partial protection against MTB infection (Werninghaus *et al.*, 2009). In this case, the activation of antigen-presenting cells (APCs) by TDM requires the immunoreceptor tyrosine-based activation motif (ITAM)-bearing adaptor protein Fc receptor γ chain. Nevertheless, TDM has recently been reported to promote MTB survival in macrophages by blocking phagosomal acidification and P–L fusion. Depletion of TDM by delipidation of MTB and the depletion of *fbpA* gene participating in the production and transport of TDM in MTB decreases MTB survival in macrophages in connection with increases in the expression of macrophage surface markers including MHCII, CD1d, CD40, CD80, and CD86 (Kan-Sutton *et al.*, 2009). Notably, TDM was found to induce suppressor cytokine signalling (SOCS) which is implicated in the suppression of certain immune signalling pathways by mycobacteria in macrophages (Fig. 12.2).

As indicated in Fig. 12.2, with respect to the interaction of mycobacterial components with TLRs, the following findings are of particular interest. First, MyD88, a common adaptor involved in TLR, IL-1, and IL-18 receptor signalling, is essential for the control of acute MTB infections. Although TLR2, TLR4, and TLR9 have been implicated in the response to mycobacteria, disruption of the genes for these TLRs impairs only the long-term control of MTB infections. While MyD88-deficient macrophages and DCs were defective in cytokine production in response to mycobacterial stimulation, IL-1R1-deficient macrophages exhibited only reduced IL-12p40 secretion with unaffected levels of proinflammatory cytokine and nitric oxide (NO) production (Fremond *et al.*, 2007). However, an aerogenic MTB infection in IL-1R1-deficient mice resulted in rapid death of infected mice with increased bacterial growth at sites of infection and necrotic pneumonia, albeit with normal T-cell responses in the lungs. Thus, it is thought that an absence of IL-1R signals leads to a marked defect in the

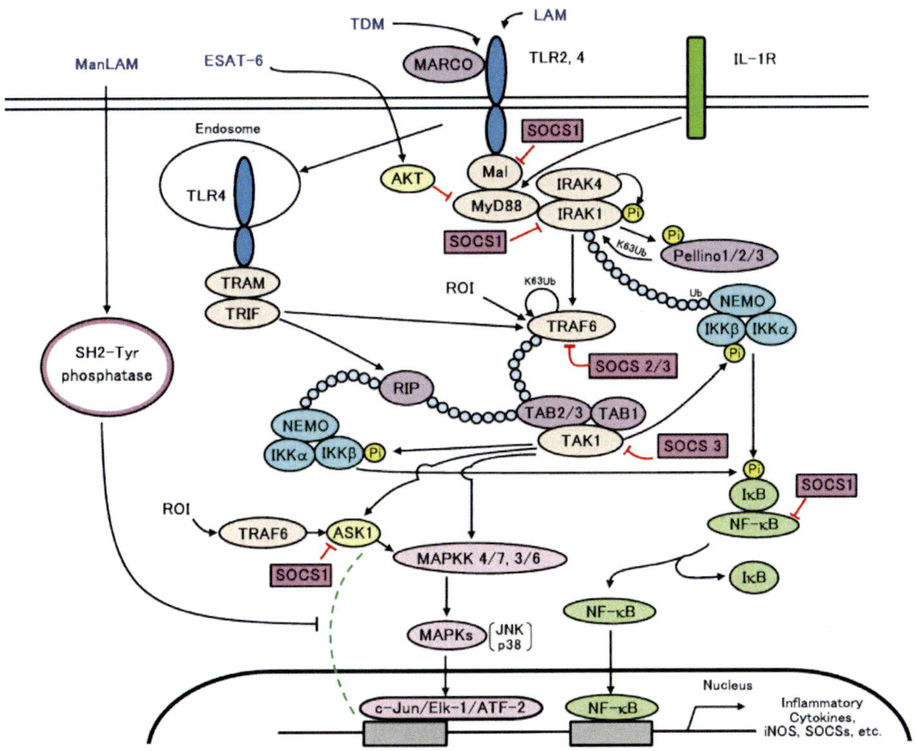

Figure 12.2 Signalling pathways triggered by TLRs, especially TLR2 and TLR4. When TLR2 and TLR4 engage bacterial ligands with PAMPs, such as LPS and PGN, on the cell surface, the adaptor proteins Mal and MyD88 are recruited to the TIR domains of the receptors. Then Mal and MyD88 recruit IRAKs and then TRAF6. After K63 polyubiquitination, TRAF6 activates TAB2/3 and subsequently TAK1. TAK1 thereafter activates an IKK complex consisting of IKKβ, IKKα, and NEMO (IKKγ). The IKK complex also receives an activating signal from a ubiquitinated IRAK1 through NEMO. The IKK complex phosphorylates IκB causing its degradation, allowing the activation and translocation of the transcription factor NF-κB into the nucleus. In addition, TAK1 simultaneously activates MAPKK4/7 and MAPKK3/6 causing the phosphorylation of MAPK (JNK and p38) and subsequently, activation of the transcription factors c-Jun, Elk-1, ATF-2, etc. Meanwhile, TLR4, which is trafficked to the endosomes, recruits the adaptors TRAM and TRIF. TRIF activates TRAF6 and subsequently TAK1, as described above. After the K63-polyubiquitination, RIP simultaneously transmits activating signals to the IKK complex through binding with NEMO. In the case of mycobacterial components with PAMPs, LAM, lipoproteins, and TDM bind to TLR2 and transmit stimulatory signals to TLR2's downstream pathways as described above. The following is worth noting: (1) the binding of TDM to TLR is facilitated by a TDM receptor MARCO; (2) TDM-mediated signalling causes the expression of SOCSs which down-regulate the activities of MyD88, IRAK1, TRAF6, NF-κB, and ASK1; (3) Man-LAM activates the SH2-containing tyrosine phosphatase 1, which subsequently inhibits the MAPK pathway; (4) ROIs, which are generated intracellularly owing to the activation of macrophages through TLR signalling pathways, subsequently activate TRAF6 and ASK1 allowing the inhibition of MAPKK; and (5) ESAT-6 activates AKT which subsequently inhibits MyD88's function. PAMPs, pathogen-associated molecular patterns; LPS, lipopolysaccharide; PGN, peptidoglycan; Mal, MyD88 adaptor-like; MyD88, myeloid differentiation primary response gene; TIR, Toll-IL-1 receptor; IRAK, IL-1-associated kinase; TRAF, TNFR-associated factor; TAB, TAK1-binding protein; TAK, TGF-β-activated kinase; IKK, IκB kinase; NEMO, NF-κB essential modulator; NF-κB, nuclear factor-κB; MAPKK, MAP kinase kinase; MAPK, mitogen-activated protein kinase; JNK, c-jun N-terminal kinase; ATF, activation transcriptional factor; TRAM, TRIF-related adapter molecule; TRIF, TIR-containing adapter inducing IFN-β; RIP, receptor interacting protein; LAM, lipoarabinomannan; TDM, trehalose dimycolate; MARCO, macrophage receptor with collagenous structure; SOCS, suppressor of cytokine signalling; ASK, apoptosis signal-regulating kinase; Man-LAM, mannosylated (mannose-capped) LAM; SH2, Src-homology 2; ROIs, reactive oxygen intermediates; ESAT-6, early secreted antigenic target protein 6; AKT, known as protein kinase B (PKB).

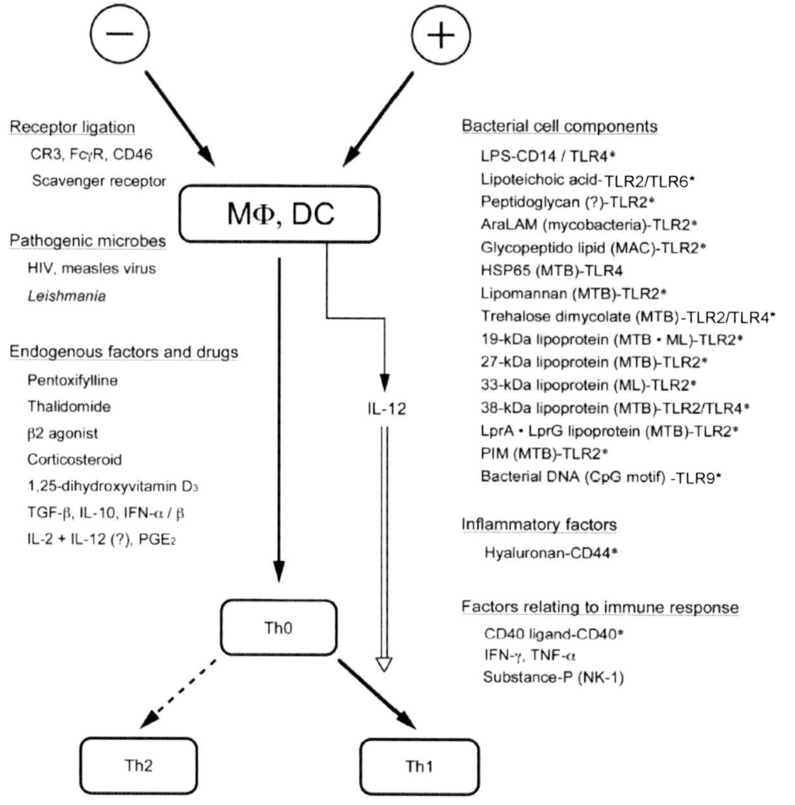

Figure 12.3 Various types of stimulatory factors able to induce the IL-12 production by macrophages (MΦs) and dendritic cells (DCs). *MΦs have been reported to produce IL-12 in response to the indicated receptors.

early control of MTB similar to that seen in the absence of MyD88. This means that IL-1 and the IL-1-induced innate response account for most of the MyD88-dependent response to an MTB infection.

Second, after an aerosol MTB infection, TLR2/4/9-deficient mice as well as wild-type mice expressed proinflammatory cytokines and expanded antigen-specific T-cell subsets which generate IFN-γ (Hölscher *et al.*, 2008). Moreover, TLR2/4/9-deficient mice expressed inducible NO synthase (iNOS) in response to IFN-γ at sites of infection. Although these events were also observed in MyD88-deficient mice, albeit in a delayed fashion, bacterial growth was not as effectively controlled in these MyD88-deficient mice as in the wild-type mice and, moreover, TLR2/4/9-deficient mice. In this context, the IFN-γ-dependent restriction of MTB replication is severely impaired in MyD88-deficient macrophages, but not in TLR2/4/9-deficient macrophages. These findings indicate that after MTB infection neither TLR2/4/9 nor MyD88 are required for the induction of adaptive T cell responses. Rather, MyD88, but not TLR2, TLR4 or TLR9, is critical for triggering macrophage effector mechanisms crucial to antimycobacterial defence. Notably, these signalling receptors are not required for the induction of adaptive T-cell responses in mice infected with MTB. In this context, BCG up-regulated the Notch homologue 1 (NOTCH1) signalling pathway, leading to the expression of suppressor of cytokine signalling 3 (SOCS3; Narayana *et al.*, 2008). The perturbation of TLR2 signalling resulted in a marked reduction in SOCS3 levels and expression of the NOTCH1 target gene *hes1*. In addition, the down-regulation of MyD88 function resulted in a significant decrease in SOCS3 expression, suggesting a role for the TLR2–MyD88 axis in

BCG-triggered signalling. In this context, cross-talk among members of the phosphoinositide 3-kinase and MAPKK cascades with NOTCH1 signalling plays a role in SOSC3 expression. It has also been reported that TDM induced expression of SOCSs (SOCS1 and SOCS3) by macrophages and inhibited IFN-γ-stimulated phosphorylation of signal transducers and activators of transcription (STAT; Imai, *et al.*, 2003). As indicated in Fig. 12.2, SOCSs inhibit TLR2/4-related signalling pathways, particularly the proteins MyD88 adaptor-like (Mal), IL-1-associated kinase 1 (IRAK1), TNFR-associated factor 6 (TRAF6), apoptosis signal-regulating kinase 1 (ASK1), and nuclear factor-κB (NF-κB), via a feedback mechanism (Yoshimura, *et al.*, 2007).

Third, the stimulation of macrophages with BCG via TLR2 and TLR4 elicits the activation of MAPKs, causing increased secretion of inflammatory cytokines including TNF-α and IL-6 and the generation of reactive oxygen intermediates (ROIs; Yuk *et al.*, 2009). In this signalling event, BCG-induced production of ROIs causes the activation of apoptosis signal-regulating kinase 1 (ASK1), a member of the MAPKKK family, upstream of c-jun-N-terminal kinase (JNK) and p38, but not ERK, via TNF receptor (TNFR)-associated factor 6 (TRAF6) signalling pathways. ASK1 activity is necessary for the effective control of intramacrophage mycobacterial survival, indicating an important role for the TLR 2/4–ROI-TRAF6-ASK1 axis in the innate immune response to mycobacteria. In this context, the following findings regarding ASK1 functions are of interest. That is, there is accumulating evidence that ROIs can be produced within cells specifically to serve as signalling molecules (Finkel, 2003). The ROI-dependent ASK1 axis is crucial for TLR4-mediated innate immunity (Matsuzawa *et al.*, 2005). In addition, Yang *et al.* (2008) recently indicated that $p47^{phox}$ (NAPDH oxidase subunits), ASK1 and p38 MAPK play essential roles in mediating the inductive effects of ROI-dependent inflammatory mediators via TLR2 during MTB infection. Stimulation with tuberculin purified protein derivative (PPD) resulted in the rapid activation of MAPKs and an early burst of ROI production in macrophages in a TLR2-dependent manner. The PPD-induced generation of ROIs leads to a robust activation of ASK1 upstream of p38 MAPK, but not JNK, via TLR2. Phosphorylation of cytosolic $p47^{phox}$ and activation of ASK1 are mutually dependent on PPD/TLR2-mediated signalling. Thus, $p47^{phox}$ is a critical mediator in TLR2/ROI-induced ASK1/p38 MAPK signalling. Notably, the generation of ROIs, phosphorylation of $p47^{phox}$ and p38 MAPK, and up-regulation of proinflammatory cytokine expression are regulated differentially in TB patients and healthy controls. In this context, it has been reported that the stimulation of macrophages with a 19 kDa lipoprotein of MTB elicited the translocation of TLR2 to lipid rafts and generation of ROIs, and consequently proinflammatory responses (Shin *et al.*, 2008). As this occurred, protein kinase Cζ (PKCζ) was recruited into detergent-resistant membrane fractions and played a role in respiratory bursts by binding with TLR2 in lipid rafts. These findings indicate that TLR2 trafficking and raft coalescence are essential for the initiation of mycobacterial lipoprotein-induced innate immune responses via TLR2 and ROI signalling.

Fourth, expression of early secreted antigenic target protein 6 (ESAT-6) by MTB is known to be associated with a weaker innate immune response to infection. As indicated in Figs. 12.2 and 12.4, in macrophages, ESAT-6 inhibits the activation of NF-κB and interferon-regulatory factors (IRFs) after TLR (TLR2, TLR7, TLR-9 but not TLR3) signalling via serine/threonine protein kinase B (PKB, also known as AKT)-mediated pathway (Pathak *et al.*, 2007). Direct binding of ESAT-6 to TLR2 does not affect the recruitment of MyD88 to TLR2, but activates AKT and consequently prevents interaction between MyD88 and the downstream kinase IL-1 receptor-associated kinase 4 (IRAK4), thereby blocking the phosphorylation and consequently the activation of IRAK4. This results in the abrogation of NF-κB activation, and in the down-regulation of IL-12p40, IL-6 and TNF-α production. The TLR-mediated inhibitory effect of ESAT-6 is dependent on its six C-terminal amino acid residues (Val90-Ala95). It has also been reported that ESAT-6 down-regulated phosphorylation of ERK1/2 in the nucleus of LPS-stimulated macrophages principally due to active dephosphorylation of

ERK1/2 by some tyrosine phosphatase (Ganguly *et al.*, 2007).

NOD-mediated signalling pathways

Nucleotide-binding oligomerization domain proteins (NODs) participate in the recognition of peptidoglycan-derived small molecules. NOD2 (CARD15) is important for host cell cytokine responses to MTB also via TLR2 signalling. The profile of NOD2-related host signalling pathways is indicated in Fig. 12.4 (Inohara *et al.*, 2005; Watanabe *et al.*, 2008; Ting *et al.*, 2010). With respect to the role of NOD2 in signalling pathways of macrophages in response to mycobacterial stimulation, the following findings have recently been reported (Fig. 12.4). First, it has been found that secretion of TNF-α in macrophages infected with mycobacteria is NOD2-dependent. Interestingly, a recent study by Coulombe *et al.* (2009) indicated that the NOD2-mediated TNF-α production is dependent on mycobacterial *N*-acyl muramic acid hydroxylase (NamH), which plays a role in the conversion of the NOD2 agonist muramyl dipeptide (MDP), a component of mycobacterial peptidoglycan, to its *N*-glycosylated form characteristic of mycobacterial organisms. Indeed, MTB (*namH* positive)-derived peptidoglycan, but not *Streptomyces* sp. (*namH* negative)-derived peptidoglycan, induced NF-κB activation. Moreover, *N*-glycosylated MDP is sensed by NOD2 and activates receptor-interacting protein 2 (RIP2), NF-κB, and JNK by stimulating NOD2 (Coulombe *et al.*, 2009). As indicated in Fig. 12.2, in the case of TLR-related signalling pathways, the polyubiquitination of certain signalling proteins such as IRAK1, TRAF6, and RIP plays important roles in the transduction of signals to downstream pathways (Wullaert *et al.*, 2006). Also in the case of NOD2-dependent signalling pathways, RIP2 interacts with its downstream protein, TAK1-binding protein (TAB2/3), through a polyubiquitin chain (Fig. 12.4). According to Coulombe *et al.* (2009), *N*-glycosyl MDP is 30-fold more effective than *N*-acyl MDP at inducing polyubiquitination of RIP2, and 100-fold more effective than *N*-acyl MDP at inducing phospho-inhibitor κB (phospho-IκB), and only *N*-glycosyl MDP could induce phosphorylation of JNK. Consequently, *N*-glycosyl MDP is more effective than *N*-acyl MDP at inducing production of proinflammatory cytokines including TNF-α and IL-6.

Second, it has recently been reported that NOD2-deficient macrophages and DCs were impaired in the production of proinflammatory cytokines and NO following infection with live, virulent MTB (Gandotra *et al.*, 2007). In this case, the NOD2 agonist MDP stimulated both TNF-α and IL-12p40 production by macrophages in the absence and presence of IFN-γ in a NOD2-dependent manner. On the other hand, mycolyl arabinogalactan peptidoglycan (PGN), the cell wall core of MTB, stimulates macrophages to generate TNF in a NOD2-dependent manner. In contrast, PGN also increased macrophage IL-12p40 production, but this was not dependent on NOD2. Thus, the mycobacterial cell wall fragments stimulate macrophages partially via NOD2 to induce cytokine production, but the degree and type of response depend on the structure and composition of the fragments used. This indicates the complexity of signalling pathways in response to MTB stimulation. Furthermore, NOD2-deficient mice and moreover NOD2- and TLR2-double deficient mice, are not more susceptible to infection with virulent MTB than wild-type mice and the replication of MTB was controlled at sites of infection. These findings imply that NOD2 participates in the recognition of MTB by APCs *in vitro* but is dispensable for the control of MTB *in vivo* infection (Gandotra *et al.*, 2007).

Third, while the recognition of microbial infections often occurs at the cell surface mainly via TLRs as described above, cytosolic recognition is also important for host cell responses to bacterial invasion (Fig. 12.4). As reported by Pandey *et al.* (2009), in an ESAT-6 secretion system 1 (ESX1; specialized protein-secreting system of MTB)-dependent manner, intraphagosomal live MTB stimulates the cytosolic NOD2 pathway that responds to bacterial MDP, eliciting the polyubiquitination of RIP2, and triggers the expression of type I IFNs in a TANK- binding kinase-1 (TBK1) kinase- and interferon regulatory factor 5 (IRF5)-dependent manner. This response is only partially impaired by the loss of Irf3, differing from the case of the pathway triggered by *Listeria monocytogenes* (Fig. 12.4). In this signalling axis, mycobacterial peptidoglycan and *N*-glycosylated

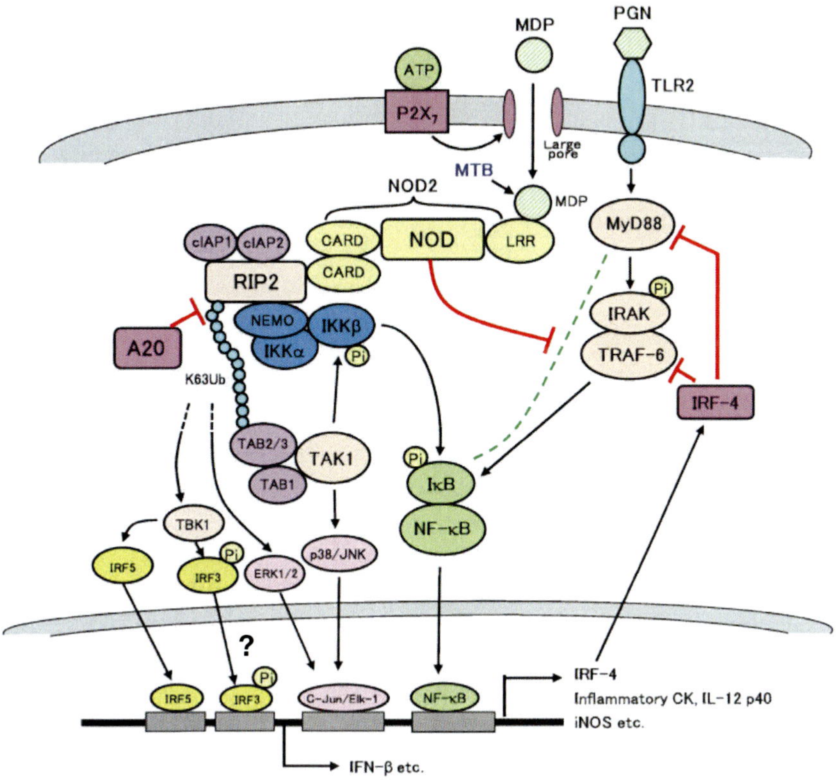

Figure 12.4 Signalling pathways related to NOD2. Intracellularly delivered MDP, one of the PGN-related small molecules, binds to the LRR domain of NOD2 consisting of a ligand-binding domain (LRR), an oligomerization domain (NOD), and two effector domains (CARDs), and activates it. Then, the CARD recruits and activates RIP2 (also called RICK), which activates the IKK complex by binding to NEMO after the recruitment of cIAP1 and cIAP2 ubiquitin ligases forming complex B. Simultaneously, K63-ubiquitinated RIP2 activates the TAK1–TAB complex by binding to TAB2/3 and the activated TAK1 subsequently up-regulates the IKK-NF-κB and MAPKK-MAPK (JNK and p38) axes, as described in Fig. 12.2. NOD-mediated activation of RICK also allows the activation of ERK1/2 and TBK1. TBK1 activates IRF3 and IRF5 and consequently up-regulates the transcription of a type-1 IFN, IFN-β. IRF-4, the expression of which is up-regulated by the NF-κB pathway, inhibits the TLR2-induced activation of MyD88 and TRAF6. Moreover, a polyubiquitin-degrading enzyme, A20, inhibits RIP2's function to activate TAB2/3. In addition, NOD itself directly inhibits the downstream signalling events of PGN-stimulated TLR2, especially the MyD88–TRAF6-NF-κB axis. The activation of the $P2X_7$ receptor with ATP causes large pores to form on the cell membrane, allowing the influx of extracellular MDP molecules into the cytosol, and the MDP activates NOD2 by binding to its LRR domain. Abbreviations are as follows. NOD2, nucleotide-binding oligomerization domain 2; MDP, muramyl dipeptide; PGN, peptidoglycan; LRR, leucine-rich repeat; CARD, caspase recruitment domain; RIP, receptor interacting protein; RICK, RIP-like interacting CLARP kinase; cIAP, CARD-like apoptotic protein; ERK, extracellular signal-regulated kinase; TBK, TANK (TNFR-associated factor family member-associated NF-κB activator) binding kinase; IRF, interferon regulatory factor. The following abbreviations are the same as in Fig. 12.2. MyD88, myeloid differentiation primary response gene; IRAK, IL-1-associated kinase; TRAF, TNFR-associated factor; TAK, TGF-β-activated kinase; TAB, TAK1-binding protein; IKK, IκB kinase; NEMO, NF-κB essential modulator; NF-κB, nuclear factor-κB; MAPKK, MAP kinase kinase; MAPK, mitogen-activated protein kinase; JNK, c-jun N-terminal kinase.

MDP act as a uniquely potent agonist of the NOD2/RIP2/IRF5 pathway. Taken together, the NOD2 system is specialized to recognize bacteria including mycobacteria that actively perturb host membranes. Indeed, some cytolysins and $P2X_7$ receptor agonists such as ATP, which cause the formation of pores in the macrophage cell membrane, promote the NOD2 pathway's activation by facilitating the release of a stimulatory mycobacterial component into the cytosol (Pandey *et al.*, 2009).

Fourth, it has been indicated that IL-1β production is induced by MTB through pathways involving TLR2/6 and NOD2 receptors but not TLR1/4/9 receptors and their downstream-pathways involve ERK, p38 and RIP2 but not JNK before transcription of pro-IL-1β (Kleinnijenhuis *et al.*, 2009). In addition, MTB stimulation is unnecessary for the caspase 1-mediated processing of pro-IL-1β to IL-1β, because monocytes constitutively express active caspase 1. Interestingly, the secretion of IL-1β is dependent on the activation of $P2X_7$ receptor-induced pathways in response to endogenously generated ATP. As described above, NOD2 mediates innate immune responses to mycobacterial infection. As reported by Divangahi *et al.* (2008), in NOD2-deficient mice given an aerosol of MTB, bacterial growth during the first 6 weeks was greater than that observed in the wild-type mice in connection with reduced histopathological responses during 4–8 weeks post infection, indicating impaired adaptive antimycobacterial immunity in NOD2-deficient mice. These phenomena are closely associated with decreased production of IL-12, the Th1 cytokine IFN-γ, the proinflammatory cytokine TNF-α and NO at the site of infection and, moreover, reduced recruitment of $CD8^+$ and $CD4^+$ T cells. In addition, the number of antigen-specific T cells was significantly lower in NOD2-deficient mice, which was associated with decreased levels of IFN-γ production, indicating that NOD2-mediated signalling serves as a bridge between innate and adaptive immune responses during mycobacterial infection. Taken together, NOD2 mediates resistance to mycobacterial infection via both innate immunity and adaptive immunity. Notably, MyD88-independent signalling pathways play important roles in the adaptive immune responses to MTB infection *in vivo*.

Other signalling pathways

With respect to new targets of anti-TB drugs, another interesting finding has been reported by Kuijl *et al.* (2007). With experiments using various kinds of kinase-inhibitors, especially the protein kinase A-inhibitor H-89 (*N*-[2-((*p*-bromo- cinnamyl)amino)ethyl]-5-isoquinolinesulfonamide) and its analogues, they identified several host kinases capable of inhibiting the intracellular (intramacrophage) growth of facultative intracellular bacteria, including *Salmonella typhimurium* and MTB. The kinases identified were clustered into one network around serine/threonine protein kinase AKT1 (also known as PKB). They proposed the following signalling pathways (Sassetti *et al.*, 2003): *S. typhimurium*-derived effector protein SopB activates AKT1. AKT1 targets p21-activated kinase 4 (PAK4), which phosphorylates GEF-H11, thus controlling RhoGTPase subfamily proteins (RhoA and Rac1) and actin. AKT1 also phosphorylates As160, a Rab14 (Ras-related protein)-GTPase activator protein (GAP). This prevents As160 from binding to the phagosome membrane, thereby causing the activation of Rab14 and consequently resulting in the inhibition of phagosomal maturation. These AKT1-mediated events promoted the intracellular survival and growth of *S. typhimurium*, especially in macrophages. Since the intramacrophage growth of MTB as well as *S. typhimurium* is inhibited by the same kinase inhibitors, such as H-89 and its analogue ETB067, it appears that AKT1 plays the same role in the signal transduction pathways for controlling phagosome–lysosome fusion in the case of MTB-infected macrophages (Sassetti *et al.*, 2003). Thus, a host cell kinase network around AKT1/PKB is a promising target for new anti-TB drugs. However, it should be noted that kinase inhibitors have generally serious selectivity issues. The employment of appropriate drug screening systems may solve this problem. In this context, the attributes of a proper quantitative structure–activity relationship (QSAR) analysis should include the ability to predict or deal with such important areas in the development of new anti-TB drugs as safety/toxicity, oral bioavailability, metabolic stability, and drug–drug interactions. In this context, it is important to study not only the drug targets in bacterial cells but also those in human cells, tissues, and organs.

Apoptosis is important for the interaction between pathogenic mycobacteria and host macrophages (Lee *et al.*, 2009). In hosts infected with MTB, innate immune responses work to eliminate infected cells by apoptosis, but virulent MTB uses strategies to prevent apoptosis. Thus, the balance between survival and apoptotic death in infected macrophages is important to the outcome of a MTB infection. In this context, it is known that caspase-8 (also called Fas-associated death domain-like IL-1β converting enzyme (FLICE))-mediated apoptosis of MTB-infected macrophages via the activation of p38 MAPK inhibits mycobacterial dissemination in hosts. Recently, Kundu *et al.* (2009) indicated that the production of TNF by macrophages in response to MTB is tightly linked to MTB-induced apoptosis in macrophages. In this case, TNF activates the ASK1-p38 MAPK-c-Abl (tyrosine kinase) axis, resulting in phosphorylation of FLICE-inhibitory proteins (FLIPs) at Ser 4, which facilitates its interaction with the E3 ubiquitin ligase c-Cbl, causing ubiquitination of FLIPs. This interaction subsequently triggers proteasomal degradation of FLIPs, which promotes activation of caspase-8 allowing this caspase to interact with the pro-apoptotic adapter protein, Fas-associated death domain protein (FADD), and initiate apoptosis. This finding is important when considering the participation of macrophage's apoptosis to the outcome of mycobacterial infection.

A recent finding by Feng *et al.* (2008) is also interesting. Mice deficient in the IFN-γ-inducible, immunity-related GTPase 1 (Irgm1) have impaired IFN-γ-dependent macrophage microbicidal activity, resulting in defective host resistance to a variety of intracellular pathogens. Interestingly, Irgm1 also regulates the survival of mature CD4 T cells by protecting them from IFN-γ-induced autophagic cell death (Feng *et al.*, 2008). Mice deficient in both IFN-γ and Irgm1 are rescued from the lymphocyte depletion after mycobacterial infection, although Irgm1 depletion alone causes enhanced T cell death, implying a feedback mechanism in Th1 response that limits the detrimental effects of IFN-γ on effector T cell survival.

Autophagy

Autophagy is a bulk protein degradation process where cytoplasmic components become enclosed by double-membrane structures termed autophagic vacuoles. These fuse with the endocytic pathway and their contents are eventually degraded in lysosomes. Autophagy is essential for cells to survive when nutrients are limited and crucial for cellular remodelling during development and differentiation (Bursch, 2001; Rich *et al.*, 2003). Concerning autophagy, the following principles are known (Deretic *et al.*, 2009). Autophagy functions as a innate defence mechanism against intracellular microbes, including mycobacteria. Autophagy is under the control of pattern recognition receptors (PRRs) including TLRs, and acts as one of the immunological output effectors of PPR and TLR signalling. Autophagy, an immune effector of Th1/Th2 cell response, is activated by Th1 cytokines and inhibited by Th2 cytokines. Recent findings by Harris *et al.* (2007) indicated the following. The Th1 cytokine IFN-γ induces autophagy in macrophages to eliminate infected MTB. The Th2 cytokines IL-4 and IL-13 abrogate autophagy-mediated killing of intracellular MTB in macrophages. Inhibition of starvation-induced autophagy by the Th2 cytokines is dependent on AKT signalling, while the inhibition of IFN-γ-induced autophagy is AKT-independent but signal transducer and activator of transcription 6 (STAT 6)-dependent.

Effects of mycobacterial components on T-cell signalling

As indicated in Fig. 12.5, very complicated signalling pathways are known for T-cell activation (Song, *et al.*, 2008; Smith-Carvin *et al.*, 2009; Nakayama and Yamashita, 2010). In the signalling events of T-cell receptor (TCR)-stimulated T cells, the following membranous or cytosolic proteins, most of which exhibit tyrosine or serine/threonine protein kinase activity, play important roles: zeta-chain-associated protein kinase 70 kDa (ZAP70), Lck, system l-amino acid transporter (LAT), SH2 domain containing leucocyte protein-76 (SLP-76), phospholipase C-γ (PLC-γ), mitogen-activated protein kinases (MAPKs), Ras, protein kinase-θ (PKC-θ), calmodulin (CaM), calcinulin (CN), IκB kinases (IKKs), nuclear factor-κB (NF-κB), activator protein-1 (AP-1), and nuclear factor of activated T cells (NFAT). Cellular components of pathogenic mycobacteria and infections are known to cause disturbances, usually the suppression of T-cell signalling events

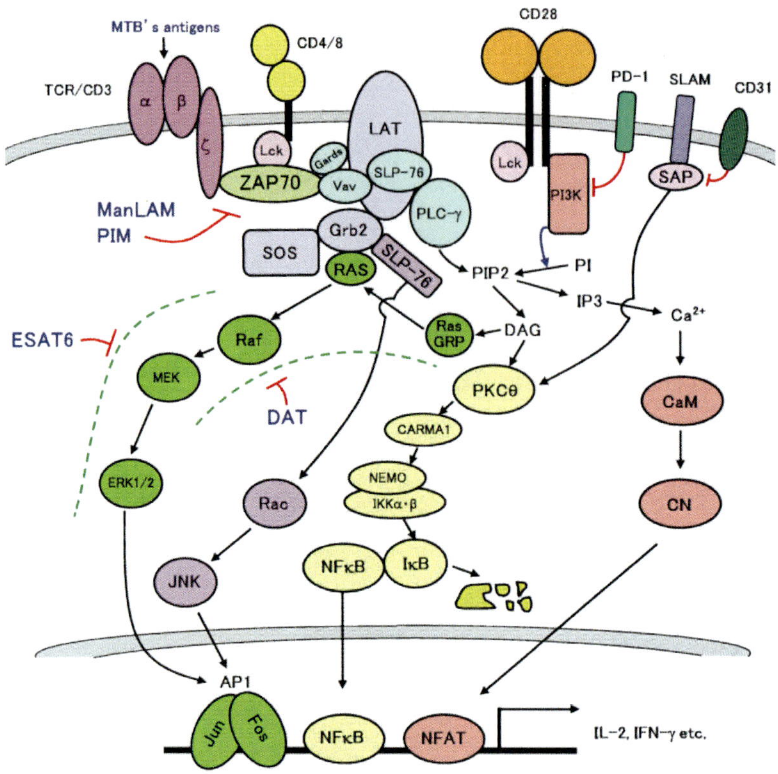

Figure 12.5 Signalling pathways participating in T-cell activation. After receiving antigen signals via TCR causing phosphorylation of the CD3ζ subunit, ZAP70 is recruited to the ζ subunit and activated on receiving stimulatory signals from CD4/8 and CD28. ZAP70 thereafter activates the LAT–Grb2-SOS-RAS axis allowing subsequent activation of Raf followed by MEK, ERK1/2, and finally the transcription factor AP1 (Jun/Fos). Grb2 also binds to SLP-76 which activates Rac and subsequently JNK and finally AP1. ZAP70 also simultaneously activates the Vav–SLP-76-PLC-γ axis, causing the PLC-γ-mediated generation of DAG and IP3 from PIP2 which is produced from PI by the enzymatic actions of PI3K recruited to the intracellular domain of CD28. PD-1's signals block the activation of PI3K. DAG activates RasGRP, which then activates Ras, and simultaneously activates PKC-θ allowing further activation of the IKK complex–NF-κB axis. In this pathway, CARMA-1 links PKC-θ to IKK. In the case of Th2 cell differentiation, the SLAM–SAP axis plays roles in promoting Th2-type responses through the activation of PKC-θ. IP3 causes the mobilization of intracellular Ca^{2+} and subsequent activation of the CaM–CN-NFAT axis. The triggering of these TCR signalling pathways finally induces the transcription of various cytokines, cytokine receptors, and CD antigens, such as IL-2, IL-2R and IFN-γ in the case of Th1 cells. Mycobacterial Man-LAM and PIM inhibit ZAP70. ESAT-6 inhibits the RAS-Raf-MAPKK-ERK1/2 pathway. DAT inhibits the RAS-Raf-MEK pathway. The major abbreviations are as follows. TCR, T cell receptor; ZAP, zeta-chain-associated protein kinase; LAT, system L-amino acid transporter; Grb, growth factor receptor-bound protein; SOS, son of sevenless; RAS, rat sarcoma; MEK, MAPK/ERK kinase; ERK, extracellular signal-regulated kinase; AP, activator protein; SLP, SH2 domain containing leukocyte protein; PLC, phospholipase C; DAG, diacylglycerol; IP3, inositol-1,4,5-trisphosphate; PIP2, phosphatidylinositol bisphosphate; PI, phosphatidylinositol; PI3K, phosphatidylinositol 3-kinase; PD-1, programmed death 1; PKC, protein kinase C; CARMA, caspase recruitment domain-containing MAGUK protein; SLAM, signalling lymphocyte activation molecule; SAP, SLAM-associated protein; CaM, calmodulin; CN, calcinulin; NFAT, nuclear factor of activated T cells; PIM, phosphatidylinositol mannoside; Lck, lymphocyte-specific protein tyrosine kinase; DAT, di-O-acyl trehalose. The following abbreviations are the same as in Fig. 12.2; IKK, IκB kinase; NEMO, NF-κB essential modulator; NF-κB, nuclear factor-κB; MAPKK, MAP kinase kinase; MAPK, mitogen-activated protein kinase; JNK, c-jun N-terminal kinase; Man-LAM, mannosylated (mannose-capped) LAM; ESAT-6, early secreted antigenic target protein 6.

induced by the stimulation of TCRs as described below.

Host cell signalling proteins and mycobacterial infection

Concerning the roles of signalling proteins of host cells in the down-regulation of T-cell functions during MTB infection, the following findings have been reported. First, as indicated in Fig. 12.5, programmed death 1 (PD-1) and its ligand PD-Ls are known to modulate T-cell activation in the regulation of IFN-γ production during TB presumably by blocking the activation of phosphoinositol 3-kinase (PI3K; Riley, 2009). Specific MTB antigen-stimulation increased numbers of $CD3^+PD\text{-}1^+$ T cells in peripheral blood of TB patients with a concomitant elevation of IFN-γ production (Jurado *et al.*, 2008). However, when T cells were stimulated with MTB, IFN-γ significantly increased the expression of PD-1 on T cells from MTB antigen-low responder TB patients (Jurado *et al.*, 2008). Therefore, it is thought that MTB-induced IFN-γ participates in the up-regulation of PD-1 expression. Blockage of PD-1 or PD-1:PD-Ls pathway results in an increase in IFN-γ-producing T cells in response to MTB, indicating that PD-1 interferes with T-cell effector functions against MTB. Moreover, it was found that costimulation through signalling lymphocytic activation molecule (SLAM) and blocking PD-1 cooperate to produce a synergic effect on IFN-γ production against MTB (Jurado *et al.*, 2008). In this context, when the role of PD-1 in regulating the antimycobacterial functions of CD8 T lymphocytes was examined in terms of T cell lytic degranulation activity, blockade of PD-1 was found to augment the T-cell activity. This indicates that a PD-1:PD-Ls pathway participates in the regulation of anti-MTB effector functions of CD8 T cells (Jurado *et al.*, 2008).

In this context, it is also known that CD31 acts as a receptor that attenuates TCR signals in MTB-infected hosts (Fig. 12.5). $CD3^+CD31^+$ T cells decrease in TB patients with a robust Th1 response to MTB (Quiroga *et al.*, 2007). Notably, IFN-γ is secreted only by $CD31^-$ T cells. In TB patients with a weak Th1 cytokine (IFN-γ) response to MTB, the level of $CD3^+CD31^+$ T cells is increased. CD31 binds to signalling lymphocyte activation molecule (SLAM)-associated protein (SAP) and the resulting complex of CD31 and SAP inhibits the IFN-γ response to MTB stimulation, indicating that CD31 has a key regulatory role in the signalling pathways leading to the IFN-γ response induced by MTB infection (Fig. 12.5). Furthermore, transmembrane signalling adaptor 12 (DAP12) also plays an important role in innate immune responses. DAP12 deletion results in pronounced proinflammatory and Th1 cytokine responses, over activation of antigen-specific T cells of a type 1 phenotype, and heightened immunopathological responses in lungs and lymphoid organs (Divangahi *et al.*, 2007). DAP12-deficient APCs display enhanced NF-κB activation and cytokine responses in response to TLR stimulation with mycobacterial cell components. These findings indicate that DAP12 functions as an critical immune regulatory molecule in controlling the level of antimicrobial type 1 T-cell activation and consequent immunopathology in response to mycobacterial stimulation.

Another interesting finding related to the suppression of antimycobacterial T-cell immune responses during the course of MTB infection is that intercellular adhesion molecule-1 (ICAM-1)-deficient mice had a pronounced reduction in $Foxp3^+$ Treg cells in the lungs compared with wild-type mice, since ICAM-1 plays a role in the induction of Foxp3 expression (Windish *et al.*, 2009). In ICAM-1-deficient mice, the Th1 cell response and inflammation were greater in the lungs, resulting in greater morbidity owing to overwhelming pathology. Therefore, ICAM-1 may play roles in the down-regulation of Th1-mediated antimycobacterial immune responses by generating Foxp 3^+Treg cells.

Mycobacterial components modulating T-cell signalling

With respect to the modulation of T-cell signalling pathways by mycobacteria and their cellular components, the following recent findings are of interest. First, as indicated in Fig. 12.5, MTB glycolipid di-O-acyl trehalose (DAT) but not free-lipid moiety trehalose was reported to inhibit CD4 T-cell proliferation and the production of cytokines including IL-2, IL-4 and IFN-γ (Palma-Nicolás *et al.*, 2009). In this case, the suppressive mechanisms target intracellular functions just after phospholipase C-γ activation because DAT inhibits T-cell proliferation in response to phorbol myristate acetate (PMA)–ionomycin stimulation. In addition, the effect of DAT involves down-regulation of diacylglycerol-dependent activation of the MAPK-ERK1/2 pathway but causes no inhibition of classical protein kinase C (PKC) activity and IκB-α phosphorylation/degradation (Fig. 12.5). Second, cell wall lipid fractions of MTB including phosphatidylinositol mannoside and Man-LAM were found to be potent inhibitors of CD4 T-cell activation in response to TCR stimulation in terms of IL-2 production, expression of activation markers (CD25, CD69), and proliferative responses (Mahon *et al.*, 2009). In this case, Man-LAM, which binds to TLRs, was 10-fold more potent as an inhibitor of IL-2 production than phosphatidylinositol mannoside (PIM), which binds to very late antigen 5 (VLA-5) on CD4 T cells. However, the MTB glycolipid-mediated inhibition due to direct interaction with target T cells was not associated with TLR/MyD88 signalling, but was related to zeta-associated protein 70 (ZAP-70) phosphorylation (Fig. 12.5). Indeed, MTB glycolipids reduced the expression of the activation-inducer molecule CD69, which is rapidly expressed upon TCR stimulation, indicating that the glycolipids affect CD4 T-cell activation through the TCR signalling pathway. In addition, MTB glycolipids did not affect PMA-ionomycin-induced IL-2 production but did inhibit IL-2 production by anti-CD3- and anti-CD28-stimulated T cells. These findings indicate that glycolipids of MTB directly inhibit CD4 T-cell activation processes by interfering with proximal TCR signalling.

ESAT-6 is known to contribute to MTB's virulence in hosts and inhibit IL-12 production by macrophages (Brodin *et al.*, 2004). Wang *et al.* (2009) recently indicated that ESAT-6 but not culture filtered protein 10 (CFP10) inhibited IFN-γ production by T cells in response to MTB and TCR stimulation with anti-CD3 and anti-CD28 antibodies at high concentrations (1.6–3.3 μM). In this case, ESAT-6 inhibited IFN-γ secretion from T cells by decreasing transcription of IFN-γ. Notably, ESAT-6 directly inhibits T-cell responses, in the absence of APCs. Although ESAT-6 is known to induce apoptosis of macrophages, it induced neither apoptotic nor necrotic responses in target T cells. ESAT-6 inhibits the expression of early activation markers, such as CD25 and CD69, without reducing the activation of ZAP70, and down-regulates the expression of transcription factors, such as activating transcription factor-2 (ATF-2) and c-Jun, causing a decrease in IFN-γ transcription, indicating that ESAT-6 directly inhibits T-cell responses to mycobacterial antigens by affecting TCR signalling pathways downstream of ZAP70 (Fig. 12.5). Notably, ESAT-6 also inhibits the production of IL-17 and TNF-α but not IL-2 and moreover suppresses the secretion of IFN-γ by T cells.

Remarks on mycobacterial components affecting cellular functions of macrophages and T cells in relation to the discovery of new drug targets

About 2.0 billion people in the world have been infected at some time with tubercle bacilli. As MTB pathogens remain dormant, existing in a latent and asymptomatic state where bacterial growth is undetectable, until the host's immune system has been compromised, prophylactic treatment with antituberculous drugs is effective at decreasing the incidence of reactivation in high-risk individuals such as those infected with HIV. MTB in persistent stages shows a metabolic shift in which bacterial glycolysis is decreased and glyoxylate shunt is up-regulated, allowing the bacteria to use the C_2 substrate generated by the β-oxidation of fatty acids. The expression of isocitrate lyase (Icl), a key enzyme in the glyoxylate cycle and essential as an anapleurotic protein for growth using acetate and certain fatty acids as a source of carbon, is up-regulated in MTB exposed to anaerobic conditions, in a stationary phase of growth, or inside macrophages (Wayne and Lin, 1982; Gould *et al.*, 2006). Indeed, an *icl* mutant of MTB reportedly replicated *in vivo* identically to wild-type organisms during the first 2 weeks of infection, but was steadily eliminated after the onset of host acquired cellular immunity to MTB antigens (McKinney *et al.*, 2000). This indicates that the *icl* gene plays a crucial role in the persistence of MTB *in vivo*. Indeed, Icl is necessary for the survival of MTB in activated macrophages (McKinney *et al.*, 2000). Since humans have no functional glyoxylate shunt, the Icl protein makes a promising target for drug treatment. In addition, a second isocitrate lyase encoded by the *aceA* gene and malate synthase (the second enzyme of the glyoxylate shunt) encoded by the *glcB* gene are thought to be required for the persistence of MTB (Smith *et al.*, 2003). Thus, these enzymes are also attractive targets for new anti-TB drugs. However, comprehensive anti-TB drug screening based on these targets will require ingenious assay systems for assessing the viability of dormant MTB. Indeed, the high-throughput screening (HTS) of a library of more than one million organic molecules for promising anti-MTB drugs, based on targets such as Icl and GlcB, met with little success, because of difficulty in whole-cell screening and for other reasons. Nevertheless, several interesting approaches have been applied on the drug targets characteristic of non-replicating MTB. Singh *et al.* (2008) recently demonstrated that nitroimidazopyran PA-824 kills non-replicating MTB by releasing reactive nitrogen intermediates (RNIs) generated from PA-824 owing to the enzymatic action of the bacterial deazaflavin-dependent nitroreductase (Ddn). Bryk *et al.* (2008) reported that a rhodanine derivative, D15707, a specific inhibitor of dihydrolipoamide acyltransferase (Dlat), selectively increased the susceptibility of non-replicating MTB to RNIs, but only when the *dlat* gene encoding Dlat had not been deleted (Bryk *et al.*, 2008). In this case, Dlat is thought to contribute to the antinitro oxidative defence of MTB by helping to detoxify peroxynitrite. These findings indicate that Ddn and Dlat are suitable as targets for antituberculous drugs active against MTB in a non-replicating state. The antimycobacterial mechanism of PA-824 is particularly unique. The MTB-sterilizing activity of PA-824 may be improved by increasing the Ddn activity and simultaneously inhibiting the Dlat activity using certain drugs. In this context, El Kasmi *et al.* (2008) demonstrated that BCG induced the expression of arginase 1 by macrophages through a STAT6-independent, TLR (TLR2)-MyD88-dependent pathway. Arginase 1, expressed in macrophages, impaired immune reactions to intramacrophage mycobacteria by competing with iNOS for the utilization of arginine as a substrate, resulting in the suppression of NO production by iNOS. Thus, appropriate drug targets may be found around TLR-mediated signalling events causing arginase 1 expression.

In any case, it is still difficult to find agents highly active against dormant- or persistent-type MTB. Consequently, it would be useful to entrust the entire sterilization process at sites of infection to the bactericidal function of host macrophages, although it is also important to continue attempts to develop novel antimicrobial drugs that are active against non-replicating MTB as described above. It would be useful to develop chemotherapeutic regimens and new antituberculous drugs that promote the bactericidal actions of macrophages. For this purpose, it is critical to potentiate and maintain the antimicrobial activity of host macrophages during the course of prolonged infection with MTB. In this context, as indicated in Fig. 12.2, some mycobacterial components, especially lipoglycans such as LAM and TDM, transmit signals via TLR2 causing the transcription of inflammatory-related genes resulting in the enhanced expression of proinflammatory mediators and effectors including IL-12, TNF-α, IL-1β, and iNOS. Fig. 12.3 lists various bacterial components that stimulate macrophages and DCs to produce IL-12. Most, including peptidoglycan, LAM, glycopeptidolipids, lipomannan, TDM, 19-kDa, 27-kDa 33-kDa, and 38-kDa lipoproteins, LprA and LprG lipoproteins, and phosphatidylinositol mannoside (PIM), transmit their stimulatory signals into the antigen-presenting cells through TLR2. It thus appears that these mycobacterial components essentially enhance host innate immunity and Th1-mediated immune responses and consequently inflammatory responses to promote host defences against mycobacterial infections. However, pathogenic mycobacteria including MTB can survive in host macrophages for long periods, principally by inhibiting phagosomal maturation and phagosome–lysosome fusion (P–L fusion). Therefore, virulent mycobacteria usually cause a prolonged infection leading to abnormally potent and sustained activation of TLR2-mediated signalling pathways. As indicated in Fig. 12.1, to prevent tissue damage from excessively activated macrophages, infected macrophages switch from the production of proinflammatory cytokines to the production of immunosuppressive and/or anti-inflammatory cytokines and factors, including TGF-β, IL-10 and prostaglandin E. These cytokines and factors in turn suppress the bactericidal and inflammatory functions of macrophages in an autocrine fashion. This homeostatic response in host macrophages and, moreover, the conversion from a Th1-type to Th2-type immune response seem to facilitate bacterial persistence and growth at sites of infection. Similarly, as shown in Fig. 12.4, MDP, a degradation product of mycobacterial peptidoglycan (PGN), also induces strong antimycobacterial inflammatory responses through NOD2-mediated signalling pathways. However, in this case, a IRF-4-mediated feedback regulation of downstream signalling pathways of NOD2 is also observed.

Consequently, for chemotherapy to be effective against mycobacterial infections, it is important to rapidly eliminate the pathogens residing within the macrophages using drugs with strong bactericidal activity. In this context, bacterial proteins (enzymes) essential for bacterial survival rather than replication are thought to be promising targets in the development of antimycobacterial drugs. In addition, the treatment of TB using appropriate antituberculous drugs in combination with TGF-β inhibitors (decorin and latency-associated peptide) and IL-10 inhibitors (anti-IL-10 receptor) is effective to some extent (Yamaguchi *et al.*, 1990; Bottinger *et al.*, 1996; Silva *et al.*, 2001). In this context, the use of two traditional Chinese medicines, Mao-Bushi-Saisin-To (MBST) and Yokuinin, in combination with antitubercular drugs is expected to improve chemotherapeutic outcome in cases of refractory TB, since these Chinese medicines mildly potentiate Th1-type responses, inhibit IL-10 expression, and moreover calm excessive inflammatory reactions in hosts during the course of refractory mycobacterial infections (Shimizu *et al.*, 1999a,b; Tomioka, 2004). In addition, the activation of TLR-related signalling pathways with mycobacterial components including TDM induces the expression of SOCSs, especially SOCS1 and SOCS3, which inhibit major signalling molecules involved in TLR-initiated events (Mal, IRAK1, TRAF6, ASK1, and NF-κB) via a feedback mechanism (Fig. 12.2). Therefore, specific inhibitors of SOCSs, such as SKF96365 (Schepers *et al.*, 2009), and inhibitors of SOCS expression may be useful to prevent SOCS-mediated inhibition of TLR-signalling pathways. Similarly, since, in the case of NOD2-related signalling pathways, NOD2 stimulation increases IRF-4 expression and IRF-4 subsequently inhibits TLR2-signalling events activating NF-κB (Fig. 12.4), IRF-4 inhibitors such as FKBP52 (Mamane *et al.*, 2000) and inhibitors of IRF-4 expression, such as 1α,25-dihydroxyvitamin D_3 (Gauzzi *et al.*, 2005), may also be useful as adjunctive agents in combination with antituberculous drugs. Thus, SOCSs and IRF-4 may be promising targets for the development of new drug regimens.

In addition, it is worth noting that ATP induces the formation of large membrane pores on macrophages by activating $P2X_7$ purinoceptor-mediated signalling pathways (Di Virgilio *et al.*, 1996; Lammas *et al.*, 1997), thereby causing an increase in the influx of extracellular MDP molecules into the macrophage cytosol. This causes the potentiation of NOD2-related signalling pathways leading to enhanced production of inflammatory cytokines, iNOS, etc. and augmentation of host antimycobacterial immune responses. In fact, ATP in combination with antimycobacterial drugs exhibited good therapeutic effects against *M. avium* complex infection induced in mice (Tomioka *et al.*, 2005). On the other hand, intramacrophage MTB escapes into the cytosol from phagosomal vesicles using cytolysin (van der Wel *et al.*, 2007; Hagedorn *et al.*, 2009). In this case, cytosolic NOD2 recognizes MDP molecules released from MTB, which resides in the cytosol of macrophages, and initiates the activation of signalling pathways, allowing the mobilization of antimycobacterial inflammatory responses by macrophages. Therefore, it is thought that, in the case of a prolonged infection of macrophages with MTB pathogens, NOD2-related signalling pathways are important to the effective expression of host defence mechanisms. Mycobacterial components, which prevent the triggering of NOD2-dependent signalling pathways, are expected to act as important virulence factors, and bacterial enzymes participating in the biosynthesis of such components may serve as good drug targets.

As indicated in Figs. 12.2 and 12.5, components of MTB, especially certain glycolipids (Man-LAM, PIM, DAT, etc.) and ESAT-6, affect host macrophage and/or T-cell signalling pathways resulting in the suppression of antimycobacterial immune responses, particularly the production of proinflammatory cytokines, Th1-inducing cytokines, and Th1 cytokines. Therefore, bacterial enzymes and membrane transporters engaged in the biosynthesis of such bacterial components may be good targets in the development of new antituberculous drugs. As shown in Table 12.2, the following enzymes may be suitable targets (Cole *et al.*, 1998; Mdluli and Spigelman, 2006; Williams and Duncan, 2007; Brennan and Crick, 2007; Tomioka, 2008). First, enzymes needed for the biosynthesis of mycolic acids (Basra and Chatterjee, 1994; Heath *et al.*, 2001; Marrakchi *et al.*, 2002; Brennan, 2003; Takayama *et al.*, 2005; Tomioka *et al.*, 2008), such as fatty acid synthetase-I (Fas) involved in the FAS-I system, acyl carrier protein (AcpM), malonyl-CoA:ACP transacylase (FabD), b-ketoacyl-acyl carrier protein synthase (FabH, KasA/B) and 2-transenoyl-ACP reductase (InhA) involved in the FAS-I to FAS-II system participating in chain elongation, cyclopropane mycolic acid synthase (CmaA1) and methyl transferases (MmaA1-A4, CmaA2, PcaA) participating in the introduction of a cyclopropane ring and a methyl group, and acyl-CoA carboxylase (AccD4/D5), fatty acyl-AMP ligase (Fad32), and polyketide synthase (Psk13) participating in Claisen-type condensation, are possible targets for anti-TB drugs. Second, bacterial enzymes and membrane transporters that play crucial roles in the biosynthesis of cell wall-associated components containing lipid moieties, including LAM, TDM, and the phthiocerol dimycocerosate (PDIM) family of complex lipids may be good targets (Brennan, 2003; Mdluli and Spigelman, 2006; Jackson *et al.*, 2007; Brennan and Crick, 2007; Tomioka *et al.*, 2008; Kaur *et al.*, 2009). These targets involve (1) enzymes needed for the biosynthesis of mycolic acids, as described above, (2) enzymes participating in the biosynthesis of LAM, such as polyprenol monophosphomannose synthase (Ppm1) and mannosyltransferase (PimB, PimF; Alderwick *et al.*, 2007), (3) enzymes playing central roles in the biosynthesis of PDIM, including phthiocerol dimycocerosyl transferase (PapA5) and moreover phthiocerol synthase A–E (PpsA–E), mycocerosic acid synthase Mas, Fad26 and FadD28 (Jackson *et al.*, 2007), (4) membrane transporter needed for the transport of PDIM through the cell membrane to the cell surface (*mmpL7* gene product; Jackson *et al.*, 2007), (5) enzymes participating in the biosynthesis of TDM and trehalose monomycolate (TMM), such as mycolyl transferase I (encoded by *Rv3802c*), mycolyl transferases II (encoded by *Rv1288*, *Rv0519* and *Rv0774c*), and mycolyl transferases called Ag85 complex (FbpA, FbpB, FbpC; Belisle *et al.*, 1997; Jackson *et al.*, 1999; Nguyen *et al.*, 2005a; Takayama *et al.*, 2005), and (6) enzymes participating in the biosynthesis of the mycolic acid-arabinogalactan-peptidoglycan (mAGP) complex, including glucose-1-phosphate thymidyltransferase (RmlA), dTDP-glucose dehydratase (RmlB), dTDP-4-dehydrorhamnose 3,5-epimerase (RmlC), dTDP-4-dehydrorhamnose reductase (RmlD), and mycolyl transferase (FbpC2; Crick *et al.*, 2001). As described above, the interaction of mycobacterial cell wall components with surface receptors of host macrophages principally induces potentiation of proinflammatory cytokines and Th1-activating cytokines primarily activating the signalling pathways of host cells. Notably, under certain conditions, some glycolipids including TDM, DAT and Man-LAM cause suppression of P–L fusion of macrophages or T-cell functions to produce Th1 cytokines, thereby attenuating host resistance to mycobacterial infection. Therefore, mycobacterial enzymes and transporters participating in the biosynthesis and secretion of these glycolipids are promising targets of new antituberculous drugs related to the signalling pathways in host immune competent cells.

ESAT-6 was initially found to be a very strong T-cell antigen principally activating Th1 cells as in the case of CFP-10 (Sørensen *et al.*, 1995; Brodin *et al.*, 2004). These proteins are secreted by ESAT-6 secretion system 1 (EXS-1) which is restricted to the genus *Mycobacterium*. Many of the genes encoding T-cell antigens are located in clusters in the RD1 region of the ESAT-6

Table 12.2 Promising drug targets in MTB related to mycobacterial cell wall components

Protein (gene)	Function	Inhibitors	References
Fatty acid synthetase-I (*fas*)[a]	Mycolic acid synthesis	–	Schwelzer (2004)
ACP (*acpM*)[a]	Mycolic acid synthesis	–	Kremer (2001)
Malonyl-CoA:ACP transacylase (*fabD*)	Mycolic acid synthesis	–	Kremer (2001)
β-Ketoacyl-ACP synthase III (*fabH*)[a]	Mycolic acid synthesis	Thiolactomycin	Choi (2000)
β-Ketoacyl-ACP synthase (*kasAB*)[a]	Mycolic acid synthesis	Thiolactomycin	Schaeffer (2001)
β-Ketoacyl-ACP reductase (*fabG1*)	Mycolic acid synthesis	–	Marrakchi (2002)
β-Ketoacyl-ACP reductase (*mabA*)[a]	Mycolic acid synthesis	Isoniazid	Cohen-Gonsaud (2002)
Enoyl ACP reductase (*fabI/inhA*)	Mycolic acid synthesis	Isoniazid Triclosan Diazaborines	Rozwarski (1999)
Enoyl ACP isomerase (*echA10?*)	Mycolic acid synthesis	–	–
S-adenosyl-methinine-dependent methyltransferase (*mmaA1-A4*)	Mycolic acid synthesis	S-adenocyl-N-decyl-amino-ethyl	Yuan (1997), Glickman (2003), Boissier (2006), Vaubourgr (2009)
Cyclopropane synthase (*pcaA*)[a]	Mycolic acid synthesis	Sinefungin Thiacetazone	Glickman (2000), Alahari (2007)
Cyclopropane mycolic acid synthase (*cmaA1,A2*)[a]	Mycolic acid synthesis	Thiacetazone	Yuan (1995), Glickman (2001), Alahari (2007)
Acyl-CoA carboxylase (*accD4, D5*)[a]	Mycolic acid synthesis	Naphthalene sulfonate	Portevin (2005), Lin (2006)
Fatty acyl-AMP ligase (*fadD32*)[a]	Mycolic acid synthesis	–	Portevin (2005)
Polyketide synthase (*pks13*)[a]	Mycolic acid synthesis	Cerulenin	Gavalda (2009)
Phosphatidylinositol mannosyltransferase (*pimA*)	mAGP complex synthesis	Phosphonate analogues	Guerin (2005); Dinev (2007)
Glucose-1-phosphate thymidyl-transferase (*rmlA*)	mAGP complex synthesis	–	Blankenfeldt (2000), Qu (2007)
dTDP-glucose dehydratase (*rmlB*)	mAGP complex synthesis	Rhodanines	Li (2006)
dTDP-keto-deoxyglucose epimerase (*rmlC*)	mAGP complex synthesis	Rhodanines	Li (2006)
dTDP-deoxyhexulose reductase (*rmlD*)	mAGP complex synthesis	Rhodanines	Ma (2001)
Polyprenol monophospho-mannose synthase (*ppm1*)[a]	LAM synthesis	–	Gurcha (2002), Alexander (2004)
Mannosyltransferase (*pimB,F*)[a]	LAM synthesis	–	Schaeffer (1999)
Phthiocerol dimycocerosyl transferase (*papA5*)[a]	PDIM (Wax) synthesis	–	Buglino (2004)
Phthiocerol synthase (*ppsA–E*)	PDIM (Wax) synthesis	–	Azad (1997)
Mycocerosic acid synthase (*mas, fad26,fadD28*)	PDIM (Wax) synthesis	–	Mathur (1992), Sirakova (2002)
Mycobacterial membrane protein large family of proteins (*mmpL7*)[a]	PDIM (Wax) synthesis	–	Domenech (2005)
Filamentation temperature-sensitive protein Z (*ftsZ*)	Septum formation	Taxanes	Rajagopalan (2005)
Mycolyltransferase I (*Rv3802c*)[a]	TDM synthesis	–	Brand (2003), Matsunaga (2008)
Mycolyltransferase II (*Rv1288, Rv518c, Rv0774c*)	TDM synthesis	–	Brand (2003)
Mycolyltransferases (Ag85 complex) (*fbpA,B,C*)	TDM synthesis	Phosphonates Trehalose analogues	Belisle (1997)

[a]Indispensable for bacterial survival and growth in macrophages and in animals.

operons as follows: *Rv3871*, *Rv3872* (PE35), *Rv3873* (PPE68), *Rv3984* (CFP-10), *Rv3872* (ESAT-6), *Rv3878* and *Rv3879c* (Okkels *et al.*, 2003; Brodin *et al.*, 2004). The RD1 region encoding ESX-1 (*Rv3868*, *Rv3869*, *Rv3870*, *Rv3872*, *Rv3877*) is simultaneously involved in enhancing mycobacterial virulence while its proteins EsxA (ESAT-6 encoded by Rv3875) and EsxB (CFP-10 encoded by Rv3874) induce Th1-mediated protective immunity (Sørensen *et al.*, 1995; Brodin *et al.*, 2004). As described by Brodin *et al.* (2004), ESX-1 may be a potential new drug target because of its dual role. Since the abolishment of bacterial production or secretion of ESAT-6 confers an attenuated phenotype on MTB, it is strongly suggested that ESAT-6 has an important role in the survival and multiplication of MTB in host macrophages. In fact, ESAT-6 suppresses macrophage functions to produce IL-12 and proinflammatory cytokines and T-cell functions to produce IFN-γ (Pathak *et al.*, 2007; Wang *et al.*, 2009). Thus, under certain conditions, ESAT-6 and the enzymes participating in its biosynthesis are thought to be good targets for the development of antituberculous drugs related to signalling pathways in host immune competent cells. In addition, Man-LAM, PIM and DAT also down-regulate TCR-initiated signalling pathways of T cells (Fig. 12.5). Therefore, the enzymes of MTB participating in the biosynthesis of these mycobacterial lipoglycans are also good targets for TB drugs.

Finally, autophagy plays an important role in host immunological defences against MTB (Deretic *et al.*, 2009). IFN-γ-induced autophagy is effective in the elimination of intraphagosomal MTB and presumably intracellular MTB that escapes from the phagosome into the cytosol (Gutierrez *et al.*, 2004). Moreover, the activation of innate immunity through the signalling pathways initiated by TLRs/MyD88 after the recognition of mycobacterial PAMPs stimulates the autophagic elimination of MTB (Delgado *et al.*, 2008). Notably, autolysosomes contain ubiquitin fragments generated by hydrolysis that act as mycobactericidal peptides (Alonso *et al.*, 2007). Such a mechanism at least partly contributes to the ATP-mediated mycobactericidal function of macrophages via $P2X_7$ receptors (Biswas *et al.*, 2008). Harris *et al.* (2007) reported that autophagy is activated by Th1 cytokines, especially IFN-γ, and inhibited by Th2 cytokines including IL-4 and IL-13. As indicated in Fig. 12.1, the differential expression of IL-12 and IL-10 by macrophages dictates the development of Th1 and Th2 cell responses. High levels of IL-10 result in decreased expression of IL-12 and a strong Th2-type response. Mycobacteria use strategies to trigger IL-10 expression in macrophages during the early stages of an infection. Nair *et al.* (2009) found that PPE18, part of the proline-proline-glutamic acid (PPE) family of mycobacterial proteins, specifically interacts with the leucine-rich repeat domain of TLR2 in macrophages, leading to an early and sustained activation of p38 MAPK, which is critical for IL-10 production. In contrast, a synthetic lipopeptide, Pam3CSK4, binds to a different site of TLR2 causing ERK1/2's activation resulting in enhanced expression of IL-12. Taken together, it is thought that the PPE18 of MTB is a promising target for the development of TB drugs which support the antimycobacterial autophagic function of host macrophages. In addition, it has been reported that LAM, especially Man-LAM, modify the signalling machinery of raft/microdomains in Th cells causing a modification of the membrane's organization that eventually leads to an enhancement of Th2 cytokine expression and the inhibition of Th1 cytokine production (Shabaana *et al.*, 2005). Thus, mycobacterial enzymes participating in the biosynthesis of Man-LAM may be good drug targets in relation to the antimycobacterial activity of macrophages based on autophagic mechanisms.

Mycobacterial protein kinases and phosphatases affecting the signalling pathways of host cells

As recently reviewed by Chao *et al.* (2009), signal transduction pathways of MTB are co-mediated by five main families of protein kinases and phosphatases as follows: (1) two-component systems consisting of a histidine kinase and a response regulator, (2) eukaryotic-like Serine/Threonine (Ser/Thr) protein kinases (STPKs), (3) Ser/Thr protein phosphatase belonging to the protein phosphatase type 2C family, (4) protein tyrosine phosphatases (PTPs), one of which interferes

staurosporin-related compounds, K-252a and K-252b, exhibit strong inhibitory activity against the kinase activity of PknB protein (Fernandez *et al.*, 2006). Moreover, an ATP-competitive inhibitor, mitoxantrone, is also capable of inhibiting PknB-mediated kinase activity (Wehenkel *et al.*, 2006). These inhibitors are reported to inhibit the growth of MTB *in vitro*. Third, in MTB, the *pknF* gene resides in an operon with an ATP-binding cassette (ABC) transporter gene, suggesting that the PknF protein participates in the regulation of the membrane transport system. In addition, the *pknH* gene also resides downstream of this ABC transporter gene and is also adjacent to the *embR* gene, which controls both *embA* and *embB* genes encoding arabinosyl transferases needed for the biosynthesis of cell wall components such as lipoarabinomannan and arabinogalactan. Thus, it is proposed that the PknH protein play some role in arabinose metabolism. The deletion of *pknH* gene in MTB resulted in decreased resistance to RNIs, suggesting that NO is a potential environmental trigger for PknH activation. However, a *pknH*-deleted mutant of MTB was found to survive and replicate more effectively in mice and in macrophages than in the wild-type strain (Papavinasasundaram *et al.*, 2005). In contrast, MTB strain with a mutation in the *pknE* gene, encoding a STPK PknE closely related to PknH, was not affected in its virulence phenotype (Papavinasasundaram *et al.*, 2005). These findings suggest that PknH plays a role in regulating MTB bacterial load in response to host signals, resulting in increased persistence in mice as a result of adaptation to the host environment. Fourth, in MTB, the *pknI* gene is located in a cluster of genes including *dacB* (Rv2911) encoding the D-amino acid hydrolase PBP5 and the *ftsY* gene participating in cell division processes. Although PknI is one of the proteins with a transmembrane domain, it is located predominantly in the bacterial cytosol (Singh *et al.*, 2006). It is thus proposed that PknI regulates mycobacterial cell division. Interestingly, PknI expression is decreased during macrophage infection with MTB, differing from the cases of PknA and PknB which are up-regulated during infection (Singh *et al.*, 2006).

With respect to the profiles of interaction of the membrane-bound STPKs including PknA, B, D, E, F, and H with cellular proteins of MTB, interesting findings have been reported by Molle *et al.* (2006).They indicated that these MTB STPKs phosphorylate some enzymes comprising the type II fatty acid synthase (FAS-II) system involved in mycolic acid biosynthesis: the β-ketoacyl acyl carrier protein (AcpM) synthases KasA and KasB, and malonyl-CoA:AcpM transacylase mtFabD. PknA exhibits the strongest kinase activity on these substrates, and multiple threonine residues (up to three) of these FAS-II enzymes are preferentially phosphorylated by these STPKs *in vitro*, and serine residues but not tyrosine residues serve as weaker substrates for the STPKs. The STPK-mediated multiple phosphorylation of KasA and KasB at threonine residues is also observed *in vivo* in *M. bovis* BCG. Notably, Ser/Thr protein phosphatase PstP of MTB (described in details in the latter section) was found to dephosphorylate *in vivo* phosphorylated KasA, indicating that PstP exerts its phosphatase activity not only on STPKs but also on the STPK-phosphorylated substrate (phosphorylatedKasA) and, consequently, phosphorylation of KasA is reversible in *M. bovis* BCG. The phosphorylation of KasA leads to a reduction in its condensation activity, whereas the phosphorylation of KasB causes an increase in its condensation activity. Thus, phosphorylation affects the enzymatic functions of KasA and KasB differently. These findings provide new insights into the regulation of mycolic acid biosynthesis, especially the elongation of acyl chains, by phosphorylation of the condensing enzymes, KasA and KasB. External signals are transduced within mycobacteria by sensor STPKs on the bacterial membrane. Autophosphorylated STPKs in response to the external signals then transfer the signal to KasA and KasB by transphosphorylation and differentially regulate the activities of these enzymes in different parts of a specialized elongation FAS-II system: E1-FAS-II comprising KasA elongates the acyl-AcpM chains and E2-FAS-II possessing KasB terminates the elongation process. Thus, STPK-mediated phosphorylation and PstP-mediated dephosphorylation of KasA and KasB are crucial for the regulation of biosynthesis of matured mycolic acid, important cell wall components closely related to mycobacterial virulence.

Soluble STPKs

The PknG protein is of particular interest, since it plays important roles in pathogenicity/virulence and survival in host macrophages (Av-Gay and Everett 2000; Wehenkel *et al.*, 2008; Chao *et al.*, 2009). Indeed, Cowley *et al.* (2004) reported that PknG-deficient MTB mutants exhibited reduced growth in target visceral organs (lung, spleen, liver) of immunocompetent host mice, which were given an intravenous bacterial challenge, and that the same PknG mutants caused delayed dissemination to the spleen after aerosol infection into the lungs (Cowley *et al.*, 2004). The PknG protein has three distinct domains, an N-terminal Trx domain (150 amino acids), a kinase domain (246 amino acids), and a C-terminal TPR domain (354 amino acids). PknG is the only STPK that has an N-terminal Trx domain, including double cysteine motifs (^{106}CXXC^{109} and ^{128}CXXC^{131}) characteristic of disulfide reductase, as the upstream region of the kinase domain as in the case of YopO of *Yersinia*. Therefore, it is proposed that PknG is capable of responding to the cellular redox status. In addition, the PknG protein contains a single tetratricopeptide sequence presumably forming a signalling domain (Sikorski *et al.*, 1990) in its C-terminal region. These findings indicate the possibility that PknG protein is activated in response to cellular redox status to regulate bacterial signalling pathways. In this context, a recent study using mutated PknG proteins by Tiwari *et al.* (2009) revealed the following.

1. The kinase domain of PknG by itself is inactive, indicating the importance of the other domains.
2. Both the Trx and TPP domains help in regulating the kinase activity of PknG.
3. Double cysteine motifs in the Trx domain are critical for modulation of PknG activity in response to the environmental redox state.
4. Threonine 63 in the Trx domain is the major autophosphorylation site in PknG protein.
5. The kinase activity is dependent on the presence of threonine 309 in the $p + 1$ loop of the activation segment. Threonine 309 plays a role in either structural stabilization of the loop or in arbitrating kinase–substrate interaction.

In MTB, the *pknG* gene resides within an operon including the *glnH* gene (Rv1411c) encoding glutamine-binding protein, thereby suggesting that PknG protein play roles in glutamine uptake in MTB. In this context, it has been indicated that inhibition of extracellular glutamine synthase attenuates bacterial growth of MTB *in vitro* and in macrophages (Harth and Horwitz, 1999), suggesting participation of PknG protein in the expression of pathogenicity and virulence by mycobacterial pathogens. The reduced growth of the *pknG* mutant is more pronounced in the stationary phase of the MTB's growth cycle and when grown in nutrient-depleted medium (Cowley *et al.*, 2004). Indeed, the deletion of *pknG* gene causes reduction in the growth of MTB in mice and delayed mortality of infected mice with a concomitant decrease in intracellular glutamate and glutamine levels (Cowley *et al.*, 2004). Thus, it is probable that PknG mediates the transfer of signals sensing nutritional stress in MTB and translates them into metabolic adaptations, especially glutamate/glutamine metabolism (Cowley *et al.*, 2004). Recent studies in a wild-type MTB strain demonstrate that the PknG phosphorylates the GarA protein, a negative regulator of NAD$^+$-dependent glutamate dehydrogenase (GDH) and α-ketoglutarate decarboxylase (KGD; Chao *et al.*, 2009). The phosphorylated GarA loses its ability to inhibit the enzymatic activity of GDH and KGD, yielding conditions for the speedy conversion of glutamate to α-ketoglutarate and further to succinic semialdehyde in the TCA cycle. This results in a down-regulation of the cellular accumulation of glutamate. Furthermore, PknG-mediated phosphorylation of GarA also causes a decrease in its activity to potentiate glutamine synthase activity (GS), thereby causing reduction of GS-mediated conversion of glutamine to glutamate. This also contributes to the down-regulation of glutamate's accumulation in cells. In this context, it should be noted that, in *M. bovis* BCG, the depletion of PknG has no effect on *in vitro* growth of the organisms and there is no detectable defect on glutamine uptake or intracellular glutamine concentrations, suggesting that modulation of glutamine metabolism does not contribute to the intracellular degradation of BCG organisms lacking PknG (Nguyen *et al.*, 2005b).

Although the physiological roles of the *pknG* gene in mycobacterial cells have not yet been elucidated, PknG protein is secreted from pathogenic mycobacteria residing within macrophages and enhances their intramacrophage survival, principally by inhibiting phagosome-lysosome (P–L) fusion (Walburger *et al.*, 2004). In this context, Tiwari *et al.* (2009) recently reported that the expression of PknG in a non-pathogenic mycobacterium, *M. smegmatis*, allowed long-term survival in the spleen of mice by blocking the fusion of bacteria-engulfing phagosomes with lysosomes. They found that the first 73 N-terminal residues, the Trx motif, and threonine 63 residue, which is the major autophosphorylation site, are independently essential for modulating PknG-mediated survival of mycobacteria in host macrophages. In addition, the study by Scherr *et al.* (2009) based on endoproteinase-digestion followed by HPLC coupled to ESI-MS indicated that threonine residues, Thr_{23}, Thr_{32} and $Thr_{63/64}$, were autophosphorylated by PknG protein. They found that PknG autophosphorylation is essential for the prevention of mycobacterial trafficking to lysosomes during infection inside macrophages, although autophosphorylation is not required for the activation of PknG kinase activity (Scherr *et al.*, 2009).

Walburger *et al.* (2004) showed that after the internalization of *M. bovis* BCG strain into host macrophages, the PknG protein is secreted within macrophage phagosomes and inhibits P–L fusion, mediating the intramacrophage survival of the mycobacteria. It is therefore thought that pathogenic mycobacteria, especially the MTB complex, possess eukaryotic cell-like signal transduction mechanisms capable of modulating host macrophage trafficking pathways. In this context, Houben *et al.* (2009) have recently reported that *M. smegmatis*, an avirulent mycobacterium, also has a pknG gene which encodes an active PknG protein performing the same function as its *M. bovis* BCG counterpart in intramacrophage survival of mycobacteria. In this case, although similar levels of the *pknG* transcript are generated in *M. bovis* BCG and *M. smegmatis*, virtually no translation product of the *pknG* gene is detected in *M. smegmatis* after *in vitro* growth or after intramacrophage infection, indicating that decreased levels of PknG in *M. smegmatis* are mainly due to a reduced translational efficiency. Therefore, PknG is an attractive target for a novel type of antituberculous drug. On the basis of this strategy, Szekely *et al.* (2008) screened agents exhibiting PknG-inhibitory activity from 1000 compounds listed in the NCL kinase inhibitory library, containing 124 kinases. They identified AX20017 (tetrahydrobenzothiophene), AX33510 and AX14585 (derivatives of AX20017) potent PknG-inhibitory agents. Indeed, the 2.4 Å X-ray crystal structure of PknG in complex with AX20017 has been reported (Scherr *et al.*, 2007). These agents were active in inducing lysosomal transfer in BCG-infected macrophages. Further studies are necessary to identify PknG's target protein in the phagosomal maturation signalling pathways shown in Fig. 12.6 (Koul *et al.*, 2004; Walburger *et al.*, 2004; Philips, 2008; Chao *et al.*, 2009; Scherr *et al.*, 2009).

Protein tyrosine kinase A

Protein tyrosine phosphorylation plays a key role in the regulation of fundamental cellular processes in eukaryotes and prokaryotes. In bacteria, signalling pathways based on tyrosine phosphorylation are involved in numerous cellular events, including cell division, biosynthesis of bacterial components and factors, and expression of pathogenicity and virulence. Protein tyrosine kinases (PTKs) are thought to play important roles in bacterial signalling pathways. Recently, a novel class of PTK (named PtkA) encoded in the open reading frame Rv2232, has been identified in MTB (Bach *et al.*, 2009). This PTK (PtkA) lacks Walker A (GXXGXGK[T/S]) and Walker B (hhhhD) motifs characteristic of PTKs of common bacteria, is capable of utilizing both ATP and GTP as phosphate donors, and its activity is not inhibited by staurosporine, a general non-specific PTK inhibitor. The *ptkA* gene is located immediately upstream of the *ptpA* gene encoding protein tyrosine phosphatase A (PtpA) in the same operon and PtkA phosphorylates tyrosine residues of PtpA, but PtkA does not serve as the substrate for dephosphorylation reaction mediated by PtpA. Although, at present, there is only limited information concerning PtkA, this MTB PTK may be promising as a target for development

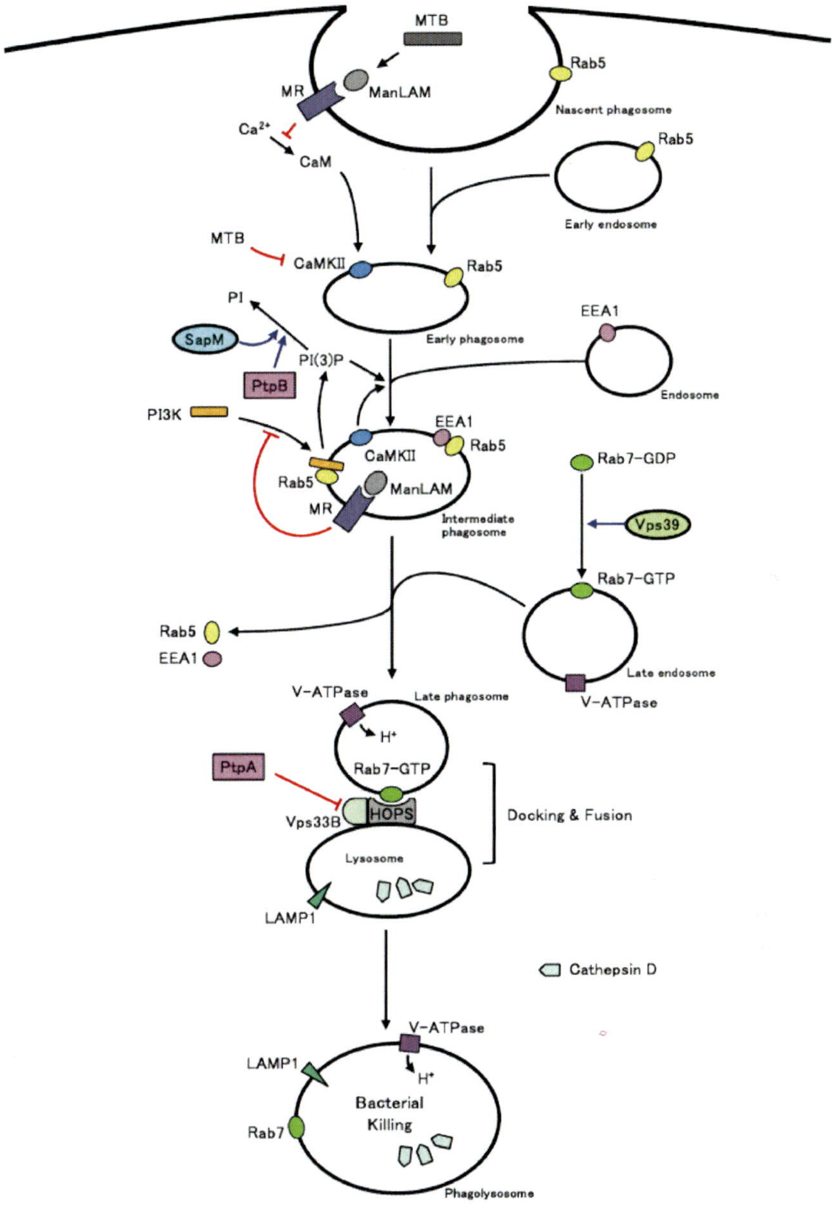

Figure 12.6 Profile of phagosomal maturation in macrophages after the phagocytosis of mycobacterial pathogens. On the phagocytosis of heat-killed mycobacteria, nascent phagosomes acquire a low molecular weight GTPase (GTP-binding protein), Rab5, specific for the control of fusion within early endosomal compartments, from the cell membrane and early endosomes. Concomitantly, a rise in the cellular Ca^{2+} concentration causes the activation of CaM which subsequently phosphorylates CaMKII, allowing its association with the phagosomal membrane. Then, Rab5 recruits PI3K, which generates PI(3)P and mediates the recruitment of EEA1 in collaboration with CaMKII. EEA1 thereafter recruits GTP-bound Rab7 (second GTPase) and V-ATPase from the late endosomes, consequently converting the intermediate phagosomes to the late phagosomes, while Rab5 and EEA1 are lost. In the cytosol, Vps39 catalyses the conversion of Rab7 from the GDP- to GTP-bound form. Rab7-GTP on the late phagosomes binds to a HOPS complex containing Vps33B, a key regulator of membrane trafficking and fusion, and HOPS binds to the lysosomal membrane (docking), allowing membranous fusion between the late phagosomes and lysosomes (P–L fusion). Via the P–L fusion, mycobacterium-engulfing phagosomes acquire LAMP1 and an antimicrobial

Figure 12.6 continued
cationic protein, cathepsin D, and the intraphagosomal pH is lowered. Pathogenic mycobacteria, especially MTB, block the phagosomal maturation pathway as follows. Man-LAM inhibits the $[Ca^{2+}]$i–CaM axis mainly via MR and consequently blocks CaMKII's activation. Other MTB-derived signals also directly inhibit CaMKII. Alternatively, Man-LAM blocks the recruitment of cytosolic PI3K to the phagosomal membrane and disrupts PI(3)P's accumulation and interaction with EEA1. Mycobacterial lipid phosphatase (SapM) converts PI(3)P to PI. The mycobacterial protein tyrosine phosphatase, PtpA, dephosphorylates Vps33B and blocks the Vps33B–HOPS-mediated fusion of late phagosomes with lysosomes. Another mycobacterial protein tyrosine phosphatase, PtpB, is also thought to convert PI(3)P to PI. Other mycobacterial proteins including PknG, LpdC, and Zmp1 are also proposed to play a role in arresting phagosome maturation pathways. LpdC plays a direct role in restricting phagosomal maturation by retaining coronin 1, an inhibitor of phagosome maturation, on the phagosomal membrane. The target protein of PknG in the signalling pathways for phagosomal maturation is not known. The major abbreviations are as follows. Rab, Ras-related protein; EEA, early endosomal antigen; CaM, calmodulin; CaMKII, calmodulin-dependent protein kinase II; PI(3)P, phosphatidylinositol 3-phosphate; V-ATPase, vacuolar ATPase; Vps, vacuolar sorting protein; HOPS, homotypic fusion and vacuole protein sorting; LAMP, lysosome-associated membrane protein; MR, mannose receptor; Ptp, protein tyrosine phosphatase; PknG, protein kinase G; LpdC, lipoamide dehydrogenase; Zmp1, Zn^{2+} metalloproteases.

of antituberculous chemotherapeutics, since PtkA-mediated tyrosine phosphorylation of MTB proteins in signalling pathways, especially Ptp proteins, is thought to greatly contribute to the regulation of MTB functions to express pathogenicity and virulence (Bach *et al.*, 2009).

Protein phosphatases

MTB possesses two protein tyrosine phosphatases (PTPs), PtpA and PtpB. These PTPs are thought to act as virulence factors, particularly enabling persistent infection of MTB organisms in host macrophages (Chao *et al.*, 2009).

PtpA has a conserved CX5R motif typical of functional PTPs and its properties are as follows.

1. PtpA activity is specific for phosphotyrosine residues.
2. PtpA expression levels are increased during infection in macrophages.
3. PtpA is secreted from the bacterial cell body into the cytoplasm of host macrophages.
4. PtpA expression in macrophages reduces phagocytosis of mycobacteria and zymosan in connection with increased actin polymerization (Castandet *et al.*, 2005).

These properties strongly indicate the intracellular interaction of PtpA with signalling pathways of host cells. Using *ptpA*-deleted mutant MTB, Bach *et al.* indicated that PtpA is essential for successful infection inside macrophages (Bach *et al.*, 2008). They found that MTB PtpA dephosphorylated vacuolar protein sorting 33B (Vps33B), a regulator of membrane fusion, as its substrate. Indeed, as indicated in Fig. 12.6, PtpA secretion combined with active-phosphorylated Vps33B inhibited P–L fusion. According to the model proposed by Philips (2008) and Chao *et al.* (2009), PtpA blocks the docking process of phagosome maturation yielding phagolysosomes by interfering with the binding between the homotypic vacuole fusion and vacuole protein sorting (HOPS) complex on the lysosomes, which contains Vps33B and Ras-related protein 7 (Rab7) protein expressed on the early endosomes, due to PtpA-mediated dephosphorylation of Vps33B (Bach *et al.*, 2008). Because the interaction of HOPS with Rab7 (a late endosome-/lysosome-associated small GTPase) is crucial for the assembly of soluble N-ethylmaleimide-sensitive fusion protein attachment receptor (SNARE) proteins on the endosomes, and this is necessary for the formation of late endosomes and subsequent fusion with lysosome membranes (Koul *et al.*, 2004; Philips, 2008; Chao *et al.*, 2009), PtpA is thought to lead to blocking of P–L fusion in MTB-infected macrophages.

PtpB, lacking human orthologues, is a secretable protein presumably specific for phosphotyrosine and phosphoserine/threonine residues. In addition, this PTP also acts on phosphoinositides, thereby suggesting its function in interfering with cellular phosphoinositide-mediated signalling events, which is important for phagosome maturation in macrophages as indicated in Fig. 12.6 (Koul *et al.*, 2004; Beresford *et al.*, 2007; Philips, 2008). Indeed, PtpB-deficient MTB exhibits reduced bacterial survival and growth in macrophages and guinea pigs (Singh *et al.*, 2003).

Both PtpA and PtpB are promising targets for the development of new antituberculosis drugs because of their roles in supporting MTB survival in hosts. Indeed, antituberculous agents, chalcone derivatives, obtained by aldolic condensation between aldehyde and acetophenones, have been reported to exhibit strong inhibitory activity against PtpA (Chiaradia *et al.*, 2008). PtpB may be a target in the development of new antituberculous drugs, since specific inhibitors against this phosphatase reduce mycobacterial survival within macrophages without showing antimicrobial activity against extracellular mycobacteria (Beresford *et al.*, 2009). Indeed, cyclic stevastelin, roseophilin and prodigiosin analogues and dimethylpyrol-1-yl benzoic acids, which were derived from natural compounds as starting points, have been reported to possess strong inhibitory activity against MTB PtpA and PtpB (Manger *et al.*, 2005). Moreover, macroline-derived compounds have recently been reported to exhibit PtpB inhibitory activity (Nören-Müller *et al.*, 2008). Thus, these protein phosphatases may be potential targets for the development of new antituberculous drug candidates.

In addition to the PTP-encoding genes, the MTB genome includes the gene encoding eukaryotic-like protein Ser/Thr phosphatase PstP, which is a membrane-anchored metalloenzyme and binds a third Mn^{2+} in the active site (Pullen *et al.*, 2004). Based on sequence motifs, Ser/Thr phosphatases are divided into the PPM and PPP families. The prototypical member of the PPM family is PP2C which negatively regulates a variety of signalling pathways including stress responses. The MTB PstP is a PP2C-family Ser/Thr protein phosphatase and is known to dephosphorylate 11 mycobacterial STPKs in MTB (Jackson and Denu, 2001). Furthermore, PstP dephosphorylates the β-ketoacyl AcpM synthase KasA, which plays an important role in the biosynthesis of mycolic acids, one of the most important virulence factors on MTB pathogens (Molle *et al.*, 2006). Therefore, PstP may be a promising target of antituberculous drugs.

Remarks on mycobacterial protein kinases and phosphatases as drug targets

Among mycobacterial enzymes participating in protein phosphorylation and dephosphorylation including STPKs, PTKs, PTPs, and PstP, the following are promising as targets for new antituberculous drugs. First, PknA, PknB, PknG, and presumably PknF and PtkA, required for the normal growth of MTB in culture, can be considered potential targets. Since these regulatory proteins participate in cellular metabolic events required for bacterial growth mostly in an indirect manner (for instance, phosphorylation of KasA and KasB participating in mycolic acid synthesis by PknA), their inhibition appears not to be critical to the replication of mycobacteria. Thus, agents screened based on inhibitory activity against these enzymes would exhibit less bactericidal activity than TB drugs which directly inhibit metabolic reactions essential to the bacterial growth and survival, such as rifamycins (inhibitors of nucleic acid synthesis), isoniazid (inhibitor of mycolic acid synthesis), and moreover, diarylquinoline TMC207 (ATP synthase inhibitor) currently under development. Second, some kinases and phosphatases, including PknG, PknH, PknI, PtpA, PtpB, and PstP, may be good drug targets, since they act as virulence factors, supporting the intramacrophage survival of MTB by inhibiting phagosomal maturation and P–L fusion (PknG and PtpB), reducing nutritional stress in macrophages (PknG), and regulating bacterial stress response.

There is a consensus that the virulence factor-mediated interference in the intramacrophage processes of phagosomal maturation leading to P–L fusion are crucial for the killing and elimination of mycobacterial pathogens in host macrophages (Koul *et al.*, 2004). In this context, among the various protein kinases and phosphatases described above, PknG is regarded as the most promising target for the development of antituberculous drugs having unique antimycobacterial mechanisms. Indeed, PknG is secreted from the MTB inside macrophages and enables the pathogen to survive in macrophages by disrupting the phagosomal maturation process in macrophages, leading to P–L fusion, although PknG's target in the signalling pathways of phagosomal maturation (Fig. 12.6) has not yet been identified (Koul *et al.*, 2004; Walburger *et al.*, 2004; Philips, 2008; Chao *et al.*, 2009; Scherr *et al.*, 2009). In fact, Szekely's study focusing on PknG as a target revealed that the PknG inhibitors AX33510 and AX14585, both tetrahydrobenzothiophenes, exhibited strong antimicrobial activity against MTB (Szekely *et al.*, 2008). As indicated in Fig. 12.6, mycobacterial protein tyrosine phosphatases, Ptps, are also thought to be a good target for antituberculous drugs, because PtpA blocks P–L fusion by dephosphorylating Vps33B, a member of the HOPS complex acting as a key regulator of membrane trafficking and fusion (Bach *et al.*, 2008; Philips, 2008; Chao *et al.*, 2009). Inactivation of Vps33B causes an arrest of the docking and fusion of the late phagosomes with lysosomes. In addition, dephosphorylation of PI(3)P by PtpB may cause the arrest of the PI(3)P-mediated signalling that is critical to the formation of intermediate phagosomes (Beresford *et al.*, 2007). Therefore, PtpB may be another target for the development of TB drugs. In any case, it is necessary to elucidate in detail the roles of mycobacterial protein kinases and phosphatases in manipulating phagosomal maturation in MTB-infected macrophages to find novel drugs that better enable host macrophages to eliminate MTB, especially non-replicating dormant and persistent MTB.

Concluding remarks

To combat intractable TB, especially MDR-TB and XDR-TB, the development of new classes of drugs exhibiting potent anti-MTB activity without cross-resistance to existing antituberculous drugs is urgently required. For this purpose, it is necessary to accelerate investigations to identify new types of anti-TB drugs acting on novel targets. It is logical to design compounds capable of interfering with specific bacterial targets involved in biosynthesis or cellular mechanisms of membrane permeability and transport, which are closely related to the manifestation of mycobacterial virulence in hosts. In this respect, critical information regarding the entire genome of MTB has been elucidated (Cole *et al.*, 1998; Fleischmann *et al.*, 2002). In conjunction with advancing knowledge on various mycobacterial virulence genes, this is advancing our efforts to identify genes that code for new potential drug targets. Employing such knowledge on the mycobacterial genome and virulence factors, drug design applying quantitative structure–activity relationship (QSAR) may play a central role in the development of antituberculous drugs in the near future (Barker, 2006). The three-dimensional (3D)-QSAR analysis of anti-mycobacterial inhibitors is an interesting area of research for the development of novel antituberculous drugs (Aparna *et al.*, 2006; Costa *et al.*, 2006). Traditional QSAR models are incapable of explaining complex structure–activity data, since the extreme specificity of biological activity is described by 3-D intermolecular forces and predicated on 3-D molecular structures. Dynamic QSAR techniques, especially the method for recognition of the common reactivity pattern (COREPA method), are currently employed in drug design and toxicology (Mekenyan, 2002). The principle assumption of the COREPA method is that biologically similar chemicals should possess a commonality in their stereoelectronic reactivity pattern. Thus, COREPA allows the visualization and interpretation of reactivity patterns. The core of the approach is the procedure for evaluating of the 3-D similarity between chemicals accounting for their conformational flexibility. In contrast to traditional methods, which compare 3-D structural patterns of fixed molecular geometries, the COREPA method compares the conformer distribution of chemicals across molecular descriptors, thereby accounting for structural differences due to conformational flexibility of the molecules. Recently, a more useful method, the comparative molecular field analysis (CoMFA), has been employed in 3-D-QSAR study for the synthesis of new antituberculous agents, in order to rationalize the design and to understand the physicochemical properties and structural parameters of the pharmacophore (Shagufta *et al.*, 2007; Hevener *et al.*, 2008; Nayyar *et al.*, 2009; Manvar *et al.*,

2010). CoMFA examines differences in targeted properties that are related to changes in the shape of the steric and electrostatic fields surrounding the molecules. The idea underlying CoMFA is that differences in biological activity are often related to differences in the magnitudes of molecular fields surrounding the receptor ligands. To employ mycobacterial proteins as useful drug targets for 3-D-QSAR analysis, a number of bacterial virulence genes have been cloned for MTB and other mycobacteria, and their structural properties and biological functions have been examined in relation to the manifestation of mycobacterial pathogenicity in host animals (Glickman and Jacobs, 2001; Smith, 2003). At present, high-resolution structures of the gene products (usually enzymes) encoded by such virulence genes are available, and structure-based inhibitor design is now under way, with the prospect of generating new classes of drugs which are potently active against mycobacteria, particularly persistent and dormant types of MTB organisms (Glickman and Jacobs, 2001; Terwilliger *et al.*, 2003; Smith *et al.*, 2004; Duncan, 2004; Kantardjieff and Rupp, 2004; Tomioka 2008; Murphy and Brown, 2008; Ioerger and Sacchettini, 2009). Such structural bioinformatics-based approaches offer a promising strategy to identify new classes of antituberculous drugs.

Proinflammatory cytokines such as IL-12, IL-23 and IL-18, which are required for the induction/activation of Th1 cells; Th1 and Th1-like cytokines including IFN-γ, IL-2, TNF-α and GM-CSF; and moreover Th17-derived inflammatory cytokines such as IL-17 with an important role in the formation of protective granulomas and IL-22; are crucial to host resistance against mycobacterial infections. On the other hand, immunosuppressive and/or macrophage-deactivating cytokines or humoral factors, such as IL-4, IL-10, IL-13, TGF-β and prostaglandin E, which are produced by Th2 cells, Th3 cells, Ts/Treg cells and M2-type macrophages, appear to play important roles in depression of manifestation of protective immunity against mycobacterial pathogens, resulting in long-term survival and multiplication of virulent mycobacteria including MTB at sites of infection in hosts. Therefore, it is important to control the host cytokine network to maintain effective expression of innate immunity and the Th1 cytokine-dominant immune responses that are crucial for hosts to combat mycobacterial pathogens, without eliciting severe inflammation and consequent bodily injury. Thus, it may be possible to identify novel targets for the development of antimycobacterial drugs, especially antituberculosis drugs, among mycobacterial genes encoding functional proteins which are capable of cross-talking with host signalling pathways and thereby affecting host cytokine network.

In this context, I described the roles of various types of mycobacterial cellular components, especially those composing the cell wall and secreting proteins, in modulating host innate and acquired immune responses related to host cytokine network. Mycobacterial components including various glycolipids, peptidoglycan and certain proteinaceous substances such as ESAT-6, affect intracellular signalling pathways via surface cellular receptors and signal proteins such as TLRs, MyD88, NOD2, MAPK and transcription factors, thereby resulting in activation or suppression of host innate and acquired immunity against mycobacterial infection by affecting cellular functions of macrophages and T cells. Therefore, mycobacterial gene products participating in the biosynthesis and transport of such bacterial immune regulatory components may be promising as the targets for novel types of antituberculous drugs which exert therapeutic effects by preventing mycobacterial manipulation of immune signalling events in host cells.

Next, I described the roles of signalling proteins of mycobacteria in the expression of their pathogenicity in hosts. Certain types of protein kinases and protein phosphatases, especially STPKs and PTPs, cross-talk with host signalling proteins such as those related to the maturation of mycobacteria-engulfing phagosomes of host macrophages, causing inhibition of P–L fusion consequently enabling intramacrophage survival of pathogenic mycobacteria. In addition, some of them are critical for bacterial growth *in vitro* and/or in host cells. Therefore, these mycobacterial signalling proteins may be good targets of the development of novel antituberculous drugs, particularly based on the inhibitors of protein kinases or phosphatases as the starting agents.

However, it should be noted that kinase inhibitors and presumably phosphatase inhibitors have generally serious selectivity issues. The employment of appropriate drug screening systems may solve this problem. In this context, the attributes of a proper QSAR analysis should include the ability to predict or deal with such important areas in the development of new anti-TB drugs as safety/toxicity, oral bioavailability, metabolic stability, and drug–drug interactions. In this context, it is important to study not only the drug targets in bacterial cells but also those in human cells, tissues, and organs.

Furthermore, the following problems should be noted. First, many studies have revealed that the high-throughput screening (HTS) of a library of organic compounds against MTB based on appropriate drug targets is not as easy as initially expected, because of the low binding affinity and electron charge of test molecules, difficulty in measuring the activity against whole mycobacterial cells, etc. Notably, the measurement of test agents regarding their inhibition of intracellular functions of target signalling proteins, such as STPKs and PTPs as mentioned above, needs *in vitro* models of macrophage infection, which are not amenable to HTS studies. Moreover, the correlation of such *in vitro* efficacy models with murine *in vivo* models is still disputable. In addition, the only truly valid corroboration is through *in vivo* studies, which have a very low throughput. Indeed, the activity of these target proteins related to MTB's virulence can be measured or assessed only in animal or macrophage cell culture models, and, therefore, early stage discovery efforts against such targets would be hindered by the need to always test new compounds in animal or macrophage models. However, it may consequently be possible to identify unique antituberculous drugs that cannot be found by screening methods based on the growth inhibition test of extracellularly growing mycobacterial organisms.

Second, bioinformatic approaches are difficult to apply to TB, since, at present, our knowledge of intracellular signal transduction and metabolic pathways of MTB organisms is still limited. Thus, further, marked efforts to comprehensively elucidate the precise roles and functions of mycobacterial gene products in intracellular pathways of MTB organisms are necessary to validate the structural bioinformatic-based approaches for the screening, identification, and chemical design of novel anti-TB drugs. In this context, as an idealistic argument, it is necessary to logically depict the sequence of biochemical metabolic events associated with target proteins (enzymes) and genes. However, at present, studies are generally undertaken by selectively focusing on one or a few target proteins. Thus, post genome studies on the detailed biological and biochemical characteristics of whole gene products of MTB are urgently required.

References

Alderwick, L.J., Birch, H.L., Mishra, A.K., Eggeling, L., and Besra, G.S. (2007). Structure, function and biosynthesis of the *Mycobacterium tuberculosis* cell wall: arabinogalactan and lipoarabinomannan assembly with a view to discovering new drug targets. Biochem. Soc. Trans. 35, 1325–1328.

Aliyu, M.H., and Salihu, H.M. (2003). Tuberculosis and HIV disease: two decades of a dual epidemic. Wien Klin. Wochenschr. *115*, 685–697.

Alonso, S., Pethe, K., Russell, D.G., and Purdy, G.E. (2007). Lysosomal killing of *Mycobacterium* mediated by ubiquitin-derived peptides is enhanced by autophagy. Proc. Natl. Acad. Sci. U.S.A. *104*, 6031–6036.

Aparna, V., Jeevan, J., Ravi, M., Desiraju, G.R., and Gopalakrishnan, B. (2006). 3-D-QSAR studies on antitubercular thymidine monophosphate kinase inhibitors based on different alignment methods. Bioorg. Med. Chem. Lett. *16*, 1014–1020.

Av-Gay, Y., and Everett, M. (2000). The eukaryotic-like Ser/Thr protein kinases of *Mycobacterium tuberculosis*. Trends Microbiol. *8*, 238–244.

Av-Gay, Y., Jamil, S., and Drews, (1999). S.J. Expression and characterization of the *Mycobacterium tuberculosis* serine/threonine protein kinase PknB. Infect. Immun. *67*, 5676–5682.

Bach, H., Papavinasasundaram, K.G., Wong, D., Hmama, Z., and Av-Gay, Y. (2008). *Mycobacterium tuberculosis* virulence is mediated by PtpA dephosphorylation of human vacuolar protein sorting 33B. Cell Host Microbe 3, 316–322.

Bach, H., Wong, D., and Av-Gay, Y. (2009). *Mycobacterium tuberculosis* PtkA is a novel protein tyrosine kinase whose substrate is PtpA. Biochem. J. *420*, 155–160.

Barker, J.J. (2006). Antibacterial drug discovery and structure-based design. Drug Discov. Today *11*, 391–404.

Basra, G.S., and Chatterjee, D. (1994). Lipids and carbohydrates of *Mycobacterium tuberculosis*. In Tuberculosis, Pathogenesis, Protection, and Control.

Bloom, B.R., ed. (ASM Press, Washington, DC), pp. 285–306.

Belisle, J.T., Vissa, V.D., Sievert, T., Takayama, K., Brennan, P.J., and Besra, G.S. (1997). Role of the major antigen of *Mycobacterium tuberculosis* in cell wall biogenesis. Science *276*, 1420–1422.

Beresford, N.J., Mulhearn, D., Szczepankiewicz, B., Liu, G., Johnson, M.E., Fordham-Skelton, A., Abad-Zapatero, C., Cavet, J.S., and Tabernero, L. (2009). Inhibition of MptpB phosphatase from *Mycobacterium tuberculosis* impairs mycobacterial survival in macrophages. J. Antimicrob. Chemother. *63*, 928–936.

Beresford, N., Patel, S., Armstrong, J., Szöor, B., Fordham-Skelton, A.P., and Tabernero, L. (2007). MptpB, a virulence factor from *Mycobacterium tuberculosis*, exhibits triple-specificity phosphatase activity. Biochem. J. *406*, 13–18.

Biswas, D., Qureshi, O.S., Lee, W.Y., Croudace, J.E., Mura, M., and Lammas, D.A. (2008). ATP-induced autophagy is associated with rapid killing of intracellular mycobacteria within human monocytes/macrophages. BMC Immunol. *9*, 35.

Blanc, D.C., Nunn, P., and Duncan, K. (2000). Incentives and disincentives for new anti-tuberculous drug development. Situational analysis. In World Health Organization 2000, (World Health Organization, Geneva Switzerland), pp.1–59.

Bottinger, E.P., Factor, V.M., Tsang, M.L., Weatherbee, J.A., Kopp, J.B., Qian, S.W., Wakefield, L.M., Roberts, A.B., Thorgeirsson, S.S., and Sporn, M.B. (1996). The recombinant proregion of transforming growth factor β1 (latency-associated peptide) inhibits active transforming growth factor β1 in transgenic mice. Proc. Natl. Acad. Sci. U.S.A. *93*,5877–5882.

Bowdish, D.M., Sakamoto, K., Kim, M.J., Kroos, M., Mukhopadhyay, S., Leifer., C.A., Tryggvason, K., Gordon, S., and Russell, D.G. (2009). MARCO, TLR2, and CD14 are required for macrophage cytokine responses to mycobacterial trehalose dimycolate and *Mycobacterium tuberculosis*. PLoS Pathog. *5*, e100047.

Brennan, P.J. (2003) Structure, function, and biogenesis of the cell wall of *Mycobacterium tuberculosis*. Tuberculosis (Edinb) *83*, 91–97.

Brennan, P.J., and Crick, D.C. (2007). The cell-wall core of *Mycobacterium tuberculosis* in the context of drug discovery. Curr. Top. Med. Chem. *7*, 475–488.

Brodin, P., Rosenkrands, I., Andersen, P., Cole, S.T., and Brosch, R. (2004). ESAT-6 proteins: protective antigens and virulence factors? Trends Microbiol. *12*, 500–508.

Bryk, R., Gold, B., Venugopal, A., Singh, J., Samy, R., Pupek, K., Cao, H., Popescu, C., Gurney, M., Hotha, S., *et al.* (2008). Selective killing of nonreplicating mycobacteria. Cell Host Microbe *3*, 137–145.

Bursch, W. (2001). The autophagosomal-lysosomal compartment in programmed cell death. Cell Death Differ. *8*, 569–581.

Canova, M.J., Veyron-Churlet, R., Zanella-Cleon, I., Cohen-Gonsaud, M., Cozzone, A.J., Becchi, M., Kremer, L., and Molle, V. (2008). The *Mycobacterium tuberculosis* serine/threonine kinase PknL phosphorylates Rv2175c: mass spectrometric profiling of the activation loop phosphorylation sites and their role in the recruitment of Rv2175c. Proteomics *8*, 521–533.

Canova, M.J., Kremer, L., and Molle, V. (2009). The *Mycobacterium tuberculosis* GroEL1 chaperone is a substrate of Ser/Thr protein kinases. J. Bacteriol. *191*, 2876–2883.

Castandet, J., Prost, J.F., Peyron, P., Astarie-Dequeker, C., Anes, E., Cozzone, A.J., Griffiths, G., and Maridonneau-Parini, I. (2005). Tyrosine phosphatase MptpA of *Mycobacterium tuberculosis* inhibits phagocytosis and increases actin polymerization in macrophages. Res. Microbiol. *156*, 1005–1013.

Chao, J., Wong, D., Zheng, X., Poirier, V., Bach, H., Hmama, Z., and Av-Gay, Y. (2010). Protein kinase and phosphatase signaling in *Mycobacterium tuberculosis* physiology and pathogenesis. Biochim. Biophys. Acta. Proteins Proteomics *1804*, 620–627.

Chiaradia, L.D., Mascarello, A., Purificação, M., Vernal, J., Cordeiro, M.N., Zenteno, M.E., Villarino, A., Nunes, R.J., Yunes, R.A., and Terenzi, H. (2008). Synthetic chalcones as efficient inhibitors of *Mycobacterium tuberculosis* protein tyrosine phosphatase PtpA. Bioorg. Med. Chem. Lett. *18*, 6227–6230.

Cole, S.T., Brosch, R., Parkhill, J., Garnier, T., Churcher, C., Harris, D., Gordon, S.V., Eiglmeier, K., Gas, S., Barry, C.E. 3rd, *et al.* (1998). Deciphering the biology of *Mycobacterium tuberculosis* from the complete genome sequence. Nature *393*, 537–544.

Costa, M.S., Boechat, N., Rangel, E.A., da Silva Fde, C., de Souza, A.M., Rodrigues, C.R., Castro, H.C., Junior, I.N., Lourenço, M.C., Wardell, S.M., *et al.* (2006). Synthesis, tuberculosis inhibitory activity, and SAR study of N-substituted-phenyl-*1*,2,3-triazole derivatives. Bioorg. Med. Chem. *14*, 8644–8653.

Coulombe, F., Divangahi, M., Veyrier, F, de Léséleuc, L., Gleason, J.L., Yang, Y., Kelliher, M.A., Pandey, A.K., Sassetti, C.M., Reed, M.B., and Behr, M.A. (2009). Increased NOD2-mediated recognition of *N*-glycolyl muramyl dipeptide. J. Exp. Med. *206*, 1709–1716.

Cowley, S., Ko, M., Pick, N., Chow, R., Downing, K.J., Gordhan, B.G., Betts, J.C., Mizrahi, V., Smith, D.A., Stokes, R.W., and Av-Gay, Y. (2004). The *Mycobacterium tuberculosis* protein serine/threonine kinase PknG is linked to cellular glutamate/glutamine levels and is important for growth *in vivo*. Mol. Microbiol. *52*, 1691–1702.

Crick, D.C., Mahapatra, S., and Brennan, P.J. (2001). Biosynthesis of the arabinogalactan–peptidoglycan complex of *Mycobacterium tuberculosis*. Glycobiology *11*, 107R-118R.

Cui, T., Zhang, L., Wang, X., and He, Z.G. (2009). Uncovering new signaling proteins and potential drug targets through the interactome analysis of *Mycobacterium tuberculosis*. BMC Genomics *10*, 118.

Delgado, M.A., Elmaoued, R.A., Davis, A.S., Kyei, G., and Deretic, V. (2008). Toll-like receptors control autophagy. EMBO J. *27*, 1110–1121.

Deretic, V., Delgado, M., Vergne, I., Master, S., De Haro, S., Ponpuak, M., and Singh, S. (2009). Autophagy in immunity against *Mycobacterium tuberculosis*: a model system to dissect immunological roles of autophagy. Curr. Top. Microbiol. Immunol. 335, 169–188.

Di Virgilio, F., Ferrari, D., Falzoni, S., Chiozzi, P., Munerati, M., Steinberg, T.H., and Baricordi, O.R. (1996). P2 purinoceptors in the immune system. Ciba Found. Symp. *198*, 290–302.

Divangahi, M., Yang, T., Kugathasan, K., McCormick, S., Takenaka, S., Gaschler, G., Ashkar, A., Stampfli, M., Gauldie, J., Bramson, J., Takai, T., Brown, E., Yokoyama, W.M., Aoki, N., and Xing, Z. (2007). Critical negative regulation of type 1 T cell immunity and immunopathology by signaling adaptor DAP12 during intracellular infection. J. Immunol. *179*, 4015–4026.

Divangahi, M., Mostowy, S., Coulombe, F., Kozak, R., Guillot, L., Veyrier, F., Kobayashi, K.S, Flavell, R.A., Gros, P., and Behr, M.A. (2008). NOD2-deficient mice have impaired resistance to *Mycobacterium tuberculosis* infection through defective innate and adaptive immunity. J. Immunol. *181*, 7157–7165.

Duncan, K. (2004). Identification and validation of novel drug targets in tuberculosis. Curr. Pharm. Des. *10*, 3185–3194.

El Kasmi, K.C., Qualls, J.E., Pesce, J.T., Smith, A.M., Thompson, R.W., Henao-Tamayo, M., Basaraba, R.J., König, T., Schleicher, U., Koo, M.S., *et al.* (2008). Toll-like receptor-induced arginase 1 in macrophages thwarts effective immunity against intracellular pathogens. Nat. Immunol. *9*, 1399–1406.

Feng, C.G., Zheng, L., Jankovic, D., Báfica, A., Cannons, J.L., Watford, W.T., Chaussabel, D., Hieny, S., Caspar, P., Schwartzberg, P.L., Lenardo, M.J., and Sher, A. (2008). The immunity-related GTPase Irgm1 promotes the expansion of activated $CD4^+$ T cell populations by preventing interferon-γ-induced cell death. Nat. Immunol. *9*, 1279–1287.

Fernandez, P., Saint-Joanis, B., Barilone, N., Jackson, M., Gicquel, B., Cole, S.T., and Alzari, P.M. (2006). The Ser/Thr protein kinase PknB is essential for sustaining mycobacterial growth. J. Bacteriol. *188*, 7778–7784.

Finkel, T. (2003). Oxidant signals and oxidative stress. Curr. Opin. Cell Biol. *15*, 247–254.

Fleischmann, R.D., Alland, D., Eisen J.A., Carpenter, L., White, O., Peterson, J., DeBoy, R., Dodson, R., Gwinn, M., Haft, D., *et al.* (2002). Whole-genome comparison of *Mycobacterium tuberculosis* clinical and laboratory strains. J. Bacteriol. *184*, 5479–5490.

Fremond, C.M., Togbe, D., Doz, E., Rose, S., Vasseur, V., Maillet, I., Jacobs, M., Ryffel, B., and Quesniaux, V.F. (2007). IL-1 receptor-mediated signal is an essential component of MyD88-dependent innate response to *Mycobacterium tuberculosis* infection. J. Immunol. *179*, 1178–1189.

Frieden, T.R., Sterling, T.R., Munsiff, S.S., Watt, C.J., and Dye, C. (2003). Tuberculosis. Lancet *362*, 887–899.

Gandotra, S., Jang, S., Murray, P.J., Salgame, P., and Ehrt, S. (2007). Nucleotide-binding oligomerization domain protein 2-deficient mice control infection with *Mycobacterium tuberculosis*. Infect. Immun. *75*, 5127–5134.

Ganguly, N., Giang, P.H., Basu, S.K., Mir, F.A., Siddiqui, I., and Sharma, P. (2007). *Mycobacterium tuberculosis* 6-kDa early secreted antigenic target (ESAT-6) protein down-regulates lipopolysaccharide induced *c-myc* expression by modulating the extracellular signal regulated kinases 1/2. BMC Immunol. *8*, 24.

Gauzzi, M.C., Purificato, C., Conti, L., Adorini, L., Belardelli, F., and Gessani, S. (2005). IRF-4 expression in the human myeloid lineage: up-regulation during dendritic cell differentiation and inhibition by 1alpha,25-dihydroxyvitamin D_3. J. Leukoc. Biol. *77*, 944–947.

Geisel, R.E., Sakamoto, K., Russell, D.G., and Rhoades, E.R. (2005). *In vivo* activity of released cell wall lipids of *Mycobacterium bovis* bacillus Calmette-Guérin is due principally to trehalose mycolates. J. Immunol. *174*, 5007–5015.

Glickman, M.S., and Jacobs, W.R. (2001). Microbial pathogenesis of *Mycobacterium tuberculosis*: Dawn of a discipline. Cell *104*, 477–485.

Gould, T.A., van de Langemheen, H., Muñoz-Elías, E.J., McKinney, J.D., and Sacchettini, J.C. (2006). Dual role of isocitrate lyase 1 in the glyoxylate and methylcitrate cycles in *Mycobacterium tuberculosis*. Mol. Microbiol. *61*, 940–947.

Greenstein, A.E., Grundner, C., Echols, N., Gay, L.M., Lombana, T.N., Miecskowski, C.A., Pullen, K.E., Sung, P.Y., and Alber, T. (2005). Structure/function studies of Ser/Thr and Tyr protein phosphorylation in *Mycobacterium tuberculosis*. J. Mol. Microbiol. Biotechnol. *9*, 167–181.

Grundner, C., Gay, L.M., and Alber, T. (2005). *Mycobacterium tuberculosis* serine/threonine kinases PknB, PknD, PknE, and PknF phosphorylate multiple FHA domains. Protein Sci. *14*, 1918–1921.

Gutierrez, M.G., Master, S.S., Singh, S.B., Taylor, G.A., Colombo, M.I., and Deretic, V. (2004). Autophagy is a defense mechanism inhibiting BCG and *Mycobacterium tuberculosis* survival in infected macrophages. Cell *119*, 753–766.

Hagedorn, M., Rohde, K.H., Russell, D.G., and Soldati, T. (2009). Infection by tubercular mycobacteria is spread by nonlytic ejection from their amoeba hosts. Science *323*, 1729–1733.

Håkansson, S., Galyov, E.E., Rosqvist, R., and Wolf-Watz, H. (1996). The *Yersinia* YpkA Ser/Thr kinase is translocated and subsequently targeted to the inner surface of the HeLa cell plasma membrane. Mol. Microbiol. *20*, 593–603.

Harris, J., De Haro, S.A., Master, S.S., Keane, J., Roberts, E.A., Delgado, M., and Deretic, V. (2007). T helper 2 cytokines inhibit autophagic control of intracellular *Mycobacterium tuberculosis*. Immunity *27*, 505–517.

Harth, G., and Horwitz, M.A. (1999). An inhibitor of exported *Mycobacterium tuberculosis* glutamine synthetase selectively blocks the growth of pathogenic mycobacteria in axenic culture and in human monocytes: extracellular proteins as potential novel drug targets. J. Exp. Med. *189*, 1425–1436.

Heath, R.J., White, S.W., and Rock, C.O. (2001). Lipid biosynthesis as a target for antibacterial agents. Prog. Lipid Res. *40*, 467–497.

Henriques, A.O., Glaser, P., Piggot, P.J., and Moran, C.P., Jr. (1998). Control of cell shape and elongation by the *rodA* gene in *Bacillus subtilis*. Mol. Microbiol. *28*, 235–247.

Hevener, K.E., Ball, D.M., Buolamwini, J.K., and Lee, R.E. (2008). Quantitative structure–activity relationship studies on nitrofuranyl anti-tubercular agents. Bioorg. Med. Chem. *16*, 8042–8453.

Hölscher, C., Reiling, N., Schaible, U.E., Hölscher, A., Bathmann, C., Korbel, D., Lenz, I., Sonntag, T., Kröger, S., Akira, S., *et al.* (2008). Containment of aerogenic *Mycobacterium tuberculosis* infection in mice does not require MyD88 adaptor function for TLR2, -4 and -9. Eur. J. Immunol. *38*, 680–694.

Houben, E.N., Walburger, A., Ferrari, G., Nguyen, L., Thompson, C.J., Miess, C., Vogel, G., Mueller, B., and Pieters, J. (2009). Differential expression of a virulence factor in pathogenic and non-pathogenic mycobacteria. Mol. Microbiol. *72*, 41–52.

Imai, K., Kurita-Ochiai, T., Ochiai, K. (2003). *Mycobacterium bovis* bacillus Calmette-Guérin infection promotes SOCS induction and inhibits IFN-γ-stimulated JAK/STAT signaling in J774 macrophages. FEMS Immunol. Med. Microbiol. *39*, 173–180.

Inohara, N. Chamaillard, M., McDonald C., and Nuñez, G. (2005). NOD-LRR proteins: role in host–microbial interactions and inflammatory disease. Annu. Rev. Biochem. *74*, 355–383.

Ioerger, T.R., and Sacchettini, J.C. (2009). Structural genomics approach to drug discovery for *Mycobacterium tuberculosis*. Curr. Opin. Microbiol. *12*, 318–325.

Jackson, M.D., and Denu, J.M. (2001). Molecular reactions of protein phosphatases-insights from structure and chemistry. Chem. Rev. *101*, 2313–2340.

Jackson, M., Raynaud, C., Lanéelle, M.A., Guilhot, C., Laurent-Winter, C., Ensergueix, D., Gicquel, B., and Daffé, M. (1999). Inactivation of the antigen 85C gene profoundly affects the mycolate content and alters the permeability of the *Mycobacterium tuberculosis* cell envelope. Mol. Microbiol. *31*, 1573–1587.

Jackson, M., Stadthagen, G., and Gicquel, B. (2007). Long-chain multiple methyl-branched fatty acid-containing lipids of *Mycobacterium tuberculosis*: biosynthesis, transport, regulation and biological activities. Tuberculosis (Edinb) *87*, 78–86.

Janin, Y.L. (2006). Antituberculosis drugs: ten years of research. Bioorg. Med. Chem. *15*, 2479–2513.

Jurado, J.O., Alvarez, I.B., Pasquinelli, V., Martínez, G.J., Quiroga, M.F., Abbate, E., Musella, R.M., Chuluyan, H.E., and García, V.E. (2008). Programmed death (PD)-*1*,PD-ligand 1/PD-ligand 2 pathway inhibits T-cell effector functions during human tuberculosis. J. Immunol. *181*, 116–125.

Kan-Sutton, C., Jagannath, C., and Hunter, R.L. Jr. (2009). Trehalose 6,6-dimycolate on the surface of *Mycobacterium tuberculosis* modulates surface marker expression for antigen presentation and costimulation in murine macrophages. Microbes Infect. *11*, 40–48.

Kantardjieff, K., and Rupp, B. (2004). Structural bioinformatic approaches to the discovery of new antimycobacterial drugs. Curr. Pharm. Des. *10*, 3195–3211.

Kaur, D., Guerin, M.E., Skovierová, H., Brennan, P.J., and Jackson, M. (2009). Chapter 2, Biogenesis of the cell wall and other glycoconjugates of *Mycobacterium tuberculosis*. Adv. Appl. Microbiol. *69*, 23–78.

Kawai, T., and Akira, S. (2007). Signaling to NF-κB by Toll-like receptors. Trends Mol. Med. *13*, 460–469.

Kawai, T., and Akira, S. (2009). The roles of TLRs, RLRs and NLRs in pathogen recognition. Int. Immunol. *21*, 317–337.

Kleinnijenhuis, J., Joosten, L.A., van de Veerdonk, F.L., Savage, N., van Crevel, R., Kullberg, B.J., van der Ven, A., Ottenhoff, T.H., Dinarello, C.A., van der Meer, J.W., *et al.* (2009). Transcriptional and inflammasome-mediated pathways for the induction of IL-1β production by *Mycobacterium tuberculosis*. Eur. J. Immunol. *39*, 1914–1922.

Koul, A., Herget T., Klebl B., and Ullrich A. (2004). Interplay between mycobacteria and host signalling pathways. Nat. Rev. Microbiol. *2*, 189–202.

Kuijl, C., Savage, N.D., Marsman, M., Tuin, A.W., Janssen, L., Egan, D.A., Ketema, M., van den Nieuwendijk, R., van den Eeden, S.J., Geluk, A., *et al.* (2007). Intracellular bacterial growth is controlled by a kinase network around PKB/AKT1. Nature *450*, 725–730.

Kumar, P., Kumar, D., Parikh, A., Rananaware, D., Gupta, M., Singh, Y., and Nandicoori, V.K. (2009). The *Mycobacterium tuberculosis* protein kinase K modulates activation of transcription from the promoter of mycobacterial monooxygenase operon through phosphorylation of the transcriptional regulator VirS. J. Biol. Chem. *284*, 11090–11099.

Kundu, M., Pathak, S.K., Kumawat, K., Basu, S., Chatterjee, G., Pathak, S., Noguchi, T., Takeda, K., Ichijo, H., Thien, C.B., Langdon, W.Y., and Basu, J. (2009). A TNF- and c-Cbl-dependent FLIP(S)-degradation pathway and its function in *Mycobacterium tuberculosis*-induced macrophage apoptosis. Nat. Immunol. *10*, 918–926.

Lammas, D.A., Stober, C., Harvey, C.J., Kendrick, N., Panchalingam, S., and Kumararatne, D.S. (1997). ATP-induced killing of mycobacteria by human macrophages is mediated by purinergic P2Z($P2X_7$) receptors. Immunity *7*, 433–444.

Lee, J., Hartman, M., and Kornfeld, H. (2009). Macrophage apoptosis in tuberculosis. Yonsei Med. J. *50*, 1–11.

Mahon, R.N., Rojas, R.E., Fulton, S.A., Franko, J.L., Harding, C.V., and Boom, W.H. (2009). *Mycobacterium tuberculosis* cell wall glycolipids directly inhibit $CD4^+$ T-cell activation by interfering with proximal T-cell-receptor signaling. Infect. Immun. *77*, 4574–4583.

Mamane, Y., S. Sharma, L. Petropoulos, R. Lin, J. Hiscott, J. (2000). Posttranslational regulation of IRF-4 activity by the immunophilin FKBP52. Immunity *12*, 129–140.

Manger, M., Scheck, M., Prinz, H., von Kries, J.P., Langer, T., Saxena, K., Schwalbe, H., Fürstner, A., Rademann, J., and Waldmann, H. (2005). Discovery of *Mycobacterium tuberculosis* protein tyrosine phosphatase A (MptpA) inhibitors based on natural products and a fragment-based approach. Chembiochem. *6*, 1749–1753.

Manvar, A.T., Pissurlenkar, R.R., Virsodia, V.R., Upadhyay, K.D., Manvar, D.R., Mishra, A.K., Acharya, H.D., Parecha, A.R., Dholakia, C.D., Shah, A.K., and Coutinho, E.C. (2010). Synthesis, *in vitro* antitubercular activity and 3-D-QSAR study of *1*,4-dihydropyridines. Mol. Divers. *14*, 285–305.

Marrakchi, H., Zhang, Y.M., and Rock, C.O. (2002). Mechanistic diversity and regulation of Type II fatty acid synthesis. Biochem. Soc. Trans. *30*, 1050–1055.

Matsuzawa, A., Saegusa, K., Noguchi, T., Sadamitsu, C., Nishitoh, H., Nagai, S., Koyasu, S., Matsumoto, K., Takeda, K., and Ichijo, H. (2005). ROS-dependent activation of the TRAF6-ASK1-p38 pathway is selectively required for TLR4-mediated innate immunity. Nat. Immunol. *6*, 587–592.

McKinney, J.D., Höner zu Bentrup, K., Muñoz-Elías, E.J. Miczak A, Chen, B., Chan, W.T., Swenson, D., Sacchettini, J.C., Jacobs, W.R.,Jr, and Russell, D.G. (2000). Persistence of *Mycobacterium tuberculosis* in macrophages and mice requires the glyoxylate shunt enzyme isocitrate lyase. Nature *406*, 735–738.

Mdluli, K., and Spigelman, M. (2006). Novel targets for tuberculosis drug discovery. Curr. Opin. Pharmacol. *6*, 459–467.

Medzhitov, R., and Janeway, C. Jr, (2000). The Toll receptor family and microbial recognition. Trends Microbiol. *8*, 452–456.

Mekenyan, O. (2002). Dynamic QSAR techniques: applications in drug design and toxicology. Curr. Pharm. Des. *8*, 1605–1621.

Molle, V., Brown, A.K., Besra, G.S., Cozzone, A.J., and Kremer, L. (2006). The condensing activities of the *Mycobacterium tuberculosis* type II fatty acid synthase are differentially regulated by phosphorylation. J. Biol. Chem. *281*, 30094–30103.

Moreno, C., Mehlert, A., and Lamb, J. (1988). The inhibitory effects of mycobacterial lipoarabinomannan and polysaccharides upon polyclonal and monoclonal human T-cell proliferation. Clin. Exp. Immunol. *74*, 206–210.

Murphy, D.J., and Brown, J.R. (2008). Novel drug target strategies against *Mycobacterium tuberculosis*. Curr. Opin. Microbiol. *11*, 422–427.

Nair, S., Ramaswamy, P.A., Ghosh, S., Joshi, D.C., Pathak, N., Siddiqui, I., Sharma, P., Hasnain, S.E., Mande, S.C., and Mukhopadhyay, S. (2009). The PPE18 of *Mycobacterium tuberculosis* interacts with TLR2 and activates IL-10 induction in macrophage. J. Immunol. *183*, 6269–6281.

Nakayama, T, and Yamashita, M. (2010). The TCR-mediated signaling pathways that control the direction of helper T-cell differentiation. Semin. Immunol. 2010 May 18 (Online).

Narayana, Y., and Balaji, K.N. (2008). NOTCH1 up-regulation and signaling involved in *Mycobacterium bovis* BCG-induced SOCS3 expression in macrophages. J. Biol. Chem. *283*, 12501–12511.

Narayan, A., Sachdeva, P., Sharma, K., Saini, A.K., Tyagi, A.K, and Singh, Y. (2007). Serine threonine protein kinases of mycobacterial genus: phylogeny to function. Physiol. Genomics *29*, 66–75.

Nayyar, A., Patel, S.R., Shaikh, M., Coutinho, E., and Jain, R. (2009). Synthesis, anti-tuberculosis activity and 3-D-QSAR study of amino acid conjugates of 4-(adamantan-1-yl) group containing quinolines. Eur. J. Med. Chem. *44*, 2017–2029.

Nguyen, L., Chinnapapagari, S., and Thompson, C.J. (2005a). FbpA-Dependent biosynthesis of trehalose dimycolate is required for the intrinsic multidrug resistance, cell wall structure, and colonial morphology of *Mycobacterium smegmatis*. J. Bacteriol. *187*, 6603–6611.

Nguyen, L., Walburger, A., Houben, E., Koul, A., Muller, S., Morbitzer, M., Klebl, B., Ferrari, G., and Pieters, J. (2005b). Role of protein kinase G in growth and glutamine metabolism of *Mycobacterium bovis* BCG. J. Bacteriol. *187*, 5852–5856.

Nigou, J., Vasselon, T., Ray, A., Constant, P., Gilleron, M., Besra, G.S., Sutcliffe, I., Tiraby, G., and Puzo, G. (2008). Mannan chain length controls lipoglycans signaling via and binding to TLR2. J. Immunol. *180*, 6696–6702.

Nören-Müller, A., Wilk, W., Saxena, K., Schwalbe, H., Kaiser, M., and Waldmann, H. (2008). Discovery of a new class of inhibitors of *Mycobacterium tuberculosis* protein tyrosine phosphatase B by biology-oriented synthesis. Angew. Chem. Int. Ed. Engl. *47*, 5973–5977.

O'Brien, R.J., and Vernon, A.A. (1998). New tuberculosis drug development. How can we do better? Am. J. Respir. Crit. Care Med. *157*, 1705–1707.

Okkels, L.M., Brock, I., Follmann, F., Agger, E.M., Arend, S.M., Ottenhoff, T.H., Oftung, F., Rosenkrands, I., and Andersen, P. (2003). PPE protein (Rv3873) from DNA segment RD1 of *Mycobacterium tuberculosis*: strong recognition of both specific T-cell epitopes and epitopes conserved within the PPE family. Infect. Immun. *71*, 6116–6123.

Palma-Nicolás, J.P., Hernández-Pando, R., Segura, E., Ibarra-Sánchez, M.J., Estrada-García, I., Zentella-Dehesa, A., and López-Marín, L.M. (2010). Mycobacterial di-O-acyl trehalose inhibits Th-1 cytokine gene expression in murine cells by down-modulation of MAPK signaling. Immunobiology *215*, 143–152.

Pandey, A.K., Yang, Y., Jiang, Z., Fortune, S.M., Coulombe, F., Behr, M.A., Fitzgerald, K.A., Sassetti, C.M., and Kelliher, M.A. (2009). NOD2, RIP2 and IRF5 play a critical role in the type I interferon response to *Mycobacterium tuberculosis*. PLoS Pathog. *5*, e100050.

Papavinasasundaram, K.G., Chan, B., Chung, J.H., Colston, M.J., Davis, E.O., and Av-Gay, Y. (2005). Deletion of the *Mycobacterium tuberculosis pknH* gene confers a higher bacillary load during the chronic phase of infection in BALB/c mice. J. Bacteriol. *187*, 5751–5760.

Parish, T., Smith, D.A., Kendall, S., Casali, N., Bancroft, G.J., and Stoker, N.G. (2003). Deletion of two-component regulatory systems increases the virulence of *Mycobacterium tuberculosis*. Infect. Immun. *71*, 1134–1140.

Parikh, A., Verma, S.K., Khan, S., Prakash, B., and Nandicoori, V.K. (2009). PknB-mediated phosphorylation of a novel substrate, *N*-acetylglucosamine-1- phosphate uridyltransferase, modulates its acetyltransferase activity. J. Mol. Biol. *386*, 451–464.

Pathak, S.K., Basu, S., Basu, K.K., Banerjee, A., Pathak, S., Bhattacharyya, A., Kaisho, T., Kundu, M., and Basu, J. (2007). Direct extracellular interaction between the early secreted antigen ESAT-6 of *Mycobacterium tuberculosis* and TLR2 inhibits TLR signaling in macrophages. Nat. Immunol. *8*, 610–618.

Philips, J.A. (2008). Mycobacterial manipulation of vacuolar sorting. Cell. Microbiol. *10*, 2408–2415.

Pullen, K.E., Ng, H.L., Sung, P.Y., Good, M.C., Smith, S.M., and Alber, T. (2004). An alternate conformation and a third metal in PstP/Ppp, the *M. tuberculosis* PP2C-Family Ser/Thr protein phosphatase. Structure *12*, 1947–1954.

Quiroga, M.F., Jurado, J.O., Martínez, G.J., Pasquinelli, V., Musella, R.M., Abbate, E., Issekutz, A.C., Bracco, M.M., Malbran, A., Sieling, P.A., *et al.* (2007). Cross-talk between CD31 and the signaling lymphocytic activation molecule-associated protein during interferon-γ production against *Mycobacterium tuberculosis*. J. Infect. Dis. *196*, 1369–1378.

Rich, K.A., Burkett, C., and Webster, P. (2003). Cytoplasmic bacteria can be targets for autophagy. Cell. Microbiol. *5*, 455–468.

Riley, J.L. (2009). PD-1 signaling in primary T cells. Immunol. Rev. *229*, 114–125.

Rook, G.A. (2007). Th2 cytokines in susceptibility to tuberculosis. Curr. Mol. Med. *7*, 327–337.

Russell, D.G., Dant, J., and Sturgill-Koszycki, S. (1996). *Mycobacterium avium*- and *Mycobacterium tuberculosis*-containing vacuoles are dynamic, fusion-competent vesicles that are accessible to glycosphingolipids from the host cell plasmalemma. J. Immunol. *156*, 4764–4773.

Sassetti, C.M., Boyd, D.H., and Rubin, E.J. (2003). Genes required for mycobacterial growth defined by high density mutagenesis. Mol. Microbiol. *48*, 77–84.

Shabaana, A.K., Kulangara, K., Semac, I., Parel, Y., Ilangumaran, S., Dharmalingam, K., Chizzolini, C., and Hoessli, DC. (2005). Mycobacterial lipoarabinomannans modulate cytokine production in human T helper cells by interfering with raft/microdomain signalling. Cell Mol. Life Sci. *62*, 179–187.

Schepers, E., Glorieux, G., Dhondt, A., Leybaert, L., and Vanholder, R. (2009). Role of symmetric dimethylarginine in vascular damage by increasing ROS via store-operated calcium influx in monocytes. Nephrol. Dial. Transplant. *24*, 1429–1435.

Scherr, N., Honnappa, S., Kunz, G., Mueller, P., Jayachandran, R., Winkler, F., Pieters, J., and Steinmetz, M.O. (2007). Structural basis for the specific inhibition of protein kinase G, a virulence factor of *Mycobacterium tuberculosis*. Proc. Natl. Acad. Sci. U.S.A. *104*, 12151–12156.

Scherr, N., Müller, P., Perisa, D., Combaluzier, B., Jenö, P., and Pieters, J. (2009). Survival of pathogenic mycobacteria in macrophages is mediated through autophosphorylation of protein kinase G. J. Bacteriol. *191*, 4546–4554.

Schreiber, M., Res, I., and Matter, A. (2009). Protein kinases as antibacterial targets. Curr. Opin. Cell Biol. *21*, 325–330.

Shagufta, Kumar, A., Panda, G., and Siddiqi, M.I. (2007). CoMFA and CoMSIA 3-D-QSAR analysis of diaryloxy-methano-phenanthrene derivatives as anti-tubercular agents. J. Mol. Model *13*, 99–109.

Shimizu, T., Tomioka, H., Sato, K., Sano, C., Akaki, T., Dekio, S., Yamada, Y., Kamei, T., Shibata, H., and Higashi, N. (1999a). Effects of the Chinese traditional medicine Mao-Bushi-Saishin-To on therapeutic efficacy of a new benzoxazinorifamycin, KRM-*1648*, against *Mycobacterium avium* infection in mice. Antimicrob. Agents. Chemother. *43*, 514–519.

Shimizu, T., Tomioka, H., Sato, K., Sano, C., Yamada, Y., Shibata, H., and Higashi., N. (1999b). Effects of Yokuinin on the therapeutic efficacy of a new benzoxazinorifamycin KRM-1648 against *Mycobacterium avium* infection. Int. J. Antimicrob. Agents *11*, 69–74.

Shin, D.M., Yang, C.S., Lee, J.Y., Lee, S.J., Choi, H.H., Lee, H.M., Yuk, J.M., Harding, C.V., and Jo, E.K. (2008). *Mycobacterium tuberculosis* lipoprotein-induced association of TLR2 with protein kinase Cζ in lipid rafts contributes to reactive oxygen species-dependent inflammatory signalling in macrophages. Cell Microbiol. *10*, 1893–1905.

Signoretto, C., Di Stefano, F., and Canepari, P. (1996). Modified peptidoglycan chemical composition in shape-altered *Escherichia coli*. Microbiology *142*, 1919–1926.

Sikorski, R.S., Boguski, M.S., Goebl, M., and Hieter, P. (1990). A repeating amino acid motif in CDC23 defines a family of proteins and a new relationship among genes required for mitosis and RNA synthesis. Cell *60*, 307–317.

Silva, R.A., Pais, T.F., and Appleberg, R. (2001). Blocking the receptor for IL-10 improves antimycobacterial chemotherapy and vaccination. J. Immunol. *167*, 1535–1541.

Singh, R., Rao, V., Shakila, H., Gupta, R., Khera, A., Dhar, N., Singh, A., Koul, A., Singh, Y., Naseema, M., Narayanan, P.R., Paramasivan, C.N., Ramanathan, V.D., and Tyagi, A.K. (2003). Disruption of *mptpB* impairs the ability of *Mycobacterium tuberculosis* to survive in guinea pigs. Mol. Microbiol. *50*, 751–762.

Singh, A., Singh, Y., Pine, R., Shi, L., Chandra, R., and Drlica, K. (2006). Protein kinase I of *Mycobacterium tuberculosis*: cellular localization and expression during infection of macrophage-like cells. Tuberculosis (Edinb) *86*, 28–33.

Part IV

Biological Agents for Antibacterial Therapy

Current Strategies for Antibacterial Vaccine Development

13

Robert G.K. Donald and Annaliesa S. Anderson*

Abstract

Prophylactic antibacterial vaccines have been responsible for a drastic reduction in global bacterial diseases. Older vaccines made from attenuated whole cells or lysates have been replaced by less reactogenic acellular vaccines made with purified components, including capsular polysaccharides and their conjugates to protein carriers, inactivated toxins (toxoids) and proteins. Examples of vaccines in each category are reviewed to illustrate underlying strategies and associated technological advances such as polysaccharide conjugation and recombinant protein expression. In addition, progress and the current status in the development of new vaccines to prevent diseases caused by *Neisseria meningitidis* serogroup B, *Staphylococcus aureus* and *Clostridium difficile* is summarized. Future progress will probably bring to the clinic passive immunotherapies based on monoclonal antibodies and new adjuvants, especially for use in vaccines against intracellular pathogens.

Introduction

The transforming effect of vaccines on global mortality is well established, exceeded only in impact by improvements in public hygiene and access to safe drinking water (Plotkin, 2008a). Broadly protective immunity elicited in vaccinated populations has led to widespread suppression or even elimination of disease. Owing to comprehensive childhood vaccicnation programmes, viral diseases such as smallpox and polio have been eliminated or nearly completely suppressed, while bacterial infections due to *Corynebacterium diphtheriae*, *Haemophilus influenzae* and *Streptococcus pneumoniae* have become much less prevalent (Roush *et al.*, 2007). For contagious diseases such as these, widespread vaccination not only protects the vaccinated from infection but also protects a vulnerable unvaccinated subpopulation by blocking transmission and reducing pathogen circulation. This principal, known as herd immunity, has become a fundamental public health control strategy to prevent epidemic outbreaks of infectious disease.

With the implementation of modern genomics and proteomics technology has come a new generation of vaccines made from well-characterized components (a previously reviewed topic; Zagursky *et al.*, 2008). The development of conjugation technologies has significantly improved immune responses to polysaccharide antigens. As a result of these advances, prophylactic vaccines based on live attenuated whole cells have largely been replaced by ones based on purified carbohydrate and protein antigens. Therapeutic vaccines based on humanized monoclonal antibodies currently in development have reinvented the strategy of emergency treatment for acute infections using sera from vaccinated animals, a practice that was abandoned with the clinical introduction of the first antibiotics. The manufacturing of commercial vaccines began with horse farms in the early 1900s, such as Lederle Laboratories in Pearl River (New York, USA) that produced immunoglobulin antitoxins

*Corresponding author

to treat diphtheria and tetanus. One hundred years later, the same site manufactures the prophylactic pneumococcal conjugate vaccines, Prevnar and Prevnar-13.

Listed in Table 13.1 are antibacterial vaccines currently used in the USA and examples of new vaccines in clinical development. They target a variety of distinct cellular components, which are broadly categorized and summarized in Fig. 13.1. Antigens include the full complement of surface antigens presented by the attenuated whole cell; secreted toxins; capsular polysaccharides (CPS); outer membrane vesicle preparations (OMVs); surface protein virulence factors; and combinations of CPS and protein antigens. Representative vaccines in each category will be discussed to illustrate the underlying technologies and associated development challenges. Another consistent theme in modern vaccinology is the important role that molecular epidemiology plays in both the initial selection of target antigens as well as in post-vaccination surveillance to monitor pathogen adaptation and vaccine coverage.

Bordetella pertussis – from whole cell to acellular antigens

Attenuated whole-cell vaccines were once the mainstay of the prophylactic antibacterial vaccine arsenal but have fallen out of favour in developed

Table 13.1 Vaccines for the prevention or treatment of bacterial diseases[1]

Species	Disease	Antigen
Currently available in the USA[1]		
Mycobacterium tuberculosis	TB	Whole cell[2]
Vibrio cholerae	Cholera	Whole cell
Salmonella enterica type Typhi	Typhoid	Whole cell or CPS
Bordetella pertussis	Pertussis (whooping cough)	Acellular (includes toxoid)
Bacillus anthracis	Anthrax	Acellular (includes PA)
Clostridium tetani	Tetanus	Toxoid
Corynebacterium diphtheriae	Diphtheria	Toxoid
Haemophilus influenzae type b	Meningitis	CPS-conjugate
Neisseria meningitidis type A, C, W_{135},Y	Meningitis	CPS or CPS-conjugate
Streptococcus pneumoniae	Pneumonia	CPS or CPS-conjugate
In development		
Bacillus anthracis	Anthrax	Protein (recombinant PA) (Emergent)
Clostridium difficile	*C. difficile*-associated disease	Toxoid (Sanofi); mAb immunotherapy[3] (Merck)
Neisseria meningitidis	Type B meningitis	Protein (Pfizer); proteins + OMV (Novartis)
Staphylococcus aureus	Sepsis, soft-tissue infections	CPS-conjugate + proteins (Pfizer, GlaxoSmithKline); protein (Merck)
Staphylococcus spp.	Sepsis	mAb immunotherapy[3] (Biosynexus)
Pseudomonas aeruginosa	VAP and sepsis	Proteins (Intercell)

[1]Vaccines approved for public use, see also http://www.cdc.gov/vaccines/vpd-vac/vaccines-list.htm.

[2]BCG vaccination is reserved for use in high risk settings (e.g. health care workers exposed to MDR strains).

[3]Therapeutic vaccine.

TB, tuberculosis; CBS, CPS, capsular polysaccharide; VAP, ventilator-associated pneumonia.

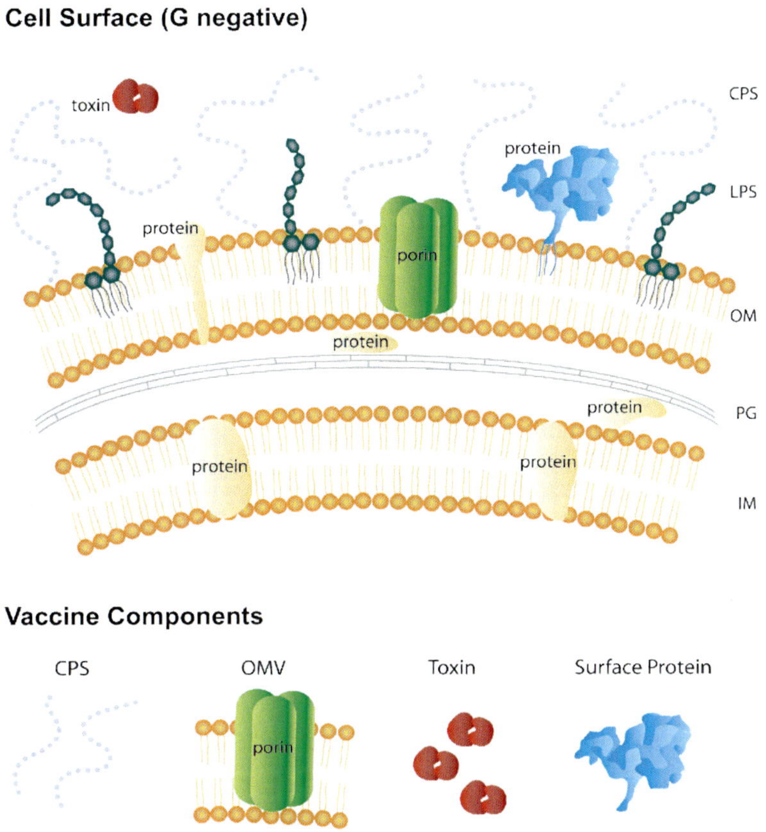

Figure 13.1 Bacterial vaccine targets. The surface of a generic Gram-negative bacterium and derived subcellular components used as vaccine antigens are illustrated. Established targets include the chemically or genetically attenuated whole cell, capsular polysaccharide (CPS), outer membrane vesicles (OMVs), secreted toxins, surface proteins or assorted combinations. Also depicted is the outer membrane (OM), peptidoglycan cell wall (PG), the inner membrane (IM), and lipopolysaccharide (LPS). Gram-positive bacteria (not shown) lack the outer membrane and do not make OMVs. They present characteristic lipid-anchored or PG-anchored teichoic acid polymers that are also under investigation as possible G^+ bacteria vaccine antigens (see text).

countries primarily owing to higher reactogenicity and safety concerns. The *Bordetella pertussis* whole-cell vaccine was discontinued in the USA in the 1990s, but the historical context that led to its replacement by acellular versions is worth mentioning. The only remaining whole-cell antibacterial vaccines still licensed in the USA are the typhus (Ty21a) and cholera (CVD 103-HgR) vaccines used by travellers to endemic areas.

Prior to the introduction of whole-cell vaccines, pertussis (whooping cough) was a major cause of infant death worldwide. *Bordetella pertussis* is a Gram-negative bacterium that infects its host by colonizing lung epithelial cells. Pertussis is found only in humans, with its severe pulmonary 'whooping cough' symptoms caused by the pertussis toxin (Pittman, 1984, 1991). Following the introduction of attenuated whole cell vaccines, the incidence of the disease declined steadily in the USA from approximately 10 deaths per 100,000 population to less than 0.01 (Cherry *et al.*, 1988). The widely used vaccine known as DPT is a combination of a bacterial suspension (killed by heat or chemical agent) and chemically inactivated toxins (or 'toxoids') from tetanus and diphtheria. Such combination vaccines were developed to minimize the number of injections given during routine infant vaccinations. In 1982 a television documentary *'DPT: Vaccine Roulette'* set off a media storm by suggesting that the vaccine caused serious neurological adverse events in children. The resulting public outcry led to lawsuits and increased liability that contributed to the reduction in vaccine research, development and manufacturing globally. A taskforce set up to review the available scientific data concluded that the benefits of pertussis immunization clearly outweighed possible risks and recommended that immunization programmes be continued (Cherry *et al.*, 1988). Subsequent epidemiology studies have consistently failed to establish a significant link between vaccine use and neurological problems. Important findings from a 2006 study appear to explain rare cases of encephalopathy (brain injury) suspected of being attributable to DPT vaccination. Investigators reported that 11 of 14 patients diagnosed with encephalopathy following DPT vaccination who met clinical criteria for severe myoclonic epilepsy of infancy (SMEI) were found to harbour mutations in the sodium channel gene *SCN1A*. As mutations in this gene are independently known to cause the condition, the authors concluded that the vaccine cannot have been responsible for these cases (Berkovic *et al.*, 2006). An important positive outcome of this otherwise unfortunate vaccine safety controversy was the passing of the US National Childhood Vaccine Injury Act (NCVIA) in 1986 to limit the financial liability of the vaccine industry owing to injury claims, and to ensure a stable market supply to protect public health. The law also established a vaccine adverse event reporting system (VAERS) as a sentinel for potential problems.

Several protein-based acellular combination vaccines have been licensed to replace whole cell-based DPT. Head-to-head comparisons of old with new in clinical trials have demonstrated equivalent (Gustafsson *et al.*, 1996; Stehr *et al.*, 1998) or superior (Greco *et al.*, 1996) efficacy in infants. The five-component pertussis vaccine comprises virulence factors including chemically or genetically inactivated pertussis toxin (or 'toxoid'), haemagglutinin (a surface adhesin), pertactin (an outer membrane transport protein) and two fimbriae (appendage proteins). Other vaccine variants contain fewer pertussis components (e.g. toxoid and haemagglutinin only) but all are used in combination with diphtheria and tetanus toxoids. The three acellular multidose infant vaccines currently licensed in the USA (known as DTaP) show comparable efficacy and contain either two pertussis components (Infanrix™, Tripedia®) or five (Acel-imune®). Available evidence suggests that the pertussis toxin causes most if not all disease symptoms and that it can induce long-term immunity in the absence of other components (Pittman, 1984; Robbins *et al.*, 1993). Nevertheless, clinical trial results suggest that acellular multicomponent pertussis vaccines are more efficacious than a single component pertussis toxin vaccine (Storsaeter *et al.*, 1990). Most significantly, the new acellular vaccines show improved safety and lack the side-effects of DPT, which commonly include injection site reactions (redness, swelling and pain) and transient fever.

Although pertussis vaccination is now an integral part of the standard childhood vaccine regimen, sporadic outbreaks of disease still occur globally, including the U.S. Waning immunity in adolescents and adults is thought to be a primary culprit. Protective immunity after vaccination is lost after 4–12 years with either whole cell or acellular vaccines (Wendelboe *et al.*, 2005). As unimmunized infants are at risk for severe disease, vaccination of adolescents and adults in close contact is expected to reduce incidence if high coverage rates can be achieved (Forsyth *et al.*, 2007; Wendelboe *et al.*, 2007). In 2005, two new single dose booster vaccines with inactivated tetanus, diphtheria, and pertussis toxin components (known as Tdap) were licensed to fulfil this objective (Plosker and Plosker, 2009). They contain the same active components but reduced doses of the diphtheria and pertussis antigens compared with the analogous infant DTaP vaccines.

Capsular polysaccharides as antigens

The capsular polysaccharide (CPS) is an essential virulence determinant for many pathogenic bacteria. It can confer resistance to phagocytosis and bacterial killing, in some cases by masking underlying antigens and by interfering with complement fixation (Merino *et al.*, 1992; Thurlow *et al.*, 2009). The recognition that natural immunity to disease is strongly correlated with antibody titres to capsular antigens has led to the development of vaccines for *Haemophilus influenza*, *Neisseria meningitidis* and *S. pneumoniae* (the pneumococcus; Plotkin, 2008b). These encapsulated pathogens are the main cause of severe and frequently lethal meningitis in infants and young children. The hyper-susceptibility of young children to these diseases is no coincidence and is now understood to be attributable to B cell immaturity (Siegrist 2007). Unlike protein antigens which induce T-cell-dependent responses, polysaccharides elicit T-cell independent immune response through direct stimulation of B cells. A weak early life B-cell response also explains why bacterial meningitis vaccines based on purified CPS are ineffective in infants but not in older children and adults.

The solution of improving the immunogenicity of polysaccharides as vaccines comes from the concept of hapten-conjugation described by Landsteiner in 1924 (Landsteiner, 1945). Coupling the hapten to a protein carrier allows the hapten to be recognized and antibodies to be formed. This hapten carrier effect is thought to derive from cooperative interactions between B and T lymphocytes, which recognize respective hapten and carrier antigens (Bartlett and Noelle, 1991). Similarly, through conjugation to a protein carrier, a weak or undeveloped B-cell mediated response to polysaccharide can be converted to a robust long-lived antibody response involving T-helper cells. Using this strategy, improved carbohydrate specificity and immunogenicity was first demonstrated by Avery and Goebel in 1929 (Avery and Goebel, 1929). Based on this principle, the first CPS–polysaccharide conjugate vaccine was developed for *H. influenzae* by Schneerson and Robbins (Schneerson and Robbins, 1980). Immunization of infants with the vaccine yielded the first demonstration of immunogenicity and protective efficacy for this population in clinical trials (Eskola *et al.*, 1985, 1987). The carrier protein in this case was the diphtheria toxoid, although tetanus toxoid or a recombinant diphtheria toxoid known as CRM_{197} have shown equivalent efficacy. *H. influenzae* vaccines utilizing these three protein carriers, produced by different manufacturers, were subsequently approved for use in infants. This breakthrough paved the way for analogous meningococcal and pneumococcal CPS conjugate vaccines that subsequently became licensed in 1999 and 2000 (Makela and Makela 2003; Pace *et al.*, 2009).

More recently, the specific cellular interactions responsible for CPS-conjugate induced immunity have been confirmed using a compartmentalized transwell *in vitro* culture system, in which responses to antigenic components of the *H. influenzae* vaccine were measured in *in vivo*-primed immune cells isolated from vaccinated patients (Breukels *et al.*, 1999). Antibody-secreting B cells induced by CPS-conjugates can be detected following direct contact between T cells (activated by exposure to the protein carrier) and polysaccharide-specific B cells. This interaction releases diffusible cytokines that stimulate B-cell differentiation and plasma cell maturation. The proposed scheme is summarized in Fig. 13.2.

Pneumococcal capsular antigen vaccines

S. pneumoniae is a leading cause of vaccine-preventable illness and death world-wide. Invasive meningitis, pneumonia and bacteraemia, or non-invasive ear-infections (acute otitis media) and sinusitis are all manifestations of pneumococcal disease. At high risk for disease are infants, the elderly and the immunocompromised. The disease is especially problematic in developing countries where based on World Health Organization (WHO) estimates approximately 1.6 million people, mostly children, are killed each year (Anonymous, 2007). The increase in resistance to penicillin among some strains makes prevention through vaccination even more important. As non-encapsulated strains do not cause disease, the capsule is considered the primary virulence determinant. Current vaccines like Prevnar B target the CPS of the most clinically relevant strains.

The unconjugated pneumococcal polysaccharide vaccine (PPSV, Pneumovax-23) was licensed in 1983 for use in children over two years of age and adults. The vaccine contains polysaccharide purified from 23 different pneumococcal capsular types. For the reasons mentioned above it is not effective in infants and it confers relatively

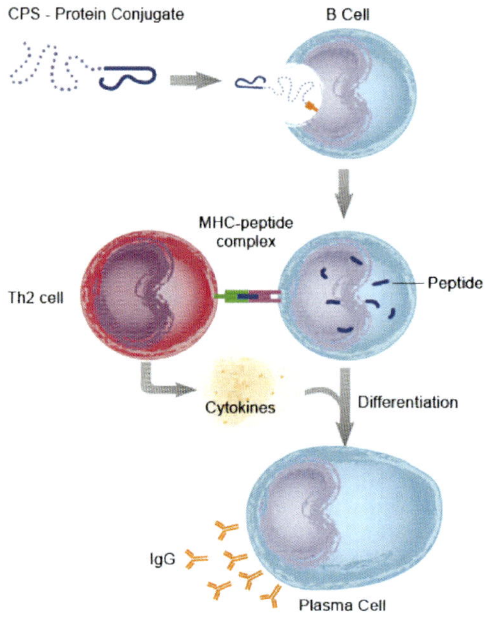

Figure 13.2 Antibody responses to capsular polysaccharide (CPS) conjugates. Polysaccharide-specific B cells internalize the CPS-conjugate, process the protein component, and present the peptides to their MHC molecules to activate matching peptide-specific helper T (Th2) cells. Activated Th2 cells then secrete cytokines (interleukins) that activate polysaccharide-specific B cells to become antibody secreting plasma cells. The resulting antibodies mediate opsonophagocytic and complement-mediated bactericidal killing of encapsulated bacteria.

short-lived immunity (five to ten years) that requires 'booster' vaccinations to extend coverage. The efficacy of this vaccine to prevent community acquired pneumonia, especially among the elderly or adults with chronic illness is controversial (Andrews *et al.*, 2009; Huss *et al.*, 2009). Nevertheless, it has served as a useful safety and immunogenicity comparator for subsequent conjugate vaccines in clinical trials.

The first pneumococcal conjugate vaccine (PCV7, Prevnar™) has had a rapid and profound impact on global public health. This 7-valent vaccine consists of CPS purified from the most problematic disease causing pneumococcal strains (serotypes 4, 6B, 9V, 14, 18C, 19F and 23F) conjugated to recombinant diphtheria toxoid CRM_{197}. Since its introduction into national infant immunization programs in nearly 30 countries, numerous studies from different geographic areas have documented substantial reductions in the incidence of invasive pneumococcal disease (IPD), pneumonia and ear-infections (Fletcher *et al.*, 2007; Dinleyici *et al.*, 2008). In the USA alone, a CDC study found that the vaccine resulted in a 77% reduction in IPD incidence for children below 5 years of age, comparing rates in 2005 with a pre-vaccination baseline of 1998–1999 (Centers for Disease and Prevention, 2008). The resulting indirect or 'herd' immunity has also undoubtedly reduced the occurrence of pneumococcal disease in older patients, although the availability of the 23-valent polysaccharide vaccine makes this hard to quantify. In a head-to-head comparison of PCV7 with Pneumovax-23 in an elderly population, the conjugate vaccine elicited improved antibacterial immune responses and immunological memory (de Roux *et al.*, 2008). Whilst the polysaccharide vaccine was incapable of inducing protective levels of antibodies when used as a priming dose, the conjugate vaccine permitted subsequent boosting with either vaccine to maintain functional levels of antipolysaccharide antibody. The conjugate vaccine is also superior to the polysaccharide vaccine for preventing life-threatening opportunistic pneumococcal coinfection in human immunodeficiency virus patients. *S. pneumoniae* is the most common bacterial respiratory pathogen in HIV patients, who are hypersusceptible. Evidence of HIV patient efficacy with the unconjugated polysaccharide vaccine is lacking and therefore is not recommended (Anonymous, 2008). In contrast, the conjugate vaccine protected HIV-infected adults from recurrent pneumococcal infection caused by vaccine serotypes (French *et al.*, 2010).

Another positive impact of PCV7 has been its effect on nasopharyngeal carriage. For the invasive meningitis causing pathogens such as *S. pneumoniae*, *N. meningitidis and H. influenzae,* colonization on the nasal mucosa is a common feature and prerequisite for promoting transmission to new hosts. The pioneering *H. influenzae* conjugate vaccines dramatically reduced carriage in children once routine vaccination became widespread (Bisgard *et al.*, 1998). The PCV7 vaccine has similarly reduced the nasal carriage of the vaccine serotypes, resulting in decreased transmission of the seven most prevalent and antibiotic resistant strains circulating prior to its introduction. In their place have come to prominence serotypes not included in the original vaccine, including type 19A which is associated with high-level penicillin and multidrug resistance (Centres for Disease and Prevention, 2008; Huang *et al.*, 2009; Reinert *et al.*, 2010). Such strain replacement creates a need for surveillance and periodic adjustment, with the introduction of new CPS-conjugates to prevent disease caused by non-vaccine strains. Despite considerable effort, alternative vaccines based on broadly distributed pneumococcal antigens have not been identified.

To address the changing epidemiology, new conjugate vaccines with expanded coverage have been developed. A 10-valent vaccine manufactured by GlaxoSmithKline (GSK) (Synflorix), expands coverage of PCV7 to three additional serotypes (1, 5 and 7F), and was approved for use in Europe in 2009. Pfizer's 13-valent successor to PCV7 (PCV13), covers an additional six serotypes (1, 3, 5, 6A, 7F and 19A) and was recently licensed in both Europe and U.S. for use in infants and young children (Centres for Disease and Prevention, 2010). The broader coverage offered by PCV13 (especially for the pivotal 19A serotype) provides an efficacy advantage.

For practical and ethical reasons demonstration of efficacy for PCV13 licensure is based on well-established protective correlates of functional activity and immunogenicity rather than on clinical efficacy data. The opsonophagocytic assay (OPA) measures the complement-dependent bactericidal killing activity of immune sera by polymorphonuclear leucocytes (PMNs) or human promyelocytic leukaemia cells (HL-60s). Immunogenicity of sera to capsular antigens is measured by standardized ELISA. The true impact of PCV13 on different populations globally will become apparent with continued surveillance of carriage and disease. Post licensure molecular epidemiological monitoring will also detect the potential for non-vaccine strains to gain a foothold.

Neisseria meningitidis (serogroup B) – the protein antigen alternative

Neisseria meningitidis is a commensal pathogen that colonizes the respiratory tract of healthy individuals. Invasive disease is rare and sporadic in industrialized countries but the symptoms are severe and debilitating. The fatality rate is ~10% even with treatment and 20% of survivors may experience brain damage, deafness and amputations (Rosenstein *et al.*, 2001). In the USA and Europe incidence per 100,000 varies from 0.35 to 4, but epidemic outbreaks in Africa yield rates as high as 10–1000 (Khatami *et al.*, 2010). Owing to rapid onset of disease antibiotic treatment often comes too late. Disease is caused by encapsulated strains and CPS-based vaccines are effective at preventing *N. meningitidis* disease caused by four of five major serogroups. As with analogous pneumococcal and *H. influenzae* vaccines, earlier unconjugated vaccines based on purified *Neisseria meningitidis* CPS antigens have been followed by highly successful CPS-conjugate vaccines that show improved immunogenicity in infants and have greater impact on carriage and transmission. A vaccination campaign in the UK with monovalent serogroup C conjugate vaccines significantly reduced the overall burden of meningococcal disease. Tetravalent conjugate vaccines that cover A,

C, Y and W serotypes are now available (Meningococcal Reference Unit *et al.*, 2006; Pace *et al.*, 2009). An inexpensive meningococcal conjugate A vaccine (MenAfriVac™) developed specifically to combat serotype A epidemics in sub-Saharan Africa is currently being distributed.

Despite the success of CPS-based meningococcal vaccines, a large gap in coverage currently exists for serogroup B, which is the predominant cause of invasive disease in developed countries where the other serogroups are covered. There is clearly an unmet need for a safe serogroup B (MenB) vaccine to protect infants and adolescents. The quest for such a vaccine has required strategies involving non-capsular antigens. The capsule of serotype B strains is poorly immunogenic, most probably due to immune tolerance induced by structural similarity of the α2–8-linked polysialic acid component and neural cell glycoproteins (Finne *et al.*, 1983). Concerns about the potential for autoimmunity and the lack of success with a modified B polysaccharide vaccine (Bruge *et al.*, 2004) have lead to the abandonment of the traditional CPS-based approach and the investigation of alternative antigens.

Outer membrane vesicles (OMVs) are produced naturally by Gram-negative bacteria during infections. This can be mimicked *in vitro*, and such vesicles can be purified from the pathogens and processed with detergents to make vaccines. These extracted vesicles are a heterogeneous mixture of lipopolysaccharide, phospholipid and integral outer membrane proteins. In the case of MenB OMVs, the dominant immune response is directed against the surface antigen porin A (PorA; Martin *et al.*, 2006). This integral membrane porin is a virulence factor that binds to complement (factor C4BP) and helps the pathogen subvert complement-mediated killing by the host (Jarva *et al.*, 2005). OMVs have achieved some success in controlling isolated MenB epidemics in Cuba (VA-MENCOG-BC) and New Zealand (MeNZB™) caused by specific clonal subtypes (Oster *et al.*, 2005; Sierra *et al.*, 1991). Unfortunately, the narrow coverage of OMV vaccines for clinically relevant serotype B strains has restricted their use. Only short loops of PorA are exposed on the cell surface, and these are highly variable in amino acid sequence resulting in an immune response that is highly specific to strain subtype.

The rational search for protein antigens as components of more broadly protective MenB vaccines has been facilitated by the availability of genomic, proteomic and genetic tools. Key desirable properties include surface accessibility, broadly cross-protective immunogenicity and role as an essential virulence factor. An attractive vaccine candidate with all of these qualities, now known as the factor H binding protein (fHBP), has been identified independently by research groups at Novartis and Wyeth (now Pfizer) (Bernfield *et al.*, 2002; Masignani *et al.*, 2003; Fletcher *et al.*, 2004). The Novartis team mined the MenB genome for signal peptide-containing surface proteins. Antigens were prioritized based on their ability to elicit antibodies that are both bactericidal and protective in a passive immunization (infant rat) model. The Wyeth/Pfizer team used a biochemical approach, identifying the protein as a component in a fractionated outer membrane preparation that was capable of inducing a broadly PorA-independent cross-reactive bactericidal response against a panel of diverse MenB strains. Flow cytometry experiments demonstrated that the candidate is accessible on the cell surface in the presence of the serogroup B capsule (Masignani *et al.*, 2003; Fletcher *et al.*, 2004; McNeil *et al.*, 2009). Its ability to bind complement factor H (fH) was shown by the ability of specific mAbs to block fH binding and by the inability of fH to bind to the surface of deletion mutants (Madico *et al.*, 2006; Schneider *et al.*, 2006). Direct evidence for high-affinity binding has come from X-ray crystallography studies (Schneider *et al.*, 2009). Knock-out mutants are hyper-susceptible to killing by whole blood or serum, consistent with a biological role for fHBP in subverting complement fixation (Welsch *et al.*, 2008). A comprehensive analysis of 1837 invasive geographically diverse MenB isolates shows the gene is ubiquitous (Murphy *et al.*, 2009). Sequence variability clusters into two phylogenetic subfamilies (A and B) with individual members capable of eliciting bactericidal antibodies that are subfamily specific. A bivalent vaccine containing representative fHBPs from both subfamilies forms the basis of the Pfizer

MenB vaccine that is currently under evaluation in clinical trials (Jiang *et al.*, 2010).

The Novartis experimental multicomponent MenB vaccine (5CVMB) contains a single fHBP B-family variant and two other antigens, neisserial heparin-binding antigen (NHBA) and NadA, an adhesion and virulence factor (Giuliani *et al.*, 2006). While NHBA is well represented across the clinical strains, its ability to elicit bactericidal antibodies with human complement is limited (Welsch *et al.*, 2003). The NadA antigen is represented in only a minority of strains. The fHBP and NHBA polypeptides are fused to two MenB accessory proteins, also identified by genome mining but which do not elicit bactericidal antibodies. The vaccine is formulated with an OMV preparation from the MeNZB™ vaccine strain, which contributes a specific PorA subtype from the New Zealand epidemic as primary immunogen (Martin *et al.*, 2006). The intent of this complex cocktail of antigens is to increase the breadth of vaccine coverage and to minimize escape mutants, although available evidence predicts that the simpler bivalent fHBP vaccine competitor is unlikely to generate escapes because of the profoundly reduced virulence of fHBP null mutants (Seib *et al.*, 2010) and the broad distribution of the fHBP antigen. Recent studies have also demonstrated that the additional antigens have little benefit over fHBP (Findlow *et al.*, 2010; Brunelli *et al.*, 2011). The possibility of gain-of-function escape mutants in which fHBP is replaced by a distinct but potentially redundant fHBP activity has recently been raised (Lewis *et al.*, 2010). In this study the surface-expressed NspA antigen was shown to bind factor H, but only in mutant strains unable to express sialated capsule or full-length lipopolysaccharide. This protein seems unlikely to contribute significantly to factor H binding in invasive encapsulated bloodstream forms but may play a role in unencapsulated forms associated with nasopharyngeal colonization and carriage

How will the effectiveness of these new MenB vaccines be measured? With low incidence of disease, the customary clinical efficacy trials are impractical for measuring efficacy. As a disease specific to humans, animal models do not accurately predict the course of disease. The only surrogate of protection against meningococcal disease for which there are supportive data from humans is *in vitro* serum bactericidal activity (SBA; Goldschneider *et al.*, 1969a,b). SBA assays measure complement-dependent killing by immune sera and are well established for existing meningococcal CPS-conjugate vaccines (serotypes A, C, Y and W135). SBA assays for serogroup B have been developed for evaluating OMV vaccines that use human rather than rabbit serum as source of complement (Borrow *et al.*, 2005). To measure vaccine coverage, the selection of MenB indicator strains used in the assay should be unbiased and representative of current isolates circulating globally. For fHBP-based vaccines, a broad survey of diverse clinical strains has shown that there is no clear correlation between susceptibility to SBA killing and clonal subtype. However, a good predictor of killing in the SBA assay is the level of surface fHBP expression (Jiang *et al.*, 2010).

Staphylococcus aureus – evaluation of capsular and protein antigens

The development of vaccines to prevent *S. aureus* infection has become a research priority. *S. aureus* causes increasingly common life-threatening hospital-acquired infections. The appearance of community acquired methicillin-resistant *S. aureus* (MRSA) strains such as USA300 that can cause severe skin and soft-tissue infections or necrotizing pneumonia in the otherwise healthy, has drawn further attention to the problem (Otto, 2010). However, despite considerable effort no vaccine has yet been licensed for use in humans. One confounding factor is the exceptional versatility of the pathogen. It expresses a myriad of surface or secreted antigens that are differentially expressed depending on both the timing and the nature of the infection. Antigens include capsular polysaccharides and a variety of adhesins, toxins, and immune system modulators. As a result, successful vaccines will probably need to be multicomponent, targeting a variety of different surface antigens. A prophylactic vaccine will most likely focus on boosting immunity during periods in which patients are most at risk for infections. Prime candidates are patients preparing for

staphylococci (CoNS). *S. epidermidis* is especially problematic in premature infants, which are deficient in functional opsogenic antibodies and prone to catheter infections that lead all too frequently to life threatening sepsis. Vaccine antigens capable of extending coverage from *S. aureus* to CoNS strains would clearly be beneficial. A promising target is teichoic acid, a cell wall-associated polymer that is ubiquitous in Gram-positive bacteria. Teichoic acid is found in two forms, either covalently cross-linked to the peptidoglycan wall (wall teichoic acid, WTA) or inserted in the cell membrane via a glycolipid anchor (lipoteichoic acid, LTA). Most teichoic acids are zwitterionic owing to the presence of negatively charged phophodiester linkages and D-alanine residues on the repeating units. They are believed to play a role in surface adherence and in protection from environmental stresses (Xia *et al.*, 2010). A *S. aureus* mutant defective in teichoic acid alanylation is compromised in its ability to adhere to endothelial cells and shows reduced virulence in animal models (Weidenmaier *et al.*, 2005). A null mutant unable to make WTA shows profound colonization defects in a cotton rat nasal colonization model (Weidenmaier *et al.*, 2008). To combat neonatal staphylococcal sepsis a humanized mAb specific for LTA has been developed (Pagibaximab®, Biosynexus). The antibody is capable of opsonizing a variety of CoNS and *S. aureus* strains and demonstrating conservation of epitope and cross-species reactivity (Weisman *et al.*, 2009). A phase IIb/III clinical trial is ongoing to assess efficacy (completed Dec 2010). For a possible prophylactic vaccine, conjugates of teichoic acid warrant investigation for preclinical proof of concept. As with other vaccine candidates, it will be necessary to assess coverage against clinically relevant staphylococcal strains and to evaluate potential masking or interference by the staphylococcal polysaccharide capsule.

Clostridium difficile – toxoids as both prophylactic and therapeutic vaccine targets

Diphtheria, pertussis and tetanus are all examples of disease where antibody-mediated neutralization of the corresponding bacterial toxin can completely abrogate disease. As mentioned earlier, diphtheria and tetanus toxoids are usually combined with pertussis toxoid to make multicomponent vaccines that are specifically formulated for use in infants or for subsequent booster dosing in adolescents and adults. *C. difficile* is a fourth bacterial pathogen following *C. diphtheriae*, *B. pertussis* and *C. tetani*, that has the potential to be controlled through vaccination with inactivated toxin (toxoid). *C. difficile* infection (CDI) is a serious healthcare problem associated with antibiotic use in hospitals that has been increasing in incidence, severity and economic impact (Song *et al.*, 2008). Most at risk are elderly patients who develop the disease following a course of broad-spectrum antibiotics, including beta-lactam drugs, clindamycin and fluoroquinolones. The resulting eradication of normal gut flora encourages *C. difficile* proliferation and toxin production that causes diarrhoea and can lead to pseudomembranous colitis, a severe and potentially lethal infection of the colon. Epidemic outbreaks have caused thousands of deaths in hospitals in the USA, UK and Canada, which have led to increased public awareness and the implementation of rigorous control and hygiene measures to block disease transmission by the heat-resistant spores. Metronidazole and vancomycin are used to treat the most serious cases, but relapses are common. Vaccines offer a much needed alternative approach for managing CDI.

Disease symptoms are mediated by two exotoxins A and B that act synergistically to induce intestinal damage when introduced intragastrically in rodent models (Lyerly *et al.*, 1985). These very large structurally related proteins (of 308 kDa and 270 kDa) enter the host cell by receptor mediated endocytosis and release an N-terminal glucosyl transferase domain that modifies cytosolic Rho subfamily GTPases leading to cytoskeleton depolymerization and apoptosis. In studies with isogenic knock-out strains of *C. difficile*, either toxin alone can cause fulminant disease in the hamster model of infection (Kuehne *et al.*, 2010). In contrast, virulence is completely attenuated in this model with a double knock-out mutant strain of *C. difficile*, suggesting that both toxins contribute to disease. Indeed, a shared role of toxin A and toxin B seems probable as it is necessary to vaccinate hamsters against both toxin A and toxin B to confer immunity (Libby *et al.*, 1982; Fernie *et al.*, 1983).

What is responsible for the dramatic increase in the incidence and severity of health care-associated CDI over the last decade? Genotyping has implicated the emergence of a previously rare hypervirulent strain characterized by its pulsed-field type (NAP1) and PCR ribotype (027) profile. This strain has spread globally and currently accounts for the majority of isolates in North American and British hospitals (Clements *et al.*, 2010). Comparative genomic analysis of PCR-ribotype 027 strains identified at least five genetic regions unique to the epidemic 027 strain that may encode new virulence factors (Stabler *et al.*, 2009). Phenotypic differences include increased sporulation rates and resistance to fluoroquinolones, the latter due to mutations in the DNA gyrase target (Drudy *et al.*, 2007; Akerlund *et al.*, 2008). With regard to the toxin antigens, minor differences in amino acid sequence between the 027 genotype and other disease causing strains cannot alone explain the hypervirulence phenotype, as the non-epidemic 027 CD196 strain has toxin sequences identical to the circulating epidemic 027 strains. Another possible contributing factor is elevated expression of toxins by the hypervirulent strain observed *in vitro*, although the *in vivo* significance of this result is controversial (Warny *et al.*, 2005; Freeman *et al.*, 2006).

Proof of concept of a bivalent toxoid A and B human vaccine was demonstrated in a small scale study in volunteers (Sougioultzis *et al.*, 2005). Partially purified toxins inactivated with formaldehyde were used to immunize patients experiencing multiple episodes of CDI. Vaccination induced immune responses to both toxins and was associated with resolution of recurrent diarrhoea. Expanded phase II trials of the vaccine (Sanofi Pasteur) also with the endpoint of reducing recurrence, are ongoing as of 2010. If successful, the broader target indication for the vaccine will be disease prevention in the elderly. Vaccination of nursing home residents and patients in long term health care facilities is expected to have a significant impact on reducing incidence of disease.

Additional validation of the toxoid vaccine approach has come from clinical trials with humanized monoclonal antibodies (mAbs) directed against the C-terminal receptor binding domains of A and B toxins (Lowy *et al.*, 2010). A single infusion of both A and B toxoid neutralizing antibodies when administered in combination with metronidazole or vancomycin reduced the incidence of second episode CDI recurrence from 25% (placebo group) to 7% (antibody group). In contrast, treatment with an anti-toxin A monoclonal antibody alone did not show any benefit (Leav *et al.*, 2010). These results are encouraging, although the combination therapy had no impact on disease severity or length of hospital stay. If approved, the *C. difficile* antitoxin monoclonal therapy would probably become the first example of a licensed mAb-based passive vaccine for treating bacterial disease, representing a major technological leap from the passive immunoglobulin therapies of the pre-antibiotic era. The first and still only immunotherapy currently licensed to treat infectious diseases, is a monoclonal antibody developed by Medimmune (AstraZeneca) to control respiratory syncytial virus (RSV; Sorrentino and Powers, 2000).

Emerging bacterial pathogens as targets for future vaccines

The emergence of multidrug resistant bacterial pathogens combined with the paucity of new antibiotics to treat them has provided new impetus for vaccine-based approaches.

This is especially the case for emerging nosocomial pathogens such vancomycin-resistant enterococci (VRE), and multidrug resistant *Acinetobacter baumanii* and *Klebsiella pneumoniae.* Increased public awareness of these hospital acquired infections and mandatory monitoring by public health agencies has created added incentive for developing new therapies. Identification of appropriate vaccine antigens will involve a systematic evaluation of the most immunodominant and broadly represented carbohydrates and proteins on the surface of these pathogens. As disease incidence is relatively rare, defining the appropriate target population and assessing the clinical impact of a vaccine on infected patients may prove challenging. These opportunistic pathogens are commonly associated with very sick patients with complicating comorbid conditions, so attributing mortality can be difficult. Often immunocompromised, such patients may not be ideal candidates for prophylactic vaccination. Instead, mAb-based immunotherapies may offer a more attractive therapeutic option.

In a separate category are new vaccines to address the threat of bioterrorism and associated security shortcomings. In response to the 2001 anthrax and September 11 attacks, the US congress passed the Project BioShield act in 2004 to provide funding and incentives for the development of new therapeutics to combat biowarfare agents. Commercial incentives under this initiative include expedited FDA approval and a provision for guaranteed governmental purchase of supplies for stockpiling. Top priority bacterial pathogens recognized for their potential impact on public health are anthrax (*Bacillus anthracis*), plague (*Yersinia pestis*), tularaemia (*Francisella tularensis*) and botulism (*Clostridium botulinum*). Since the bill's passage, governmental contracts have been awarded for the manufacturing of prophylactic and therapeutic anthrax vaccines (PharmAthene, HGS), as well as hyperimmune horse sera to neutralize the botulism toxin (Cangene). The history and development of anthrax vaccines has been previously reviewed (Grabenstein, 2008). The currently licensed anthrax vaccine used by the military is considered unsuitable for civilian use, as it requires six doses over a period of 18 months to afford protection. The vaccine is a partially purified filtrate from anthrax cultures and contains secreted virulence factors including protective antigen (PA). Recombinant PA is the target antigen of both the new investigational prophylactic vaccine as well as a therapeutic monoclonal antibody therapy. Antibodies against PA prevent disease by interfering with its role in mediating the cellular uptake of anthrax lethal factor and oedema factor toxins. Also at a preclinical stage of development are a recombinant plague vaccine based on secreted F1 and V virulence protein antigens (DynPort), and live attenuated whole cell vaccines for tularaemia (DynPort and US Army).

This chapter has focused primarily on the development of vaccines against extracellular bacterial pathogens where protective immunity is dependent on the induction of antibody responses to surface antigens. The antibodies preventing these diseases are capable of neutralizing toxins, or facilitating complement-mediated or opsonophagocytic killing. For intracellular pathogens, which are physically shielded from antibody attack, a broader repertoire of T lymphocyte-based immune responses is required. For viral pathogens these responses can be effectively induced by live attenuated vaccines, such as those currently used to treat measles, mumps and rubella. A live attenuated vaccine for the intracellular bacterial pathogen *S. enterica* Typhi is still licensed, but general concerns about safety (from regulatory agencies) encourage the development of subunit vaccines derived from purified components. The live attenuated *Bacillus* Calmette–Guerin (BCG) vaccine used to prevent tuberculosis (TB) caused by *M. tuberculosis* is no longer recommended for widespread use in the US. Factors contributing to its disfavour include injection site reactogenicity, the rare incidence of TB meningitis in infants in developed countries and its limited efficacy against pulmonary TB disease. While incidence of TB is stable or falling in wealthy countries, it is the leading opportunistic infection and cause of death for those infected with HIV/AIDS in impoverished countries. A surge in drug-resistant strains has set back efforts to control the disease by drug therapy and the search for a more effective vaccine has become urgent.

A non-profit product development partnership known as the Aeres foundation has been established to address the global TB problem (http://www.aeras.org/). It is a consortium of experienced industrial, governmental and academic partners with the goal of developing, licensing, manufacturing and distributing affordable new TB vaccines. The overall strategy is to find both a safer alternative to BCG for priming infants and a follow-up booster vaccine based on recombinant TB antigens. The antigens will be expressed from attenuated viral vectors or adjuvanted to induce cell-mediated immune responses. The most advanced of these exploratory booster vaccines, used in combination with BCG, is currently being evaluated for safety and efficacy in clinical trials in South Africa (Nicol *et al.*, 2010). This vaccine is a live attenuated modified vaccinia virus expressing the immunodominant TB antigen 85A.

Chlamydia trachomatis is another example of an important intracellular pathogen for which there is an unmet medical need for a vaccine. Chlamydia is the most frequently reported bacterial sexually transmitted disease in the United States and can cause irreversible female infertility. The disease is considered a prime candidate for intervention by a prophylactic vaccine as damage can occur asymptomatically before antibiotics can be administered. As chlamydia is an obligate intracellular pathogen, any vaccine based on chlamydial surface antigens will need to be formulated to elicit the requisite Th1 (cellular) immune response. Despite high regulatory barriers to approval, there is substantial investment and interest in implementing new adjuvant technologies for intracellular pathogens, as evidenced by various ongoing clinical studies for vaccine to treat high-impact viral, parasitic and bacterial diseases such as HIV, malaria and TB.

Future challenges

Challenges to the advancement of new vaccines are not only technological in nature but also include social, economic and regulatory factors. A prevailing viewpoint is that governmental regulatory agencies have become increasingly risk averse, fuelled to a significant extent by highly publicized and often unscientifically informed controversies over vaccine safety, such as the now discredited link with autism (Gerber and Offit 2009). Commercial vaccine development must carefully weigh the probability of success against development costs and regulatory constraints. New vaccines for MenB and *C. difficile* that can be readily consolidated into existing schedules for infants, adolescents and the elderly and that have a high probability of success are more easily justified. The success of a vaccine for *S. aureus* is harder to predict but the scope of the public health problem encourages continued investment. Increased product liability and more stringent regulatory requirements have reduced the number of Pharmaceutical companies engaged in vaccine development. To encourage future investment in vaccines a number of changes have been proposed, including governmental action to incentivize research and development, strengthening protection against scientifically unfounded litigation and bolstering public health systems to assure vaccine delivery (Offit 2005). In addition, the emergence of collaborative public–private partnerships such as the Aeres foundation, PATH Malaria Vaccine Initiative (MVI) and international AIDS vaccine initiative (AIVI) has created a new paradigm for the development of low cost vaccines for neglected diseases in developing countries. By funding preclinical basic research and clinical trials, these organizations remove risk from the process, encouraging innovative approaches to vaccine development that are likely to advance the field in the future. For example, experience from these programmes with novel antigen expression systems and adjuvant formulations is likely to accelerate their acceptance by regulatory agencies and encourage their broader use.

Acknowledgements

We'd like to thank Dawn DeThomas for illustrations and our many Pfizer colleagues who helped shape the chapter through discussions and suggestions. We'd also like to thank Kathrin Jansen for her critical review and leadership in the development of vaccines to treat infectious diseases.

References

Akerlund, T., Persson, I., Unemo, M., Noren, T., Svenungsson, B., Wullt, M., Burman, L.G., Akerlund, T., Persson, I., Unemo, M., *et al.* (2008). Increased sporulation rate of epidemic *Clostridium difficile* Type 027/NAP1. J. Clin. Microbiol. *46*, 1530–1533.

Andrews, R., Moberley, S.A., Andrews, R., and Moberley, S.A. (2009). The controversy over the efficacy of pneumococcal vaccine. CMAJ Canadian Med. Assoc. J. *180*, 18–19.

Anonymous. (2007). Pneumococcal conjugate vaccine for childhood immunization–WHO position paper. Wkly Epidemiol. Rec. *82*, 93–104.

Anonymous. (2008). 23-valent pneumococcal polysaccharide vaccine. WHO position paper. Wkly Epidemiol. Rec. *83*, 373–384.

Avery, O.T., and Goebel, W.F. (1929). Chemo-immunological studies on conjugated carbohydrate-proteins. J. Exp. Med. *50*, 533–550.

Bartlett, W.C., and Noelle, R.J. (1991). T cell-dependent B-cell activation. In T-cell Dependent and Independent B-cell Activation, Snow, E.C., ed. (CRC Press, Boca Raton), pp. 111–131.

Berkovic, S.F., Harkin, L., McMahon, J.M., Pelekanos, J.T., Zuberi, S.M., Wirrell, E.C., Gill, D.S., Iona, X., Mulley,

J.C., Scheffer, I.E., *et al.* (2006). De-novo mutations of the sodium channel gene SCN1A in alleged vaccine encephalopathy: a retrospective study. Lancet Neurol. *5*, 488–492.

Bernfield, L., Fletcher, L.D., Howell, A., Farley, J.E., Zagursky, R., Knauf, M., and Zlotnick, G. (2002). Identification of a novel vaccine candidate for group B *Neisseria meningitidis*. Thirteenth International Pathogenic *Neisseria* Conference. Norwegian Institute of Public Health, Oslo, Norway; 1–6 September 2002.

Bisgard, K.M., Kao, A., Leake, J., Strebel, P.M., Perkins, B.A., and Wharton, M. (1998). *Haemophilus influenzae* invasive disease in the United States, 1994–1995, near disappearance of a vaccine-preventable childhood disease. Emerg. Infect. Dis. *4*, 229–237.

Borrow, R., Aaberge, I.S., Santos, G.F., Eudey, T.L., Oster, P., Glennie, A., Findlow, J., Hoiby, E.A., Rosenqvist, E., Balmer, P., *et al.* (2005). Interlaboratory standardization of the measurement of serum bactericidal activity by using human complement against meningococcal serogroup b, strain 44/76-SL, before and after vaccination with the Norwegian MenBvac outer membrane vesicle vaccine. Clin. Diagn. Lab. Immunol. *12*, 970–976.

Breukels, M.A., Rijkers, G.T., Voorhorst-Ogink, M.M., and Zegers, B.J. (1999). Regulatory T cells in the antibody response to *Haemophilus influenzae* type b polysaccharide. Infect. Immun. *67*, 789–793.

Bruge, J., Bouveret-Le Cam, N., Danve, B., Rougon, G., Schulz, D., Bruge, J., Bouveret-Le Cam, N., Danve, B., Rougon, G., and Schulz, D. (2004). Clinical evaluation of a group B meningococcal N-propionylated polysaccharide conjugate vaccine in adult, male volunteers. Vaccine *22*, 1087–1096.

Brunelli, B., Del Tordello, E., Palumbo, E., Biolchi, A., Bambini, S., Comanducci, M., Muzzi, A., Pizza, M., Rappuoli, R., Donnelly, J.J., *et al.* (2011). Influence of sequence variability on bactericidal activity sera induced by Factor H binding protein variant 1.1. Vaccine *29*, 1072–1081.

Cheng, A.G., Kim, H.K., Burts, M.L., Krausz, T., Schneewind, O., and Missiakas, D.M. (2009). Genetic requirements for *Staphylococcus aureus* abscess formation and persistence in host tissues. FASEB J. *23*, 3393–3404.

Centers for Disease and Prevention. (2008). Invasive pneumococcal disease in children 5 years after conjugate vaccine introduction–eight states, 1998–2005. MMWR – Morbid. Mortal. Wkly Rep. *57*, 144–148.

Centers for Disease and Prevention. (2010). Licensure of a 13-valent pneumococcal conjugate vaccine (PCV13) and recommendations for use among children – Advisory Committee on Immunization Practices (ACIP), 2010. MMWR – Morbid. Mortal. Wkly Rep.*59*, 258–261.

Cherry, J.D., Brunell, P.A., Golden, G.S., and Karzon, D.T. (1988). Report of the task force on *Pertussis* and *Pertussis* immunization – 1988. Pediatrics *81*, 933–984.

Clements, A.C., Magalhaes, R.J., Tatem, A.J., Paterson, D.L., Riley, T.V., Clements, A.C.A., Magalhaes, R.J.S., Tatem, A.J., Paterson, D.L., and Riley, T.V. (2010). *Clostridium difficile* PCR ribotype *027*, assessing the risks of further worldwide spread. Lancet Infect. Dis. *10*, 395–404.

Creech, C.B., 2nd, Johnson, B.G., Alsentzer, A.R., Hohenboken, M., Edwards, K.M., Talbot, T.R., 3rd, Creech, C.B., 2nd, Johnson, B.G., Alsentzer, A.R., Hohenboken, M., *et al.* (2009). Vaccination as infection control: a pilot study to determine the impact of *Staphylococcus aureus* vaccination on nasal carriage. Vaccine *28*, 256–260.

Dinleyici, E.C., Yargic, Z.A., Dinleyici, E.C., and Yargic, Z.A. (2008). Pneumococcal conjugated vaccines: impact of PCV-7 and new achievements in the postvaccine era. Exp. Rev. Vaccines *7*, 1367–1394.

Drudy, D., Kyne, L., O'Mahony, R., Fanning, S., Drudy, D., Kyne, L., O'Mahony, R., and Fanning, S. (2007). gyrA mutations in fluoroquinolone-resistant *Clostridium difficile* PCR-027. Emerg. Infect. Dis. *13*, 504–505.

Eskola, J., Kayhty, H., Peltola, H., Karanko, V., Makela, P.H., Samuelson, J., and Gordon, L.K. (1985). Antibody levels achieved in infants by course of *Haemophilus influenzae* type B polysaccharide/diphtheria toxoid conjugate vaccine. Lancet *1*, 1184–1186.

Eskola, J., Peltola, H., Takala, A.K., Kayhty, H., Hakulinen, M., Karanko, V., Kela, E., Rekola, P., Ronnberg, P.R., Samuelson, J.S., *et al.* (1987). Efficacy of *Haemophilus influenzae* type b polysaccharide-diphtheria toxoid conjugate vaccine in infancy. N. England J. Med. *317*, 717–722.

Etz, H., Minh, D.B., Henics, T., Dryla, A., Winkler, B., Triska, C., Boyd, A.P., Sollner, J., Schmidt, W., von Ahsen, U., *et al.* (2002). Identification of *in vivo* expressed vaccine candidate antigens from *Staphylococcus aureus*. Proc. Natl. Acad. Sci. U.S.A. *99*, 6573–6578.

Fattom, A.I., Horwith, G., Fuller, S., Propst, M., Naso, R., Fattom, A.I., Horwith, G., Fuller, S., Propst, M., and Naso, R. (2004). Development of StaphVAX, a polysaccharide conjugate vaccine against *S. aureus* infection: from the lab bench to phase III clinical trials. Vaccine *22*, 880–887.

Fernie, D.S., Thomson, R.O., Batty, I., and Walker, P.D. (1983). Active and passive immunization to protect against antibiotic associated caecitis in hamsters. Dev. Biol. Standardization *53*, 325–332.

Findlow, J., Borrow, R., Snape, M.D., Dawson, T., Holland, A., John, T.M., Evans, A., Telford, K.L., Ypma, E., Toneatto, D., *et al.* (2010). Multicenter, open-label, randomized phase II controlled trial of an investigational recombinant Meningococcal serogroup B vaccine with and without outer membrane vesicles, administered in infancy. Clin. Infect. Dis. *51*, 1127–1137.

Finne, J., Leinonen, M., and Makela, P.H. (1983). Antigenic similarities between brain components and bacteria causing meningitis. Implications for vaccine development and pathogenesis. Lancet *2*, 355–357.

Fletcher, L.D., Bernfield, L., Barniak, V., Farley, J.E., Howell, A., Knauf, M., Ooi, P., Smith, R.P., Weise,

P., Wetherell, M., *et al.* (2004). Vaccine potential of the Neisseria meningitidis 2086 lipoprotein. Infect. Immun. *72*, 2088–2100.

Fletcher, M.A., Fritzell, B., Fletcher, M.A., and Fritzell, B. (2007). Brief review of the clinical effectiveness of PREVENAR against otitis media. Vaccine *25*, 2507–2512.

Forsyth, K.D., Wirsing von Konig, C.-H., Tan, T., Caro, J., and Plotkin, S. (2007). Prevention of pertussis: recommendations derived from the second Global Pertussis Initiative roundtable meeting. Vaccine *25*, 2634–2642.

Freeman, J., Fawley, W., Baines, S., Wilcox, M., Freeman, J., Fawley, W., Baines, S., and Wilcox, M. (2006). Measurement of toxin production by Clostridium difficile. Lancet *367*, 982–983; author reply 983–984.

French, N., Gordon, S.B., Mwalukomo, T., White, S.A., Mwafulirwa, G., Longwe, H., Mwaiponya, M., Zijlstra, E.E., Molyneux, M.E., Gilks, C.F., *et al.* (2010). A trial of a 7-valent pneumococcal conjugate vaccine in HIV-infected adults. N. Engl. J. Med. *362*, 812–822.

Gerber, J.S., and Offit, P.A. (2009). Vaccines and autism: a tale of shifting hypotheses. Clin. Infect. Dis. *48*, 456–461.

Giuliani, M.M., Adu-Bobie, J., Comanducci, M., Arico, B., Savino, S., Santini, L., Brunelli, B., Bambini, S., Biolchi, A., Capecchi, B., *et al.* (2006). A universal vaccine for serogroup B meningococcus. Proc. Natl. Acad. Sci. U.S.A. *103*, 10834–10839.

Goldschneider, I., Gotschlich, E.C., and Artenstein, M.S. (1969a). Human immunity to the meningococcus. I. The role of humoral antibodies. J. Exp. Med. *129*, 1307–1326.

Goldschneider, I., Gotschlich, E.C., and Artenstein, M.S. (1969b). Human immunity to the meningococcus. II. Development of natural immunity. J. Exp. Med. *129*, 1327–1348.

Grabenstein, J.D. (2008). Vaccines: countering anthrax: vaccines and immunoglobulins. Clin. Infect. Dis. *46*, 129–136.

Greco, D., Salmaso, S., Mastrantonio, P., Giuliano, M., Tozzi, A.E., Anemona, A., Ciofi degli Atti, M.L., Giammanco, A., Panei, P., Blackwelder, W.C., *et al.* (1996). A controlled trial of two acellular vaccines and one whole-cell vaccine against pertussis. Progetto Pertosse Working Group. N. Engl. J. Med. *334*, 341–348.

Gustafsson, L., Hallander, H.O., Olin, P., Reizenstein, E., and Storsaeter, J. (1996). A controlled trial of a two-component acellular, a five-component acellular, and a whole-cell pertussis vaccine. N. Engl. J. Med. *334*, 349–355. Erratum appears in N. Engl. J. Med. 1996: *334*, 1207.

Hair, P.S., Echague, C.G., Sholl, A.M., Watkins, J.A., Geoghegan, J.A., Foster, T.J., Cunnion, K.M., Hair, P.S., Echague, C.G., Sholl, A.M., *et al.* (2010). Clumping factor A interaction with complement factor I increases C3b cleavage on the bacterial surface of *Staphylococcus aureus* and decreases complement-mediated phagocytosis. Infect. Immun. *78*, 1717–1727.

Hall, A.E., Domanski, P.J., Patel, P.R., Vernachio, J.H., Syribeys, P.J., Gorovits, E.L., Johnson, M.A., Ross, J.M., Hutchins, J.T., and Patti, J.M. (2003). Characterization of a protective monoclonal antibody recognizing *Staphylococcus aureus* MSCRAMM protein clumping factor A. Infect. Immun. *71*, 6864–6870.

Hidron, A.I., Edwards, J.R., Patel, J., Horan, T.C., Sievert, D.M., Pollock, D.A., Fridkin, S.K., National Healthcare Safety Network, Participating National Healthcare Safety Network, Hidron, A.I., *et al.* (2008). NHSN annual update: antimicrobial-resistant pathogens associated with healthcare-associated infections: annual summary of data reported to the National Healthcare Safety Network at the Centers for Disease Control and Prevention, 2006–2007. Infect. Control Hosp. Epidemiol. *29*, 996–1011.

Huang, S.S., Hinrichsen, V.L., Stevenson, A.E., Rifas-Shiman, S.L., Kleinman, K., Pelton, S.I., Lipsitch, M., Hanage, W.P., Lee, G.M., Finkelstein, J.A., *et al.* (2009). Continued impact of pneumococcal conjugate vaccine on carriage in young children. Pediatrics *124*, e1–11.

Huss, A., Scott, P., Stuck, A.E., Trotter, C., Egger, M., Huss, A., Scott, P., Stuck, A.E., Trotter, C., and Egger, M. (2009). Efficacy of pneumococcal vaccination in adults: a meta-analysis. CMAJ Can. Med. Assoc. J. *180*, 48–58. Erratum appears in CMAJ 2009: *180*, 1038

Jarva, H., Ram, S., Vogel, U., Blom, A.M., Meri, S., Jarva, H., Ram, S., Vogel, U., Blom, A.M., and Meri, S. (2005). Binding of the complement inhibitor C4bp to serogroup B Neisseria meningitidis. J. Immunol. *174*, 6299–6307.

Jiang, H.-Q., Hoiseth, S., Harris, S., McNeil, L., Zhu D, Tan, C., Scott, A.A., Alexander, K., Mason, K., Miller, L., *et al.* (2010). Broad vaccine coverage predicted for a bivalent recombinant factor H binding protein based vaccine to prevent serogroup B meningococcal disease. Vaccine *28*, 6086–6093.

Josefsson, E., Hartford, O., O'Brien, L., Patti, J.M., and Foster, T. (2001). Protection against experimental *Staphylococcus aureus* arthritis by vaccination with clumping factor A, a novel virulence determinant. J. Infect. Dis. *184*, 1572–1580.

Kampen, A.H., Tollersrud, T., Lund, A., Kampen, A.H., Tollersrud, T., and Lund, A. (2005). *Staphylococcus aureus* capsular polysaccharide types 5 and 8 reduce killing by bovine neutrophils *in vitro*. Infect. Immun. *73*, 1578–1583.

Khatami, A., Pollard, A.J., Khatami, A., and Pollard, A.J. (2010). The epidemiology of meningococcal disease and the impact of vaccines. Exp. Rev. Vaccines *9*, 285–298.

Kuehne, S.A., Cartman, S.T., Heap, J.T., Kelly, M.L., Cockayne, A., and Minton, N.P. (2010). The role of toxin A and toxin B in *Clostridium difficile* infection. Nature *467*, 711–713.

Kuklin, N.A., Clark, D.J., Secore, S., Cook, J., Cope, L.D., McNeely, T., Noble, L., Brown, M.J., Zorman, J.K., Wang, X.M., *et al.* (2006). A novel *Staphylococcus aureus* vaccine: iron surface determinant B induces rapid antibody responses in rhesus macaques and specific increased survival in a murine *S. aureus* sepsis model. Infect. Immun. *74*, 2215–2223.

Landsteiner, K. (1945). The Specificity of Serological Reactions (Harvard University Press, USA).

Leav, B.A., Blair, B., Leney, M., Knauber, M., Reilly, C., Lowy, I., Gerding, D.N., Kelly, C.P., Katchar, K., Baxter, R., *et al.* (2010). Serum anti-toxin B antibody correlates with protection from recurrent *Clostridium difficile* infection (CDI). Vaccine *28*, 965–969.

Lewis, L.A., Ngampasutadol, J., Wallace, R., Reid, J.E., Vogel, U., Ram, S., Lewis, L.A., Ngampasutadol, J., Wallace, R., Reid, J.E.A., *et al.* (2010). The meningococcal vaccine candidate neisserial surface protein A (NspA) binds to factor H and enhances meningococcal resistance to complement. PLoS Pathog. *6*, e100102.

Libby, J.M., Jortner, B.S., and Wilkins, T.D. (1982). Effects of the two toxins of *Clostridium difficile* in antibiotic-associated cecitis in hamsters. Infect. Immun. *36*, 822–829.

Lowy, I., Molrine, D.C., Leav, B.A., Blair, B.M., Baxter, R., Gerding, D.N., Nichol, G., Thomas, W.D., Jr., Leney, M., Sloan, S., *et al.* (2010). Treatment with monoclonal antibodies against *Clostridium difficile* toxins. N. Engl. J. Med. *362*, 197–205.

Luby, C.D., Middleton, J.R., Ma, J., Rinehart, C.L., Bucklin, S., Kohler, C., and Tyler, J.W. (2007). Characterization of the antibody isotype response in serum and milk of heifers vaccinated with a *Staphylococcus aureus* bacterin (Lysigin). J. Dairy Res. *74*, 239–246.

Lyerly, D.M., Saum, K.E., MacDonald, D.K., and Wilkins, T.D. (1985). Effects of *Clostridium difficile* toxins given intragastrically to animals. Infect. Immun. *47*, 349–352.

Ma, J., Cocchiaro, J., and Lee, J.C. (2004). Evaluation of serotypes of *Staphylococcus aureus* strains used in the production of a bovine mastitis bacterin. J. Dairy Sci. *87*, 178–182.

McNeil, L.K., Murphy, E., Zhao, X.J., Guttmann, S., Harris, S.L., Scott, A.A., Tan, C., Mack, M., DaSilva, I., Alexander, K., *et al.* (2009). Detection of LP2086 on the cell surface of *Neisseria meningitidis* and its accessibility in the presence of serogroup B capsular polysaccharide. Vaccine *27*, 3417–3421.

Madico, G., Welsch, J.A., Lewis, L.A., McNaughton, A., Perlman, D.H., Costello, C.E., Ngampasutadol, J., Vogel, U., Granoff, D.M., Ram, S., *et al.* (2006). The meningococcal vaccine candidate GNA1870 binds the complement regulatory protein factor H and enhances serum resistance. J. Immunol. *177*, 501–510.

Makela, P.H., and Makela, P.H. (2003). Conjugate vaccines–a breakthrough in vaccine development. Southeast Asian J. Trop. Med. Public Health *34*, 249–253.

Martin, D.R., Ruijne, N., McCallum, L., O'Hallahan, J., and Oster, P. (2006). The VR2 epitope on the PorA P1.7–2,4 protein is the major target for the immune response elicited by the strain-specific group B meningococcal vaccine MeNZB. Clin.Vaccine Immunol. CVI *13*, 486–491.

Masignani, V., Comanducci, M., Giuliani, M.M., Bambini, S., Adu-Bobie, J., Arico, B., Brunelli, B., Pieri, A., Santini, L., Savino, S., *et al.* (2003). Vaccination against *Neisseria meningitidis* using three variants of the lipoprotein GNA1870. J. Exp. Med. *197*, 789–799.

Meningococcal Reference Unit, U., Gray, S.J., Trotter, C.L., Ramsay, M.E., Guiver, M., Fox, A.J., Borrow, R., Mallard, R.H., Kaczmarski, E.B., Gray, S.J., *et al.* (2006). Epidemiology of meningococcal disease in England and Wales 1993/94 to 2003/04, contribution and experiences of the Meningococcal Reference Unit. J. Med. Microbiol. *55*, 887–896.

Merino, S., Camprubi, S., Alberti, S., Benedi, V.J., and Tomas, J.M. (1992). Mechanisms of *Klebsiella pneumoniae* resistance to complement-mediated killing. Infect. Immun. *60*, 2529–2535.

Middleton, J.R., Ma, J., Rinehart, C.L., Taylor, V.N., Luby, C.D., and Steevens, B.J. (2006). Efficacy of different Lysigin formulations in the prevention of *Staphylococcus aureus* intramammary infection in dairy heifers. J. Dairy Res. *73*, 10–19.

Middleton, J.R., Luby, C.D., Adams, D.S., Middleton, J.R., Luby, C.D., and Adams, D.S. (2009). Efficacy of vaccination against staphylococcal mastitis: a review and new data. Vet. Microbiol. *134*, 192–198.

Murphy, E., Andrew, L., Lee, K.L., Dilts, D.A., Nunez, L., Fink, P.S., Ambrose, K., Borrow, R., Findlow, J., Taha, M.K., *et al.* (2009). Sequence diversity of the factor H binding protein vaccine candidate in epidemiologically relevant strains of serogroup B *Neisseria meningitidis*. J. Infect. Dis. *200*, 379–389.

Nanra, J.S., Timofeyeva, Y., Buitrago, S.M., Sellman, B.R., Dilts, D.A., Fink, P., Nunez, L., Hagen, M., Matsuka, Y.V., Mininni, T., *et al.* (2009). Heterogeneous *in vivo* expression of clumping factor A and capsular polysaccharide by *Staphylococcus aureus*: implications for vaccine design. Vaccine *27*, 3276–3280.

Nicol, M.P., Grobler, L.A., Nicol, M.P., and Grobler, L.A. (2010). MVA-85A, a novel candidate booster vaccine for the prevention of tuberculosis in children and adults. Curr. Opin. Mol. Therapeutics *12*, 124–134.

Offit, P.A. (2005). Why are pharmaceutical companies gradually abandoning vaccines? Health Affairs *24*, 622–630.

Oster, P., Lennon, D., O'Hallahan, J., Mulholland, K., Reid, S., Martin, D., Oster, P., Lennon, D., O'Hallahan, J., Mulholland, K., *et al.* (2005). MeNZB: a safe and highly immunogenic tailor-made vaccine against the New Zealand Neisseria meningitidis serogroup B disease epidemic strain. Vaccine 23, 2191–2196.

Otto, M. (2010). Looking toward basic science for potential drug discovery targets againts community-associated MRSA. Med. Res. Rev. *30*, 1–22.

Pace, D., Pollard, A.J., Messonier, N.E., Pace, D., Pollard, A.J., and Messonier, N.E. (2009). Quadrivalent meningococcal conjugate vaccines. Vaccine *27 Suppl* 2, B30–41.

Pittman, M. (1984). The concept of pertussis as a toxin-mediated disease. Pediatr. Infect. Dis. 3, 467–486.

Pittman, M. (1991). History of the development of pertussis vaccine. Dev. Biol. Standardization *73*, 13–29.

Plosker, G.L., and Plosker, G.L. (2009). Combined, reduced-antigen content tetanus, diphtheria, and acellular pertussis vaccine (Boostrix): a review of its use as a single-dose booster immunization in individuals aged 10–64 years in the US. Biodrugs *23*, 253–267.

Plotkin, S. (2008a). A short history of vaccination. In Vaccines, Plotkin, S., Orenstein, W.A., Offit, P.A., eds. (Saunders Elsevier, Philadelphia).

Plotkin, S.A. (2008b). Vaccines: correlates of vaccine-induced immunity. Clin. Infect. Dis. *47*, 401–409.

Reinert, R., Jacobs, M.R., Kaplan, S.L., Reinert, R., Jacobs, M.R., and Kaplan, S.L. (2010). Pneumococcal disease caused by serotype 19A: review of the literature and implications for future vaccine development. Vaccine *28*, 4249–4259.

Robbins, J.B., Pittman, M., Trollfors, B., Lagergard, T.A., Taranger, J., and Schneerson, R. (1993). Primum non nocere: a pharmacologically inert pertussis toxoid alone should be the next pertussis vaccine. Pediatr. Infect. Dis. J. *12*, 795–807.

Rosenstein, N.E., Perkins, B.A., Stephens, D.S., Popovic, T., and Hughes, J.M. (2001). Meningococcal disease. N. Engl. J. Med. *344*, 1378–1388.

Roush, S.W., Murphy, T.V., and Vaccine-Preventable Disease Table Working. (2007). Historical comparisons of morbidity and mortality for vaccine-preventable diseases in the United States. JAMA *298*, 2155–2163.

de Roux, A., Schmole-Thoma, B., Siber, G.R., Hackell, J.G., Kuhnke, A., Ahlers, N., Baker, S.A., Razmpour, A., Emini, E.A., Fernsten, P.D., *et al.* (2008). Comparison of pneumococcal conjugate polysaccharide and free polysaccharide vaccines in elderly adults: conjugate vaccine elicits improved antibacterial immune responses and immunological memory. Clin. Infect. Dis. *46*, 1015–1023.

Schneerson, R., Barrera, O., Sutton, A., and Robbins, J.B. (1980). Preparation, characterization, and immunogenicity of *Haemophilus influenzae* type b polysaccharide–protein conjugates. J. Exp. Med. *152*, 361–376.

Schneider, M.C., Exley, R.M., Chan, H., Feavers, I., Kang, Y.H., Sim, R.B., and Tang, C.M. (2006). Functional significance of factor H binding to *Neisseria meningitidis*. J. Immunol. *176*, 7566–7575.

Schneider, M.C., Prosser, B.E., Caesar, J.J., Kugelberg, E., Li, S., Zhang, Q., Quoraishi, S., Lovett, J.E., Deane, J.E., Sim, R.B., *et al.* (2009). *Neisseria meningitidis* recruits factor H using protein mimicry of host carbohydrates. Nature *458*, 890–893.

Seib, K.L., Oriente, F., Adu-Bobie, J., Montanari, P., Ferlicca, F., Giuliani, M.M., Rappuoli, R., Pizza, M., Delany, I., Seib, K.L., *et al.* (2010). Influence of serogroup B meningococcal vaccine antigens on growth and survival of the meningococcus *in vitro* and in *ex vivo* and *in vivo* models of infection. Vaccine *28*, 2416–2427.

Siegrist, C.A. (2007). The challenges of vaccine responses in early life: selected examples. J. Comp. Pathol. *137 (Suppl 1)*, S4–9.

Sierra, G.V., Campa, H.C., Varcacel, N.M., Garcia, I.L., Izquierdo, P.L., Sotolongo, P.F., Casanueva, G.V., Rico, C.O., Rodriguez, C.R., and Terry, M.H. (1991). Vaccine against group B Neisseria meningitidis: protection trial and mass vaccination results in Cuba. NIPH Annals *14*, 195–207; discussion 208–210.

Song, X., Bartlett, J.G., Speck, K., Naegeli, A., Carroll, K., and Perl, T.M. (2008). Rising economic impact of clostridium difficile-associated disease in adult hospitalized patient population. Infect. Control Hosp. Epidemiol. *29*, 823–828.

Sorrentino, M., and Powers, T. (2000). Effectiveness of palivizumab: evaluation of outcomes from the 1998 to 1999 respiratory syncytial virus season. The Palivizumab Outcomes Study Group. Pediatr. Infect. Dis. J. *19*, 1068–1071.

Sougioultzis, S., Kyne, L., Drudy, D., Keates, S., Maroo, S., Pothoulakis, C., Giannasca, P.J., Lee, C.K., Warny, M., Monath, T.P., *et al.* (2005). *Clostridium difficile* toxoid vaccine in recurrent *C. difficile*-associated diarrhoea . Gastroenterology *128*, 764–770.

Stabler, R.A., He, M., Dawson, L., Martin, M., Valiente, E., Corton, C., Lawley, T.D., Sebaihia, M., Quail, M.A., Rose, G., *et al.* (2009). Comparative genome and phenotypic analysis of *Clostridium difficile* 027 strains provides insight into the evolution of a hypervirulent bacterium. Genome Biol. *10*, R102.

Stehr, K., Cherry, J.D., Heininger, U., Schmitt-Grohe, S., Uberall, M., Laussucq, S., Eckhardt, T., Meyer, M., Engelhardt, R., Christenson, P., *et al.* (1998). A comparative efficacy trial in Germany in infants who received either the lederle/takeda acellular pertussis component DTP (DTaP) vaccine, the Lederle whole-cell component DTP vaccine, or DT vaccine. Pediatrics *101*, 1–11.

Storsaeter, J., Hallander, H., Farrington, C.P., Olin, P., Mollby, R., and Miller, E. (1990). Secondary analyses of the efficacy of two acellular pertussis vaccines evaluated in a Swedish phase III trial. Vaccine *8*, 457–461.

Stranger-Jones, Y.K., Bae, T., Schneewind, O., Stranger-Jones, Y.K., Bae, T., and Schneewind, O. (2006). Vaccine assembly from surface proteins of *Staphylococcus aureus*. Proc. Natl. Acad. Sci. U.S.A. *103*, 16942–16947.

Thurlow, L.R., Thomas, V.C., Fleming, S.D., and Hancock, L.E. (2009). Enterococcus faecalis capsular polysaccharide serotypes C and D and their contributions to host innate immune evasion. Infect. Immun. *77*, 5551–5557.

Warny, M., Pepin, J., Fang, A., Killgore, G., Thompson, A., Brazier, J., Frost, E., McDonald, L.C., Warny, M., Pepin, J., *et al.* (2005). Toxin production by an emerging strain of *Clostridium difficile* associated with outbreaks of severe disease in North America and Europe. Lancet *366*, 1079–1084.

Weidenmaier, C., Peschel, A., Kempf, V.A., Lucindo, N., Yeaman, M.R., and Bayer, A.S. (2005). DltABCD- and

MprF-mediated cell envelope modifications of *Staphylococcus aureus* confer resistance to platelet microbicidal proteins and contribute to virulence in a rabbit endocarditis model. Infect. Immun. *73*, 8033–8038.

Weidenmaier, C., Kokai-Kun, J.F., Kulauzovic, E., Kohler, T., Thumm, G., Stoll, H., Gotz, F., and Peschel, A. (2008). Differential roles of sortase-anchored surface proteins and wall teichoic acid in *Staphylococcus aureus* nasal colonization. Int. J. Med. Microbiol. *298*, 505–513.

Weisman, L.E., Thackray, H.M., Garcia-Prats, J.A., Nesin, M., Schneider, J.H., Fretz, J., Kokai-Kun, J.F., Mond, J.J., Kramer, W.G., Fischer, G.W., *et al.* (2009). Phase 1/2 double-blind, placebo-controlled, dose escalation, safety, and pharmacokinetic study of pagibaximab (BSYX-A110), an antistaphylococcal monoclonal antibody for the prevention of staphylococcal bloodstream infections, in very-low-birth-weight neonates. Antimicrob. Agents Chemother. *53*, 2879–2886.

Welsch, J.A., Moe, G.R., Rossi, R., Adu-Bobie, J., Rappuoli, R., Granoff, D.M., Welsch, J.A., Moe, G.R., Rossi, R., Adu-Bobie, J., *et al.* (2003). Antibody to genome-derived neisserial antigen 2132, a Neisseria meningitidis candidate vaccine, confers protection against bacteremia in the absence of complement-mediated bactericidal activity. J. Infect. Dis. *188*, 1730–1740.

Welsch, J.A., Ram, S., Koeberling, O., and Granoff, D.M. (2008). Complement-dependent synergistic bactericidal activity of antibodies against factor H-binding protein, a sparsely distributed meningococcal vaccine antigen. J. Infect. Dis. *197*, 1053–1061.

Wendelboe, A.M., Van Rie, A., Salmaso, S., Englund, J.A., Wendelboe, A.M., Van Rie, A., Salmaso, S., and Englund, J.A. (2005). Duration of immunity against pertussis after natural infection or vaccination. Pediatr. Infect. Dis. J. *24*, S58–61.

Wendelboe, A.M., Njamkepo, E., Bourillon, A., Floret, D.D., Gaudelus, J., Gerber, M., Grimprel, E., Greenberg, D., Halperin, S., Liese, J., *et al.* (2007). Transmission of Bordetella pertussis to young infants. Pediatr. Infect. Dis. J. *26*, 293–299.

Xia, G., Kohler, T., and Peschel, A. (2010). The wall teichoic acid and lipoteichoic acid polymers of *Staphylococcus aureus*. Int. J. Med. Microbiol. *300*, 148–154.

Zagursky, R.J., Anderson, A.S., Zagursky, R.J., and Anderson, A.S. (2008). Application of genomics in bacterial vaccine discovery: a decade in review. Curr. Opin. Pharmacol. *8*, 632–638.

Recent Advances in Vaccine Adjuvants

14

Risini D. Weeratna* and Michael J. McCluskie

Abstract

Infectious disease remains one of the main causes of mortality and morbidity worldwide. Vaccination has had the greatest impact of any medical intervention technique in controlling infectious diseases. Most notably, eradication of smallpox was achieved through concerted and rigorous mass vaccination programs, and the incidence of diphtheria, pertussis, polio and other childhood diseases have been significantly reduced through routine infant immunization.

However, with a move away from whole-killed vaccines for safety reasons, a key challenge in realizing the full potential of vaccination has been the lack of immunogenicity of many novel vaccines especially in certain populations such as the elderly and the immunocompromised. Adjuvants are a key component in enhancing immunogenicity of vaccines. Furthermore, adjuvants can play a vital role in facilitating the induction of the appropriate type of immunity that is required to either prevent, such as in prophylactic vaccines, or to treat, such as in therapeutic vaccines. Therefore, careful consideration of the choice of adjuvants becomes quintessential for developing an effective vaccine. This chapter focuses on the importance of choosing the correct adjuvant or adjuvant combination to induce the appropriate immune responses to control the target pathogen.

Introduction

Vaccination has had a tremendous impact worldwide; decreasing morbidity and mortality associated with infectious disease. Well planned and orchestrated mass vaccination programs have resulted in the eradication of smallpox and near eradication of polio; two diseases that have resulted in significant morbidity and mortality throughout the world. In the early 1950s, there were an estimated 50 million new cases of smallpox worldwide each year. The World Health Organization launched its eradication plan in 1967. At that time, an estimated 60% of the world's population was threatened by smallpox, a disease that had a 25% mortality rate and left many survivors scarred or blinded with no effective treatment. The immunization programme was so successful that there has not been a single natural case since 1977 (http://www.who.int/mediacentre/factsheets/smallpox/en/index.html; 2010).

However, the battle to control infectious disease is far from over with numerous infectious agents still rampant in the globe including respiratory, enteric, sexually transmitted, vector-borne viral, zoonotic and parasitic pathogens, many of which have become targets for vaccine development. Despite the growth of interest in infectious disease vaccines, progress has been slow such that infectious disease still remains the number one cause of morbidity and mortality worldwide.

The top success stories in vaccinology, namely smallpox and polio vaccines are whole killed or live attenuated vaccines. Whole-cell vaccines are inherently immunogenic as they contain the necessary 'danger signals' that function to appropriately activate the innate immune cells that have a crucial role in immune induction, without the

*Corresponding author

need for additional extraneous agents as adjuvants. However, there are safety risks associated with whole killed or live attenuated vaccines, especially in the very young, the elderly and the immunocompromised. These risks have created a shift away from using whole cell vaccines to using more pure forms of recombinant subunit vaccines. Although there are fewer safety concerns associated with recombinant subunit vaccines, a key challenge in realizing their full potential has been their lack of immunogenicity, especially in the populations stated above where the whole cell vaccines pose a potential safety risk. Despite the tremendous effort that has been made to discover and develop new adjuvants in recent decades, their successful use in human vaccines has been far from satisfactory. This chapter will discuss some of the recent advances in the field of vaccine adjuvants and address the factors which must be considered when choosing appropriate adjuvants to enhance vaccine efficacy.

Types of adjuvants

By simple definition, adjuvants are agents incorporated into vaccine formulations to enhance the immunogenicity of the vaccine. The word 'adjuvant' comes from the Latin word *adjuvare*, meaning to help or aid.

Use of an adjuvant can be beneficial in a number of different ways. Adjuvants can (i) enhance low immunogenicity of pure antigens; (ii) help promote the optimal quality of immune response (e.g. promote desired type of immune response; Th1 versus Th2, humoral versus cell mediated immunity etc.); (iii) help overcome poor immune responses in certain populations (e.g. elderly, infants and immunocompromised individuals); (iv) enable vaccine delivery through non-parenteral routes (e.g. mucosal, transcutaneous) which may be less invasive and potentially safer by eliminating needles or by promoting a broader range of immune responses; and (v) add socio-economic advantages through antigen dose sparing, for example, reducing the number of boosts required for induction of protective immunity.

Vaccine adjuvants function through influencing or re-directing three major components of the immune response: (1) they can enhance the awareness of the immune system to the vaccine antigen through its innate components [dendritic cells (DC), macrophages, natural cells, polymorphonuclear (PMN) cells, mast cells etc.]; (2) they enhance the strength and duration of antigenic stimulation through the T- and B-cell receptors; and, most importantly, (3) adjuvants can help direct the profile of the antigen-specific immune response.

Adjuvants can be broadly divided into two groups based on their dominant mechanism of action. Immune stimulants enhance immune responses by acting as replacements for 'danger signals' normally provided by infectious agents. Delivery systems help induce productive immune responses by helping optimize delivery of antigen(s) and/or other adjuvants to appropriate immune cells. Some have dual properties acting both as immune stimulants and delivery vehicles (Fig. 14.1).

While many of the adjuvants have strong immune stimulatory properties when used alone, an emerging paradigm in vaccinology is that multiple adjuvant combinations may be more effective than individual adjuvants/delivery systems in enhancing immune responses to vaccine antigens. While an immune stimulant may activate innate immunity in an antigen independent fashion, they may need appropriate delivery systems to direct them to specific immune compartments in order to achieve optimal benefit. One example of such combinations is ISCOMs (immunostimulatory complexes) which consists of a delivery system (liposome) and an immune stimulant (saponin), to which antigen can be added (Pearse and Drane, 2005). GlaxoSmithKline (GSK) has also developed a number of adjuvant systems combining immune stimulants and delivery systems (e.g. AS01, AS02, AS04 and AS15; Garcon *et al.*, 2007), and some are described in length below.

Another approach to maximizing the adjuvant effect in vaccination is to combine multiple immune stimulants that may act through different immune cells and/or cell signalling pathways. Indeed this approach has been used in Fendrix® (a hepatitis B vaccine) and Cervarix® (a human papillomavirus vaccine), which contain the Toll-like receptor 4 (TLR4) agonist in combination with

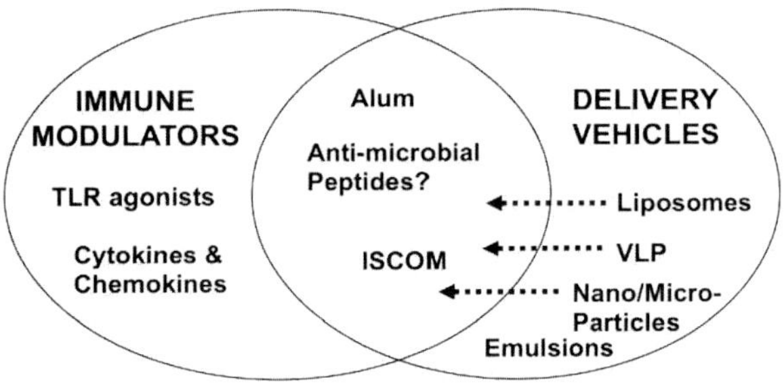

Figure 14.1 Simplified classification of basic types of adjuvants and delivery systems.

aluminium hydroxide as both delivery vehicle and co-stimulant (Kundi, 2007; Schwarz, 2009). However, as with any novel approach, safety and reactogenicity must be carefully considered when developing combinational approaches. This is especially true when dealing with agents for which the mechanism of action has not yet been elucidated.

There are numerous adjuvants and delivery systems and combinations thereof that are being investigated at various stages of development as discussed below.

Immune modulatory agents as vaccine adjuvants

Immune modulatory agents are agents that activate the innate immune system through interaction with their respective receptors which in turn activate various immune signalling pathways. There are a number of molecules that fall into this category; for example Toll-like receptors (TLR), nucleotide-binding oligomerization domain-like receptors (NLR), and retinoic acid-inducible gene (RIG)-1-like receptors (RLR), cytokines, chemokines, etc. Of these, those that have been most widely investigated in the realm of vaccine adjuvants are the TLRs, cytokines and chemokines.

Cytokines/chemokines as vaccine adjuvants

Cytokines and chemokines are proteins secreted by immune cells that play a vital role in regulating immune responses to various immunological stimuli and, as such, can modulate adaptive immune responses to vaccine antigens. Numerous cytokines/chemokines including but not limited to interleukin (IL)-1, -2, -3, -6, -12, interferon gamma (IFN-γ), tumour necrosis factor alpha (TNF-α) and granulocyte-monocyte-colony stimulating factor (GM-CSF) have been tested as vaccine adjuvants. However, as cytokines are relatively small molecules, they rapidly distribute throughout the body and away from site of injection. This may not only result in unwanted systemic toxicities but may also result in undesirable inflammation at distal sites causing deviation of the immune response from site of immunization and ultimately no adjuvant benefit. To overcome this, a number of strategies have been tested such as fusion of the cytokine with antigen or the use of liposomes or microparticles as delivery agents for cytokines.

In clinical settings, cytokines have most commonly been evaluated as adjuvants with tumour vaccines. However, despite the promising preclinical data, clinical success has been limited. This may be due to the fact that most preclinical studies are conducted in a prophylactic setting, with evaluation of tumour rejection in vaccinated animals, whereas clinically, tumour vaccines are used as therapeutic agents to treat patients with advanced disease, frequently with compromised immune systems due to a heavy tumour burden and/or prior cytoreductive therapies. However, cytokines have been used in several late stage clinical trials. For example, Melacine® is a melanoma vaccine derived from human melanoma cell lysate and adjuvanted with Detox (*Salmonella*-derived monophosphoryl lipid-A and mycobacterial cell wall components), which has been approved in Canada for treatment of late stage melanoma. Melacine® is currently being evaluated in combination with low-dose interferon alpha-2b (Intron® A) in patients with lymph node positive stage III melanoma (Ding and Wei, 2007; Mitchell *et al.*, 2007b) and data published to date show similar long relapse free survival (five years) and overall survival with this combination therapy compared with that with high dose Intron® A alone (Mitchell *et al.*, 2007a). Therefore, this approach may result in similar clinical efficacy but with a lower incidence of the side-effects that have typically been associated with high dose interferon alpha therapy (Enudi and Udolu, 2009; Navines *et al.*, 2009).

Another cytokine, GM-CSF has also been evaluated in late stage clinical trials with encouraging data. Sipuleucel-T (Provenge®), a dendritic cell vaccine that utilizes autologous dendritic cells treated with recombinant fusion protein consisting of GM-CSF linked to prostatic acid phosphatase (an antigen expressed in 95% prostate cancers) increased overall survival compared with placebo controls (Higano *et al.*, 2009). Although the specific contribution made by GM-CSF to vaccine efficacy cannot be determined; it is highly likely that GM-CSF plays a vital role in activation and maturation of dendritic cells which are necessary for vaccine efficacy. Provenge® was approved by the US Food and Drug Administration (FDA) for treatment of advanced prostate cancer in April 2010, making it the first therapeutic vaccine approved in the USA. Use of GM-CSF as an adjuvant for peptide-, protein-, or viral-based tumour vaccines has also been evaluated in clinical studies. A number of these studies have reported enhanced anti-tumour immune responses with some showing improved clinical outcome (Dranoff, 2002; Berinstein, 2007; Heimberger and Sampson, 2009). However, it has been suggested that prolonged sustained release of low-dose GM-CSF may be needed for maximal effect and that high doses of GM-CSF may be detrimental owing to expansion of myeloid suppressor cells (Parmiani *et al.*, 2007).

Cytokines have also been evaluated in clinical settings with infectious disease vaccines. For example use of GM-CSF as an adjuvant to HBV vaccination in patients with end-stage renal disease enhances sero-protection rates (Fabrizi *et al.*, 2006b). This is of particular interest since chronic renal dialysis patients often fail to respond to vaccination against hepatitis B virus (HBV; Fabrizi *et al.*, 2006a). Use of IL-12 and IL-15 in DNA vaccination against human immunodeficiency virus (HIV) or simian immunodeficiency virus SIV has also been shown to enhance vaccine efficacy by augmenting cell mediated immune responses (Morrow and Weiner, 2008). In addition, IL-12 has been shown to be an effective adjuvant with *Leishmania* vaccines in animal models (Scott and Trinchieri, 1997), and a heat-killed vaccine containing autoclaved *Leishmania* antigen with recombinant human IL-12 and aluminium hydroxide gel as adjuvants is currently in phase I/II clinical trials.

TLR agonists as vaccine adjuvants

Toll-like receptors (TLRs) are pattern recognition receptors that can recognize pathogens via pathogen-associated molecular patterns (PAMPs; Medzhitov and Janeway, Jr., 1997). TLRs play a crucial role in both innate and adaptive immunity (Akira *et al.*, 2001). Cells of the innate immune system can recognize invading pathogens as non-self through the TLRs leading to activation, maturation of certain immune cells and in turn the induction of pro-inflammatory cytokines and other anti-microbial compounds (Medzhitov and Janeway, Jr., 1997). Immature DCs resident in peripheral tissues possess numerous TLRs which recognize invading pathogens. This leads to the activation, maturation and trafficking of the DCs to local lymph nodes and presentation of antigens to naïve T cells leading to the induction of adaptive immunity. Furthermore, DCs can also regulate T-cell differentiation (Th1 vs. Th2) based on the particular pro-inflammatory cytokines that are produced by them, which in turn may depend on the TLR:PAMP interaction (Morse *et al.*, 1999; Viney, 1999; Kaisho and Akira, 2000, 2001; Tanaka *et al.*, 2000). For example, different subsets of DCs respond to different TLR ligands depending on their differential expression of TLR repertoire (Morse *et al.*, 1999; Viney, 1999; Kaisho and Akira, 2000, 2001; Tanaka *et al.*, 2000; Boonstra *et al.*, 2003; Mazzoni and Segal, 2004). Therefore, the cytokines that are secreted by these DCs, which in turn influence T-cell differentiation, may also vary depending on the subset of DCs that are activated. In addition, TLRs can also induce the expression of co-stimulatory molecules that are necessary for sustained activation of adaptive immunity (Kaisho and Akira, 2002; Akira *et al.*, 2001). Therefore, use of TLR agonists as adjuvants in vaccines is a promising approach to accelerate and enhance vaccine-specific immune responses.

However, differential expression of TLRs within immune cells and between species, differential cell signalling pathways used by different TLRs and more importantly the cellular localization of TLRs should be taken into consideration when choosing an appropriate TLR agonist for a particular vaccine of interest. For example TLR1, 2, 4, 5 and 6 are thought to be expressed on the cell surface of certain innate immune cells, where they can detect components of extracellular pathogens, whereas TLR3, 7, 8 and 9 are expressed within the endosomal compartments of certain innate immune cells, where they can detect components of intracellular pathogens such as nucleic acids (Kaisho and Akira, 2002). Therefore, it is critical that the TLR agonist to be employed is selected based on the disease target. It is also noteworthy that expression of TLRs differs between humans and mouse (the most commonly used laboratory species) which must be considered when interpreting preclinical data with these TLR agonists. Ten functional TLRs have been identified in humans and twelve in mice. TLRs 1–9 are conserved in both species, while TLR10 is not functional in mice because of a retrovirus insertion and TLR11, 12 and 13 have been lost from the human genome (Kawai and Akira, 2010). In addition, until recently TLR8 was thought to be non-functional in murine cells yet is expressed on monocytes, macrophages and B cells in humans (Jurk *et al.*, 2002). However, recent studies have suggested that TLR8 may indeed have some functionality in mice as well (Gorden *et al.*, 2006; Martinez *et al.*, 2010).

TLR signalling pathways can generally be classified as being either MyD88-dependent or TRIF dependent (Kawai and Akira, 2010). MyD88 signalling leads to induction of inflammatory cytokines whereas TRIF-dependent signalling leads to induction of both inflammatory cytokines and type-1 IFNs. All TLRs, with the exception of TLR3, signal via MyD88 leading to activation of transcription factor NF-κB and mitogen-activated protein kinases (MAPK) resulting in induction of inflammatory cytokines. In contrast, TLR3 signals via TRIF, leading to activation of transcription factors IRF3 and NF-κB resulting in induction of Type 1 IFNs and inflammatory cytokines, whereas TLR4 is capable of signalling via both MyD88 and TRIF pathways (Kawai and Akira, 2010).

The natural ligands for most of the ten human TLRs have been identified (Rock *et al.*, 1998; Takeuchi *et al.*, 1999; Akira *et al.*, 2001; Chuang and Ulevitch, 2000, 2001; Table 14.1). However, only a few TLRs including TLR3, 4, 5, 7/8 and 9 are currently under clinical investigation as vaccine adjuvants. Only one TLR agonist, monophosphoryl lipid A (TLR4 agonist) is currently approved for use in two human vaccines; Fendrix® (HBV) and Cervarix® (HPV) both marketed by GSK and used in combination with aluminium salt (Kundi, 2007; Schwarz, 2009).

TLR3 agonists as vaccine adjuvants

TLR3 is the pattern recognition receptor that recognizes double-stranded RNA (dsRNA), a PAMP-associated with viral infection. In humans, TLR3 expression has been reported in myeloid DCs, natural killer (NK) cells and in intestinal epithelial cells (Cario and Podolsky, 2000; Muzio *et al.*, 2000; Schmidt *et al.*, 2004). Among all TLRs, TLR3 is the only TLR that does not use MyD88 as the adapter protein to mediate cell signalling (Akira *et al.*, 2006). Instead, TLR3 mediates cell signals using the adaptor protein TRIF (Oshiumi *et al.*, 2003b; Yamamoto *et al.*, 2003), which leads to the activation of transcription factors IRF-3, NF-κB and AP-1, resulting in the induction of type I IFN, cytokine/chemokine production (Alexopoulou *et al.*, 2001; Diebold *et al.*, 2003; Oshiumi *et al.*, 2003a; Okahira *et al.*, 2005), up-regulation of co-stimulatory molecules on DCs (Bochtler *et al.*, 2008; Groschel *et al.*, 2008) and the promotion of antigen cross-presentation (Datta *et al.*, 2003; Schulz *et al.*, 2005). Polyriboinosinic:polyribocytidylic acid (poly I:C), a synthetic double stranded RNA, has been shown to promote the induction of IFN-α and enhance anti-viral and adjuvant activity in mice (Alexopoulou *et al.*, 2001). However,

of clinical importance. In another study, repeated immunization of mixed cancer patients with recombinant cancer-testis antigen (NY-ESO-1) using CPG 7909 and a proprietary water in oil emulsion (Montanide ISA-51) resulted in early induction of $CD4^+$ responses and Ab responses in most patients, followed by later development of CD8+ responses in 50% of the patients (Valmori *et al.*, 2007). A more recent study demonstrated that vaccination of cancer patients with NY-ESO-1 peptide P157–165 using CPG 7909 and Montanide ISA-51 resulted in induction of high numbers of antigen-specific and tumour-reactive T $CD8^+$ T cells in the majority of vaccinated patients and that a proportion of patients with vaccine-induced immune responses experienced increased survival times following vaccination (Karbach *et al.*, 2010). These results confirm the preclinical results and demonstrate the potency of CpG ODN to promote strong antigen-specific anti-tumour $CD8^+$ T-cell responses in humans with advanced cancer.

An adjuvant system (IC31®) consisting of an immunostimulatory oligonucleotide with natural phosphodiester backbone consisting of repeats of deoxyinosine and deoxycytosine dinucleotides (ODN1a) in combination with an antimicrobial peptide $KLKL_5KLK$ has been shown to activate antigen-presenting cells and promote Th1-biased immune responses. Despite the fact that ODN1a lacks immunostimulatory CpG motifs, this has been shown to stimulate innate immune activation through the TLR9 (Lingnau *et al.*, 2007). IC31® has been shown to augment immune responses against a number of different vaccine antigens in preclinical models (Kamath *et al.*, 2008a,b; Riedl *et al.*, 2008; Aagaard *et al.*, 2009; Olafsdottir *et al.*, 2009). In a clinical setting, IC31® has been shown to elicit strong antigen-specific T-cell responses against a mycobacterial vaccine antigen, Ag85B-ESAT-6 which persisted through 2 years and 6 months of follow-up (van Dissel *et al.*, 2010).

Delivery systems

While the immunomodulatory agents described above generally enhance immune responses by acting as replacements for the 'danger signals' normally provided by infectious agents, delivery systems enhance immune responses by optimizing delivery of antigen(s) and/or other adjuvants to the appropriate immune cells. It should be noted, however, that certain delivery systems also contain inherent immunostimulatory activity as discussed below.

A large number of different delivery systems have been tested preclinically in efforts to enhance vaccine efficacy. These include mineral salts, emulsions, surface active agents, synthetic microparticles, liposomes, virus-like particles, virosomes etc. For many delivery systems, the exact mode of action has not yet been fully elucidated, however, in general, delivery systems are thought to physically protect antigens such that they are degraded and/or eliminated less rapidly, increase their size or epitope density such that antigens are more readily detected by the immune system, retain associated antigens at a particular target site or deliver the antigens and any immune modulatory agents associated with the vaccine to appropriate immune cells for efficient presentation and immune induction.

Mineral salts

Both aluminium and calcium based salts have been tested as vaccine adjuvants. Aluminium salts, collectively known as alum, are suspensions of aluminium hydroxide or phosphate which have been used in human vaccines for nearly 70 years now and are the most widely used adjuvant in both human and veterinary vaccines worldwide. Alum was the only approved adjuvant for human use in the Unites States until the approval in late 2009 of Cervarix® which uses aluminium hydroxide in combination with the TLR4 agonist MPL as adjuvant. Although alum has been used safely for many years, and has been shown to enhance efficacy of prophylactic vaccines which require strong antibody responses to be effective, alum is not capable of generating effective immune responses against intracellular pathogens owing to its propensity to promote strongly Th2-biased immune responses (Lindblad, 2004; Petrovsky and Aguilar, 2004).

Until recently alum was thought to confer adjuvant activity solely by having an antigen depot effect and therefore enhancing uptake of aluminium-adsorbed antigen by antigen-presenting cells. More recently, alum has been shown to induce secretion of pro- inflammatory cytokines through the activation of the inflammasome complex and thus possess immune stimulatory properties as well (Eisenbarth *et al.*, 2008; Li *et al.*, 2008).

Emulsions

Emulsions made from water and oil (e.g. water-in-oil, oil-in-water, water-in-oil-in-water emulsions) can be used as adjuvants in vaccines. The adjuvant potential of emulsions was first shown in 1930s by Freund with his discovery that immunogenicity of antigens can be enhanced by emulsifying them in mineral oils. Mineral oil containing killed mycobacteria, termed Freund's complete adjuvant, induced potent immune responses consisting of both strong humoral and cell mediated immunity. In contrast, its potent inflammatory effects which may result in the formation of granulomas at the site of injection make it unacceptable for human use. However, Freund's incomplete adjuvant, which contains mineral oil without the mycobacteria, has been tested as an adjuvant in human vaccines for many decades: with prophylactic vaccines such as polio and influenza vaccines in the 1950s and currently with therapeutic vaccines against cancer and chronic viral infections. A new generation of water-in-oil emulsions such as Montanide ISA-51 and Montanide ISA-720 are also being evaluated with a number of human vaccines. Montanide ISA-51 contains a proprietary refined emulsifier from the mannide monoleate family in a natural metabolizable oil solution (SEPPIC Inc., Paris, France) which has been demonstrated to be non-carcinogenic, teratogenic or mutagenic. It uses a surfactant; a mannide monoleate which is rapidly metabolized and eliminated following administrations (Aucouturier *et al.*, 2006). Montanide ISA-720 contains squalene, a non-mineral oil (Aucouturier *et al.*, 2006). Both are in currently in phase II/III clinical in a number of vaccines for cancer and chronic infectious diseases.

A number of commercial vaccines contain oil-in-water emulsions. These can be used as single agents for example MF-59; used in Fluad®, an influenza vaccine marketed by Novartis and given to over one million people in Europe so far; or AS03 used in GSK's pandemic influenza vaccine or in combination with immunostimulatory adjuvants, for example AS02 (MPL, QS-21, oil in water emulsion); an adjuvant system developed by GSK.

Liposomes

In recent decades there have been a large number of studies conducted to evaluate liposomes, in particular cationic liposomes, as adjuvants and delivery systems. However, cationic liposomes are generally not sufficiently immunostimulatory on their own and so are frequently used to deliver other more potent immunomodulators. For example, certain saponin immunostimulants isolated from the bark of the *Quillaja saponaria* tree have been incorporated into lipid particles composed of cholesterol, phospholipids and cell membrane antigens to yield immunostimulating complexes or ISCOMs. A liposome based vaccine composed of cationic liposomes in combination with MPL and QS21 (AS01) has shown great promise in protecting infants from malaria (Cummings *et al.*, 2009; Owusu-Agyei *et al.*, 2009). To date more than 8000 doses have been given to more than 3000 children (Bojang *et al.*, 2005; Cummings *et al.*, 2009; Owusu-Agyei *et al.*, 2009; Kester *et al.*, 2008, 2009). A phase III anti-malarial vaccine trial using AS01 has recently been launched by GSK which will evaluate this vaccine in 16,000 children in seven different African countries.

Cationic liposomes composed of dimethyldioctadecylammonium (DDA) and trehalose 6,6′-dibehenate (TDB) have been shown in a murine chlamydia model to give better protection than CpG ODN or ISCOMs (Yu *et al.*, 2010). Cationic liposomes have also been extensively used to deliver nucleic acids for gene therapy purposes and also in numerous studies when combined with plasmid DNA as a vaccine delivery mechanism (Duan *et al.*, 2009). In addition, recent studies have demonstrated the ability of cationic liposomes to deliver TLRs, including TLR3, TLR7/8 and TLR9, highlighting their potential in a combinational adjuvant approach to enhance vaccine efficacy (Dow, 2008).

A number of liposomal vaccines based on viral membrane proteins (virosomes) without the use of additional immunopotentiators have been developed. For example, a virosome-based influenza vaccine (InflexalV®) has been used since 1997 in Europe, is currently approved in 38 countries and has sold more than 40 million doses. This vaccine is composed of reconstituted influenza virus envelopes without inner core or genetic information, and is believed to more closely mimic a natural viral infection. During vaccine production, the influenza virus surface antigens neuraminidase and haemagglutinin are integrated into phosphatidylcholine bilayer liposomes, yielding unilamellar virosomes with diameter of ~150 nm (Herzog *et al.*, 2009). A similar approach has been used to produce a licensed hepatitis A vaccine (Epaxal®) which was introduced in Europe in 1994 (Bovier, 2008).

Microparticles

A number of different degradable and non-degradable micro- and nanoparticles have been described for drug delivery, although for vaccine delivery, only degradable microparticles are commonly used. Microparticles can be used to simultaneously deliver antigens or antigens plus immunomodulators to immune cells whether encapsulated or absorbed.

The biodegradable polyesters, polylactide-co-glycolides (PLG) are the most commonly used components of microparticles (O'Hagan and Singh, 2003), with their well-established safety (from years of use in resorbable sutures) making them an attractive approach. Typically, antigens are encapsulated within microparticles, and generally enhanced immunogenicity is associated with smaller particles (O'Hagan and Singh, 2003). The immunogenicity of microparticles can be significantly enhanced by the co-encapsulation or delivery of additional adjuvants, such as the TLR9 agonist CpG ODN (Beaudette *et al.*, 2009; Malyala *et al.*, 2009), or TLR4 agonists MPL and the synthetic LPS mimetic RC529 (Kazzaz *et al.*, 2006).

In addition, several studies have reported absorption of antigen to the surface of pre-formed anionic PLG microparticles (Kazzaz *et al.*, 2000; Otten *et al.*, 2003; Singh *et al.*, 2004a,b). This approach has been shown in some cases to better preserve antigenic structure compared with encapsulation within particles (Singh *et al.*, 2004a). An alternative approach has been to adsorb immunomodulators onto PLG microparticles and co-administer with existing vaccines. For example, co-administration of CpG ODN-coated PLG microparticles with BioThrax® (Anthrax Vaccine Adsorbed, AVA, the licensed human anthrax vaccine) was shown to elicit a stronger and faster IgG response against the protective antigen of anthrax than AVA alone (Xie *et al.*, 2005).

While PLG microparticles remain by far the most commonly used microparticles, additional studies have used alternative biodegradable polymers including chitosan, polyanhydrides, polyorthoesters, hyaluronic acid and starch (Gupta *et al.*, 1998). In some cases, some of these polymers (chitosan and trimethyl chitosan) have also been used to surface coat PLG microparticles, for example to enhance nasal delivery (Pawar *et al.*, 2010).

Mucosal adjuvants

The vast majority of infectious diseases are transmitted through the mucosal surfaces of the gastrointestinal, genitourinary and respiratory tracts, where the primary source of protection is the mucosal immune system. The most effective way to stimulate the mucosal immune system is by immunizing at mucosal surfaces (i.e. via oral, nasal, rectal, vaginal or ocular routes), since while mucosal immunization can induce both mucosal and systemic responses, parenteral immunization (SC, IM) will generally only induce systemic immune responses. Mucosal vaccines, for example the Sabin oral polio vaccine, are also more practical to administer, are also generally preferred by both children and parents and may have safety advantages particularly in developing countries since injection is avoided. One of the major impediments to the development of mucosal vaccines has been the lack of safe and effective adjuvants and in recent years there has been considerable focus on developing mucosal adjuvants and/or delivery vehicles. These include derivatives of bacterial toxins (e.g. cholera toxin B subunit and genetically detoxified mutants of the *E. coli* heat-labile enteroxin, LT), microparticles, immunostimulatory complexes (ISCOMs), lipid A derivatives, liposomes, virus-like particles as well as various TLR agonists (McCluskie and

Weeratna 2001; Vajdy *et al.*, 2004). However, while mucosal adjuvants have shown great success in animal models, at present, relatively few mucosal vaccines are suitable for use in humans. These include the oral polio vaccine, oral cholera vaccines (live-attenuated or whole killed plus cholera toxin B subunit), an oral live-attenuated rotavirus vaccine, an oral BCG vaccine licensed in Brazil and a live attenuated influenza vaccine (FluMist®). A nasally delivered influenza vaccine (Nasalflu®) composed of virosomes containing influenza haemagglutinin and neuramidase antigens, adjuvanted with a low dose of LT, was initially approved in several European countries, but had subsequently been withdrawn due to a potential association with Bell's palsy (Mutsch *et al.*, 2004). In two subsequent phase I studies, Bell's palsy was also reported despite the use of a genetically detoxified mutant of LT (LTK63; Lewis *et al.*, 2009). This vaccine highlights the potential safety concerns associated with mucosal delivery of vaccines and it is probably due to regulatory concerns that parenteral immunization will remain the route of choice for the foreseeable future.

Conclusions

There is a clear need for adjuvants to enhance immunogenicity of new and existing vaccines. Not only may they result in a broadening of antibody responses (for example Cervarix®), but also may enhance T-cell responses critical for certain diseases (for example TB, HIV). They may allow antigen dose sparing (for example in pandemic flu vaccines to enable a larger population to be immunized with a limited supply of vaccine antigens), vaccine dosage sparing (for example in Fendrix®) and may overcome immune senescence. It is probable that adjuvants will also play a critical role in therapeutic vaccines (for example in oncology, chronic infectious diseases, Alzheimer's etc.).

While in recent years a number of vaccines have been approved which contain novel adjuvants (e.g. Cervarix and Fendrix), this has also served to highlight the regulatory hurdles which must be overcome in the approval of such vaccines. For the development and widespread use of novel adjuvants, it is critical that a better understanding of their mechanism of action be achieved not only so that the appropriate adjuvant and adjuvant formulation can be selected for a particular antigen and disease indication but also so that a greater confidence in their safety can be obtained. This will probably require the development of better non-clinical safety models and technologies such that concerns around hypersensitivity reactions, autoimmunity and the induction of tolerance can be addressed. A better understanding of the mechanism of action and non-clinical safety of adjuvants will help in the design of earlier and more relevant clinical safety endpoints which may help in designing better clinical trials, in return expediting regulatory approval for novel vaccines.

Of the many adjuvants currently being developed, it is probable that for optimal immune enhancement, a combinatorial approach will be required. The choice of adjuvants will depend on whether they possess the capacity to enhance the desired immune response, have a clearly defined mechanism of action and an acceptable safety profile. In addition, ease and reproducibility of manufacturing, overall cost of goods and commercial availability are also key determinants in whether an adjuvanted vaccine will be further developed.

References

Aagaard, C., Hoang, T.T., Izzo, A., Billeskov, R., Troudt, J., Arnett, K., Keyser, A., Elvang, T., Andersen, P., and Dietrich, J. (2009). Protection and polyfunctional T cells induced by Ag85B-TB10.4/IC31 against *Mycobacterium tuberculosis* is highly dependent on the antigen dose. PLoS ONE *4*, e5930.

Ahmad-Nejad, P., Hacker, H., Rutz, M., Bauer, S., Vabulas, R.M., and Wagner, H. (2002). Bacterial CpG-DNA and lipopolysaccharides activate Toll-like receptors at distinct cellular compartments. Eur. J. Immunol. 32, 1958–1968.

Akira, S., Takeda, K., and Kaisho, T. (2001). Toll-like receptors: critical proteins linking innate and acquired immunity. Nat. Immunol. *2*, 675–680.

Akira, S., Uematsu, S., and Takeuchi, O. (2006). Pathogen recognition and innate immunity. Cell *124*, 783–801.

Alexopoulou, L., Holt, A.C., Medzhitov, R., and Flavell, R.A. (2001). Recognition of double-stranded RNA and activation of NF-kappaB by Toll-like receptor 3. Nature *413*, 732–738.

An, H., Yu, Y., Zhang, M., Xu, H., Qi, R., Yan, X., Liu, S., Wang, W., Guo, Z., Guo, J., Qin, Z., and Cao, X. (2002). Involvement of ERK, p38 and NF-kappaB signal transduction in regulation of TLR2, TLR4 and

TLR9 gene expression induced by lipopolysaccharide in mouse dendritic cells. Immunology *106*, 38–45.

Angel, J.B., Cooper, C.L., Clinch, J., Young, C.D., Chenier, A., Parato, K.G., Lautru, M., Davis, H., and Cameron, D.W. (2008). CpG increases vaccine antigen-specific cell-mediated immunity when administered with hepatitis B vaccine in HIV infection. J. Immune. Ther. Vaccines *6*, 4.

Aucouturier, J., Ascarateil, S., and Dupuis, L. (2006). The use of oil adjuvants in therapeutic vaccines. Vaccine *24* (Suppl. 2), S2–S5.

Baldridge, J.R., McGowan, P., Evans, J.T., Cluff, C., Mossman, S., Johnson, D., and Persing, D. (2004). Taking a Toll on human disease: Toll-like receptor 4 agonists as vaccine adjuvants and monotherapeutic agents. Expert Opin. Biol. Ther. *4*, 1129–1138.

Bargieri, D.Y., Rosa, D.S., Braga, C.J., Carvalho, B.O., Costa, F.T., Espindola, N.M., Vaz, A.J., Soares, I.S., Ferreira, L.C., and Rodrigues, M.M. (2008). New malaria vaccine candidates based on the plasmodium vivax merozoite surface protein-1 and the TLR-5 agonist *Salmonella typhimurium* FliC flagellin. Vaccine *26*, 6132–6142.

Beaudette, T.T., Bachelder, E.M., Cohen, J.A., Obermeyer, A.C., Broaders, K.E., Frechet, J.M., Kang, E.S., Mende, I., Tseng, W.W., Davidson, M.G., and Engleman, E.G. (2009). *In vivo* studies on the effect of co-encapsulation of CpG DNA and antigen in acid-degradable microparticle vaccines. Mol. Pharm. *6*, 1160–1169.

Berinstein, N.L. (2007). Enhancing cancer vaccines with immunomodulators. Vaccine *25* (*Suppl* 2), B72-B88.

Bochtler, P., Kroger, A., Schirmbeck, R., and Reimann, J. (2008). Type I IFN-induced, NKT cell-mediated negative control of CD8 T cell priming by dendritic cells. J. Immunol. *181*, 1633–1643.

Bojang, K.A., Olodude, F., Pinder, M., Ofori-Anyinam, O., Vigneron, L., Fitzpatrick, S., Njie, F., Kassanga, A., Leach, A., Milman, J., *et al.* (2005). Safety and immunogenicty of RTS, S/AS02A candidate malaria vaccine in Gambian children. Vaccine 23, 4148–4157.

Boonstra, A., Asselin-Paturel, C., Gilliet, M., Crain, C., Trinchieri, G., Liu, Y.J., and O'Garra, A. (2003). Flexibility of mouse classical and plasmacytoid-derived dendritic cells in directing T helper type 1 and 2 cell development: dependency on antigen dose and differential toll-like receptor ligation. J. Exp. Med. *197*, 101–109.

Bovier, P.A. (2008). Epaxal: a virosomal vaccine to prevent hepatitis A infection. Expert. Rev. Vaccines *7*, 1141–1150.

Brazolot Millan, C.L., Weeratna, R., Krieg, A.M., Siegrist, C.A., and Davis, H.L. (1998). CpG DNA can induce strong Th1 humoral and cell-mediated immune responses against hepatitis B surface antigen in young mice. Proc. Natl. Acad. Sci. U.S.A. *95*, 15553–15558.

Buwitt-Beckmann, U., Heine, H., Wiesmuller, K.H., Jung, G., Brock, R., Akira, S., and Ulmer, A.J. (2005). Toll-like receptor 6-independent signaling by diacylated lipopeptides. Eur. J. Immunol. *35*, 282–289.

Buwitt-Beckmann, U., Heine, H., Wiesmuller, K.H., Jung, G., Brock, R., Akira, S., and Ulmer, A.J. (2006). J. Biol. Chem. *281*, 9049–9057.

Campodonico, V.L., Llosa, N.J., Grout, M., Doring, G., Maira-Litran, T., and Pier, G.B. (2010). Evaluation of flagella and flagellin of *Pseudomonas aeruginosa* as vaccines. Infect. Immun. *78*, 746–755.

Cario, E. and Podolsky, D.K. (2000). Differential alteration in intestinal epithelial cell expression of toll-like receptor 3 (TLR3) and TLR4 in inflammatory bowel disease. Infect. Immun. *68*, 7010–7017.

Chow, J.C., Young, D.W., Golenbock, D.T., Christ, W.J., and Gusovsky, F. (1999). Toll-like receptor-4 mediates lipopolysaccharide-induced signal transduction. J. Biol. Chem. *274*, 10689–10692.

Chuang, T.H., and Ulevitch, R.J. (2000). Cloning and characterization of a sub-family of human toll-like receptors: hTLR7, hTLR8 and hTLR9. Eur. Cytokine Netw. *11*, 372–378.

Chuang, T., and Ulevitch, R.J. (2001). Identification of hTLR*10*, a novel human Toll-like receptor preferentially expressed in immune cells. Biochim. Biophys. Acta *1518*, 157–161.

Cluff, C.W., Baldridge, J.R., Stover, A.G., Evans, J.T., Johnson, D.A., Lacy, M.J., Clawson, V.G., Yorgensen, V.M., Johnson, C.L., Livesay, M.T., *et al.* (2005). Synthetic toll-like receptor 4 agonists stimulate innate resistance to infectious challenge. Infect. Immun. 73, 3044–3052.

Cooper, C.L., Davis, H., and Cameron, D.W. (2004a). Influenza vaccination with 1/10th the full dose. N. Engl. J Med. *351*, 2339–2340.

Cooper, C.L., Davis, H.L., Morris, M.L., Efler, S.M., Adhami, M.A., Krieg, A.M., Cameron, D.W., and Heathcote, J. (2004b). CPG *7909*, an immunostimulatory TLR9 agonist oligodeoxynucleotide, as adjuvant to engerix-B((R)) HBV vaccine in healthy adults: A double-blind Phase I/II study. J. Clin. Immunol *24*, 693–701.

Cooper, C.L., Davis, H.L., Morris, M.L., Efler, S.M., Krieg, A.M., Li, Y., Laframboise, C., Al Adhami, M.J., Khaliq, Y., Seguin, I., *et al.* (2004c). Safety and immunogenicity of CPG 7909 injection as an adjuvant to Fluarix influenza vaccine. Vaccine *22*, 3136–3143.

Cooper, C.L., Davis, H.L., Angel, J.B., Morris, M.L., Elfer, S.M., Seguin, I., Krieg, A.M., and Cameron, D.W. (2005). CPG 7909 adjuvant improves hepatitis B virus vaccine seroprotection in antiretroviral-treated HIV-infected adults. Aids *19*, 1473–1479.

Cooper, C.L., Angel, J.B., Seguin, I., Davis, H.L., and Cameron, D.W. (2008). CPG 7909 adjuvant plus hepatitis B virus vaccination in HIV-infected adults achieves long-term seroprotection for up to 5 years. Clin. Infect. Dis. *46*, 1310–1314.

Crompton, P.D., Mircetic, M., Weiss, G., Baughman, A., Huang, C.Y., Topham, D.J., Treanor, J.J., Sanz, I., Lee, F.E., Durbin, A.P., *et al.* (2009). The TLR9 ligand CpG promotes the acquisition of *Plasmodium falciparum*-specific memory B cells in malaria-naive individuals. J. Immunol. *182*, 3318–3326.

Cummings, J.F., Spring, M.D., Schwenk, R.J., Ockenhouse, C.F., Kester, K.E., Polhemus, M.E.,

Walsh, D.S., Yoon, I.K., Prosperi, C., Juompan, L.Y., *et al.* (2009). Recombinant liver stage antigen-1 (LSA-1) formulated with AS01 or AS02 is safe, elicits high titer antibody and induces IFN-gamma/IL-2 CD4+ T cells but does not protect against experimental *Plasmodium falciparum* infection. Vaccine *28*, 5135–5144.

Datta, S.K., Redecke, V., Prilliman, K.R., Takabayashi, K., Corr, M., Tallant, T., DiDonato, J., Dziarski, R., Akira, S., Schoenberger, S.P., *et al.* (2003). A subset of Toll-like receptor ligands induces cross-presentation by bone marrow-derived dendritic cells. J. Immunol. *170*, 4102–4110.

Davis, H.L., Suparto, I.I., Weeratna, R.R., Jumintarto, Iskandriati, D.D., Chamzah, S.S., Ma'ruf, A.A., Nente, C.C., Pawitri, D.D., Krieg, A.M., *et al.* (2000). CpG DNA overcomes hyporesponsiveness to hepatitis B vaccine in orangutans. Vaccine *18*, 1920–1924.

Davis, H.L., Weeratna, R., Waldschmidt, T.J., Tygrett, L., Schorr, J., Krieg, A.M., and Weeratna, R. (1998). CpG DNA is a potent enhancer of specific immunity in mice immunized with recombinant hepatitis B surface antigen J. Immunol. *160*, 870–876. Erratum in J. Immunol. 1999; *162*, 3103.

Diebold, S.S., Montoya, M., Unger, H., Alexopoulou, L., Roy, P., Haswell, L.E., Al-Shamkhani, A., Flavell, R., Borrow, P., and Reis e Sousa (2003). Viral infection switches non-plasmacytoid dendritic cells into high interferon producers. Nature *424*, 324–328.

Ding, Z. and Wei, Y. (2007). Therapeutic vaccine for melanoma: the untouched grail? Expert. Rev. Vaccines. *6*, 907–911.

van Dissel, J.T., Arend, S.M., Prins, C., Bang, P., Tingskov, P.N., Lingnau, K., Nouta, J., Klein, M.R., Rosenkrands, I., Ottenhoff, T.H., *et al.* (2010). Ag85B-ESAT-6 adjuvanted with IC31 promotes strong and long-lived *Mycobacterium tuberculosis* specific T-cell responses in naive human volunteers. Vaccine *28*, 3571–3581.

Dow, S. (2008). Liposome-nucleic acid immunotherapeutics. Expert. Opin. Drug Deliv. *5*, 11–24.

Dranoff, G. (2002). GM-CSF-based cancer vaccines. Immunol. Rev. *188*, 147–154.

Duan, Y., Zhang, S., Wang, B., Yang, B., and Zhi, D. (2009). The biological routes of gene delivery mediated by lipid-based non-viral vectors. Expert. Opin. Drug Deliv. *6*, 1351–1361.

Eisenbarth, S.C., Colegio, O.R., O'Connor, W., Sutterwala, F.S., and Flavell, R.A. (2008). Crucial role for the Nalp3 inflammasome in the immunostimulatory properties of aluminium adjuvants. Nature *453*, 1122–1126.

Enudi, W. and Udolu, O.J. (2009). Neuropsychiatric complications associated with interferon-alpha-2b treatment of malignant melanoma. Afr. J. Psychiatry (Johannesbg.) *12*, 227–228.

Fabrizi, F., Ganeshan, S.V., Dixit, V., and Martin, P. (2006a). Meta-analysis: the adjuvant role of granulocyte macrophage-colony stimulating factor on immunological response to hepatitis B virus vaccine in end-stage renal disease. Aliment. Pharmacol. Ther. *24*, 789–796.

Fabrizi, F., Ganeshan, S.V., Dixit, V., and Martin, P. (2006b). Meta-analysis: the adjuvant role of granulocyte macrophage-colony stimulating factor on immunological response to hepatitis B virus vaccine in end-stage renal disease. Aliment. Pharmacol. Ther. *24*, 789–796.

Garcon, N., Chomez, P., and Van, M.M. (2007). GlaxoSmithKline Adjuvant Systems in vaccines: concepts, achievements and perspectives. Expert. Rev. Vaccines. *6*, 723–739.

Gawchik, S.M. and Saccar, C.L. (2009). Pollinex Quattro Tree: allergy vaccine. Expert. Opin. Biol. Ther. *9*, 377–382.

Gewirtz, A.T., Navas, T.A., Lyons, S., Godowski, P.J., and Madara, J.L. (2001). Cutting edge: bacterial flagellin activates basolaterally expressed TLR5 to induce epithelial proinflammatory gene expression. J. Immunol. *167*, 1882–1885.

Gibson, S.J., Lindh, J.M., Riter, T.R., Gleason, R.M., Rogers, L.M., Fuller, A.E., Oesterich, J.L., Gorden, K.B., Qiu, X., McKane, S.W., *et al.* (2002). Plasmacytoid dendritic cells produce cytokines and mature in response to the TLR7 agonists, imiquimod and resiquimod. Cell Immunol. *218*, 74–86.

Gorden, K.K., Qiu, X.X., Binsfeld, C.C., Vasilakos, J.P., and Alkan, S.S. (2006). Cutting edge: activation of murine TLR8 by a combination of imidazoquinoline immune response modifiers and polyT oligodeoxynucleotides. J. Immunol. *177*, 6584–6587.

Groschel, S., Piggott, K.D., Vaglio, A., Ma-Krupa, W., Singh, K., Goronzy, J.J., and Weyand, C.M. (2008). TLR-mediated induction of negative regulatory ligands on dendritic cells. J. Mol. Med. *86*, 443–455.

Gupta, R.K., Singh, M., and O'Hagan, D.T. (1998). Poly(lactide-co-glycolide) microparticles for the development of single-dose controlled-release vaccines. Adv. Drug Deliv. Rev. *32*, 225–246.

Halperin, S.A., Dobson, S., McNeil, S., Langley, J.M., Smith, B., McCall-Sani, R., Levitt, D., Nest, G.V., Gennevois, D., and Eiden, J.J. (2006). Comparison of the safety and immunogenicity of hepatitis B virus surface antigen co-administered with an immunostimulatory phosphorothioate oligonucleotide and a licensed hepatitis B vaccine in healthy young adults. Vaccine *24*, 20–26.

Halpern, M.D., Kurlander, R.J., and Pisetsky, D.S. (1996). Bacterial DNA induces murine interferon-gamma production by stimulation of interleukin-12 and tumor necrosis factor-alpha. Cell Immunol. *167*, 72–78.

Hasan, U., Chaffois, C., Gaillard, C., Saulnier, V., Merck, E., Tancredi, S., Guiet, C., Briere, F., Vlach, J., Lebecque, S., Trinchieri, G., and Bates, E.E. (2005). Human TLR10 is a functional receptor, expressed by B cells and plasmacytoid dendritic cells, which activates gene transcription through MyD88. J. Immunol. *174*, 2942–2950.

Hayashi, F., Smith, K.D., Ozinsky, A., Hawn, T.R., Yi, E.C., Goodlett, D.R., Eng, J.K., Akira, S., Underhill, D.M., and Aderem, A. (2001). The innate immune response to bacterial flagellin is mediated by Toll-like receptor 5. Nature *410*, 1099–1103.

Heil, F., Hemmi, H., Hochrein, H., Ampenberger, F., Kirschning, C., Akira, S., Lipford, G., Wagner, H., and Bauer, S. (2004). Species-specific recognition of single-stranded RNA via Toll-like receptor 7 and 8. Science *303*, 1526–1529.

Heimberger, A.B. and Sampson, J.H. (2009). The PEPvIII-KLH (CDX-110) vaccine in glioblastoma multiforme patients. Expert. Opin. Biol. Ther. *9*, 1087–1098.

Hemmi, H., Takeuchi, O., Kawai, T., Kaisho, T., Sato, S., Sanjo, H., Matsumoto, M., Hoshino, K., Wagner, H., Takeda, K., and Akira, S. (2000). A Toll-like receptor recognizes bacterial DNA. Nature *408*, 740–745.

Hemmi, H., Kaisho, T., Takeuchi, O., Sato, S., Sanjo, H., Hoshino, K., Horiuchi, T., Tomizawa, H., Takeda, K., and Akira, S. (2002). Small anti-viral compounds activate immune cells via the TLR7 MyD88-dependent signaling pathway. Nat. Immunol. 3, 196–200.

Herzog, C., Hartmann, K., Kunzi, V., Kursteiner, O., Mischler, R., Lazar, H., and Gluck, R. (2009). Eleven years of Inflexal V-a virosomal adjuvanted influenza vaccine. Vaccine *27*, 4381–4387.

Higano, C.S., Schellhammer, P.F., Small, E.J., Burch, P.A., Nemunaitis, J., Yuh, L., Provost, N., and Frohlich, M.W. (2009). Integrated data from 2 randomized, double-blind, placebo-controlled, phase 3 trials of active cellular immunotherapy with sipuleucel-T in advanced prostate cancer. Cancer *115*, 3670–3679.

Huleatt, J.W., Jacobs, A.R., Tang, J., Desai, P., Kopp, E.B., Huang, Y., Song, L., Nakaar, V., and Powell, T.J. (2007). Vaccination with recombinant fusion proteins incorporating Toll-like receptor ligands induces rapid cellular and humoral immunity. Vaccine *25*, 763–775.

Huleatt, J.W., Nakaar, V., Desai, P., Huang, Y., Hewitt, D., Jacobs, A., Tang, J., McDonald, W., Song, L., Evans, R.K., *et al.* (2008). Potent immunogenicity and efficacy of a universal influenza vaccine candidate comprising a recombinant fusion protein linking influenza M2e to the TLR5 ligand flagellin. Vaccine *26*, 201–214.

Jurk, M., Heil, F., Vollmer, J., Schetter, C., Krieg, A.M., Wagner, H., Lipford, G., and Bauer, S. (2002). Human TLR7 or TLR8 independently confer responsiveness to the antiviral compound R-848. Nat. Immunol. 3, 499.

Kagan, J.C., and Medzhitov, R. (2006). Phosphoinositide-mediated adaptor recruitment controls Toll-like receptor signaling. Cell *125*, 943–955.

Kagan, J.C., Su, T., Horng, T., Chow, A., Akira, S., and Medzhitov, R. (2008). TRAM couples endocytosis of Toll-like receptor 4 to the induction of interferon-beta. Nat. Immunol. *9*, 361–368.

Kaisho, T., and Akira, S. (2000). Critical roles of Toll-like receptors in host defense. Crit. Rev. Immunol. *20*, 393–405.

Kaisho, T., and Akira, S. (2001). Toll-like receptors and their signaling mechanism in innate immunity. Acta Odontol. Scand. *59*, 124–130.

Kaisho, T., and Akira, S. (2002). Toll-like receptors as adjuvant receptors. Biochim. Biophys. Acta *1589*, 1–13.

Kamath, A.T., Rochat, A.F., Valenti, M.P., Agger, E.M., Lingnau, K., Andersen, P., Lambert, P.H., and Siegrist, C.A. (2008a). Adult-like anti-mycobacterial T cell and *in vivo* dendritic cell responses following neonatal immunization with Ag85B-ESAT-6 in the IC31 adjuvant. PLoS ONE *3*, e3683.

Kamath, A.T., Valenti, M.P., Rochat, A.F., Agger, E.M., Lingnau, K., von, G.A., Andersen, P., Lambert, P.H., and Siegrist, C.A. (2008b). Protective anti-mycobacterial T-cell responses through exquisite *in vivo* activation of vaccine-targeted dendritic cells. Eur. J. Immunol. *38*, 1247–1256.

Karbach, J., Gnjatic, S., Bender, A., Neumann, A., Weidmann, E., Yuan, J., Ferrara, C.A., Hoffmann, E., Old, L.J., Altorki, N.K., and Jager, E. (2010). Tumor-reactive CD8+ T-cell responses after vaccination with NY-ESO-1 peptide, CpG 7909 and Montanide ISA-*51*, association with survival. Int. J. Cancer *126*, 909–918.

Kawai, T., and Akira, S. (2010). The role of pattern-recognition receptors in innate immunity: update on Toll-like receptors. Nat. Immunol. *11*, 373–384.

Kazzaz, J., Neidleman, J., Singh, M., Ott, G., and O'Hagan, D.T. (2000). Novel anionic microparticles are a potent adjuvant for the induction of cytotoxic T lymphocytes against recombinant p55 gag from HIV-1. J. Control Release *67*, 347–356.

Kazzaz, J., Singh, M., Ugozzoli, M., Chesko, J., Soenawan, E., and O'Hagan, D.T. (2006). Encapsulation of the immune potentiators MPL and RC529 in PLG microparticles enhances their potency. J. Control Release *110*, 566–573.

Kester, K.E., Cummings, J.F., Ockenhouse, C.F., Nielsen, R., Hall, B.T., Gordon, D.M., Schwenk, R.J., Krzych, U., Holland, C.A., Richmond, G., *et al.* (2008). Phase 2a trial of *0*, *1*, and 3 month and *0*, *7*, and 28 day immunization schedules of malaria vaccine RTS, S/AS02 in malaria-naive adults at the Walter Reed Army Institute of Research. Vaccine *26*, 2191–2202.

Kester, K.E., Cummings, J.F., Ofori-Anyinam, O., Ockenhouse, C.F., Krzych, U., Moris, P., Schwenk, R., Nielsen, R.A., Debebe, Z., Pinelis, E., *et al.* (2009). Randomized, double-blind, phase 2a trial of falciparum malaria vaccines RTS, S/AS01B and RTS, S/AS02A in malaria-naive adults: safety, efficacy, and immunologic associates of protection. J. Infect. Dis. *200*, 337–346.

Klinman, D.M. (2006). CpG oligonucleotides accelerate and boost the immune response elicited by AVA, the licensed anthrax vaccine. Expert Rev. Vaccines *5*, 365–369.

Klinman, D.M., Yi, A.K., Beaucage, S.L., Conover, J., and Krieg, A.M. (1996). CpG motifs present in bacteria DNA rapidly induce lymphocytes to secrete interleukin *6*, interleukin *12*, and interferon gamma. Proc. Natl. Acad. Sci. U.S.A. *93*, 2879–2883.

Klinman, D.M., Currie, D., Gursel, I., and Verthelyi, D. (2004). Use of CpG oligodeoxynucleotides as immune adjuvants. Immunol. Rev. *199*, 201–216.

Klinman, D.M., Xie, H., and Ivins, B.E. (2006). CpG oligonucleotides improve the protective immune response induced by the licensed anthrax vaccine. Ann. N. Y. Acad. Sci. *1082*, 137–150.

Kobayashi, H., Horner, A.A., Takabayashi, K., Nguyen, M.D., Huang, E., Cinman, N., and Raz, E. (1999). Immunostimulatory DNA pre-priming: a novel approach for prolonged Th1- biased immunity. Cell Immunol. *198*, 69–75.

Krieg, A.M. (2002). CpG motifs in bacterial DNA and their immune effects. Annu. Rev. Immunol. *20*, 709–760.

Krieg, A.M. (2003). CpG motifs: the active ingredient in bacterial extracts? Nat. Med. *9*, 831–835.

Krieg, A.M., Yi, A.K., Matson, S., Waldschmidt, T.J., Bishop, G.A., Teasdale, R., Koretzky, G.A., and Klinman, D.M. (1995). CpG motifs in bacterial DNA trigger direct B-cell activation. Nature *374*, 546–549.

Krieg, A.M., Efler, S.M., Wittpoth, M., Al Adhami, M.J., and Davis, H.L. (2004). Induction of systemic TH1-like innate immunity in normal volunteers following subcutaneous but not intravenous administration of CPG *7909*, a synthetic B-class CpG oligodeoxynucleotide TLR9 agonist. J. Immunother. *27*, 460–471.

Krug, A., Towarowski, A., Britsch, S., Rothenfusser, S., Hornung, V., Bals, R., Giese, T., Engelmann, H., Endres, S., Krieg, A.M., and Hartmann, G. (2001). Toll-like receptor expression reveals CpG DNA as a unique microbial stimulus for plasmacytoid dendritic cells which synergizes with CD40 ligand to induce high amounts of IL-12. Eur. J. Immunol. *31*, 3026–3037.

Kundi, M. (2007). New hepatitis B vaccine formulated with an improved adjuvant system. Expert. Rev. Vaccines. *6*, 133–140.

Lee, S.E., Kim, S.Y., Jeong, B.C., Kim, Y.R., Bae, S.J., Ahn, O.S., Lee, J.J., Song, H.C., Kim, J.M., Choy, H.E., *et al.* (2006). A bacterial flagellin, Vibrio vulnificus FlaB, has a strong mucosal adjuvant activity to induce protective immunity. Infect. Immun. *74*, 694–702.

Lewis, D.J., Huo, Z., Barnett, S., Kromann, I., Giemza, R., Galiza, E., Woodrow, M., Thierry-Carstensen, B., Andersen, P., Novicki, D., *et al.* (2009). Transient facial nerve paralysis (Bell's palsy) following intranasal delivery of a genetically detoxified mutant of *Escherichia coli* heat labile toxin. PLoS ONE *4*, e6999.

Li, H., Willingham, S.B., Ting, J.P., and Re, F. (2008). Cutting edge: inflammasome activation by alum and alum's adjuvant effect are mediated by NLRP3. J. Immunol. *181*, 17–21.

Lien, E., Means, T.K., Heine, H., Yoshimura, A., Kusumoto, S., Fukase, K., Fenton, M.J., Oikawa, M., Qureshi, N., Monks, B., *et al.* (2000). Toll-like receptor 4 imparts ligand-specific recognition of bacterial lipopolysaccharide. J. Clin. Invest. *105*, 497–504.

Lien, E., Sellati, T.J., Yoshimura, A., Flo, T.H., Rawadi, G., Finberg, R.W., Carroll, J.D., Espevik, T., Ingalls, R.R., Radolf, J.D., and Golenbock, D.T. (1999). Toll-like receptor 2 functions as a pattern recognition receptor for diverse bacterial products. J. Biol. Chem. *274*, 33419–33425.

Lindblad, E.B. (2004). Aluminium compounds for use in vaccines. Immunol. Cell Biol. *82*, 497–505.

Lingnau, K., Riedl, K., and von, G.A. (2007). IC31 and IC30, novel types of vaccine adjuvant based on peptide delivery systems. Expert. Rev. Vaccines. *6*, 741–746.

Lipford, G.B., Sparwasser, T., Zimmermann, S., Heeg, K., and Wagner, H. (2000). CpG-DNA-mediated transient lymphadenopathy is associated with a state of Th1 predisposition to antigen-driven responses. J. Immunol. *165*, 1228–1235.

McCluskie, M.J., and Davis, H.L. (2000). Oral, intrarectal and intranasal immunizations using CpG and non-CpG oligodeoxynucleotides as adjuvants. Vaccine *19*, 413–422.

McCluskie, M.J., and Krieg, A.M. (2006). Enhancement of infectious disease vaccines through TLR9-dependent recognition of CpG DNA. Curr. Top. Microbiol. Immunol *311*, 155–178.

McCluskie, M.J., and Weeratna, R.D. (2001). Novel adjuvant systems. Curr. Drug Targets. Infect. Disord. *1*, 263–271.

McCluskie, M.J., Weeratna, R.D., and Davis, H.L. (2000a). Intranasal immunization of mice with CpG DNA induces strong systemic and mucosal responses that are influenced by other mucosal adjuvants and antigen distribution. Mol. Med. *6*, 867–877.

McCluskie, M.J., Weeratna, R.D., Krieg, A.M., and Davis, H.L. (2000b). CpG DNA is an effective oral adjuvant to protein antigens in mice. Vaccine *19*, 950–957.

McCluskie, M.J., Weeratna, R.D., Clements, J.D., and Davis, H.L. (2001a). Mucosal immunization of mice using CpG DNA and/or mutants of the heat- labile enterotoxin of *Escherichia coli* as adjuvants. Vaccine *19*, 3759–3768.

McCluskie, M.J., Weeratna, R.D., Payette, P.J., and Davis, H.L. (2001b). The potential of CpG oligodeoxynucleotides as mucosal adjuvants. Crit Rev. Immunol. *21*, 103–120.

McDonald, W.F., Huleatt, J.W., Foellmer, H.G., Hewitt, D., Tang, J., Desai, P., Price, A., Jacobs, A., Takahashi, V.N., Huang, Y., *et al.* (2007). A West Nile virus recombinant protein vaccine that coactivates innate and adaptive immunity. J. Infect. Dis. *195*, 1607–1617.

McNeilly, T.N., Mitchell, M.C., Nisbet, A.J., McAteer, S., Erridge, C., Inglis, N.F., Smith, D.G., Low, J.C., Gally, D.L., Huntley, J.F., and Mahajan, A. (2010). IgA and IgG antibody responses following systemic immunization of cattle with native H7 flagellin differ in epitope recognition and capacity to neutralise TLR5 signalling. Vaccine *28*, 1412–1421.

Malyala, P., O'Hagan, D.T., and Singh, M. (2009). Enhancing the therapeutic efficacy of CpG oligonucleotides using biodegradable microparticles. Adv. Drug Deliv. Rev. *61*, 218–225.

Martinez, J., Huang, X., and Yang, Y. (2010). Toll-like receptor 8-mediated activation of murine plasmacytoid dendritic cells by vaccinia viral DNA. Proc. Natl. Acad. Sci. U.S.A. *107*, 6442–6447.

Mazzoni, A., and Segal, D.M. (2004). Controlling the Toll road to dendritic cell polarization. J. Leukoc. Biol. *75*, 721–730.

Medzhitov, R., and Janeway, C.A., Jr. (1997). Innate immunity: the virtues of a nonclonal system of recognition. Cell *91*, 295–298.

Mitchell, M.S., Abrams, J., Thompson, J.A., Kashani-Sabet, M., DeConti, R.C., Hwu, W.J., Atkins, M.B., Whitman, E., Ernstoff, M.S., Haluska, F.G., *et al.* (2007a). Randomized trial of an allogeneic melanoma lysate vaccine with low-dose interferon Alfa-2b compared with high-dose interferon Alfa-2b for Resected stage III cutaneous melanoma. J. Clin. Oncol. *25*, 2078–2085.

Mitchell, M.S., Abrams, J., Thompson, J.A., Kashani-Sabet, M., DeConti, R.C., Hwu, W.J., Atkins, M.B., Whitman, E., Ernstoff, M.S., Haluska, F.G., *et al.* (2007b). Randomized trial of an allogeneic melanoma lysate vaccine with low-dose interferon Alfa-2b compared with high-dose interferon Alfa-2b for Resected stage III cutaneous melanoma. J. Clin. Oncol. *25*, 2078–2085.

Montoya, C.J., Jie, H.B., Al-Harthi, L., Mulder, C., Patino, P.J., Rugeles, M.T., Krieg, A.M., Landay, A.L., and Wilson, S.B. (2006). Activation of plasmacytoid dendritic cells with TLR9 agonists initiates invariant NKT cell-mediated cross-talk with myeloid dendritic cells. J. Immunol. *177*, 1028–1039.

Morrow, M.P., and Weiner, D.B. (2008). Cytokines as adjuvants for improving anti-HIV responses. AIDS *22*, 333–338.

Morse, M.A., Vredenburgh, J.J., and Lyerly, H.K. (1999). A comparative study of the generation of dendritic cells from mobilized peripheral blood progenitor cells of patients undergoing high-dose chemotherapy. J. Hematother. Stem Cell Res. *8*, 577–84.

Moseman, E.A., Liang, X., Dawson, A.J., Panoskaltsis-Mortari, A., Krieg, A.M., Liu, Y.J., Blazar, B.R., and Chen, W. (2004). Human plasmacytoid dendritic cells activated by CpG oligodeoxynucleotides induce the generation of CD4+CD25+ regulatory T cells. J. Immunol. *173*, 4433–4442.

Mutsch, M., Zhou, W., Rhodes, P., Bopp, M., Chen, R.T., Linder, T., Spyr, C., and Steffen, R. (2004). Use of the inactivated intranasal influenza vaccine and the risk of Bell's palsy in Switzerland. N. Engl. J. Med. *350*, 896–903.

Muzio, M., Bosisio, D., Polentarutti, N., D'amico, G., Stoppacciaro, A., Mancinelli, R., van't, V.C., Penton-Rol, G., Ruco, L.P., Allavena, P., *et al.* (2000). Differential expression and regulation of toll-like receptors (TLR) in human leukocytes: selective expression of TLR3 in dendritic cells. J. Immunol. *164*, 5998–6004.

Navines, R., Gomez-Gil, E., Puig, S., Baeza, I., De, P.J., and Martin-Santos, R. (2009). Depression in hospitalized patients with malignant melanoma treated with interferon-alpha-2b: primary to induced disorders. Eur. J. Dermatol. *19*, 611–615.

O'Hagan, D.T., and Singh, M. (2003). Microparticles as vaccine adjuvants and delivery systems. Expert. Rev. Vaccines. *2*, 269–283.

Okahira, S., Nishikawa, F., Nishikawa, S., Akazawa, T., Seya, T., and Matsumoto, M. (2005). Interferon-beta induction through toll-like receptor 3 depends on double-stranded RNA structure. DNA Cell Biol. *24*, 614–623.

Olafsdottir, T.A., Lingnau, K., Nagy, E., and Jonsdottir, I. (2009). IC*31*, a two-component novel adjuvant mixed with a conjugate vaccine enhances protective immunity against pneumococcal disease in neonatal mice. Scand. J. Immunol. *69*, 194–202.

Oshiumi, H., Matsumoto, M., Funami, K., Akazawa, T., and Seya, T. (2003a). TICAM-*1*, an adaptor molecule that participates in Toll-like receptor 3-mediated interferon-beta induction. Nat. Immunol. *4*, 161–167.

Oshiumi, H., Matsumoto, M., Funami, K., Akazawa, T., and Seya, T. (2003b). TICAM-*1*, an adaptor molecule that participates in Toll-like receptor 3-mediated interferon-beta induction. Nat. Immunol. *4*, 161–167.

Otten, G., Schaefer, M., Greer, C., Calderon-Cacia, M., Coit, D., Kazzaz, J., Medina-Selby, A., Selby, M., Singh, M., Ugozzoli, M., *et al.* (2003). Induction of broad and potent anti-human immunodeficiency virus immune responses in rhesus macaques by priming with a DNA vaccine and boosting with protein-adsorbed polylactide coglycolide microparticles. J. Virol. *77*, 6087–6092.

Owusu-Agyei, S., Ansong, D., Asante, K., Kwarteng, O.S., Owusu, R., Wireko Brobby, N.A., Dosoo, D., Osei, A.A., Osei-Kwakye, K., Adjei, E.A., *et al.* (2009). Randomized controlled trial of RTS, S/AS02D and RTS, S/AS01E malaria candidate vaccines given according to different schedules in Ghanaian children. PLoS ONE *4*, e7302.

Parmiani, G., Castelli, C., Pilla, L., Santinami, M., Colombo, M.P., and Rivoltini, L. (2007). Opposite immune functions of GM-CSF administered as vaccine adjuvant in cancer patients. Ann. Oncol. *18*, 226–232.

Pawar, D., Goyal, A.K., Mangal, S., Mishra, N., Vaidya, B., Tiwari, S., Jain, A.K., and Vyas, S.P. (2010). Evaluation of mucoadhesive PLGA microparticles for nasal immunization. AAPS J. *12*, 130–137.

Pearse, M.J., and Drane, D. (2005). ISCOMATRIX adjuvant for antigen delivery. Adv. Drug Deliv. Rev. *57*, 465–474.

Petrovsky, N. and Aguilar, J.C. (2004). Vaccine adjuvants: current state and future trends. Immunol. Cell Biol. *82*, 488–496.

Poltorak, A., He, X., Smirnova, I., Liu, M.Y., Huffel, C.V., Du, X., Birdwell, D., Alejos, E., Silva, M., Galanos, C., *et al.* (1998). Defective LPS signaling in C3H/HeJ and C57BL/10ScCr mice: mutations in Tlr4 gene. Science *282*, 2085–2088.

Poltorak, A., Ricciardi-Castagnoli, P., Citterio, S., and Beutler, B. (2000). Physical contact between lipopolysaccharide and toll-like receptor 4 revealed by genetic complementation. Proc. Natl. Acad. Sci. U.S.A. *97*, 2163–2167.

Rhee, S.H., and Hwang, D. (2000). Murine TOLL-like receptor 4 confers lipopolysaccharide responsiveness as determined by activation of NF kappa B and expression of the inducible cyclooxygenase. J. Biol. Chem. *275*, 34035–34040.

Riedl, K., Riedl, R., von, G.A., Nagy, E., and Lingnau, K. (2008). The novel adjuvant IC31 strongly improves influenza vaccine-specific cellular and humoral immune responses in young adult and aged mice. Vaccine *26*, 3461–3468.

Rock, F.L., Hardiman, G., Timans, J.C., Kastelein, R.A., and Bazan, J.F. (1998). A family of human receptors structurally related to *Drosophila* Toll. Proc. Natl. Acad. Sci. U.S.A. *95*, 588–593.

Rowe, D.C., McGettrick, A.F., Latz, E., Monks, B.G., Gay, N.J., Yamamoto, M., Akira, S., O'Neill, L.A., Fitzgerald, K.A., and Golenbock, D.T. (2006). The myristoylation of TRIF-related adaptor molecule is essential for Toll-like receptor 4 signal transduction. Proc. Natl. Acad. Sci. U.S.A. *103*, 6299–6304.

Salman, H.H., Irache, J.M., and Gamazo, C. (2009). Immunoadjuvant capacity of flagellin and mannosamine-coated poly(anhydride) nanoparticles in oral vaccination. Vaccine *27*, 4784–4790.

Schmidt, K.N., Leung, B., Kwong, M., Zarember, K.A., Satyal, S., Navas, T.A., Wang, F., and Godowski, P.J. (2004). APC-independent activation of NK cells by the Toll-like receptor 3 agonist double-stranded RNA. J. Immunol. *172*, 138–143.

Schulz, O., Diebold, S.S., Chen, M., Naslund, T.I., Nolte, M.A., Alexopoulou, L., Azuma, Y.T., Flavell, R.A., Liljestrom, P., and Reis e Sousa, C. (2005). Toll-like receptor 3 promotes cross-priming to virus-infected cells. Nature *433*, 887–892.

Schwandner, R., Dziarski, R., Wesche, H., Rothe, M., and Kirschning, C.J. (1999). Peptidoglycan- and lipoteichoic acid-induced cell activation is mediated by toll-like receptor 2. J. Biol. Chem. *274*, 17406–17409.

Schwarz, T.F. (2009). Clinical update of the AS04-adjuvanted human papillomavirus-16/18 cervical cancer vaccine, Cervarix. Adv. Ther. *26*, 983–998.

Scott, P., and Trinchieri, G. (1997). IL-12 as an adjuvant for cell-mediated immunity. Semin. Immunol. *9*, 285–291.

Siegrist, C.A., Pihlgren, M., Tougne, C., Efler, S.M., Morris, M.L., AlAdhami, M.J., Cameron, D.W., Cooper, C.L., Heathcote, J., Davis, H.L., *et al.* (2004). Co-administration of CpG oligonucleotides enhances the late affinity maturation process of human anti-hepatitis B vaccine response. Vaccine *23*, 615–622.

Singh, M., Chesko, J., Kazzaz, J., Ugozzoli, M., Kan, E., Srivastava, I., and O'Hagan, D.T. (2004a). Adsorption of a novel recombinant glycoprotein from HIV (Env gp120dV2 SF162) to anionic PLG microparticles retains the structural integrity of the protein, whereas encapsulation in PLG microparticles does not. Pharm. Res. *21*, 2148–2152.

Singh, M., Kazzaz, J., Chesko, J., Soenawan, E., Ugozzoli, M., Giuliani, M., Pizza, M., Rappouli, R., and O'Hagan, D.T. (2004b). Anionic microparticles are a potent delivery system for recombinant antigens from *Neisseria meningitidis* serotype B. J. Pharm. Sci. *93*, 273–282.

Song, L., Zhang, Y., Yun, N.E., Poussard, A.L., Smith, J.N., Smith, J.K., Borisevich, V., Linde, J.J., Zacks, M.A., Li, H., *et al.* (2009). Superior efficacy of a recombinant flagellin:H5N1 HA globular head vaccine is determined by the placement of the globular head within flagellin. Vaccine *27*, 5875–5884.

Sparwasser, T., Koch, E.S., Vabulas, R.M., Heeg, K., Lipford, G.B., Ellwart, J.W., and Wagner, H. (1998). Bacterial DNA and immunostimulatory CpG oligonucleotides trigger maturation and activation of murine dendritic cells. Eur. J. Immunol. *28*, 2045–2054.

Speiser, D.E., Lienard, D., Rufer, N., Rubio-Godoy, V., Rimoldi, D., Lejeune, F., Krieg, A.M., Cerottini, J.C., and Romero, P. (2005). Rapid and strong human CD8(+) T-cell responses to vaccination with peptide, IFA, and CpG oligodeoxynucleotide 7909. J. Clin. Invest. *115*, 739–746.

Stephen, E.L., Sammons, M.L., Pannier, W.L., Baron, S., Spertzel, R.O., and Levy, H.B. (1977). Effect of a nuclease-resistant derivative of polyriboinosinic-polyribocytidylic acid complex on yellow fever in rhesus monkeys (*Macaca mulatta*). J. Infect. Dis. *136*, 122–126.

Sun, C.M., Deriaud, E., Leclerc, C., and Lo-Man, R. (2005). Upon TLR9 signaling, CD5+ B cells control the IL-12-dependent Th1-priming capacity of neonatal DCs. Immunity. *22*, 467–477.

Takeda, K., Takeuchi, O., and Akira, S. (2002). Recognition of lipopeptides by Toll-like receptors. J. Endotoxin. Res. *8*, 459–463.

Takeshita, F., Leifer, C.A., Gursel, I., Ishii, K.J., Takeshita, S., Gursel, M., and Klinman, D.M. (2001). Cutting edge: Role of Toll-like receptor 9 in CpG DNA-induced activation of human cells. J. Immunol. *167*, 3555–3558.

Takeuchi, O., Kawai, T., Sanjo, H., Copeland, N.G., Gilbert, D.J., Jenkins, N.A., Takeda, K., and Akira, S. (1999). TLR6, A novel member of an expanding toll-like receptor family. Gene *231*, 59–65.

Takeuchi, O., Sato, S., Horiuchi, T., Hoshino, K., Takeda, K., Dong, Z., Modlin, R.L., and Akira, S. (2002). Cutting edge: role of Toll-like receptor 1 in mediating immune response to microbial lipoproteins. J. Immunol. *169*, 10–14.

Tanaka, H., Demeure, C.E., Rubio, M., Delespesse, G., and Sarfati, M. (2000). Human monocyte-derived dendritic cells induce naive T-cell differentiation into T helper cell type 2 (Th2) or Th1/Th2 effectors. Role of stimulator/responder ratio. J. Exp. Med. *192*, 405–412.

Tanimura, N., Saitoh, S., Matsumoto, F., kashi-Takamura, S., and Miyake, K. (2008). Roles for LPS-dependent interaction and relocation of TLR4 and TRAM in TRIF-signaling. Biochem. biophys. Res. Commun. *368*, 94–99.

Ulrich, J.T., and Myers, K.R. (1995). Monophosphoryl lipid A as an adjuvant. Past experiences and new directions. Pharm. Biotechnol. *6*, 495–524.

Vajdy, M., Srivastava, I., Polo, J., Donnelly, J., O'Hagan, D., and Singh, M. (2004). Mucosal adjuvants and delivery systems for protein-, DNA- and RNA-based vaccines. Immunol. Cell Biol. *82*, 617–627.

Valmori, D., Souleimanian, N.E., Tosello, V., Bhardwaj, N., Adams, S., O'Neill, D., Pavlick, A., Escalon, J.B., Cruz, C.M., Angiulli, A., *et al.* (2007). Vaccination with NY-ESO-1 protein and CpG in Montanide induces integrated antibody/Th1 responses and CD8 T cells through cross-priming. Proc. Natl. Acad. Sci. U.S.A. *104*, 8947–8952.

Viney, J.L. (1999). Dendritic cell subsets: the ultimate T-cell differentiation decision makers? Gut *45*, 640–641.

Vollmer, J. and Krieg, A.M. (2009). Immunotherapeutic applications of CpG oligodeoxynucleotide TLR9 agonists. Adv. Drug Deliv. Rev. *61*, 195–204.

Weeratna, R.D., McCluskie, M.J., Xu, Y., and Davis, H.L. (2000). CpG DNA induces stronger immune responses with less toxicity than other adjuvants. Vaccine *18*, 1755–1762.

Weeratna, R.D., Brazolot Millan, C.L., McCluskie, M.J., and Davis, H.L. (2001a). CpG ODN can re-direct the Th bias of established Th2 immune responses in adult and young mice. FEMS Immunol. Med. Microbiol. *32*, 65–71.

Weeratna, R.D., Brazolot Millan, C.L., McCluskie, M.J., Siegrist, C.A., and Davis, H.L. (2001b). Priming of immune responses to hepatitis B surface antigen in young mice immunized in the presence of maternally derived antibodies. FEMS Immunol. Med. Microbiol. *30*, 241–247.

Weeratna, R.D., Comanita, L., and Davis, H.L. (2003). CPG ODN allows lower dose of antigen against hepatitis B surface antigen in BALB/c mice. Immunol. Cell Biol. *81*, 59–62.

Weeratna, R.D., Davis H.L, Medynski, L., and Krieg, A.M. (2006). Potential use of CpG ODN for cancer immunotherapy. Update Cancer Ther. 49–58.

WHO Media centre. (2010). Ref Type: Internet Communication.

Xie, H., Gursel, I., Ivins, B.E., Singh, M., O'Hagan, D.T., Ulmer, J.B., and Klinman, D.M. (2005). CpG oligodeoxynucleotides adsorbed onto polylactide-co-glycolide microparticles improve the immunogenicity and protective activity of the licensed anthrax vaccine. Infect. Immun. *73*, 828–833.

Yamamoto, M., Sato, S., Hemmi, H., Hoshino, K., Kaisho, T., Sanjo, H., Takeuchi, O., Sugiyama, M., Okabe, M., Takeda, K., and Akira, S. (2003). Role of adaptor TRIF in the MyD88-independent toll-like receptor signaling pathway. Science *301*, 640–643.

Yi, A.K., Peckham, D.W., Ashman, R.F., and Krieg, A.M. (1999). CpG DNA rescues B cells from apoptosis by activating NFkappaB and preventing mitochondrial membrane potential disruption via a chloroquine-sensitive pathway. Int. Immunol. *11*, 2015–2024.

Yoshimura, A., Lien, E., Ingalls, R.R., Tuomanen, E., Dziarski, R., and Golenbock, D. (1999). Cutting edge: recognition of Gram-positive bacterial cell wall components by the innate immune system occurs via Toll-like receptor 2. J. Immunol. *163*, 1–5.

Yu, H., Jiang, X., Shen, C., Karunakaran, K.P., Jiang, J., Rosin, N.L., and Brunham, R.C. (2010). *Chlamydia muridarum* T-cell antigens formulated with the adjuvant DDA/TDB induce immunity against infection that correlates with a high frequency of gamma interferon (IFN-gamma)/tumor necrosis factor alpha and IFN-gamma/interleukin-17 double-positive CD4+ T cells. Infect. Immun. *78*, 2272–2282.

Host Defence Peptides

15

Olga M. Pena*, John D. F. Hale* and Robert E.W. Hancock

Abstract

The increasing problem of resistance to antimicrobial agents, combined with the limited development of novel agents to treat infectious diseases is a serious threat to human morbidity and mortality around the world. Among the available strategies to create new therapeutic agents is the enhancement of the multifunctional properties of the natural anti-infectives, cationic host defence (antimicrobial) peptides (HDPs). This chapter will provide a summary of our current understanding of the different types of HDPs including natural and synthetic peptides and their antimicrobial and immunomodulatory modes of action. Additionally, we will describe new approaches to peptide design and discuss both the therapeutic potential and prospective challenges in the utilization of peptides for antibacterial therapy.

Introduction

Host defence peptides (HDPs; also known as antimicrobial peptides – AMPs), are produced as the first line of defence by a wide variety of organisms, from single celled microorganisms, insects and other invertebrates, to amphibians, birds, fish and mammals including humans (Wang *et al.*, 2009). They are generally described as relatively short (12–50 amino acids), amphiphilic, and having at least two positively charged residues such as arginine or lysine. They can be broadly distributed into four classes based on structure: α-helical, β-sheet, loop and extended structures (Jenssen *et al.*, 2006). The important role of HDPs is supported by the existence of the hundreds of examples identified to date and the initial observation of their broad functions as direct antimicrobials capable of destabilizing bacterial, fungal or viral membranes. In addition, more recent work has shown that they have a diverse range of functions in modulating the immune response to prevent or combat infections (Easton *et al.*, 2009). Some of these functions include: the neutralization of the septic effects of bacterial components like lipopolysaccharide (LPS), also known as endotoxin; promoting the recruitment of immune cells to the site of infection; inducing cell differentiation; adjuvant activity enhancing the adaptive immune response to antigens; wound healing properties; and enhancing cell survival. HDPs can be constitutive or inducible depending on the species, tissue type, cellular lineage, or differentiation state of the cells. Their production can be triggered by infectious and/or inflammatory stimuli, like LPS, or pro-inflammatory cytokines, and usually leads to the expression of a pro-peptide which requires proteolytic cleavage for final production of a mature (biologically active) HDP (Brown and Hancock, 2006). Many of them are potent enough to be considered natural antibiotics with direct microbicidal activity; however it has been shown that at physiologically relevant salt concentrations, many mammalian HDPs have reduced microbicidal activity at normally available peptide concentrations and under such conditions their functions as immunomodulators are foremost

*Corresponding authors

(Mookherjee and Hancock, 2007). The name *host defence peptides* has been given to this class of molecule to reflect their distinctive role in reinforcing hosts defences and enhancing clearance of pathogen, highlighting that their function is much broader than the antimicrobial activity initially observed (Hancock and Sahl 2006). This chapter will describe the different aspects of HDPs, such as their classifications, modes of action, advantages and possible limitations for their prospective therapeutic use.

Structural classes

Despite their ubiquity and lack of sequence similarity, HDPs can be classified into four major themes based on their secondary structures: (A) α-helical; (B) β-stranded, owing to the presence of two or more disulfide bonds; (C) β-hairpin or loop, owing to the presence of a single disulfide bond and/or cyclization of the peptide chain; and (D) extended. Fig. 15.1 shows examples of HDPs that are characteristic of each of the structural classes and these will be discussed in the next section. It is important to clarify that many of these peptides are unstructured in free solution and fold into their final configuration upon insertion into lipid bilayer membranes.

The α-helical HDPs are highly abundant in nature and are present in a range of organisms that are evolutionarily quite distant including insects, amphibians and humans. The contexts in which they function, and anatomical locations of their production, are quite different. For example, in insects they are produced in the haemolymph, while in amphibians they are secreted at the skin, gastric and mucosal epithelial surfaces. They are often described as potent and broad antimicrobials, however their activities and spectra can vary substantially. Regarding amino acid sequence, they vary substantially (Tossi *et al.*, 2000). Magainins are one of the common examples of α-helical HDPs present in frogs. They have been isolated from the skin of the African clawed frog *Xenopus laevi* and found to have a broad spectrum of antimicrobial activities against Gram-positive and Gram-negative bacteria, as well as fungi (Zasloff, 1987). Additionally, magainins have been found to have spermicidal activity whereby *in vitro* and *in vivo* monkey studies demonstrated that 1 mg of magainin-A was sufficient to arrest sperm motility, preventing pregnancy without affecting the ovulation cycle (Zairi *et al.*, 2009). In humans, LL-37 from the cathelicidin family is the most studied α-helical HDP. It is produced mainly by phagocytic and epithelial cells, as the result of a proteolytic cleavage from the C-terminus of the cathelicidin antimicrobial precursor protein, hCAP18 (Nijnik and Hancock, 2009), but its broad spectrum antimicrobial activities are

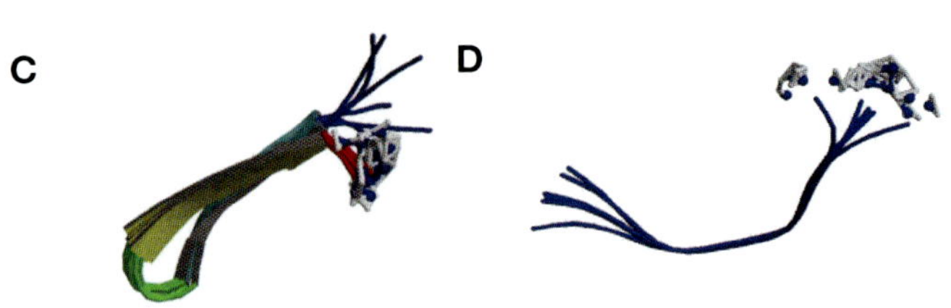

Figure 15.1 Structural classes of host defence peptides. (A) α-helical, LL-37 (PDB ID: 2K6O Wang *et al.*, 2008). (B) β-stranded, human neutrophil protein-3 (HNP-3) also known as α defensin 3 (PDB ID: 1DFN. Hill *et al.*, 1991); (C) looped, protegrin 1 (PDB ID: 1PG1. Fahrner *et al.*, 1996); (D) extended, indolicidin (PDB 1G89. Rozek, *et al.*, 2000).

substantially diminished by physiological action and glycosaminoglycan concentrations (Bowdish *et al.*, 2005a), indicating that its immunomodulatory properties may be more physiologically relevant.

The most well characterized β-stranded HDPs are the defensins, which are found in fungi, insects and mammals. They are a family of small cationic peptides from 30–50 amino acids in size, with three intramolecular disulfide bonds between six conserved cysteine residues: they are often expressed by various epithelial cells such as respiratory and epidermal cells (Oren, 2005). They have antimicrobial activity not only against Gram-negative and Gram-positive bacteria but also against mycoplasma such as *Mycoplasma pneumoniae* (Kuwano *et al.*, 2006). The β-stranded HDP structure is attributed to the presence of two or more disulfide bonds which provide an important role in stabilising peptide structure. For example, it was found that while human β-defensin-3 (HβD3) requires correct disulfide bond formation for inducing chemotactic activity, this was not necessary for its antimicrobial activity (Wu *et al.*, 2003). Conversely, the disulfide bond which forms the loop of the bovine bactenecin was required for activity against wild type Gram-negative organisms (but actually inhibited activity against Gram-positive bacteria) and is thought to play a role in outer membrane permeabilization (Wu and Hancock, 1999b).

Thanatin, an example of the rarer loop HDPs, has both bactericidal and fungicidal activity. It was first discovered in the insect *Podisus maculiventris* and is a 21-residue peptide with similar sequence homology to frog HDPs, such as the Brevinin family of peptides. The role of the C-terminal loop in the biological activity of this peptide was confirmed using synthetic truncated isoforms (Fehlbaum *et al.*, 1996).

Indolicidin is a typical model for an extended HDP. With a length of only 13 amino acids, and a composition consisting of 39% tryptophan and 23% proline, it is present in the cytoplasmic granules of bovine neutrophils. Its structure is distinct from the α-helical and β-stranded HDPs and it is too small to span the lipid bilayer of biological membranes. Thus, it probably adopts an extended structure as suggested by circular dichroism measurements. In Gram-negative bacteria, indolicidin permeabilizes the outer membrane and, although originally suspected to disrupt the cytoplasmic membrane by channel formation (Falla *et al.*, 1996), is now considered to enter cells and act on cytoplasmic targets (Zhang *et al.*, 2001).

Despite their different structures, HDPs have two main functions, antimicrobial and immunomodulatory, involving diverse modes of action which will be discussed in the following sections.

Antimicrobial mode of action

All bacteria, with certain exceptions, e.g. *Burkholderia* sp. and sometimes *Serratia* and *Proteus* spp. are susceptible to HDPs (Hancock and Patrzykat 2002). Disappointingly, despite the growing number (>1000) of antibacterial HDPs that have now been identified (Wang *et al.*, 2009) or synthesized, the mode of action is understood in depth for only a few peptides. Initially, HDPs were thought to kill bacteria solely through their interaction with the bacterial cell cytoplasmic membrane. However, as shown with the winter flounder (*Pleuronectes americanus*) HDP pleuricidin, the mode of action can differ in a concentration-dependent manner (Patrzykat *et al.*, 2002). Assays using elevated concentrations of the HDP (above its minimal inhibitory concentration; MIC), demonstrated a bias towards lytic membrane interactions whereas at the MIC no membrane perforation could be demonstrated and instead inhibition of macromolecular synthesis was evident. Given this issue, recent studies have tended to focus on the antimicrobial effect of HDPs at their MIC to gain a more accurate picture of their mechanism of action.

In contrast to classical antibiotics, HDPs have a uniquely complex mode of action. A growing body of evidence indicates that HDPs may work through simultaneous interactions with multiple bacterial targets and as a result have been termed 'dirty drugs' (Peschel and Sahl, 2006). Certainly, their promiscuous action can help explain why resistance against HDPs is much rarer than resistance against classical antibiotics. The current view on how HDPs function is still debatable and it would appear to be very much dependent on the individual HDP and the bacterial target. However, it is agreed that owing to their amphipathic properties, all HDPs are able to interact with the cytoplasmic membrane. Regardless, antibacterial HDPs can be loosely classified as membrane-acting or alternative target molecules (Hale and Hancock, 2007). New technology and advances in biological tools have accelerated our understanding of bacteria and assisted us in learning more about the behaviour of HDPs. The use of microarrays is well established and can provide a snapshot of bacterial cell transcriptional responses to an HDP. This technology has been used to increase our understanding of selected HDPs, including those produced by humans such as HβD3 (Sass *et al.*, 2008) and LL37 (Overhage *et al.*, 2008) as well as from animals, e.g. temporin L, dermaseptin K4-S4(1–16) and ovispirin-1 (Pietiainen *et al.*, 2009) and insect cecropins (Hong *et al.*, 2003). Other techniques such as bacterial mutant libraries and essential gene libraries (Hale and Hancock, unpublished data) are providing an insight into how HDPs affect bacterial cells through possible direct interaction with protein targets. Now that HDPs are having a renaissance with regards to their therapeutic potential, a detailed understanding of their mode of action would greatly assist us in predicting possible resistance mechanisms as well as strategies to optimize their potency. A summary of some known mechanisms of action are listed below and reviewed in Fig. 15.2.

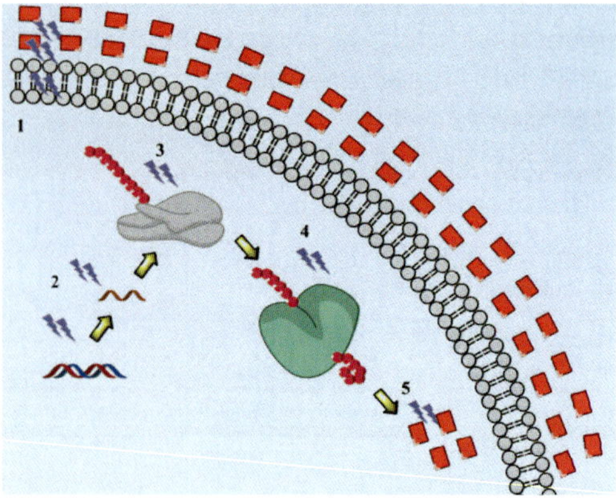

Figure 15.2 Currently defined possible antimicrobial host defence peptide (HDP) target sites. HDPs must passage across the cell wall (red blocks) (or initially the outer membrane in Gram-negative bacteria) reaching the cytoplasmic membrane (CM). HDPs exert killing at the following locations: (1) through insertion and disruption of CM integrity; (2) by binding and disrupting DNA and RNA synthesis; (3) interfering with protein synthesis, (4) protein folding and (5) cell wall synthesis.

Membrane interactive mode of action

Regardless of bacterial cell, type cationic HDPs are attracted to the polyanionic surface of bacterial cells owing to the presence of polyanionic LPS on the outer leaflet of the outer membrane in Gram-negative bacteria or teichoic acids in the Gram-positive cell wall. To reach the cytoplasmic membrane of the bacterial cells, HDPs must translocate across the thick cell wall for Gram-positive bacteria or the outer membrane and cell wall peptidoglycan of Gram-negative bacteria. The interaction of cationic HDPs with the outer membrane of Gram-negative bacteria is reasonably well understood and has been termed 'self-promoted uptake' (Hancock and Chapple, 1999; Brogden, 2005). Following initial binding, the outer membrane is disrupted by the high net positive charge and amphipathicity of the peptides and self-promoted peptide uptake into the cell occurs through the disrupted outer membrane. Once a HDP reaches the cytoplasmic membrane, it associates parallel to the plane of the membrane with the phospholipid heads of the outer leaflet (Jenssen *et al.*, 2006). As more HDP molecules associate, above a critical level, the peptides molecules are proposed to aggregate and/or insert into lipid bilayer. This can lead to membrane disruption, interaction with bacterial cell wall synthesizing enzymes in the membrane (Peschel and Sahl, 2006) or to passage of the HDP to the cytoplasm where it can interact with cytoplasmic targets and cause bacterial cell death. Death by cell disruption has been proposed to result from the formation of ion channels, transmembrane pores or extensive structural damage of the membrane (Brogden, 2005). Four basic models as to how membrane disruption might occur have been described by different authors and include: the barrel-stave, carpet, toroidal pore and aggregate models (Hale and Hancock, 2007). Currently no single model gives a more complete explanation than others and it appears likely that different HDPs might have different interaction mechanisms (Wu *et al.*, 1999c; Zhang *et al.*, 2001). However, of the four models, only the aggregate model explains easily how HDPs can translocate across the membrane to the cytoplasm. This model is consistent with the demonstrated inhibitory actions of HDPs on intracellular components.

Membrane non-lytic (alternative) mode of action

There has been a recent escalation in our understanding of alternative modes of action of HDPs. The initial inability to identify non-membrane targets might be attributed to a bias towards observing membrane interaction owing to the use of elevated concentrations of HDPs in previous assays and also a possible 'needle in a haystack' concept whereby there were too many possible targets to predict any one accurately. The nature of HDPs (amphipathic molecules with an overall cationic charge) suggests that it is possible to have multiple simultaneous targets (that would be correspondingly hydrophobic and/or negatively charged) and the list of HDPs that interfere with alternative targets which display no activity against the membrane, is growing (Hale and Hancock, 2008) [see specific examples of HβD3 (Sass *et al.*, 2008) and PR39 (Boman *et al.*, 1993)]. It is difficult to identify all possible targets, but with new genomic techniques this is becoming more feasible. Included in the current list of alternative targets are nucleic acids and their synthesis (both RNA and DNA, e.g. buforin II; Park *et al.*, 1998)), proteins and enzymes including chaperones (e.g. chaperone DnaK is targeted by HDPs such as pyrrhocoricin and its analogues; Otvos *et al.*, 2000) and cell wall synthesis enzymes, including glutamate racemase (Hale and Hancock, unpublished) and others (Peschel and Sahl, 2006). Multiple targets have been implicated for a range of peptides including pig PR-39 (Boman, *et al.*, 1993; Falla *et al.*, 1996; Subbalakshmi and Sitaram, 1998; Marchand, *et al.*, 2006), bacterial nisin (Brotz *et al.*, 1998) various synthetic peptides and flounder pleurocidin (Patrzykat *et al.*, 2002).

Immunomodulatory mode of action

The innate immune response is the most primitive form of the immune system. It is highly conserved amongst animals and is especially important for host defences in invertebrates that lack an adaptive response. The effectiveness of the innate immune response originates from many different processes that work actively and simultaneously or consecutively to eradicate pathogens. One of the main strategies of the innate immune response is the production of HDPs. Their importance is supported by *in vivo* evidence in rodent models in which the injection of HDPs, such as cathelicidins can reduce bacterial load, at physiological concentrations (Scott *et al.*, 2002; Bowdish *et al.*, 2005a; Cirioni *et al.*, 2006). Conversely many HDP lose their direct antimicrobial activities at physiological salt or serum concentrations (Bowdish *et al.*, 2005a; Ciornei *et al.*, 2005), indicating that under these conditions the immunomodulatory properties of HDP are a more likely explanation for their protective activities *in vivo*. Additionally, transgenic mice lacking β-defensin-1 and cathelicidin-related antimicrobial peptide (CRAMP) fail to clear infections which is either due to decreased direct killing or to a dysregulated and inefficient innate immune response, leading to an increased susceptibility to infection (Moser *et al.*, 2002; Iimura *et al.*, 2005; Huang *et al.*, 2007). Similar patterns have been seen in humans, where patients with primary immunodeficiencies such as neutrophil-granule deficient syndrome thus lacking α-defensins, present with frequent skin infections that persist for months. The same scenario is seen for Morbus–Kostmann syndrome patients lacking LL-37 and α-defensin-3, who suffer from frequent oral bacterial infections and severe periodontal disease (Pütsep *et al.*, 2002). However, the lack of any particular HDP in these human syndromes may not be the sole cause of the observed symptoms in these patients, but reflect the importance of the combined presence of many different elements of innate immunity in orchestrating an efficient immune response and fighting infections.

In general, the immunomodulatory functions of HDPs are not restricted to boosting the immune response to attack and destroy the pathogen rapidly, but also to regulate and selectively activate the immune response, prevent the overwhelming effects of excessive inflammation, and smoothing the transition to adaptive immunity while promoting wound healing (Fig. 15.3). The different immunomodulatory functions of HDPs include the various activities described below.

Anti-endotoxin activity

The detection by immune cells of microbes or microbe-associated signature molecules (sometimes termed pathogen-associated molecular patterns – PAMPs) such as endotoxin (LPS), triggers inflammatory responses that are beneficial for fighting and eradicating infections. However, it is essential that this process occurs in a well balanced manner, since excessive inflammation can lead to host tissue damage and deadly syndromes like sepsis. Many different HDPs have the ability to suppress the activity of endotoxin-induced pro-inflammatory mediators such as tumour necrosis factor alpha (TNF-α; Scott *et al.*, 2000; Bowdish and Hancock, 2005b). It has been suggested that these actions can occur through blocking or modulation of the intracellular signalling cascades downstream of TLRs or through direct binding to LPS, thus preventing its detection by Toll-like receptors (TLR).

Chemotactic activity

Upon encounter with pathogens, local tissues produce chemokines that attract other immune effector cells. HDPs have been found to selectively induce the gene expression of chemokines such as CXCL8/IL8 and CCL2/MCP-1 in a large variety of cell types including monocytes, mast cells and T cells. While induction of chemokine expression occurs at physiological concentrations, it has been demonstrated that at slightly higher but pathologically relevant, concentrations HDPs

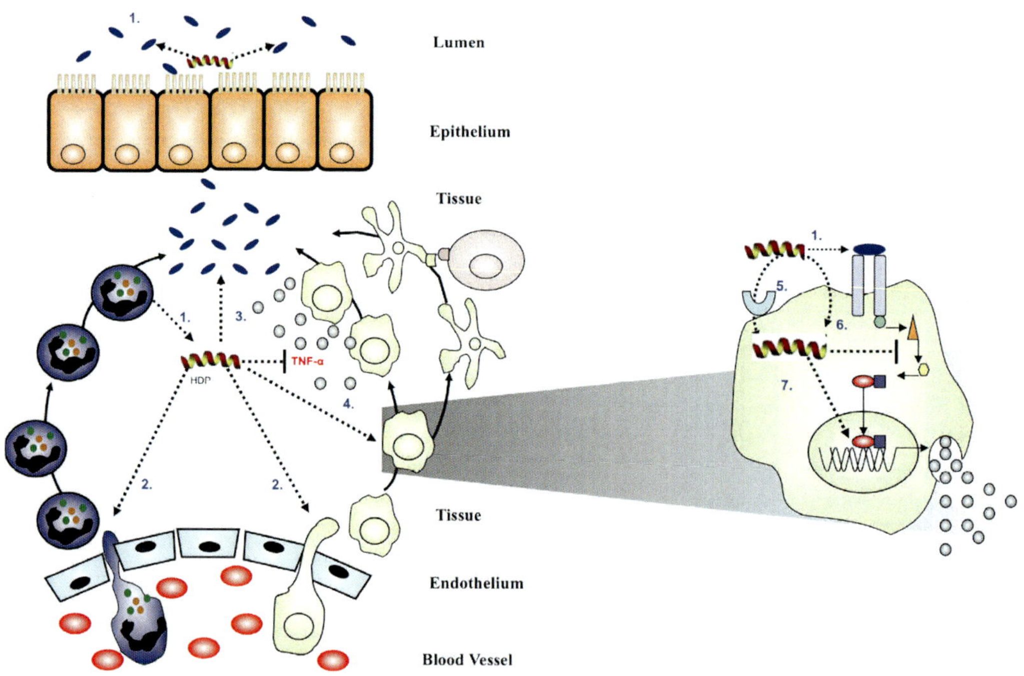

Figure 15.3 Immunomodulatory activities of host defence peptides (HDPs) (e.g. LL-37). The figure presents the different immune functions of HDPs at the systemic level, and in an individual cell. At the systemic level, HDPs released by epithelial cells and immune cells such as neutrophils can provide direct anti-microbial activity (1), stimulate either directly or indirectly the recruitment of immune cells (2), block pro-inflammatory mediators such as TNF-α (3), and induce dendritic cell differentiation and activation leading to the connection of adaptive immunity (4). In a single cell, HDPs may enter the cells by directly translocating across the membrane and/or interact with a variety of different cell receptors (5), inhibiting different intracellular signalling pathways (6) and generating specific gene transcription profiles (7).

can exhibit direct chemotactic activity (Durr and Peschel 2002; Tjabringaa *et al.*, 2006).

Immune cell differentiation

Correct and effective cellular differentiation of dendritic cells (DC), macrophages or lymphocytes is necessary for a successful immune response. Some studies suggest that HDPs can induce cellular differentiation, for example LL-37 was found to be a potent modifier of dendritic cell differentiation, leading to significantly up-regulated endocytic capacity, co-stimulatory molecule expression, enhanced secretion of Th1 inducing cytokines, and promotion of Th1 responses *in vitro* (Davidson *et al.*, 2004).

Cell survival

Apoptosis is a non-inflammatory natural mechanism for eliminating old, unnecessary and unhealthy cells. Immune cells are normally rapidly eliminated from the body through apoptosis. Lengthening the normal half-life of immune cells can, under some circumstances, allow the host to respond in a more efficient manner. Particular HDPs, such as LL-37, have demonstrated potent anti-apoptotic properties in enhancing the normal lifespan of neutrophils. This event appears to involve signalling through the $P2X_7$ receptor, modulation of Mcl-1 expression and the inhibition of BCL-2 interacting domain and procaspase-3 cleavage (Barlow *et al.*, 2006). In contrast LL-37 actually stimulates apoptosis in epithelial cells and T lymphocytes (Lau *et al.*, 2005; Mader *et al.*, 2009).

Initiation and polarization of adaptive immunity

By directing the two arms of the immune system (innate and adaptive) through their chemotactic effects, HDPs can influence the initiation of the adaptive immune response. They can also act as adjuvants by interacting through various receptors on or in DCs and T cells, influencing the polarization of antigen-specific adaptive immunity. Their functions are extended to enhancing cellular proliferation and cytokine responses of CD4 T cells, such as IL-10 and IL-6 production and the expression of co-stimulatory molecules, and the production of antigen-specific serum antibodies by B cells (Yang *et al.*, 1999; Brogden *et al.*, 2003).

Wound healing and angiogenesis

Angiogenesis, the formation of new blood vessels, is a requirement for tissue repair and wound healing. Some HDPs have demonstrated roles in angiogenesis by acting directly on endothelial cells and inducing the growth of small capillaries. This has also been confirmed by observing the opposing scenario, whereby a lack of HDPs was associated with impaired re-epithelialization of chronic wounds and thus failure of wound healing (Heilborn *et al.*, 2003; Koczulla *et al.*, 2003). Other mechanisms may be involved in the stimulation of wound repair including chemokine production as described above, leading to the influx of healthy cells and enhanced proliferation of epithelial cells (Shaykhiev *et al.*, 2005).

The complex mechanisms through which peptides perform the above-mentioned functions are slowly being unravelled. Emerging evidence shows that peptides may act through multiple points of intervention, which may vary depending on the cell type. For example, LL-37 has been found to act through multiple interactors, such as the $P2X_7$ receptor, formyl peptide receptor like-1 (FPRL-1) and other unknown Gi-protein coupled receptors, and an intracellular receptor GAPDH (Tomasinsig *et al.*, 2008; Coffelt *et al.*, 2009; Mookherjee *et al.*, 2009). Key signalling pathways such as those involving the mitogen-activated protein kinases (MAPK) p38, JNK, and extracellular signal-regulated kinase-1/2 (ERK1/2) as well as Src-family kinases, NFκB (transiently) and the PI3 kinase pathways, are activated by stimulation with HDPs, and are essential for downstream functions. Such mechanisms have been demonstrated in mast cells, keratinocytes and monocytes by β-defensins and LL-37. In addition, systems biology studies have shown a large number of transcriptional changes that HDPs can induce, demonstrating the complexity of these immunomodulatory responses (Mookherjee *et al.*, 2006).

Overall, both the antimicrobial properties and immunomodulatory activities make HDPs excellent candidates for beneficial therapeutic applications in a range of disease scenarios.

Table 15.2 Peptides in clinical trials

Drug name	Company name	Function	Use	Progress	References
IMX942 (5 amino acid peptide)	Inimex	Immunomodulatory	Nosocomial infections, Febrile neutropenia	Post-phase IA	http://www.inimexpharma.com/documents/PressRelease_FirstClinicalStudy_Apr2709.pdf
EA-230 (oligopeptide fragment from β-hCG)	Exponential Biotherapies	Immunomodulatory	Sepsis	phase II	http://www.expobio.com/clinical-trials/index.php
RDP58 (semisynthetic D-amino acid decapeptide derived from HLA class I B2702)	Genzyme	Immunomodulatory	Inflammatory bowel disease	Post-phase II	http://www.genzyme.com/corp/licensing/genz_p_rdp58_login.asp
PMX-30063 (defensin structural mimetic, non-peptide, small molecule/copolymer)	PolyMedix	Antimicrobial	Antibiotic	Phase IB	http://www.polymedix.com/pipeline.php
PAC-113 (synthetic cationic host defence peptide (12-mer), histatin derivative)	Pacgen Biopharmaceuticals	Antimicrobial	Antifungal	Phase II	NCT0065991[a]
hLF1–11 (Cationic peptide, human lactoferricin – amino acid fragment 1–11)	AM-Pharma	Immunomodulatory/antimicrobial	Bacteraemia and fungal infections	Phase I/II	NCT0050998[a]
Opebacan (rBPI21) (21-amino-acid peptide derivative of bactericidal/permeability-increasing protein)	Xoma	Immunomodulatory/antimicrobial	Endotoxaemia in haematopoietic stem cell transplant recipients	Phase I/II	[a]NCT0045415
Pexiganan acetate (MSI-78) (synthetic cationic host defence peptide) (22-mer), magainin derivative)	MacroChem	Antimicrobial/immunomodulatory	Topical antibiotic	Phase III	[a]NCT0056343
Omiganan (MX-226) (synthetic cationic host defence peptide) (12-mer), indolicidin derivative)	MIGENIX	Antimicrobial/immunomodulatory	Topical antiseptic, *Acne vulgaris*, papulopustular rosacea	Phase III	[a]NCT0000045
Iseganan (IB-367) (synthetic protegrin-1 derivative) (17 amino acids)	Ardea Biosciences	Antimicrobial/immunomodulatory	Oral mucositis in radiation therapy patients	Phase III	[a]NCT0002233

[a]Registered clinical trial number http://clinicaltrials.gov/ct2/home.

are major pathogens in our society (Devine and Hancock, 2002); indeed by and large the most recalcitrant multidrug resistant bacteria appear to be major potential targets for HDPs. A range of specific resistance mechanisms have been identified in bacteria, but in contrast to resistance to classical antibiotics, the genetic determinants associated with HDP resistance are usually not located on transferable elements such as transposons or plasmids (Kraus and Peschel, 2006). Also, in our experience, natural resistance to the most potent antimicrobial peptides is quite unusual. Similarly *in vitro* studies have experienced difficulty in selecting for resistance by serial passage (Devine and Hancock, 2002). A range of the better characterized resistance mechanisms are presented below.

Mutational resistance: modification of the cell's net charge

The majority of HDPs have a net cationic charge, which allows for the initial attraction of the peptide to the negatively charged cell surface. Thus it is perhaps no surprise that the most successful resistance mechanisms of bacteria employ molecular mechanisms that alter the cell's net charge, effectively reducing the affinity that HDPs have with the cell (Ernst *et al.*, 2001; Peschel 2002).

The Gram-positive bacterial cell wall has a strong polyanionic charge due to phosphate groups found in the repeating glycerophosphate units of teichoic acid polymers (Neuhaus and Baddiley, 2003). Many Gram-positive bacteria including *S. aureus* (Peschel *et al.*, 1999), *Streptococcus agalactiae* (Poyart *et al.*, 2003) and *Streptococcus pyogenes* (Kristian *et al.*, 2005) can reduce this negative charge by the addition of cationic D-alanine residues effectively decreasing the action of HDPs. This resistance mechanism has been found to be encoded by the *dltABCD* operon (Peschel *et al.*, 1999) and is regulated by a three component antimicrobial peptide sensor system that comprises a sensor (*apsS*), response regulator (*apsR*) and a third protein of unknown function (*apsX*; Li *et al.*, 2007a,b).

Though very different structurally, Gram-negative bacteria employ similar mechanisms of resistance to those used by Gram-positive species described above. The outer leaflet of the outer membrane comprises the molecule LPS, which comprises three segments: the conserved anionic Lipid A anchor molecule; the variable core oligosaccharide; and the repeating O antigen oligosaccharide. Interaction of cationic peptides with the Lipid A phosphates in particular is the basis of self-promoted uptake into Gram-negative bacteria. A reduction the negative charge of Lipid A can result in reduced binding by HDPs. Reduction in charge can be achieved through the addition of cationic molecules such as aminoarabinose and is achieved through the effects of the *pmr* operon (*pmrHFIJKLM*; also termed *arnBCADTEF*) which has now been identified in a multitude of Gram-negative species (including *E. coli*, *S. typhi* and *P. aeruginosa*) indicating that this is probably a common resistance mechanism (Devine and Hancock, 2002; Miller *et al.*, 2005). Regulation of these operons is under the control of two-component regulatory systems: *pmrAB* and *phoPQ*. Under Mg^{2+} limiting conditions or in the presence of peptide that displaces Mg^{2+}, two membrane-bound histidine kinase molecules (PhoQ, which in *Salmonella* but not *Pseudomonas* is the actual peptide sensor, and PmrB) autophosphorylate and activate their respective cognate response regulators PhoP and PmrA. Activation of these regulators causes the transcriptional up-regulation of their own operons as well as the Lipid A modifying *arnBCADEFTEF* operon (Gooderham and Hancock, 2009). The outcome is the reduced interaction of HDPs with the outer membrane, due to the reduced anionic charge causing a loss in AMPs ability to promote their own uptake across the outer membrane. Interestingly not all peptides are equally able to induce these peptide resistance mechanisms and it was demonstrated that the MIC of *P. aeruginosa* was inversely proportional to HDPs ability to induce the *pmrAB* and *arn* operons (McPhee *et al.*, 2003).

Modification of the cytoplasmic membrane charge can also affect resistance to HDPs. Homologues of a large membrane bound protein MprF (multiple resistance factor) have been identified in many bacterial species such as *S. aureus* (Peschel *et al.*, 2001), *Listeria monocytogenes* (Thedieck *et al.*, 2006) and *Bacillus anthracis* (Samant *et al.*, 2009) indicating that it has evolved as a universal resistance mechanism (Weidenmaier *et al.*, 2003;

Kraus and Peschel, 2006). MprF modifies anionic phospholipid phosphatidylglycerol with L-lysine to reduce the net anionic charge of the membrane.

Mutational resistance: degradation by proteases

The most direct method to inactivate HDPs is by degrading them. There are many examples, although not all well characterized, of bacteria from both major classes that have evolved to express HDP-degrading proteases. These proteases tend to be selective for specific peptides. For example, the α-helical HDPs are often prone to proteolysis due to their simpler structure (Peschel and Sahl, 2006). The introduction of disulfide bonds, as seen in looped and β-sheet type HDPs increases resistance to proteases (Rozek *et al.*, 2003; Campopiano *et al.*, 2004). Despite the advantage afforded by the inclusion of disulfide bonds, α-helical HDPs are conserved throughout different species in nature (Peschel and Sahl, 2006).

The omptins are a class of outer membrane proteases, produced by many Gram-negative species including *E. coli* (Stumpe *et al.*, 1998), *Salmonella enterica* (Guina *et al.*, 2000b) and *Yersinia pestis* (Sodeinde and Goguen, 1989), that have been implicated in bacterial pathogenesis (Hritonenko and Stathopoulos, 2007). Despite sharing sequence similarity, representative omptins vary in their roles. OmpT, the prototype omptin produced by *E. coli* degrades the HDPs colicin (Cavard and Lazdunski, 1990) and protamine (Stumpe *et al.*, 1998). Another omptin, PgtE produced by *S. enterica* has been shown to degrade a range of HDPs, including the human cathelcidin LL-37, its mouse equivalent CRAMP, and other α-helical HDPs (Guina *et al.*, 2000a). Omptins are thought to work by cleaving peptides between two paired basic amino acids or between one basic amino acids and a following non-polar amino acid (Sugimura and Nishihara, 1988), thus it is possible to engineer peptides for decreased susceptibility to these proteases.

Gram-positive bacteria also produce proteases. *S. aureus* has been shown to produce many different proteases including the metalloprotease aureolysin (Sieprawska-Lupa *et al.*, 2004) and the serine protease V8 (Sieprawska-Lupa, 2004). Other bacterial species shown to produce proteases that can cleave LL-37 include *Enterococcus faecalis*, *Porphymonas gingivalis* and *Prevotella* spp. (Schmidtchen *et al.*, 2002).

Mutational resistance: capsule

Some species of bacteria are able to produce an extracellular capsule that is thought to help protect the cell from phagocytosis by the host immune system. The capsule in *Neisseria meningitidis* (*Jones et al., 2009; Spinosa et al., 2007*) and *Klebsiella pneumoniae* (Campos *et al.*, 2004) plays an important role in HDP resistance with capsule-deficient mutants being more sensitive to a variety of HDPs than the wild-type strain. It is proposed that this resistance is achieved by the ability of the anionic capsule to bind HDPs and limit their access into the cell (Llobet *et al.*, 2008). However, the capsule may not act alone as a resistance mechanism since other HDP resistance genes [such as those for efflux pumps (Tzeng *et al.*, 2005; Spinosa *et al.*, 2007)] were also encoded by some of the same bacteria in these studies.

Conclusions

There is an urgent need to develop new antimicrobial therapies to stem the rise of infections caused by multidrug resistant bacteria. HDPs provide a fresh alternative for new possible therapeutic options. Initial peptide research showed that HDPs are produced naturally in all living organisms, and function either directly as antimicrobials against microbes or through immunomodulation of the host immune defences. However, their development towards clinical use has been slow because of issues such as cost and potential toxicities. The past decade has seen many major advances in peptide research both as antimicrobials and immune modulators, in part bolstered by the significant reduction in cost of synthesis paired with new assay technologies. This has led to the identification of lead candidates, some of which are now in clinical trials. Recent development of cost-effective synthesis procedures has reduced the financial burden required of screening large libraries of peptides as well as allowed for the production of large batches of peptides for research use. Furthermore, peptides can provide a structural backbone for the synthesis of a myriad of new classes of peptides.

Acknowledgements

The cationic peptide research of the authors is supported by grants from the Grand Challenges in Global Health Research programme, the Canadian Institutes for Health Research (CIHR) and the Advanced Foods and Materials Network. OMP is the recipient of studentships from University of British Colombia fellowship and the CIHR-Vanier Canada Graduate Scholarships, and REWH holds a Canada Research Chair. We wish to thank Jelena Pistolic and Donna Easton for their assistance in proof reading this chapter.

References

Angus, D.C.,Linde-Zwirble, W.T., Lidicker, J., Clermont, G., Carcillo, J., and Pinsky, M.R. (2001). Epidemiology of severe sepsis in the United States: analysis of incidence, outcome, and associated costs of care. Crit. Care Med. *29*, 1303–1310.

Barlow, P.G., Li, Y., Wilkinson, T.S., Bowdish, D.M.E., Lau, Y.E., Cosseau, C., Haslett, C., Simpson, A.J., Hancock, R.E.W., and Davidson, D.J. (2006). The human cationic host defense peptide LL-37 mediates contrasting effects on apoptotic pathways in different primary cells of the innate immune system. J. Leukoc. Biol. *80*, 509–520.

Batista, C.V., Scaloni, A., Rigden, D.J., Silva, L.R., Rodrigues Romero, A., Dukor, R., Sebben, A., Talamo, F., and Bloch, C. (2001). A novel heterodimeric antimicrobial peptide from the tree-frog *Phyllomedusa distincta*. FEBS Lett. *494*, 85–89.

Bierbaum, G., and Sahl, H.G. (1987). Autolytic system of *Staphylococcus simulans 22*, influence of cationic peptides on activity of N-acetylmuramoyl-L-alanine amidase. J. Bacteriol. *169*, 5452–5458.

Blondelle, S.E., and Houghten, R.A. (1996). Novel antimicrobial compounds identified using synthetic combinatorial library technology. Trends Biotechnol. *14*, 60–65.

Blondelle, S. E., Takahashi, E., Weber, P. A., and Houghten, R. A. (1994). Identification of antimicrobial peptides by using combinatorial libraries made up of unnatural amino acids. Antimicrob. Agents Chemother. *38*, 2280–2286

Boman, H.G., Agerberth, B., and Boman, A. (1993). Mechanisms of action on *Escherichia coli* of cecropin P1 and PR-39, two antibacterial peptides from pig intestine. Infect. Immun. *61*, 2978–2984.

Bowdish, D., and Hancock, R.E.W. (2005b). Anti-endotoxin properties of cationic host defence peptides and proteins. J. Endotoxin Res. *11*, 230–236.

Bowdish, D., Davidson, D.J., Lau, Y.E., Lee, K., Scott M.G., and Hancock, R.E.W. (2005a). Impact of LL-37 on anti-infective immunity. J. Leukocyte Biol. *77*, 451–459.

Brogden, K.A. (2005). Antimicrobial peptides: pore formers or metabolic inhibitors in bacteria. Nat. Rev. Microbiol. 3, 238–250.

Brogden, K.A., Heidari, M., Sacco, R.E., Palmquist, D., Guthmiller, J.M., Johnson, G.K., Jia, H.P., Tack, B.F., and P. B. McCray, J. (2003). Defensin-induced adaptive immunity in mice and its potential in preventing periodontal disease. Oral Microbiol. Immunol. *18*, 95–99.

Brotman, Y., Makovitzki, A., Shai, Y., Chet, I., and Viterbo, A. (2009). Synthetic ultrashort cationic lipopeptides induce systemic plant defense responses against bacterial and fungal pathogens. Appl. Environ. Microbiol. 75, 5373–5379

Brotz, H., Bierbaum, G., Reynolds, P.E., and Sahl, H.G. (1997). The lantibiotic mersacidin inhibits peptidoglycan biosynthesis at the level of transglycosylation. Eur. J. Biochem. *246*, 193–199.

Brotz, J., Josten, M., Wiedemann, I., Schneider, U., Götz, F., Bierbaum, G., Sahl, H.G. (1998). Role of lipid bound peptidoglycan precursors in the formation of pores by nisin, epidermin and other lantibiotics. *30*, 317–327.

Brouwer, C.P., Wulferink, M., and Welling, M.M. (2008). The pharmacology of radiolabeled cationic antimicrobial peptides. J. Pharm. Sci. *97*, 1633–1651.

Brown, K.L., and Hancock, R.E.W. (2006). Cationic host defense (antimicrobial) peptides. Curr. Opin. Immunol. *18*, 24–30.

Campopiano, D.J., Clarke, D.J., Polfer, N.C., Barran, P.E., Langley, R.J., Govan, J.R., Maxwell, A., and Dorin, J.R. (2004). Structure–activity relationships in defensin dimers: a novel link between beta-defensin tertiary structure and antimicrobial activity. J. Biol. Chem. *279*, 48671–48679.

Campos, M.A., Vargas, M.A., Regueiro, V., Llompart, C.M., Alberti, S., and Bengoechea, J.A. (2004). Capsule polysaccharide mediates bacterial resistance to antimicrobial peptides. Infect. Immun. *72*, 7107–7114.

Cavard, D., and Lazdunski, C. (1990). Colicin cleavage by OmpT protease during both entry into and release from *Escherichia coli* cells. J. Bacteriol. *172*, 648–652.

Chongsiriwatana, N.P., Patch, J.A., Czyzewski, A.M., Dohm, M.T., Ivankin, A., Gidalevitz, D., Zuckermann, R.N., and Barron, A.E. (2008). Peptoids that mimic the structure, function, and mechanism of helical antimicrobial peptides. Proc. Natl. Acad. Sci. U.S.A. *105*, 2794–2799.

Ciornei, C.D., Sigurdardóttir, T., Schmidtchen, A., and Bodelsson, M. (2005). Antimicrobial and chemoattractant activity, lipopolysaccharide neutralization, cytotoxicity, and inhibition by serum of analogs of human cathelicidin LL-37. Antimicrob Agents Chemother. *49*, 2845–50.

Cirioni, O., Giacometti, A., Ghiselli, R., Bergnach, C., Orlando, F., Silvestri, C., Mocchegiani, F., Licci, A., Skerlavaj, B., Rocchi, M., *et al.* (2006). LL-37 protects rats against lethal sepsis caused by Gram-negative bacteria. Antimicrob. Agents Chemother. *50*, 1672–9.

Coffelt, S., Tomchuck, S., Zwezdaryk, K., Danka, E., and Scandurro, A. (2009). Leucine leucine-37 uses formyl peptide receptor-like 1 to activate signal transduction pathways, stimulate oncogenic gene expression, and enhance the invasiveness of ovarian cancer cells. Mol. Cancer Res. *7*, 907–915.

Cudic, M., Condie, B. A., Weiner, D. J., Lysenko, E. S., Xiang, Z. Q., Insug, O., Bulet, P., and Otvos, L., Jr. (2002). Development of novel antibacterial peptides that kill resistant isolates. Peptides *23*, 2071–2083

Davidson, D.J., Currie, A.J., Reid, G.S.D., Bowdish, D.M.E., MacDonald, K.L., Ma, R.C., Hancock, R.E.W., and Speert, D.P. (2004). The cationic antimicrobial peptide LL-37 modulates dendritic cell differentiation and dendritic cell-induced T cell polarization. J. Immunol. *172*, 1146–1156.

Devine, D.A., and Hancock, R.E.W. (2002). Cationic peptides: distribution and mechanisms of resistance. Curr. Pharm. Des. *8*, 703–714.

Dewan, P. C., Anantharaman, A., Chauhan, V. S., and Sahal, D. (2009). Antimicrobial action of prototypic amphipathic cationic decapeptides and their branched dimers. Biochemistry *48*, 5642–5657

Durr, M., and Peschel, A. (2002). Chemokines meet defensins: the merging concepts of chemoattractants and antimicrobial peptides in host defense. Infect. Immun. *70*, 6515–6517.

Easton, D.M., Nijnik, A., Mayer, M.L., and Hancock, R.E.W. (2009). Potential of immunomodulatory host defense peptides as novel anti-infectives. Trends Biotechnol. *27*, 582–590.

Eisenstein, B.I. (2004). Lipopeptides, focusing on daptomycin, for the treatment of Gram-positive infections. Expert Opin. Investig. Drugs *13*, 1159–1169.

Ernst, R.K., Guina, T., and Miller, S.I. (2001). *Salmonella typhimurium* outer membrane remodelling: role in resistance to host innate immunity. Microbes Infect. *3*, 1327–1334.

Fahrner, R.L., Dieckmann, T., Harwig, S.S., Lehrer, R.I., Eisenberg, D., and Feigon, J. (1996) Solution structure of protegrin-*1*, a broad-spectrum antimicrobial peptide from porcine leukocytes. Chem. Biol. 3, 543–50.

Falagas, M.E., and Kasiakou, S.K. (2005). Colistin: the revival of polymyxins for the management of multidrug-resistant Gram-negative bacterial infections. Clin. Infect. Dis. *40*, 1333–1341.

Falla, T.J., Karunaratne, D.N., and Hancock, R.E.W. (1996). Mode of action of the antimicrobial peptide indolicidin. J. Biol. Chem. *271*, 19298–19303.

Fehlbaum, P., Bulet, P., Chernysh, S., Briand, J.P., Roussel, J.P., Letellier, L., Hetru, C., and A, H.J. (1996). Structure–activity analysis of thanatin, a 21-residue inducible insect defense peptide with sequence homology to frog skin antimicrobial peptides. Proc. Natl. Acad. Sci. *93*, 1221–1225.

Fjell, C.D., Jenssen, H., Hilpert, K., Cheung, W.A., Panté, N., Hancock, R.E.W., and Cherkasov, A. (2009). Identification of novel antibacterial peptides by chemoinformatics and machine learning. J. Med. Chem. *52*, 2006–2015.

Fowler, S.A., and Blackwell, H.E. (2009). Structure-function relationships in peptoids: recent advances toward deciphering the structural requirements for biological function. Org. Biomol. Chem. *7*, 1508–1524.

Gooderham, W.J., and Hancock, R.E.W. (2009). Regulation of virulence and antibiotic resistance by two-component regulatory systems in *Pseudomonas aeruginosa*. FEMS Microbiol. Rev. *33*, 279–294.

Goodson, B., Ehrhardt, A., Ng, S., Nuss, J., Johnson, K., Giedlin, M., Yamamoto, R., Moos, W.H., Krebber, A., Ladner, M., *et al.* (1999). Characterization of novel antimicrobial peptoids. Antimicrob. Agents Chemother. *43*, 1429–1434.

Guina, T., Yi, E.C., Wang, H., Hackett, M., and Miller, S.I. (2000a). A PhoP-regulated outer membrane protease of *Salmonella enterica* serovar *typhimurium* promotes resistance to alpha-helical antimicrobial peptides. J. Bacteriol. *182*, 4077–4086.

Guina, T., Yi, E.C., Wang, H.L., Hackett, M., and Miller, S.I. (2000b). A PhoP-regulated outer membrane protease of *Salmonella enterica* serovar typhimurium promotes resistance to alpha-helical antimicrobial peptides. J. Bacteriol. *182*, 4077–4086.

Giuliani, A., Pirri, G., Bozzi, A., Di Giulio, A., Aschi, M., and Rinaldi, A. C. (2008). Antimicrobial peptides: natural templates for synthetic membrane-active compounds. Cell Mol. Life Sci. *65*, 2450–2460.

Hale, J.D., and Hancock, R.E.W. (2007). Alternative mechanisms of action of cationic antimicrobial peptides on bacteria. Expert Rev. Anti Infect. Ther. *5*, 951–959.

Hamill, P., Brown, K., Jenssen, H., and Hancock, R.E.W. (2008). Novel anti-infectives: is host defence the answer. Curr.Opin. Biotechnol. *19*, 628–636.

Hancock, R.E., and Chapple, D.S. (1999). Peptide antibiotics. Antimicrob. Agents Chemother. *43*, 1317–1323.

Hancock, R.E.W., and Patrzykat, A. (2002). Clinical development of cationic antimicrobial peptides: from natural to novel antibiotics. Curr. Drug Targets Infect. Disord. *2*, 79–83.

Hancock, R.E.W., and Sahl, H.-G. (2006). Antimicrobial and host-defense peptides as new anti-infective therapeutic strategies. Nat. Biotech. *24*, 1551–1557.

Heilborn, J.D., Nilsson, M.F., Kratz, G., Weber, G., Sorensen, O., Borregaard, N., and Stahle-Backdahl, M. (2003). The cathelicidin anti-microbial peptide LL-37 is involved in re-epithelialization of human skin wounds and is lacking in chronic ulcer epithelium. J. Investig. Dermatol. *120*, 379–389.

Hilpert, K., Volkmer-Engert, R., Walter, T., and Hancock, R.E. (2005). High-throughput generation of small antibacterial peptides with improved activity. Nat Biotechnol 23, 1008–1012.

Hilpert, K., Elliott, M.R., Volkmer-Engert, R., Henklein, P., Donini, O., Zhou, Q., Winkler, D.F., and Hancock, R.E. (2006). Sequence requirements and an optimization strategy for short antimicrobial peptides. Chem. Biol. *13*, 1101–1107.

Hill, C.P., Yee, J., Selsted, M.E., and Eisenberg, D. (1991) Crystal structure of defensin HNP-3, an amphiphilic dimer: mechanisms of membrane permeabilization. Science *22*, 1481–1485

Hong, R.W., Shchepetov, M., Weiser, J.N., and Axelsen, P.H. (2003). Transcriptional profile of the *Escherichia coli* response to the antimicrobial insect peptide cecropin A. Antimicrob. Agents Chemother. *47*, 1–6.

Hornef, M.W., Pütsep, K., Karlsson, J., Refai E., and Andersson M. (2004). Increased diversity of intestinal antimicrobial peptides by covalent dimer formation. Nat. Immunol. *5*, 836–843.

Hritonenko, V., and Stathopoulos, C. (2007). Omptin proteins: an expanding family of outer membrane proteases in Gram-negative Enterobacteriaceae. Mol. Membr. Biol. *24*, 395–406.

Humet, M., Carbonell, T., Masip, I., Sanchez-Baeza, F., Mora, P., Canton, E., Gobernado, M., Abad, C., Perez-Paya, E., and Messeguer, A. (2003). A positional scanning combinatorial library of peptoids as a source of biological active molecules: identification of antimicrobials. J. Comb. Chem. *5*, 597–605.

Huang LC, Reins RY, Gallo RL, and McDermott AM. (2007). Cathelicidin-deficient (Cnlp -/-) mice show increased susceptibility to *Pseudomonas aeruginosa* keratitis. Invest. Ophthalmol. Vis. Sci. *48*, 4498–508.

Iimura, M., Gallo, R.L., Hase, K., Miyamoto, Y., Eckmann, L., and Kagnoff, M.F. (2005). Cathelicidin Mediates Innate Intestinal Defense against Colonization with Epithelial Adherent Bacterial Pathogens. J. Immunol. *174*, 4901–4907.

Jenssen, H., Hamill, P., and Hancock, R.E.W. (2006). Peptide antimicrobial agents. Clin. Microbiol. Rev. *19*, 491–511.

Jerala, R. (2007). Synthetic lipopeptides: a novel class of anti-infectives. Expert Opin. Investig. Drugs *16*, 1159–1169.

Jones, A., Georg, M., Maudsdotter, L., and Jonsson, A.B. (2009). Endotoxin, capsule, and bacterial attachment contribute to *Neisseria meningitidis* resistance to the human antimicrobial peptide LL-37. J. Bacteriol. *191*, 3861–3868.

Jung, D., Rozek, A., Okon, M., and Hancock, R. E. (2004). Structural transitions as determinants of the action of the calcium-dependent antibiotic daptomycin. Chem. Biol. *11*, 949–957.

Koczulla, R., von Degenfeld, G., Kupatt, C., Krotz, F., Zahler, S., Gloe, T., Issbrucker, K., Unterberger, P., Zaiou, M., Lebherz, C., *et al.* (2003). An angiogenic role for the human peptide antibiotic LL-37/hCAP-18. J. Clin. Invest. *111*, 1665–1672.

Kraus, D., and Peschel, A. (2006). Molecular mechanisms of bacterial resistance to antimicrobial peptides. Curr. Top. Microbiol. Immunol .*306*, 231–250.

Kristian, S.A., Datta, V., Weidenmaier, C., Kansal, R., Fedtke, I., Peschel, A., Gallo, R.L., and Nizet, V. (2005). D-alanylation of teichoic acids promotes Group A Streptococcus antimicrobial peptide resistance, neutrophil survival, and epithelial cell invasion. J. Bacteriol. *187*, 6719–6725.

Kruszewska, D., Sahl, H.G., Bierbaum, G., Pag, U., Hynes, S.O., and Ljungh, A. (2004). Mersacidin eradicates methicillin-resistant *Staphylococcus aureus* (MRSA) in a mouse rhinitis model. J. Antimicrob. Chemother. *54*, 648–653.

Kuwano, K., Tanaka, N., Shimizu, T., and Kida, Y. (2006). Antimicrobial activity of inducible human β defensin-2 against mycoplasma pneumoniae. Curr. Microbiol. *52*, 435–438.

Lawyer, C., Pai, S., Watabe, M., Borgia, P., Mashimo, T., Eagleton, L., and Watabe, K. (1996). Antimicrobial activity of a 13 amino acid tryptophan-rich peptide derived from a putative porcine precursor protein of a novel family of antibacterial peptides. FEBS Lett. *390*, 95–98.

Lau, Y.E., Rozek, A., Scott, M.G., Goosney, D.L., Davidson, D.J., and Hancock, R.E.W. (2005). Interaction and cellular localization of the human host defence peptide, LL-37, with lung epithelial cells. Infect. Immun. *73*, 583–591

Lee, J.Y., Yang, S.T., Lee, S.K., Jung, H.H., Shin, S.Y., Hahm, K.S., and Kim, J.I. (2008). Salt-resistant homodimeric bactenecin, a cathelicidin-derived antimicrobial peptide. FEBS J. *275*, 3911–3920.

Li, J., Nation, R.L., Milne, R.W., Turnidge, J.D., and Coulthard, K. (2005). Evaluation of colistin as an agent against multiresistant Gram-negative bacteria. Int. J. Antimicrob. Agents 25, 11–25.

Li, M., Cha, D.J., Lai, Y., Villaruz, A.E., Sturdevant, D.E., and Otto, M. (2007a). The antimicrobial peptide-sensing system aps of *Staphylococcus aureus*. Mol. Microbiol. *66*, 1136–1147.

Li, M., Lai, Y., Villaruz, A.E., Cha, D.J., Sturdevant, D.E., and Otto, M. (2007b). Gram-positive three-component antimicrobial peptide-sensing system. Proc. Natl. Acad. Sci. U.S.A. *104*, 9469–9474.

Lynn, D.J., Winsor, G.L., Chan, C., Richard, N., Laird, M.R., Barsky, A., Gardy, J.L., Roche, F.M., Chan, T.H., Shah, N., *et al.* (2008). InnateDB: facilitating systems-level analyses of the mammalian innate immune response. Mol. Syst. Biol. *4*, 218–222.

Llobet, E., Tomas, J.M., and Bengoechea, J.A. (2008). Capsule polysaccharide is a bacterial decoy for antimicrobial peptides. Microbiology *154*, 3877–3886.

McPhee, J.B., Lewenza, S., and Hancock, R.E.W.(2003). Cationic antimicrobial peptides activate a two-component regulatory system, PmrA-PmrB, that regulates resistance to polymyxin B and cationic antimicrobial peptides in *Pseudomonas aeruginosa*. Mol. Microbiol. *50*, 205–217.

Mader, J.S., Mookherjee, N., Hancock, R.E.W., and Bleackley, R.C. (2009). The human host defense peptide LL-37 induces apoptosis in a calpain- and apoptosis-inducing factor dependent manner involving Bax activity. Mol. Cancer Res. 7, 689–702.

Marchand, C., Kajewski, K., Lee, H.F., Antony, S., Johnson, Amin, R., Roller, P., Kvaratskhelia, M., and Pommier, Y. (2006). Covalent binding of the natural antimicrobial peptide to DNA abasic sites. Nucleic Acids Res. *34*, 5157–65.

Malina, A., and Shai, Y. (2005). Conjugation of fatty acids with different lengths modulates the antibacterial and antifungal activity of a cationic biologically inactive peptide. Biochem. J. *390*, 695–702.

Makovitzki, A., Avrahami, D., and Shai, Y. (2006). Ultrashort antibacterial and antifungal lipopeptides. Proc. Natl. Acad. Sci. U.S.A. *103*, 15997–16002.

Miller, S.I., Ernst, R.K., and Bader, M.W. (2005). LPS, TLR4 and infectious disease diversity. Nat. Rev. Microbiol. 3, 36–46.

Mookherjee, N., and Hancock, R.E.W. (2007). Cationic host defence peptides: Innate immune regulatory peptides as a novel approach for treating infections. Cell. Mol. Life Sci. *64*, 922–933.

Mookherjee, N., Lippert, D.N.D., Hamill, P., Falsafi, R., Nijnik, A., Kindrachuk, J., Pistolic, J., Gardy, J., Miri, P., Naseer, M., *et al.* (2009). Intracellular receptor for human host defense peptide LL-37 in monocytes. J. Immunol. *183*, 2688–2696.

Mookherjee, N., Wilson, H.L., Doria, S., Popowych, Y., Falsafi, R., Yu, J., Li, Y., Veatch, S., Roche, F.M., Brown, K.L., *et al.* (2006). Bovine and human cathelicidin cationic host defense peptides similarly suppress transcriptional responses to bacterial lipopolysaccharide. J. Leukoc. Biol. *80*, 1563–1574.

Moser, C., Weiner, D.J., Lysenko, E., Bals, R., Weiser, J.N., and Wilson, J.M. (2002). {beta}-defensin 1 contributes to pulmonary innate immunity in mice. Infect. Immun. *70*, 3068–3072.

Neuhaus, F.C., and Baddiley, J. (2003). A continuum of anionic charge: structures and functions of D-alanyl-teichoic acids in Gram-positive bacteria. Microbiol. Mol. Biol. Rev. *67*, 686–723.

Ng, S., Goodson, B., Ehrhardt, A., Moos, W.H., Siani, M., and Winter, J. (1999). Combinatorial discovery process yields antimicrobial peptoids. Bioorg. Med. Chem. 7, 1781–1785.

Nijnik, A., and Hancock, R.E.W. (2009). The roles of cathelicidin LL-37 in immune defences and novel clinical applications. Curr. Opini. Hematol. 16,41–47 *16*, 41–47.

Nijnik A., Madera, L., Ma, S., Waldbrook, M., Elliott, M., Mullaly, S.C., Kindrachuk, J., Jenssen, H., and Hancock, R.E. (2010). Synthetic cationic peptide IDR-1002 provides protection against bacterial infections through chemokine induction and enhanced leukocyte recruitment. J. Immunol. *184*, 2539–2550.

Oh, H.S., Ko, S.S., Cho, H., and Lee, K.H. (2005). Design and synthesis of antibacterial pseudopeptides with a potent antibacterial activity and more improved stability from a short cationic antibacterial peptide. Bull. Korean Chem. Soc. *26*, 161–164.

Oren, F. (2005). Regulation of mammalian defensin expression by Toll-like receptor-dependent and independent signalling pathways. Cell Microbiol. *7*, 1387–1397.

Otvos, L., Jr., O, I., Rogers, M.E., Consolvo, P.J., Condie, B.A., Lovas, S., Bulet, P., and Blaszczyk-Thurin, M. (2000). Interaction between heat shock proteins and antimicrobial peptides. Biochemistry 39, 14150–14159.

Overhage, J., Campisano, A., Bains, M., Torfs, E.C., Rehm, B.H., and Hancock, R.E.W. (2008). Human host defense peptide LL-37 prevents bacterial biofilm formation. Infect. Immun. *76*, 4176–4182.

Park, C.B., Kim, H.S., and Kim, S.C. (1998). Mechanism of action of the antimicrobial peptide buforin II: buforin II kills microorganisms by penetrating the cell membrane and inhibiting cellular functions. Biochem. Biophys. Res. Commun. *244*, 253–257.

Patch, J.A., and Barron, A.E. (2003). Helical peptoid mimics of magainin-2 amide. J. Am. Chem. Soc. *125*, 12092–12093.

Patrzykat, A., Friedrich, C.L., Zhang, L., Mendoza, V., and Hancock, R.E.W. (2002). Sublethal concentrations of pleurocidin-derived antimicrobial peptides inhibit macromolecular synthesis in *Escherichia coli*. Antimicrob. Agents Chemother. *46*, 605–614.

Park, C.B., Kim, H.S., and Kim, S.C. (1998) Mechanism of action of the antimicrobial peptide buforin II: buforin II kills microorganisms by penetrating the cell membrane and inhibiting cellular functions. Biochem. Biophys. Res. Commun. *244*, 253–257.

Perron, G., Zasloff, M., and Bell, G. (2006). Experimental evolution of resistance to an antimicrobial peptide. Pro. Biol. Sci. *273*, 251–256.

Peschel, A. (2002). How do bacteria resist human antimicrobial peptides. Trends Microbiol. *10*, 179–186.

Peschel, A., and Sahl, H.G. (2006). The co-evolution of host cationic antimicrobial peptides and microbial resistance. Nat. Rev. Microbiol. *4*, 529–536.

Peschel A., Otto M., Jack, R.W., Kalbach, H., Jung, G., and Gotz, F. (1999). Inactivation of the *dlt* operon in *Staphylococcus aureus* confers sensitivity to defensins, protegrins, and other antimicrobial peptides. J. Biol. Chem. *274*, 8405–8410.

Peschel, A., Otto, M., Jack, R.W., Kalbacher, H., Jung, G., and Gotz, F. (1999). Inactivation of the dlt operon in *Staphylococcus aureus* confers sensitivity to defensins, protegrins, and other antimicrobial peptides. J. Biol. Chem. *274*, 8405–8410.

Peschel, A., Jack, R.W., Otto, M., Collins, L.V., Staubitz, P., Nicholson, G., Kalbacher, H., Nieuwenhuizen, W.F., Jung, G., Tarkowski, A., *et al.* (2001). *Staphylococcus aureus* resistance to human defensins and evasion of neutrophil killing via the novel virulence factor MprF is based on modification of membrane lipids with l-lysine. J. Exp. Med. *193*, 1067–1076.

Pietiainen, M., Francois, P., Hyyrylainen, H.L., Tangomo, M., Sass, V., Sahl, H.G., Schrenzel, J., and Kontinen, V.P. (2009). Transcriptome analysis of the responses of *Staphylococcus aureus* to antimicrobial peptides and characterization of the roles of *vraDE* and *vraSR* in antimicrobial resistance. BMC Genomics *10*, 429.

Poyart, C., Pellegrini, E., Marceau, M., Baptista, M., Jaubert, F., Lamy, M.C., and Trieu-Cuot, P. (2003). Attenuated virulence of *Streptococcus agalactiae* deficient in D-alanyl-lipoteichoic acid is due to an increased susceptibility to defensins and phagocytic cells. Mol. Microbiol. *49*, 1615–1625.

Powers, J.P., Martin, M.M., Goosney D.L., and Hancock, R.E.W. (2006). The antimicrobial peptide polyphemusin localizes to the cytoplasm of *Escherichia coli* following treatment. Antimicrob. Agents Chemother. 50:1522–1524.

Pütsep, K., Carlsson, G., Boman, H.G., and Andersson, M. (2002). Deficiency of antibacterial peptides in patients with morbus Kostmann: an observation study. Lancet *360*, 1144–1149.

Romeo, D., Skerlavaj, B., Bolognesi, M., and Gennaro, R. (1988). Structure and bactericidal activity of an antibiotic dodecapeptide purified from bovine neutrophils. J. Biol. Chem. *263*, 9573–9575.

Rozek, A., Friedrich, C.L., Hancock, R.E.W. (2000) Structure of the bovine antimicrobial peptide indolicidin bound to dodecylphosphocholine and sodium dodecyl sulfate micelles. Biochemistry *39*, 15765–74.

Rozek, A., Powers, J.P., Friedrich, C.L., and Hancock, R.E.W. (2003). Structure-based design of an indolicidin peptide analogue with increased protease stability. Biochemistry *42*, 14130–14138.

Salomon, R.A., and Farias, R.N. (1992). Microcin 25, a novel antimicrobial peptide produced by *Escherichia coli*. J. Bacteriol. *174*, 7428–7435.

Samant, S., Hsu, F.F., Neyfakh, A.A., and Lee, H. (2009). The *Bacillus anthracis* protein MprF is required for synthesis of lysylphosphatidylglycerols and for resistance to cationic antimicrobial peptides. J. Bacteriol. *191*, 1311–1319.

Sandgren, S., Wittrup, A., Cheng, F., Jonsson, M., Eklund, E., Busch, S., and Belting, M. (2004). The human antimicrobial peptide LL-37 transfers extracellular DNA plasmid to the nuclear compartment of mammalian cells via lipid rafts and proteoglycan-dependent endocytosis. J. Biol. Chem. *279*, 17951–17956.

Sass, V., Pag, U., Tossi, A., Bierbaum, G., and Sahl, H.G. (2008). Mode of action of human beta-defensin 3 against *Staphylococcus aureus* and transcriptional analysis of responses to defensin challenge. Int. J. Med. Microbiol. *298*, 619–633.

Schmidtchen, A., Frick, I.M., Andersson, E., Tapper, H., and Bjorck, L. (2002). Proteinases of common pathogenic bacteria degrade and inactivate the antibacterial peptide LL-37. Mol. Microbiol. *46*, 157–168.

Scott, M.G., Rosenberger, C.M., Gold, M.R., Finlay, B.B., and Hancock, R.E.W. (2000). An α-helical cationic antimicrobial peptide selectively modulates macrophage responses to lipopolysaccharide and directly alters macrophage gene expression. J. Immunol. *165*, 3358–3365.

Scott, M.G., Davidson, D.J., Gold, M.R., Bowdish, D., and Hancock R.E.W. (2002) The human antimicrobial peptide LL-37 is a multifunctional modulator of innate immune responses. J. Immunol. *169*, 3883–91.

Scott, M.G., Dullaghan, E., Mookherjee, N., Glavas, N., Waldbrook, M., Thompson, A., Wang, A., Lee, K., Doria, S., Hamill, P., *et al.* (2007). An anti-infective peptide that selectively modulates the innate immune response. Nat. Biotech. *25*, 465–472.

Sebbage, V. (2009). Cell-penetrating peptides and their therapeutic applications. Biosci. Horizons *1*, 64–72.

Shai, Y., Makovitzky, A., and Avrahami, D. (2006). Host defense peptides and lipopeptides: modes of action and potential candidates for the treatment of bacterial and fungal infections. Curr. Protein Pept. Sci. *7*, 479–486.

Shankaramma, S.C., Athanassiou, Z., Zerbe, O., Moehle, K., Mouton, C., Bernardini, F., Vrijbloed, J.W., Obrecht, D., and Robinson, J.A. (2002). Macrocyclic hairpin mimetics of the cationic antimicrobial peptide protegrin I: a new family of broad-spectrum antibiotics. Chembiochem 3, 1126–1133.

Shankaramma, S.C., Moehle, K., James, S., Vrijbloed, J.W., Obrecht, D., and Robinson, J.A. (2003). A family of macrocyclic antibiotics with a mixed peptide-peptoid beta-hairpin backbone conformation. Chem. Commun. (Camb), 1842–1843.

Shaykhiev R., Beißwenger C., Kändler K., Senske J., Püchner A., Damm T., Behr J., and Bals R. (2005). Human endogenous antibiotic LL-37 stimulates airway epithelial cell proliferation and wound closure. Am. J. Physiol. Lung Cell Mol. Physiol. *289*, L842–L848.

Sieprawska-Lupa, M.P., Krawczyk, K., Wójcik, K., Puklo, M., Lupa, B., Suder, P., Silberring, J., Reed, M., Pohl, J., Shafer, W., *et al.* (2004). Degradation of human antimicrobial peptide LL-37 by *Staphylococcus aureus*-derived proteinases. J. Antimicrob. Chemother. *8*, 4673–4679.

Sieprawska-Lupa, M., Mydel, P., Krawczyk, K., Wojcik, K., Puklo, M., Lupa, B., Suder, P., Silberring, J., Reed, M., Pohl, J., *et al.* (2004). Degradation of human antimicrobial peptide LL-37 by *Staphylococcus aureus*-derived proteinases. Antimicrob. Agents Chemother. *48*, 4673–4679.

Simon, R.J., Kania, R.S., Zuckermann, R.N., Huebner, V.D., Jewell, D.A., Banville, S., Ng, S., Wang, L., Rosenberg, S., Marlowe, C.K., *et al.* (1992). Peptoids: a modular approach to drug discovery. Proc. Natl. Acad. Sci. U.S.A. *89*, 9367–9371.

Sodeinde, O.A., and Goguen, J.D. (1989). Nucleotide sequence of the plasminogen activator gene of *Yersinia pestis*: relationship to *ompT* of *Escherichia coli* and gene E of *Salmonella typhimurium*. Infect. Immun. 57, 1517–1523.

Spinosa, M.R., Progida, C., Tala, A., Cogli, L., Alifano, P., and Bucci, C. (2007). The *Neisseria meningitidis* capsule is important for intracellular survival in human cells. Infect. Immun. 75, 3594–3603.

Staubitz, P., Peschel, A., Nieuwenhuizen, W.F., Otto, M., Gotz, F., Jung, G., and Jack, R.W. (2001). Structure–function relationships in the tryptophan-rich, antimicrobial peptide indolicidin. J. Pept. Sci. *7*, 552–564.

Storici, P., Tossi, A., Lenarcic, B., and Romeo, D. (1996). Purification and structural characterization of bovine cathelicidins, precursors of antimicrobial peptides. Eur. J. Biochem. *238*, 769–776.

Stumpe, S., Schmid, R., Stephens, D.L., Georgiou, G., and Bakker, E.P. (1998). Identification of OmpT as the protease that hydrolyzes the antimicrobial peptide protamine before it enters growing cells of *Escherichia coli*. J. Bacteriol. *180*, 4002–4006.

Subbalakshmi, C., and Sitaram, N. (1998). Mechanism of antimicrobial action of indolicidin. FEMS Microbiol. Lett. *160*, 91–96.

Sugimura, K., and Nishihara, T. (1988). Purification, characterization, and primary structure of *Escherichia coli* protease VII with specificity for paired basic residues: identity of protease VII and OmpT. J. Bacteriol. *170*, 5625–5632.

Tally, F.P., and DeBruin, M.F. (2000). Development of daptomycin for Gram-positive infections. J. Antimicrob. Chemother. *46*, 523–526.

Tedesco, K.L., and Rybak, M.J. (2004). Daptomycin. Pharmacotherapy *24*, 41–57.

Tencza, S.B., Creighton, D.J., Yuan, T., Vogel, H.J., Montelaro, R.C., and Mietzner. T.A. (1999). Lentivirus-derived antimicrobial peptides: increased potency by sequence engineering and dimerization. J. Antimicrob. Chemother. *44*, 33–41.

Thennarasu, S., Lee, D. K., Tan, A., Prasad Kari, U., and Ramamoorthy, A. (2005). Antimicrobial activity and membrane selective interactions of a synthetic lipopeptide MSI-843. Biochim. Biophys. Acta *1711*, 49–58

Thedieck, K., Hain, T., Mohamed, W., Tindall, B.J., Nimtz, M., Chakraborty, T., Wehland, J., and Jansch, L. (2006). The MprF protein is required for lysinylation of phospholipids in listerial membranes and confers resistance to cationic antimicrobial peptides (CAMPs) on *Listeria monocytogenes*. Mol. Microbiol. *62*, 1325–1339.

Tjabringaa, G.S., Ninabera, D.K., Drijfhoutb, J.W., Rabea, K.F., and Hiemstraa, P.S. (2006). Human cathelicidin ll-37 is a chemoattractant for eosinophils and neutrophils that acts via formyl-peptide Receptors. Int. Arch. Allergy Immunol. *140*, 103–112.

Tomasinsig, L., Pizzirani, C., Skerlavaj, B., Pellegatti, P., Gulinelli, S., Tossi, A., Di Virgilio, F., and Zanetti, M. (2008). The human cathelicidin LL-37 modulates the activities of the P2X7 receptor in a structure-dependent manner. J. Biol. Chem. *283*, 30471–30481.

Tossi, A., Sandri, L., and Giangaspero, A. (2000). Amphipathic, alpha-helical antimicrobial peptides. Peptide Sci. *55*, 4–30.

Tzeng, Y.L., Ambrose, K.D., Zughaier, S., Zhou, X., Miller, Y.K., Shafer, W.M., and Stephens, D.S. (2005). Cationic antimicrobial peptide resistance in *Neisseria meningitidis*. J. Bacteriol. *187*, 5387–5396.

Wang G. (2008). Structures of human host defense cathelicidin LL-37 and its smallest antimicrobial peptide KR-12 in lipid micelles. J. Biol. Chem. *21*, 32637–32643

Wang, G., Li, X., and Wang, Z. (2009b). APD2, the updated antimicrobial peptide database and its application in peptide design. Nucl. Acids Res. *37*, D933–937.

Weidenmaier, C., Kristian, S.A., and Peschel, A. (2003). Bacterial resistance to antimicrobial host defenses–an emerging target for novel antiinfective strategies. Curr. Drug Targets *4*, 643–649.

Whitmore, L., Chugh, J.K., Snook, C.F., and Wallace, B.A. (2003). The peptaibol database: a sequence and structure resource. J. Pept. Sci. *9*, 663–665.

WHO. (2009). World Health Statistics. Available at: http://www.who.int/whosis/whostat/2009/en/index.html.

Wiest, A., Grzegorski, D., Xu, B.W., Goulard, C., Rebuffat, S., Ebbole, D.J., Bodo, B., and Kenerley, C. (2002). Identification of peptaibols from *Trichoderma virens* and cloning of a peptaibol synthetase. J. Biol. Chem. *277*, 20862–20868.

Wu, M., and Hancock, R.E.W. (1999a). Improved derivatives of bactenecin, a cyclic dodecameric antimicrobial cationic peptide. Antimicrob. Agents Chemother. *43*, 1274–1276.

Wu, M., and Hancock, R.E.W. (1999b). Interaction of the cyclic antimicrobial cationic peptide bactenecin with the outer and cytoplasmic membrane. J. Biol. Chem. *274*, 29–35.

Wu, M., Maier, E., Benz, R., and Hancock, R.E.W. (1999c). Mechanism of interaction of different classes of cationic antimicrobial peptides with planar bilayers and with the cytoplasmic membrane of *Escherichia coli*. Biochemistry *38*, 7235–7242.

Wu, Z., Hoover, D.M., Yang, D., Boulègue, C., Santamaria, F., Oppenheim, J.J., Lubkowski, J., and Lu, W. (2003). Engineering disulfide bridges to dissect antimicrobial and chemotactic activities of human β-defensin 3. Proc. Natl. Acad. Sci. U.S.A. *100*, 8880–8885.

Yang, D., Chertov, O., Bykovskaia, S.N., Chen, Q., Buffo, M.J., Shogan, J., Anderson, M., der Schrouml, J.M., *et al.* (1999). -Defensins: linking innate and adaptive immunity through dendritic and T cell CCR6. Science *286*, 525–528.

Yang, S.T., Kim, J.I., and Shin, S.Y. (2009). Effect of dimerization of a beta-turn antimicrobial peptide, PST13-RK, on antimicrobial activity and mammalian cell toxicity. Biotechnol. Lett. *31*, 233–237.

Zairi, A., Tangy, F., Bouassida, K., and Hani, K. (2009). Dermaseptins and Magainins: Antimicrobial Peptides from Frogs' Skin–New Sources for a Promising Spermicides Microbicides–A Mini Review. J. Biomed. Biotechnol. *2009*, 1–8.

Zhang, L., Rozek, A., and Hancock, R.E.W. (2001). Interaction of cationic antimicrobial peptides with model membranes. J. Biol. Chem. *276*, 35714–35722.

Zasloff, M. (1987). Magainins, a class of antimicrobial peptides from *Xenopus* skin: isolation, characterization of two active forms, and partial cDNA sequence of a precursor. Proc. Natl. Acad. Sci. *84*, 5449–5443.

Antibodies for Antibacterials

16

Bret R. Sellman and C. Ken Stover*

Abstract

Prior to the use of antibiotics, antibody (or serum) therapy was used with some success to treat bacterial infections. Antibiotics almost completely replaced the use of antibody therapies for bacterial disease with few exceptions. Based upon the information available at the time, this was an obvious progression given the broader spectrum activity of antibiotics. Antibiotics revolutionized medicine and the approach to treating infectious disease. In addition to their broad spectrum, they exhibited few side-effects relative to the potential for serum sickness (following the administration of equine immune serum) and they were inexpensive. But bacterial resistance to antibiotics became evident in the decades to follow, and we are now faced with a shortage of effective antibiotics and a need for alternative approaches to stand-alone antibiotic therapy. One such approach which could supplement antibiotic use, thereby removing some of the selective pressure from antibiotics, is monoclonal antibody therapy or prophylaxis. Recent advances in monoclonal antibody technology and discovery strategies and the ability to make a fully human antibody have led to the marketing of ~30 recombinant antibodies and Fc fusion proteins to treat a variety of human diseases. Although this technology has yet to yield an antibacterial product, many clinical and preclinical programmes are under way to explore varied and novel approaches to monoclonal antibody-based anti-infectives.

Introduction

With the development of horse serum antitoxin for diphtheria and tetanus in the late 1800s, the use of antibodies and passive immunization in the prevention and treatment of bacterial disease actually predate the antibiotic era. While antitoxin immunoglobulin against diphtheria, tetanus and botulinum toxin are still in limited use for the treatment of acute infections with these pathogens, the development of antibiotics in the 1940s rapidly supplanted the use of passive immunization for other diseases such as pneumococcal pneumonia because chemotherapy was found to be safer, more cost effective and efficacious in comparison to horse serum therapy. In addition, and perhaps more importantly, was the broader-spectrum coverage afforded by antibiotics which enabled physicians to practice empiric treatment of suspected bacterial infections. However, the ongoing battle against antibiotic resistant bacteria has suffered greatly in the last two decades with the relative lack of productivity in the discovery of new antibiotics with novel mechanisms of action and the rise of resistance to existing broader-spectrum antibiotics. For this reason, there is an increasing trend towards thinking that more specific and targeted therapies are ultimately needed to slow the progression of multidrug-resistant bacteria. Given the serious nature of multidrug-resistant bacterial infections, particularly in the hospital environment, coupled with the rapid

*Corresponding author

development of antibody-based discovery and production technologies, the consideration of returning to antibody-based therapy or prophylaxis to supplement antibiotic standard of care has become more attractive and economically feasible, at least in some clinical contexts.

More specific antibody-based antibacterial strategies offer a considerable number of potential advantages over small molecule-based, less selective antibiotics. Firstly, specific antibodies directed at pathogen targets are less likely to be toxic than small molecule inhibitors of enzymes which may have off-target effects on related enzymes. Nature has evolved the specific antibody mechanism to be very narrow in spectrum. Because of this specificity and accompanying safety, narrow spectrum antibodies are much less likely to impact commensal organisms and the delicate balance of beneficial normal flora populations within the human host. While small molecule antibiotics may have an advantage in their general ability to penetrate some tissues or intracellular locations in comparison to antibodies, this potential advantage can also be disadvantageous, particularly in the gastrointestinal tract. Indeed, the use of broader-spectrum antibiotics has led to the unintended emergence of serious infections with enteric bacterial pathogens such as *Clostridium difficile* and drug-resistant enterococci, which are normally held in check by beneficial non-pathogenic flora (Top *et al.*, 2008; Rupnik *et al.*, 2009). Furthermore, small-molecule antibiotics targeting a broader-spectrum bacterial mechanism are not always optimally active on all the susceptible bacteria in that broad spectrum, or present in sufficient concentrations long enough to sufficiently kill the bacteria, whether they be beneficial flora or the targeted pathogen. This subinhibitory concentration scenario is optimal for stepwise development of resistance to antibiotics, which can potentially be laterally transferred between natural, beneficial flora and undesirable pathogens. With regard to pharmacokinetics, antibody-based biological agents are also likely to have longer half-lives, thereby reducing timing issues and offering the potential to simplify dosing regimens and improve compliance. This is particularly advantageous for prophylactic and pre-emptive applications. Biological products are also less likely to have drug–drug interactions or influence small molecule metabolic clearance mechanisms which can affect the safety and efficacy of other drugs. This is particularly important in the chronic or critical care environment where patients are often treated with multiple drugs and the potential for drug–drug safety and metabolic clearance interactions are high.

In spite of the potential advantages for antibody-based antibacterial agents, at the time of this writing, there are no monoclonal antibody-based drugs approved for antibacterial therapy or prevention, although there are more than a few in various stages of clinical and preclinical development (Table 16.1). The progression in the technologies for the discovery and development of monoclonal antibody-based drugs has led to the approval of many drugs for diseases in other non-infectious disease therapeutic areas (Reichert, 2010). To date, however, the only monoclonal antibody-based therapy targeting an infectious agent is Synagis® (palivizumab), targeting respiratory syncytial virus (Pai *et al.*, 2009). While there have been a substantial number of monoclonal antibody efforts aimed at bacterial targets, these were mostly started in the absence of significant genomic and systems biology information and pursued before the technology revolution that has occurred in antibody discovery in recent years and they have largely been driven by smaller resource-limited efforts. Nevertheless, the single greatest impediment to returning or moving to more specific antibody-based pre-emptive and therapeutic strategies may be the requirement for a considerable cultural and economic evolution in the practice of infectious disease specialists, hospitals and health care payers away from empiric and seemingly less expensive broad-spectrum approaches which may in reality be more expensive in terms of morbidity, mortality and longer term total downstream health care costs. This would also require the progression towards the development of acceptable rapid diagnostics for the selection of a more specific therapy. While new and powerful diagnostic technology platforms and capabilities do exist to make more specific therapy practicable, the current standards of practice for treating infectious disease do not demand higher resolution diagnostics, hence

Table 16.1 Selected antibacterial antibody-based preparations at various stages of clinical and preclinical development

Bacteria/disease	Target	Product	Antibody class	Stage[a]	Originator/sponsor	References
Gram-negative/sepsis	LPS core (undefined)	Donor anti-J5 *E. coli* Hyperimmune globulin	Polyclonal	III	–	Anonymous (1992a,b)
Gram-negative/sepsis	Lipid A	HA-1	IgM	III	Senecor	Ziegler *et al.*, (1991), McCloskey *et al.*, (1994)
Gram- negative/sepsis	Lipid A	E5	IgM	III	Xoma	Greenman *et al.* (1991),; Bone *et al.* (1995)
P. aeruginosa	PcrV T3SS	KB001	PEGylated FAB	Phase I/II	Kalobios Pharmaceuticals	Baer *et al.* (2009)
P. aeruginosa	LPS Outer Core	–	IgM MAb (mix)	–	Sumitomo Pharmaceuticals	Terashima *et al.* (1991), Yokota *et al.* (1992), Harrison *et al.* (1997)
P. aeruginosa	Alginate	F429 (Aerucin™)	IgG MAb	Preclinical	Brigham and Women's Aridis Pharmaceuticals	Zaidi and Pier (2008)
P. aeruginosa	O- antigen	KBPA101	IgM	II	Kenta Biotech	Lazar *et al.* (2009)
Shiga toxin-producing *E. coli* (STEC)	Stx 1 and 2	Shigamabs	IgG	I	Thallion Pharmaceuticals	Bitzan *et al.* (2009)
Shiga toxin producing *E. coli* (STEC)	Stx 2	Urotxazomab	IgG	I	Teijin Pharma	Lopez *et al.* (2010)
A. baumannii	Undisclosed	KBAB401	Undisclosed	Preclinical	Kenta Biotech	www.kentabiotech.com
Multiple species	dPNAG	F598	IgG MAb	Preclinical	Alopexx/Sanofi Aventis	Cerca *et al.* (2007)
C. difficile	TcdA, TcdB	CDA-1, CDB-1	IgG	Phase II	Medarex/Merck	Lowy *et al.* (2010)
B. anthracis	Anthrax toxin (PA)	Raxibacumab (Abthrax™)	IgG	Pending FDA approval	Human Genome Sciences	Mazumdar (2009), Migone *et al.* (2009)
B. anthracis	Anthrax toxin (PA)	Valortim®	IgG	I	Medarex/PharmAthene	Vitale *et al.* (2006)
B. anthracis	Anthrax toxin (PA)	Anthim™	IgG	I	Elusys Therapeutics	www.elusys.com
S. aureus	Enriched ClfA SdrG	INH-A21	Polyclonal	III – failed endpoint	Inhibitex	DeJonge *et al.* (2007)
S. aureus	ClfA	Tefibazumab (Aurexis®)	IgG	II	Inhibitex	Weems *et al.* (2006)
S. aureus	ABC transporter GrfA	Aurograb®	scFv	III – discontinued	Neutec/Novartis	Burnie *et al.* (2000), www.neutecpharma.com

Table 16.1 continued

Bacteria/disease	Target	Product	Antibody class	Stage[a]	Originator/sponsor	References
S. aureus	LTA	Pagibaximab	IgG	Phase II/III	Biosynexus	Weisman *et al.* (2009)
S. aureus	Protein A	ETI-211	–	Preclinical	Elusys Therapeutics	www.elusys.com
S. aureus	IsdB	–	IgG	Preclinical	Merck	Brown *et al.* (2009)
S. aureus	Undisclosed	KBSA301–302	–	Preclinical	Kenta Biotech	www.kentabiotech.com
S. pneumoniae	Undisclosed	–	Undisclosed	Preclinical	Intercell/Kyowa Hakko Kirin	www.intercell.com
Group A streptococci	Undisclosed	–	Undisclosed	Preclinical	Intercell	www.intercell.com
Group B streptococci	Undisclosed	–	undisclosed	Preclinical	Intercell	www.intercell.com

[a]Last reported.

the need for the sea change in infectious disease medical practice.

Antitoxins: a proven approach

Bacterial toxins are well validated targets for use of antibodies as therapeutics against bacterial disease. Successful treatment of infections with a biologic targeting a bacterial toxin (horse and sheep antitoxins) was pioneered over 100 years ago (Pope, 1963). One drawback to such therapy is the incidence of serum sickness resulting from a hypersensitivity reaction in the patient to antibodies from another species. Even so, similar antitoxins are used with success in the clinic today to treat bacterial infections or to alleviate the symptoms of the toxins alone (i.e. diphtheria, tetanus and botulinum toxins; Shapiro *et al.*, 1998; Arnon *et al.*, 2006; Underwood *et al.*, 2007; Wagner *et al.*, 2009). One such example is the human-derived botulism antitoxin, human botulism immune globulin intravenous (BIG-IV, BabyBig®), developed for the treatment of infant botulism. In a clinical trial, BIG-IV was shown to significantly reduce the mean length of hospital stay, duration of intensive care and hospital charges per patient in infants with suspected botulism (Arnon *et al.*, 2006; Long, 2007). Such toxins are ideal targets for a biologic approach to therapy because a single human antibody or antibody fragments (i.e. Fab, $F(ab')_2$ or scFv) can neutralize a toxin with little concern for serum sickness or toxicity (Maynard *et al.*, 2002; Mohamed *et al.*, 2005a; Jones *et al.*, 2006; Chen *et al.*, 2009b; Migone *et al.*, 2009; Ragle and Bubeck Wardenburg, 2009). Antitoxins are amenable to the use of antibody fragments because toxin neutralization does not rely on Fc effector function for direct interaction with the host immune system. This is in contrast to an antibody in which the desired effect is to fix complement and target the bacterial cell for opsonophagocytic killing. Antitoxin therapy does not target bacteria directly but neutralizes a toxin. In doing so, it can alleviate the symptoms of disease resulting from intoxication or compromise the ability of bacteria to cause disease thereby providing the host with an advantage which could ultimately enhance clearance of the infection. Use of an antitoxin in therapy or prophylaxis will, however, only be successful if the toxin itself causes the disease symptoms or if it plays a critical role in the pathogenesis of the infection. Similar to other specific biological approaches to treat bacterial infections, the utility of an antitoxin as a therapeutic is limited by the accuracy and timing of diagnosis. Infections caused by organisms such as *S. aureus* are more difficult to diagnose than infections for which antitoxins are already in use, such as infant botulism. In prophylaxis, accurate diagnosis is not in question; however, identification of an appropriate at-risk patient population is often difficult. Nonetheless, past successes with antitoxin therapies have paved the way for several clinical and preclinical programs currently under way exploring toxin neutralizing monoclonal antibodies (mAbs) for prophylaxis and treatment of bacterial infections.

Programmes that have entered phase I or II clinical trials include mAbs against Shiga-like toxin (STX) produced by Shiga toxin-producing *Escherichia coli* (STEC), the protective antigen (PA) subunit of anthrax toxin produced by *Bacillus anthracis* and the TcdA and TcdB toxins produced by *Clostridium difficile*. STEC is an important foodborne pathogen and the etiologic agent of haemorrhagic colitis and haemolytic uraemic syndrome (HUS; Serna and Boedeker, 2008; Tarr *et al.*, 2005). Approximately 15% of STEC infections progress to HUS, but the percentage can be as high as 50% if treated with antibiotics. STEC typically express one of two STX molecules, STX-1 or STX-2, which are reported to play a major role in the disease and ultimately development of HUS and kidney failure (Tarr *et al.*, 2005). At present, there is no treatment for HUS, and antibiotic therapy is contraindicated at any stage of an STEC infection due to the association with an increased risk of developing HUS. As an alternative, STX neutralizing mAbs specific for STX-1 (13C4) and STX-2 (11E10) have been identified using hybridoma technology in mice and converted into human chimeras by genetically fusing the mouse variable region sequences with a human kappa light chain constant domain sequence and a human IgG1 heavy chain sequence (Strockbine *et al.*, 1985; Perera *et al.*, 1988; Sauter *et al.*, 2008; Bitzan *et al.*, 2009). 13C4 and 11E10 both neutralize their respective toxins *in vitro*, and 11E10 was shown to protect mice from STX-induced HUS (Sauter *et al.*, 2008). In two independent phase I clinical trials, a single antibody or a cocktail of the two mAbs was deemed safe and tolerable in humans with a half-life of between 8 and 10 hours. If successful, these antibodies would be the only available treatment, aside from supportive therapy, for STEC infections and ultimately for the prevention and treatment of HUS.

Following the anthrax attacks of 2001 there was renewed interest (and funding) in alternative approaches to therapy and prophylaxis against anthrax infections. Since anthrax toxin is viewed as the major virulence determinant of anthrax, a concerted effort was undertaken in the scientific community to identify novel inhibitors of toxin function as potential therapeutics for treatment and prophylaxis. Anthrax toxin is an A-B type bacterial toxin in which the B moiety (i.e. protective antigen, PA) is responsible for the delivery of two different enzymatic A moieties (lethal factor, LF and oedema factor, EF) into the cytosol of the cell. LF and EF in combination with PA form lethal toxin (LF + PA, LTx) and oedema toxin (EF + PA, ETx), respectively. Lethal toxin is thought to be responsible for the symptoms of shock that occur prior to death from anthrax. These efforts have in some cases pushed the envelope and expanded the possible biologic approaches to neutralize toxins from more traditional inhibitory mAbs to a polyvalent peptide inhibitor and dominant negative mutants, all of which could be used as potent inhibitors of toxin function (Mourez *et al.*, 2001; Sellman *et al.*, 2001; Mohamed *et al.*, 2005a; Vitale *et al.*, 2006; Chen *et al.*, 2009a,b). The knowledge gained from these studies will probably be applied not only to other antitoxins in the future but possibly to other biological approaches to combat infectious disease. Of all the approaches studied thus far, two investigational mAbs, raxibacumab and Valortim®, have progressed to clinical trials (Vitale *et al.*, 2006; Migone *et al.*, 2009). Raxibacumab and Valortrim® are human IgGs directed against PA that inhibit either binding of PA to its cellular receptor or cell surface assembly, respectively (Subramanian *et al.*, 2005; Vitale *et al.*, 2006; Migone *et al.*, 2009). As a possible testament to its potential utility, raxibacumab was shown to inhibit LTx *in vivo* in a rat model of intoxication, even after the animals exhibited symptoms of shock (Cui *et al.*, 2005). This ability to neutralize the effects of a toxin after the bacteria have been killed by antibiotics is a key feature of an antitoxin. In trials of symptomatic inhalational anthrax in monkeys or rabbits, a single administration of Raxibacumab or Valortrim alone improved survival when administered either prophylactically or therapeutically. No information is available whether either provides a benefit (additive or synergistic) in combination with an antibiotic. Raxibacumab's half-life of 20–22 days in humans could possibly provide benefit over antibiotic alone for the recommended 60-day postexposure prophylaxis, particularly since compliance is always in question (Inglesby *et al.*, 2002; Subramanian *et al.*, 2005; Migone *et al.*, 2009). This potential pharmacokinetic benefit could be further enhanced by including a mutation in the mAb's Fc portion that results in increased half-life (Dall'Acqua *et al.*, 2006b). Owing to the nature of anthrax infections, it is not possible to test the actual efficacy of raxibacumab and Valortim® in a human population; however, their performance thus far in rabbits and non-human primates has been promising, leading the US government to purchase supplies of raxibacumab (Mazumdar, 2009).

The antibiotic era is largely responsible for the emergence of pseudomembranous colitis caused by *C. difficile* (Rupnik *et al.*, 2009). This type of infection typically occurs in hospitalized individuals following broad spectrum antibiotic therapy and is believed to result from an antibiotic-mediated imbalance in the normal intestinal microflora. Two toxins produced by *C. difficile*, TcdA and TcdB, are implicated as the major virulence determinants responsible for *C. difficile* disease, and high serum titres of anti-TcdB and anti-TcdA antibodies are reported to correlate with a reduced incidence of infection and recurrence (Kyne *et al.*, 2001; Voth and Ballard, 2005; Leav *et al.*, 2009). Fully human mAbs were generated against both TcdA (CDA1) and TcdB (MDX1388) in HuMAb mice transgenic for human immunoglobulin genes (Babcock *et al.*, 2006). Both mAbs neutralize the effect of the target toxins *in vitro*, however, only CDA1 inhibited toxin-mediated death in mice. Similarly, in a hamster model of infection, CDA1 but not MDX1388 significantly reduced both death and relapse following challenge with *C. difficile*. In hamsters, a combination of MDX1388 and CDA1 provided added benefit over CDA1 alone, suggesting the effect of MDX1388 alone is not detectable in these models but its inhibition of TcdB in combination with inhibition of TcdA by CDA1 might improve the clinical outcome. CDA1 was tested in a dose escalation phase I clinical trial and deemed safe and tolerable but provided no benefit over the standard of care in a phase II trial (Taylor *et al.*, 2008; Leav *et al.*, 2009). The investigators did, however, collect data suggesting that disease outcome and recurrence is dependent on the levels of both anti-TcdA and anti-TcdB serum antibody titres thus providing a rationale to test a combination of anti-TcdA and anti-TcdB mAbs further (Leav *et al.*, 2009). In a follow-up phase II placebo-controlled trial, CDA1 was coadministered with an anti-TcdB IgG (CDB1) along with either metronidazole or vancomycin to patients who had diarrhoea associated with a positive stool test for a *C. difficile* toxin (Lowy *et al.*, 2010). The administration of these mAbs in combination with an antibiotic achieved the primary endpoint by reducing the recurrence of *C. difficile* infection compared with antibiotics provided in combination with a placebo. These results are exciting and warrant further clinical trials in a larger patient population for confirmation. Furthermore, these results demonstrate a potential for the augmentation of antibiotic therapy by combination therapy with an antibody-based product.

Another toxin for which monoclonal antibodies are being explored as possible therapies to prevent or treat bacterial infections is alpha toxin (AT) produced by *Staphylococcus aureus*. Unlike the aforementioned bacteria, *S. aureus* causes a wide array of diseases ranging from mild skin and soft tissue infections (i.e. boils) to more serious and invasive infections such as endocarditis, pneumonia and osteomyelitis. As a consequence, *S. aureus* employs a multitude of virulence determinants in pathogenesis (Foster, 2005). Despite this wide array of virulence determinants, AT has been demonstrated to play a key role in multiple animal models of infection including a murine model of staphylococcal pneumonia (Patel *et al.*, 1987; Bramley *et al.*, 1989; Callegan *et al.*, 1994; Bubeck Wardenburg *et al.*, 2007; Bubeck Wardenburg and Schneewind, 2008). AT is a pore-forming cytolytic toxin hypothesized to potentiate immune evasion and dysregulation, thereby providing a means for the bacteria to exploit the host immune system for benefit (Craven *et al.*, 2009). AT binds to a surface receptor and assembles into a membrane-inserted heptameric pore leading to lysis of the affected cell (Bhakdi and Tranum-Jensen, 1991; Song *et al.*, 1996). Murine mAbs were identified that neutralize toxin function by preventing the toxin from assembling into the heptameric ring (Ragle and Bubeck Wardenburg, 2009). Prophylactic administration of the individual mAbs resulted in significant reductions in bacterial numbers and associated mortality following intranasal bacterial challenge. These data suggest that targeting a toxin like AT can tilt the balance of infection thus enabling the host to clear the bacteria more readily and limiting disease.

All of the antitoxin mAbs above look promising; however, success of these mAbs or other biologics in the clinic is not yet clear. Antitoxins have had successes in the past and testing of these mAbs in a human population will pave the way for future studies testing antibacterial mAbs, antitoxin or otherwise, in treatment of and prophylaxis against serious bacterial infections.

Other methods targeting virulence

Targeting toxins as a means to treat or prevent bacterial disease is a proven approach for some infections. However, many bacterial species do not secrete a toxin as the primary virulence determinant, but, have evolved alternative mechanisms to assist in the pathogenesis of disease (e.g. type III secretion or quorum sensing). Targeting a virulence mechanism with a therapeutic agent is another antibacterial approach where an antibody could have utility. Similar to an antitoxin, an antivirulence approach to therapy does not target the bacteria directly, and the mAb would not require direct interaction with the host immune system to be functional. However, the immune system would ultimately be responsible for the clearance of the pathogen. An antivirulence mAb would function by providing the infecting bacteria with a deficit resulting in an increased sensitivity to antibiotic therapy, enhanced clearance by the host or both. One such target is the type III secretion system (T3SS) of Gram-negative bacteria. The T3SS utilizes a 'needle-like' apparatus on the surface of the bacteria to contact host cells and deliver bacterial effector proteins into the cytosol of the cell where they elicit cellular damage and in some cases provide a mechanism for the bacteria to evade the host innate immune system (Galan and Wolf-Watz, 2006; Hauser, 2009). Inhibition of the T3SS with a mAb or mAb fragment might, therefore, prevent the bacteria from causing disease, improve clinical outcomes or enhance antibiotic therapy. An anti-T3SS approach in phase II clinical testing is a PEGylated Humaneered^{TM} Fab' directed against PcrV on the *P. aeruginosa* T3SS (Bebbington and Yarranton, 2008; Baer *et al.*, 2009; Hauser, 2009). *P. aeruginosa* is a leading cause of hospital acquired pneumonia and a common pathogen associated with lung infections in cystic fibrosis patients (Park, 2005; Hassett *et al.*, 2009). Expression of the T3SS in *P. aeruginosa* is reported to be important in numerous animal infection models and to be associated with poor clinical outcomes and death in humans (Roy-Burman *et al.*, 2001; Hauser *et al.*, 2002, 2009). PcrV is located at the needle tip and is required for translocation by the T3SS (Sawa *et al.*, 1999; Hauser, 2009). Antibodies directed against PcrV (either rabbit polyclonal IgG or mAb 166) inhibit T3SS-mediated cytotoxicity *in vitro* and prophylactic administration of either mAb 166 or rabbit IgG reduced both lethality and bacterial numbers in a murine acute pneumonia model (Sawa *et al.*, 1999; Shime *et al.*, 2001; Frank *et al.*, 2002; Baer *et al.*, 2009). Similar results were seen following administration of mAb 166 or rabbit IgG Fab fragments indicating that Fc effector function is not necessary to inhibit T3SS and facilitate bacterial clearance (Shime *et al.*, 2001; Baer *et al.*, 2009). These data demonstrate that inhibition of a T3SS has an impact on disease probably through an indirect influence on the immune system, which allows for the host to more readily clear the invading pathogen. An approach to inhibit the T3SS holds promise for use in a human infection not only to prevent the bacteria from damaging the host but to assist the host immune system in bacterial clearance. One potential drawback of using a T3SS as a target is the kinetics of expression. A T3SS probably plays a role in the establishment of an infection and is a valid target in prophylaxis or pre-emptive therapy. The role of T3SS in the maintenance of chronic infection is unclear bringing into question its utility in therapy; however, it may provide value in combination with an antibiotic to prevent spread of the infection.

Quorum sensing (QS) is a mechanism employed by Gram-positive and -negative bacteria to sense local population density and to coordinate behaviour in response to changes in cell density (Bassler and Losick, 2006). QS is mediated by small molecules called autoinducers that are produced and recognized by the bacteria. Once the autoinducer reaches a threshold concentration the bacteria detect its presence which affects a coordinated population-wide alteration in gene expression influencing the expression of hundreds of genes. These changes in gene expression can impact virulence mechanisms including virulence gene expression, motility and biofilm formation (de Kievit and Iglewski, 2000). As mentioned above, inhibition of a virulence mechanism important for disease pathogenesis (i.e. QS) will probably benefit the host. Quenching of QS with mAbs that recognize and scavenge autoinducers as a means to attenuate virulence is being explored for the treatment of infections caused by both *P. aeruginosa* and *S. aureus*. Unfortunately the autoinducers produced by these two pathogens are structurally dissimilar and cannot be targeted by a single mAb (Bassler and Losick, 2006; Kaufmann *et al.*, 2008a). *P. aeruginosa* produces two distinct acyl homoserine lactone (AHL) autoinducers, N-(3-oxododecanoyl) homoserine lactone (3-oxo-C_{12}-HSL) and N-butyrylhomoserine lactone (C_4-HSL). At sufficiently high concentrations these molecules bind to their cognate receptors and lead to alterations in gene expression which promote virulence (de Kievit and Iglewski, 2000; Fuqua and Greenberg, 2002). In *P. aeruginosa*, these QS systems are reported to regulate the expression of up to 600 genes some of which play a role in disease pathogenesis (Schuster *et al.*, 2003; Wagner *et al.*, 2003). Mutants in which either of these QS pathways have been genetically inactivated exhibit reduced virulence in animal models of infection (Rumbaugh *et al.*, 2000).

In addition to its role in QS, 3-oxo-C_{12}-HSL has been reported to be toxic for both neutrophils and macrophages (Tateda *et al.*, 2003). A mAb that scavenges free AHLs may attenuate the virulence of Pseudomonas infections and possibly protect phagocytic cells from the toxic effect of AHL. An anti-3-oxoC_{12}-HSL mAb (RS2–1G9) was generated using hybridoma technology in mice vaccinated with a synthetic 3-oxo-C_{12}-HSL analogue conjugated to KLH (Kaufmann *et al.*, 2006). RS2–1G9 bound the AHL *in vitro* by ELISA, inhibited 3-oxo-C_{12}-HSL-mediated QS in cell culture and protected macrophages from its toxic effects (Kaufmann *et al.*, 2008b). These *in vitro* results show promise, however, a role for this mAb *in vivo* has yet to be reported.

S. aureus is another nosocomial pathogen which employs QS to regulate virulence determinants. The accessory gene regulatory (agr) system is the QS system in *S. aureus*. It is a two-component signal transduction system which utilizes cyclic oligopeptides termed autoinducing peptides (AIP) as the signalling molecules. Thus far, four different AIPs with varying degrees of sequence similarity have been identified (Novick and Geisinger, 2008). Each peptide comprises 7–9 amino acids and a thiolactone or a lactone ring. Like the AHL of *P. aeruginosa*, these peptides are secreted from the bacteria, and when they reach a threshold in concentration they engage a two-component signalling system to alter bacterial gene expression. This complex regulatory system impacts a large portion of *S. aureus* virulence determinants (Novick and Geisinger, 2008). Genetic knockouts and inhibitors of the *agr* system have been used to demonstrate that QS is important for virulence in animal models of infection and raise the hypothesis that inhibition of the *agr* system is a valid approach for antistaphylococcal therapy or prophylaxis (Abdelnour *et al.*, 1993; Cheung *et al.*, 1994; Gillaspy *et al.*, 1995; Mayville *et al.*, 1999). One such approach generated a mAb specific for an AIP and tested the ability of the mAb to inhibit *agr in vitro* and alter disease progression *in vivo*. Mice were vaccinated with a synthetic analogue of AIP-4 conjugated to KLH and one mAb was identified, AP4-24H11, by hybridoma technology. AP4-24H11 bound AIP-4 with a high affinity but displayed little cross-reactivity to either of the other three AIPs including AIP-1, which differs only by a single amino acid (Park *et al.*, 2007). Not only did this mAb inhibit QS signalling *in vitro* but it also reduced abscess formation when admixed with *S. aureus* in a murine infection model. These data suggest that a mAb-mediated anti-QS approach for *S. aureus* holds promise and should be further explored. There is however some doubt concerning the validity of *agr* as a target for antistaphylococcal therapies (Yarwood and Schlievert 2003; Otto, 2004; Harraghy *et al.*, 2007). Expression of *agr* up-regulates exotoxin production and down-regulates expression of surface adhesins and factors involved in biofilm formation (Yarwood and Schlievert 2003; Novick and Geisinger, 2008). Consequently, inhibition of *agr* could increase factors necessary for attachment and biofilm formation resulting in a shift from an acute infection to a chronic and persistent infection. This hypothesis is supported by reports which indicate that mutants in *agr* are present in staphylococcal isolates from serious biofilm-associated infections and that constructed mutations in *agr* increase biofilm formation (Vuong *et al.*, 2000, 2004; Schwan *et al.*, 2003). Although *agr* is important for virulence in animal models, and it does play a role in the regulation of a large number of virulence determinants, more work is warranted to fully understand the impact inhibition of *agr* will have on staphylococcal disease. Characterization and development of antibodies like AP4-24H11 will be invaluable tools in the evaluation of *agr* as a target for antibacterial therapy.

Antibodies recognizing bacterial surface

Targeting bacteria directly with a mAb that recognizes an epitope on the bacterial surface is another antibacterial approach being explored. Antibodies that bind the surface of the bacteria may act by preventing functions important for disease pathogenesis (i.e. attachment and colonization), targeting the bacterial cell for killing by the host innate immune system or both. Consequently, the target antigen must be expressed *in vivo* (preferably over a time course) and accessible to antibody binding in the context of the bacterial surface architecture (i.e. capsular polysaccharide) and the presence of host factors attached to the surface of the bacteria (Risley *et al.*, 2007; Sellman *et al.*, 2008; Nanra *et al.*, 2009). Current approaches typically utilize IgG and to a lesser extent IgM and rely on the effector functions of the Fc domain for activity. One of the important effector functions of the Fc domain is the fixation and activation of the complement cascade which can lead to the direct lysis of some Gram-negative bacteria and the deposition of opsonins (i.e. C3b, C4b) which act in concert with the antibody Fc to engage receptors on the surface of phagocytic cells to trigger phagocytosis. Following phagocytosis, bacteria are exposed to antibacterial mechanisms within the phagosome including reactive oxygen species, antibacterial peptides, low pH, etc. Targeting bacteria directly with an antibody therefore requires functional phagocytes and an active complement cascade for the mAb to be effective. As a means to assist the host immune system, IgG with Fc enhanced for increased complement binding or optimized for interaction with immune effector cells are being explored (Mohamed *et al.*, 2005b; Dall'Acqua *et al.*, 2006a; Strohl, 2009).

An opsonic antibody approach for prophylaxis or treatment of bacterial infections has yet to be validated in humans, however, there are some promising results following prophylactic or therapeutic administration of mAbs reactive with the cell surface of *S. aureus* or *P. aeruginosa* in animal models of infection. Liptoteichoic acid (LTA) is a cell surface glycolipid present on the surface of Gram-positive bacteria which is thought to share some of the pathophysiological properties of LPS (Ginsburg, 2002). LTA is proinflammatory and believed to play a role in the development of Gram-positive septic shock and other immune sequelae subsequent to bacterial infection. A humanized mouse chimeric IgG1, pagibaximab (Biosynexus), specific for staphylococcal LTA is presently under evaluation in a phase II clinical trial for the prevention of staphylococcal sepsis in premature infants (Weisman *et al.*, 2009). The primary role of pagibaximab and other cell surface binding mAbs *in vivo* is to promote opsonophagocytic killing, ultimately leading to bacterial clearance. Pagibaximab may also alleviate the symptoms of Gram-positive septic shock by neutralizing staphylococcal LTA (Ginsburg, 2002). While this antibody has been demonstrated to bind and promote killing of *S. epidermidis* and *S. aureus* by neutrophils in an *in vitro* opsonophagocytic assay, animal model data are limited and it is unclear if LTA is sufficiently accessible to antibody *in vivo* during an infection (Weisman, 2007, 2009; Burkhardt *et al.*, 2008). In a phase I/II double-blind placebo controlled trial in very low birth weight (VLBW) infants, pagibaximab exhibited a safety profile that supports continued development with no adverse event attributed to the antibody. Although pagibaxumab is not fully human, there was no evidence of anti-drug antibody responses (ADA) and pagibaximab exhibited a serum half-life of 20.5 days which is typical for a human IgG (Weisman *et al.*, 2009). The long half-life of a mAb is valuable in the prevention of infections during the hospitalization typical for premature infants. Pagibaximab is currently being tested in a phase IIb/III trial to evaluate safety and efficacy in VLBW infants (www.clinicaltrials.gov; Weisman *et al.*, 2009).

A second cell surface antigen on *S. aureus* being pursued as a target for an antibacterial mAb is clumping factor A (ClfA). ClfA is 97 kDa fibrinogen binding protein covalently linked to the cell wall of *S. aureus* (Patti, 2004). Attachment of staphylococci to fibrinogen coating an indwelling device or present at a site of tissue damage is thought to play a role in staphylococcal colonization. A mAb specific for ClfA could have the dual function of targeting the bacteria for opsonophagocytic killing and limiting colonization by inhibiting binding to fibrinogen. 12–9, a murine mAb specific for ClfA exhibits some of these characteristics. 12–9 was shown to block adherence of ClfA and *S. aureus* to fibrinogen and reduce the incidence of death in a murine model of staphylococcal sepsis following prophylactic administration (Hall *et al.*, 2003). The humanized version of 12–9, tefibazumab (Aurexis®, Inhibitex), was reported to target beads coated with ClfA for phagocytic uptake by PMNs, yet there is no report that either tefibazumab or 12–9 are opsonic for bacteria (Domanski *et al.*, 2005). Prophylactic administration of tefibazumab in a rabbit model of infectious endocarditis significantly reduced bacterial numbers recovered from cardiac vegetations, blood and kidneys at both 10 mg/kg and 30 mg/kg doses. Following therapeutic administration in combination with vancomycin in the rabbit endocarditis model, tefibazumab resulted in a decrease in bacterial numbers in the cardiac vegetations and kidneys relative to vancomycin alone (Domanski *et al.*, 2005). Although this was a trend in the right direction, the results failed to meet statistical significance. Tefibazumab progressed to a phase II clinical trial in patients with *S. aureus* bacteraemia in combination with antibiotics and was shown to have a safety profile similar to other mAbs along with a half-life of 18 days (Weems *et al.*, 2006).

A recent phase III clinical trial with a polyclonal IgG preparation (INH-A21) enriched for antibodies against ClfA failed to reach its primary endpoint of 50% reduction in late onset *S. aureus* bacteraemia in VLBW infants (DeJonge *et al.*, 2007). This was after a favourable result in a phase II trial. The reason for the failure in the phase III trial is unclear and could be related to the impaired innate immune response in VLBW infants (Levy, 2007). It is also unclear if INH-A21 is opsonic, as there are no publications describing its activity in an *in vitro* opsonophagocytic assay with live bacteria. Additionally, a recent report describes the heterogeneous expression of ClfA *in vivo* suggesting that ClfA alone may not be an ideal target (Nanra *et al.*, 2009). The exact reason for the failure of INH-A21 is unknown and warrants further understanding as tefibazumab and other opsonic mAbs are developed.

An opsonic mAb specific for a cell-surface carbohydrate, alginate, on *P. aeruginosa* is under development. A hybridoma expressing an alginate specific IgG1, F429, was isolated from a human volunteer vaccinated with alginate. F429 binds to and targets numerous clinical isolates for opsonophagocytic killing *in vitro* independent of serotype (Pier *et al.*, 2004). Alginate is an *in vivo* expressed antigen and F429 binds in the context of the host during an infection. Prophylactic administration of F429 reduced disease severity and bacterial numbers in murine models of pneumonia and keratitis (Pier *et al.*, 2004; Zaidi and Pier, 2008). F429 was also shown to reduce bacterial numbers and disease severity when administered therapeutically via a systemic or topical route (Zaidi and Pier, 2008). Whether therapeutic administration of F429 with the current standard of care antibiotic would improve disease outcome is unknown. F429 is an attractive candidate for both prophylaxis and therapy of *P. aeruginosa* infections.

Most efforts are being directed at the identification of opsonic monoclonal IgG however there are also programs exploring the use of monoclonal IgM as well. One clear advantage to using an opsonic IgM mAb is that IgM molecules are potent activators of the classical complement cascade. In fact, a single IgM molecule bound to a red blood cell can trigger complement mediated lysis of the red blood cell while 1000 IgG molecules are required (Boes, 2000). Such enhanced complement deposition could increase both opsonization of the invading pathogen and infiltration of effector cells, resulting in more effective bacterial clearance. IgM are the first antibodies expressed upon exposure to an invading pathogen and are effective at combating infection (Boes, 2000). Because IgM lack somatic mutations in their V segments they tend to exhibit low binding affinity. IgM molecules are, however, pentameric resulting in an increased avidity which will help compensate for the reduced affinity. Modern techniques of antibody engineering may also help to overcome the low affinity of IgM through either affinity optimization or placing the antigen binding domain of a high affinity IgG in an IgM molecule (Wu *et al.*, 2007). However, IgM mAbs do have some drawbacks relative to IgG. These molecules are large and consequently may have reduced tissue availability. The size also may make production more difficult however technology is currently available to make IgM on a scale necessary to produce clinical trial material and it is probable that the commercial potential for an IgM as a candidate drug will drive technological advances in antibody production and purification (Lazar *et al.*, 2009). Also, IgM molecules have a serum half-life of ~5 days, which is shorter than the ~ 20 day half-life of IgG-1 but longer than most antibiotics (Mix *et al.*, 2006).

At present, there is a fully human IgM, KBPA-101, in clinical trials for the treatment of *P. aeruginosa* infections possibly in combination with antibiotics (Lazar *et al.*, 2009). KBPA-101 is serotype specific and binds the O-polysaccharide on serotype O11 *P. aeruginosa*. IgM may be the ideal mAb format to target a polysaccharide of repeating subunits such as O-polysaccharide. The avidity of the pentavalent IgM coupled with the high density repeating subunits of O-polysaccharide would probably act to increase the relative affinity of the mAb for its target. However, the specificity for a single serotype limits the use of KBPA-101 to infections with an O11 serotype *P. aeruginosa* and necessitates the co-development of a diagnostic tool. In preclinical studies, it was reported to exhibit protection in murine burn wound and pneumonia models although the data are yet to be published (Lazar *et al.*, 2009). KBPA-101 was deemed safe and tolerable with a half-life of 70–95 hours in a phase I clinical trial. Further clinical testing is warranted and will provide valuable information regarding the use of IgM to treat and prevent infectious diseases.

The research programs discussed thus far using mAbs that bind the surface of bacteria have focused on antibodies that rely on individuals with fully functional neutrophils and complement cascade for killing of the targeted bacteria. This approach may not be optimal in patients with low levels of complement or impaired neutrophil function such as VLBW infants or end stage renal disease patients (Lucchi *et al.*, 1989; Porter *et al.*, 1997; Levy, 2007). A radically different antibody-based approach for the treatment of infectious disease utilizes heteropolymer (HP) technology, which takes advantage of the natural mechanism of erythrocyte immune complex-clearing or immune adherence (IA; Mohamed *et al.*, 2005b). During an infection the HP binds both the pathogen and CR1 on red blood cells. In theory, the pathogen is then transported on the red blood cell to the liver where it is released and killed by resident macrophages with the end result being complement independent opsonization and phagocytosis. If this were effective it would be of value in both immunocompetent patients and patients with defects in the complement cascade or neutrophil function. Experiments demonstrating the complement independent attachment of *P. aeruginosa*, *Escherichia coli* or *S. aureus* to red blood cells and subsequent delivery to macrophages support this hypothesis (Kuhn *et al.*, 1998; Lindorfer *et al.*, 2001; Gyimesi *et al.*, 2004). One concern of this approach is the half-life *in vivo*. If the CR-1 specific mAb binds rapidly to RBCs following injection the HP could be cleared from circulation prior to interaction with the invading pathogen necessitating multiple injections. Another potential drawback is the construction of the bispecific HP through a chemical conjugation reaction (Taylor *et al.*, 1997). Chemical cross-linking is a process which can be difficult to control and may result in significant lot-to-lot variation. It is also possible that the HP bound to the red blood cells could activate the complement cascade leading to lysis of the circulating red blood cells. Nonetheless, this general approach is an exciting new possibility as a platform for antibody-based treatment of bloodstream infections.

An intriguing approach to antibacterial therapy that ultimately failed in the clinic is Aurograb®, an scFv[1] specific for the cell surface ABC transporter, GrfA, on *S. aureus* which reportedly enhanced the sensitivity of the bacteria to treatment with vancomycin (www.neutecpharma.com). This cell-surface target shares a high degree of sequence similarity with ABC transporters involved with cell wall biosynthesis. One could envisage that scFv-mediated inhibition of this transporter would impact the integrity of the cell wall itself and the effect would be amplified in the presence of a cell wall active antibiotic such as vancomycin (Burnie *et al.*, 2000). In early phase I and phase

[1] An scFv is a small chain fragment of the immunoglobulin variable region. It is comprises the variable domains of the light and heavy chains separated by a linker. It is considered to be a minimal form of a functional antibody (Hagemeyer *et al.*, 2009).

II trials, Aurograb® was reported to have some success when administered in combination with vancomycin yet development was ultimately ceased following the results from the phase II clinical trial (www.novartis.com). Unfortunately, information available about this molecule and its mechanism of action is quite limited as there are no publications describing its activity *in vitro* or *in vivo* (in animal models or in humans). That being said, the concept of potentiating the activity of an antibiotic with a mAb is attractive and merits future exploration.

Targeting the LPS endotoxin of Gram-negative bacteria

Lipopolysaccharide (LPS) or endotoxin is the most accessible component of the Gram-negative outer membrane and it plays a vital role in bacterial cell-wall physiology and the virulence of Gram-negative infections. These two functions (structural component and endotoxin virulence factor) provide different and complementary methods to the targeting of Gram-negative bacteria infections and LPS endotoxaemia and resulting sepsis using an antibody mediated approach. Well before endotoxin was known to be a strong pattern recognition molecule for the innate immune system, it was recognized that LPS serotype specific antibodies could be protective in animal models of infection (Cryz *et al.*, 1985; Eisenstein, 1975; Fisher, 1977; Plant *et al.*, 1980; Sawada *et al.*, 1987; Svenson and Lindberg, 1981). It was also established that localized Gram-negative infections were accompanied by endotoxaemia and that the systemic symptoms associated with these infections were similar to those observed in animal models following LPS administration, suggesting that it was important to neutralize freely circulating endotoxin in addition to targeting the bacteria directly (BRAUDE *et al.*, 1963). These finding also explained why antibiotic therapy did not always prevent and sometimes hastened progression to Gram-negative sepsis. However, at that time, and even today, an antibody based therapy for every possible Gram-negative LPS O-antigen serotype would have been impractical and untenable. But early studies employing immunization with heat-killed 'rough' mutants of *E. coli* (J5) and Salmonella (*S. minnesota* R595) that do not express serotype specific O-antigen, but only express the common core glycolipid (CGL) of these enteric bacteria, demonstrated promising serotype and even Gram-negative species cross protection (McCabe 1972; Braude *et al.*, 1977, 1981; Bruins *et al.*, 1977; McCabe *et al.*, 1977). These findings, coupled with data using extracted CGL as immunogens and data from patients showing a correlation between antibody titres to the CGL of LPS at the onset of sepsis and a positive outcome, encouraged efforts targeting CGL as an important antigen (Pollack *et al.*, 1983; Zinner and McCabe, 1976).

Extensive work with *E. coli* J5 and *S. minnesota* R595 strains, O-antigen minus, rough mutants expressing lipid A anchored CGL, eventually led to passive immunization clinical trials in the early 1980s. Patients treated with hyperimmune antisera from volunteers immunized with a boiled *E. coli* J5-based vaccine had significant improved survival (22% mortality) compared with those receiving non-immune control serum (39% mortality; Ziegler *et al.*, 1982). Results seen in patients with shock were the most significant as lower mortality rates were observed in patients receiving the J5 immune sera (46%) versus patients receiving the control non-immune sera (76%, $p<0.009$). This clinical study concluded that the J5 antiserum provided about 50% protection from mortality in Gram-negative infected patients, and the protection was apparent even with the best antibiotic therapy and surgical management available at that time. A follow-up prophylactic clinical trial in high-risk surgical patients also demonstrated significant potential for the polyclonal anti-J5 sera mediated protection against serious consequences of Gram-negative infection. Although the J5 antibody study group did not exhibit significant differences in incidence of infection versus control, control patients were more than twice as likely to progress to septic shock and mortality rates in J5 passively immunized patients progressing to shock were half that of control patients (Baumgartner *et al.*, 1985).

While subsequent clinical studies based on polyclonal anti-CGL IgG immune sera generated with a variety of CGL immunogen sources were not as encouraging (Calandra *et al.*, 1988; Anonymous,

1992a,b), the initial promising results and the recognition that large scale production of polyclonal sera from immunized donors was impractical, led to work focused on targeting lipid A with IgM monoclonal antibodies because lipid A is the most toxic and most highly conserved component of CGL and because it was also believed that the protective isotype was IgM. Although neither antibody demonstrated an overall benefit in phase III clinical trials, both seemed to have afforded some benefit in Gram-negative subgroups. The HA-1A antibody appeared to improve survival in patients with Gram-negative bacteraemia and E5 in patients with Gram-negative infections that were not in shock (Greenman *et al.*, 1991; Ziegler *et al.*, 1991). However, subsequent clinical studies did not confirm the subgroup results of both HA-1A and E5 and indicated a higher mortality rate in HA-1A treated patients without Gram-negative bacteraemia (Anonymous, 1994; McCloskey *et al.*, 1994; Bone *et al.*, 1995; Angus *et al.*, 2000)

The lack of a clear overall benefit with the anti-lipid A monoclonal antibodies brought into question the entire concept of targeting therapy on the conserved epitopes of LPS CGL even though previous data suggested that targeting lipid A and not other regions of CGL was a questionable approach. In retrospect, it can be questioned whether the concept of targeting CGL and lipid A with antibodies has actually been adequately tested and the modern tenets of translational medicine satisfied in these failed studies (Baumgartner, 1991; Cross *et al.*, 1999). Firstly, these efforts were undertaken despite the fact that the protective anti-CGL epitope specificity in the positive studies had not been localized to the lipid A portion of CGL and that anti-lipid A specific mAbs had not been convincingly shown to neutralize endotoxin or effectively treat or prevent sepsis in animal models (Baumgartner, 1991; Baumgartner and Glauser, 1993). Even if they had possessed LPS neutralization activity, it is also possible that the amount and frequency of antibody administration tested in the clinical trials may have been grossly inadequate for a fulminant condition such as sepsis (Cross *et al.*, 1999). But, additional studies after the E5 and HA-1A clinical trials have also revealed that these mAbs were relatively non-specific and weak in their affinity for lipid A and have confirmed that even in large molar excess, neither antibody neutralized LPS in mitogenic assays measuring release of inflammatory cytokines (Warren *et al.*, 1993; Marra *et al.*, 1994; Rietschel *et al.*, 1996). The inability of the HA-1A and E5 antibodies to neutralize LPS *in vitro* and *in vivo* is probably due to the fact that the epitopes these mAbs recognize are not exposed on intact bacteria or even free LPS multimeric aggregates.

While the validity of the previous clinical efforts targeting CGL and in particular lipid A for sepsis have been critically questioned, the dismal failures for the anti-lipid A antibodies made any approach targeting bacterial LPS generally less attractive to the pharmaceutical industry. Nevertheless, it is clear that serotype specific antibodies to O-antigen and non-serotype-specific cross-reactive antibodies to LPS core and the non-lipid A portion of CGL, can in some cases be protective in animal models by targeting either the bacteria for opsonophagocytic killing or by enhancing Fc-dependent detoxification and clearance of the free CGL (Nnalue *et al.*, 1992; Di Padova *et al.*, 1993; Rietschel *et al.*, 1996). While much more is known today about innate immunity, pattern recognition and the interaction of lipid A with TLR-4, it is clear that freely circulating endotoxin 'debris' can be cleared by Fc mediated mechanisms with antibodies targeting CGL even if they do not directly block interaction with lipid A pattern recognition receptors in the innate immune system (Burd *et al.*, 1993a,b). Perhaps the best approach to targeting bacterial infections via LPS would be with antibodies that mediate both opsonophagocytosis and endotoxin neutralization. Whether single antibodies with suitable cross-reactivity, affinity and characteristics can be discovered remains to be seen, but, now that more is known about LPS structure in the major Gram-negative pathogens, a more rational and targeted approach may be taken to discover and develop protective antibodies against CGL. LPS structural information suggests that single antibodies to CGL that cross-protect against all major Gram-negative pathogens are highly unlikely given the structural differences in CGL that might be exposed or accessible to antibodies *in vivo*, particularly between the *Klebsiella*,

Pseudomonas and the enterobacterial species (Muller-Loennies *et al.*, 2007). Therefore, more targeted species-specific anti-CGL approaches may have a greater potential for success. However, work also continues with CGL-based immunogens lacking lipid A and antibodies binding other regions of the LPS core and data still suggests the potential for significant cross-species activity with anti CGL antibodies (Terashima *et al.*, 1991; Harrison *et al.*, 1997; Muller-Loennies *et al.*, 2003; Cross *et al.*, 2004; Muller-Loennies *et al.*, 2007). Because healthy humans normally have measurable titres of anti-endotoxin antibodies which have been correlated with reduced incidence or severity of sepsis, the old approach of using high-dose intravenously administered immunoglobulin (IVIG), particularly IgM enriched IVIG, is also still being studied and used for the treatment of sepsis with encouraging results (Neilson *et al.*, 2005; Rodriguez *et al.*, 2005).

Antibody antimicrobial peptide hybrid approaches

Another interesting approach is the utilization of antimicrobial peptides (AMPs) linked to antibodies as a means of targeting the bacterial cell surface. To date, this approach has focused on Gram-negative bacterial applications by taking advantage of the outer membrane, LPS binding, endotoxin neutralization and antibacterial properties of the AMP, while providing a longer half life and the antibody-dependent clearance effecter function afforded by the Fc domain of the antibody. Fletcher *et al.* (1997) demonstrated a proof of concept for this idea with AMP CAP18 (aka LL-37) chemically conjugated to a non-specific IgG fraction (Fletcher *et al.*, 1997; Ogata *et al.*, 1997; Warren *et al.*, 2003). Similarly, Drabick *et al.* (1998) chemically conjugated polymixin to a non-specific IgG fraction (Drabick *et al.*, 1998). Both efforts demonstrated retention of the AMP's antimicrobial activity with the AMP antibody conjugate and successful prevention but not treatment of sepsis in mouse models. Because these efforts did not combine the AMP with a specific antibody, the studies were aimed at extending the half-life of the relatively unstable and short lived AMPs, enhancing clearance of free endotoxin and reducing the toxicity of the AMPs. While these studies were apparently successful in this regard, a more recent and interesting approach, albeit non-bacterial, comes from a 2004 study with an antifungal peptide (AFP), genetically fused to a chicken derived single chain variable fragment specific for the surface of the *Fusarium* plant pathogen (Peschen *et al.*, 2004). In this study, it was clearly demonstrated that the AFP–scFv fusion was more potent than a simple mixture of the AFP or the anti-*Fusarium* scFv *in vitro*. The study went on to show that when the gene for the AFP–scFv fusion was inserted into the simple plant *Arabidopsis thaliana*, it conferred high levels of protection against *Fusarium* spore challenge in comparison to *A. thaliana* expressing either the AFP or the scFv alone, which exhibited only moderate resistance. Despite this interesting and successful study in a plant system, a similar study with a bacterial pathogen specific antibody and AMP in an animal infection model has yet to be published.

Conclusion

A successful monoclonal antibody-based antibacterial product has yet to be realized for clinical use, although a number of mAbs have been tested. In retrospective analysis, there are logical explanations for the past failures that are not necessarily due to the nature of antibodies. As discussed above, in multiple cases including anti-ClfA and anti-lipid A approaches, the antibodies or targets may not have been fully characterized or adequately validated for use in a clinical setting. In other cases, an inappropriate or difficult patient population may have been chosen for the clinical trial. Some production and quality control issues may have also contributed to some failures. In addition to the knowledge gained from previously failed trials about the importance of target selection, *in vivo* expression profile of target antigens, appropriate dosing levels and the selection of an appropriate patient population, significant advances have also been made in antibody engineering to provide potential for optimized antibody function. Such advances include affinity optimization, enhanced Fc effector function, enhanced Fcγ receptor binding, or the potential use of IgM to improve mAb directed complement fixation (Boes, 2000; Dall'Acqua *et al.*, 2006a; Wu *et al.*, 2007; Strohl 2009). The development of heterofunctional multivalent antibodies or antibody-antimicrobial peptide hybrids may also hold significant promise (Warren *et al.*, 2003; Mohamed *et al.*, 2005b). Additionally, advances in antibody manufacturability to reduce the cost of goods have made mAb use in the clinic more attainable from an economic standpoint (Shire, 2009). Finally, the principles of translational medicine have evolved significantly with knowledge gained from numerous antibody based efforts in other therapeutic areas and much more can be done prior to entering the clinic to help limit the risk of clinical efforts. Lack of success does not necessarily predict the future, but, the scientific and medical community must approach this strategy with an open mind that will require new insights and innovation along with assimilation of the lessons learned from failed attempts in order to capitalize on one of nature's most diverse and useful molecules.

While antibody based approaches to antibacterial strategies may hold considerable promise, it is not likely that antibodies would actually supplant antibiotic use, at least in the foreseeable future. Given their ability to rapidly kill without reliance on host immune capabilities, antibiotics could still be needed to do the heavy lifting in bacterial clearance, particularly in immune compromised patients. Indeed, more specific antibiotics with novel mechanisms of action are also needed. It is more likely, and perhaps appropriate, that antibody-based strategies will be employed in special high risk and critical care settings, used prophylactically and pre-emptively to prevent serious disease progression and adjunctively with antibiotics to neutralize bacterial factors that elicit hyperinflammatory responses leading to bacterial sepsis or reduce the potential for recurrence. Thus, specific antibody-based antibacterials could provide another useful weapon in the armamentarium of infectious disease physicians, thereby taking some of the pressure off antibiotics and potentially preserving their utility for longer periods of time in the face of ever increasing antibiotic resistance.

Perhaps the greatest impediment to the development of antibody-based antibacterial strategies is the comfort and familiarity with the status quo of broader spectrum infectious disease care and not a technology or learning gap. Acceptance of novel therapeutic modalities requires a paradigm and cultural shift in both the prevention and treatment of infectious disease. This could be facilitated by the development of adequate, rapid and specific diagnostics which could be used to indicate a more selective therapy. While technically achievable today, the driver for rapid and specific diagnostics also requires the specific therapies that would be indicated by the diagnostic. While this 'Catch 22' conundrum and required sea change in mind set may at times seem impossible to achieve, the trend towards more specific antibacterial therapy is unlikely to deviate as there has never been more motivation and pressure from the spectre of antibiotic resistance than there is today.

References

Anonymous. (1992a). Prophylactic intravenous administration of standard immune globulin as compared with core-lipopolysaccharide immune globulin in patients at high risk of postsurgical infection. The Intravenous Immunoglobulin Collaborative Study Group. N. Engl. J. Med. *327*, 234–240.

Anonymous. (1992b). Treatment of severe infectious purpura in children with human plasma from donors immunized with *Escherichia coli* J5, a prospective double-blind study. J5 study Group. J. Infect. Dis. *165*, 695–701.

Anonymous. (1994). The French National Registry of HA-1A (Centoxin) in septic shock. A cohort study of 600 patients. The National Committee for the Evaluation of Centoxin. Arch. Intern. Med. *154*, 2484–2491.

Abdelnour, A., Arvidson, S., Bremell, T., Ryden, C., and Tarkowski, A. (1993). The accessory gene regulator (agr) controls *Staphylococcus aureus* virulence in a murine arthritis model. Infect. Immun. *61*, 3879–3885.

Angus, D.C., Birmingham, M.C., Balk, R.A., Scannon, P.J., Collins, D., Kruse, J.A., Graham, D.R., Dedhia, H.V., Homann, S., and MacIntyre, N. (2000). E5 murine monoclonal antiendotoxin antibody in Gram-negative sepsis: a randomized controlled trial. E5 Study Investigators. JAMA *283*, 1723–1730.

Babcock, G.J., Broering, T.J., Hernandez, H.J., Mandell, R.B., Donahue, K., Boatright, N., Stack, A.M., Lowy, I., Graziano, R., Molrine, D., *et al.* (2006). Human monoclonal antibodies directed against toxins A and B prevent Clostridium difficile-induced mortality in hamsters. Infect. Immun. *74*, 6339–6347.

Baer, M., Sawa, T., Flynn, P., Luehrsen, K., Martinez, D., Wiener-Kronish, J.P., Yarranton, G., and Bebbington, C. (2009). An engineered human antibody fab fragment specific for *Pseudomonas aeruginosa* PcrV antigen has potent antibacterial activity. Infect. Immun. *77*, 1083–1090.

Bassler, B.L., and Losick, R. (2006). Bacterially speaking. Cell *125*, 237–246.

Baumgartner, J.D. (1991). Immunotherapy with antibodies to core lipopolysaccharide: a critical appraisal. Infect. Dis. Clin. North Am. *5*, 915–927.

Baumgartner, J.D., and Glauser, M.P. (1993). Immunotherapy of endotoxemia and septicemia. Immunobiology *187*, 464–477.

Baumgartner, J.D., Glauser, M.P., McCutchan, J.A., Ziegler, E.J., van Melle, G., Klauber, M.R., Vogt, M., Muehlen, E., Luethy, R., and Chiolero, R. (1985). Prevention of Gram-negative shock and death in surgical patients by antibody to endotoxin core glycolipid. Lancet *2*, 59–63.

Bebbington, C., and Yarranton, G. (2008). Antibodies for the treatment of bacterial infections: current experience and future prospects. Curr. Opin. Biotechnol. *19*, 613–619.

Bhakdi, S., and Tranum-Jensen, J. (1991). Alpha-toxin of *Staphylococcus aureus*. Microbiol. Rev. *55*, 733–751.

Bitzan, M., Poole, R., Mehran, M., Sicard, E., Brockus, C., Thuning-Roberson, C., and Riviere, M. (2009). Safety and pharmacokinetics of chimeric anti-Shiga toxin 1 and anti-Shiga toxin 2 monoclonal antibodies in healthy volunteers. Antimicrob. Agents Chemother. *53*, 3081–3087.

Boes, M. (2000). Role of natural and immune IgM antibodies in immune responses. Mol. Immunol. *37*, 1141–1149.

Bone, R.C., Balk, R.A., Fein, A.M., Perl, T.M., Wenzel, R.P., Reines, H.D., Quenzer, R.W., Iberti, T.J., Macintyre, N., and Schein, R.M. (1995). A second large controlled clinical study of E5, a monoclonal antibody to endotoxin: results of a prospective, multicenter, randomized, controlled trial. The E5 Sepsis Study Group. Crit. Care Med. *23*, 994–1006.

Bramley, A.J., Patel, A.H., O'Reilly, M., Foster, R., and Foster, T.J. (1989). Roles of alpha-toxin and beta-toxin in virulence of *Staphylococcus aureus* for the mouse mammary gland. Infect. Immun. *57*, 2489–2494.

Braude, A.I., Jones, J.L., and Douglas, H. (1963). The behavior of *Escherichia coli* endotoxin (somatic antigen) during infectious arthritis. J. Immunol. *90*, 297–311.

Braude, A.I., Ziegler, E.J., Douglas, H., and McCutchan, J.A. (1977). Antibody to cell wall glycolipid of Gram-negative bacteria: induction of immunity to bacteremia and endotoxemia. J. Infect. Dis. *136 Suppl*, S167–73.

Braude, A.I., Ziegler, E.J., McCutchan, J.A., and Douglas, H. (1981). Immunization against nosocomial infection. Am. J. Med. *70*, 463–466.

Brown, M., Kowalski, R., Zorman, J., Wang, X.M., Towne, V., Zhao, Q., Secore, S., Finnefrock, A.C., Ebert, T., Pancari, G., *et al.* (2009). Selection and characterization of murine monoclonal antibodies to *Staphylococcus aureus* iron-regulated surface determinant B with functional activity *in vitro* and *in vivo*. Clin. Vaccine Immunol. *16*, 1095–1104.

Bruins, S.C., Stumacher, R., Johns, M.A., and McCabe, W.R. (1977). Immunization with R mutants of Salmonella minnesota. II. Comparison of the protective effect of immunization with lipid A and the Re mutant. Infect. Immun. *17*, 16–20.

Bubeck Wardenburg, J., Bae, T., Otto, M., Deleo, F.R., and Schneewind, O. (2007). Poring over pores: alpha-hemolysin and Panton-Valentine leukocidin in *Staphylococcus aureus* pneumonia. Nat. Med. *13*, 1405–1406.

Bubeck Wardenburg, J., and Schneewind, O. (2008). Vaccine protection against *Staphylococcus aureus* pneumonia. J. Exp. Med. *205*, 287–294.

Burd, R.S., Battafarano, R.J., Cody, C.S., Farber, M.S., Ratz, C.A., and Dunn, D.L. (1993a). Anti-endotoxin monoclonal antibodies inhibit secretion of tumor necrosis factor-alpha by two distinct mechanisms. Ann. Surg. *218*, 250–259; discussion 259–261.

Burd, R.S., Cody, C.S., Raymond, C.S., and Dunn, D.L. (1993b). Anti-endotoxin monoclonal antibodies protect by enhancing bacterial and endotoxin clearance. Arch. Surg. *128*, 145–150; discussion 150–151.

Burkhardt, M., Sei, C., Lees, A., and Mond, J. (2008). A rapid opsonic assay for measuring killing of bioluminescent *Staphylococcus epidermidis*. Hybridoma (Larchmt) *27*, 487–491.

Burnie, J.P., Matthews, R.C., Carter, T., Beaulieu, E., Donohoe, M., Chapman, C., Williamson, P., and Hodgetts, S.J. (2000). Identification of an immunodominant ABC transporter in methicillin-resistant *Staphylococcus aureus* infections. Infect. Immun. *68*, 3200–3209.

Calandra, T., Glauser, M.P., Schellekens, J., and Verhoef, J. (1988). Treatment of Gram-negative septic shock with human IgG antibody to *Escherichia coli* J5, a prospective, double-blind, randomized trial. J. Infect. Dis. *158*, 312–319.

Callegan, M.C., Engel, L.S., Hill, J.M., and O'Callaghan, R.J. (1994). Corneal virulence of *Staphylococcus aureus*: roles of alpha-toxin and protein A in pathogenesis. Infect. Immun. *62*, 2478–2482.

Cerca, N., Maira-Litran, T., Jefferson, K.K., Grout, M., Goldmann, D.A., and Pier, G.B. (2007). Protection against *Escherichia coli* infection by antibody to the *Staphylococcus aureus* poly N-acetylglucosamine surface polysaccharide. Proc. Natl. Acad. Sci. U.S.A. *104*, 7528–7533.

Chen, Z., Moayeri, M., Crown, D., Emerson, S., Gorshkova, I., Schuck, P., Leppla, S.H., and Purcell, R.H. (2009a). Novel chimpanzee/human monoclonal antibodies that neutralize anthrax lethal factor, and evidence for possible synergy with anti-protective antigen antibody. Infect. Immun. *77*, 3902–3908.

Chen, Z., Moayeri, M., Zhao, H., Crown, D., Leppla, S.H., and Purcell, R.H. (2009b). Potent neutralization of anthrax edema toxin by a humanized monoclonal antibody that competes with calmodulin for edema factor binding. Proc. Natl. Acad. Sci. U.S.A. *106*, 13487–13492.

Cheung, A.L., Eberhardt, K.J., Chung, E., Yeaman, M.R., Sullam, P.M., Ramos, M., and Bayer, A.S. (1994). Diminished virulence of a sar-/agr- mutant of *Staphylococcus aureus* in the rabbit model of endocarditis. J. Clin. Invest. *94*, 1815–1822.

Craven, R.R., Gao, X., Allen, I.C., Gris, D., Bubeck Wardenburg, J., McElvania-Tekippe, E., Ting, J.P., and Duncan, J.A. (2009). *Staphylococcus aureus* alpha-hemolysin activates the NLRP3-inflammasome in human and mouse monocytic cells. PLoS One *4*, e7446.

Cross, A.S., Opal, S.M., Bhattacharjee, A.K., Donta, S.T., Peduzzi, P.N., Furer, E., Que, J.U., and Cryz, S.J. (1999). Immunotherapy of sepsis: flawed concept or faulty implementation? Vaccine *17 Suppl 2*, S13–21.

Cross, A.S., Opal, S., Cook, P., Drabick, J., and Bhattacharjee, A. (2004). Development of an anti-core lipopolysaccharide vaccine for the prevention and treatment of sepsis. Vaccine *22*, 812–817.

Cryz, S.J.,Jr, Meadow, P.M., Furer, E., and Germanier, R. (1985). Protection against fatal *Pseudomonas aeruginosa* sepsis by immunization with smooth and rough lipopolysaccharides. Eur. J. Clin. Microbiol. *4*, 180–185.

Cui, X., Li, Y., Moayeri, M., Choi, G.H., Subramanian, G.M., Li, X., Haley, M., Fitz, Y., Feng, J., Banks, S.M., Leppla, S.H., and Eichacker, P.Q. (2005). Late treatment with a protective antigen-directed monoclonal antibody improves hemodynamic function and survival in a lethal toxin-infused rat model of anthrax sepsis. J. Infect. Dis. *191*, 422–434.

Dall'Acqua, W.F., Cook, K.E., Damschroder, M.M., Woods, R.M., and Wu, H. (2006a). Modulation of the effector functions of a human IgG1 through engineering of its hinge region. J. Immunol. *177*, 1129–1138.

Dall'Acqua, W.F., Kiener, P.A., and Wu, H. (2006b). Properties of human IgG1s engineered for enhanced binding to the neonatal Fc receptor (FcRn). J. Biol. Chem. *281*, 23514–23524.

DeJonge, M., Burchfield, D., Bloom, B., Duenas, M., Walker, W., Polak, M., Jung, E., Millard, D., Schelonka, R., Eyal, F., *et al.* (2007). Clinical trial of safety and efficacy of INH-A21 for the prevention of nosocomial staphylococcal bloodstream infection in premature infants. J. Pediatr. *151*, 260–5, 265.e1.

Di Padova, F.E., Brade, H., Barclay, G.R., Poxton, I.R., Liehl, E., Schuetze, E., Kocher, H.P., Ramsay, G., Schreier, M.H., and McClelland, D.B. (1993). A broadly cross-protective monoclonal antibody binding to *Escherichia coli* and *Salmonella* lipopolysaccharides. Infect. Immun. *61*, 3863–3872.

Domanski, P.J., Patel, P.R., Bayer, A.S., Zhang, L., Hall, A.E., Syribeys, P.J., Gorovits, E.L., Bryant, D., Vernachio, J.H., Hutchins, J.T., and Patti, J.M. (2005). Characterization of a humanized monoclonal antibody recognizing clumping factor A expressed by *Staphylococcus aureus*. Infect. Immun. 73, 5229–5232.

Drabick, J.J., Bhattacharjee, A.K., Hoover, D.L., Siber, G.E., Morales, V.E., Young, L.D., Brown, S.L., and Cross, A.S. (1998). Covalent polymyxin B conjugate with human immunoglobulin G as an antiendotoxin reagent. Antimicrob. Agents Chemother. *42*, 583–588.

Eisenstein, T.K. (1975). Evidence for O antigens as the antigenic determinants in 'ribosomal' vaccines prepared from *Salmonella*. Infect. Immun. *12*, 364–377.

Fisher, M.W. (1977). A polyvalent human gamma-globulin immune to *Pseudomonas aeruginosa*: passive protection of mice against lethal infection. J. Infect. Dis. *136 (Suppl)* S181–185.

Fletcher, M.A., Kloczewiak, M.A., Loiselle, P.M., Ogata, M., Vermeulen, M.W., Zanzot, E.M., and Warren, H.S. (1997). A novel peptide–IgG conjugate, CAP18(106–138)-IgG, that binds and neutralizes endotoxin and kills Gram-negative bacteria. J. Infect. Dis. *175*, 621–632.

Foster, T.J. (2005). Immune evasion by staphylococci. Nat. Rev. Microbiol. *3*, 948–958.

Frank, D.W., Vallis, A., Wiener-Kronish, J.P., Roy-Burman, A., Spack, E.G., Mullaney, B.P., Megdoud, M., Marks, J.D., Fritz, R., and Sawa, T. (2002). Generation and characterization of a protective monoclonal antibody to *Pseudomonas aeruginosa* PcrV. J. Infect. Dis. *186*, 64–73.

Fuqua, C., and Greenberg, E.P. (2002). Listening in on bacteria: acyl-homoserine lactone signalling. Nat. Rev. Mol. Cell Biol. 3, 685–695.

Galan, J.E., and Wolf-Watz, H. (2006). Protein delivery into eukaryotic cells by type III secretion machines. Nature *444*, 567–573.

Gillaspy, A.F., Hickmon, S.G., Skinner, R.A., Thomas, J.R., Nelson, C.L., and Smeltzer, M.S. (1995). Role of the accessory gene regulator (agr) in pathogenesis of staphylococcal osteomyelitis. Infect. Immun. *63*, 3373–3380.

Ginsburg, I. (2002). Role of lipoteichoic acid in infection and inflammation. Lancet Infect. Dis. 2, 171–179.

Greenman, R.L., Schein, R.M., Martin, M.A., Wenzel, R.P., MacIntyre, N.R., Emmanuel, G., Chmel, H., Kohler, R.B., McCarthy, M., and Plouffe, J. (1991). A controlled clinical trial of E5 murine monoclonal IgM antibody to endotoxin in the treatment of Gram-negative sepsis. The XOMA Sepsis Study Group. JAMA *266*, 1097–1102.

Gyimesi, E., Bankovich, A.J., Schuman, T.A., Goldberg, J.B., Lindorfer, M.A., and Taylor, R.P. (2004). *Staphylococcus aureus* bound to complement receptor 1 on human erythrocytes by bispecific monoclonal antibodies is phagocytosed by acceptor macrophages. Immunol. Lett. *95*, 185–192.

Hagemeyer, C.E., von Zur Muhlen, C., von Elverfeldt, D., and Peter, K. (2009). Single-chain antibodies as diagnostic tools and therapeutic agents. Thromb. Haemost. *101*, 1012–1019.

Hall, A.E., Domanski, P.J., Patel, P.R., Vernachio, J.H., Syribeys, P.J., Gorovits, E.L., Johnson, M.A., Ross, J.M., Hutchins, J.T., and Patti, J.M. (2003). Characterization of a protective monoclonal antibody recognizing *Staphylococcus aureus* MSCRAMM protein clumping factor A. Infect. Immun. *71*, 6864–6870.

Harraghy, N., Kerdudou, S., and Herrmann, M. (2007). Quorum-sensing systems in staphylococci as therapeutic targets. Anal. Bioanal Chem. *387*, 437–444.

Harrison, F.J., Rohm, D., Kohzuki, T., and Noguchi, H. (1997). Pharmacokinetics, tolerability, and preliminary efficacy of human anti-*Pseudomonas aeruginosa* monoclonal antibodies in pneumonia and burn infection patients. Hybridoma *16*, 413–420.

Hassett, D.J., Sutton, M.D., Schurr, M.J., Herr, A.B., Caldwell, C.C., and Matu, J.O. (2009). *Pseudomonas aeruginosa* hypoxic or anaerobic biofilm infections within cystic fibrosis airways. Trends Microbiol. *17*, 130–138.

Hauser, A.R. (2009). The type III secretion system of *Pseudomonas aeruginosa*: infection by injection. Nat. Rev. Microbiol. 7, 654–665.

Hauser, A.R., Cobb, E., Bodi, M., Mariscal, D., Valles, J., Engel, J.N., and Rello, J. (2002). Type III protein secretion is associated with poor clinical outcomes in patients with ventilator-associated pneumonia caused by *Pseudomonas aeruginosa*. Crit. Care Med. *30*, 521–528.

Inglesby, T.V., O'Toole, T., Henderson, D.A., Bartlett, J.G., Ascher, M.S., Eitzen, E., Friedlander, A.M., Gerberding, J., Hauer, J., Hughes, J., *et al.* (2002). Anthrax as a biological weapon, *2002*, updated recommendations for management. JAMA *287*, 2236–2252.

Jones, R.G., Alsop, T.A., Hull, R., Tierney, R., Rigsby, P., Holley, J., and Sesardic, D. (2006). Botulinum type A toxin neutralisation by specific IgG and its fragments: a comparison of mouse systemic toxicity and local flaccid paralysis assays. Toxicon *48*, 246–254.

Kaufmann, G.F., Park, J., and Janda, K.D. (2008a). Bacterial quorum sensing: a new target for anti-infective immunotherapy. Expert Opin. Biol. Ther. *8*, 719–724.

Kaufmann, G.F., Park, J., Mee, J.M., Ulevitch, R.J., and Janda, K.D. (2008b). The quorum quenching antibody RS2-1G9 protects macrophages from the cytotoxic effects of the *Pseudomonas aeruginosa* quorum sensing signalling molecule N-3-oxo-dodecanoyl-homoserine lactone. Mol. Immunol. 45, 2710–2714.

Kaufmann, G.F., Sartorio, R., Lee, S.H., Mee, J.M., Altobell, L.J.,3rd, Kujawa, D.P., Jeffries, E., Clapham, B., Meijler, M.M., and Janda, K.D. (2006). Antibody interference with N-acyl homoserine lactone-mediated bacterial quorum sensing. J. Am. Chem. Soc. *128*, 2802–2803.

de Kievit, T.R., and Iglewski, B.H. (2000). Bacterial quorum sensing in pathogenic relationships. Infect. Immun. *68*, 4839–4849.

Kuhn, S.E., Nardin, A., Klebba, P.E., and Taylor, R.P. (1998). *Escherichia coli* bound to the primate erythrocyte complement receptor via bispecific monoclonal antibodies are transferred to and phagocytosed by human monocytes in an *in vitro* model. J. Immunol. *160*, 5088–5097.

Kyne, L., Warny, M., Qamar, A., and Kelly, C.P. (2001). Association between antibody response to toxin A and protection against recurrent *Clostridium difficile* diarrhoea. Lancet 357, 189–193.

Lazar, H., Horn, M.P., Zuercher, A.W., Imboden, M.A., Durrer, P., Seiberling, M., Pokorny, R., Hammer, C., and Lang, A.B. (2009). Pharmacokinetics and safety profile of the human anti-*Pseudomonas aeruginosa* serotype O11 immunoglobulin M monoclonal antibody KBPA-101 in healthy volunteers. Antimicrob. Agents Chemother. 53, 3442–3446.

Leav, B.A., Blair, B., Leney, M., Knauber, M., Reilly, C., Lowy, I., Gerding, D.N., Kelly, C.P., Katchar, K., Baxter, R., Ambrosino, D., and Molrine, D. (2009). Serum anti-toxin B antibody correlates with protection from recurrent *Clostridium difficile* infection (CDI). Vaccine

Levy, O. (2007). Innate immunity of the newborn: basic mechanisms and clinical correlates. Nat. Rev. Immunol. 7, 379–390.

Lindorfer, M.A., Nardin, A., Foley, P.L., Solga, M.D., Bankovich, A.J., Martin, E.N., Henderson, A.L., Price, C.W., Gyimesi, E., Wozencraft, C.P., *et al.* (2001). Targeting of *Pseudomonas aeruginosa* in the bloodstream with bispecific monoclonal antibodies. J. Immunol. *167*, 2240–2249.

Long, S.S. (2007). Infant botulism and treatment with BIG-IV (BabyBIG). Pediatr. Infect. Dis. J. *26*, 261–262.

Lopez, E.L., Contrini, M.M., Glatstein, E., Gonzalez Ayala, S., Santoro, R., Allende, D., Ezcurra, G., Teplitz, E., Koyama, T., Matsumoto, Y., *et al.* (2010). Safety and pharmacokinetics of urtoxazumab, a humanized monoclonal antibody, against Shiga-like toxin 2 in healthy adults and in pediatric patients infected with Shiga-like toxin-producing *Escherichia coli*. Antimicrob. Agents Chemother. *54*, 239–243.

Lowy, I., Molrine, D.C., Leav, B.A., Blair, B.M., Baxter, R., Gerding, D.N., Nichol, G., Thomas, W.D.,Jr, Leney, M., Sloan, S., Hay, C.A., and Ambrosino, D.M. (2010). Treatment with monoclonal antibodies against *Clostridium difficile* toxins. N. Engl. J. Med. *362*, 197–205.

Lucchi, L., Cappelli, G., Acerbi, M.A., Spattini, A., and Lusvarghi, E. (1989). Oxidative metabolism of polymorphonuclear leukocytes and serum opsonic activity in chronic renal failure. Nephron *51*, 44–50.

McCabe, W.R. (1972). Immunization with R mutants of *S. minnesota*. I. Protection against challenge with heterologous Gram-negative bacilli. J. Immunol. *108*, 601–610.

McCabe, W.R., Bruins, S.C., Craven, D.E., and Johns, M. (1977). Cross-reactive antigens: their potential for immunization-induced immunity to Gram-negative bacteria. J. Infect. Dis. *136 (Suppl)*, S161–166.

McCloskey, R.V., Straube, R.C., Sanders, C., Smith, S.M., and Smith, C.R. (1994). Treatment of septic shock with human monoclonal antibody HA-1A. A randomized, double-blind, placebo-controlled trial. CHESS Trial Study Group. Ann. Intern. Med. *121*, 1–5.

Marra, M.N., Thornton, M.B., Snable, J.L., Wilde, C.G., and Scott, R.W. (1994). Endotoxin-binding and -neutralizing properties of recombinant bactericidal/permeability-increasing protein and monoclonal antibodies HA-1A and E5. Crit. Care Med. *22*, 559–565.

Maynard, J.A., Maassen, C.B., Leppla, S.H., Brasky, K., Patterson, J.L., Iverson, B.L., and Georgiou, G. (2002). Protection against anthrax toxin by recombinant antibody fragments correlates with antigen affinity. Nat. Biotechnol. *20*, 597–601.

Mayville, P., Ji, G., Beavis, R., Yang, H., Goger, M., Novick, R.P., and Muir, T.W. (1999). Structure–activity analysis of synthetic autoinducing thiolactone peptides from *Staphylococcus aureus* responsible for virulence. Proc. Natl. Acad. Sci. U.S.A. *96*, 1218–1223.

Mazumdar, S. (2009). Raxibacumab. MAbs *1*, 531–538.

Migone, T.S., Subramanian, G.M., Zhong, J., Healey, L.M., Corey, A., Devalaraja, M., Lo, L., Ullrich, S., Zimmerman, J., Chen, A., *et al.* (2009). Raxibacumab for the treatment of inhalational anthrax. N. Engl. J. Med. *361*, 135–144.

Mix, E., Goertsches, R., and Zett, U.K. (2006). Immunoglobulins–basic considerations. J. Neurol. 253 *(Suppl 5)*, V9–17.

Mohamed, N., Clagett, M., Li, J., Jones, S., Pincus, S., D'Alia, G., Nardone, L., Babin, M., Spitalny, G., and Casey, L. (2005a). A high-affinity monoclonal antibody to anthrax protective antigen passively protects rabbits before and after aerosolized Bacillus anthracis spore challenge. Infect. Immun. 73, 795–802.

Mohamed, N., Jones, S.M., Casey, L.S., Pincus, S.E., and Spitalny, G.L. (2005b). Heteropolymers: a novel technology against blood-borne infections. Curr. Opin. Mol. Ther. 7, 144–150.

Mourez, M., Kane, R.S., Mogridge, J., Metallo, S., Deschatelets, P., Sellman, B.R., Whitesides, G.M., and Collier, R.J. (2001). Designing a polyvalent inhibitor of anthrax toxin. Nat. Biotechnol. *19*, 958–961.

.Muller-Loennies, S., Brade, L., MacKenzie, C.R., Di Padova, F.E., and Brade, H. (2003). Identification of a cross-reactive epitope widely present in lipopolysaccharide from enterobacteria and recognized by the cross-protective monoclonal antibody WN1 222–5. J. Biol. Chem. *278*, 25618–25627.

Muller-Loennies, S., Brade, L., and Brade, H. (2007). Neutralizing and cross-reactive antibodies against enterobacterial lipopolysaccharide. Int. J. Med. Microbiol. 297, 321–340.

Nanra, J.S., Timofeyeva, Y., Buitrago, S.M., Sellman, B.R., Dilts, D.A., Fink, P., Nunez, L., Hagen, M., Matsuka, Y.V., Mininni, T., *et al.* (2009). Heterogeneous *in vivo* expression of clumping factor A and capsular polysaccharide by *Staphylococcus aureus*: implications for vaccine design. Vaccine 27, 3276–3280.

Neilson, A.R., Burchardi, H., and Schneider, H. (2005). Cost-effectiveness of immunoglobulin M-enriched immunoglobulin (Pentaglobin) in the treatment of severe sepsis and septic shock. J. Crit. Care *20*, 239–249.

Nnalue, N.A., Lind, S.M., and Lindberg, A.A. (1992). The disaccharide L-alpha-D-heptose1->7-L-alpha-D-heptose1->of the inner core domain of Salmonella lipopolysaccharide is accessible to antibody and is the epitope of a broadly reactive monoclonal antibody. J. Immunol. *149*, 2722–2728.

Novick, R.P., and Geisinger, E. (2008). Quorum sensing in staphylococci. Annu. Rev. Genet. 42, 541–564.

Ogata, M., Fletcher, M.F., Kloczewiak, M., Loiselle, P.M., Zanzot, E.M., Vermeulen, M.W., and Warren, H.S. (1997). Effect of anticoagulants on binding and neutralization of lipopolysaccharide by the peptide immunoglobulin conjugate CAP18(106–138)-immunoglobulin G in whole blood. Infect. Immun. *65*, 2160–2167.

Otto, M. (2004). Quorum-sensing control in Staphylococci – a target for antimicrobial drug therapy? FEMS Microbiol. Lett. *241*, 135–141.

Pai, J.C., Sutherland, J.N., and Maynard, J.A. (2009). Progress towards recombinant anti-infective antibodies. Recent. Pat. Antiinfect Drug Discov. *4*, 1–17.

Park, D.R. (2005). The microbiology of ventilator-associated pneumonia. Respir. Care *50*, 742–63; discussion 763–5.

Park, J., Jagasia, R., Kaufmann, G.F., Mathison, J.C., Ruiz, D.I., Moss, J.A., Meijler, M.M., Ulevitch, R.J., and Janda, K.D. (2007). Infection control by antibody disruption of bacterial quorum sensing signaling. Chem. Biol. *14*, 1119–1127.

Patel, A.H., Nowlan, P., Weavers, E.D., and Foster, T. (1987). Virulence of protein A-deficient and alpha-toxin-deficient mutants of *Staphylococcus aureus* isolated by allele replacement. Infect. Immun. 55, 3103–3110.

Patti, J.M. (2004). A humanized monoclonal antibody targeting *Staphylococcus aureus*. Vaccine *22 Suppl 1*, S39–43.

Perera, L.P., Marques, L.R., and O'Brien, A.D. (1988). Isolation and characterization of monoclonal antibodies to Shiga-like toxin II of enterohemorrhagic *Escherichia coli* and use of the monoclonal antibodies in a colony enzyme-linked immunosorbent assay. J. Clin. Microbiol. *26*, 2127–2131.

Peschen, D., Li, H.P., Fischer, R., Kreuzaler, F., and Liao, Y.C. (2004). Fusion proteins comprising a Fusarium-specific antibody linked to antifungal peptides protect plants against a fungal pathogen. Nat. Biotechnol. *22*, 732–738.

Pier, G.B., Boyer, D., Preston, M., Coleman, F.T., Llosa, N., Mueschenborn-Koglin, S., Theilacker, C., Goldenberg, H., Uchin, J., Priebe, G.P., *et al.* (2004). Human monoclonal antibodies to *Pseudomonas aeruginosa* alginate that protect against infection by both mucoid and nonmucoid strains. J. Immunol. *173*, 5671–5678.

Plant, J.E., Glynn, A.A., and Valtonen, M.V. (1980). O-antigenic specificity of the protective supernatant factor from *Salmonella typhimurium* effective in S. typhimurium-infected mice. Parasite Immunol. *2*, 293–302.

Pollack, M., Huang, A.I., Prescott, R.K., Young, L.S., Hunter, K.W., Cruess, D.F., and Tsai, C.M. (1983). Enhanced survival in *Pseudomonas aeruginosa* septicemia associated with high levels of circulating antibody to *Escherichia coli* endotoxin core. J. Clin. Invest. 72, 1874–1881.

Pope, C.G. (1963). Development of knowledge of antitoxins. Br. Med. Bull. *19*, 230–234.

Porter, C.J., Burden, R.P., Morgan, A.G., Daniels, I., and Fletcher, J. (1997). Impaired bacterial killing and hydrogen peroxide production by polymorphonuclear neutrophils in end-stage renal failure. Nephron *77*, 479–481.

Ragle, B.E., and Bubeck Wardenburg, J. (2009). Anti-alpha-hemolysin monoclonal antibodies mediate protection against *Staphylococcus aureus* pneumonia. Infect. Immun. *77*, 2712–2718.

Reichert, J.M. (2010). Antibodies to watch in 2010. MAbs *2*, 84–100.

Rietschel, E.T., Brade, H., Holst, O., Brade, L., Muller-Loennies, S., Mamat, U., Zahringer, U., Beckmann, F., Seydel, U., Brandenburg, K., *et al.* (1996). Bacterial endotoxin: Chemical constitution, biological recognition, host response, and immunological detoxification. Curr. Top. Microbiol. Immunol. *216*, 39–81.

Risley, A.L., Loughman, A., Cywes-Bentley, C., Foster, T.J., and Lee, J.C. (2007). Capsular polysaccharide masks clumping factor A-mediated adherence of *Staphylococcus aureus* to fibrinogen and platelets. J. Infect. Dis. *196*, 919–927.

Rodriguez, A., Rello, J., Neira, J., Maskin, B., Ceraso, D., Vasta, L., and Palizas, F. (2005). Effects of high-dose of intravenous immunoglobulin and antibiotics on survival for severe sepsis undergoing surgery. Shock *23*, 298–304.

Roy-Burman, A., Savel, R.H., Racine, S., Swanson, B.L., Revadigar, N.S., Fujimoto, J., Sawa, T., Frank, D.W., and Wiener-Kronish, J.P. (2001). Type III protein secretion is associated with death in lower respiratory and systemic *Pseudomonas aeruginosa* infections. J. Infect. Dis. *183*, 1767–1774.

Rumbaugh, K.P., Griswold, J.A., and Hamood, A.N. (2000). The role of quorum sensing in the *in vivo* virulence of *Pseudomonas aeruginosa*. Microbes Infect. *2*, 1721–1731.

Rupnik, M., Wilcox, M.H., and Gerding, D.N. (2009). *Clostridium difficile* infection: new developments in epidemiology and pathogenesis. Nat. Rev. Microbiol. *7*, 526–536.

Sauter, K.A., Melton-Celsa, A.R., Larkin, K., Troxell, M.L., O'Brien, A.D., and Magun, B.E. (2008). Mouse model of hemolytic-uremic syndrome caused by endotoxin-free Shiga toxin 2 (Stx2) and protection from lethal outcome by anti-Stx2 antibody. Infect. Immun. *76*, 4469–4478.

Sawa, T., Yahr, T.L., Ohara, M., Kurahashi, K., Gropper, M.A., Wiener-Kronish, J.P., and Frank, D.W. (1999). Active and passive immunization with the Pseudomonas V antigen protects against type III intoxication and lung injury. Nat. Med. *5*, 392–398.

Sawada, S., Kawamura, T., and Masuho, Y. (1987). Immunoprotective human monoclonal antibodies against five major serotypes of *Pseudomonas aeruginosa*. J. Gen. Microbiol. *133*, 3581–3590.

Schuster, M., Lostroh, C.P., Ogi, T., and Greenberg, E.P. (2003). Identification, timing, and signal specificity of *Pseudomonas aeruginosa* quorum-controlled genes: a transcriptome analysis. J. Bacteriol. *185*, 2066–2079.

Schwan, W.R., Langhorne, M.H., Ritchie, H.D., and Stover, C.K. (2003). Loss of hemolysin expression in *Staphylococcus aureus* agr mutants correlates with selective survival during mixed infections in murine abscesses and wounds. FEMS Immunol. Med. Microbiol. *38*, 23–28.

Sellman, B.R., Mourez, M., and Collier, R.J. (2001). Dominant-negative mutants of a toxin subunit: an approach to therapy of anthrax. Science *292*, 695–697.

Sellman, B.R., Timofeyeva, Y., Nanra, J., Scott, A., Fulginiti, J.P., Matsuka, Y.V., and Baker, S.M. (2008). Expression of Staphylococcus epidermidis SdrG increases following exposure to an *in vivo* environment. Infect. Immun. *76*, 2950–2957.

Serna, A.,4th, and Boedeker, E.C. (2008). Pathogenesis and treatment of Shiga toxin-producing *Escherichia coli* infections. Curr. Opin. Gastroenterol. *24*, 38–47.

Shapiro, R.L., Hatheway, C., and Swerdlow, D.L. (1998). Botulism in the United States: a clinical and epidemiologic review. Ann. Intern. Med. *129*, 221–228.

Shime, N., Sawa, T., Fujimoto, J., Faure, K., Allmond, L.R., Karaca, T., Swanson, B.L., Spack, E.G., and Wiener-Kronish, J.P. (2001). Therapeutic administration of anti-PcrV F(ab')(2) in sepsis associated with *Pseudomonas aeruginosa*. J. Immunol. *167*, 5880–5886.

Shire, S.J. (2009). Formulation and manufacturability of biologics. Curr. Opin. Biotechnol. *20*, 708–714.

Song, L., Hobaugh, M.R., Shustak, C., Cheley, S., Bayley, H., and Gouaux, J.E. (1996). Structure of staphylococcal alpha-hemolysin, a heptameric transmembrane pore. Science *274*, 1859–1866.

Strockbine, N.A., Marques, L.R., Holmes, R.K., and O'Brien, A.D. (1985). Characterization of monoclonal antibodies against Shiga-like toxin from *Escherichia coli*. Infect. Immun. *50*, 695–700.

Strohl, W.R. (2009). Optimization of Fc-mediated effector functions of monoclonal antibodies. Curr. Opin. Biotechnol. *20*, 685–691.

Subramanian, G.M., Cronin, P.W., Poley, G., Weinstein, A., Stoughton, S.M., Zhong, J., Ou, Y., Zmuda, J.F., Osborn, B.L., and Freimuth, W.W. (2005). A phase 1 study of PAmAb, a fully human monoclonal antibody against Bacillus anthracis protective antigen, in healthy volunteers. Clin. Infect. Dis. *41*, 12–20.

Svenson, S.B., and Lindberg, A.A. (1981). Artificial Salmonella vaccines: *Salmonella typhimurium* O-antigen-specific oligosaccharide–protein conjugates elicit protective antibodies in rabbits and mice. Infect. Immun. *32*, 490–496.

Tarr, P.I., Gordon, C.A., and Chandler, W.L. (2005). Shiga-toxin-producing *Escherichia coli* and haemolytic uraemic syndrome. Lancet *365*, 1073–1086.

Tateda, K., Ishii, Y., Horikawa, M., Matsumoto, T., Miyairi, S., Pechere, J.C., Standiford, T.J., Ishiguro, M., and Yamaguchi, K. (2003). The *Pseudomonas aeruginosa* autoinducer N-3-oxododecanoyl homoserine lactone accelerates apoptosis in macrophages and neutrophils. Infect. Immun. *71*, 5785–5793.

Taylor, C.P., Tummala, S., Molrine, D., Davidson, L., Farrell, R.J., Lembo, A., Hibberd, P.L., Lowy, I., and Kelly, C.P. (2008). Open-label, dose escalation phase I study in healthy volunteers to evaluate the safety and pharmacokinetics of a human monoclonal antibody to *Clostridium difficile* toxin A. Vaccine *26*, 3404–3409.

Taylor, R.P., Sutherland, W.M., Martin, E.N., Ferguson, P.J., Reinagel, M.L., Gilbert, E., Lopez, K., Incardona, N.L., and Ochs, H.D. (1997). Bispecific monoclonal antibody complexes bound to primate erythrocyte complement receptor 1 facilitate virus clearance in a monkey model. J. Immunol. *158*, 842–850.

Terashima, M., Uezumi, I., Tomio, T., Kato, M., Irie, K., Okuda, T., Yokota, S., and Noguchi, H. (1991). A protective human monoclonal antibody directed to the outer core region of *Pseudomonas aeruginosa* lipopolysaccharide. Infect. Immun. *59*, 1–6.

Top, J., Willems, R., and Bonten, M. (2008). Emergence of CC17 *Enterococcus faecium*: from commensal to hospital-adapted pathogen. FEMS Immunol. Med. Microbiol. *52*, 297–308.

Underwood, K., Rubin, S., Deakers, T., and Newth, C. (2007). Infant botulism: a 30-year experience spanning the introduction of botulism immune globulin intravenous in the intensive care unit at Childrens Hospital Los Angeles. Pediatrics *120*, e1380–5.

Vitale, L., Blanset, D., Lowy, I., O'Neill, T., Goldstein, J., Little, S.F., Andrews, G.P., Dorough, G., Taylor, R.K., and Keler, T. (2006). Prophylaxis and therapy of inhalational anthrax by a novel monoclonal antibody to protective antigen that mimics vaccine-induced immunity. Infect. Immun. *74*, 5840–5847.

Voth, D.E., and Ballard, J.D. (2005). *Clostridium difficile* toxins: mechanism of action and role in disease. Clin. Microbiol. Rev. *18*, 247–263.

Vuong, C., Saenz, H.L., Gotz, F., and Otto, M. (2000). Impact of the agr quorum-sensing system on adherence to polystyrene in *Staphylococcus aureus*. J. Infect. Dis. *182*, 1688–1693.

Vuong, C., Kocianova, S., Yao, Y., Carmody, A.B., and Otto, M. (2004). Increased colonization of indwelling medical devices by quorum-sensing mutants of Staphylococcus epidermidis *in vivo*. J. Infect. Dis. *190*, 1498–1505.

Wagner, V.E., Bushnell, D., Passador, L., Brooks, A.I., and Iglewski, B.H. (2003). Microarray analysis of *Pseudomonas aeruginosa* quorum-sensing regulons: effects of growth phase and environment. J. Bacteriol. *185*, 2080–2095.

Wagner, K.S., Stickings, P., White, J.M., Neal, S., Crowcroft, N.S., Sesardic, D., and Efstratiou, A. (2009). A review of the international issues surrounding the availability and demand for diphtheria antitoxin for therapeutic use. Vaccine *28*, 14–20.

Warren, H.S., Amato, S.F., Fitting, C., Black, K.M., Loiselle, P.M., Pasternack, M.S., and Cavaillon, J.M. (1993). Assessment of ability of murine and human anti-lipid A monoclonal antibodies to bind and neutralize lipopolysaccharide. J. Exp. Med. *177*, 89–97.

Warren, H.S., Matyal, R., Allaire, J.E., Yarmush, D., Loiselle, P., Hellman, J., Paton, B.G., and Fink, M.P. (2003). Protective efficacy of CAP18106-138-immunoglobulin G in sepsis. J. Infect. Dis. *188*, 1382–1393.

Weems, J.J., Jr, Steinberg, J.P., Filler, S., Baddley, J.W., Corey, G.R., Sampathkumar, P., Winston, L., John, J.F., Kubin, C.J., Talwani, R., *et al.* (2006). Phase II, randomized, double-blind, multicenter study comparing the safety and pharmacokinetics of tefibazumab to placebo for treatment of *Staphylococcus aureus* bacteremia. Antimicrob. Agents Chemother. *50*, 2751–2755.

Weisman, L.E. (2007). Antibody for the prevention of neonatal noscocomial staphylococcal infection: a review of the literature. Arch. Pediatr. *14 Suppl 1*, S31–4.

Weisman, L.E., Thackray, H.M., Garcia-Prats, J.A., Nesin, M., Schneider, J.H., Fretz, J., Kokai-Kun, J.F., Mond, J.J., Kramer, W.G., and Fischer, G.W. (2009). Phase 1/2 double-blind, placebo-controlled, dose escalation, safety, and pharmacokinetic study of pagibaximab (BSYX-A110), an antistaphylococcal monoclonal antibody for the prevention of staphylococcal bloodstream infections, in very-low-birth-weight neonates. Antimicrob. Agents Chemother. *53*, 2879–2886.

Wu, H., Pfarr, D.S., Johnson, S., Brewah, Y.A., Woods, R.M., Patel, N.K., White, W.I., Young, J.F., and Kiener, P.A. (2007). Development of motavizumab, an ultra-potent antibody for the prevention of respiratory syncytial virus infection in the upper and lower respiratory tract. J. Mol. Biol. *368*, 652–665.

Yarwood, J.M., and Schlievert, P.M. (2003). Quorum sensing in *Staphylococcus* infections. J. Clin. Invest. *112*, 1620–1625.

Yokota, S., Terashima, M., Chiba, J., and Noguchi, H. (1992). Variable cross-reactivity of *Pseudomonas aeruginosa* lipopolysaccharide-code-specific monoclonal antibodies and its possible relationship with serotype. J. Gen. Microbiol. *138*, 289–296.

Zaidi, T., and Pier, G.B. (2008). Prophylactic and therapeutic efficacy of a fully human immunoglobulin G1 monoclonal antibody to *Pseudomonas aeruginosa* alginate in murine keratitis infection. Infect. Immun. *76*, 4720–4725.

Ziegler, E.J., McCutchan, J.A., Fierer, J., Glauser, M.P., Sadoff, J.C., Douglas, H., and Braude, A.I. (1982). Treatment of Gram-negative bacteremia and shock with human antiserum to a mutant *Escherichia coli*. N. Engl. J. Med. *307*, 1225–1230.

Ziegler, E.J., Fisher, C.J., Jr., Sprung, C.L., Straube, R.C., Sadoff, J.C., Foulke, G.E., Wortel, C.H., Fink, M.P., Dellinger, R.P., and Teng, N.N. (1991). Treatment of Gram-negative bacteremia and septic shock with HA-1A human monoclonal antibody against endotoxin. A randomized, double-blind, placebo-controlled trial. The HA-1A Sepsis Study Group. N. Engl. J. Med. *324*, 429–436.

Zinner, S.H., and McCabe, W.R. (1976). Effects of IgM and IgG antibody in patients with bacteremia due to Gram-negative bacilli. J. Infect. Dis. *133*, 37–45.

Therapeutic Applications of Phage Biology: History, Practice and Recommendations

17

Jason Gill and Ryland F. Young, III*

Abstract

A bacteriophage, or 'phage', is a virus that infects bacteria. This chapter is aimed at assessing the record and potential of the use of phage and phage-derived molecules in antibacterial therapeutics and prophylactics. Unlike other areas of current biomedicine, phage therapy has a long history that pre-dates even the basics of modern biology, and even the development of phage biology itself. Thus it is important to reflect on the historical record to establish a context before considering the more recent literature and, finally, the prospects and obstacles facing phage therapy at the current time. In addition, although the study of phage was vibrant through the mid-1970s, the last decades of the twentieth and the first decade of the twenty-first centuries witnessed a drastic contraction in the number of phage biology laboratories. This has led now to an odd situation where interest and activity in phage research are outstripping the available expertise. Accordingly, a section of this chapter is devoted to a summary of the fundamental characteristics of bacteriophage that would be important to the prospective phage therapist. Next, we present a review and meta-review of the recent phage therapy literature and then summarize the current practices in the field. Finally, we consider the future, in terms of what should be done, according to our perspective.

Please note that throughout this text, we define terminology for elements and concepts important to phage biology and its practical applications. We have done this in an overt attempt to simplify the text, but in some cases we admit to promoting what we think is better and less confusing terminology than that currently in general use. To this end, a glossary is provided at the end of the chapter.

Phage therapy and the development of phage biology

d'Herelle and the rise and fall of phage therapy

It was the work of Felix d'Herelle, a self-educated French Canadian, that definitively established the existence of phages, beginning during World War I at the Pasteur Institute. d'Herelle discovered that, in a ward of French *poilus* suffering from shigellosis dysentery, the faeces of patients whose symptoms had suddenly improved contained agents that caused cleared zones within *Shigella* lawns grown on agar plates. These zones, or 'plaques', became the basis of the bio-assay used to this day to titre phage suspensions, which are quantified in terms of 'pfu' (plaque-forming units) on a lawn of a particular *indicator* strain (see below 'Isolating phage from the environment: phage hunts'). Plaques could be used to inoculate *Shigella* cultures which, in general, were entirely lysed after overnight growth; again, to this day, flask or tube 'lysates' are the normal way that phage are grown in the laboratory. d'Herelle named the agent *bacteriophage* (Dublanchet and Bourne, 2007) and largely devoted the rest of his life to promulgating phage biology and, true

*Corresponding author

to the Pasteurian 'bench to bedside' tradition, its applications against bacterial disease (Summers, 1999). This focus, combined with d'Herelle's brilliance and powerful personality, as well as the general lack of other antibacterial agents, led to the widespread implementation of phage therapy around the world in the 1920–30s. A reflection of the excitement that d'Herelle's discoveries generated can be seen in the novel *Arrowsmith*, which helped win Sinclair Lewis Pulitzer and Nobel Prizes with its story of a young doctor who discovers bacteriophage and becomes consumed with its development as a 'magic bullet', only to find that d'Herelle has already described and exploited the same phenomenon (Lewis, 1925).

Although he published and travelled extensively, d'Herelle's most publicized work reported the treatment of bubonic plague in Alexandria (Summers, 1999). Four patients in the acute phase of the disease were treated with direct injections of phage preparations in the buboes, leading in all four cases to rapid and dramatic recoveries. Although anecdotal by today's standards, this success gave d'Herelle access to international venues and resources, which he exploited assiduously to influence a generation of medical scientists. This led, for example, to the use of oral doses of vibriophage in successful large scale interventions against cholera epidemics in India during the late 1920s.

Nevertheless, by the beginning of the next world war, d'Herelle and phage therapy had risen and fallen. There is no need here to recount the events in detail, especially in view of excellent accounts in both short review and book-length formats (Summers, 1999, 2001). These works constitute an essential resource if the reader wants an integrated understanding of the origins of phage biology, pre-antibiotic era phage therapy and how intertwined these subjects were with d'Herelle's rather tragic trajectory. The fundamental problem was that the science underlying phage therapy was immature, since it was decades before the definition of the modern concept of DNA and the gene. d'Herelle engaged in sometimes acrimonious debates with Nobel Laureates Bordet and Northrop and others about numerous issues regarding the essential nature of phage: whether they were viral or enzymatic, whether each phage was unique or was generated to match the bacterial prey, and whether phages existed in the environment or were generated by the infected tissues. Even d'Herelle's thesis that phages were particulate virions was not finally accepted until they were imaged by the newly developed electron microscope in 1942 (Luria and Anderson, 1942). On the practical side, the primitive state of microbiology meant that there was not even a semblance of quality control for phage preparations or significant useful knowledge about effective ways to employ them. In fact, d'Herelle himself tested a number of the phage preparations used in many of the failed clinical applications and reported that few, if any, actually contained any viable phage particles. This led him to establish the Laboratory for Bacteriophage in Paris, which strove to produce legitimate phage 'cocktails'. Labelled descriptively, as in *intestiphage* and *pyophage*, these cocktails were pools of multiple phages that were isolated against a range of pathogenic bacteria involved in a particular syndrome (Summers, 2001). These efforts were largely too late, because, in 1934, a comprehensive study commissioned by the American Medical Association reported in depth on the unproven record of efficacy of phage therapy (Eaton and Baynes-Jones, 1934). Although often cited as a conclusive rejection of phage therapeutics, in fact this report concluded, properly, that largely due to the lack of accepted standards for purity, dosage and potency, its efficacy could not be assessed with the current knowledge base and technology. Concomitant with the unsettled state of the basic science, the other main factor in the decline of interest in phage therapeutics was, in the years leading up to World War II, the advent of small-molecule antibiotics. During this period, a wide variety of chemical antibiotics became available to treat virtually all forms of bacterial infections. These drugs were effective, easily formulated and administered, and available technology made their pharmacology straightforward to establish. While certain bacteria, like *M. tuberculosis*, were known to possess some innate resistance to chemical antibiotics, acquired resistance was not a significant clinical problem and acquired resistance to multiple drugs was virtually unknown. Thus, there was minimal clinical demand for non-antibiotic approaches for the treatment of bacterial infections. Ultimately, phage therapy was largely extinct in the West by the mid-1940s, although phage preparations continued to be marketed in the U.S. by some major pharmaceutical companies even through the 1950s (Summers, 2001).

Phage therapy continues in Georgia and Poland

Despite its disappearance in the West, phage therapy against human disease survived principally in two institutions, the Eliava Institute in Tbilisi, Georgia, and in the Hirszfeld Institute in Wrocław, Poland, as recently documented in detail by Kutter *et al.* (2009) and Gorski *et al.* (2009). George Eliava was a Pasteur colleague of d'Herelle, and it was through the Eliava–d'Herelle partnership, and the support of the dictator Stalin (a native Georgian) that the Institute in Tbilisi was established. Through the efforts of Eliava scientists and physicians, phage preparations were developed and produced in large quantities throughout the Soviet era and were in routine use in the Red Army. After the fall of the Soviet Union, the Institute lost its state support and nearly disappeared, but the recent resurgent interest in phage therapy, and some grant funding from the West, have given it new visibility. The major focus at Eliava has been to use two cocktails, Pyophage™ and Intestiphage™, derived from the original preparations produced by d'Herelle's laboratory. Each cocktail contains multiple phages targeting a wide range of bacteria: pyophage has phages against numerous Gram-positive and Gram-negative bacteria that cause purulent infections, whereas Intestiphage has phages against enteric bacteria. These preparations are widely used in Georgia at tertiary care centres and wound/burn facilities. Unfortunately, there has been little or nothing in the peer-reviewed western literature to document the Eliava record, other than a few anecdotal retrospectives (Kutateladze and Adamia, 2008). Moreover, owing to the exigencies of operating under the Soviet system and then surviving in the chaotic post-Soviet era in Central Asia, the Institute was until recently largely isolated from the rapid technological advances of modern molecular biology. Consequently, until recently, little of the work done at Eliava was documented at the molecular level. Nevertheless, Eliava remains a significant source of experience with phage-based therapeutics and tested phage preparations.

The approach at Hirszfeld has been different and also has been much more thoroughly documented in English-language literature. In the Wrocław model, extensive libraries of phages against important pathogenic bacteria are maintained. Patients are accepted after the failure of standard antibiotic therapies. The bacteria responsible for the pathology are identified and tested against the library; treatment occurs only if the bacterial isolates are shown to be susceptible to specific phages present in the library. In addition, reviews describing the experience at Hirszfeld have been published in journals accessible to Western scientists. Impressive success rates have been reported, especially considering the infections being treated have been pre-selected for antibiotic resistance. As discussed later, these results suggest that phage therapy should at least be accorded the status of emergency treatment for cases where antibiotic treatment has failed.

Delbrück and the phage group

It is a historical quirk that in the 1940s, as the applied side of phage biology withered, basic research in the field began to flourish, with the establishment of the Phage Group, by the German physicist Max Delbrück and colleagues (Cairns *et al.*, 2000). Using an annual summer course taught at Cold Spring Harbor Laboratory on Long Island, Delbrück recruited an illustrious cohort of young scientists, mostly trained in the physical sciences, to use 'quantitative biology' to uncover the secrets of heredity. There was a strict requirement, derived from the reductionist philosophy of physics, that experiments be done with bacteriophages, chosen both because of their facile experimental character and because Delbrück decided that they were the simplest possible genetic system. Furthermore, the phages to be studied were restricted to a set of seven 'well-behaved' phages of *E. coli*, designated T1 through T7. This focus, although gradually relaxed to include a few other phages of *E. coli* and related bacteria, notably the paradigm temperate phage lambda, proved to be immensely successful. Indeed, most of the fundamental principles of modern molecular biology were initially established using these phage systems, as was recognized by several Nobel Prizes. One of the original experimental systems was to use phage to select for *resistant* mutants of bacteria, i.e. for bacteria that could not be recognized by the particular phage. The major purpose of these experiments was to get at spontaneous mutation rates and to discredit the still prevalent idea that adaptive mutations could be induced by a selective agent (Newcombe, 1949). A collateral effect was to make it dogma within the phage community that phage-based therapeutics was doomed to failure, since every culture of bacteria would always contain a few phage-resistant mutants that would rapidly multiply after clinical application of a phage (Stent, 1963). Thus the remarkable scientific success of phage biology contributed to the demise of phage therapy, rather than reinvigorating it. Nevertheless, as a result of the enterprise, there is an incredibly detailed understanding of many aspects of the biology of the few intensely studied coliphages, especially phages T4, T7 and lambda, which are today the best characterized genetic entities at any level. In addition, and less obviously, the spectacular successes of phage biology triggered another consequence of reductionism: the rapid abandonment of basic phage biology by many of its practitioners, who moved on in large numbers to other 'higher' model systems. That this happened just before the advent of modern high-throughput DNA and protein technology and high-resolution structural biology is particularly frustrating, considering the momentum of the genetic, physiological and biochemical studies that were abandoned.

The phage therapist must understand this recent history and realize both the depth and the limitations of the experimentally established knowledge base of phage biology, before extrapolating too much to the very different phage systems of pathogenic bacteria. There are also glaring gaps even in the classic systems, especially in issues dealing with the bacterial surface and membrane envelope, which were largely ignored during the decades of using phages to work out the principles of replication, transcription, translation and regulation. For example, no one has any real idea how lambda DNA gets into the host cytoplasm, despite the paradigm status of the phage. Of course the focus on the classical systems also meant that the biology of phages of the great variety of bacterial pathogens has been, until recently, relatively unexplored, and the ecology and behaviour of phages outside of closely constrained laboratory environments were a complete mystery. Interestingly, the most important exceptions stem from the dairy industry, where, owing to the sensitivity of large volume bacterial fermentations to catastrophic viral infections, significant effort was placed on basic phage biology (Brussow, 2001).

Phage biology for phage therapists

To paraphrase the foregoing section, phage biology is 'a mile deep and an inch wide'. Not only does this mean that we may make important mistakes in generalizing for practical applications, but it is probably true that many fundamental, and useful, features of phage are yet to be discovered, as is strongly indicated by the recent discoveries of the diversity-generating modules in *Bordetella* phages and the CRISPR defence systems in streptococcal dairy strains (see below). Nevertheless, there is a body of knowledge about how phages work, and it is useful here to summarize those aspects that may be important for phage therapy.

Phages are everywhere

No consideration of the potential for phage therapy can be accomplished without an appreciation for the place and role of bacteriophage in the biosphere. Phages are ubiquitous, outnumbering bacteria 10–100-fold in unbiased environmental samples (Fuhrman, 1999; Wommack and Colwell, 2000; Suttle, 2005), easily the dominant life form on the planet not just in terms of numbers but also in terms of DNA sequence diversity. A number of reviews highlight the important role played by bacteriophage in ecology and evolution (Ochman *et al.*, 2000; Weinbauer and Rassoulzadegan, 2004; Abedon, 2009b; Thurber, 2009). However, for our purposes here, it is important to stress that phages are an integral part of the human biome. Metagenomic analysis of the total viral content of human faecal material revealed the presence of thousands of different phages (Breitbart *et al.*, 2003). This fact alone should have an impact on the regulatory situation for the use of phages in therapeutic applications, at least for enteric bacteria, a consideration we shall revisit later.

Phage structure: the *Caudovirales*

For the purposes of phage therapy, bacteriophages of eubacteria fall into two categories: the *Caudovirales*, or 'tailed' phages, and everything else. The reason for this is implicit in the definition: the existence in *Caudovirales* of a *tail* structure that allows high efficiency adsorption to the host cell and high efficiency transfer of the DNA from the phage particle into the host cytoplasm. Other solutions are simply too inefficient or depend on the host providing, unreliably, a complex structure like a conjugative pilus to permit the penetration of the viral nucleic acid. The tailed phages come in three basic shapes: myophages, siphophages and podophages (Fig. 17.1; Ackermann, 2006). All have a single, linear, double-stranded DNA chromosome, ranging in size from ~15 kbp to >500 kbp, packaged into a *head* structure. Phage heads have fundamental icosahedral symmetry, although they are often elongated into prolate structures. The tail always emanates from one of the 12 icosahedral vertices, called the *portal*, and distinguishes the three types: (1) *myophages* have rather inflexible, contractile tails, ending with some sort of flat base plate; (2) *siphophages* have flexible, non-contractile tails, ending in a distinct structure, sometimes conical, with a central spike; and (3) *podophages* have stubby, indistinct protrusions, often surrounded by spikes. Phages often have long, jointed tail fibres that are attached by knob-in-socket joints around the periphery of whatever structures exist at the end of the tail in myophages and siphophages, or around the portal protrusion of a podophage. For the purposes of this chapter, these structures are designated as long (or lateral) tail fibres (LTFs). Depending on the phage, there are either three, four or six of these LTFs, thought to flail about randomly, sweeping rapidly through the space surrounding the virion. Consequently, the LTFs are expected make the first contact with the surface of the bacterial cell. Accordingly there are always host-specificity domains which recognize and bind to host surface features near the distal tips of LTFs. Note, however, phages usually have another host-recognition determinant, much more variable in structure. For example, in the classic siphophages of *E. coli*, which have conical tail tips, there is a central tail fibre, which we will abbreviate as CTF for our purposes here, emanating from the cone structure. The tip of the CTF also has a cell surface binding domain. If what we know from detailed

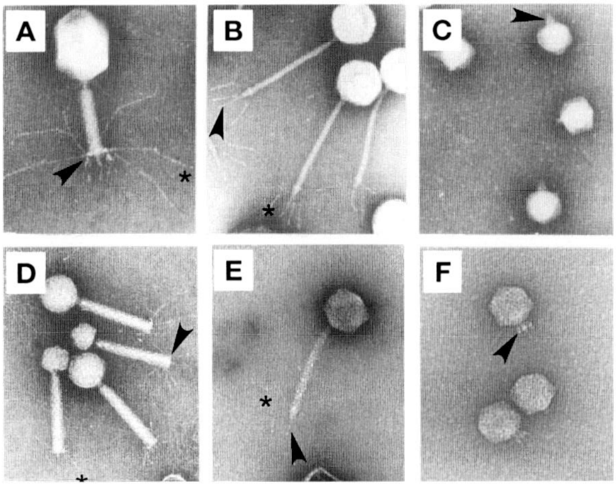

Figure 17.1 Morphotypes of the Caudovirales. Paradigm examples of the tailed phages of *E. coli* (A–E) and *Salmonella* (F). A. Myophage T4; B. Siphophage T5; C. Podophage T7; D. Myophage P2 and satellite myophage P4 (small head). E. Siphophage lambda; F. Podophage P22. Asterisks mark distal end of lateral tail fibres (LTF). Arrows mark: base plate of myophages (A and D); distal end of central tail fibre (CTF) of siphophages (B and E); stubby tail in podophage T7 (C) and tail spike assemblies in podophage P22 (E). The six T7 LTFs attached at the top of the stubby tail are not visible in panel C. Negative stain, ×155,000 magnification (adapted from Calendar *et al.*, 1988); panel E is unpublished data from R. Duda and R. Hendrix.

studies of coliphage T4 can be generalized, the base plates of myophages have six short tail fibres (STFs), which also recognize the host surface. Some podophages, like the *Salmonella* phage P22, lack LTFs but have a set of six tail spikes, which are essentially STF-equivalents; in P22, the tail spikes recognize the lipopolysaccharide O-antigen repeat of the host. In other cases, polysaccharide depolymerase enzymes are found incorporated onto the tail structures of phages and have been shown to degrade capsular material, which otherwise would block recognition of the host (Scholl *et al.*, 2005; Leiman *et al.*, 2007). Since these enzymes are stable and catalyse 'slime' degradation even if not mounted on the virion, they are being considered for use in dispersing biofilms (Hughes *et al.*, 1998; Mushtaq *et al.*, 2005; Verma *et al.*, 2010).

Hopefully, this brief discussion of Caudovirales structure will entice the reader to review some of the spectacular new structural studies on several of the classical phages, generated by a powerful leveraging of X-ray crystallography and cryo-electron microscopy (Rossmann *et al.*, 2004; Lander *et al.*, 2006). The combination of the resulting images with the extensive molecular biology of these phages is bringing about a new level of understanding of the phage infection process and helping to rekindle general interest in phage systems. Also, before closing, we want to re-emphasize our terminology is designed for clarity in the context of this chapter and does not reflect consensus in the classical phage biology community, which is largely fragmented into phage-specific dialects. For example, the term 'tail fibre' has been and continues to be used promiscuously and indiscriminately, not only by genomicists but also by professional phage biologists. For example, in the extensive literature on the siphophages lambda and T5, LTF and CTF are called 'side tail fibre ' (Stf) and 'tail fibre' in the former and 'L-shaped tail fibre' (LTF) and 'straight tail fibre' (STF) in the latter.

Phage adsorption, specificity and resistance

From the point of view of phage therapy, an important issue is what determines the specificity of the phage for a given bacterium. It turns out that all myophages and siphophages and many podophages have two different determinants, one defined by the LTF and the other by whatever structure is at the end of the tail itself (CTF or STF). For example, in phage lambda, the CTF recognizes and binds to surface loops of LamB (Gehring *et al.*, 1987; Wang *et al.*, 1998), a minor outer membrane porin of *E. coli*, and its LTF recognizes OmpC (Hendrix and Duda, 1992), a major outer membrane porin protein in all Gram-negative enteric bacteria. Similarly, the T4 LTF recognizes a diglucose sequence in the lipopolysaccharide (LPS) of *E. coli* B, whereas the STF is thought to bind LPS non-specifically (Henning *et al.*, 1994). Sometimes the specificity binding domain is polyvalent; for example, the binding domain of the T4 LTF is also able to recognize the OmpC protein in *E. coli* K-12. One or both of these interactions are absolutely required for successful adsorption; the host feature involved in the most specific interaction is usually designated as the *receptor*. Most phages of Gram-negative hosts use either an outer membrane protein, usually a porin, or a specific sugar moiety of the LPS. Less is known about phages of Gram-positive hosts, but in some cases, phages recognize peptidoglycan or teichoic acid moieties (Wendlinger *et al.*, 1996), or cell wall proteins (Sao-Jose *et al.*, 2006).

Herein is a key issue for phage therapy. First, if the bacterium changes or loses the receptor through mutation, it becomes *resistant* to the phage. It should be noted that resistance is just one form of insensitivity to the phage (for a comprehensive review, see Hyman and Abedon, 2010) but, as noted above, it is the one most cited as a fatal shortcoming of the concept of phage therapeutics. In fact, however, the ability of phage to overcome resistance is the one of the principal strengths of the approach. First, tail fibres often have more than one receptor binding domain (RBD), as noted above for the LTF of T4. Second, unlike antibiotics, phages have DNA and a mutation rate at least an order of magnitude greater than that of bacteria (Santos and Drake, 1994), not to mention a huge advantage in generation time and progeny count. This makes it relatively easy for a phage to achieve new specificities by mutation of the RBD, which in several cases have been shown to be mutationally plastic and able to switch to new receptors of either proteinaceous or carbohydrate character (Henning *et al.*, 1994; Qimron *et al.*, 2006). In classic phage biology, these would be called '*h*' for host range mutations. Some phages even have tail fibre genes served by arrays of reading frames encoding different C-terminal RBD domains (Fig. 17.2). In these genes, different RBDs can be attached to the tail fibre by site-specific recombination events catalysed by a phage enzyme (Koch, 1987; Sandmeier *et al.*, 1991; Summer *et al.*, 2004). Recently, perhaps the best example of the unequal nature of this contest (and of the scientific gold mines awaiting us in non-*E. coli* phage biology) was found in some phages of *Bordetella*. In these phages, the DNA segment encoding the RBD is explicitly mutagenized by a phage-encoded HIV-like reverse transcriptase (Liu *et al.*, 2002; Doulatov *et al.*, 2004)! The function of this diversity-generating retroelement (DGR) results in mutation rates many orders of magnitude higher than normal and allows the phage to keep ahead of a host that constantly changes its surface protein array. Moreover, if a phage targets a surface feature that is important for virulence, resistant strains may have compromised pathogenicity; while this is not a universal phenotype, examples of loss of virulence associated with phage resistance have been found in both Gram-negative (Smith and Huggins, 1982) and Gram-positive (Capparelli *et al.*, 2010b) organisms. Resistance is thus an issue but not a disqualifying consideration for phage therapy. We will return to this issue later but in passing it is worth noting that the analogous 'resistance' to a chemical antibiotic molecule often requires the change of only a single amino acid in the target enzyme and that a chemical synthesis response to this could be prohibitively difficult.

In general, quantitative aspects of the interplay between the receptors and the receptor-binding domains are the reason why phage and bacteria can coexist in a particular domain of the environment. Phage infection is a two-body process, and every phage-receptor system has an intrinsic affinity constant that reflects the biochemical nature of the interaction. If phage and bacteria are present at low levels, they simply will not collide very often. Even in simple laboratory conditions, the mathematics of phage–host interactions are complex, because many if not most phage have at least one mode of low affinity, reversible interaction before becoming affixed irreversibly to the host surface (Stent and Wollman, 1952; Watanabe *et al.*, 1977; Moldovan *et al.*, 2007). Moreover, there is evidence that the level of host receptors may be stochastically determined within a population of isogenic host cells, perhaps as an evolutionary tactic to avoid over-predation (Chapman-McQuiston and Wu,

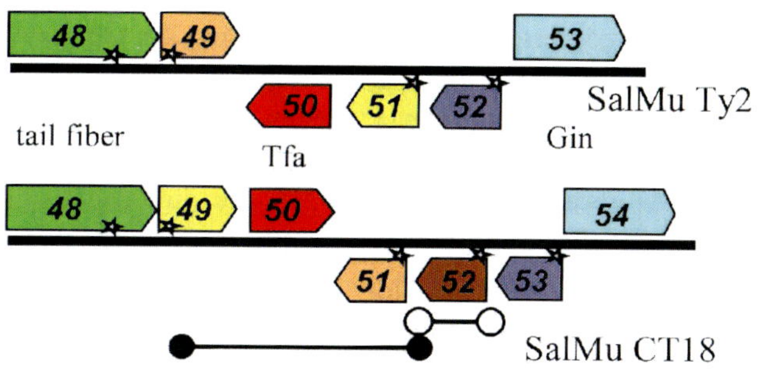

Figure 17.2 Combinatorial tail fibre gene arrays. The tail fibre-encoding regions of the genomes of two Mu-like myophages, SalMuTy2 and SalMuCT18 are shown. Genes *48* and *50* encode the LTF protein and the tail fibre assembly chaperone, respectively. In these genomes, separate modules encoding alternate C-terminal receptor-binding domains are encoded by the reading frames *49*, *51*, *52* and (in SalMuCT18) *53*. Black stars mark homologous site-specific recombination sites. Genes *53* in SalMuTy2 and *54* in SalMuCT18 encode a recombinase (Gin) which can catalyse inversion or loop-out recombination events between any two of these sites. Thus, SalMuTy2 can rearrange to replace the receptor-binding domain of its gp48 tail fibre with the domains in modules *49*, *51*, and *52*. Similar events can occur in SalMut CT18 to generate four alternate receptor binding specificities. The arrangement in SalMu Ty2 may have been derived from SalMu CT18 by site-specific recombination events resulting in a loop-out (indicated by bar with open circles) and an inversion (indicated by bar with filled circles).

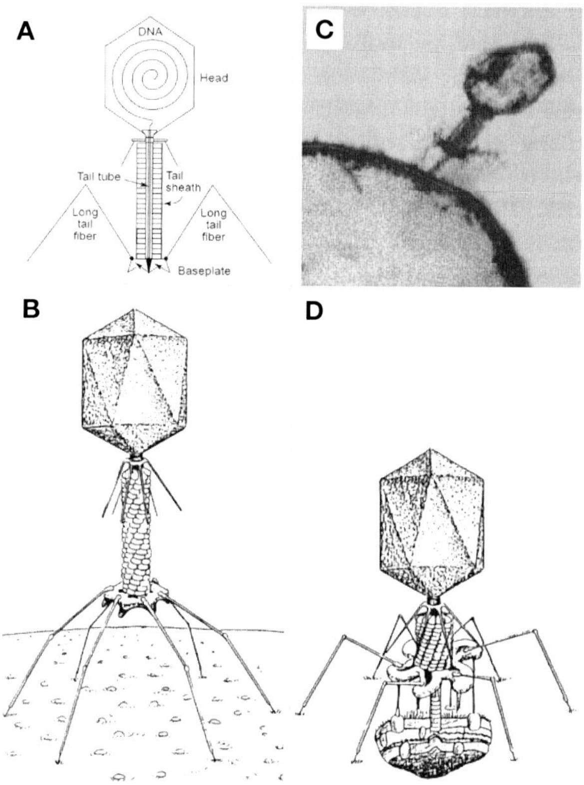

Figure 17.3 Model for T4 adsorption and DNA ejection. The T4 infection process begins when the LTFs (A) adsorb to the receptor on the cell surface, which in *E. coli* K-12 is the porin OmpC (B). This induces a conformational change originating in the base plate and propagating rapidly up the tail sheath. The final result is the contraction of the tail sheath and the deployment of the short tail fibres, driving the tail tube, capped with a 'puncturing device' that includes a lysozyme activity, through the outer membrane. Panels C–D show an electron micrograph and a cartoon interpretation, respectively, of the myophage in its contracted state. The lysozyme causes a local degradation of the murein, allowing the tube to approach the surface of the cytoplasmic membrane. A temporary hole is made in the membrane, allowing the DNA to pass into the cell. This is shown in cross-section in panel D. Panels B and D coutesy of F. Eiserling, UCLAY (http://www.mimg.ucla.edu/faculty/eiserling/fac_eiserling.html). Panel C courtesy of L. Simon, Rutgers University (http://genetics.rutgers.edu/?page=faculty/faculty_details&id=16.

contraction of the tail sheath, which drives the central tail tube through the outer membrane. There is also a cell-wall degrading activity associated with the penetration apparatus, so a local degradation of the peptidoglycan is effected. All this activity delivers the tail tube to the surface of the cytoplasmic membrane. From this point on the process is not understood. Somehow, a temporary channel is opened in the membrane; in fact, the cell nearly 'bleeds out' through this hole, in the sense that a substantial fraction of the cytoplasmic potassium ions are released and the membrane becomes briefly de-energized (Letellier and Boulanger, 1989; Letellier *et al.*, 2004). However, something associated with the passage of the huge dsDNA molecule apparently triggers closure of the channel, after which the membrane is re-energized and macromolecular biosynthesis resumes. Something like this temporary hole occurs with siphophages and many podophages too, but little is known about how these morphotypes gain access to the cytoplasmic membrane, much less about penetration by the DNA.

It should also be noted that for many phages, especially the larger ones, many protein molecules are also ejected into the cytoplasm with the DNA (Goldberg *et al.*, 1994; Bair and Black, 2007; Bair *et al.*, 2007). Most of these have to do with protecting the incoming DNA from host enzymes but some have other functions, including redirection of the cellular transcription system (Wilkens and Ruger, 1994).

Tailocins

Many bacteria have taken advantage of the cellular leakage associated with the phage DNA ejection process by evolving 'tailocins', a term we propose here for what were originally called 'high-molecular weight pyocins' when they were first identified in *Pseudomonas* strains (Michel-Briand and Baysse, 2002). Basically, these appear to be the products of degenerate prophage elements which lack head genes; when the tailocin determinants are expressed, headless phage particles are produced that appear to kill the target cells by the same hole-forming process used for phage DNA ejection, except that no DNA ever goes through to cause re-sealing of the hole. The new term tailocin is justified because the same entities have now been found in many other bacterial species, and the continued use of the pyocin term would be misleading. Extending our terminology, tailocins have been identified in both myotailocin and siphotailocin form, originally called 'R-type' and 'F-type'. The simplest notion is that the tailocins appear to be simply headless phage particles. Indeed, when headless mutants of the temperate *Pseudomonas* myophage PS17, which is antigenically related to common myotailocins, were constructed and induced, complete phage tails were produced that acted as single-hit killing complexes and collapsed the membrane potential of the target cells, exactly as found for *Pseudomonas* myotailocins. However, the specific killing activity of the purified PS17 tails was ~50-fold less than the native myotailocin; thus it may be that some additional mutational change might be required for robust tailocin activity (Shinomiya and Shiga, 1979). The potential of tailocins for therapeutic use is briefly discussed below.

The infection cycle: replication, morphogenesis and lysis

Classically, the ensuing infection cycle is divided into an 'early' programme, mostly concerned with establishing DNA replication, and a 'late' programme, where the virion components are synthesized and assembled into head, tail and tail fibres and lysis proteins are deployed into the envelope (Fig. 17.4). DNA replication reaches its maximum rate before late genes are turned on. Rough numerology is instructive for considering the tasks faced by a phage. Typically bacteria have ~4–6 Mbp chromosomes that must be replicated once per generation, and the DNA replication cycle takes ~ 1 hour, so the dNTP substrates have to be supplied at ~10^5 molecules/min. Typically, virulent *Caudovirales* have 'burst sizes' (average yield of viable virions per infected cell) of ~200 in laboratory medium and package ~50% of the phage DNA produced in a cycle. A typical virulent phage infection cycle takes ~30 minutes and most of the DNA is made in the first part of the cycle. The larger phages, with genomes exceeding ~100 kbp, have to make >4 Mb of DNA in ~15 minutes, several times faster than the host normally requires. So it is not surprising that the classical virulent coliphage T4 (170 kb), for example, expresses a full complement of enzymes needed for making dNTPs, not to mention degrading and recycling the host chromosome and converting much of the host RNA to dNTPs as well (Greenberg *et al.*, 1994). In most cases, the end-products of DNA replication are long concatemers of the phage genome.

Morphogenesis of phage virions consists of up to three independent, parallel pathways, resulting in the DNA-filled, mature head, the tail, and the LTFs. The heads are first assembled into a prohead of icosahedral symmetry, constructed of many copies of the major capsid ('coat') protein, typically 45 kDa; the assembly is based on an internal scaffold rooted on a portal ring that defines one icosahedral vertex. The smallest completely functional *Caudovirales* have >200 coat proteins per capsid, and the largest have several thousand. After the scaffold is removed either by disassembly or proteolysis, the mature prohead is filled with the linear dsDNA by the terminase enzyme (Ter), which both cuts the long, concatemeric phage DNA intermediate into packageable unit lengths and also pumps the DNA into the head, using ATP hydrolysis for energy.

An important detail is how Ter determines the packageable length (Fig. 17.5). For some phages, Ter cuts at a specific site called *cos*, so every phage

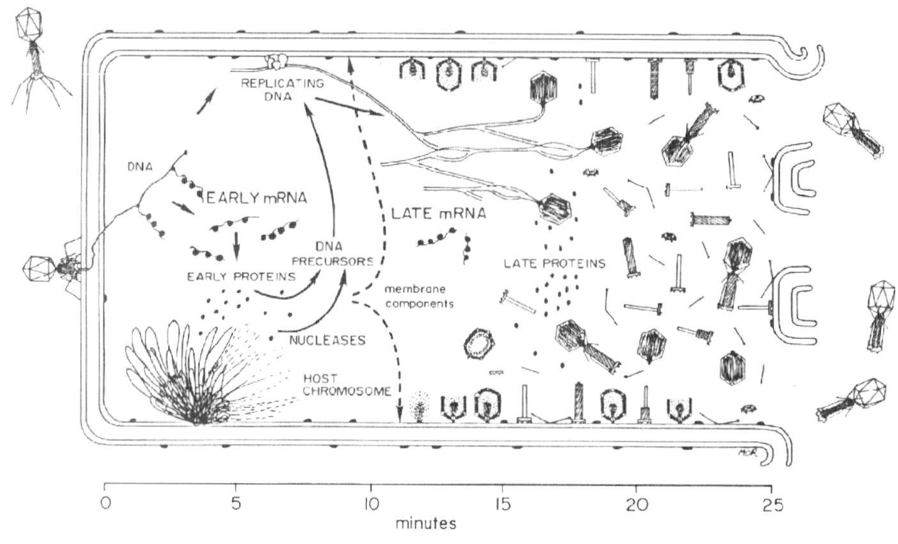

Figure 17.4 The life cycle of T4. This cartoon model by F.A. Eiserling summarizes the important features of the infection cycle of the virulent myophage T4. (See Fig. 17.3 for adsorption and DNA ejection steps.) Following DNA ejection, early gene expression leads to destruction of the host chromosome and synthesis of the T4 DNA biosynthesis and recombination apparatus, generating highly branched, replicating T4 DNA molecules. Late gene expression leads to assembly of procapsids, tails and LTFs, in membrane-attached intermediates. The T4 DNA is debranched and packaged in headful amounts, resulting in filled heads, which then join the tails. The LTFs are assembled originally in an undeployed conformation, with the 'knee' attached to whiskers emanating from the head–tail junction. Lysis occurs at about 25 minutes. Adapted from Mathews (1994), with permission.

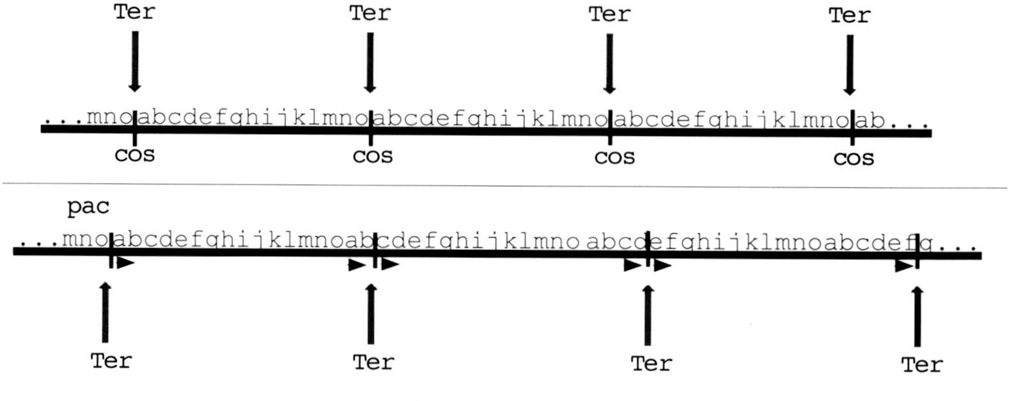

Figure 17.5 *pac* versus *cos* packaging. Most phages generate concatemeric DNA during replication, usually by rolling circle replication and/or promiscuous recombination. In *cos* phages, the terminase enzyme (Ter) recognizes and cleaves a specific site, *cos*, and pumps the DNA into the prohead from that cleavage. The sites are called 'cos', short for cohesive ends, because Ter makes staggered cuts, with the result that the two ends of the phage DNA are complementary ('sticky ends'). In *cos* phages (like lambda), every particle has the same DNA sequence. In *pac* phages, usually Ter still recognizes a specific sequence, *pac*. However, the initiating cleavage event is usually not precisely determined, although it is near *pac*. Moreover, packaging continues until the head is full of DNA. The capacity of the head is always greater than one unit length genome, so each virion has a terminal redundancy (arrowheads) and a different DNA sequence. In general, *pac* phages have a tendency to package host DNA at some frequency, and so are capable of generalized transduction. This does not happen in *cos* phages because of the requirement for the site-specific cleavage event for each virion (Casjens and Gilcrease, 2009).

head gets filled with the same sequence of phage DNA, between cos sites on the concatemeric intermediate. In other phages, Ter is 'sloppier'. In general, a site called *pac* is recognized, near which an initial cut is made and the packaging process initiates, but the subsequent cut occurs when the head is physically full of DNA. At this point, another prohead is filled with another headful, and so on. In these *pac* or 'headful-packaging' phages, the amount of DNA packaged in the head is always more than the unit genome length. For example, in T4, the head capacity is ~2% longer than the unit genome, so each particle (a) has a unique sequence and (b) has ~3 kb of terminal redundancy. Importantly, phages that use the *pac* strategy are usually also capable of generalized transduction, which means that the Ter enzyme is so sloppy that some heads, typically 0.1% to as high as 5%, get filled with randomly chosen host DNA. In nature, generalized transduction by phage is recognized as an important mode for lateral transmission of DNA between bacteria (Breitbart *et al.*, 2005). Obviously, phages that support generalized transduction are to be avoided in phage therapeutics, although the transmission of virulence factors by this mode is basically passive and inadvertent. Recently, a form of specialized transduction, which means the directed transduction of a specific DNA sequence, has been discovered as a major route of the spread of pathogenicity in *S. aureus* (Novick and Subedi, 2007). SaPIs (for S. aureus pathogenicity islands), are small prophage-like genetic elements encoding virulence factors and proteins that, upon infection of the host cell by certain temperate phages, actively subvert the morphogenesis and DNA packaging pathway, causing the formation of virions with small heads containing only the SaPI DNA. Thus the altered viral progeny are basically unit-efficiency vectors for the spread of the SaPI virulence factor.

Tail assembly has common themes for siphophages and myophages. For siphophages, construction starts at the tail tip and, typically, >200 copies of the major tail protein are assembled into a tube, as stacks of trimeric or hexameric subunits. The number of discs, and thus the length of the tail, is determined by the length of the tape-measure protein, which is protein H in phage lambda (Katsura and Hendrix, 1984). H, present in six copies, spans the entire length of the flexible tail in alpha-helical conformation and, after adsorption to LamB, is thought to be involved in lambda DNA penetration into the cytoplasm. Myophage tails are similarly constituted, except that an identical number of tail sheath proteins arranged around the central tube are also required (Coombs *et al.*, 1994). The point of this numerology is that the cell is essentially turned into a 'natural' overexpression system for the major virion proteins. For example, a T4-infected cell has to make ~300 heads, each with ~1000 copies of a 48-kDa major capsid protein, in about 20 minutes. Because virulent phages usually shut down host gene expression, typically the bulk of the protein synthesis capacity, and the machinery for protein folding, of the infected cell is devoted to the major proteins of the capsid and tail. Other features of the virion, including the LTFs, tail spikes, portals, etc., are made in much lower amounts, reflecting the lower numbers of these molecules per virion. Oddly, these pathways are independent; i.e. there does not appear to be any feedback regulation to ensure that the right numbers of each protein are made or that heads, tails and LTFs are made in appropriate amounts. One can knock out production of the major capsid protein by mutation, for example, and although no heads are made, the normal amount of LTFs and tails accumulate. For completion of morphogenesis in myophages and siphophages, heads are joined to tails and LTFs are attached to the sockets on the tails. In podophages, the tail-equivalents like the ring of tail spikes in P22 are assembled directly onto the filled heads. In phages which express them, the final step is probably the assembly of polysaccharide depolymerase molecules onto the virion, although this has not yet been established rigorously.

The final step in the infection cycle is host lysis, which, counter-intuitively, occurs via a completely independent pathway (Young and Wang, 2006). That is, the lysis proteins are made, deployed to various sites in the host cell envelope, and function to achieve a temporally programmed, efficient destruction of the cell, irrespective of whether any virus particles are actually synthesized, with no feedback control at all. Lysis is initiated by the holin, a small membrane protein which suddenly permeabilizes the cytoplasmic membrane at its predetermined time. Besides bringing metabolism and macromolecular synthesis to a sudden halt, (i.e. causing physiological death), the holin function allows the phage endolysin to degrade the peptidoglycan. Finally, in phages of Gram-negative hosts, a third type of protein, the spanins, disrupt the outer membrane (Summer *et al.*, 2007; Berry *et al.*, 2008). In mycobacteria and other Gram-positive hosts that have a waxy outer membrane-equivalent, phages also produce a lysis protein that hydrolyses the surface mycolic acid esters (Gil *et al.*; Gil *et al.*, 2008; Payne *et al.*, 2009).

Specialists versus generalists

The preceding overview is meant to highlight common themes in the incredibly diverse pathways used by phages in their respective infection cycles. It does not remotely do justice to the detailed molecular information that is known for the best-studied phage systems. One issue of particular interest, from the point of view of phage therapy, is the relative dependence on specific host factors for any part of the infection cycle. That is, given a solution to the receptor-RBD problem, can 'generalist' phages be found or developed that can grow effectively in a wide range of hosts? Because nearly all the molecular-level studies on phage were done after the demise of phage therapy and were focused on a few model phages, little has been done relevant to this issue. Nevertheless, the classical studies certainly reveal a wide diversity among phages in this respect. For example, lambda redirects the entire host DNA replisome to its own chromosomal origin, relies on phage-encoded positive control factors to redirect the host RNA polymerase to the phage promoters, requires the host chaperonin for folding of the portal protein and effects regulation of transcriptional termination with complexes involving multiple bacterial proteins (Hendrix and Casjens, 2006). Thus lambda, which interacts with host proteins at many steps in the lytic pathway, is restricted to a relatively narrow host range. In contrast T7, even though it has a smaller genome, deploys its own RNA polymerase for middle and late gene expression and its DNA replisome is entirely phage encoded, except for the ingenious exploitation of a single small and highly conserved host protein (thioredoxin) as a processivity clamp for replication (Molineux, 2006b). Indeed 'T7-like' virulent podophage appear to be one of the most successful phage types in every environmental niche and, because of their simplicity, may be good candidates for intensive development of 'superphages' for therapeutic use (see below).

Obstacles to phage predation

From the point of view of phage therapy, it is most important to know what can go wrong with the phage infection cycle, which provides, in principle, the unique advantage of a potentially exponential agent against the target bacteria. The mechanism and consequences of the co-evolutionary 'arms-race' between phage and bacteria have been extensively examined (Lenski, 1984; Jessup and Forde 2008; Vos *et al.*, 2009). Indeed, recent experiments have indicated that bacterial mutation frequencies increase dramatically when forced to co-exist with phage in continuous culture (Pal *et al.*, 2007). So from first principles, one would expect that bacteria are constantly evolving ways to protect themselves against phages and phages are constantly evolving to overcome these defences. Here, we will consider some of the more important ways that this might affect the utility of a phage chosen for a therapeutic application.

Resistance

As noted above, true *resistance* occurs when the host receptor is altered or eliminated by mutation. In this case, the simplest response is to select phage *h* mutants by plating on the resistant host. Of course, having a 'cocktail' with multiple phages would be ideal. Assuming the phages have independent receptors, it would then probably require multiple host mutations to achieve resistance to all of the phages in the cocktail. Operationally, this could involve the 'low tech' expedient of selecting resistant mutants using each phage and then checking which of your other phages still plate on the selected resistant strains. Ultimately, it seems probable that regulatory agencies may insist on knowing what the receptors are for any phage proposed for therapeutic use. Modern high-throughput genomics, proteomics and glycomics may soon make the precise identification of the phage receptor easily accessible. It should be noted that most phage 'cocktails' that have been put together by experimentalists for combating bacteria are not defined in this way. Instead, these cocktails are often a collection of phages that, together, attack a high percentage of the clinical strains of the target bacteria, as judged by spot-plating tests using lawns of each host. Thus only a single host mutation could at least temporarily defeat the entire cocktail. Of course, this could also happen if the host mutations affect multiple features of the envelope, perhaps due to changes in important global regulators or severe disruption of envelope integrity. However, it is likely that such profound changes would significantly reduce the fitness of the pathogenic bacterium. Moreover, the selection for mutations conferring phage resistance can be turned into a therapeutic strategy by deliberately selecting phages that target bacterial surface features required for virulence. One way to do this is to assemble a panel of isogenic pathogen strains with knockouts of various genes encoding surface virulence features, and conduct the phage hunt looking for phages that grow on the parental strain but not on at least one of the knockouts.

Abortive infection morons

It turns out that there is another, even more complex war going on between phages. Temperate phages often carry morons that specify 'abortive infection' (AI) suicide functions that are activated if certain other phages infect the lysogenic cell (Molineux, 1991; Hendrix *et al.*, 2003; Chopin *et al.*, 2005; Hyman and Abedon, 2010). Basically when the cell is infected by the 'enemy' phage, the moron in the prophage causes the cell to die and thus squelch the infection cycle. The bad news here is that a phage chosen for a therapeutic application might suddenly stop growing on the target population because a temperate phage carrying an AI module has spread throughout that population. Our knowledge of what is being detected by these AI systems is sketchy. In general, a virulent phage that runs into this problem cannot get around it very easily in the short term; it appears to require the acquisition of more genes to block the AI suicide function, as opposed to what would seem to be the simpler strategy of eliminating the molecular signal involved. The most famous example is in phage T4. There is an AI moron called *rexAB* in the lambda prophage which interferes with the infection cycle of many, very different bacteriophages, including the classic phages T1, T4, T5, T7, and P22, a temperate lambdoid podophage of *Salmonella* (Molineux, 1991). However, at some point during evolution, T4 acquired the *rIIAB* genes, which completely block the lethal AI event imposed by the RexAB proteins. Although the *rex-rII* system was described genetically before the last mid-century, we still do not know what Rex detects that is in common to the disparate phages sensitive to it, how Rex kills the infected cell or how the RII proteins block Rex killing. Moreover, genetic studies indicate that an *rII*-independent solution to the Rex block is not readily accessible to T4 by mutation. It would not be possible in the laboratory to replicate this evolutionary development, so if your favourite phage ran into an AI block, even the remarkable genetic flexibility of phages might be stymied. Fortunately, the AI morons appear to have been around for a long time, because many phages found in the environment have already acquired AI-counteracting genes, as in the case of T4.

Restriction systems

Many biological scientists are unaware that the restriction enzymes they use every day were evolved primarily for defence against bacteriophage. In fact, 'restriction' means restricting the growth of phage on a host carrying the restriction system (Bertani and Weigle 1953; Arber and Dussoix, 1962). Most bacteria isolated in the wild have at least one active restriction system. In a nutshell, restriction systems operate by being able to distinguish phage DNA, recently injected into the cytoplasm, from the host DNA. In general, this happens because the restriction system chemically modifies the host DNA as it is being replicated, usually by methylation of a base within the restriction site sequence, wherever it occurs in the chromosome (Tock and Dryden, 2005). Usually, restriction sequences are palindromes of four to six bases, so a typical phage of 30–100 kb would be expected to have 5–50 sites of any particular sequence. This means that the phage DNA is cleaved into many pieces when it is injected into the host cytoplasm, effectively halting the phage infection. Restriction differs from AI because the host cell actually survives. Unfortunately for bacteria, most restriction systems represent a classic 'eggshell' defence. If a phage somehow gets past the initial scan by the restriction enzyme system and gets its DNA replicated (and modified) just once inside the cell, it will now be essentially invisible to the defence system. All the progeny will be modified and the entire population of bacteria is thus defenceless. The frequency that a phage can escape a restriction system in this way depends on many things but typically is 10^{-3}–10^{-8} per infection. From a phage therapy point of view, this is easy; you just plate out your favourite phage on the restricting host and isolate the rare plaques that emerge, exactly as was done by W. Arber when he first studied phage restriction systems (Arber and Dussoix, 1962). The 'escapees', of course, will have lost whatever modification had occurred in the previous host, which may have had its own restriction-modification system. Nevertheless, overcoming restriction systems in the target bacteria is conceptually straight-forward. Of course, if a host has more than one restriction system, the problem of getting escapees gets multiplicatively worse.

Phages have also developed other strategies to avoid restriction. The simplest, if slowest, way is to eliminate the restriction sequence from the genome by point mutations, on the basis of the fewer such sites, the more likely to become an escapee. The most amazing example of this is phage K, one of the few known virulent phages of *S. aureus*; in its 127 kb genome, there should be hundreds of instances of the sequence GATC, the target sequence of a common *S. aureus* restriction enzyme, yet there are none, suggesting how powerful the evolutionary pressure must be (O'Flaherty *et al.*, 2004). Many phages, especially the large ones, modify their own DNA in various ways, to confound restriction systems. The most successful of these strategies documented to date is the replacement of all the cytosines in T-even phage DNA with glycosylated hydroxymethylcytosine, rendering such DNA insensitive to enzymes with C-containing restriction sites. During evolution, bacteria have countered with restriction systems that recognize and cleave modified C or A bases (Carlson *et al.*, 1994). In turn, phages have acquired genes for proteins that get injected into the cytoplasm at the same time as the phage DNA and inhibit the host restriction enzymes directly or by reducing the levels of restriction enzyme cofactors (Bair and Black, 2007; Bair *et al.*, 2007). The most stunning example of this is Ocr protein of phage T7, which actually folds to mimic double-stranded DNA and blocks the active site of the restriction enzyme (Walkinshaw *et al.*, 2002). Some phages take a different approach by initially injecting only a small part of their DNA, as noted above for T5 and T7, thereby reducing the likelihood that a restriction site would be present. Typically, proteins that inhibit restriction enzymes are made from this 'toe in the water' fragment of the DNA, allowing the rest of the chromosome to be internalized in safety. Again, from the point of view of phage therapy, the simplest approach is to isolate phages from the wild that already have these adaptations.

CRISPR systems

In recent years, researchers in dairy microbiology documented a novel way for bacteria to prevent phage predation, mediated by the CRISPR loci (clustered regularly interspaced short palindromic repeats; Barrangou *et al.*, 2007; Horvath *et al.*, 2008; Sorek *et al.*, 2008). These appear to be 'acquired immunity' systems in which short DNA sequences homologous to previously encountered foreign DNA, usually from phages or plasmids, are stored as 'spacers' in a repetitive array. If a phage infects a host with a CRISPR spacer that matches its DNA exactly, somehow the phage infection cycle is blocked, possibly by cleavage of the phage DNA at the matching site. Nothing is known about how new CRISPR spacers are acquired, except that, at least in Streptococci, persistent exposure to a virulent phage seems to select for generation of a new spacer specific to that phage at the promoter proximal position in the CRISPR array. In any case, from a phage therapy point of view, CRISPR interference is easy to overcome, since even a single mismatch between the phage DNA and the CRISPR spacer is sufficient to confer sensitivity.

Physiological insensitivity, nooks and crannies

Aside from the above described systems, many of which probably evolved specifically to protect their hosts from phages, there are a number of less well-defined, transient states which may affect the ability of phages to attack their targets. A number of pathogenic bacteria are known to propagate within the eukaryotic cell during infection, including *S. aureus*, *L. monocytogenes*, *S. enterica* Typhimurium, and, perhaps most famously, *M. tuberculosis*. These intracellular states may be transient or permanent, but as long as a bacterium resides inside a host cell it is expected to be well-protected from phage, as phages are not able to diffuse through biological membranes. The physiological state of the host cell has been shown to influence phage infections and, as is the case with many other aspects of 'non-standard' phage biology, research in this area is patchy but illustrative cases are available. For example, *E. coli* phage T4 is unable to complete its life cycle in stationary phase host cells (Kutter *et al.*, 1994), and *E. coli* grown in nutrient-poor conditions adsorbs phage T4 more slowly and produces a smaller burst size than cells grown in rich medium (Hadas *et al.*, 1997). Conversely, *E. coli* phage T7 is able to infect and produce progeny phage within stationary phase host cells (Schrader *et al.*, 1997), indicating that this phenomenon is at least partially phage specific. Bacterial biofilms have been shown to present a barrier to phage attack (Doolittle *et al.*, 1996), a finding which is especially relevant when considering that many types of wound-colonizing bacteria reside in biofilms (Costerton *et al.*, 1999; Wolcott and Rhoads, 2008). A study of the *in vivo* effect of phage on non-pathogenic *E. coli* in the mouse gut showed that, when established as part of the normal gut flora, the bacterium was largely resistant to phage attack, even though it was sensitive to the phage *in vitro* (Chibani-Chennoufi *et al.*, 2004). It was presumed that the bacterium avoided attack by occupying protected niches within the gut or by being physiologically insensitive to the phage. Some bacteria may be phenotypically altered during infections in ways that cannot be predicted from their behaviour in standard lab culture. The production of capsule by *S. aureus*, for example, is variable and affected by its environment (Dassy and Fournier, 1996; Herbert *et al.*, 2001), and capsule has been shown to inhibit phage adsorption in some systems (Scholl *et al.*, 2005). *S. aureus* also appears to accumulate host proteins on its surface which hinder the ability of phage to adsorb when grown in *ex vivo* medium (O'Flaherty *et al.*, 2005; Gill *et al.*, 2006b).

These few examples can give some idea of the types of barriers therapeutic phages must overcome *in vivo*, and these types of interactions are an area which has received little focused research attention to date. Considering that the CRISPR system was completely unknown until less than five years ago, there are undoubtedly novel mechanisms by which bacteria may resist phage attack that remain to be discovered. Obviously, many phages are perfectly able to overcome these barriers, as demonstrated by the many successful experimental phage therapy trials described below. Better understanding of the interaction between phages and their hosts *in vivo*, however, will always be useful for predicting situations where phages may be less effective, for interpreting cases which may be refractory to phage treatment, and for designing remedies that circumvent these obstacles. In the case of intracellular bacteria, for example, the avirulent *Mycobacterium smegmatis* can be used as a 'Trojan horse' to deliver phage into cells harbouring pathogenic *Mycobacterium* (Broxmeyer *et al.*, 2002; Danelishvili *et al.*, 2006).

The development of therapeutic phage

The most important component of phage therapy is obviously the phages themselves. Below we summarize how phages are isolated, characterized in terms of therapeutic potential, and produced for laboratory and clinical applications.

Isolating phage from the environment: phage hunts

The usual goal of phage hunts is to find virulent phages that are able to grow on the widest possible collection of strains of the targeted bacterial genus. In general, these phages are isolated from environmental samples that contain the target bacterium. In untreated sewage, for example, samples plated directly on lawns of laboratory

E. coli invariably give plaques. Usually, however, the environmental abundance of phages against any particular pathogenic bacterium is low, so the direct plating method can require collection of large sample volumes followed by laborious concentration and filtration steps. Consequently, phages are more commonly selected by a technique called enrichment. In this method, an early logarithmic culture of the target bacterium is mixed with a filter-sterilized environmental sample and then allowed to grow for some period of time, usually overnight, allowing any phage present to amplify to high levels. At the end of the enrichment period, the bacteria are removed from the culture either by filtration or treatment with chloroform, which kills bacteria but not, in general, *Caudovirales*. After being cleared of cells and debris, the lysate is plated on a lawn of the target. In reconstruction experiments, one can mix a 10-ml sample containing a single particle of phage T7 with 100 ml of early log phase *E. coli* and obtain, after overnight growth, a visibly viscous lysate with titres exceeding 10^9 pfu/ml.

The enrichment method, while generally successful, has its limitations. First, it selects for certain types of phages; typically only a single phage will overrun the enrichment culture, probably one that grows robustly on high-density cultures and has short infection cycles. This may not be optimal, since overnight broth culture is not necessarily a good model for the conditions of therapeutic usage. In addition, it obviously selects for specialist phages that do exceptionally well on the single strain of target host used for the enrichment. There is a report, cited in several reviews, that claims virulent phages able to bridge large evolutionary divides were isolated using the enrichment technique with pairwise combinations of different bacteria (Jensen *et al.*, 1998). This report has profound implications for phage therapy and is thus worth considering carefully. Two of the strains were well-characterized laboratory strains of *E. coli* and *Pseudomonas aeruginosa*, while the third was the sheathed beta-proteobacterium *Sphaerotilus natans*. However, recent attempts to reproduce the results with three of these phages indicated instead that the phages were absolutely limited to their propagation hosts (C. F. Gonzalez, unpublished data). Moreover, the titres reported for some of the phages are significantly beyond expectations for *Caudovirales* and suggest methodological defects were involved. Nevertheless, it is clear that phages run the gamut between extreme specificity and extreme promiscuity in terms of host range. For example, the coliphage P1, a large temperate myophage that has been the standard tool for genetic transduction for *E. coli*, can adsorb to and transfer its DNA into the cytoplasm of a wide range of Gram-negative hosts, including *Proteus*, *Pseudomonas*, *Flavobacterium*, *Acetobacter*, and even *Myxococcus* spp. Its ability to propagate efficiently is restricted to the enterics, but low-level phage production can be achieved without overt engineering in *Proteus*, *Pseudomonas* and *Flavobacterium*. P1 was isolated by chance as one of three prophages, including another well-studied phage, P2, in a clinical *E. coli* isolate (Bertani, 1951). Clearly, the properties of P1 show that generalist phages exist and that there is scope for engineering phages to very broad host ranges.

A brief note on phage quantification and purification is in order here. As mentioned above, the standard method for titering phage preparations since the time of d'Herelle has been the plaque assay. It is important to remember that the choice of indicator bacteria for the lawn is critical. That is, the absence of plaques on a lawn does not mean there are no viable phages in the sample; as we have discussed, a bacterial cell can be insensitive to a particular phage for many different reasons. Moreover, if plaques are generated, the resultant pfu titre, usually expressed as pfu/ml, does not necessarily reflect the actual number of virions. Every phage has a distinct efficiency of plating (EOP) on a particular indicator strain, where EOP is rigorously defined as plaques per virion. To obtain a true EOP, the investigator must have an independent measure of the concentration of virions. Often this is derived from biochemical measurements. *Caudovirales* have a unique density of ~1.5 g/cm^3, because of their molecular composition of ~50% protein/50% DNA. Phage biologists traditionally exploit this property to purify phage particle away from the myriad contaminants in a phage lysate using isopycnic density gradient centrifugation, usually with the salt CsCl. With such purified stocks, one can accurately determine the virion concentration by spectrophotometric analysis, using the relation that A_{260} = 50 μg/ml of native dsDNA. A useful rule of thumb is that 10^{10} virions/ml equals 1 μg/ml dsDNA if the phage has a ~100 kbp chromosome. Real EOPs approaching 1.0 can be obtained for some of the classic coliphages on standard *E. coli* indicators, especially the robust T-even phages. However, this rigorous methodology is problematic for a number of reasons, including the fact that the high salt environment of the isopycnic gradient can damage phage particles. It is far more common for investigators to determine relative EOPs, comparing the ability of a particular phage preparation to make plaques on different lawns compared with a standard laboratory indicator or, in phage hunts, compared with the lawn on which the phage was originally isolated. The EOPs so derived are important characteristics when one is trying to assemble a panel of phages to cover a large variety of pathogenic bacterial strains.

From plaque picking to phage preparation

Many companies, organizations and research laboratories are currently hunting for candidate therapeutic phages. Invariably, the practice is to pick the largest, clearest plaques obtained after an enrichment procedure. The underlying assumptions are that the size of the plaque bespeaks desirable traits like fast adsorption and rapid replication and the lack of turbidity within the plaque indicates a virulent lifestyle. Neither assumption is universally valid. Plaques are, after all, artefacts, and the rate of growth and the morphology are derived from a combination of complex parameters, including growth phase of the host, diffusion of the virions in the agar and inactivation of the virions by cellular debris (Abedon and Yin, 2009). The standard laboratory version of lambda, better known as λ_{PaPa}, makes large plaques that are distinctly turbid, as befits a temperate phage. This property made lambda ideal for studying the molecular basis of lysogeny, because mutants defective in the establishment or maintenance of the lysogenic state could be easily discerned as clear plaques. However, in this case the large plaque phenotype is due to a mutation that destroys the lambda LTF (see Fig. 17.1). The original lambda, now known as λ_{Ur}, with the four functional LTFs, make much smaller plaques, although it adsorbs to the *E. coli* cell ~10-fold faster (Hendrix and Duda, 1992). Another example are the T-even phages, like T4, which make small, indistinct fuzzy plaques, even though they adsorb much faster than even λ_{Ur}, and have a completely virulent infection cycle which produces more than twice as many, much larger virions in one-half the time of the lambda infection cycle. Moreover, T4 has a much broader host range than lambda and, as is noted above, is largely insensitive to restriction systems because of its highly modified DNA. The fact that it makes small plaques is largely due to a phenomenon called lysis inhibition, where infected cells at the edges of the growing plaque postpone lysis indefinitely and instead fill up with large numbers of virions (Abedon, 1994). So, the little fuzzy plaques of T4 have huge numbers of phage particles in them, far more than the much larger lambda plaque.

Not only can plaque size and morphology be misleading, but it is also true that many of the phages that make the largest and most robust plaques are not useful for other reasons. For example, the ssRNA and ssDNA phages make very large plaques, mostly because they are small and thus diffuse rapidly, but they are not good candidates for therapeutics because they either need an easily lost host organelle, like a pilus, for adsorption or are simply highly inefficient at infection. The moral is that phage hunts, seemingly simple, are tricky propositions, not well suited to generalized optimization, but instead will have to be set up and optimized for each particular target bacterium.

In any case, typically, a phage hunt is aimed at a particular bacterial pathogen, and the hunters will assemble a panel of clinical isolates. Typically the pathogen of choice is isolated from a clinical sample, which is then also processed, after some sort of sterilization, for phages by enrichment with the new bacterial isolate, making a 'pickate' by punching out a plaque from a lawn and transferring to a small volume of buffer, and then making minilysates by inoculating the pickate into a small culture of the enrichment strain. Next, the phages are checked for the ability to attack other members of a panel of clinical isolates, usually by spot tests with a few microliters of each new lysate. In general, phages are chosen that have different host range patterns across the panel, based on the simple notion that different host ranges implies different receptors, which, in view of some of the considerations noted above, is not necessarily true. It is rare that any effort is made to identify the host receptor, despite the obvious importance to the potential usefulness of the phage.

Most of the studies discussed in the next section were done with phage collections that were largely uncharacterized beyond this level. As standards evolve and technology progresses, it seems probable that it will become more common to determine a few basic distinguishing features for candidate phages, including determining crude virion structure, using simple negative-stain electron microscopy, as well as preliminary characterization of the DNA. Good commercial DNA mini-preparation kits can be used on a few ml of the crude phage lysate to generate DNA samples that can be characterized by a few standard restriction enzymes and sized by pulsed-field gel electrophoresis. As discussed below, large-scale production of phages, and the methods of purification appropriate for use as therapeutics, is an area that is ripe for invention and intellectual property.

Modern phage therapy

Following the abandonment of phage therapy in the 1940s, research in this area remained essentially frozen in the West for many years. By 1963, the general scientific consensus within the microbiology research community was that, although 'the medical use of bacteriophages still persists in some out-of-the-way places… The strange bacteriophage… therapy chapter of the history of medicine may now be fairly considered as closed' (Stent, 1963). This is hardly surprising considering the state of the science during this period, and the availability and efficacy of chemical antibiotics. Phage therapy began the slow process of its 'modern' resurgence some 28 years ago, with the publication of a series of well-designed studies examining the use of phage for the treatment of *E. coli* diarrhoea in calves (Smith and Huggins, 1982, 1983). These studies were followed by work published by Soothill studying the ability of phage to treat *P. aeruginosa*, *A. baumanii* and *S. aureus* infections in mice (Soothill, 1992) and *P. aeruginosa* infections in guinea pigs (Soothill, 1994). In 1996, Merril *et al.* found that it was a relatively simple procedure to mutationally alter the phage capsid to allow it to circulate for extended periods in the bloodstream of mice, and thereby enhance its therapeutic efficacy. The field again lay fallow until, shortly after the turn of this century, research interest in evaluating phage therapy in various animal models expanded greatly, with over 30 primary research articles detailing such experimental studies published between 2002 and May of 2010. We will not comprehensively review this work here; numerous excellent reviews currently exist in the literature (Matsuzaki *et al.*, 2005; Clark and March, 2006; Kropinski, 2006). In fact, the field of phage therapy is currently one of the most – if not the most – extensively reviewed research topics in biology; as of this writing, there are almost as many review articles published on phage therapy as there are experimental studies. Instead, we have compiled the results of some 38 phage therapy efficacy studies conducted in animal models in order to evaluate the current state of the art, which are shown in Table 17.1. Overall, the efficacy of phage therapy across a variety of experimental systems has been demonstrated, with 35/38 studies reporting some beneficial effect of phage treatment, supporting the oft-stated promise of phages as potential antimicrobials. Further analysis of this data can be used to indicate future directions for research and development of phage therapy.

Treatment of acute and chronic infections

Almost all experimental phage studies to date have used a challenge model of infection, wherein a known amount of a given pathogen strain is administered to a healthy animal, followed by administration of phage. A common type of phage therapy efficacy study is developed as follows: novel phages which lyse the desired pathogen strain are procured from the environment and briefly characterized. One or more of these phages are selected, propagated and purified. Healthy model animals, typically mice, are challenged with a large pathogen load which is selected to cause 100% mortality within several days of challenge. This bacterial challenge is then followed by a dose

Table 17.1 Tabular summary of modern phage therapy experimental studies published in the peer-reviewed literature from 1982 (the studies conducted by Smith *et al.* in cattle) to May 2010)

Study author	Pathogen	System	Disease model	Type	Number of phages used (phage names)	Evaluation criteria
Smith and Huggins (1982)	*E. coli*	Mouse	Bacteraemia	C	9 (B, C, D, F G, K, L, P, R), mixed and separately	Survival
Smith and Huggins (1983)	*E. coli*	Calf, lamb, piglet	Diarrhoea	C	6 (B44/1, B44/2, B44/3, S13, P433/1, P433/2), in pairs or separately	Symptoms, survival
Soothill (1992)	*P. aeruginosa*, *A. baumanii*, *S. aureus*	Mouse	Peritonitis	C	3 (BS24, BS46, phi-131), used separately	Survival
Soothill (1994)	*P. aeruginosa*	Guinea pigs	Skin graft	C	1 (BS24)	Graft healing
Merril *et al.* (1996)	*E. coli*	Mouse	Bacteraemia	C	3 (λW60, Argo1, Argo2), used separately	Symptoms, survival
Biswas *et al.* (2002)	*E. faecium*	Mouse	Bacteraemia	C	1 (ENB6)	Survival
Cerveny *et al.* (2002)	*V. vulnificus*	Mouse	Systemic infection	C	3 (CK-2, 153A-5, 153A-7), used separately	Lesion severity
Bach *et al.* (2003)	*E. coli* O157:H7	Sheep	Carriage	C	1 (DC22)	Bacterial enumeration
Matsuzaki *et al.* (2003)	*S. aureus*	Mouse	Peritonitis	C	1 (MR11)	Survival
Park and Nakai (2003)	*P. plecoglossicida*	Fish	Haemorrhagic ascites	C, N	2 (PPpW-3, PPpW-4), mixed and separately	Survival
Loc Carrillo *et al.* (2005)	*C. jejuni*	Poultry	Carriage	C	2 (CP34, CP8), used separately	Bacterial enumeration
Tanji *et al.* (2005)	*E. coli* O157:H7	Mouse	Carriage	C	3 (SP15, SP21, SP22), mixture	Bacterial enumeration

Table 17.1 continued

Study author	Pathogen	System	Disease model	Type	Number of phages used (phage names)	Evaluation criteria
Vinodkumar *et al.* (2005)	*K. pneumoniae*	Mouse	Sepsis	C	1 (CSV-29)	Survival
Danelishvili *et al.* (2006)	*M. avium*	Mouse	Bacteraemia	C	1 (TM4)	Bacterial enumeration
Gill *et al.* (2006a)	*S. aureus*	Cattle	Mastitis	N	1 (K)	Cure, bacterial enumeration
Raya *et al.* (2006)	*E. coli* O157:H7	Sheep	Carriage	C	1 (CEV1)	Bacterial enumeration
Sheng *et al.* (2006)	*E. coli* O157:H7	Sheep, mouse, cattle	Carriage	C	2 (KH1, SH1) mixed and separately	Bacterial enumeration
Wang *et al.* (2006)	*P. aeruginosa*	Mouse	Peritonitis	C	1 (A392)	Survival
Atterbury *et al.* (2007)	*S. enterica* Enteriditis, Typhimurium, Hadar	Poultry	Carriage	C	3 (phi151, phi25, phi10), used separately	Bacterial enumeration
Capparelli *et al.* (2007)	*S. aureus*	Mouse	Bacteraemia, Abscess	C	1 (Msa)	Survival, bacterial enumeration, abscess severity
McVay *et al.* (2007)	*P. aeruginosa*	Mouse	Burn wound	C	3 (Pa1, Pa2, Pa11), mixture	Survival
Watanabe *et al.* (2007)	*P. aeruginosa*	Mouse	Sepsis	C	1 (KPP10)	Survival, bacterial enumeration
Callaway *et al.* (2008)	*E. coli* O157:H7	Sheep	Carriage	C	8-phage cocktail	Bacterial enumeration
Chhibber *et al.* (2008)	*K. pneumoniae*	Mouse	Pneumonia	C	1 (SS)	Bacterial enumeration
Nishikawa *et al.* (2008)	*E. coli*	Mouse	Urinary tract infections	C	2 (T4, KEP10), used separately	Survival
Uchiyama *et al.*, (2008)	*E. faecalis*	Mouse	Sepsis	C	1 (phiEF24c)	Survival
Vinodkumar *et al.* (2008)	*P. aeruginosa*	Mouse	Bacteraemia	C	1 (CSV-31)	Survival

Zimecki *et al.* (2008)	*E. coli*, *S. aureus*	Mouse	Bacteraemia	C	2 (T4, A5/L), used separately	Bacterial enumeration
El-Shibiny *et al.* (2009)	*C. jejuni*, *C. coli*	Poultry	Carriage	C	1 (CP220)	Bacterial enumeration
Heo *et al.* (2009)	*P. aeruginosa*	Mouse	Peritonitis	C	2 (MPK1, MPK6), used separately	Survival, bacterial enumeration
Rozema *et al.* (2009)	*E. coli* O157:H7	Cattle	Carriage	C	4 (rV5, wV7, wV8, Wv11), mixture	Bacterial enumeration
Wright *et al.* (2009)	*P. aeruginosa*	Human	Otitis	N	6 (Biophage-PA product)	Symptoms, bacterial enumeration
Zimecki *et al.* (2009)	*S. aureus*	Mouse	Bacteraemia	C	1 (A5/L)	Bacterial enumeration
Capparelli *et al.* (2010a)	*S. enterica* Paratyphi	Mouse	Bacteraemia	C	1 (phi1)	Survival
Carmody *et al.* (2010)	*B. cenocepacia*	Mouse	Pneumonia	C	1 (BcepIL02)	Bacterial enumeration
Debarbieux *et al.* (2010)	*P. aeruginosa*	Mouse	Pneumonia	C	1 (PAK-P1)	Survival
Miller *et al.* (2010)	*C. perfringens*	Poultry	Enteritis	C	5 (CPAS-7, CPAS-12, CPAS-15, CPAS-16, CPLV-42), mixture	Weight gain, feed conversion, mortality
(Wall *et al.*, 2010)	*Salmonella*	Swine	Carriage	C	14 (PEW1 – PEW14), mixture	Bacterial enumeration

Inclusion criteria for this survey: primary research papers (not reviews) published in English, and must include an *in vivo* experimental component wherein a bacterial infection in an animal is treated with one or more phages. Studies examining the dynamics of the phages themselves in an animal model were not included, nor were individual case reports.

Type, type of infection; C, challenge; N, natural.

of phage or placebo treatment, and mortality among the treatment groups is recorded over the following days. Other markers of disease progress (inflammatory cytokines, bacterial loads present in the spleen, etc.) are also sometimes recorded.

The infection model described above represents a medical worst-case scenario such as acute peritonitis or sepsis, conditions which are often fatal if left untreated. However, this study design serves only as proof-of-concept for phage therapy, rather than as a demonstration of clinical feasibility. The interval from challenge to treatment is tightly controlled in these types of experimental studies, and even modest delays in treatment can result in markedly reduced efficacy (e.g. Smith and Huggins, 1982; Cerveny *et al.*, 2002; Matsuzaki *et al.*, 2003). Often the phage is administered immediately following bacterial challenge. While these studies do indicate that phage may be able to treat acute sepsis or peritonitis in humans, the very severe pathogen challenges used are not typical of normally acquired infections.

Other disease models which are designed to mimic more chronic types of infections have also been evaluated. In these models, a sublethal pathogen dose is administered followed by phage treatment after an interval of 1 day to 1 week, and recovery from the infection is monitored by clinical observation and often by enumeration of bacteria remaining in the affected organs. These models are meant to represent a more natural infection state, with the pathogen allowed some time to become established in the host, and therefore they presumably reflect a slightly more 'realistic' treatment condition. Phage have been shown to reduce such chronic or sub-clinical infections in a number of mouse model systems, including *S. aureus* abscesses (Capparelli *et al.*, 2007), *S. enterica* Paratyphi bacteraemia (Capparelli *et al.*, 2010a) and *B. cenocepacia* pneumonia (Carmody *et al.*, 2010). Phage therapy studies using infections in agricultural animals also tend to use this chronic infection model and are discussed below.

Choice of animal model

The experimental efficacy of phage therapy has most often been evaluated in a mouse model, with agriculturally important animals such as poultry, cattle and sheep being next most common animal systems. An interesting and important distinction to note here is that the experiments conducted in mouse models tend to be designed to examine acute, often lethal bacterial infections and typically evaluate treatment success in terms of animal survival over the course of several days. These models are, for the most part, used as proxies for equivalent bacterial infections in humans. In contrast, phage therapy studies in agricultural animals typically study infections which are of concern in the actual system under evaluation: carriage of *E. coli* O157:H7 by cattle, for example, or colonization of poultry by *Salmonella* or *Campylobacter*. Such infections are often asymptomatic and create problems in terms of animal production rates or pose potential downstream food safety risks for humans, rather than causing acute and rapidly fatal illnesses. In these studies, phage therapy can be evaluated under conditions which closely mimic the 'real world' production setting in which they could be used, against pathogens which commonly infect the study population under natural conditions, and as such these studies are perhaps more reliable indicators of the potential utility of phage therapy in practice.

In poultry studies, efficacy has been observed in treatment of *Salmonella* (Atterbury *et al.*, 2007), *C. jejuni* (Loc Carrillo *et al.*, 2005; El-Shibiny *et al.*, 2009), and *C. perfringens* (Miller *et al.*, 2010). Phage treatment has also been found to reduce colonization by *E. coli* O157:H7 in cattle (Sheng *et al.*, 2006) and sheep (Raya *et al.*, 2006; Callaway *et al.*, 2008), and to be effective against pathogens in fish (Park and Nakai, 2003). Other studies, however, have shown phages to be ineffective in the treatment of animal pathogens, again including *E. coli* O157:H7 in sheep (Bach *et al.*, 2003; Sheng *et al.*, 2006) and cattle (Rozema *et al.*, 2009), and also *S. aureus* in cattle (Gill *et al.*, 2006a). Of all the experimental animal studies surveyed, only two attempted to treat naturally acquired infections. Park and Nakai (2003) reported a 60% reduction in mortality following treatment of fish with a mixture of two phages against *P. plecoglossicida*, while Gill *et al.* (2006a) reported no efficacy in treatment of subclinical mastitis caused by *S. aureus* in lactating dairy cattle.

Choice of phage dose

The amount of phage required to achieve efficacy in experimental infections is difficult to predict and therefore there is no 'standard dose' of phage, for example on a mg/kg or pfu/kg basis. Each phage isolate used experimentally is a unique biological entity, with its own adsorption rate, latent period and burst size, factors which are also undoubtedly affected by the location and physiological state of the pathogen as well. Different phages may also be presumed to possess varying rates of removal from the host circulation, depending on the initial route of administration.

Experimental studies of phage therapy often use high doses of phage, on the order of 10^8–10^{11} pfu per dose. This is especially high when one considers such a dose in a mouse model, in which the entire animal typically weighs only 15–20 g. In many cases such high doses appear to be required for treatment efficacy, as in several studies a clear dose–response effect is observed in which higher phage doses result in greater protection (Soothill, 1992; Merril *et al.*, 1996; Biswas *et al.*, 2002; Matsuzaki *et al.*, 2003; Capparelli *et al.*, 2007; Nishikawa *et al.*, 2008; Debarbieux *et al.*, 2010). These observations are in contrast to other experimental and clinical experience, in which lower phage doses are not only effective but appear to be more effective than higher doses. While these dose effects appear to be somewhat contradictory, the reported experience in human clinical work has used relatively low concentrations of phage (on the order of 10^5–10^7 pfu) with reportedly good efficacy (Kutter *et al.*, 2009; Wright *et al.*, 2009). A term from classical phage biology is useful here: multiplicity of infection (MOI), which is the ratio of phage to bacteria in a system. In animal studies, Callaway *et al.* (2008) reported that a low phage dose (reported as an MOI of 1) produced greater reductions in *E. coli* O157:H7 colonization than 10 or 100-fold greater doses in sheep. Similarly, a 10^7 pfu dose of phage was more effective than a 10^9 pfu dose in reducing colonization by *C. jejuni* in poultry (Loc Carrillo *et al.*, 2005).

The observations may be explicable by theoretical considerations of phage replication, and the distinction between active and passive phage therapy. Passive therapy, also called inundative therapy, supplies an initial phage concentration that is great enough to infect all of the cells in the pathogen population at the onset of treatment, such that phage-mediated killing of the pathogen is accomplished without the need for further rounds of phage replication. Passive therapy can be accomplished by treatment with high initial doses of phage (MOIs of 10 or greater) in a system where most or all of the pathogen population is accessible and susceptible to phage attack. In such circumstances, the vast majority of bacterial cells will initially adsorb at least one phage, and thus be killed as the phage completes its lytic programme. In contrast, in active therapy, a relatively lower dose of phage is applied initially (MOIs of one or less) and successive rounds of phage replication using the pathogen population as hosts is required to achieve local phage populations great enough to infect and lyse the entire population. In this situation, the final bactericidal effect is expected to be delayed compared with inundative control due to the time required for the phage to replicate and locate their new hosts. There is also an important caveat to bear in mind when discussing phage dosing in terms of MOI: the MOI calculated from the ratio of phages and bacterial cells (the input MOI) is not necessarily the same as the actual MOI which will be experienced by the bacterial population. Because time is required for phage to diffuse to and adsorb to their host cells, and, in the context of an animal model, many phage may be sequestered or degraded prior to even reaching the site of infection, the actual ratio of phage to host cells in the pathogen population is expected to be functionally lower than the expectation based on calculation alone (Kasman *et al.*, 2002). Owing to these considerations, it is not straightforward to determine which mode of treatment, active or inundative, is actually in effect *in vivo*. Because the actual MOI may be considerably lower than the input MOI, doses of phage administered at high MOI, which are expected to provide inundative control, may in reality translate into low actual MOI at the site of infection, resulting in an active control model being responsible for the treatment effects seen. Other theoretical concerns surrounding phage therapy dosages and the distinction between inundative and active control have been discussed thoroughly elsewhere (Payne and Jansen, 2001, 2003; Gill 2008; Abedon, 2009a).

In the acute infection treatments described above, very high doses of phage were required for efficacy, suggesting that such models are best treated by an inundative control strategy. Such a scenario seems plausible considering the high bacterial doses used in these experiments, and the importance of rapid bacterial killing; the administration of lower phage doses may shift the treatment strategy from an inundative to an active model, and time required for the phage to replicate to a high enough concentration to kill the pathogen population reduces overall efficacy. In contrast, in a chronic infection model where the death of the animal is not imminent, a certain luxury of time may be afforded for an active therapy strategy to play itself out.

Phage selection

As noted above, the majority of phage therapy studies begin with the *de novo* isolation of phages from the environment which will attack the desired pathogen. In some cases this is a purely practical requirement, as virulent phages are simply not available for many pathogenic bacteria, and there is certainly no standardized phage panel which is generally available in the West for selection of therapeutic phages. Well-studied laboratory phages such as T2, for example, while perfectly able to infect *E. coli* K-12, will not infect pathogenic *E. coli* strains such as O157:H7 (Yoichi *et al.*, 2005). This basic requirement that a therapeutic phage include the bacterial challenge strain within its host range is one reason for researchers to seek out novel phages. Experimental phage therapy studies also have tended to be carried out by research groups as 'one shot' experiments, with the publication documenting the trial itself being the apparent alpha and omega of the entire project. This lack of sustained effort also contributes to the fragmentary nature of the field, with each group apparently starting from scratch in the study of a given pathogen system. It will be important that more research groups emerge as sustained centres of experimental phage research with a focus on phage applications and therapeutics.

A side-effect of the one-shot tendency, as illustrated in Table 17.1, is that virtually no phage therapy study has used the same phage twice, even in cases where the pathogen under study has been treated by phages in previously published work. The case of *P. aeruginosa* is an example of this phenomenon. One of the earliest modern phage therapy studies (Soothill 1992) treated experimental *P. aeruginosa* infections in mice using a phage called BS24, and reported 100% efficacy at a dose of 3×10^8 pfu. In six subsequently published studies on phage therapy against *P. aeruginosa*, however, none used phage BS24 but instead all opted for the isolation and use of novel phages. A seventh study, a clinical trial of phages for treatment of otitis in humans, used a commercial six-phage cocktail of undisclosed composition. It is possible, of course, that BS24 would have been unusable in these subsequent trials owing to considerations of host range or other factors, but it is not clear if the use of this phage was ever attempted.

This repetition of phage isolation can also be seen in studies examining pathogenic *E. coli* infections, particularly *E. coli* O157:H7. Of six published trials, none utilized a common phage; each group used one or more phages which they had isolated themselves for the purposes of their own experiment. This practice is especially puzzling in the case of *E. coli* O157:H7, as isolates of this pathogen are related to one another and there is a high likelihood that the phages used in one trial would efficiently infect the challenge strains used in the others. The major drawback of this fragmented approach to experimental therapeutics is that the use of different phages in every trial makes it difficult or impossible to determine if differences in efficacy between trials are caused by factors such as phage dose, timing, or route of administration, or are due to phage-to-phage variations.

Phage proteins and components as antibacterials

Phage lytic enzymes

Every *Caudovirales* produces at least one bacteriolytic enzyme, the endolysin, as an essential element of the host lysis process. It has long been recognized that phage endolysins are capable of destroying Gram-positive bacteria from the outside (Ralston *et al.*, 1955; Maxted, 1957) and standard assays for phage endolysins typically used whole cells as substrates (Raina, 1981). Recently, the use of the endolysins from phages of Gram-positive hosts as therapeutic agents against numerous pathogenic bacteria, including *B. anthracis*, *Streptococcus* spp., *Listeria*, and *S. aureus*, has been explored in laboratory conditions and in animal models and shown to have great promise, as recently reviewed by Fischetti (2010). These endolysins generally have two domains: a C-terminal domain that has specific binding affinity for the cell wall of the host bacterium, and an N-terminal catalytic domain. Several different types of catalytic domains have been identified, specific for the peptide, amide or glycosidic bonds of the peptidoglycan. In each case, the cell wall-binding domain exhibits extremely high affinity and specificity. This molecular organization provides several important features to these endolysins, in terms of their potential as anti-infectives. First, resistance, in the sense of mutations that can prevent the binding of the enzyme to the cell wall, has not yet been observed for any of the endolysins, indicating that the molecular features recognized by the binding domain are essential to the structure of the cell wall and, typically, common to a whole genus of bacteria. In addition, new chimeras that reassort the catalytic activities with different cell wall binding domains can and have been generated. Quantitative enzymology of endolysins is difficult because of the inherent difficulties of quantifying of the peptidoglycan substrate, but practical readouts in terms of killing rate in liquid suspensions are very impressive; for example, nanogram amounts of a Streptococcal phage endolysin are able to reduce viable cell counts of *S. pyogenes* by six logs within seconds. Endolysin killing can also be combined with antibiotic therapy, to some advantage.

Animal models of both challenge and established infection modes have been conducted, in some cases with attention paid to the immunological and inflammatory responses. Again, results have been generally very encouraging; it may even turn out that modes of treatment can be devised that would minimize or eliminate the production of peptidoglycan degradation products that would lead to enhanced cytokine response. Moreover, in some cases at least endolysins appear to be largely resistant to neutralization by antibodies, possibly because the major epitopes are distant from the active sites.

Besides the obvious advantages in terms of spontaneous resistance and broad intra-generic host range, endolysins offer advantages over whole phages in several other respects, such as a more defined regulatory and intellectual property framework, GMP production practices and a potential for avoiding immunological interference. Consequently, multiple laboratories and some industrial interests are currently pursuing endolysin-based strategies for the treatment of Gram-positive pathogens. Another approach is derived from the bacteriolytic activities associated with the tail structures of many *Caudovirales*, used to perforate the peptidoglycan layer and allow the ejection apparatus access to the cytoplasmic membrane (http://www.gangagen.com/staph-tame.asp). Since these 'tail-associated muralytic enzymes' are designed to persist in the environment (as part of the virion), at least in principle they might be expected to have robust shelf-lives.

Tailocins as potential therapeutics

As mentioned above, myotailocins and siphotailocins are bactericidal assemblages derived from temperate phages. Obviously, tailocins cannot replicate and thus lack the exponential growth feature of whole phage. However, as protein-only assemblages, they do offer the advantage of defined dose and, since they do not disseminate DNA into the environment, are less problematic in terms of potential environmental regulation (Scholl and Martin, 2008; Williams *et al.*, 2008). Moreover, since they are carried as prophages, they can be readily engineered by high-efficiency chromosomal 'recombineering' methodologies, which would allow the tailoring of host range and other properties. By contrast, virulent phages are much more difficult to engineer from a technical standpoint, since they must be propagated as viable phages. The tailocin concept may be particularly attractive in systems like *S. aureus* and *Mycobacterium* spp., where very few virulent phages have been identified. Although phage therapists must avoid the use of temperate phages, there is no reason to avoid the use of tailocins that could be derived from those same temperate phages. Further research is needed to see if all it takes to make a tailocin is to inactivate the head genes of a temperate myophage or siphophage.

Phage lethal agents

Some studies have explored the use of genetically modified phages, which are able to infect their hosts and cause cell death but do not lyse the cell and do not produce any progeny phage in the process. These phage lethal agents are, like lysins and tailocins, easier to manage in terms of dosing because they are incapable of replicating on their own and can only operate under an inundative treatment mode. These agents start out as normal phages, the genomes of which are modified to carry a 'killer' gene such as a restriction enzyme (Hagens *et al.*, 2004) or intracellular toxins such as *gef* or *chpBK* (Westwater *et al.*, 2003). A form of phage lethal agent in development by Phico Therapeutics, dubbed SASPject, uses modified phage to deliver genes encoding small acid-soluble spore (SASP) proteins (Fairhead, 2009); a SASPject product designed to target *S. aureus* is currently in early clinical evaluation. Upon infection of a target cell, the phage genome is ejected into the host cytoplasm and the toxic gene is expressed, resulting in the death of the cell, in essence trading away the ability of the phage to produce progeny in order to gain more control over the agent dose and the bactericidal mechanism. In principle, the 'phage genome' in these agents could comprise nothing but toxin determinants, depending on how aggressively the agent is engineered. One possible advantage of this system is that the use of temperate phage or even filamentous phages becomes possible for this application, since lysogen formation and transduction are no longer issues in the absence of progeny phage, and this factor may be useful against pathogens where useful virulent phages are difficult to isolate. Additionally, the alternative bactericidal mechanism engineered into the agent typically means that the killed cells are not lysed, which may reduce the amount of bacterial endotoxin released during the course of therapy (Hagens *et al.*, 2004), but it is worth noting that such lysis-associated toxicity is not commonly reported in experimental phage therapy studies and is of less importance in the treatment of Gram-positive pathogens. Unlike tailocins, phage lethal agents still contain a viral genome and therefore would be considered as genetically modified organisms.

IP and the regulatory barrier

As illustrated by the number of successful phage therapy trials in animal models, plus the extensive clinical experience of phage therapy in humans reported from regions such as Georgia and Poland, phages certainly represent a promising antimicrobial strategy, particularly in combating antibiotic-resistant bacterial infections for which other treatments are unavailable. Why, then, are phages not more widely available for medical use?

At this time we believe that the greatest barrier to medical use of phage therapy in the West is caused by the interaction of two components of the traditional drug development process: the requirement of conducting large scale (i.e. phase III) clinical trials in human subjects prior to regulatory approval, and the requirements of the companies which typically fund such trials to possess intellectual property (IP) of the therapeutic agent they are testing.

Limited early steps in the regulatory approval process have been undertaken by individual small companies and research groups in the USA, UK and EU. Limited phase I trials, designed to demonstrate the general safety of a proposed treatment, have been conducted for a small number of phage-based therapeutics (Verbeken *et al.*, 2007; Rhoads *et al.*, 2009; Wright *et al.*, 2009). Additionally, a significant amount of work in humans and animals has been conducted in Poland (Weber-Dabrowska *et al.*, 2000, 2002; Miedzybrodzki *et al.*, 2009) and Georgia (Kutateladze and Adamia, 2008) demonstrating phage safety and some aspects of phage pharmacology in humans. At least one phase II trial, testing the efficacy of phage against *P. aeruginosa* otitis in humans, was recently completed and reported significant clinical improvements following a single phage treatment in 12 volunteers (Wright *et al.*, 2009). Undoubtedly other phase II trials, which tend to be limited in scale and can be financially supported by a small company or well-supported research group, will be completed over the next few years. The final regulatory barrier is the completion of a multicentre phase III trial, which typically enrols hundreds to thousands of patients over the course of one to three years and can incur out-of-pocket costs in the range of $80 million or more (Adams and Brantner, 2006). In the case of a traditional, small-molecule drug, the immense costs involved in drug development and approval are recouped during the lifetime of the drug's intellectual property protection, which grants a period of exclusivity to the IP holder. Phages, however, are not traditional small-molecule drugs and the normal paradigm of pharmaceutical IP protection does not strictly apply.

There are many forms of IP protection which may be used to maintain exclusive control over a therapeutic product. The most commonly employed of these in the development of pharmaceuticals is the patent, which allows public disclosure of the agent while still protecting the patent assignee from unauthorized distribution of the product. The state of phage therapy patent practices has been recently reviewed (Gill *et al.*, 2007). In the case of phage therapy, this form of protection is difficult to implement in the traditional manner. Patenting the 'invention' of phage therapy has been attempted by several parties, under so-called 'use patents' as they patent the use of phages in a given application rather than the phage themselves. However, it is not clear how well these patents could be defended given the extensive prior art, dating from the 1920s, around the practice of phage therapy. Individual phages or phage mixtures can – and have been – patented, a protection which prevents would-be competitors from simply isolating the phages from the marketed product, propagating them and re-selling them. However, a patent on a specific phage does not prevent a competitor from going into the environment and isolating a new phage which has the same or similar antibacterial properties of the original phage, thereby 'inventing around' the patent on the original phage. Various accessory practices surrounding phage therapy, including production methods, storage and formulation techniques, and so on, present better targets for direct IP protection, either in the form of patents or possibly as trade secrets, depending on the nature of the technology. These protections do not provide exclusivity for the phage itself, which is the entity that must justify the massive costs currently associated with the regulatory approvals process.

How, then, can this conundrum be resolved? A phage-based therapeutic must be shepherded through the regulatory framework and still provide a return on this investment. There is no clear answer to this question, although we can provide some possible scenarios for its resolution.

Models for industrial adoption of phage therapy and the influence of regulation

There are two broad frameworks which can describe how phage-based therapeutics may be developed in the future: a high-throughput, low-investment strategy which is essentially the model used in current phage therapy practice, and a high-investment, few-phage model which is conceptually closer to the current small-molecule antimicrobial paradigm. In the high-throughput model, phages which infect and kill the desired pathogen strains are isolated from the environment and undergo a relatively brief characterization process prior to deployment either alone or as part of a phage cocktail. This is the system used in clinical practice in Georgia and Poland, and it is also similar in principle to the model used for many of the experimental phage therapy studies described in Table 17.1. It has the advantages which are commonly listed when discussing phage therapy as an alternative or adjunct to small-molecule antibiotics: low up-front development costs and rapid ability to overcome bacterial resistance. However, this model also provides much more limited IP protection for the producer of the phage product, for the reasons described above.

In the high-investment model, extensive pre-clinical investment is directed towards the isolation or engineering of what can essentially be described as 'superphage' which possess distinctly enhanced therapeutic potential, such as a broadened host range within or across pathogen species, the ability to withstand clearance by the immune system of the patient, greater receptor affinity for its target, or ease of production and purification. As noted above, the patenting of individual phages collected from the environment (as though they were new molecular entities) is a strategy unlikely to provide a great deal of exclusivity. However, a phage which is unique, and therefore unlikely or impossible to be re-isolated from the environment, could be harder to replicate (i.e. 'invent around') by a competitor and may be more securely protected by a patent. Such a strategy would diminish one of the innate advantages of phage therapy, which is the ease with which novel phages may be isolated from the environment when the need arises. On the other hand, an engineered 'superphage' could contain features which allow greater clinical efficacy than would a wild-type phage: altered or broadened host range and lower rates of bacterial resistance, for example, achieved by modification of the phage receptor binding domains, or addition of genes which allow the phage to overcome abortive infection or restriction systems in the target bacteria. The phage itself may be recruited to perform 'double duty' by modification of the capsid to carry vaccine or other antigens to the site of infection via protein fusions. Engineering of phage to become resistant to clearance by the mammalian innate immune system may also increase clinical efficacy; the long-circulating phage described in Merril *et al.* (1996) was shown to possess a single amino acid change in its major capsid protein, resulting in a net change in the surface charge of the virion. If such charge differences eventually prove to be a paradigm for reducing phage clearance, these may be deliberately engineered into therapeutic phages.

Such engineered phages would require a significant amount of pre-clinical R&D investment in order to determine which features are most desirable, and then implementing strategies which fulfil them. Furthermore, such an entity would by necessity become a genetically modified organism which would subject it to further regulatory restrictions and also possibly more rigorous phase I and II clinical evaluation. The other phage-derived antibacterials described in this chapter – lysins, phage lethal agents and probably most tailocins – would almost by definition fall into this second high-investment model, owing to the extensive pre-clinical engineering required for their development.

The paradigm under which phage therapy proceeds will be influenced on a fundamental level by the regulatory framework which governs the approval of phage therapeutics. Currently, it is presumed that at least the first phage-based therapeutic products, which are generally expected to contain natural phage isolated from the environment in a method which more closely resembles the low-investment, high-throughput model, will be required to undergo the standard three-phase regulatory pipeline prior to approval. However, is not yet clear what the regulatory process for phage therapeutics will resemble in the long-term. The ease with which new phages may be approved for use as a part of existing approved phage mixtures remains unknown and will need to be determined by the regulatory agencies as such issues are presented to them. For example, a common practice in phage therapy in the Georgian and Polish experience – and a virtual necessity for the success of the low-investment model – has been to routinely reformulate the phage preparations used, tailoring their phage content to maintain efficacy as new strains of pathogenic bacteria are encountered. If a three-phage preparation for the treatment of a specific pathogen is produced and approved under the current framework, what will happen if the producer determines that a new phage should be added to maintain clinical efficacy? If this new formulation is required to undergo new phase I, II and III clinical trials, several of the advantages of phage therapy in this model, namely the ability to rapidly overcome bacterial resistance and adapt to the appearance of new pathogen strains, will be lost. Furthermore, the lack of strong IP protection in this model provides a strong disincentive for any company to pursue such repeated, costly trials for the sake of an antibacterial that requires constant reformulation and re-approval. Such a regulatory regime would tend to favour the alternative, high-investment model, in which additional significant initial R&D resources are incurred in the development of phages which are more likely to retain long-term utility and possess stronger IP protectability.

Alternatively, a regulatory framework which allows an accelerated or simplified approval process for novel phages would greatly favour the currently practiced low-investment model. Such a regime would require the implicit assumption that a newly isolated phage is not a fundamentally novel agent and could be added to or substituted for other phages in a pre-existing, approved phage-based antimicrobial, provided that the phage could be shown to meet a set of basic safety and efficacy criteria. This system may be made to resemble the current licensing process for influenza vaccines, which are reformulated annually and are subject to an accelerated approval process (Wood and Levandowski, 2003). Under such a regulatory system, concerns of IP protection become somewhat less relevant owing to the removal of the high cost barrier associated with phase III clinical trials, coupled with the relatively low costs required for pre-clinical R&D if environmentally isolated, non-engineered phages are to be used therapeutically. Such a system may resemble the current market for the production of generic antibiotics, which have already received regulatory approval, are unprotected by IP, and the major costs are incurred in the manufacture and distribution of the products.

Criteria for approval of a therapeutic phage

Production

As mentioned previously, only a handful of phages have completed any form of regulatory process and been formally approved for human consumption in the West. Based on this experience so far, it has already become clear that the common use of sterilized crude phage lysates, standard in the Georgian clinical practice (Kutter *et al.*, 2009) will not be readily accepted by Western regulators. A recent publication details attempts to produce a defined, purified therapeutic phage preparation under conditions designed to pass a stringent EU regulatory process (Merabishvili *et al.*, 2009), and this work may serve as a model for future efforts, particularly if the low-cost, high-throughput phage therapy model is adopted as general practice. In this effort, phages were isolated and tested to determine their host range among clinical bacterial isolates. Three phages with a suitably broad host range were selected, propagated by a classic plate lysate method using animal-protein-free bacteriological medium and filter-sterilized to remove bacteria and large bacterial debris. These lysates were then passed over a commercially available column to remove bacterial endotoxin from the preparation. This production method is quite simple to perform and is actually less stringent than that conducted for the execution of many experimental animal trials, which often used highly purified phage preparations (e.g. Merril *et al.*, 1996; Matsuzaki *et al.*, 2003; Gill *et al.*, 2006a; Carmody *et al.*, 2010; Debarbieux *et al.*, 2010). The phage preparation was then subjected to a quality control process determining pH, pyrogenicity, cytotoxicity and sterility. The phages themselves were also characterized by whole genome sequencing to determine the presence of known toxin- or virulence factor-encoding genes, and to help determine if the phages were temperate.

This approach, which incorporated GMP production methods and quality control analysis by recognized, accredited laboratories, generated a phage-based antimicrobial product which satisfied the regulatory requirements to be tested in a human clinical trial. Even so, this preparation approach is expected to yield a final product that contains various contaminants remaining from the phage culturing step, including medium components, small cellular debris and soluble proteins of bacterial origin. In the case described above, these contaminants did not appear to possess any pyrogenic or cytotoxic properties but this may not always be the case, depending on the host bacterium used for production of the phage. While bacterial endotoxin may be removed by a number of commercially proven technologies, a wide variety of soluble protein toxins, including haemolysins, lipases, cytolysins and enterotoxins are produced by various pathogen species, and their presence in the phage lysate may require additional purification steps in producing final phage preparations that satisfy quality control requirements. Additionally, the presence of functional endogenous prophages in the bacterial strains used for phage production may result in low-level contamination of the final product with viable, temperate phage, which may be able to lysogenize (and possibly lysogenically convert) the new hosts that they encounter.

A solution to this issue may be to require that, in addition to the phages themselves, the bacterial strains used for phage production be sequenced to determine the presence of endogenous prophages or critical virulence determinants. Such data could then be used to direct modifications to the bacterial propagation strain, using established molecular biology techniques to delete all suspected prophage regions and virulence factors. Such a strategy could not only result in a more well-defined end product, but also streamline downstream purification steps. Merabishvili *et al.* (2009) screened their bacterial propagation hosts for endogenous prophages using induction followed by plating of the culture supernatant on an indicator host. While we applaud this consideration in the phage production process, it is difficult to confirm a negative result using this technique as there is always the possibility that the host prophage is not inducible or that the indicator host is not sensitive to the phage(s) produced, if any. Complete sequencing of the propagation strain essentially guarantees the absence of prophage, and logistically this requirement continues to become less burdensome as we move further into the post-genomic era.

Phage characterization

As mentioned previously, phages selected for therapeutic use must meet several criteria, some of which are important for practical considerations of efficacy, while others are important for the avoidance of possible adverse effects. Phage host range, life style (virulent vs. temperate), adsorption, stability and the ability to infect and lyse pathogenic bacteria *in vivo* at the site of infection fall under the former category. The capacity of a phage for generalized transduction or lysogeny falls under the latter category. Temperate phages, owing to considerations of lysogeny and superinfection immunity, are obviously poor choices as therapeutic phages: they are essentially guaranteed to give rise to a population of prophage-carrying lysogens that will be immune to further phage attack, and also may express additional virulence factors if they are encoded in morons contained in the phage genome. However, clear plaques and the inability to readily form lysogens is not a guaranteed indicator of a virulent lifestyle of a given phage. The phenotypes of 'clear' and 'turbid' are subjective, with no obvious binary distinction between the two states. The ability to easily recover lysogens from a phage-infected bacterial culture is a simple and useful screen but again a negative result remains inconclusive; only 18% of the bacteria surviving challenge with *Burkholderia cenocepacia* phage KS9 were reportedly lysogens, with the remainder of the bacterial population presumably becoming insensitive to KS9 via some other mechanism (Lynch *et al.*, 2010), suggesting that some temperate phage form lysogens at a low frequency. Additionally, some apparently virulent phages are able to transduce bacterial DNA with considerable efficiency, such as *Caulobacter crescentus* phage CR30 (Bender, 1981).

Therefore, empirical characterization of candidate therapeutic phages may be required, including general tests for lysogeny and transduction. While the acquisition of the complete phage genome has already become a *de facto* standard for therapeutic phages, this only allows inferred understanding of the lifestyle and virulence determinant content of the phage based on its relationships to other known phages. The sequence of phage TerL proteins may be used to indicate the DNA packaging strategy of the phage as either *pac* or *cos* type (Casjens *et al.*, 2005), and cos-type phages are expected to be at low risk of generalized transduction. Molecular-based methods for determining the generalized transducing ability of a novel phage have been proposed (Sander and Schmieger, 2001), and such methods may be adaptable for routine use in the characterization of novel phages. From a purely regulatory standpoint, exactly how low a transduction frequency should be considered acceptable in a therapeutic phage remains to be determined. One benchmark may be to keep the transduction frequency below the natural rate at which the target bacterium takes up DNA from the environment, which could be considered as the 'background level' of genetic exchange for the organism.

Another useful criterion for confirming phage virulence may be the determination of host DNA degradation immediately following phage infection. Several known virulent phages such as *E. coli* phages T4 (Mosig and Eiserling, 2006) and *S. aureus* phage K (Rees and Fry, 1981) degrade the bacterial chromosome shortly after infection, a strategy which has two important implications. First, a phage cannot lysogenize a host lacking a chromosome, so phages that immediately degrade the chromosome are more likely to be completely incapable of a temperate lifestyle. Second, the degradation of the host chromosome means that, at the time of phage DNA packaging, there is little or no bacterial DNA left in the cell which could be erroneously packaged into a phage capsid. Therefore the ability of virulent phages to transduce host DNA, even if it uses a low-specificity *pac*-type Ter, is expected to be low.

The immediate future of phage therapy

Despite its promise and over ten years of Western research interest, phage therapy in humans remains a highly experimental procedure, generally unavailable in the West outside of a tiny handful of clinical trials. As these trials proceed and specific clinical applications for phage therapy become more apparent, the regulatory, IP and financial frameworks surrounding these treatments will become more defined. At this time it is difficult to foresee if phage therapy will eventually move into general use as an antimicrobial, such as is now practiced in Georgian medicine; whether it will become a specialty niche treatment confined to specific ailments, or, perhaps, whether phage therapy will prove to be unworkable in the Western medical framework, and abandoned again in favour of a new, small-molecule antibiotic approach which has yet to be discovered.

Compassionate emergency use

At the Hirszfeld Institute in Poland, despite a long history of the use of phage as therapeutics, this treatment is still considered as an experimental procedure and is typically reserved for patients which have failed standard (i.e. antibiotic) treatment (Kutter *et al.*, 2009). This restriction, however, has still allowed the Institute to treat many hundreds of patients with phage therapy (Weber-Dabrowska *et al.*, 2000), undoubtedly saving lives. Given its general safety record and the fact that it is already currently available from some practitioners in the USA, we can see no compelling reason why phage therapy cannot be made more widely available on compassionate-use grounds to patients afflicted with serious bacterial infections which are refractory to standard treatments. A Hirszfeld Institute model for such treatment could be implemented in one or more centres, preferably within hospitals with a strong research mandate, with phages used as a 'treatment of last resort' in patients which have not responded to standard care regimens. As in Poland, the availability of such a treatment would not only save lives, but allow for the collection of clinical data under well-documented clinical conditions within the USA. In keeping with this Hirszfeld model, a standardized phage bank, containing virulent phages with a documented history of treatment success, could be established and maintained at one or more sites and serve as a repository for therapeutic phages. A number of sources for such phages already exist, including the phage collections maintained at the Hirszfeld Institute in Poland, the Eliava Institute in Georgia, and at the universities around the world which have conducted experimental animal trials of phage therapy.

Among major pathogens, *S. aureus* is a prime candidate for compassionate-use phage therapy for a number of reasons. *S. aureus* poses a significant health threat to humans, and drug-resistant strains (e.g. MRSA) can be extremely difficult to treat by conventional antibiotics. *S. aureus* is also somewhat unique in that a broad range of isolates of this bacterium are susceptible to a closely related group of virulent myophages related to phage Twort. This group includes phages K, 812 (Pantucek *et al.*, 1998), G1 (isolated from a Georgian therapeutic preparation; Kwan *et al.*, 2005) and ISP, which is currently under clinical evaluation (Merabishvili *et al.*, 2009). These phages tend to exhibit a broad host range in *S. aureus*, typically infecting 80–90% of the *S. aureus* strains tested and often possessing an ability to infect some coagulase-negative staphylococci as well. Given this background, a relatively small library of these Twort-like phages could be utilized to treat a broad range of staphylococcal infections.

Veterinary and food applications

While phage therapy to treat life-threatening infections in humans is the stated or implied goal of much of the Western experimental work described here, the use of phages in the treatment of agricultural, veterinary and food-borne pathogens has also received considerable attention and has the added benefit of being subject to less stringent regulatory requirements. In general, the regulatory focus in these settings is on the safety of the product, and efficacy does not need to be rigorously demonstrated in a phase III-style clinical trial in order for its use to be approved. Companies within the USA such as OmniLytics have been marketing phages for the treatment of plant diseases for several years, and phages against *L. monocytogenes* have already been approved for use as an additive to foods [Food and Drug Administration (FDA), 2006]. Veterinary, agricultural and food applications represent a potential arena for the early adoption of phage therapy, and these also have the potential to provide extremely large-scale clinical data on phage efficacy: consider that a moderately sized poultry operation may house 50,000 birds or more, providing an extremely large potential sample size.

These applications, however, represent a double-edged sword for the prospects of the eventual adoption of phage therapy in human medicine. On one hand, the lower regulatory barrier for the use of phage in these settings allows for their early adoption and evaluation in real-world situations. On the other, the lack of stringent regulation opens up these markets to many of the same problems which plagued the early days of phage therapy: poor standardization of products, lax quality control, and improper use of phages owing to a poor understanding of phage biology. Reckless or inappropriate use of phages in non-human medical settings may have the effect of poisoning attitudes towards the practice, if their misuse leads to a reputation of ineffectiveness. Therefore, it is crucial that properly controlled, large-scale trials of phage therapy be conducted in these settings in order to determine efficacy (or lack thereof), an honest appraisal of the practice and determination of which systems may be most benefitted by phages for the control of their bacterial burdens.

Recommendations

We hope we have left the reader with an enhanced perspective on the history and current state of phage therapy. The research efforts in this area have been scattered both in time and space. Thus, inevitably, and regrettably, we will have missed some important points, slighted some research achievements and cut some corners in our effort to simplify and summarize for interested but non-expert readers. Awareness of these shortcomings, however, does not inhibit us from, in closing this chapter, relating a set of specific recommendations that could be useful in moving the field forward.

1. Bacteriophages which meet basic criteria: virulent lifestyle, low transducing potential, absence of virulence determinants in the genome, and which have been prepared by prescribed methods (to be determined) should be granted generally recognized as safe (GRAS) status for use as antimicrobials in humans. Such a designation could accelerate the introduction of phage therapy into clinical evaluation. This is clearly reasonable for phages of enteric bacteria, given the high level and great diversity of phage particles normally present in the human gut.
2. We also recommend that steps be taken to clarify the regulatory framework for phage-based therapeutics, under the assumption that at least some will be shown to have clinical efficacy in some situations. In particular, issues surrounding the requirements for reformulated phage products (e.g. the periodic updating of phage cocktails with new phages and the removal of old ones) should be discussed now, rather than when the need becomes urgent. This kind of undertaking is rightfully the province of the FDA and USDA in the USA, perhaps as a joint effort with the equivalent agencies in other countries.
3. We recommend the establishment of compassionate-use dispensation of phage therapy under the Polish Hirszfeld Institute model in one or more major treatment centres. Under this model, a standardized phage bank containing well-characterized phages active against important pathogens would be established and the phages dispensed on compassionate-use grounds in cases where standard antimicrobial treatments have failed.

4 The community involved in this area straddles academic, clinical and corporate institutions, as well as disparate international scientific and medical cultures. It has no regular focused convocation where ideas can be exchanged nor an international body that could set standards. Major international organizations concerned with research in microbiological and infectious disease, like the American Society for Microbiology and the Federation of European Microbiological Societies, should collaborate to convene a regular conference and a committee to establish standards.

5 Major funding agencies should consider establishing a proposal market for basic and applied research that would address the large gaps in the knowledge base that underlies phage therapy. A late 2010 survey of the NIH Report grants database using the words 'bacteriophage' or 'phage' as a search term was revealing. After filtering out grants involving phage display, there were 42 funded projects by the standard R01 mechanism. However, most (31) of these grants concerned the detailed structure and function of proteins involved in transcription and DNA metabolism of classical phage systems, including 13 grants on the single subject of DNA packaging. These projects have been very fruitful in terms of fundamental molecular biology but few of them have much current connection to phage-based therapeutics. By contrast, there were only 10 grants concerning phage biology outside of the classic laboratory systems, and only a few of these focused on phage in pathogenic bacteria in any respect. Clearly the basic phage biology of bacteria outside the classical laboratory standards deserves much more attention. For important pathogenic species, some high-profile phage systems, e.g. the Twort-like phages of *S. aureus*, should be developed in some depth. Considering the powerful tools that are now available in addition to the intrinsic experimental facility of phage genetics, acquiring a working knowledge of the phage biology of, for example, *Enterococcus* or *Clostridium* could be achieved with modest investment. The scientific payoff should be handsome, considering the immensely powerful tools that the study of the classical coliphage systems has provided for modern biology. It would also help foster a new generation of phage biologists with the expertise and tools to address the many serious questions facing phage therapeutics.

In the same search, the terms 'bacteria' and 'antibiotic' generated over 3500 and 1000 hits respectively, whereas there were only two R01s exploring the use of phage endolysins as therapeutics and not a single one explicitly aimed at phage therapy. This seems disproportionate, by any standard, especially considering the world-wide increase in general antibiotic-resistance in pathogens and the clear potential for phage as an alternative approach.

Glossary

Active therapy A model of phage therapy characterized by lower initial phage dosages, requiring active replication of the phage at the site of infection to generate enough phage to eradicate the bacterial population. In contrast to inundative therapy.

Bacteriophage A virus which infects bacteria.

Capsid The proteinaceous shell of a virus, which protects the viral genome and allows the virus to adsorb to new host cells. In *Caudovirales*, the capsid has two major components: the head and the tail.

Caudovirales An order of viral taxonomy that includes the tailed phages (Myoviridae, Siphoviridae and Podoviridae). These phages all contain double-stranded DNA genomes.

Central tail fibre (CTF) In the siphophage lambda, the CTF is visible as a spike projecting from the conical tail tip and is composed of three copies of the J protein, with the C-terminus at the distal end. The CTF makes specific interactions with the LamB porin which are necessary and sufficient for irreversible adsorption of the lambda virion to the surface of the bacterial cell. In the other well-studied siphophage, T5, the CTF is composed of gp5, which makes an essential interaction with the porin FhuA as its receptor.

Cryptic prophage An apparently non-functional or defective prophage remnant which is still carried as part of a bacterial genome. These phages may be made to be functional again if complemented with compatible phage genes from another source.

Delbrück, Max Trained as a physicist under Niels Bohr, Delbrück turned to bacteriophage biology in the late 1930s and, with Salvador Luria and Alfred Hershey, founded the phage group at Cold Spring Harbour laboratory during the 1940s. The three shared the Nobel Prize in Medicine in 1960. Delbrück is widely regarded as the father of modern molecular biology.

Eliava Institute Eliava Institute of Bacteriophages, Microbiology and Virology, located in Tbilisi, Republic of Georgia.

EOP (efficiency of plating) Rigorously, this is the ratio of pfu/ml, on a specified indicator lawn, to virions/ml in a phage sample. In practice, the physical concentration of virions is rarely known. Instead, EOP is usually the ratio of pfu/ml on any particular host compared with a standard indicator host or compared with the host on which the phage was isolated.

Head The hollow, icosahedral proteinaceous shell which contains the packaged phage DNA.

Hirzsfeld Institute Hirszfeld Institute of Immunology and Experimental Therapy located in Wrøclaw, Poland.

Immunity Bacterial insensitivity to a phage caused by lysogeny of the cell by the same or related phage; this term only relates to temperate phages.

Indicator An indicator is the bacterial strain used to form lawns to detect bacteriophage plaques.

Inovirus Filamentous phage of the *Inoviridae* have a single-stranded circular DNA chromosome and require a conjugative pilus for adsorption and penetration of the viral DNA into the host cytoplasm. The temperate Inovirus CTXphi encodes the cholera exotoxin.

Insensitivity Blanket term for bacteria which are unable to support infection by a given phage. May be the result of lysogeny, receptor absence (see resistance), bacterial restriction, abortive infection mechanisms, or other mechanisms.

Inundative therapy A model of phage therapy wherein a large phage dose is initially applied, providing enough phage to immediately infect most or all of the target bacterial population. Replication of the phage is not required for success. In contrast to active therapy.

Lambda Along with myophages P1 and P2, discovered by G. Bertani, the siphophage lambda was the first temperate phage to be described, in 1951. Originally identified and characterized as an inducible prophage in *E. coli* K-12 by Esther and Joshua Lederberg, lambda became the focus of intense study to elucidate the 'genetic switch' between the lytic and lysogenic lifestyles. Overall, the siphophage lambda is indisputably the best understood genetic entity.

Lateral tail fibre (LTF) Most Caudovirales encode LTFs, which are generally long and flexible and bind to a specific receptor on the surface of the host cell. LTFs are usually attached to the top of the conical tail tip assembly of siphophages and to the base-plate of myophages, and in some podophages, to the top of the 'stubby' tail. LTFs are assembled by trimerization of protomers, generally catalysed by a specific tail fibre-assembly chaperone which, in some cases, remains associated with the mature tail fibre. In some phages, the entire LTF is composed of a single tail fibre protein, with the N-terminal domains involved in binding to the virion and the receptor-binding domain located at the distal (C-terminus). In more complex phages, like T4, the LTFs are composed of multiple proteins. LTFs have at least one 'knee', where the LTF is bent. It is thought that individual LTFs bind and release rapidly, allowing the virion to 'walk' around the surface of the cell, until irreversible adsorption occurs.

LPS (lipopolysaccharide) The major component of the outer leaflet of the outer bacterial membrane in Gram-negative bacteria. LPS is composed of three domains: the highly conserved Lipid A, which is embedded in the outer membrane; the core, a semi-conserved oligosaccharide domain typically containing heptose sugars and KDO; and the O-antigen, a repeating polysaccharide chain which is variable at the strain level. LPS is a commonly used phage receptor.

MOI (multiplicity of infection) MOI can be considered in two ways. Classically, the MOI is the ratio of phages to bacteria in a given system (a culture containing five phages for each cell would have an MOI of 5). This simplistic definition (also called the input MOI) does not account for the time required for phages to locate and adsorb to their hosts; in reality, not every phage will infect a host during a given time period, so the ratio of adsorbed phages to total bacteria (termed actual MOI) is expected to be lower than the input MOI. Additionally, due to the random nature of phage adsorption, some cells will adsorb more phages than others as described by a Poisson distribution (e.g. with an actual MOI of 1, about 1/3 of cells will adsorb no phages, 1/3 will adsorb one phage, and 1/3 will adsorb more than one phage).

Myophage A family of *Caudovirales* containing phages with contractile tails.

Omp Gram-negative bacteria generally have $>10^5$ outer membrane proteins (Omps), nearly all of which are porins, which span the bilayer with beta-barrel architecture. Many phages use Omps as receptors for irreversible or reversible binding.

pfu (plaque-forming unit) Analogous to the concept of a colony forming unit (CFU) in bacteriology. As a single virion initiates the formation of a single plaque, enumeration of pfu allows quantification of the number of phages in a given sample. Note that, as is the case with CFU, the number of pfu reflects the number of phages able to form plaques in a sample, not the total number of virions.

Phage See *bacteriophage*.

Phage cocktail A mixture of two or more different phages in a single suspension for the purposes of eliminating a bacterial population. Phage cocktails are typically formulated to contain phages with different host ranges, such that a given target bacterial strain will be infected by at least one of the phages in the cocktail.

Plaque A clearing in a confluent lawn of bacteria grown on semi-solid substrate medium which is the result of successive rounds of phage infection, lysis and diffusion. A single plaque is the result of an initial infection by a single virion, therefore all of the phages in an isolated plaque are expected to be clonal.

Podophage A family of *Caudovirales* containing phages with short, non-contractile tails.

Portal Part of the viral capsid, the portal is a 12-member protein ring which serves as the entry point for DNA into the phage head during DNA packaging.

Receptor A component of the bacterial surface that is recognized by the receptor binding domains of the phage. Surface-exposed carbohydrates, membrane proteins, or surface organelles such as flagella or pili may be used as receptors by various phages.

Receptor binding domain A specific receptor-binding domain is generally located on the LTFs, CTFs, or other component of the phage tail. In general, siphophages have two receptor binding domains, one on the LTF and one on the CTF. In contrast, the domain is located at the distal end of the LTFs in myophages and in some podophages.

Resistance Bacterial insensitivity to a phage caused by alteration of the outer bacterial surface resulting in the phage being unable to adsorb to the cell. The most common mechanism of resistance is the mutational loss or alteration of the bacterial receptor recognized by the phage.

Short tail fibre Myophages have short tail fibres attached at the vertices of the baseplate. Detailed studies of phage T4 indicate that these short tail fibres deploy as part of the irreversible adsorption event, after the LTFs have positioned the virion above the surface of the host cell.

Siphophage A family of *Caudovirales* containing phages with long, non-contractile tails.

T1 through T7 A set of seven virulent phages of *E. coli*, designated by Delbrück as the only legitimate subjects for experimentation by the phage group. This decision, the Phage Treaty of 1944 and enforced in the summer Phage Courses by which many scientists received their initial training in phage biology, resulted in intense focus on the properties of the members of the set and, eventually, elevated the myophage T4 and podophage T7 to paradigm status in molecular biology.

Tail (bacteriophage) The part of the viral capsid in members of the *Caudovirales* which contains the specificity determinants that allow the phage to adsorb to and introduce its DNA genome into the host.

Tail fibre Part of the phage tail which carries the receptor binding domains at its distal end.

Tail spike In some podophages, the portal vertex of the virion is crowned with a ring of six trimeric tail spikes. In the best studied case, the *Salmonella* phage P22, the tail spikes have enzymatic activity that can degrade the O-antigen polymer.

References

Abedon, S.T. (1994). Lysis and the interaction between free phages and infected cells. In Molecular Biology of Bacteriophage T4, Karam, J.D., ed. (American Society for Microbiology, Washington D.C.), pp. 397–405.

Abedon, S.T. (2009a). Kinetics of phage-mediated biocontrol of bacteria. Foodborne Pathog. Dis. *6*, 807–815.

Abedon, S.T. (2009b). Phage evolution and ecology. Adv. Appl. Microbiol. *67*, 1–45.

Abedon, S.T., and Yin, J. (2009). Bacteriophage plaques: theory and analysis. Methods Mol. Biol. *501*, 161–174.

Ackermann, H.-W. (2006). In The Bacteriophages, Calendar, R., ed. (Oxford: Oxford University Press), pp. 8–16.

Adams, C.P., and Brantner, V.V. (2006). Estimating the cost of new drug development: is it really 802 million dollars? Health Aff. (Millwood) 25, 420–428.

Arber, W., and Dussoix, D. (1962). Host specificity of DNA produced by *Escherichia coli*. I. Host controlled modification of bacteriophage lambda. J. Mol. Biol. *5*, 18–36.

Atterbury, R.J., Van Bergen, M.A., Ortiz, F., Lovell, M.A., Harris, J.A., De Boer, A., Wagenaar, J.A., Allen, V.M., and Barrow, P.A. (2007). Bacteriophage therapy to reduce salmonella colonization of broiler chickens. Appl. Environ. Microbiol. 73, 4543–4549.

Bach, S.J., McAllister, T.A., Veira, D.M., Gannon, V.P.J., and Holley, R.A. (2003). Effect of bacteriophage DC22 on *Escherichia coli O157*,H7 in an artificial rumen system (Rusitec) and inoculated sheep. Anim. Res. 52, 89–101.

Bair, C.L., and Black, L.W. (2007). A type IV modification dependent restriction nuclease that targets glucosylated hydroxymethyl cytosine modified DNAs. J. Mol. Biol. *366*, 768–778.

Bair, C.L., Rifat, D., and Black, L.W. (2007). Exclusion of glucosyl-hydroxymethylcytosine DNA containing bacteriophages is overcome by the injected protein inhibitor IPI*. J. Mol. Biol. *366*, 779–789.

Barondess, J.J., and Beckwith, J. (1990). A bacterial virulence determinant encoded by lysogenic coliphage lambda. Nature *346*, 871–874.

Barondess, J.J., and Beckwith, J. (1995). bor gene of phage lambdagk, involved in serum resistance, encodes a widely conserved outer membrane lipoprotein. J. Bacteriol. *177*, 1247–1253.

Barrangou, R., Fremaux, C., Deveau, H., Richards, M., Boyaval, P., Moineau, S., Romero, D.A., and Horvath, P. (2007). CRISPR provides acquired resistance against viruses in prokaryotes. Science *315*, 1709–1712.

Bender, R.A. (1981). Improved generalized transducing bacteriophage for *Caulobacter crescentus*. J. Bacteriol. *148*, 734–735.

Berry, J., Summer, E.J., Struck, D.K., and Young, R. (2008). The final step in the phage infection cycle: the Rz and Rz1 lysis proteins link the inner and outer membranes. Mol. Microbiol. *70*, 341–351.

Bertani, G. (1951). Studies on lysogenesis. I. The mode of phage liberation by lysogenic *Escherichia coli*. J. Bacteriol. *62*, 293–300.

Bertani, G., and Weigle, J.J. (1953). Host controlled variation in bacterial viruses. J. Bacteriol. *65*, 113–121.

Biswas, B., Adhya, S., Washart, P., Paul, B., Trostel, A.N., Powell, B., Carlton, R., and Merril, C.R. (2002). Bacteriophage therapy rescues mice bacteremic from a clinical isolate of vancomycin-resistant *Enterococcus faecium*. Infect. Immun. *70*, 204–210.

Breitbart, M., Hewson, I., Felts, B., Mahaffy, J.M., Nulton, J., Salamon, P., and Rohwer, F. (2003). Metagenomic analyses of an uncultured viral community from human feces. J. Bacteriol. *185*, 6220–6223.

Breitbart, M., Rohwer, F., Abedon, S.T. (2005). Phage ecology and bacterial pathogenesis. In Phages: Their Role in Bacterial Pathogenesis and Biotechnology, Waldor, M.K., Friedman, D.I., Adhya, S.L., eds. (ASM Press, Washington, DC), pp. 66–91.

Broxmeyer, L., Sosnowska, D., Miltner, E., Chacon, O., Wagner, D., McGarvey, J., Barletta, R.G., and Bermudez, L.E. (2002). Killing of *Mycobacterium avium* and *Mycobacterium tuberculosis* by a mycobacteriophage delivered by a nonvirulent mycobacterium: a model for phage therapy of intracellular bacterial pathogens. J. Infect. Dis. *186*, 1155–1160.

Brussow, H. (2001). Phages of dairy bacteria. Annu. Rev. Microbiol. *55*, 283–303.

Brussow, H., Canchaya, C., and Hardt, W.D. (2004). Phages and the evolution of bacterial pathogens: from genomic rearrangements to lysogenic conversion. Microbiol. Mol. Biol. Rev. *68*, 560–602.

Cairns, J., Stent, G.S., and Watson, J.D. (2000). Phage and the ORIgins of Molecular Biology (Cold Spring Harbor Laboratory Press, Cold Spring Harbor, NY).

Calendar, R. (1988). The Bacteriophages (New York and London: Kluwer Academic/Plenum Publishers).

Callaway, T.R., Edrington, T.S., Brabban, A.D., Anderson, R.C., Rossman, M.L., Engler, M.J., Carr, M.A., Genovese, K.J., Keen, J.E., Looper, M.L., *et al.* (2008). Bacteriophage isolated from feedlot cattle can reduce *Escherichia coli O157*,H7 populations in ruminant gastrointestinal tracts. Foodborne Pathog. Dis. *5*, 183–191.

Canchaya, C., Fournous, G., Chibani-Chennoufi, S., Dillmann, M.L., and Brussow, H. (2003a). Phage as agents of lateral gene transfer. Curr. Opin. Microbiol. *6*, 417–424.

Canchaya, C., Proux, C., Fournous, G., Bruttin, A., and Brussow, H. (2003b). Prophage genomics. Microbiol. Mol. Biol. Rev. *67*, 238–276.

Capparelli, R., Nocerino, N., Iannaccone, M., Ercolini, D., Parlato, M., Chiara, M., and Iannelli, D. (2010a). Bacteriophage therapy of Salmonella enterica: a fresh appraisal of bacteriophage therapy. J. Infect. Dis. *201*, 52–61.

Capparelli, R., Nocerino, N., Lanzetta, R., Silipo, A., Amoresano, A., Giangrande, C., Becker, K., Blaiotta, G., Evidente, A., Cimmino, A., *et al.* (2010b). Bacteriophage-resistant *Staphylococcus aureus* mutant confers broad immunity against staphylococcal infection in mice. PLoS ONE *5*, e11720.

Capparelli, R., Parlato, M., Borriello, G., Salvatore, P., and Iannelli, D. (2007). Experimental phage therapy against *Staphylococcus aureus* in mice. Antimicrob. Agents Chemother. *51*, 2765–2773.

Carlson, K., Raleigh, E.A., and Hattman, S. (1994). Restriction and modification. In Molecular Biology of Bacteriophage T4, Karam, J.D., ed. (ASM Press, Washington D.C.), pp. 369–381.

Carmody, L.A., Gill, J.J., Summer, E.J., Sajjan, U.S., Gonzalez, C.F., Young, R.F., and LiPuma, J.J. (2010). Efficacy of bacteriophage therapy in a model of Burkholderia cenocepacia pulmonary infection. J. Infect. Dis. *201*, 264–271.

Casjens, S. (2003). Prophages and bacterial genomics: what have we learned so far? Mol. Microbiol. *49*, 277–300.

Casjens, S.R., and Gilcrease, E.B. (2009). Determining DNA packaging strategy by analysis of the termini of the chromosomes in tailed-bacteriophage virions. Methods Mol. Biol. *502*, 91–111.

Casjens, S.R., Gilcrease, E.B., Winn-Stapley, D.A., Schicklmaier, P., Schmieger, H., Pedulla, M.L., Ford, M.E., Houtz, J.M., Hatfull, G.F., and Hendrix, R.W. (2005). The generalized transducing Salmonella bacteriophage ES*18*, complete genome sequence and DNA packaging strategy. J. Bacteriol. *187*, 1091–1104.

Cerveny, K.E., DePaola, A., Duckworth, D.H., and Gulig, P.A. (2002). Phage therapy of local and systemic disease caused by *Vibrio vulnificus* in iron-dextran-treated mice. Infect. Immun. *70*, 6251–6262.

Chapman-McQuiston, E., and Wu, X.L. (2008). Stochastic receptor expression allows sensitive bacteria to evade phage attack. Part I: experiments. Biophys. J. *94*, 4525–4536.

Chhibber, S., Kaur, S., and Kumari, S. (2008). Therapeutic potential of bacteriophage in treating *Klebsiella pneumoniae* B5055-mediated lobar pneumonia in mice. J. Med. Microbiol. *57*, 1508–1513.

Chibani-Chennoufi, S., Sidoti, J., Bruttin, A., Kutter, E., Sarker, S., and Brussow, H. (2004). *In vitro* and *in vivo* bacteriolytic activities of *Escherichia coli* phages: implications for phage therapy. Antimicrob. Agents Chemother. *48*, 2558–2569.

Chopin, M.C., Chopin, A., and Bidnenko, E. (2005). Phage abortive infection in lactococci: variations on a theme. Curr. Opin. Microbiol. *8*, 473–479.

Clark, J.R., and March, J.B. (2006). Bacteriophages and biotechnology: vaccines, gene therapy and antibacterials. Trends Biotechnol. *24*, 212–218.

Coombs, D., and Arisaka, F. (1994). T4 tail structure and function. In Molecular Biology of Bacteriophage T4, Karam I.D., ed. (American Society for Microbiology, Washington D.C.), pp. 209–300.

Costerton, J.W., Stewart, P.S., and Greenberg, E.P. (1999). Bacterial biofilms: a common cause of persistent infections. Science *284*, 1318–1322.

Danelishvili, L., Young, L.S., and Bermudez, L.E. (2006). *In vivo* efficacy of phage therapy for Mycobacterium avium infection as delivered by a nonvirulent mycobacterium. Microb. Drug Resist. *12*, 1–6.

Dassy, B., and Fournier, J.M. (1996). Respiratory activity is essential for post-exponential-phase production of type 5 capsular polysaccharide by *Staphylococcus aureus*. Infect. Immun. *64*, 2408–2414.

Debarbieux, L., Leduc, D., Maura, D., Morello, E., Criscuolo, A., Grossi, O., Balloy, V., and Touqui, L. (2010). Bacteriophages can treat and prevent *Pseudomonas aeruginosa* lung infections. J. Infect. Dis. *201*, 1096–1104.

Doolittle, M.M., Cooney, J.J., and Caldwell, D.E. (1996). Tracing the interaction of bacteriophage with bacterial biofilms using fluorescent and chromogenic probes. J. Ind. Microbiol. *16*, 331–341.

Doulatov, S., Hodes, A., Dai, L., Mandhana, N., Liu, M., Deora, R., Simons, R.W., and Miller, J.F. (2004). Tropism switching in Bordetella bacteriophage defines a family of diversity-generating retroelements. Nature *431*, 476–481.

Dublanchet, A., and Bourne, S. (2007). The epic of phage therapy. Can. J. Infect. Dis. Med. Microbiol. *18*, 15–18.

Eaton, M.D., and Baynes-Jones, S. (1934). Bacteriophage therapy. JAMA *103*, 1769–1939.

El-Shibiny, A., Scott, A., Timms, A., Metawea, Y., Connerton, P., and Connerton, I. (2009). Application of a group II Campylobacter bacteriophage to reduce strains of Campylobacter jejuni and Campylobacter coli colonizing broiler chickens. J. Food Prot. *72*, 733–740.

Fairhead, H. (2009). SASP gene delivery: a novel antibacterial approach. Drug News Perspect. *22*, 197–203.

FDA (2006). Food Additives Permitted For Direct Addition To Food For Human Consumption; Bacteriophage Preparation. U.S.A Federal Register *71*, 47729–47732.

Fischetti, V.A. (2010). Bacteriophage endolysins: A novel anti-infective to control Gram-positive pathogens. Int. J. Med. Microbiol. *300*, 357–362.

Fuhrman, J.A. (1999). Marine viruses and their biogeochemical and ecological effects. Nature *399*, 541–548.

Gehring, K., Charbit, A., Brissaud, E., and Hofnung, M. (1987). Bacteriophage lambda receptor site on the

Escherichia coli K-12 LamB protein. J. Bacteriol. *169*, 2103–2106.

Gil, F., Catalao, M.J., Moniz-Pereira, J., Leandro, P., McNeil, M., and Pimentel, M. (2008). The lytic cassette of mycobacteriophage Ms6 encodes an enzyme with lipolytic activity. Microbiology *154*, 1364–1371.

Gil, F., Grzegorzewicz, A.E., Catalao, M.J., Vital, J., McNeil, M.R., and Pimentel, M. (2010). Mycobacteriophage Ms6 LysB specifically targets the outer membrane of Mycobacterium smegmatis. Microbiology *156*, 1497–1504.

Gill, J.J. (2008). Modelling of bacteriophage therapy. In Bacteriophage Ecology: Population Growth, Evolution, and Impact of Bacterial Viruses, Abedon, S.T. ed. (Cambridge University Press, Cambridge), p. 508.

Gill, J.J., Hollyer, T., and Sabour, P.M. (2007). Bacteriophages and phage-derived products as antibacterial therapeutics. Expert Opin. Ther. Patents *17*, 1341–1350.

Gill, J.J., Pacan, J.C., Carson, M.E., Leslie, K.E., Griffiths, M.W., and Sabour, P.M. (2006a). Efficacy and pharmacokinetics of bacteriophage therapy in the treatment of subclinical *Staphylococcus aureus* mastitis in lactating dairy cattle. Antimicrob. Agents Chemother. *50*, 2912–2918.

Gill, J.J., Sabour, P.M., Leslie, K.E., and Griffiths, M.W. (2006b). Bovine whey proteins inhibit the interaction of *Staphylococcus aureus* and bacteriophage K. J. Appl. Microbiol. *101*, 377–386.

Goldberg, E.B., Grinius, L., and Letellier, L. (1994). Recognition, attachment and injection. In Molecular Biology of Bacteriophage T4, Karam, J.D., ed. (American Society for Microbiology, Washington D.C.), pp. 347–357.

Gorski, A., Miedzybrodzki, R., Borysowski, J., Weber-Dabrowska, B., Lobocka, M., Fortuna, W., Letkiewicz, S., Zimecki, M., and Filby, G. (2009). Bacteriophage therapy for the treatment of infections. Curr. Opin. Investig. Drugs *10*, 766–774.

Greenberg, G.R., He, P., Hilfinger, J., and Tseng, M.-J. (1994). Deoxyribonucleoside triphosphate synthesis and phage T4 DNA replication. In Molecular Biology of Bacteriophage T4, Karam, J.D., ed. (American Society for Microbiology, Washington D.C.), pp. 14–27.

Hadas, H., Einav, M., Fishov, I., and Zaritsky, A. (1997). Bacteriophage T4 development depends on the physiology of its host *Escherichia coli*. Microbiology *143*,179–185.

Hagens, S., Habel, A., von Ahsen, U., von Gabain, A., and Blasi, U. (2004). Therapy of experimental pseudomonas infections with a nonreplicating genetically modified phage. Antimicrob. Agents Chemother. *48*, 3817–3822.

Hendrix, R.W., and Casjens, S. (2006). Bacteriophage l and its genetic neighborhood. In The Bacteriophages, Calendar, R., ed. (Oxford University Press, Oxford), pp. 409–447.

Hendrix, R.W., and Duda, R.L. (1992). Bacteriophage lambda PaPa: not the mother of all lambda phages. Science *258*, 1145–1148.

Hendrix, R.W., Hatfull, G.F., and Smith, M.C. (2003). Bacteriophages with tails: chasing their origins and evolution. Res. Microbiol. *154*, 253–257.

Henning, U., Hashemolhosseini, S., Karam, J.D., Drake, J.W., Kreuzer, K.N., Mosig, G., Hall, D.H., Eiserling, F.A., Black, L.W., Spicer, E.K., *et al.* (1994). Receptor recognition by T-even type coliphages. In Molecular Biology of Bacteriophage T4, Karam, J.D., ed. (ASM Press, Washington D.C.), pp. 291–298.

Heo, Y.J., Lee, Y.R., Jung, H.H., Lee, J., Ko, G., and Cho, Y.H. (2009). Antibacterial efficacy of phages against *Pseudomonas aeruginosa* infections in mice and Drosophila melanogaster. Antimicrob. Agents Chemother. 53, 2469–2474.

Herbert, S., Newell, S.W., Lee, C., Wieland, K.P., Dassy, B., Fournier, J.M., Wolz, C., and Doring, G. (2001). Regulation of *Staphylococcus aureus* type 5 and type 8 capsular polysaccharides by CO(2). J. Bacteriol. *183*, 4609–4613.

Horvath, P., Romero, D.A., Coute-Monvoisin, A.C., Richards, M., Deveau, H., Moineau, S., Boyaval, P., Fremaux, C., and Barrangou, R. (2008). Diversity, activity, and evolution of CRISPR loci in Streptococcus thermophilus. J. Bacteriol. *190*, 1401–1412.

Hughes, K.A., Sutherland, I.W., and Jones, M.V. (1998). Biofilm susceptibility to bacteriophage attack: the role of phage-borne polysaccharide depolymerase. Microbiology *144*, 3039–3047.

Hyman, P., and Abedon, S.T. (2010). Bacteriophage host range and bacterial resistance. Adv. Appl. Microbiol. *70*, 217–248.

Jensen, E.C., Schrader, H.S., Rieland, B., Thompson, T.L., Lee, K.W., Nickerson, K.W., and Kokjohn, T.A. (1998). Prevalence of broad-host-range lytic bacteriophages of *Sphaerotilus natans*, *Escherichia coli*, and *Pseudomonas aeruginosa*. Appl. Environ. Microbiol. *64*, 575–580.

Jessup, C.M., and Forde, S.E. (2008). Ecology and evolution in microbial systems: the generation and maintenance of diversity in phage–host interactions. Res. Microbiol. *159*, 382–389.

Juhala, R.J., Ford, M.E., Duda, R.L., Youlton, A., Hatfull, G.F., and Hendrix, R.W. (2000). Genomic sequences of bacteriophages HK97 and HK022, pervasive genetic mosaicism in the lambdoid bacteriophages. J. Mol. Biol. *299*, 27–51.

Kasman, L.M., Kasman, A., Westwater, C., Dolan, J., Schmidt, M.G., and Norris, J.S. (2002). Overcoming the phage replication threshold: a mathematical model with implications for phage therapy. J. Virol. *76*, 5557–5564.

Katsura, I., and Hendrix, R.W. (1984). Length determination in bacteriophage lambda tails. Cell *39*, 691–698.

Kobiler, O., Rokney, A., Friedman, N., Court, D.L., Stavans, J., and Oppenheim, A.B. (2005). Quantitative kinetic analysis of the bacteriophage lambda genetic network. Proc. Natl. Acad. Sci. U.S.A. *102*, 4470–4475.

Koch, C.M., G., Rudt, F., Kahmann, R., Kanaar, R., Plasterk, R.H.A., Van de Putte, P., Sandulache, R., and Kamp, D. (1987). The invertible G segment. In Phage Mu, Symonds, N.T., Van de Putte, A.P, Howe, M.H., ed. (Cold Spring Harbor Laboratory, Cold Spring Harbor), pp. 75–92.

Kropinski, A.M. (2006). Phage therapy – everything old is new again. Can. J. Infect. Dis. Med. Microbiol. *17*, 297–306.

Kutateladze, M., and Adamia, R. (2008). Phage therapy experience at the Eliava Institute. Med. Mal. Infect. *38*, 426–430.

Kutter, E., De Vos, D., Gvasalia, G., Alavidze, Z., Gogokhia, L., Kuhl, S., and Abedon, S.T. (2009). Phage therapy in clinical practice: treatment of human infections. Curr. Pharm. Biotechnol. *11*, 69–86.

Kutter, E., Kellenberger, E., Carlson, K., Eddy, S., Neitzel, J., Messinger, L., North, J., and Guttman, B. (1994). Effects of bacterial growth conditions and physiology on T4 infection. In Molecular Biology of Bacteriophage T4, Karam, J.D., ed. (ASM Press, Washington, DC), pp. 406–418.

Kwan, T., Liu, J., DuBow, M., Gros, P., and Pelletier, J. (2005). The complete genomes and proteomes of 27 *Staphylococcus aureus* bacteriophages. Proc. Natl. Acad. Sci. U.S.A. *102*, 5174–5179.

Lander, G.C., Tang, L., Casjens, S.R., Gilcrease, E.B., Prevelige, P., Poliakov, A., Potter, C.S., Carragher, B., and Johnson, J.E. (2006). The structure of an infectious P22 virion shows the signal for headful DNA packaging. Science *312*, 1791–1795.

Leiman, P.G., Chipman, P.R., Kostyuchenko, V.A., Mesyanzhinov, V.V., and Rossmann, M.G. (2004). Three-dimensional rearrangement of proteins in the tail of bacteriophage T4 on infection of its host. Cell *118*, 419–429.

Leiman, P.G., Battisti, A.J., Bowman, V.D., Stummeyer, K., Muhlenhoff, M., Gerardy-Schahn, R., Scholl, D., and Molineux, I.J. (2007). The structures of bacteriophages K1E and K1-5 explain processive degradation of polysaccharide capsules and evolution of new host specificities. J. Mol. Biol. *371*, 836–849.

Lenski, R.E. (1984). Coevolution of bacteria and phage: are there endless cycles of bacterial defenses and phage counterdefenses? J. Theor. Biol. *108*, 319–325.

Letellier, L., and Boulanger, P. (1989). Involvement of ion channels in the transport of phage DNA through the cytoplasmic membrane of *E. coli*. Biochimie *71*, 167–174.

Letellier, L., Boulanger, P., Plancon, L., Jacquot, P., and Santamaria, M. (2004). Main features on tailed phage, host recognition and DNA uptake. Front Biosci. *9*, 1228–1339.

Lewis, S. (1925). Arrowsmith (Grosset and Dunlap, New York).

Little, J.W. (2005). Lysogeny, prophage induction and lysogenic conversion. In Phages: Their Role in Bacterial Pathogenesis and Biotechnology, Waldor, M.K., Friedman, D.I., and Adhya, S.L., eds. (ASM Press, Washington, D.C.), pp. 37–54.

Liu, M., Deora, R., Doulatov, S.R., Gingery, M., Eiserling, F.A., Preston, A., Maskell, D.J., Simons, R.W., Cotter, P.A., Parkhill, J., *et al.* (2002). Reverse transcriptase-mediated tropism switching in Bordetella bacteriophage. Science *295*, 2091–2094.

Loc Carrillo, C., Atterbury, R.J., el-Shibiny, A., Connerton, P.L., Dillon, E., Scott, A., and Connerton, I.F. (2005). Bacteriophage therapy to reduce Campylobacter jejuni colonization of broiler chickens. Appl. Environ. Microbiol. *71*, 6554–6563.

Luria, S.E., and Anderson, T.F. (1942). The identification and characterization of bacteriophages with the electron microscope. Proc. Natl. Acad. Sci. U.S.A. *28*, 127–130 121.

Lynch, K.H., Seed, K.D., Stothard, P., and Dennis, J.J. (2010). Inactivation of Burkholderia cepacia complex phage KS9 gp41 identifies the phage repressor and generates lytic virions. J. Virol. *84*, 1276–1288.

Mathews, C.K. (1994). An overview of the T4 developmental program. In Molecular Biology of Bacteriophage T4, Karam, J.D., ed. (American Society for Microbiology Press, Washington, D.C.), pp. 1–8.

Matsuzaki, S., Rashel, M., Uchiyama, J., Sakurai, S., Ujihara, T., Kuroda, M., Ikeuchi, M., Tani, T., Fujieda, M., Wakiguchi, H., *et al.* (2005). Bacteriophage therapy: a revitalized therapy against bacterial infectious diseases. J. Infect. Chemother. *11*, 211–219.

Matsuzaki, S., Yasuda, M., Nishikawa, H., Kuroda, M., Ujihara, T., Shuin, T., Shen, Y., Jin, Z., Fujimoto, S., Nasimuzzaman, M.D., *et al.* (2003). Experimental protection of mice against lethal *Staphylococcus aureus* infection by novel bacteriophage phi MR11. J. Infect. Dis. *187*, 613–624.

Maxted, W.R. (1957). The active agent in nascent phage lysis of streptococci. J. Gen. Microbiol. *16*, 584–595.

McVay, C.S., Velasquez, M., and Fralick, J.A. (2007). Phage therapy of *Pseudomonas aeruginosa* infection in a mouse burn wound model. Antimicrob. Agents Chemother. *51*, 1934–1938.

Merabishvili, M., Pirnay, J.P., Verbeken, G., Chanishvili, N., Tediashvili, M., Lashkhi, N., Glonti, T., Krylov, V., Mast, J., Van Parys, L., *et al.* (2009). Quality-controlled small-scale production of a well-defined bacteriophage cocktail for use in human clinical trials. PLoS ONE *4*, e4944.

Merril, C.R., Biswas, B., Carlton, R., Jensen, N.C., Creed, G.J., Zullo, S., and Adhya, S. (1996). Long-circulating bacteriophage as antibacterial agents. Proc. Natl. Acad. Sci. U.S.A. *93*, 3188–3192.

Michel-Briand, Y., and Baysse, C. (2002). The pyocins of *Pseudomonas aeruginosa*. Biochimie *84*, 499–510.

Miedzybrodzki, R., Fortuna, W., Weber-Dabrowska, B., and Gorski, A. (2009). A retrospective analysis of changes in inflammatory markers in patients treated with bacterial viruses. Clin. Exp. Med. *9*, 303–312.

Miller, R.W., Skinner, E.J., Sulakvelidze, A., Mathis, G.F., and Hofacre, C.L. (2010). Bacteriophage therapy for control of necrotic enteritis of broiler chickens experimentally infected with *Clostridium perfringens*. Avian Dis. *54*, 33–40.

bacteriophage T7, a protein that mimics B-form DNA. Mol. Cell *9*, 187–194.

Wall, S.K., Zhang, J., Rostagno, M.H., and Ebner, P.D. (2010). Phage therapy to reduce preprocessing Salmonella infections in market-weight swine. Appl. Environ. Microbiol. *76*, 48–53.

Wang, J., Hu, B., Xu, M., Yan, Q., Liu, S., Zhu, X., Sun, Z., Reed, E., Ding, L., Gong, J., *et al.* (2006). Use of bacteriophage in the treatment of experimental animal bacteremia from imipenem-resistant *Pseudomonas aeruginosa*. Int. J. Mol. Med. *17*, 309–317.

Wang, J., Michel, V., Hofnung, M., and Charbit, A. (1998). Cloning of the J gene of bacteriophage lambda, expression and solubilization of the J protein: first *in vitro* studies on the interactions between J and LamB, its cell surface receptor. Res. Microbiol. *149*, 611–624.

Watanabe, K., Takesue, S., and Ishibashi, K. (1977). Reversibility of the adsorption of bacteriophage PL-1 to the cell walls isolated from *Lactobacillus casei*. J. Gen. Virol .*34*, 189–194.

Watanabe, R., Matsumoto, T., Sano, G., Ishii, Y., Tateda, K., Sumiyama, Y., Uchiyama, J., Sakurai, S., Matsuzaki, S., Imai, S., *et al.* (2007). Efficacy of bacteriophage therapy against gut-derived sepsis caused by *Pseudomonas aeruginosa* in mice. Antimicrob. Agents Chemother. *51*, 446–452.

Weber-Dabrowska, B., Mulczyk, M., and Gorski, A. (2000). Bacteriophage therapy of bacterial infections: an update of our institute's experience. Arch. Immunol. Ther. Exp. (Warsz) *48*, 547–551.

Weber-Dabrowska, B., Zimecki, M., Mulczyk, M., and Gorski, A. (2002). Effect of phage therapy on the turnover and function of peripheral neutrophils. FEMS Immunol. Med. Microbiol. *34*, 135–138.

Weinbauer, M.G., and Rassoulzadegan, F. (2004). Are viruses driving microbial diversification and diversity? Environ. Microbiol. *6*, 1–11.

Wendlinger, G., Loessner, M.J., and Scherer, S. (1996). Bacteriophage receptors on Listeria monocytogenes cells are the N-acetylglucosamine and rhamnose substituents of teichoic acids or the peptidoglycan itself. Microbiology *142*, 985–992.

Westwater, C., Kasman, L.M., Schofield, D.A., Werner, P.A., Dolan, J.W., Schmidt, M.G., and Norris, J.S. (2003). Use of genetically engineered phage to deliver antimicrobial agents to bacteria: an alternative therapy for treatment of bacterial infections. Antimicrob. Agents Chemother. *47*, 1301–1307.

Wilkens, K., and Ruger, W. (1994). Transcription from early promoters. In Molecular Biology of Bacteriophage T4, Karam, J.D., ed. (American Society for Microbiology, Washington D.C.), pp. 132–141.

Williams, S.R., Gebhart, D., Martin, D.W., and Scholl, D. (2008). Retargeting R-type pyocins to generate novel bactericidal protein complexes. Appl. Environ. Microbiol. *74*, 3868–3876.

Wolcott, R.D., and Rhoads, D.D. (2008). A study of biofilm-based wound management in subjects with critical limb ischaemia. J. Wound Care *17*, 145–*148*, 150–*142*, 154–145.

Wommack, K.E., and Colwell, R.R. (2000). Virioplankton: viruses in aquatic ecosystems. Microbiol. Mol. Biol. Rev. *64*, 69–114.

Wood, J.M., and Levandowski, R.A. (2003). The influenza vaccine licensing process. Vaccine *21*, 1786–1788.

Wright, A., Hawkins, C.H., Anggard, E.E., and Harper, D.R. (2009). A controlled clinical trial of a therapeutic bacteriophage preparation in chronic otitis due to antibiotic-resistant *Pseudomonas aeruginosa*; a preliminary report of efficacy. Clin. Otolaryngol. *34*, 349–357.

Yoichi, M., Abe, M., Miyanaga, K., Unno, H., and Tanji, Y. (2005). Alteration of tail fiber protein gp38 enables T2 phage to infect *Escherichia coli* O*157*,H7. J. Biotechnol. *115*, 101–107.

Young, R., and Wang, I.N. (2006). Phage lysis. In The Bacteriophages, Calendar, R., ed. (Oxford University Press, Oxford), pp. 104–126.

Zhang, X., McDaniel, A.D., Wolf, L.E., Keusch, G.T., Waldor, M.K., and Acheson, D.W. (2000). Quinolone antibiotics induce Shiga toxin-encoding bacteriophages, toxin production, and death in mice. J. Infect. Dis. *181*, 664–670.

Zimecki, M., Artym, J., Kocieba, M., Weber-Dabrowska, B., Borysowski, J., and Gorski, A. (2009). Effects of prophylactic administration of bacteriophages to immunosuppressed mice infected with *Staphylococcus aureus*. BMC Microbiol. *9*, 169.

Zimecki, M., Artym, J., Kocieba, M., Weber-Dabrowska, B., Lusiak-Szelachowska, M., and Gorski, A. (2008). The concerted action of lactoferrin and bacteriophages in the clearance of bacteria in sublethally infected mice. Postepy. Hig. Med. Dosw. (Online) *62*, 42–46.

Part V

Enabling Antibacterial Discovery

Role of Within-host Models in Target Identification, Compound Optimization and Dose Selection

18

Juilee Thakar* and Eric T. Harvill

Abstract

Integrated pharmacokinetic–pharmacodynamic models are commonly used to study the *in vivo* dynamics of antimicrobial agents and bacterial pathogens. These models are extremely useful for understanding the properties of antimicrobial agents such as absorption, transport, rate of binding, etc. However, they fail to consider within-host aspects of the infectious process that are likely to affect the bacteria–host interactions. For example, immune-mediated mechanisms to contain bacteria or limit their access to nutrients can also affect the access of a drug to its bacterial target. Alternatively, pathogens have various strategies to sequester themselves from host immune mechanisms that can also affect the access of therapeutic agents. The search for new antibacterial agents that will be effective *in vivo* can be substantially informed by an understanding of the within-host dynamics of bacterial pathogens. Mathematical modelling of immune responses can assist in this process by providing new predictions, by offering mechanistic understanding and by revealing the gaps in our current understanding. Such models are based on experiments that reveal the components of the immune system that play important roles during infections. But knowing the components alone usually provides only a static picture of bacterium–host interactions. Mathematical models aim to use the information obtained from experiments to construct the interactions and dependencies between various components. Thus mathematical models offer a mechanistic understanding of the interplay between various immunological processes and simulations of these models give a dynamic view of the entire process. In this chapter we will first provide an overview of pharmacokinetic and pharmacodynamic models followed by a review of some of the immunological processes involved in bacterial infections which are generally ignored in pharmacodynamic models but are likely to affect access or activity of treatments. We will then discuss the development of mathematical models by different approaches. We will end the chapter by exploring implications of these models in the discovery of new antibacterial agents.

Introduction

In recent years pharmacokinetic–pharmacodynamic (PK–PD) models have been commonly used to quantitatively characterize the distribution of the administered drugs *in vivo* and their effects on health and disease. Pharmacokinetic models aim to computationally track the changes in drug concentration and, when integrated with pharmacodynamic models, focus on the access and activity of the drugs without necessarily considering how the disease process may affect various parameters. Within-host dynamic (WHD) models study the dynamics arising from the interactions between host components, especially immune components and the pathogen factors. These models can be developed at varying levels of complexity depending on the biological

*Corresponding author

knowledge about the system under construction. The models can give a dynamic insight into the static view generally obtained from experimental results. Depending on the specific models they can offer a mechanistic understanding of the interplay between various immunological processes, their interactions with the pathogens and quantitative rate estimates. The objectives of both PK–PD models and WHD models converge since they aim to understand aspects of dynamics within the host. Unlike PK–PD models, WHD models incorporate some of the known bacterium–host interactions that are likely to affect the outcomes.

Specifically, the outcome of antimicrobial treatments on pathogenesis of bacterial infections are affected by at least three key factors that WHD models can incorporate, allowing their effects to be modelled and studied; first, the pathogen strategies that might sequester them from host immune mechanisms, second, immune-mediated processes that might affect the density and distribution of a drug relative to its target and third, intercellular and intracellular crosstalk within the immune system that can be affected by any treatment. Antibiotic treatment of systemic bacterial infection comes with some risks. The sudden release of the components of bacteria killed by antibiotic treatment can be more deadly than the live bacteria. However, similar antibiotic treatment of bacteria sequestered within a granuloma might have no effect on bacterial numbers or the long term health of the patient. Understanding and considering the within-host dynamics of the bacterium–host interactions are therefore critical to the design and discovery of therapeutics that will be most effective. WHD models can dissect each of these parameters to reveal the processes affecting drug action and its direct and indirect effect. By breaking these processes into their key components and identifying the critical steps, mathematical modelling that includes consideration of immune responses can, for example, identify key targets for potential interventions. Importantly, WHD models can provide new predictions for the effects of specific interventions, guiding new experiments to either identify new targets or reveal gaps in our current understanding.

In this chapter we will first provide an overview of PK–PD models followed by a review of important immunological processes which are generally ignored in PK–PD models. We will then discuss the development of mathematical models by different approaches. We will end the chapter by exploring the implications of these models in the discovery of new antibacterial agents.

Pharmacokinetic–pharmacodynamic (PK–PD) modelling

Integrated pharmacokinetic and pharmacodynamic (PK–PD) modelling quantitatively characterizes the causal path between drug administration and its effect (Fig. 18.1). PK-PD modelling of phase I data (biomarker response) is often the driver for phase II study design. Recently, PK–PD modelling has been used for the optimization of the dosing regimen and the delivery profile of drugs in phase II and III clinical trials (Breimer and Danhof, 1997; Peck *et al.*, 1994). Pharmacokinetic modelling analyses the effects of the body on the drug including absorption, distribution, metabolism and elimination. There are many factors affecting individual PK, including but not limited to, age, sex, disease state, medications, genetics, ethnicity etc. Pharmacodynamics refers to the target binding and the effects of the drug on the biological system considering specific properties such as time-dependent transduction mechanisms and homeostatic feedback mechanisms. Transduction mechanisms include the

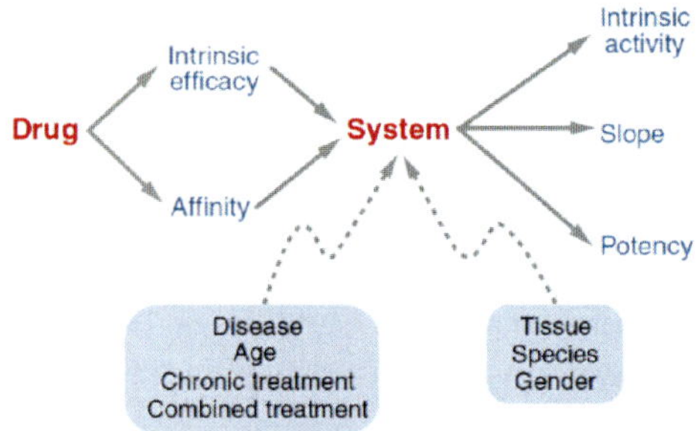

Figure 18.1 The shape and location of *in vivo* drug concentration-effect relationships is determined by drug-specific (affinity, intrinsic efficacy) and biological system-specific (transducer function relating receptor activation to effect) properties. Reproduced with permission from Van der Graaf and Danhof (1997).

interactions between a drug and its target and in cases when multiple targets are available, the pathways activated by those targets are also studied. Pharmacodynamic modelling also has important contributions to the study of synergistic, additive or antagonistic effects of one drug on another, or more complex combinations.

'Within host dynamic' models (WHDs) will substantially influence our understanding of pharmacodynamics, particularly in regards to the process of target activation leading to pharmacological responses. Typically, the binding of a drug to its target activates a cascade of electrophysiological and/or biochemical events resulting in the observable biological response. Targets consist of messenger molecules such as diacylglycerol and can mediate regulation of the expression of second messengers. Simple models (Dayneka *et al.*, 1993) have been used to study the effect of drugs on the physiological entity,

$$\frac{dR}{dt} = k_{in} - k_{out}R \tag{18.1}$$

where R is a physiological entity that is constantly being produced and eliminated in time, k_{in} is the rate constant for the production of the physiological entity and in its simplest form it is independent of the concentration of the drug. Thus in this case we assume that the production of R is constitutive. k_{out} is the rate constant for the loss of R, which is dependent on its own concentration. The drug effect is described as stimulation or inhibition of the factors controlling either the input or the dissipation of drug response in a direct concentration-dependent manner. As an example, let us evaluate two cases in which a drug can affect k_{in} or k_{out}. First, the production of R is activated by the drug when R in turn can inhibit the pathogen. This could be incorporated in the equations as follows:

$$\frac{dR}{dt} = k_{in}d - k_{out}R \tag{18.2}$$

where d is the *in vivo* concentration of the drug. Second, the drug can also induce the removal of R by directly interacting with it, which can be given as

$$\frac{dR}{dt} = k_{in}d - k_{out}Rd \tag{18.3}$$

WHD models are particularly useful in understanding pharmacodynamics. To exemplify this we can use the above example in which the interaction between pathogen (p) and R can be incorporated as

$$\frac{dR}{dt} = k_{in}p - k_{out}Rd \tag{18.4}$$

Such simple models can be used to model complex mechanisms by incorporating series of differential equations to describe the cascade of events across receptor activation and final response (Sun and Jusko, 1998).

Commonly for antibacterial agents, PK–PD models have been developed to study time-kill curves (Nielsen *et al.*, 2007; Tam *et al.*, 2007) which assess the influence of pharmacokinetic parameters on the growth curves of bacteria. In PK–PD and WHD, bacterial growth dynamics are given by exponential or logistic form (Fig. 18.2),

$$\text{exponential: } \frac{dB}{dt} = rB$$
$$\text{logistic: } \frac{dB}{dt} = rB\left(1 - \frac{B}{K}\right) \tag{18.5}$$

where $B(t)$ is bacterial concentration at time t and the equations describe the change in the concentration over small time unit dt. r is a rate of proliferation and it stays the same in both exponential and logistic forms. Exponential function usually cannot simulate the long term dynamics of bacteria since the actual growth curves of bacteria have four phases, namely lag, growth, stationary and decay. Hence the logistic equation has the term carrying capacity K which encompasses the various elements such as resource limitation affecting the growth of the population.

Now in equation 18.2 the effect of inhibition of bacteria by drug can be incorporated by introducing a function g of the drug d;

$$\frac{dB(t)}{dt} = f[B] - g[d]B \tag{18.6}$$

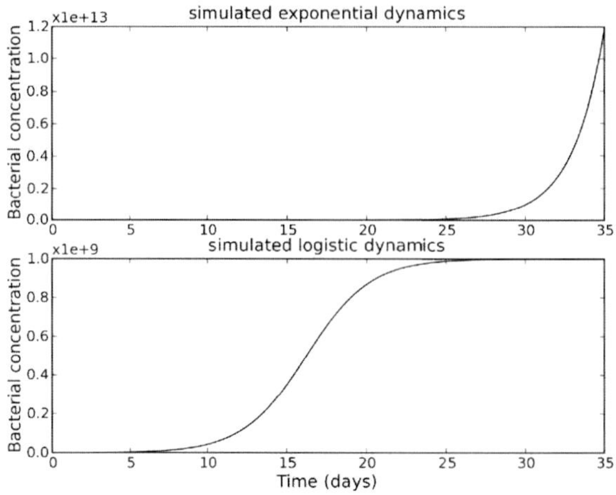

Figure 18.2 Simulated exponential (upper panel) and logistic (lower panel) functions used to represent *in vivo* bacterial dynamics. Note the limit of 10^9 on the bacterial numbers in logistic function representing carrying capacity.

Here f is either an exponential or logistic function and g is usually a function similar to Michaelis–Menten. Pharmacodynamic modelling estimates parameters such as bacterial growth rate constant, maximum killing rate constant and concentration for half-maximum effect. It has been shown that these parameters provide a rational basis for determination of a more optimal dosing regimen in terms of dose and dosing interval. Because antibacterial agents are used such that they work in combination with host defence mechanisms, the difference between the minimal inhibitory concentrations (MIC) and minimal bactericidal concentrations (MBC) determined *in vitro* becomes important. Bacteriostatic concentrations are best measured as the MIC, which is generally lower than the MBC. When host defences are active and effective bacteriostatic agents can be given in lower doses and for shorter time, allowing bacterial clearance by host defence mechanisms (Levison and Levison, 2009). However, when host defence mechanisms are ineffective the MBC must be maintained for long enough to eliminate all of the bacteria. Thus better descriptions of host defence mechanisms can substantially improve modelling and design of drug treatment regimens.

The increasing inclination to construct mechanism-based pharmacodynamic model decreases the gap between WHD and PK–PD models. A recent model by Bulitta *et al.* (2010) not only describe the receptor occupancy, drug (colistin) effect and inoculum effect but also track the concentrations of viable bacteria in susceptible cells, cells.

Overview of mathematical modelling of 'within host dynamics'

Some bacterial infections have received considerable attention from the modelling world, particularly infections by *Helicobacter pylori*, *Mycobacterium tuberculosis* and *Neisseria gonorrhoeae* (Blaser and Parsonnet 1994; Blower *et al.*, 1995; Blaser and Kirschner, 1999). Though not a main focus of this chapter, models of infection epidemics pre-date WHD models. These models have defined critical aspects such as the notion of a core population in sexually transmitted diseases (Hethcote and Yorke, 1984) as well as approaches to determine herd immunity levels within populations, all important findings that currently guide vaccination policies (Anderson, 1982). Regarding bacterial infections, issues such as drug resistance, transmission rates, epidemic trends, and the effects of treatment (Kirschner and Marino, 2005) and vaccinations have also been addressed by epidemiological models.

Dynamic models can be constructed to study both temporal and spatial dynamics. Dynamic models have as input information (i) the interactions and regulatory relationships among components (i.e. the interaction network), (ii) the manner in which the strength of the interactions depends on the state of the interacting components (i.e. the transfer functions, including kinetic parameters for each interaction) and (iii) the initial state of each component in the system. Given these inputs, the model will output the time evolution of the state of the system, for example the system's response to the presence or absence of a given signal. Temporal dynamics are simpler and are often studied without the additional complexities of spatial information owing to the greater conceptual and computational efforts required for the latter. However, approximations at various levels of details are available that allow for the inclusion of some spatial information.

In partial differential equations, intracellular space is divided into elements that can be specified by their volumes. Molecular species in different volume elements are treated like different species, with diffusion between volume elements being equivalent to interspecies transitions. Thus, these models are used to examine spatial dynamics at the molecular level, such as diffusion of signals through time and space (Meier-Schellersheim *et al.*, 2006) or receptor complex formation (Slepchenko *et al.*, 2003), and at the cellular level, such as age-structured dynamics of infections (Althaus *et al.*, 2009).

Spatial aspects can also be added in *finite state automata* models. These models focus on cell state transitions and their consequences for intercellular communication as opposed to details of intracellular biochemistry. They consist of a set of states and rules specifying how each state reacts to input signals, for instance, by switching from one state to another. To add spatial aspects, components of the system are assumed to be on a grid so that coordinates are assigned for the components. Generalizations of the simple structure of cellular automata treat the single 'cells' as agents that can carry their states with them as they move on the grid representing extracellular space. Such agent-based models have been used for simulations of multicellular systems such as the adaptive immune system (Celada and Seiden, 1992; Efroni *et al.*, 2007; Mata and Cohn, 2007), excitable tissues (Bartocci *et al.*, 2008), or populations of migrating cells (Hatzikirou and Deutsch, 2008), whereas others aim to provide generic platforms for studying how tissue, organ and organism-level biological phenomena may emerge from the behaviour of simple, locally interacting cells.

The modelling methods for studying temporal dynamics are roughly categorized in Fig. 18.3. Continuous and deterministic models are formulated as differential equations based on mass-action or Michaelis–Menten kinetics for the production and decay of all components (Irvine and Savageau, 1990). With sufficiently thorough knowledge of the biochemical interactions comprising a system, i.e. a compilation of all pairwise interactions among the system's components, measurements or estimates of kinetic parameters such as dissociation constants and half-lives, and a known initial state for the system, it is possible to quite accurately reproduce the dynamic behaviour of a complex biological system (von Dassow *et al.*, 2000; Rao *et al.*, 2002) by describing the constituent interactions as coupled (usually, nonlinear) differential equations. For example, using a *continuous deterministic model*, von Dassow *et al.* (2000) reproduced the expression patterns of segment polarity genes in *D. melanogaster*, and demonstrated that these patterns are remarkably resistant to changes in the kinetic parameters governing the biochemical reactions that result in gene expression.

Boolean models assume binary states for network nodes and are formulated as logical rules describing the change in state of each regulated node as a function of its regulators. The utility of Boolean dynamic models lies in their ability to predict large-scale dynamic trends *in the absence of* detailed kinetic parameters. For example, Boolean models successfully described the wild type or mutant expression patterns of segment polarity genes in *D. melanogaster* and predicted a significant error correcting ability for the gene regulatory network (Albert and Othmer, 2003; Chaves *et al.*, 2005). Boolean modelling can also reproduce observed gene expression patterns and mutant behaviour of the floral organs of *Arabidopsis thaliana* (Mendoza and Alvarez-Buylla, 1998). Recently, asynchronous Boolean modelling was implemented to study the sequence of events in bacterial infections by different *Bordetella* species

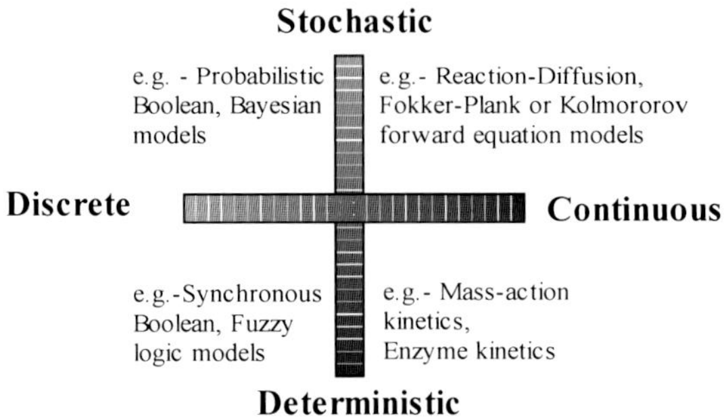

Figure 18.3 Classification of mechanistic methods for modelling dynamics of biological systems. In each quadrant we have given some of the common examples of modelling methods used in that category. Most of the methods offer flexibility to incorporate resolution required for the specific questions under examination (Thakar and Albert, 2009).

(Thakar *et al.*, 2007). These models have not only reproduced the known observations but have also generated novel predictions which have been experimentally validated. The fact that simple Boolean methods can capture the same (broad) dynamic trends and predict the same asymptotic, biologically viable behaviours as their far more detailed differential equation-based counterparts often makes them a more feasible approach for studying a system's dynamics. Moreover, many observations support the use of Boolean formulations. First, the input-output curves of many regulatory relationships are strongly sigmoidal and can be approximated by step functions (von Dassow *et al.*, 2000). Second, regulatory networks maintain their function even when faced with fluctuations in components and reaction rates (Alon *et al.*, 1999; Eldar *et al.*, 2002; Csete and Doyle, 2004; Conant and Wagner, 2004). Thus, various qualitative and quantitative approaches can be developed depending on the availability of the data to study within host dynamics.

Hybrid dynamic models meld Boolean logical functions with continuous synthesis and decay (McAdams and Shapiro, 1995; Yuh *et al.*, 2001). In this approach, the rates of change in concentration of effector molecules are expressed as differential equations, and combinatorial regulation is described by Boolean functions. Genes that are known to have a clear activation threshold within a narrow range of effector concentrations are assumed to act as threshold-dependent ON/OFF switches, while the expression levels of genes for which activation can occur over a much broader range of effector concentrations are instead described as continuous, graded functions (McAdams and Shapiro, 1995; Yuh *et al.*, 2001). For example, dynamics of the cis-regulatory system governing the embryonic expression of the Endo16 sea urchin gene has been successfully modelled with a hybrid model, demonstrating that combinatorial transcriptional regulation is essentially logical and hard-wired into an organism's DNA (Yuh *et al.*, 2001). This type of hybrid approach is appealing in that its continuous features can incorporate a great deal of quantitative detail while its discrete features allow potential uncertainty in interactions (Chaves *et al.*, 2006). Being less computationally intensive than strict deterministic models, and requiring fewer initial estimates and less initial input, hybrid approaches can be applied to larger networks than is possible with pure continuous models.

Continuous deterministic models have been successful in estimating rates of killing by cytotoxic T lymphocytes (Regoes *et al.*, 2007) and understanding T-cell differentiation and development (Borghans *et al.*, 1999; de Boer, 2002). Most of these models have been developed to analyse viral dynamics (Clark *et al.*, 1999; Dahari *et al.*, 2009; Handel *et al.*, 2009) and there are only limited numbers of models of host-bacteria infections (Marino and Kirschner, 2004; Kirschner and Marino, 2005; Thakar *et al.*, 2007, 2009). Some of the interesting features of bacterial infections such as persistence in the host (exemplified by mycobacterial infections; Antia *et al.*, 1996; Ganusov *et al.*, 2002), and invasion of gut flora and plasmid transfer among bacteria in the large intestine have been successfully studied by modelling. Discrete dynamic models have also been constructed to study the intracellular (Franke *et al.*, 2008) and cellular interactions (Thakar *et al.*, 2007, 2009) involved in bacterial infection. These models reveal connectivity between various immune components and its effect on the dynamics.

Though we do not discuss the stochastic within-host models any further it is important to note that there exist a number of approaches aimed at incorporating stochasticity and discrete events into dynamic models of cellular systems (for a review, see Andrews and Arkin, 2006). In general, stochastic models either formulate a *master equation* that follows the time evolution of the probabilities of each of a system's possible configurational states (the probability density function), or they append stochastic (noise) terms to differential equations. The former method usually uses a Monte Carlo (Gillespie) algorithm to select a state of the system, compatible with the master equation, at each time step. The sequence of these states will form one time course for the system, and the results of multiple iterations will then be interpreted statistically. The time-evolution of a system's probability density function can also be estimated from stochastic differential equations by using a Monte Carlo algorithm to select (multiple times) the noise addendums to the differential equations; statistics are then used to interpret the results in terms of probability densities (Morton-Firth and Bray, 1998; Rao *et al.*, 2002). This type of approach adequately describe observed noise, variability and heterogeneity of biological systems over a range of scales or biological organizations (Wilkinson, 2009). Their use has been increased recently with the availability of powerful tools for computation particularly at the single cell level (Hodgkin, 2007).

Characteristic within-host infection dynamics of bacterial infections

Bacteria make a wide variety of products that activate cascades of highly coordinated immune responses resulting in a complex interplay during the infection. Innate immune responses triggered upon initial recognition of pathogen-associated molecular patterns lead to the early induction of inflammatory responses. If this initial response is not sufficient to clear bacteria a long-term adaptive immune response involving T cells and antibodies is generated. Pathogens can evade and subvert the immune responses, allowing them to colonize and persist. Various cellular and intra-cellular pathways are affected by pathogen modulations. As predicted by WHD modelling, key central signalling components of host immunity are common targets for bacterial manipulations. For example, the intra-cellular signalling molecules NF-κB and mitogen-activated protein (MAP) kinase pathways are common targets. Some of the organisms modulating these pathways are species of the genus *Yersinia* (YopJ/P proteins; Monack *et al.*, 1997; Orth *et al.*, 1999), *Salmonella* (AvrA, SpvC; Collier-Hyams *et al.*, 2002; Jones *et al.*, 2008; Mazurkiewicz *et al.*, 2008) and *Shigella* (OspF). Various components of intracellular signalling pathways modulated by bacteria result in modulation of cellular behaviour. For example, recruitment of neutrophils is inhibited by pertussis toxin (Kirimanjeswara *et al.*, 2005a), while the tyrosine phosphatase YopH blocks macrophage and neutrophil phagocytosis and also interferes with the ability of B cells to up-regulate MHC class II and co-stimulatory molecules. Importantly, although substantial information about specific downstream effects may be available, the complexity of the system makes it increasingly difficult to intuitively predict the effects of disrupting particular key signalling steps, increasing the value of WHD models that can incorporate multiple downstream effects.

In general, bacterial infections can be classified as either acute or persistent. The bacteria causing these infections have different strategies; the acute pathogens aim to replicate rapidly and to optimize transmission whereas chronic pathogens aim for long-term residence of a niche within the host. These different strategies can involve different strategies for the modulation of host immunity, although often targeting similar sets of key signalling molecules such as NF-κB and MAP-kinases, reflecting the relative importance of these in the function of the immune system.

Recently it has been suggested that the properties of signalling networks explain the choice of cellular targets of pathogen factors. Many independent studies have revealed that biological systems show properties of scale free networks (Jeong *et al.*, 2000) which are characterized by few highly connected nodes termed as 'hubs' and many nodes with lower connectivity (Fig. 18.4). Thus, in scale-free networks the likelihood that any given node has a specified number of connections is inversely related to that number of connections. Brodsky and Medzhitov (2009) propose that pathogens responsible for causing acute infection tend to target central components that are hubs of cellular signalling pathways, causing global disruption of the host response. On the contrary, pathogens that cause chronic or persistent infections tend to target more peripheral signalling network components to promote persistence, often resulting in more specific effects on a smaller set of immune functions. Thus acute infections can lead to transient generalized immune modulation, whereas chronic infections tend to lead to much more limited and specific immune modulation that lasts much longer.

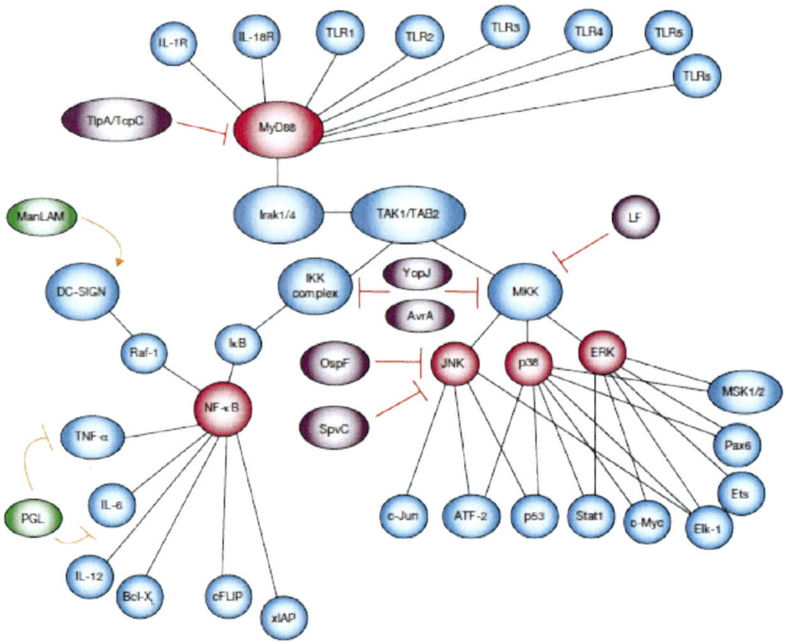

Figure 18.4 Innate immune network components (blue and pink ovals) and their targeting by bacterial virulence factors. Virulence factors of acute pathogens (purple ovals) preferentially target central hubs (pink ovals) or essential regulators of hub proteins in the signalling network. A large body of work from several laboratories has contributed to this understanding of how acute pathogens disrupt immune defence networks. Lethal factor (LF) of *B. anthracis* targets proteins of the MAPK kinase (MKK) family by cleaving and inactivating them (Duesbery *et al.*, 1998), whereas the YopJ/AvrA family of virulence factors targets both MKK and IKK proteins (Collier-Hyams *et al.*, 2002; Jones *et al.*, 2008; Orth *et al.*, 1999). SpvC and OspF are members of a newly identified class of bacterial enzymes, phospho-threonine lyases, which inhibit MAPK signalling by preventing their phosphorylation (Mazurkiewicz *et al.*, 2008). TlpA and TcpC, on the other hand, are bacterial TIR-domain-containing proteins that interfere with TLR signalling by targeting MyD88 (Newman *et al.*, 2006; Cirl *et al.*, 2008). Inhibition of these targets leads to major disruption of transcriptional programs required for robust immune responses. By contrast, how persistent pathogens selectively modulate specific aspects of the immune response is only beginning to be understood. A number of mycobacterial glycolipids, such as ManLAM and PGL (green ovals), modulate the innate immune response either by interacting with the surface protein DC-SIGN on dendritic cells (ManLAM; Geijtenbeek *et al.*, 2003), or inhibiting production of pro-inflammatory cytokines by macrophages (PGL; Constant *et al.*, 2002). However, PGL expression seems to be selected against in Mycobacteria, suggesting selective pressure against virulence factors that might act to globally disrupt the immune response. Figure reproduced from Brodsky and Medzhitov (2009).

Examples of within-host bacterial infection models

Mathematical models have been developed to analyse the co-evolution of host and bacteria in acute and persistent bacterial infections. Antia *et al.* wrote two consecutive models examining simplified immune response to acute and persistent infections (Antia *et al.*, 1996; Ganusov *et al.*, 2002). In the model of acute infections the virulence of the pathogens was assumed to be proportional to its growth rate in the host. Their results indicated that optimal transmission of the parasites would result if the 'parasite' e.g. bacteria had an intermediate rate of growth. This would result in the evolution and maintenance of an intermediate level of parasite virulence. In the second paper, the authors modelled the dynamics of persistent mycobacterial infections. This model predicted that the initial persistence of the parasite may be achieved by very-slow-growing parasites or by parasites having a refuge that is inaccessible to the immune response. They also suggested that escape from immune control at a later time might be a consequence of processes that limit the generation or persistence of active immune functions. In their scenario, the persistent parasite antigens prevent the generation of new parasite-specific cells from the thymus and the existing parasite-specific cells are eventually eliminated. Although they specifically propose antigen deletion of T cells in the thymus and the presence of a limit to the maximal number of divisions a T cell can undergo, it is important to point out that these are simply examples. Other mechanisms for modulating immunity over time (e.g. Treg cells) have been described, and many more are likely to be discovered.

In addition to the rather generic treatment of persistent and acute bacterial infections, substantial recent work has involved the modelling of the WHD of specific pathogens, as described below.

Mathematical models of *Mycobacterium tuberculosis* infections

Mycobacterium tuberculosis (TB) is one of the oldest human pathogens and is still amongst the most successful, infecting nearly a third of the human population worldwide. Although experimental data are available to establish reasonable values for some of the parameters required for modelling, much of these data come from the very difficult non-human primate infection model. Although tuberculosis is persistent in almost 90% of human cases, only about 5% of the infected individuals have active, infectious disease, the rest remaining dormant until reactivation occurs. Understanding the emergence of disease is particularly important and difficult. In the absence of reliable markers of the status of *M. tuberculosis* infections in humans, mathematical modelling has been crucial in guiding the field. Kirschner's group has developed several mathematical models to capture global dynamics regarding trafficking between the lung and its associated draining hilar lymph node (Marino and Kirschner, 2004; Marino *et al.*, 2004) and local dynamics occurring within a cell, within a granuloma (Gammack *et al.*, 2004) or within lungs of infected individuals (Wigginton and Kirschner, 2001).

In these models bacterial load has been used as the marker of TB progression. The models indicate that latency is a state whereby bacterial numbers in the lung are low and relatively stable; by contrast, during active TB, bacterial burdens increase exponentially (Fig. 18.5).

To evaluate the role of cell-mediated immunity on different stages of TB a continuous deterministic mathematical model using a set of ordinary differential equations has been developed. This model makes specific predictions concerning the roles of IL-10, IL-12, IFN-γ, and IL-4 and describes key elements of cell-mediated immunity that lead to latency or active disease (Wigginton and Kirschner, 2001). The mathematical model highlighted the factors determinative in triggering and enhancing an efficient adaptive T-cell response, as follows: (1) cellular factors, including activation of resting macrophages and T-cell killing of infected macrophages, and (2) environmental cues, such as trafficking and recruitment induced by different chemokine profiles and cell maturation. Another model has indicated that delaying the dendritic cell trafficking and migration into lymph nodes leads to altered cell-mediated immunity leading to active TB (Marino *et al.*, 2004).

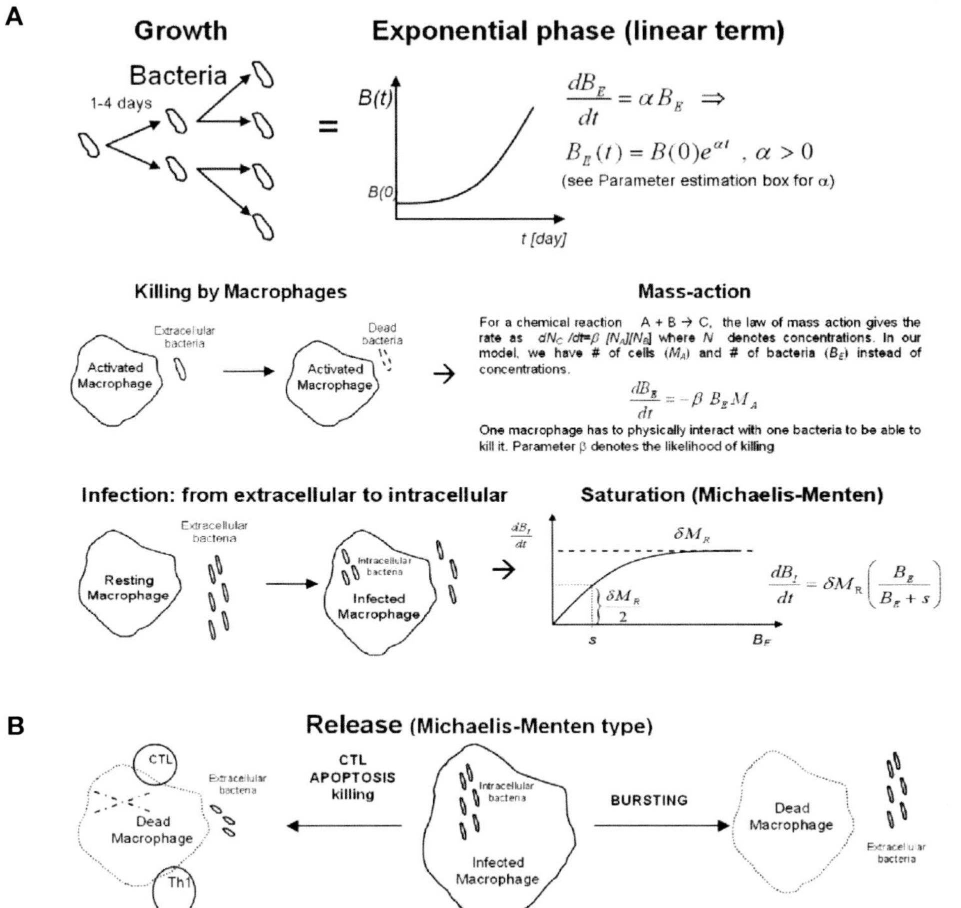

$$
\begin{aligned}
\frac{dB_E}{dt} = & +\alpha B_E && \text{(growth)} \\
& - \beta_1 M_R B_E - \beta_2 M_A B_E && \text{(killing due to macrophages)} \\
& - \beta_3 \, IDC \; B_E && \text{(killing due to dendritic cells)} \\
& - \delta M_R \left(\frac{B_E}{B_E + s_1} \right) && \text{(uptake by macrophages)} \\
& + \gamma M_I \left(\frac{R_1}{R_1 + s_2} \right) && \text{(release due to CTL/apoptosis)} \\
& + \varphi M_I \left(\frac{B_I^2}{B_I^2 + (X_I)^2} \right) && \text{(release due to bursting)}
\end{aligned}
$$

C

LEGEND

$\frac{dB_E}{dt}$: variation in # of B_E over time
B_E: extracellular bacteria
B_I: intracellular bacteria
M_R: resting macrophages
M_A: activated macrophages
M_I: infected macrophages
IDC: immature dendritic cells
s_i: half-saturation terms (i=1,2,3)
β_i: rate of interaction (i=1,2,3)
δM_R: max uptake
γM_I: max release (apoptosis)
T_1: T helper Type 1
$R_1 = \frac{T_1}{M_I}$ (Th1-M_I ratio)
CTL: Cytotoxic T cells
φM_I: max release (bursting)
X_I: average # of bacteria released

Figure 18.5 Extracellular bacteria dynamics: an example of the modelling process. (A) From biology to math: an example of how biological concepts are translated into mathematical equations. Parameter estimation of the growth rate a of *Mycobacterium tuberculosis*: *in vitro* estimates for doubling times of H37Rv laboratory strain within macrophages ranged from 28 to 96 hours. In mouse lung tissue, H37Rv was estimated to have a doubling time of 63.2 hours. We can estimate the intracellular versus extracellular growth rates from these values: rateZln2/doubling time/aBIZ (0.007, 0.024), aBEZ0.011 (per day). (B) Differential equation for extracellular bacteria, BE. Each equation of the model represents the incremental variation of a certain quantity over time (days): pg/ml (×10^6 cells) for cytokine concentrations, cell/cm^3 of tissue for cellular variables and bacteria/cm^3 of tissue for *M. tuberculosis*. (C) Table of symbols (reproduced with permission from Kirschner and Marino (2005).

Granuloma formation during TB has been studied by an agent-based model which captures spatial aspects of the process (Segovia-Juarez *et al.*, 2004). Increased recruitment of resting macrophages to the infection site is positively correlated with bacterial load, most likely by providing additional cells for productive infection. This suggests the paradoxical conclusion that inflammation might be detrimental, and further, that unless macrophages become activated, they serve to propagate infection rather than halt it. This model also showed that there is a strong correlation between the intracellular growth rate of *M. tuberculosis* and granuloma size. Using sensitivity analysis they showed that in the first 12 days of infection higher growth rates are significantly more favourable for granuloma growth, but between one to three months, granulomas grow larger when growth rates are slowest. Finally, after three months, the correlation is again positive, where slightly higher growth rates favour granuloma growth. In TB there is a balance regarding necrosis and this model predicts that the necrotic centre is a site within which extracellular bacteria reside (Kirschner and Marino, 2005).

Mathematical models of infections by *Bordetella* species

Acute infections by the bordetellae offer good examples of the interplay between host's immune components and bacterial factors since these bacteria modulate the immune responses so that they can establish themselves in the host but eventually healthy hosts can clear these infections. The roles of toll like receptors, T helper cells and cytokines such as IL-10 and IFN-γ have been well characterized (Mann *et al.*, 2004; Kirimanjeswara *et al.*, 2005b; Pilione and Harvill, 2006) in bordetellae infections. The role of some species-specific factors such as type III secretion system and pertussis toxin have also been revealed.

In the absence of quantitative information we have developed a qualitative Boolean model to study the sequence of events in the pathogenesis of *Bordetella bronchiseptica* and *B. pertussis* (Thakar *et al.*, 2007). Some quantitative information such as the time scales of immune interactions, a threshold for phagocytosis and the decay of cytokines and bacterial virulence factors was incorporated.

The model was then used to study the pathogenesis of wild-type infections and the effect of perturbations of the host and pathogen nodes. The simulations reveal (Fig. 18.6) three phases corresponding to the activation of innate immune responses, followed by the generation of Th2 and B cell responses and then Th1-related responses. The definition of three infection phases is preferable to using exact timing because specific processes can have different durations in different hosts, and timing results obtained in model organisms might not be directly applicable to the pathogens' natural hosts. The model described above reproduces the effect of the mutation of specific bacterial genes and hosts deficiency in B cells and T cells. The model's novel predictions about the secondary infections and the adoptive transfer of antibodies

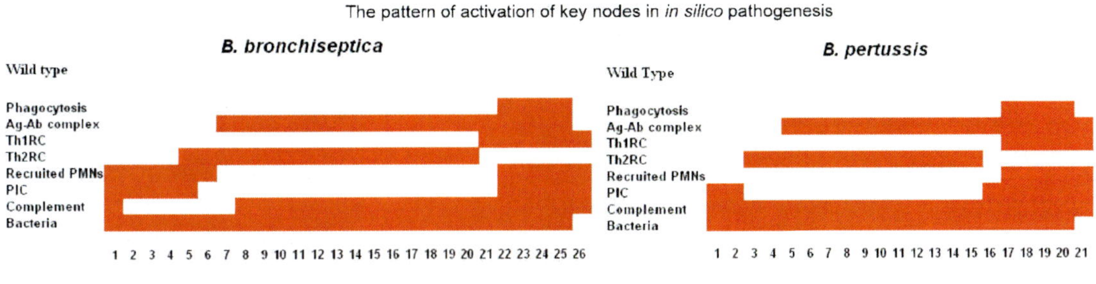

Figure 18.6 Activation patterns indicating the activity of the immune components at each time-step in two different models of infection by two different *Bordetella* species. The two panels correspond to the two bacterial strains modelled. Coloured squares represent the ON state of the given node, while boxes mean the OFF state of the node. The box in the lower right corner of each pattern represents the clearance of bacteria on the last time-step. The figure is adapted from Thakar *et al.* (2007).

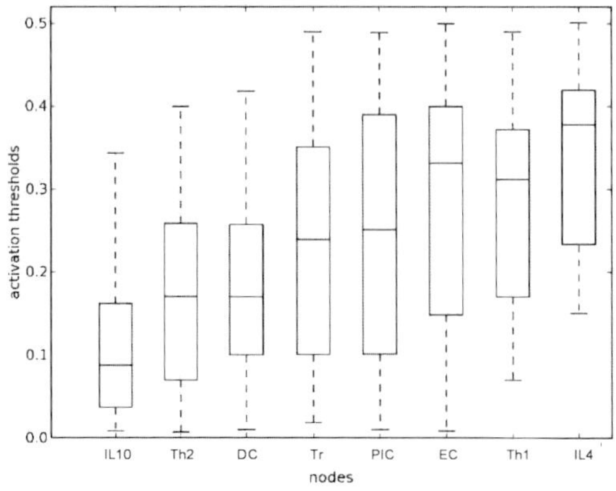

Figure 18.7 Parameters which can reproduce the experimentally observed results show that the distribution of the activation thresholds of various nodes have non-uniform distributions. IL-10, interleukin 10; Th2, T helper subtype 2; DC, dendritic cells; Tr, T regulatory cells; PIC, pro-inflammatory cytokines; EC, epithelial cells; Th1, T helper subtype 1; IL-4, interleukin 4.

were validated. The model also predicts that the decay rate of the effects of certain host and bacterial factors such as Th2 related cytokines, TTSS and FHA/ACT are similar.

Network models such as these can contribute insight into how virulence factors change the course of innate and adaptive immune responses, and allow for the comparison of species-specific virulence factors. For example, both the *B. bronchiseptica* O-antigen and the *B. pertussis* pertussis toxin (PTX) inhibit early immune responses, assisting bacterial multiplication and survival (Burns *et al.*, 2003; Kirimanjeswara *et al.*, 2005a). The model suggests that the O-antigen interacts with host components earlier than does PTX, as it inhibits bacterial recognition, while PTX disables the recruitment of immune cells that recognize bacteria. However, the inhibitory effect of PTX is ultimately more effective since it gives *B. pertussis* a means to resist the effects of serum antibodies. Comparing the way in which the *B. bronchiseptica* type three secretion system (TTSS) and the *B. pertussis* filamentous haemagluttinin (FHA) and adenylate cyclase toxin (ACT) modulate T helper cell balance gives insight into why the TTSS genes may not be expressed in *B. pertussis*. The TTSS causes necrosis of phagocytes which is possibly a side-effect of the evolution of the TTSS's primary role to modulate cytokines, and which ultimately leads to a stronger immune response. Unlike *B. bronchiseptica*, the milder pathology during *B. pertussis* infection allows it to modulate immunity silently.

More recently, a continuous-discrete (hybrid) formulation incorporating a Boolean approach with a detailed quantitative formulation for critical nodes such as bacteria and phagocytosis reproduced the observed characteristics of the known cytokine and bacterial time courses (Thakar *et al.*, 2009).This model revealed minimum numbers of inhibitory interactions required between the nodes representing IL-10, IL-4, IL-12 and IFN-γ to correctly reproduce the experimental observations. The dynamic simulations of this network show that the regulation is context-dependent. The characterization of the time courses show that they can have multiple activation peaks, depending on the interplay between the half-life of the immune components and the duration of the infection. The parameters selection and following statistical analysis (Fig. 18.7) showed the components which were highly regulated and revealed that in the mutual inhibition between IL-12 and IL-4, IL-4 is dominantly inhibited in infections involving the bordetellae. The model makes predictions about the time scales of each process, the activity thresholds of each node and novel regulatory interactions. For example, the model predicts that the activity threshold of IL-4 is higher (Fig. 18.7, IL4II) than that of IL-12 and that pro-inflammatory cytokines influence the activity of Th2 cells. Some of these predictions are supported by the literature, and many can serve as targets of future experiments. Correlation analysis revealed the strongly dependent parameter, highly regulated nodes and the expression profiles during perturbations.

Conclusion: applications of 'within host models' to antibacterial discovery

There are a variety of ways in which mathematical models can improve our understanding of the current antibacterial agents and assist in the increasingly urgent search for new antibacterial agents. Mathematical models can help guide bench experiments by identifying where our understanding of the system, incorporated as parameters in the model, does not exactly match experimental outcomes. They can also explore the probable outcomes of experiments not presently accessible in the laboratory setting. For example, multiple simultaneous virtual depletion experiments can be performed predicting results for gene knockout murine models that do not currently exist. Alternatively, models can predict the effects of multiple different drugs and therapeutics that are likely to interact *in vivo*.

Collaboration between continuous deterministic models and the PK–PD models

Models, especially those validated by experimental data, can be useful for testing the interactions which have the most influence on pathogenesis. For example, in the models of TB mentioned above there are multiple paths by which active and latent infections can be achieved by manipulation of parameters. The differential equations can be introduced or modified by incorporating a functional form depicting the mechanistic response of the treatment. The global consequence can be analysed through uncertainty analysis of the parameters. Knowing kinetic parameters of within-host interactions will also allow us to determine the optimum dose of the antibacterial agents by assessing the dose required for the control of the infection, thereby allowing for the most efficient administration of the minimum effective dose.

The parameters of drug absorption and elimination estimated from PK–PD models can be used to better construct the WHD model of drug effect. PK–PD models often simplify the spatially separated compartments. WHD models capturing the spatial dynamics can inform the parameters of drug delivery to the PK–PD models. For example, a detailed approach used by Gammack *et al.* (Gammack *et al.*, 2004, 2005) describes granuloma formation in *M. tuberculosis* infections with a reaction-diffusion model using partial differential equations and the movement of innate immune cells towards a focal point of Mtb infection with a coarse-grid spatial formulation.

New targets, insights from Boolean models

Boolean models can simulate the sequence of events arising during bacterial infections from the interplay between host immune components and pathogen factors. The networks of interactions can either be inferred (Christensen *et al.*, 2007) or constructed by using data mining tools or by manual curation. These models can assist in discovering new drugs by performing perturbation analysis. Perturbations of the network can be directly compared with knock-out or overexpression experiments. For instance in the network of cellular interactions representing regulatory relations between cells and cytokines; simulations could be run by turning off particular cytokines to examine their effects on pathogenesis. The systematic analysis of such perturbations will allow us to select particular components which have the most effect on reducing the period of infection or the intensity of the pathogenesis (Thakar *et al.*, 2007, 2009).

These network models also give us insight into the active regulatory motifs at certain stages of pathogenesis. Dynamic roles of motifs like feed-forward loops and feed-back loops can guide us to detect the critical players of the dynamics (Rosenfeld *et al.*, 2007; Kaplan *et al.*, 2008). It has been shown that various two component systems in bacterial pathogens lead to a robust response to the inputs (Shinar *et al.*, 2007). Network motifs leading to such robust response can be studied by systematic analysis of Boolean operators involved in the regulation of downstream components. Moreover, the analysis of information processing through biological networks described by continuous and binary models has revealed characteristic network topologies leading to reproducible behaviour in the presence of noise (Klemm and Bornholdt, 2005). Although the challenges and the complexities are great, the rapid progress in these areas are strong evidence that modelling approaches to the study of pathogens and the dynamics of their interactions with the host will increasingly contribute to and guide various approaches to the study of, and search for, new anti-microbial agents.

References

Albert, R., and Othmer, H.G. (2003). The topology of the regulatory interactions predicts the expression pattern of the segment polarity genes in Drosophila melanogaster. J. Theor. Biol. *223*, 1–18.

Alon, U., Surette, M.G., Barkai, N., and Leibler, S. (1999). Robustness in bacterial chemotaxis. Nature *397*, 168–171.

Althaus, C.L., De Vos, A.S., and De Boer, R.J. (2009). Reassessing the human immunodeficiency virus type 1 life cycle through age-structured modelling: life span of infected cells, viral generation time, and basic reproductive number, R0. J. Virol. *83*, 7659–7667.

Anderson, R.M. (1982). Transmission Dynamics and Control of Infectious Disease Agents. (Springer-Verlag, Berlin, Germany), pp. 149–176.

Andrews, S.S., and Arkin, A.P. (2006). Simulating cell biology. Curr. Biol. *16*, R523–527.

Antia, R., Koella, J.C., and Perrot, V. (1996). Models of the within-host dynamics of persistent mycobacterial infections. Proc. Biol. Sci. *263*, 257–263.

Bartocci, E., Corradini, F., Entcheva, E., Grosu, R., and Smolka, S.A. (2008). CellExcite: an efficient simulation environment for excitable cells. BMC Bioinformatics *9 (Suppl 2)*, S3.

Blaser, M.J., and Kirschner, D. (1999). Dynamics of *Helicobacter pylori* colonization in relation to the host response. Proc. Natl. Acad. Sci. U.S.A. *96*, 8359–8364.

Blaser, M.J., and Parsonnet, J. (1994). Parasitism by the 'slow' bacterium *Helicobacter pylori* leads to altered gastric homeostasis and neoplasia. J. Clin. Invest. *94*, 4–8.

Blower, S.M., McLean, A.R., Porco, T.C., Small, P.M., Hopewell, P.C., Sanchez, M.A., and Moss, A.R. (1995). The intrinsic transmission dynamics of tuberculosis epidemics. Nat. Med. *1*, 815–821. Borghans, J.A., Noest, A.J., and De Boer, R.J. (1999). How specific should immunological memory be? J. Immunol. *163*, 569–575.

Breimer, D.D., and Danhof, M. (1997). Relevance of the application of pharmacokinetic-pharmacodynamic modelling concepts in drug development. The 'wooden shoe' paradigm. Clin. Pharmacokinet. *32*, 259–267.

Brodsky, I.E., and Medzhitov, R. (2009). Targeting of immune signalling networks by bacterial pathogens. Nat. Cell Biol. *11*, 521–526.

Bulitta, J.B., Yang, J.C., Yohonn, L., Ly, N.S., Brown, S.V., D'Hondt, R.E., Jusko, W.J., Forrest, A., and Tsuji, B.T. (2010). Attenuation of colistin bactericidal activity by high inoculum of *Pseudomonas aeruginosa* characterized by a new mechanism-based population pharmacodynamic model. Antimicrob. Agents Chemother. *54*, 2051–2062.

Burns, V.C., Pishko, E.J., Preston, A., Maskell, D.J., and Harvill, E.T. (2003). Role of *Bordetella* O antigen in respiratory tract infection. Infect. Immun. *71*, 86–94.

Celada, F., and Seiden, P.E. (1992). A computer model of cellular interactions in the immune system. Immunol. Today *13*, 56–62.

Chaves, M., Albert, R., and Sontag, E.D. (2005). Robustness and fragility of Boolean models for genetic regulatory networks. J. Theor. Biol. *235*, 431–449.

Chaves, M., Sontag, E.D., and Albert, R. (2006). Methods of robustness analysis for Boolean models of gene control networks. Syst. Biol. (Stevenage) *153*, 154–167.

Christensen, C., Thakar, J., and Albert, R. (2007). Systems-level insights into cellular regulation: inferring, analysing, and modelling intracellular networks. IET Syst. Biol. *1*, 61–77.

Cirl, C., Wieser, A., Yadav, M., Duerr, S., Schubert, S., Fischer, H., Stappert, D., Wantia, N., Rodriguez, N., Wagner, H., *et al.* (2008). Subversion of Toll-like receptor signaling by a unique family of bacterial Toll/interleukin-1 receptor domain-containing proteins. Nat. Med. *14*, 399–406.

Clark, D.R., de Boer, R.J., Wolthers, K.C., and Miedema, F. (1999). T cell dynamics in HIV-1 infection. Adv. Immunol. *73*, 301–327.

Collier-Hyams, L.S., Zeng, H., Sun, J., Tomlinson, A.D., Bao, Z.Q., Chen, H., Madara, J.L., Orth, K., and Neish, A.S. (2002). Cutting edge: *Salmonella* AvrA effector inhibits the key proinflammatory, anti-apoptotic NF-kappa B pathway. J. Immunol. *169*, 2846–2850.

Conant, G.C., and Wagner, A. (2004). Duplicate genes and robustness to transient gene knock-downs in *Caenorhabditis elegans*. Proc. Biol. Sci *271*, 89–96.

Constant, P., Perez, E., Malaga, W., Laneelle, M.A., Saurel, O., Daffe, M., and Guilhot, C. (2002). Role of the pks15/1 gene in the biosynthesis of phenolglycolipids in the *Mycobacterium tuberculosis* complex. Evidence that all strains synthesize glycosylated *p*-hydroxybenzoic methyl esters and that strains devoid of phenolglycolipids harbor a frameshift mutation in the pks15/1 gene. J. Biol. Chem. *277*, 38148–38158.

Csete, M., and Doyle, J. (2004). Bow ties, metabolism and disease. Trends Biotechnol. *22*, 446–450.

Dahari, H., Shudo, E., Cotler, S.J., Layden, T.J., and Perelson, A.S. (2009). Modelling hepatitis C virus kinetics: the relationship between the infected cell loss rate and the final slope of viral decay. Antivir. Ther. *14*, 459–464.

von Dassow, G., Meir, E., Munro, E.M., and Odell, G.M. (2000). The segment polarity network is a robust developmental module. Nature *406*, 188–192.

Dayneka, N.L., Garg, V., and Jusko, W.J. (1993). Comparison of four basic models of indirect pharmacodynamic responses. J. Pharmacokinet. Biopharm. *21*, 457–478.

de Boer, R.J. (2002). Mathematical models of human CD4+ T-cell population kinetics. Neth. J. Med. *60*, 17–26; discussion 26.

Duesbery, N.S., Webb, C.P., Leppla, S.H., Gordon, V.M., Klimpel, K.R., Copeland, T.D., Ahn, N.G., Oskarsson, M.K., Fukasawa, K., Paull, K.D., *et al.* (1998). Proteolytic inactivation of MAP-kinase-kinase by anthrax lethal factor. Science *280*, 734–737.

Efroni, S., Harel, D., and Cohen, I.R. (2007). Emergent dynamics of thymocyte development and lineage determination. PLoS Comput. Biol. *3*, e13.

Eldar, A., Dorfman, R., Weiss, D., Ashe, H., Shilo, B.Z., and Barkai, N. (2002). Robustness of the BMP morphogen gradient in *Drosophila* embryonic patterning. Nature *419*, 304–308.

Franke, R., Muller, M., Wundrack, N., Gilles, E.D., Klamt, S., Kahne, T., and Naumann, M. (2008). Host-pathogen systems biology: logical modelling of hepatocyte growth factor and *Helicobacter pylori* induced c-Met signal transduction. BMC Syst. Biol. *2*, 4.

Gammack, D., Doering, C.R., and Kirschner, D.E. (2004). Macrophage response to *Mycobacterium tuberculosis* infection. J. Math. Biol. *48*, 218–242.

Gammack, D., Ganguli, S., Marino, S., Segovia-Juarez, J., and Kirschner, D.E. (2005). Understanding the immune response in tuberculosis using different mathematical models and biological scales. SIAM J. Multiscale Model. Simulation *3*, 312–345.

Ganusov, V.V., Bergstrom, C.T., and Antia, R. (2002). Within-host population dynamics and the evolution of microparasites in a heterogeneous host population. Evolution *56*, 213–223.

Geijtenbeek, T.B., Van Vliet, S.J., Koppel, E.A., Sanchez-Hernandez, M., Vandenbroucke-Grauls, C.M., Appelmelk, B., and Van Kooyk, Y. (2003). Mycobacteria target DC-SIGN to suppress dendritic cell function. J. Exp. Med. *197*, 7–17.

Handel, A., Margolis, E., and Levin, B.R. (2009). Exploring the role of the immune response in preventing antibiotic resistance. J. Theor. Biol. *256*, 655–662.

Hatzikirou, H., and Deutsch, A. (2008). Cellular automata as microscopic models of cell migration in heterogeneous environments. Curr. Top. Dev. Biol. *81*, 401–434.

Hethcote, H.W., and Yorke, J.A. (1984). Gonorrhea: Transmission, Dynamics and Control. (Springer-Verlag, Berlin, Germany).

Hodgkin, P.D. (2007). A probabilistic view of immunology: drawing parallels with physics. Immunol. Cell Biol. *85*, 295–299.

Irvine, D.H., and Savageau, M.A. (1990). Efficient solution of nonlinear ODE's expressed in S-system canonical form. SIAM J. Num. Anal. *27*, 704–735.

Jeong, H., Tombor, B., Albert, R., Oltvai, Z.N., and Barabasi, A.L. (2000). The large-scale organization of metabolic networks. Nature *407*, 651–654.

Jones, R.M., Wu, H., Wentworth, C., Luo, L., Collier-Hyams, L., and Neish, A.S. (2008). *Salmonella* AvrA coordinates suppression of host immune and apoptotic defenses via JNK pathway blockade. Cell Host Microbe. *3*, 233–244.

Kaplan, S., Bren, A., Dekel, E., and Alon, U. (2008). The incoherent feed-forward loop can generate non-monotonic input functions for genes. Mol. Syst. Biol. *4*, 203.

Kirimanjeswara, G.S., Agosto, L.M., Kennett, M.J., Bjornstad, O.N., and Harvill, E.T. (2005a). Pertussis toxin inhibits neutrophil recruitment to delay antibody-mediated clearance of Bordetella pertussis. J. Clin. Invest. *115*, 3594–3601.

Kirimanjeswara, G.S., Mann, P.B., Pilione, M., Kennett, M.J., and Harvill, E.T. (2005b). The complex mechanism of antibody-mediated clearance of Bordetella from the lungs requires TLR4. J. Immunol. *175*, 7504–7511.

Kirschner, D., and Marino, S. (2005). *Mycobacterium tuberculosis* as viewed through a computer. Trends Microbiol. *13*, 206–211.

Klemm, K., and Bornholdt, S. (2005). Topology of biological networks and reliability of information processing. Proc. Natl. Acad. Sci. U.S.A. *102*, 18414–18419.

Levison, M.E., and Levison, J.H. (2009). Pharmacokinetics and pharmacodynamics of antibacterial agents. Infect. Dis. Clin. North Am. *23*, 791–*815*, vii.

Mann, P.B., Elder, K.D., Kennett, M.J., and Harvill, E.T. (2004). Toll-like receptor 4-dependent early elicited tumor necrosis factor alpha expression is critical for innate host defense against *Bordetella bronchiseptica*. Infect. Immun. *72*, 6650–6658.

Marino, S., and Kirschner, D.E. (2004). The human immune response to *Mycobacterium tuberculosis* in lung and lymph node. J. Theor. Biol. *227*, 463–486.

Marino, S., Pawar, S., Fuller, C.L., Reinhart, T.A., Flynn, J.L., and Kirschner, D.E. (2004). Dendritic cell trafficking and antigen presentation in the human immune response to *Mycobacterium tuberculosis*. J. Immunol. *173*, 494–506.

Mata, J., and Cohn, M. (2007). Cellular automata-based modelling program: synthetic immune system. Immunol. Rev. *216*, 198–212.

Mazurkiewicz, P., Thomas, J., Thompson, J.A., Liu, M., Arbibe, L., Sansonetti, P., and Holden, D.W. (2008). SpvC is a *Salmonella* effector with phosphothreonine lyase activity on host mitogen-activated protein kinases. Mol. Microbiol. *67*, 1371–1383.

McAdams, H.H., and Shapiro, L. (1995). Circuit simulation of genetic networks. Science *269*, 650–656.

Mehrotra, N., Gupta, M., Kovar, A., and Meibohm, B. (2007). The role of pharmacokinetics and pharmacodynamics in phosphodiesterase-5 inhibitor therapy. Int. J. Impot. Res. *19*, 253–264.

Meier-Schellersheim, M., Xu, X., Angermann, B., Kunkel, E.J., Jin, T., and Germain, R.N. (2006). Key role of local regulation in chemosensing revealed by a new molecular interaction-based modelling method. PLoS Comput. Biol. *2*, e82.

Mendoza, L., and Alvarez-Buylla, E.R. (1998). Dynamics of the genetic regulatory network for Arabidopsis thaliana flower morphogenesis. J. Theor. Biol. *193*, 307–319.

Monack, D.M., Mecsas, J., Ghori, N., and Falkow, S. (1997). Yersinia signals macrophages to undergo apoptosis and YopJ is necessary for this cell death. Proc. Natl. Acad. Sci. U.S.A. *94*, 10385–10390.

Morton-Firth, C.J., and Bray, D. (1998). Predicting temporal fluctuations in an intracellular signalling pathway. J. Theor. Biol. *192*, 117–128.

Newman, R.M., Salunkhe, P., Godzik, A., and Reed, J.C. (2006). Identification and characterization of a novel bacterial virulence factor that shares homology with mammalian Toll/interleukin-1 receptor family proteins. Infect. Immun. *74*, 594–601.

Nielsen, E.I., Viberg, A., Lowdin, E., Cars, O., Karlsson, M.O., and Sandstrom, M. (2007). Semimechanistic pharmacokinetic/pharmacodynamic model for assessment of activity of antibacterial agents from time-kill curve experiments. Antimicrob. Agents Chemother. *51*, 128–136.

Orth, K., Palmer, L.E., Bao, Z.Q., Stewart, S., Rudolph, A.E., Bliska, J.B., and Dixon, J.E. (1999). Inhibition of the mitogen-activated protein kinase kinase superfamily by a Yersinia effector. Science *285*, 1920–1923.

Peck, C.C., Barr, W.H., Benet, L.Z., Collins, J., Desjardins, R.E., Furst, D.E., Harter, J.G., Levy, G., Ludden, T., Rodman, J.H., *et al.* (1994). Opportunities for integration of pharmacokinetics, pharmacodynamics, and toxicokinetics in rational drug development. J. Clin. Pharmacol. *34*, 111–119.

Pilione, M.R., and Harvill, E.T. (2006). The Bordetella bronchiseptica type III secretion system inhibits gamma interferon production that is required for efficient antibody-mediated bacterial clearance. Infect. Immun. *74*, 1043–1049.

Rao, C.V., Wolf, D.M., and Arkin, A.P. (2002). Control, exploitation and tolerance of intracellular noise. Nature *420*, 231–237.

Regoes, R.R., Barber, D.L., Ahmed, R., and Antia, R. (2007). Estimation of the rate of killing by cytotoxic T lymphocytes *in vivo*. Proc. Natl. Acad. Sci. U.S.A. *104*, 1599–1603.

Rosenfeld, N., Young, J.W., Alon, U., Swain, P.S., and Elowitz, M.B. (2007). Accurate prediction of gene feedback circuit behavior from component properties. Mol. Syst. Biol. *3*, 143.

Segovia-Juarez, J.L., Ganguli, S., and Kirschner, D. (2004). Identifying control mechanisms of granuloma formation during *M. tuberculosis* infection using an agent-based model. J. Theor. Biol. *231*, 357–376.

Shinar, G., Milo, R., Martinez, M.R., and Alon, U. (2007). Input output robustness in simple bacterial signaling systems. Proc. Natl. Acad. Sci. U.S.A. *104*, 19931–19935.

Slepchenko, B.M., Schaff, J.C., Macara, I., and Loew, L.M. (2003). Quantitative cell biology with the virtual cell. Trends Cell Biol. *13*, 570–576.

Sun, Y.N., and Jusko, W.J. (1998). Transit compartments versus gamma distribution function to model signal transduction processes in pharmacodynamics. J. Pharm. Sci. *87*, 732–737.

Tam, V.H., Schilling, A.N., Poole, K., and Nikolaou, M. (2007). Mathematical modelling response of *Pseudomonas aeruginosa* to meropenem. J. Antimicrob. Chemother. *60*, 1302–1309.

Thakar, J., Christensen, C., and Albert, R. Boolean networks in inference and dynamic modelling of biological systems at the molecular and physiological level. In Handbook of Biological Networks, Boccaletti, S., Latora, V., Moreno Y., eds. (World Scientific Publishing).

Thakar, J., Pilione, M., Kirimanjeswara, G., Harvill, E.T., and Albert, R. (2007). Modelling systems-level regulation of host immune responses. PLoS Comput Biol. *3*, e109.

Thakar, J., Saadatpour, A., Harvill, E.T., and Albert, R. (2009). Constraint based network model of pathogen-immune system interactions. J. R. Soc. Interface *doi:10.1098/rsif.2008.0363*.

Van der Graaf, P.H., and Danhof, M. (1997). Analysis of drug-receptor interactions *in vivo*: a new approach in pharmacokinetic-pharmacodynamic modelling. Int. J. Clin. Pharmacol.Ther. *35*, 442–446.

Wigginton, J.E., and Kirschner, D. (2001). A model to predict cell-mediated immune regulatory mechanisms during human infection with *Mycobacterium tuberculosis*. J. Immunol. *166*, 1951–1967.

Wilkinson, D.J. (2009). Stochastic modelling for quantitative description of heterogeneous biological systems. Nat. Rev. Genet. *10*, 122–133.

Yuh, C.H., Bolouri, H., and Davidson, E.H. (2001). Cis-regulatory logic in the endo16 gene: switching from a specification to a differentiation mode of control. Development *128*, 617–629.

Advances and Applications in Diagnostic Microbiology and Antibacterial Discovery

19

Audrey N. Schuetz and Yi-Wei Tang*

Abstract

Despite the rising numbers of multidrug-resistant pathogens, and their continuously emerging resistance patterns, few novel antibacterial agents have been approved or released recently. In order to combat this problem, efforts are being made to extend the utility of existing antibiotics as long as possible, while attempting to develop new drugs. The clinical practice of evidence-based therapy, based on diagnosing early and narrowing antimicrobial coverage, with timely administration of an antibiotic, may help alleviate the problem. Diagnostic procedures optimized for accuracy and turn-around time further improve patient therapy. We review techniques currently in use in diagnostic microbiology, such as direct microscopic examination, rapid biochemical and antigen testing, microorganism culture, serological diagnosis, and a variety of molecular diagnostic techniques. In addition, we introduce various emerging diagnostic techniques, which show promise in their application towards a more exact antibacterial practice. Such emerging technologies include ultra high-throughput sequencing, microarray science, quantum dots, PCR electrospray ionization mass spectrometry, atomic force microscopy, and carbon nanotubes. Point-of-care testing devices are also reviewed. As diagnostic methods have changed over the years, the novel applications of these technologies hold promise in their rapidity and accuracy, while showing potential application in drug target testing and drug discovery.

Introduction

In a 2004 policy report, 'Bad Bugs, No Drugs: As Antibiotic R&D Stagnates, a Public Health Crisis Brews', the Infectious Diseases Society of America first viewed with concern the contrast between the increasing numbers of multidrug-resistant microbial pathogens and the decreasing investment by major pharmaceutical companies in antimicrobial research and development in the United States and throughout the world (Diekema *et al.*, 2004; Talbot *et al.*, 2006). Since the published report, infections caused by antibiotic-resistant gram-positive and gram-negative bacterial pathogens have continued to be a challenge in the hospital and in the community (Giske *et al.*, 2008; Rice 2008; Arias and Murray, 2009; Boucher *et al.*, 2009). For instance, the 'ESKAPE' pathogens, which include *Enterococcus faecium*, *Staphylococcus aureus*, *Klebsiella pneumoniae*, *Acinetobacter baumanii*, *Pseudomonas aeruginosa*, and *Enterobacter* species, currently cause the majority of hospital infections in the United States (Rice 2008; Spellberg *et al.*, 2008; Boucher *et al.*, 2009). With the addition of *Clostridium difficile*, *K. pneumoniae* and *Enterobacter* species, these pathogens effectively 'escape' the effects of antibacterial drugs (Peterson, 2009). Currently, more people die of methicillin-resistant *S. aureus* (MRSA) infections in United States hospitals than of HIV/AIDS and tuberculosis combined (Boucher and Corey, 2008; Arias and Murray, 2009).

*Corresponding author

The emerging resistance trend is being driven by an increase in resistant pathogens and a decline in the development of new antimicrobial agents with which to treat them (Pakyz *et al.*, 2008). Despite increasing frequency and severity of antimicrobial resistance, the future development of new anti-infective agents is now threatened by a decrease in research in this field by many major pharmaceutical companies (Wenzel, 2004; Norrby *et al.*, 2005). The steady discovery of novel antibiotics that occurred between 1940 and 1980 has not been sustained. The discovery and development of new antimicrobials is both expensive and time-consuming, requiring a long-term commitment in order to maintain a substantial and sophisticated infrastructure (Payne *et al.*, 2007). Only a handful of new antibacterials have been approved in recent years, and the number of new antibacterial drugs released for marketing in the United States continues to decrease (Boucher *et al.*, 2009; Payne *et al.*, 2007; Rice, 2008). Fuelled by the use, overuse, and misuse of antibiotic therapy, the dawn of a troubling post-antibiotic era probably is on the horizon, during which there will be few antimicrobials available to treat multidrug-resistant pathogens such as the 'ESKAPE' pathogens (Falagas and Bliziotis, 2007; Rice, 2008; Boucher *et al.*, 2009; Peterson, 2009).

What can be done, as the current 'bad bugs–no drugs' status will not improve soon? This growing problem requires a multipronged strategy that includes adherence to infection control principles, parsimonious and rational use of current antimicrobial agents, and development of new agents active against multidrug-resistant pathogens. More efforts need to be made to preserve the activity of available antibiotics, or at least to expand as much as possible their useful period. Meanwhile, intense research efforts should focus on the development and introduction into clinical practice of new antimicrobial agents. In clinical practice, one potential proposal is to change the current practice of empiric antibiotic therapy, which is usually based on expensive, multidrug anti-infective regimens to provide adequate coverage of important pathogens (Ritchie *et al.*, 2009). Physicians should practice evidence-based, single antimicrobial therapy with narrow but specific pathogen coverage. Timely provision of adequate antimicrobial coverage in an initial anti-infective treatment regimen results in optimal outcomes for bacterial and fungal infections. Diagnostic procedures with improved test turnaround times and accuracies will enable the practice of evidence-based therapy. We will review diagnostic techniques and introduce advanced technologies in the field of diagnostic microbiology, as well as novel and emerging applications of these techniques in antibacterial practice.

Current technologies in diagnostic microbiology

The purpose of diagnostic microbiology is to determine whether suspected pathogenic microorganisms are present in specimens collected from human beings, animals, or the environment. In medical practice, diagnostic microbiology aids in defining infectious processes and elucidating treatment options through detection, quantification and characterization of specific pathogens, including antimicrobial susceptibility profiles of the identified microorganisms. Directly related to the scope of this book, advanced diagnostic techniques can also provide tools for antibacterial agent discovery and evaluation. Technological revolutions in microbiology, immunology and molecular biology have significantly expanded and improved the capabilities of diagnostic microbiology. A microorganism from a test sample can be detected and identified in any of five possible ways: (i) direct microscopic examination, (ii) rapid microbial antigen testing, (iii) cultivation of microorganisms using artificial media or living hosts, (iv) measurement of microorganism-induced host immune responses, and (v) amplification and detection of microorganism-specific nucleic acids (Tang and Persing, 2009). Certain characteristics of these techniques are summarized in Table 19.1, and the following sections will discuss each technique separately.

Direct microscopic examination

Although very simple, direct examination by microscopy is still just as important today as it was many years ago in evaluating a specimen for the presence of pathogenic microbes. A wet mount, prepared by applying specimens directly onto the

Table 19.1 Methods used for microorganism diagnosis

Test	Ease of performance	Turn-around time	Result interpretation	Advantages	Disadvantages	Examples (references)
Direct examination and antigen testing	Performed in routine clinical lab and in nurse station	1–3 hours	Direct if correlated with symptoms	Rapid	Variable sensitivity and specificity; special skills are needed for interpretation	Dark field examination of Borrelia burgdorferi (Wittenbrink et al., 1994; Pejchalova et al., 2007); Cryptococcus neoformans antigen testing (Kiska et al., 1994; Tanner et al., 1994); urea breath test for *Helicobacter pylori* infection (Blecker et al., 1993; Marchildon et al., 1996)
Cell culture	Performed in sophisticated clinical lab and in research lab	2–14 days	Definite	Phenotypic drug susceptibility testing is possible	Time-consuming; poor sensitivity; limited microorganisms are culturable	Chromogenic media for culture of MRSA (Merlino et al., 2000; Flayhart et al., 2005); mixed cell lines for enterovirus culture (Buck et al., 2002)
Serology	Performed in larger and sophisticated clinical lab	4–6 hours	Indirect	Automation	Results are generally retrospective; immunosuppressed host may be unable to mount a response	Serology diagnosis of HIV infection (Jackson et al., 1997; Weber et al., 2002; Campbell and Fedoriw, 2009); serology diagnosis of ehrlichiosis and anaplasmosis (Paddock and Childs, 2003; Eshoo et al., 2010); IFN-γ release assays for tuberculosis diagnosis (Chee et al., 2008)
Molecular diagnostics	Performed in only a few very sophisticated research and clinical lab	2.5 hours –2 days	Direct without knowing microbial viability	High sensitivity and specificity	Facility requirement (pre- and post-amplification rooms); possible false-positive results due to carry-over contamination and false-negative results due to inhibitors in specimen; most expensive and highly complex of approaches listed in this table	Rapid identification of MRSA (Oliveira et al., 2002, 2003; Chapin and Musgnug, 2003); detection of a panel of respiratory viruses (Li et al., 2007; Mahony et al., 2007); antiretroviral therapy monitoring by HIV quantification (O'Brien et al., 1996; Saag et al., 1996; Mellors et al., 1997; Ford et al., 2009)

MRSA, methicillin-resistant *Staphylococcus aureus*; IFN-γ, interferon-gamma.

surface of a slide, is often used to detect motile trophozoites of parasites such as *Giardia lamblia* in stools and *Trichomonas vaginalis* in vaginal discharges. Certain bacteria are so thin that they cannot be observed in direct preparations. Their characteristic motility, however, is an important feature of presumptive identification. Dark field examination, during which light is reflected or refracted off the surface of the object, is used to identify these bacteria, primarily spirochetes such as *Borrelia burgdorferi*, the microorganism which causes Lyme disease (Wittenbrink *et al.*, 1994; Pejchalova *et al.*, 2007). Identification of characteristic cytoplasmic morulae in peripheral blood leucocytes confirms *Ehrlichia/Anaplasma* infection. Detection of viral inclusions in smears or tissues has been the traditional method to directly demonstrate viral infections. Electron microscopy has been used directly from stool specimens for detection of the viral causes of gastroenteritis, such as rotaviruses.

Examination of stained material, either direct test specimens or samples of growth from cultures, is the most useful method for presumptive identification of several microorganisms. The Gram stain, which depends on the amount of peptidoglycan in the bacterium cell wall, remains the most fundamental and reliable technique in diagnostic microbiology. A positive Gram stain on concentrated cerebrospinal fluid (CSF) allows for a presumptive diagnosis of bacterial meningitis. Gram stain on signal-raised blood culture broth media directs further subculture and identification. Another useful stain is the acid-fast stain, by which mycobacterial species including *Mycobacterium tuberculosis* are identified. The fluorescent auramine–rhodamine acid-fast stain allows mycobacteria to be visualized with a fluorescent microscope. Similarly, a modified acid-fast stain has been widely used to identify certain organisms, such as the bacterial pathogen *Nocardia* and parasitic pathogens *Cryptosporidium* and *Cyclospora*. Another stain, the calcofluor white/potassium hydroxide stain, allows for direct fungal detection, because it binds to the cell walls of fungi. *Pneumocystis carinii/jiroveci*, a common opportunistic infection in AIDS and other immunocompromised hosts, can be detected by this method.

Rapid biochemical and antigen testing

Immuno-based devices for the detection of microorganism antigens remain important tools for the diagnosis and management of microbial diseases. The use of microorganism-specific, high binding affinity monoclonal antibodies and peptides has increased the availability of high quality reagents with improved reproducibility, sensitivity and specificity in detection of microbial antigens. Latex particle agglutination, immunofluorescence assays (IFA), and enzyme immunoassays (EIA) are the techniques most frequently used for immunologic detection of microbial antigens. Agglutination is produced by the binding of antibody to a specific antigen. This technique has been extremely useful in the detection of *Cryptococcus neoformans* in bodily fluid specimens, such as plasma, serum and cerebrospinal fluid (Kiska *et al.*, 1994; Tanner *et al.*, 1994). Direct immunofluorescence assays have been combined with the shell vial technique for rapid detection of virus from culture (Matthey *et al.*, 1992). A lateral flow chromatographic immunoassay (CIA) has recently been introduced and has performed well in rapidly detecting microbe-specific antigens (Henrickson and Hall, 2007).

Microorganisms undergo metabolism and produce by-products that can be targeted as detection and identification tools in diagnostic microbiology. For instance, *Helicobacter pylori* produces large amounts of extracellular urease which hydrolyses urea to form ammonia and soluble carbon dioxide, which can be detected within hours in human exhaled breath specimens. The non-invasive urea breath test (Meretek Diagnostics, Lafayette, CO, USA) enables the direct detection of *H. pylori* without the need for endoscopy (Blecker *et al.*, 1993; Marchildon *et al.*, 1996). For Enterobacteriaceae, sugar metabolism, indole reactions, and the presence of urease are important biochemical reactions when differentiating between genera and species. Gas–liquid chromatography (GLC) has been used widely for anaerobic microorganism identification by targeting the volatile metabolic materials. *Staphylococcus aureus* is differentiated from other staphylococcal species by the use of the coagulase test. Matrix-assisted laser desorption ionization-time-of-flight mass spectrometry (MALDI-TOF MS) has recently been reported for rapid and accurate microbial identification and characterization (Camara and Hays, 2007; Mellmann *et al.*, 2008, 2009). In antibiotic discovery, the MALDI-TOF MS has revolutionized protein biochemistry and has proven to be a superior tool for the identification and detailed analysis of peptides and proteins (Pusch and Kostrzewa, 2005). For example, the MALDI-TOF MS has been used to identify antifungal biosynthesis genes of *Bacillus* species isolated from different microhabitats (Athukorala *et al.*, 2009). A system named BioTyper based on the MALDI-TOF MS is commercially available from Bruker Daltonics (Leipzig, Germany).

Microorganism and virus culture

Culture remains the most definitive method for establishing the presence of a particular pathogen in a suspected specimen. A pure isolate is essential for performing subsequent phenotypic antimicrobial susceptibility tests. Bacteria, mycobacteria, mycoplasma, and fungi may be cultured in liquid or on solid artificial media. Isolation of microorganisms from blood specimens using automatic, continuous monitoring systems has been a major innovation which has greatly improved diagnostic microbiology. Multiple media types are utilized on each culture sample, in order to provide differential growth information on different substrates. When compounds such as antimicrobial agents that inhibit indigenous flora are incorporated into culture media, the media may selectively permit the growth of specific microorganisms resistant to these inhibitors. Growth patterns on selective media are extremely important for the isolation and identification of diarrhoeagenic pathogenic microorganisms in stool specimens, such as *Salmonella*, *Shigella* and *E. coli* O157:H7. In addition, several chromogenic media are commercially available for surveillance cultures of MRSA (Merlino *et al.*, 2000; Flayhart *et al.*, 2005).

Many *Chlamydia* spp. and viruses can be isolated and identified using eukaryotic cell cultures. With this traditional cell culture technique, the clinical samples are first inoculated onto several types of cell monolayer cultures, covered with growth media and subsequently incubated. If a viable organism is present in the sample, the cell monolayer becomes infected. The cells change their shape, become distorted, and eventually will lyse. Initial changes in the infected cell monolayer can initially be identified and differentiated by the type of cytopathic effect (CPE) observed, the speed with which the CPE is evident after inoculation, and the type of cell culture lines within which the pathogen replicates. Mixed cell lines stably transfected with host cofactors involved in the binding and entry of some enteroviruses (Super E-Mix, Diagnostic Hybrids, Athens, OH, USA) have been shown to be more sensitive than conventional cell culture for viral recovery from clinical samples (Buck *et al.*, 2002). Shell vial isolation, which combines cell culture and antigen detection techniques, has been adapted as a relatively rapid means of identifying agents. This technique is especially useful for the detection of cytomegalovirus (CMV) infections, since the appearance of CMV-induced CPE presents relatively late.

Serological diagnosis

To examine patients' immunological response to organisms, several methods may be utilized. Serological assays evaluate the presence of IgM and IgG antibodies expressed in response to microorganisms. IgM is an indicator of a new and on-going infection. For IgG testing, it is often necessary to obtain paired serum specimens, the first drawn during the acute phase of the disease and the latter drawn during convalescence, to assess for a rise in antibody titres. A 4-fold increase indicates recent, active infection. Detection of specific anti-organism antibodies in the patient is useful for organisms that are not easily cultivated such as HIV, hepatitis viruses (A, B, C), *Mycoplasma* spp., *Chlamydia* spp. and syphilis. Cultivatable viruses are also tested by this method [herpes simplex virus (HSV), varicella zoster virus (VZV), CMV, and Epstein–Barr virus (EBV)]. Retuned antibody assays, which evaluate the avidity of an antibody by dilution or by treating tested serum specimens, can be helpful in making the distinction between recent and remote infections. This technique has been used primarily in HIV epidemiology surveillance. In addition to antibody responses, other host responses have been useful in diagnostic microbiology. The

QuantiFERON-TB Gold (Cellestis, Carnegie, Australia) and the T-SPOT.TB (Oxford Immunotec, Oxford, UK) tests measure host cell-mediated immune reactivity to *M. tuberculosis* as an aid in detecting latent and active tuberculosis infection (Chee *et al.*, 2008). However, for all serological assays, it is important to note that serology may be limited in immunocompromised hosts owing to the impaired ability to mount an effective immune response.

Several conventional serological techniques continue to be widely used in the diagnostic microbiology field. Serology remains a common method for the diagnosis of several tick-borne diseases, such as ehrlichiosis and anaplasmosis (Paddock and Childs, 2003; Eshoo *et al.*, 2010). Agglutination and complement fixation tests are still used in the serodiagnosis of several fungal infections. IFA remains the primary choice for diagnosis of several viral infections, including EBV and measles. The application of EIA serological techniques for the diagnosis of viral and rickettsial infections has expanded significantly over the past several years. The EIA assay is most frequently used in screening for immune status, such as rubella antibody testing, but EIAs are also used as primary screening tests for infections due to HIV, hepatitis C virus (HCV), hepatitis B virus (HBV), and human T-cell lymphotrophic virus. Immunoblotting assays, which allow the discrimination of specific and non-specific reactions, have become a mainstay in confirming infection with HIV, HCV, and *B. burgdorferi* after initial positive EIA results (Jackson *et al.*, 1997; Weber *et al.*, 2002; Campbell and Fedoriw, 2009). In addition, new techniques and formats in antibody testing are emerging. A multianalyte profile (xMAP) allows simultaneous detection and differentiation of a panel of antibodies in a single reaction by using xMAP multiplexed technology (Luminex, Austin, TX, USA), which is highly useful in clinical laboratories in providing a wide range of differential diagnoses (Biagini *et al.*, 2004).

Molecular diagnosis

Nucleic acid amplification techniques have recently become another cornerstone in the diagnostic microbiology field, especially for detection, quantitation and monitoring of viral infections. As the front component, nucleic acid extraction techniques have advanced significantly. An optimal nucleic acid extraction method comprises efficient target recovery, establishment of the integrity of nucleic acid targets, optimal removal of amplification inhibitors, elimination of components which affect other enzymatic substrates, and sterilization of potentially hazardous organisms. Among a variety of extraction methods, the Boom technique (Boom *et al.*, 1990) has been widely applied in both manual and automatic formats. Numerous commercial kits are available for nucleic acid extraction from a variety of patient specimen types. For example, QIAamp MinElute virus kit (Qiagen Inc., Valencia, CA, USA) provides a manual, low-throughput tool to efficiently extract total viral nucleic acids from respiratory specimens and cerebrospinal fluid (Smalling *et al.*, 2002). The easyMAG platform (bioMérieux, Durham, NC, USA), which automatically processes 1–24 different types of specimens each run, has become widely used, because it provides high-quality nucleic acids in high yield (Tang *et al.*, 2005).

Nucleic acid probes detect unique nucleotide sequences within the DNA or RNA of a microorganism; the presence of these unique nucleotide 'signatures' is a surrogate for the presence of the organism itself. Several diagnostic devices which utilize non-amplified nucleic acid probes are commercially available. One such device is the peptide nucleic acid (PNA) fluorescence in situ hybridization tool (FISH, AdvanDx, Woburn, MA, USA; Oliveira *et al.*, 2002, 2003; Chapin and Musgnug, 2003). This assay was first approved for rapid identification of MRSA and recently has been extended to identify other bacterial and fungal pathogens (Sogaard *et al.*, 2005; Shepard *et al.*, 2008). Gen-Probe Inc. (Dan Diego, CA, USA) and Becton Dickinson (Sparks, MD, USA) offer several nucleic acid probes for identification of fungal, mycobacterial and bacterial organisms from culture (Reisner *et al.*, 1994; Della-Latta and Whittier, 1998; Alcaide *et al.*, 2000). Although these commercial molecular diagnostic products are more expensive than conventional approaches, the decrease in turnaround time has the potential to improve patient outcome and reduce overall health care costs.

In vitro nucleic acid amplification techniques belong to one of three broad categories: target, probe or signal amplification (Tang and Persing, 2009). Examples of target amplification systems, in which many copies of the nucleic acid target are generated, include PCR, transcription-mediated amplification (TMA), nucleic acid sequence-based amplification (NASBA) and strand displacement amplification (SDA). Numerous diagnostic devices are commercially available from Roche (Pleasanton, CA, USA), Abbott Molecular (Chicago, IL, USA), Siemens (Tarrytown, NY), bioMérieux (Durham, NC, USA), Gen-Probe Inc., and Becton Dickinson for detection of HIV, HCV, HBV, CMV, *Mycobacterium tuberculosis*, *Chlamydia trachomatis, Neisseria gonorrhoeae*, and enteroviruses. A probe amplification system, such as ligase chain reaction (LCR) and cycling probe technology (CPT), amplifies many copies of the probe, which itself detects the target nucleic acids (Wiedmann *et al.*, 1994). Since LCR-based devices were recalled, no FDA-cleared LCR products are available currently. Finally, signal amplification systems function by increasing the concentration of label attached to the target nucleic acid. Currently, diagnostic devices for HIV and HCV quantification and human papilloma virus (HPV) detection are commercially available from Qiagen, Siemens and Hologic, Inc. (Bedford, MA, USA). The measurement of HIV RNA viral load in plasma, in conjunction with the CD4 T-lymphocyte cell counts, are used routinely as laboratory markers to guide the initiation of antiretroviral therapy as well as to monitor treatment effectiveness and determine the probability of clinical progression (O'Brien *et al.*, 1996; Saag *et al.*, 1996; Mellors *et al.*, 1997; Ford *et al.*, 2009).

Once initial nucleic acid amplification is completed, detection and identification of amplification products can be performed by several different methods. Microwell plate detection methods, which offer the advantages of simplicity, high throughput and convenience, are produced by several commercial companies, such as Roche, Becton Dickinson, Argene (North Massapequa, NY, USA) and Prodesse (Waukesha, WI, USA). Suspension array, which has allowed for the adaptation of highly multiplex assays to diagnostic clinical microbiology laboratories, has been widely used in detection and identification of panels of respiratory viruses (Li *et al.*, 2007; Mahony *et al.*, 2007). Amplicon sequencing provides an accurate, rapid and simple way of detection and identification of products. Several newer sequencing technologies, including pyrosequencing and deep sequencing (Ronaghi, 2001; Balwierz *et al.*, 2009), provide more rapid and higher throughput results, as compared with the traditional Sanger sequencing technique which was developed over 30 years ago (see below). Several real-time PCR-based identification devices are available from Roche, Qiagen, Gen-Probe, Abbott and Cepheid (Sunnyvale, CA, USA).

Some molecular assays combine hybridization and amplification techniques. One such assay is the Verigene® nanosphere system (Nanosphere Inc., Northbrook, IL, USA), which is a bench-top molecular diagnostics system, which uses gold nanoparticles linked to oligonucleotides. The oligonucleotides show high affinity for complementary DNA, allowing for efficient hybridization. Recently, the Verigene Respiratory Virus Nucleic Acid $Test_{SP}$ (VRNAT) has been shown to detect virus-specific sequences for respiratory syncytial virus (RSV) and influenza A and B (Jannetto *et al.*, 2009). Results are available within 3 hours for the VRNAT assay, and testing takes place within a single unit.

Emerging diagnostic techniques

Although currently available diagnostic techniques have proven to provide generally accurate and rapid results, there is nonetheless a growing demand to develop faster, easier and more sensitive diagnostic methods (Table 19.2). Several research groups have attempted to improve the detection capabilities of traditional molecular assays and shorten overall response times. Ultrafast PCR methods have allowed investigators to detect as few as five colony-forming units within 20 minutes (Belgrader *et al.*, 1999). However, successes with ultrafast PCR methods have been variable. Investigators have also demonstrated that the use of a DNA-based nanobarcode may identify certain pathogens or groups of pathogens, such as bioterrorism agents (Li *et al.*, 2005). Although

Table 19.2 Advantages and disadvantages of emerging technologies in diagnostic microbiology

Technology	Advantages	Disadvantages
Deep sequencing	• Measures absolute abundance of nucleic acids • Wider dynamic range than microarrays • Allows discovery of novel microRNAs • More complete genetic data than offered by microarray	• Sequence peaks somewhat more variable than microarray • Expensive, especially start-up costs
Microarray science	• Analysis of multiple and complex combinations of nucleic acids at one time • Adaptability for new organisms or drugs	• Allows measurement of the relative abundance of nucleic acids • Background and cross-hybridization problems • Expensive, especially start-up costs
Quantum dots	• High number of sharp fluorescent signals • Potential for multiplexing • Photostable compounds, allowing for longer term storage, in comparison to traditional fluorescent compounds • Genes or proteins can be examined with this technology	• Usually composed of heavy metals, which could lead to long-term toxicity
PCR electrospray ionization mass spectrometry	• Discriminates individual PCR products within complex mixtures of amplicons • Useful for identification of organisms and strain typing in epidemic surveillance • Offers high resolution and substantial automation	• Difficulty of introduction and engineering of mass spectrophotometer into the laboratory • Requires continuous database updating with new genomic sequences
Atomic force microscopy	• Monitoring of intact organism without fixation artefacts • Cell changes due to antimicrobial treatment may be monitored • High-throughput analysis is possible	• Lower resolution than electron microscopy or X-ray crystallography
Carbon nanotubes	• Can detect a wide range of molecular targets	• Currently in experimental stages

these methods are promising, the techniques are presently cumbersome and often require time-consuming preparation.

Many other novel diagnostic techniques to detect and quantify bacterial cells or other targets have proven useful in enabling timely diagnosis of infection with particular organisms, which may lead to earlier, targeted antimicrobial therapy (Wu and Tang, 2009). Although available for many years, molecular assays, particularly in the area of sequencing, continue to improve. In addition, several novel diagnostic techniques are emerging, particularly within the field of nanotechnology, which builds on data gathered from bacterial genome sequencing. Various methods will be reviewed. The advantages and disadvantages of each method, as they pertain to clinical microbiology applications, are summarized in Table 19.2.

Ultra-high-throughput sequencing methods

As compared with traditional nucleic acid amplification methods, sequencing has allowed for better target identification, while allowing for characterization of genetic pathways which are shared between organisms. In addition, characterization of the population diversity within species by sequencing may also allow for more targeted antibacterial practice. Gene and capillary sequencing is highly automated and widely used throughout microbiology laboratories (Fox *et al.*, 2009). However, new ultra-high-throughput sequencing instruments, known as next generation or massively parallel sequencers, can sequence parallel DNA fragments on a single chip, resulting in millions of DNA sequence reads in a single run, as compared with conventional capillary-based sequencing (Coppee, 2008; Droege and Hill, 2008; Morozova and Marra, 2008). Deep sequencing technology enables the simultaneous sequencing of millions of DNA or RNA molecules, allowing for massive parallel sequencing (Table 19.2). This technology can also lead to the discovery of novel RNA or DNA associated with a pathogen, an antibacterial, or an infectious state. Some investigators have used deep sequencing to profile the small RNA fraction of microRNAs (Creighton *et al.*, 2009). Other areas of development include potential selectivity of targets for bacterial, rather than mammalian, cells. Smith and colleagues describe application of deep sequencing to genome profiling, a method termed 'barcode analysis by sequencing,' which can quantitatively identify drug targets (Smith *et al.*, 2009). In this application of a chemogenomic assay, certain organisms are placed in a competitive growth environment with certain antibiotics. The barcode readout aids in identifying genes which are important for growth in the presence of the compounds.

Several ultra-high-throughput sequencing systems are now available, including the Roche/454 pyrosequencing (http://www.454.com/), Illumina (formerly Solexa; http://illumina.com/), ABI sequencing by ligation (ABI SOLiD: http://www.appliedbiosystems.com), and Helicos BioSciences (http://www.helicosbio.com/; Droege and Hill, 2008; Morozova and Marra, 2008). A key benefit of all three approaches over Sanger sequencing is that there is no need to clone and propagate the DNA in bacteria. The clone-free approach has multiple benefits, such as little or no bias in sequence representation and decreased time and cost for library construction. The 454 pyrosequencing system generates reads up to 200–300 bp, but it is presently more expensive per base than Illumina or SOLiD microread approaches. Owing to the relatively longer read lengths of the 454 system, it is currently the best of the three platforms for *de novo* genome sequencing. To date, these technologies have been applied in a variety of contexts, including whole-genome sequencing, targeted resequencing, discovery of transcription factor binding sites, and non-coding RNA expression profiling. In the diagnostic field, these systems have been used to generate microorganism full genome sequencing in a timely manner, providing rapid diagnosis and monitoring information (Harris *et al.*, 2008; Holt *et al.*, 2008; Hoper *et al.*, 2009; Margeridon-Thermet *et al.*, 2009; Simen *et al.*, 2009). These systems have also been used for discovery and characterization of emerging viral pathogens (Kapoor *et al.*, 2008; Victoria *et al.*, 2008; Blinkova *et al.*, 2009; Li *et al.*, 2009) and detection of low-level HIV drug-resistant viral variants in treatment-experienced persons (Wang *et al.*, 2007; Le *et al.*, 2009; Margeridon-Thermet *et al.*, 2009; Simen *et al.*, 2009).

Microarray science

DNA microarrays, or 'DNA chips,' enable analysis of global patterns of gene expression in either microorganisms or host cells, allowing for improved diagnosis and molecular resistance testing. In general, arrays are defined as a group of more than two known 'probes' which are immobilized onto a solid support or substrate (Schena *et al.*, 1995; Anthony *et al.*, 2001). Arrays can conceptually be divided into macroarrays, which have fewer than 500 features/cm^2, and microarrays, which possess more than 500 features/cm^2 (Miller and Tang, 2009). The line probe assay, a type of macroarray, consists of a series of probes which are deposited onto a membrane; an example is the INNO-LiPA Mycobacteria reverse line blot (InnoGenetics, Ghent, Belgium). On the other hand, DNA microarrays are higher density arrays, consisting of small portions of DNA or other probes immobilized onto a solid support, which allow for examination of over 100,000 DNA sequences at one time. Investigators have used an oligonucleotide microarray with 16S rRNA, *katG*, and *rpoB* sequences for species identification of mycobacteria, as well as mycobacterial resistance testing (Troesch *et al.*, 1999). Microarrays have proven useful in the diagnosis of drug resistance, particularly for *M. tuberculosis* (Antonova *et al.*, 2008; Gegia *et al.*, 2008; Sun *et al.*, 2009). Another microarray, the MultiPathogen Identification assay (developed by researchers at Lawrence Livermore National Laboratory, Livermore, CA,

USA), contains 53,660 oligonucleotide probes and detects 18 pathogens, including viruses and bacteria (Wilson *et al.*, 2002). Other investigators have designed a microarray platform named the Virochip, which has been used to detect all known viruses, including novel viruses (Wang *et al.*, 2002; Chiu *et al.*, 2008). The Virochip consists of approximately 22,000 oligonucleotide probes and has been tested in tissue culture cells infected by viruses as well as in patients (Wang *et al.*, 2002; Kistler *et al.*, 2007). In a recent study, Chiu and colleagues demonstrated that the Virochip performed with similar sensitivity and specificity to virus-specific PCR in a cohort of children (Chiu *et al.*, 2008). In addition, the Virochip detected viruses which were missed by PCR or direct fluorescent antibody testing.

Another microarray, the Akonni TruArray® system (Akonni Diagnostics, Frederick, Maryland, USA), is an integrated microarray system based on microfluidic technology, that automates the process of gel-drop, analysing a specific biochemical test in each drop. Therefore, a small volume from a single clinical sample may be analysed for many disease markers simultaneously. The technique has been shown to be cost-effective, with a rapid turn-around time of approximately two hours. The TruArray MDR-TB (multidrug resistant tuberculosis) test is designed to diagnose *Mycobacterium tuberculosis* and document resistance to isoniazid and rifampin (http://www.akonni.com/products/mdx-test-for-mdrtb.html).

Suspension bead-based liquid xMAP microarrays have been developed by Luminex Corporation (Fig. 19.1). xMAP microarrays are essentially three-dimensional arrays, which combine the use of microscopic polystyrene beads as the solid support and flow cytometry for bead and target detection (Horan and Wheeless, 1977). Numerous studies have described the use of xMAP technology for the detection

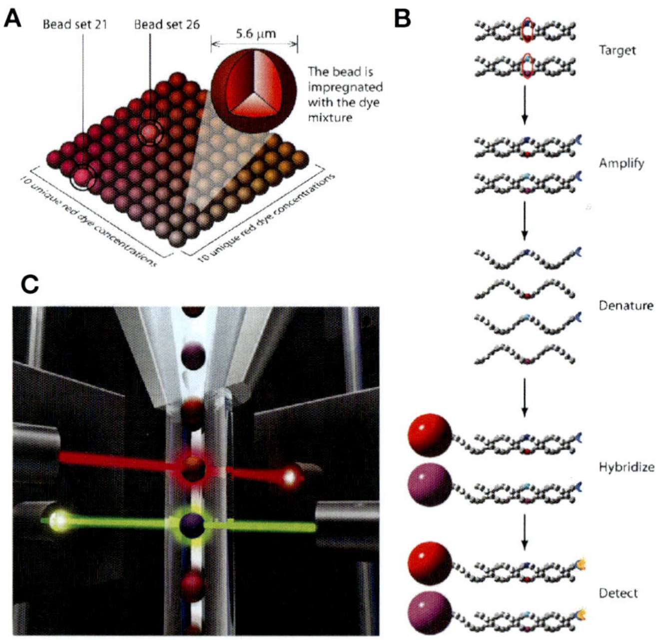

Figure 19.1 Luminex xMAP suspension bead array. (A) Microspheres 5.6 μm in diameter are filled with different relative concentrations of an infrared dye and a red dye to create 100 beads, each with a unique spectral identity. (B) Potential targets are amplified using a biotinylated primer and then denatured and hybridized to microspheres tagged with target-specific sequence probes. Probe–target hybridization is measured using a streptavidin-bound green fluorophore. (C) Flow cytometry is used to analyse the microsphere suspension. A red laser is used to determine the spectral identity of the bead and, therefore, the probe being analysed. The reporter fluorochrome is excited by a green laser, which quantifies the probe-target reaction on the microsphere surface. Adapted from Miller and Tang (2009) with permission.

and differentiation of nucleic acid sequences of microbial pathogens, including enteric bacteria, viruses, mycobacteria, fungi and protozoa (McNamara *et al.*, 2004; Brunstein and Thomas, 2006; Han *et al.*, 2006; Schmitt *et al.*, 2006; Dunbar and Jacobson, 2007; Tang *et al.*, 2007; Tarr *et al.*, 2007; Benson *et al.*, 2008; Gegia *et al.*, 2008; Podzorski *et al.*, 2008; Deng *et al.*, 2009). A molecular typing method incorporating the xMAP suspension array was used to characterize and type influenza A viruses, including H5N1 (Zou *et al.*, 2007). The Luminex suspension array has been incorporated into several commercial devices, such as a detection platform to support the laboratory differential diagnosis of common respiratory viral pathogens. Such devices include the xTAG Respiratory Viral Panel from Luminex Molecular Diagnostics (Mahony *et al.*, 2007; Merante *et al.*, 2007; Pabbaraju *et al.*, 2008), the ResPlex II assay from Qiagen (Brunstein and Thomas, 2006; Li *et al.*, 2007; Brunstein *et al.*, 2008), and the MultiCode-PLx RVP assay from EraGen Biosciences (Lee *et al.*, 2007; Nolte *et al.*, 2007).

In the antibiotic development and research area, the Luminex protein array has been reported to detect residual antibiotics in the environment, including in raw milk (de Keizer *et al.*, 2008). The suspension array system exhibits rapid hybridization kinetics, flexibility in assay design and format, and relatively low costs, making it the most practical microarray platform for clinical diagnostic applications at this time.

DNA microarrays have been proven to be expandable and adaptable, as new viruses or organisms are identified. Although the prospect of DNA microarrays for diagnosis of infection and possible drug targets is promising, there are several challenges with this technology. First, reproducibility and consistency have proven to be an issue. In addition, the turnaround time is generally 24 hours or longer. These problems, coupled with the large initial start-up costs of the microarray and oligonucleotides, can present barriers to use of this technology in some non-research laboratories. In addition, although microarrays allow analysis of multiple and complex combinations of nucleic acids at one time, the data provided by direct sequencing is typically more complete than that provided by DNA microarrays. Nevertheless, microarray hybridization has proven to be a useful tool in pathogen discovery and drug mutation detection and is expected to expand in the direction of antibacterial discovery as well.

Quantum dots

Applicable to many areas of laboratory testing, quantum dots (Qdots) are colloidal semiconductor nanocrystals which possess quantum properties. Upon excitation by incident light, electrons in the Qdots undergo changes in their energy levels in discrete steps, rendering discrete wavelengths of fluorescent light with minimal spectral overlap. In contrast to the organic fluorochromes which are currently the standard of use in fluorescent testing, Qdots offer an increased number of fluorescent signals with a potentially higher order of multiplexing, as well as improved signal brightness. In addition, Qdots are more photostable than organic fluorochromes. Both genes and proteins may be attached to Qdots, allowing for a wider range of analysis of genetic function. Water-soluble Qdots are also available, resulting not only in higher yield but also a longer shelf-life (Chan and Nie, 1998).

Investigators have utilized Qdots to visualize infecting organisms. Bentzen and colleagues demonstrated that Qdots served as immunofluorescent labels for key RSV proteins F and G, using confocal laser scanning microscopy of monolayer cell cultures of HEp-2 cells infected with RSV (Bentzen *et al.*, 2005). The investigators demonstrated the progression of viral infection over time by following the expression of these proteins and comparing Qdot expression to alternative, older techniques, such as Western blot analysis of proteins and real-time quantitative RT-PCR. They showed that the low multiplicity of infection viewed with Qdots approached the limit of detection for standard RT-PCR protocols, demonstrating that Qdots were at least as sensitive as RT-PCR methods. Other investigators have also demonstrated the use of Qdots in identifying RSV (Agrawal *et al.*, 2005; Tripp *et al.*, 2007).

Analysis of Qdot emission using hyperspectral imaging (also known as multispectral imaging), which provides spectral images for each pixel of a screen, may extend the range of Qdot colours, enhancing diagnostic capability (True and Gao, 2007). Hyperspectral imaging has been reported to remove background autofluorescence during imaging with Qdots (Xing *et al.*, 2007). Other investigators have explored the area of traceable drug delivery and targeted therapy using Qdots. Qi and Gao reported on the use of Qdots to study siRNA delivery into cells using amphipols (Qi and Gao, 2008). Amphipols are nanomaterial, linear polymers with alternating hydrophilic and hydrophobic side chains, which solubilize integral membrane proteins. In contrast to micelles, amphipols do not cause disruption in the cell membrane during delivery. As conventional siRNA delivery agents are suboptimal, the investigators demonstrated that amphipols formed stable complexes with Qdots to deliver siRNA molecules into the cytoplasm, protecting the siRNA from enzymatic degradation. The new siRNA delivery technology which integrates Qdots and amphipols may allow for a wider range of gene therapy and drug delivery to appropriate targets.

Traceable therapeutics are important not only for drug delivery and validation but also for drug discovery. Qdots will probably prove to be an important system with which to study drug delivery. Currently, high quality Qdots are primarily composed of heavy metals with possible long-term toxicity effects. Although they have as yet been applied only to cells or small animals as drug carriers, Qdots may serve as discovery tools for drug screening in humans.

PCR electrospray ionization mass spectrometry

The rapidity and high-resolution of MALDI-TOF MS provides another powerful platform for the detection and characterization of nucleic acid amplification products. The technology is premised on the capacity of MALDI-TOF MS to discriminate individual PCR products contained in complex amplicon mixtures according to nucleotide base composition (Muddiman *et al.*, 1997). The deconvolution algorithm allows base composition of PCR products to be deduced from mass spectrometrically measured molecular weights and the complementary nature of DNA, leading to organism identification. Early studies successfully used this technique to directly detect amplification products from PCR (Hurst *et al.*, 1996) and LCR (Jurinke *et al.*, 1996). Soon thereafter, the MALDI-TOF MS platform was linked to PCR amplification for genotypic analysis of HCV (Kim *et al.*, 2005) and HPV (Hong *et al.*, 2008; Soderlund-Strand *et al.*, 2008). Detection of human herpesviruses from clinical specimens was performed using MALDI-TOF MS following multiplex PCR amplification (Sjoholm *et al.*, 2008). A MALDI-TOF MS-based genotyping assay has been described that monitors development of HBV polymerase YMDD mutant genotypes during lamivudine treatment (Hong *et al.*, 2004; Lee *et al.*, 2006).

An integrated system, the Ibis T5000 Biosensor, now marketed as Plex-ID (Abbott Molecular, Des Plaines, IL, USA), has been developed to couple broad-range nucleic acid amplification to high-performance electrospray ionization (ESI) MS and base-composition analysis. The system enables the identification and quantification of a broad set of pathogens, including all known bacteria, all major groups of pathogenic fungi and the major families of viruses that cause disease in humans and animals, along with the detection of virulence factors and antibiotic resistance markers (Ecker *et al.*, 2008). The system has been used for rapid identification and strain-typing of respiratory bacterial pathogens for epidemic surveillance (Ecker *et al.*, 2005), identification and genotyping of *Acinetobacter baumannii* strains in an outbreak associated with war trauma (Ecker *et al.*, 2006; Wortmann *et al.*, 2008), determination of quinolone resistance in *Acinetobacter* species (Hannis *et al.*, 2008; Hujer *et al.*, 2009), genotyping of *Campylobacter* species (Hannis *et al.*, 2008; Hujer *et al.*, 2009), and rapid genotyping and clonal complex assignment of *Staphylococcus aureus* isolates (Hall *et al.*, 2009). The system has also been used to detect *Ehrlichia, Anaplasma,* and *Rickettsia* pathogens directly from blood specimens for diagnosis of tick-borne sepsis (Eshoo *et al.*, 2010). The main advantages of the Plex-ID system are high resolution, speed, and a substantial degree of automation. The main disadvantages include the engineering difficulty of MS device miniaturization and the need for continuous enrichment of databases with new genomic sequences, as new pathogens are being identified.

Atomic force microscopy

Atomic force microscopy (AFM) is becoming an important new tool in the biological sciences (Huff *et al.*, 2004; Dufrene, 2008). AFM provides high-resolution topographical imaging of organisms, using a sharp micron-scale tip which scans and amplifies surface features. Principally applied to the field of virology at this time, AFM is a non-destructive method of viewing an intact virus (or other organism), in liquid or other medium, without the artefacts caused by fixation and/or staining. Although AFM does not provide the high resolution of electron microscopy or X-ray crystallography, it is better suited than electron microscopy for high-throughput analysis. AFM allows for examination of changes in cell surface morphology resulting from treatment with antimicrobials, which can be examined in real time (Dufrene, 2002).

Carbon nanotubes

Researchers have also recently shown that carbon nanotubes, equipped with aptamers as bacteria-specific binding sites, can detect extremely low concentrations of bacteria (Zelada-Guillen *et al.*, 2009). Aptamers, which are short, synthetic strands of DNA or RNA, can be anchored to the carbon nanotubes and deposited onto an electrode in a thin layer. Any bacteria of interest may bind to the aptamers, which, upon binding, cause a change in electrical voltage. Researchers in Spain have shown that electrochemical detection techniques using carbon nanotubes detected *Salmonella typhi* bacteria in liquid phosphate buffer solution (Zelada-Guillen *et al.*, 2009). The sensors were highly selective for *Salmonella* and detected the organisms in real time. Other researchers have demonstrated electrical changes due to the binding and unbinding of single influenza A viruses to carbon nanowires, or nanotubes (Patolsky *et al.*, 2004).

In conclusion, there are many novel and emerging assays in the field of molecular and nanotechnology, which have improved our abilities to not only diagnose infection at very low levels, but also to open avenues for discovery of targeted antimicrobial therapy. Genomic analysis can now supply information on targets which may address the potential treatment spectrum of an antimicrobial agent. Although these techniques appear promising, many are in the developmental stage in industry or in research laboratories. Ultra-high-throughput sequencing, liquid microarrays, and PCR electrospray ionization mass spectrometry are perhaps the closest of the techniques to clinical testing.

On-demand, near-patient techniques

Point-of-care testing (POCT) is performed in homes, workplaces, pharmacies, physicians' offices, outpatient clinics, emergency rooms, disaster sites and patient bedsides. The need of POCT has been driven by the practitioner's need to provide a rapid, evidence-based, therapeutic intervention at or near the site of patient care. The design of POCT devices is simple enough for direct use by healthcare providers. POCT devices are usually designated as waived tests by Clinical Laboratory Improvement Amendments (CLIA; Overturf, 2008).

Table 19.3 lists various POCT commercial diagnostic microbiology tests which are currently available (Overturf, 2008). Most of the devices are immunoassays, which incorporate specific antigen–antibody binding into simple and rapid lateral flow, strip-based test formats. Among them, rapid group A streptococcus POCT has become a prominent method of diagnosis for streptococcal pharyngitis in clinicians' offices. In addition to rapid diagnosis, such testing may have a role in guiding antiviral therapy and avoiding the use of antibacterial therapy in children (Trevino and Weissfeld, 2005; Overturf, 2008). The availability of rapid HIV tests has the potential to increase the access of underserved populations to serological testing for HIV and allow prompt diagnosis for initiation of critical prophylactic regimens (Roberts *et al.*, 2007). However, most rapid antigen tests based on immunoassays are relatively insensitive, including the monospot test, and the RSV and influenza A and B tests. Negative tests are recommended to be retested by a more sensitive method, usually by culture

Table 19.3 CLIA-waived infectious diseases tests

Laboratory area	Tests
Infectious diseases	Group A Streptococcus
	Helicobacter pylori antibody
	HIV-1 and HIV-2 antibodies
	Influenza A and B
	Lyme antibody screen
	Monospot
	Respiratory syncytial virus
	Respiratory syncytial virus antibody
Hepatic function/inflammation[a]	Alanine aminotransferase, aspartate aminotransferase, microalbumin
Renal function and urinary tract infection[a]	Dipstick urinalysis, creatinine
Sepsis[a]	Lactate
Haematology[a]	Erythrocyte sedimentation rate, faecal/gastric blood haematocrit, haemoglobin prothrombin time

Adapted from Overturf (2008), with permission.

CLIA, Clinical Laboratory Improvement Act.

[a]Additional waived tests of interest in infectious disease diagnosis.

or molecular methods (Trevino and Weissfeld, 2005; Overturf, 2008).

Application of *in vitro* nucleic acid amplification into the field of POCT would significantly improve test sensitivities and reduce turn-around time, thereby allowing real-time decisions about patient management (Campos *et al.*, 2008). Owing to technical complexity and regulatory issues, molecular-based POCT infectious disease diagnostic devices are not generally available. However, several on-demand, near-patient molecular diagnostic devices techniques are commercially available. The Cepheid GeneXpert System (Sunnyvale, CA, USA), a single-use sample processing cartridge system with integrated multicolour real-time PCR capacity (Raja *et al.*, 2005), has the potential to greatly simplify nucleic acid amplification tests. Several devices are available for the screening and detection of MRSA, methicillin-susceptible *S. aureus*, vancomycin-resistant enterococci, enteroviruses, group B streptococcus and *C. difficile* (Gavino and Wang, 2007; Kost *et al.*, 2007; Rossney *et al.*, 2008). Recently, this system was reported to employ a novel six-colour dye set to detect *M. tuberculosis* and identify rifampin-resistance as a surrogate for multidrug resistance directly from patient sputum in less than 2 hours (Helb *et al.*, 2009). While most of these assays have been designated by the US Food and Drug Administration as moderately complex, current versions of devices need additional improvement and optimization, particularly those run by medical technologists in clinical laboratories (Sefers *et al.*, 2009).

Miniaturized PCR 'POCT' devices have been used for microbial agent detection and identification. Several lab-on-a-chip devices have been described for detection of viral pathogens, including HIV (Lee *et al.*, 2008). BioHelix (Beverly, MA, USA) developed an instrument-free and cross-contamination-proof device, which consists of isothermal helicase-dependent nucleic acid amplification in a lateral-flow-based disposable format (Fig. 19.2). The entire procedure, starting at specimen processing and ending with result readout, can be completed in less than 1.5 hours and has the potential for use as a simple and instrument-free POCT (Goldmeyer *et al.*, 2008). The system has been used successfully for detection and identification of several microbial pathogens including HIV-1, *C. difficile* and *S. aureus* (Chow *et al.*, 2008; Goldmeyer *et al.*, 2008; Tang *et al.*, 2010). DxNA's GeneSTAT (St. George, UT, USA) POCT for H1N1 influenza virus is a new platform which enables rapid diagnosis from nasal swabs; results are available within 45 minutes

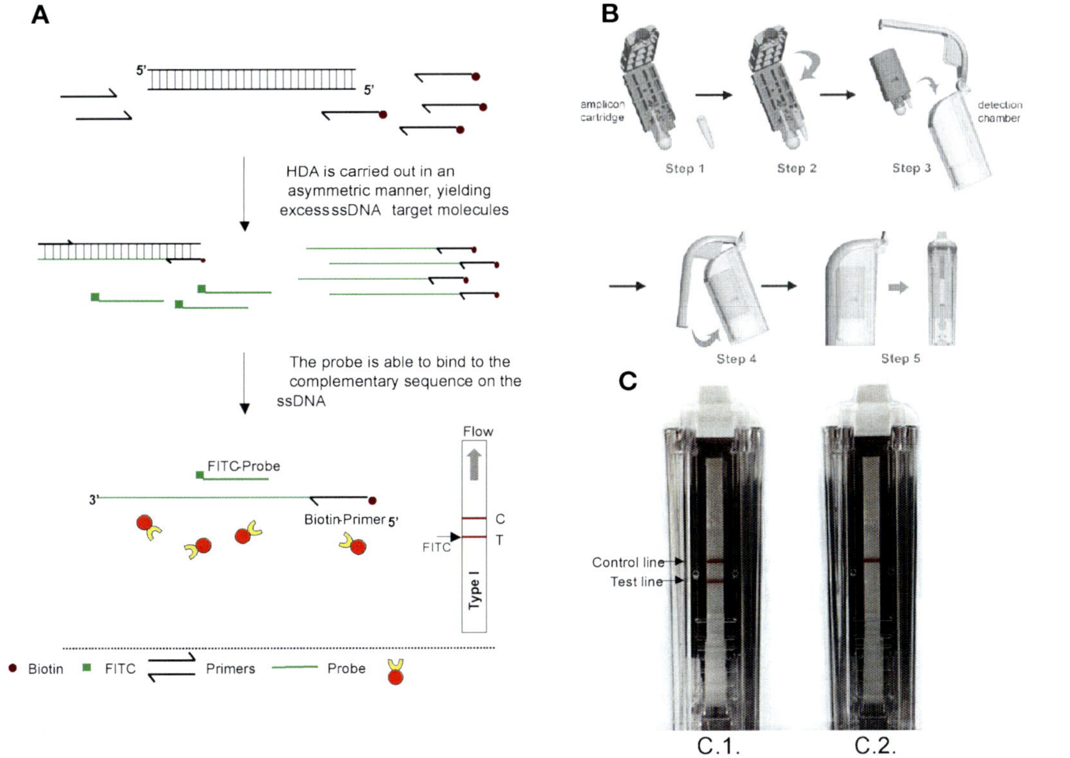

Figure 19.2 Molecular IsoAmp® HIV assay for detection and identification of methicillin resistant *S. aureus* from positive blood culture media. (A) First, a single-stranded DNA capable of binding the specific probe is generated by asymmetric RT-helicase-dependent amplification. The probe is bound by the cassette insert, and the biotin moiety binds the chromogen, resulting in the red test line. (B) Next, the amplification reaction vessel is placed in an amplicon cartridge (step 1); the cartridge is closed to immobilize the reaction vessel (step 2); the amplicon cartridge is inserted into the detection chamber (step 3); the handle of the detection chamber is closed to seal the vessel into the chamber and to cut open the running buffer reservoir and the reaction vessel (step 4); finally, after 10 minutes the detection window of the chamber is read by eye to score the assay result (step 5). (C) The readout of the type 1 BESt (BioHelix Express strip) Cassette, C.1. is a positive reaction and C.2. is a negative reaction. Adapted from Goldmeyer *et al.* (2008) with permission.

(http://www.dxna.com/products-overview.htm). With continuous technique improvements, microfluidics- and array-based systems, combined with incorporation of micro/nano fabrications/crystals (e.g. quantum dots), will enable the development of more feasible immunological and molecular tests for the POCT (Kopp *et al.*, 1998; Michalet *et al.*, 2005; Edgar *et al.*, 2006; Liu *et al.*, 2006; Mokkapati *et al.*, 2007; Sista *et al.*, 2008).

Concluding remarks

Non-culture-based methods for diagnosis of infections have changed from primarily serological methods to pathogen detection and will probably continue to include more point-of-care and multianalyte testing in the future. Although emerging technologies offer some advantages over more established techniques, such as traditional sequencing, most are still in development. However, these novel techniques hold promise for increased sensitivity and more rapid turn-around time. In addition, researchers are beginning to apply several of these techniques to the areas of drug targeting and discovery.

As the field of diagnostic microbiology continues to grow, researchers are attempting to apply novel technologies to antibacterial drug discovery. Such areas include selection of gene targets and resistance monitoring, some of which have been discussed in this chapter. These technologies can also be used to assess host susceptibility to infection by pre-screening the patient before treatment. In addition, real-time monitoring of antimicrobial outcome and side-effects may potentially be achieved, which will add to our growing knowledge of antibacterial practice and lead to evidence-based interventions for antimicrobial therapy. The future also holds promise for improved robotics and automation which can decrease the turn-around time and potential for user error of these potentially complicated techniques.

References

Agrawal, A., Tripp, R.A., Anderson, L.J., and Nie, S. (2005). Real-time detection of virus particles and viral protein expression with two-color nanoparticle probes. J. Virol. *79*, 8625–8628.

Alcaide, F., Benitez, M.A., Escriba, J.M., and Martin, R. (2000). Evaluation of the BACTEC MGIT 960 and the MB/BacT systems for recovery of mycobacteria from clinical specimens and for species identification by DNA AccuProbe. J. Clin. Microbiol. *38*, 398–401.

Anthony, R.M., Brown, T.J., and French, G.L. (2001). DNA array technology and diagnostic microbiology. Expert Rev. Mol. Diagn. *1*, 30–38.

Antonova, O.V., Gryadunov, D.A., Lapa, S.A., Kuz'min, A.V., Larionova, E.E., Smirnova, T.G., Nosova, E.Y., Skotnikova, O.I., Chernousova, L.N., Moroz, A.M., *et al.* (2008). Detection of mutations in *Mycobacterium tuberculosis* genome determining resistance to fluoroquinolones by hybridization on biological microchips. Bull. Exp. Biol. Med. *145*, 108–113.

Arias, C.A., and Murray, B.E. (2009). Antibiotic-resistant bugs in the 21st century–a clinical super-challenge. N. Engl. J. Med. *360*, 439–443.

Athukorala, S.N., Fernando, W.G., and Rashid, K.Y. (2009). Identification of antifungal antibiotics of *Bacillus* species isolated from different microhabitats using polymerase chain reaction and MALDI-TOF mass spectrometry. Can. J. Microbiol. *55*, 1021–1032.

Balwierz, P.J., Carninci, P., Daub, C.O., Kawai, J., Hayashizaki, Y., Van Belle, W., Beisel, C., and van Nimwegen, E. (2009). Methods for analyzing deep sequencing expression data: constructing the human and mouse promoterome with deepCAGE data. Genome Biol. *10*, R79.

Belgrader, P., Benett, W., Hadley, D., Richards, J., Stratton, P., Mariella, R., Jr., and Milanovich, F. (1999). PCR detection of bacteria in seven minutes. Science *284*, 449–450.

Benson, R., Tondella, M.L., Bhatnagar, J., Carvalho Mda, G., Sampson, J.S., Talkington, D.F., Whitney, A.M., Mothershed, E., McGee, L., Carlone, G., *et al.* (2008). Development and evaluation of a novel multiplex PCR technology for molecular differential detection of bacterial respiratory disease pathogens. J. Clin. Microbiol. *46*, 2074–2077.

Bentzen, E.L., House, F., Utley, T.J., Crowe, J.E., Jr., and Wright, D.W. (2005). Progression of respiratory syncytial virus infection monitored by fluorescent quantum dot probes. Nano Lett. *5*, 591–595.

Biagini, R.E., Sammons, D.L., Smith, J.P., MacKenzie, B.A., Striley, C.A., Semenova, V., Steward-Clark, E., Stamey, K., Freeman, A.E., Quinn, C.P., *et al.* (2004). Comparison of a multiplexed fluorescent covalent microsphere immunoassay and an enzyme-linked immunosorbent assay for measurement of human immunoglobulin G antibodies to anthrax toxins. Clin. Diagn. Lab. Immunol. *11*, 50–55.

Blecker, U., Lanciers, S., Hauser, B., and Vandenplas, Y. (1993). Diagnosis of *Helicobacter pylori* infection in adults and children by using the Malakit *Helicobacter pylori*, a commercially available enzyme-linked immunosorbent assay. J. Clin. Microbiol. *31*, 1770–1773.

Blinkova, O., Kapoor, A., Victoria, J., Jones, M., Wolfe, N., Naeem, A., Shaukat, S., Sharif, S., Alam, M.M., Angez, M., *et al.* (2009). Cardioviruses are genetically diverse and cause common enteric infections in South Asian children. J. Virol. *83*, 4631–4641.

Boom, R., Sol, C.J., Salimans, M.M., Jansen, C.L., Wertheim-van Dillen, P.M., and van der Noordaa, J. (1990). Rapid and simple method for purification of nucleic acids. J. Clin. Microbiol. *28*, 495–503.

Boucher, H.W., and Corey, G.R. (2008). Epidemiology of methicillin-resistant *Staphylococcus aureus*. Clin. Infect. Dis. *46 (*Suppl. 5*)*, S344–349.

Boucher, H.W., Talbot, G.H., Bradley, J.S., Edwards, J.E., Gilbert, D., Rice, L.B., Scheld, M., Spellberg, B., and Bartlett, J. (2009). Bad bugs, no drugs: no ESKAPE! An update from the Infectious Diseases Society of America. Clin. Infect. Dis. *48*, 1–12.

Brunstein, J., and Thomas, E. (2006). Direct screening of clinical specimens for multiple respiratory pathogens using the Genaco Respiratory Panels 1 and 2. Diagn. Mol. Pathol. *15*, 169–173.

Brunstein, J.D., Cline, C.L., McKinney, S., and Thomas, E. (2008). Evidence from multiplex molecular assays for complex multipathogen interactions in acute respiratory infections. J. Clin. Microbiol. *46*, 97–102.

Buck, G.E., Wiesemann, M., and Stewart, L. (2002). Comparison of mixed cell culture containing genetically engineered BGMK and CaCo-2 cells (Super E-Mix) with RT-PCR and conventional cell culture for the diagnosis of enterovirus meningitis. J. Clin. Virol. *25 (Suppl 1)*, S13–18.

Camara, J.E., and Hays, F.A. (2007). Discrimination between wild-type and ampicillin-resistant *Escherichia coli* by matrix-assisted laser desorption/ionization time-of-flight mass spectrometry. Anal. Bioanal. Chem. *389*, 1633–1638.

Campbell, S., and Fedoriw, Y. (2009). HIV testing near the patient: changing the face of HIV testing. Clin. Lab. Med. *29*, 491–501.

Campos, M., Quartin, A., Mendes, E., Abreu, A., Gurevich, S., Echarte, L., Ferreira, T., Cleary, T., Hollender,

E., and Ashkin, D. (2008). Feasibility of shortening respiratory isolation with a single sputum nucleic acid amplification test. Am. J. Respir. Crit. Care Med. *178*, 300–305.

Chan, W.C., and Nie, S. (1998). Quantum dot bioconjugates for ultrasensitive nonisotopic detection. Science *281*, 2016–2018.

Chapin, K., and Musgnug, M. (2003). Evaluation of three rapid methods for the direct identification of *Staphylococcus aureus* from positive blood cultures. J. Clin. Microbiol. *41*, 4324–4327.

Chee, C.B., Gan, S.H., Khinmar, K.W., Barkham, T.M., Koh, C.K., Liang, S., and Wang, Y.T. (2008). Comparison of sensitivities of two commercial gamma interferon release assays for pulmonary tuberculosis. J. Clin. Microbiol. *46*, 1935–1940.

Chiu, C.Y., Urisman, A., Greenhow, T.L., Rouskin, S., Yagi, S., Schnurr, D., Wright, C., Drew, W.L., Wang, D., Weintrub, P.S., *et al.* (2008). Utility of DNA microarrays for detection of viruses in acute respiratory tract infections in children. J. Pediatr. *153*, 76–83.

Chow, W.H., McCloskey, C., Tong, Y., Hu, L., You, Q., Kelly, C.P., Kong, H., Tang, Y.W., and Tang, W. (2008). Application of isothermal helicase-dependent amplification with a disposable detection device in a simple sensitive stool test for toxigenic *Clostridium difficile*. J. Mol. Diagn. *10*, 452–458.

Coppee, J.Y. (2008). Do DNA microarrays have their future behind them? Microbes Infect. *10*, 1067–1071.

Creighton, C.J., Reid, J.G., and Gunaratne, P.H. (2009). Expression profiling of microRNAs by deep sequencing. Brief Bioinform. *10*, 490–497.

Della-Latta, P., and Whittier, S. (1998). Comprehensive evaluation of performance, laboratory application, and clinical usefulness of two direct amplification technologies for the detection of *Mycobacterium tuberculosis* complex. Am. J. Clin. Pathol. *110*, 301–310.

Deng, J., Zheng, Y., Zhao, R., Wright, P.F., Stratton, C.W., and Tang, Y.W. (2009). Culture versus polymerase chain reaction for the etiologic diagnosis of community-acquired pneumonia in antibiotic-pretreated pediatric patients. Pediatr. Infect. Dis. J. *28*, 53–55.

Diekema, D.J., BootsMiller, B.J., Vaughn, T.E., Woolson, R.F., Yankey, J.W., Ernst, E.J., Flach, S.D., Ward, M.M., Franciscus, C.L., Pfaller, M.A., *et al.* (2004). Antimicrobial resistance trends and outbreak frequency in United States hospitals. Clin. Infect. Dis. *38*, 78–85.

Droege, M., and Hill, B. (2008). The Genome Sequencer FLX System–longer reads, more applications, straight forward bioinformatics and more complete data sets. J. Biotechnol. *136*, 3–10.

Dufrene, Y.F. (2002). Atomic force microscopy, a powerful tool in microbiology. J. Bacteriol. *184*, 5205–5213.

Dufrene, Y.F. (2008). Towards nanomicrobiology using atomic force microscopy. Nat. Rev. Microbiol. *6*, 674–680.

Dunbar, S.A., and Jacobson, J.W. (2007). Quantitative, multiplexed detection of *Salmonella* and other pathogens by Luminex xMAP suspension array. Methods Mol. Biol. *394*, 1–19.

Ecker, D.J., Sampath, R., Blyn, L.B., Eshoo, M.W., Ivy, C., Ecker, J.A., Libby, B., Samant, V., Sannes-Lowery, K.A., Melton, R.E., *et al.* (2005). Rapid identification and strain-typing of respiratory pathogens for epidemic surveillance. Proc. Natl. Acad. Sci. U.S.A. *102*, 8012–8017.

Ecker, J.A., Massire, C., Hall, T.A., Ranken, R., Pennella, T.T., Agasino Ivy, C., Blyn, L.B., Hofstadler, S.A., Endy, T.P., Scott, P.T., *et al.* (2006). Identification of Acinetobacter species and genotyping of *Acinetobacter baumannii* by multilocus PCR and mass spectrometry. J. Clin. Microbiol. *44*, 2921–2932.

Ecker, D.J., Sampath, R., Massire, C., Blyn, L.B., Hall, T.A., Eshoo, M.W., and Hofstadler, S.A. (2008). Ibis T*5000*, a universal biosensor approach for microbiology. Nat. Rev. Microbiol. *6*, 553–558.

Edgar, R., McKinstry, M., Hwang, J., Oppenheim, A.B., Fekete, R.A., Giulian, G., Merril, C., Nagashima, K., and Adhya, S. (2006). High-sensitivity bacterial detection using biotin-tagged phage and quantum-dot nanocomplexes. Proc. Natl. Acad. Sci. U.S.A. *103*, 4841–4845.

Eshoo, M.W., Crowder, C.D., Li, H., Matthews, H.E., Meng, S., Sefers, S.E., Sampath, R., Stratton, C.W., Blyn, L.B., Ecker, D.J., *et al.* (2010). Detection and identification of *Ehrlichia* species in blood using PCR and electrospray ionization mass spectrometry. J. Clin. Microbiol. *48*, 472–478.

Falagas, M.E., and Bliziotis, I.A. (2007). Pandrug-resistant Gram-negative bacteria: the dawn of the post-antibiotic era? Int. J. Antimicrob. Agents *29*, 630–636.

Flayhart, D., Hindler, J.F., Bruckner, D.A., Hall, G., Shrestha, R.K., Vogel, S.A., Richter, S.S., Howard, W., Walther, R., and Carroll, K.C. (2005). Multicenter evaluation of BBL CHROMagar MRSA medium for direct detection of methicillin-resistant *Staphylococcus aureus* from surveillance cultures of the anterior nares. J. Clin. Microbiol. *43*, 5536–5540.

Ford, N., Nachega, J.B., Engel, M.E., and Mills, E.J. (2009). Directly observed antiretroviral therapy: a systematic review and meta-analysis of randomised clinical trials. Lancet *374*, 2064–2071.

Fox, S., Filichkin, S., and Mockler, T.C. (2009). Applications of ultra-high-throughput sequencing. Methods Mol. Biol. *553*, 79–108.

Gavino, M., and Wang, E. (2007). A comparison of a new rapid real-time polymerase chain reaction system to traditional culture in determining group B streptococcus colonization. Am. J. Obstet. Gynecol. *197*, 388 e381–384.

Gegia, M., Mdivani, N., Mendes, R.E., Li, H., Akhalaia, M., Han, J., Khechinashvili, G., and Tang, Y.W. (2008). Prevalence of and molecular basis for tuberculosis drug resistance in the Republic of Georgia: validation of a QIAplex system for detection of drug resistance-related mutations. Antimicrob. Agents Chemother. *52*, 725–729.

Giske, C.G., Monnet, D.L., Cars, O., and Carmeli, Y. (2008). Clinical and economic impact of common multidrug-resistant Gram-negative bacilli. Antimicrob. Agents Chemother. *52*, 813–821.

Goldmeyer, J., Li, H., McCormac, M., Cook, S., Stratton, C., Lemieux, B., Kong, H., Tang, W., and Tang, Y.W. (2008). Identification of *Staphylococcus aureus* and determination of methicillin resistance directly from positive blood cultures by isothermal amplification and a disposable detection device. J. Clin. Microbiol. *46*, 1534–1536.

Hall, T.A., Sampath, R., Blyn, L.B., Ranken, R., Ivy, C., Melton, R., Matthews, H., White, N., Li, F., Harpin, V., *et al.* (2009). Rapid molecular genotyping and clonal complex assignment of *S. aureus* isolates by PCR/ESI-MS. J. Clin. Microbiol. *47*, 1733–1741.

Han, J., Swan, D.C., Smith, S.J., Lum, S.H., Sefers, S.E., Unger, E.R., and Tang, Y.W. (2006). Simultaneous amplification and identification of 25 human papillomavirus types with Templex technology. J. Clin. Microbiol. *44*, 4157–4162.

Hannis, J.C., Manalili, S.M., Hall, T.A., Ranken, R., White, N., Sampath, R., Blyn, L.B., Ecker, D.J., Mandrell, R.E., Fagerquist, C.K., *et al.* (2008). High-resolution genotyping of *Campylobacter* species by use of PCR and high-throughput mass spectrometry. J. Clin. Microbiol. *46*, 1220–1225.

Harris, T.D., Buzby, P.R., Babcock, H., Beer, E., Bowers, J., Braslavsky, I., Causey, M., Colonell, J., Dimeo, J., Efcavitch, J.W., *et al.* (2008). Single-molecule DNA sequencing of a viral genome. Science *320*, 106–109.

Helb, D., Jones, M., Story, E., Boehme, C., Wallace, E., Ho, K., Kop, J., Owens, M.R., Rodgers, R., Banada, P., *et al.* (2010). Rapid detection of *Mycobacterium tuberculosis* and rifampin-resistance using on-demand, near patient technology. J. Clin. Microbiol. *48*, 229–237.

Henrickson, K.J., and Hall, C.B. (2007). Diagnostic assays for respiratory syncytial virus disease. Pediatr. Infect. Dis. J. *26*, S36–40.

Holt, K.E., Parkhill, J., Mazzoni, C.J., Roumagnac, P., Weill, F.X., Goodhead, I., Rance, R., Baker, S., Maskell, D.J., Wain, J., *et al.* (2008). High-throughput sequencing provides insights into genome variation and evolution in *Salmonella typhi*. Nat. Genet. *40*, 987–993.

Hong, S.P., Kim, N.K., Hwang, S.G., Chung, H.J., Kim, S., Han, J.H., Kim, H.T., Rim, K.S., Kang, M.S., Yoo, W., *et al.* (2004). Detection of hepatitis B virus YMDD variants using mass spectrometric analysis of oligonucleotide fragments. J. Hepatol. *40*, 837–844.

Hong, S.P., Shin, S.K., Lee, E.H., Kim, E.O., Ji, S.I., Chung, H.J., Park, S.N., Yoo, W., Folk, W.R., and Kim, S.O. (2008). High-resolution human papillomavirus genotyping by MALDI-TOF mass spectrometry. Nat. Protoc. *3*, 1476–1484.

Hoper, D., Hoffmann, B., and Beer, M. (2009). Simple, sensitive, and swift sequencing of complete H5N1 avian influenza virus genomes. J. Clin. Microbiol. *47*, 674–679.

Horan, P.K., and Wheeless, L.L., Jr. (1977). Quantitative single cell analysis and sorting. Science *198*, 149–157.

Huff, J.L., Lynch, M.P., Nettikadan, S., Johnson, J.C., Vengasandra, S., and Henderson, E. (2004). Label-free protein and pathogen detection using the atomic force microscope. J. Biomol. Screen *9*, 491–497.

Hujer, K.M., Hujer, A.M., Endimiani, A., Thomson, J.M., Adams, M.D., Goglin, K., Rather, P.N., Pennella, T.T., Massire, C., Eshoo, M.W., *et al.* (2009). Rapid determination of quinolone resistance in Acinetobacter spp. J. Clin. Microbiol. *47*, 1436–1442.

Hurst, G.B., Doktycz, M.J., Vass, A.A., and Buchanan, M.V. (1996). Detection of bacterial DNA polymerase chain reaction products by matrix-assisted laser desorption/ionization mass spectrometry. Rapid Commun. Mass Spectrom. *10*, 377–382.

Jackson, J.B., Parsons, J.S., Nichols, L.S., Knoble, N., Kennedy, S., and Piwowar, E.M. (1997). Detection of human immunodeficiency virus type 1 (HIV-1) antibody by western blotting and HIV-1 DNA by PCR in patients with AIDS. J. Clin. Microbiol. *35*, 1118–1121.

Jannetto, P.J., Vaughan, K., Prokhorova, A., Brener, D., Lonosky, P.M., Sonpatki, V.M., Marla, S.S., Cork, W.H., N., G.-A., Pathak, N., *et al.* (2009). Nanoparticle probes for the detection and discrimination of influenza A/B and RSV A/B in respiratory specimens. Paper presented at: 25th Clinical Virology Symposium (Daytona Beach, FL).

Jurinke, C., van den Boom, D., Jacob, A., Tang, K., Worl, R., and Koster, H. (1996). Analysis of ligase chain reaction products via matrix-assisted laser desorption/ionization time-of-flight-mass spectrometry. Anal. Biochem. *237*, 174–181.

Kapoor, A., Victoria, J., Simmonds, P., Slikas, E., Chieochansin, T., Naeem, A., Shaukat, S., Sharif, S., Alam, M.M., Angez, M., *et al.* (2008). A highly prevalent and genetically diversified Picornaviridae genus in South Asian children. Proc. Natl. Acad. Sci. U.S.A. *105*, 20482–20487.

de Keizer, W., Bienenmann-Ploum, M.E., Bergwerff, A.A., and Haasnoot, W. (2008). Flow cytometric immunoassay for sulfonamides in raw milk. Anal. Chim. Acta *620*, 142–149.

Kim, Y.J., Kim, S.O., Chung, H.J., Jee, M.S., Kim, B.G., Kim, K.M., Yoon, J.H., Lee, H.S., Kim, C.Y., Kim, S., *et al.* (2005). Population genotyping of hepatitis C virus by matrix-assisted laser desorption/ionization time-of-flight mass spectrometry analysis of short DNA fragments. Clin. Chem. *51*, 1123–1131.

Kiska, D.L., Orkiszewski, D.R., Howell, D., and Gilligan, P.H. (1994). Evaluation of new monoclonal antibody-based latex agglutination test for detection of cryptococcal polysaccharide antigen in serum and cerebrospinal fluid. J. Clin. Microbiol. *32*, 2309–2311.

Kistler, A., Avila, P.C., Rouskin, S., Wang, D., Ward, T., Yagi, S., Schnurr, D., Ganem, D., DeRisi, J.L., and Boushey, H.A. (2007). Pan-viral screening of respiratory tract infections in adults with and without asthma reveals unexpected human coronavirus and human rhinovirus diversity. J. Infect. Dis. *196*, 817–825.

Kopp, M.U., Mello, A.J., and Manz, A. (1998). Chemical amplification: continuous-flow PCR on a chip. Science *280*, 1046–1048.

Kost, C.B., Rogers, B., Oberste, M.S., Robinson, C., Eaves, B.L., Leos, K., Danielson, S., Satya, M., Weir, F., and Nolte, F.S. (2007). Multicenter beta trial of the GeneXpert enterovirus assay. J. Clin. Microbiol. *45*, 1081–1086.

Le, T., Chiarella, J., Simen, B.B., Hanczaruk, B., Egholm, M., Landry, M.L., Dieckhaus, K., Rosen, M.I., and Kozal, M.J. (2009). Low-abundance HIV drug-resistant viral variants in treatment-experienced persons correlate with historical antiretroviral use. PLoS One *4*, e6079.

Lee, C.H., Kim, S.O., Byun, K.S., Moon, M.S., Kim, E.O., Yeon, J.E., Yoo, W., and Hong, S.P. (2006). Predominance of hepatitis B virus YMDD mutants is prognostic of viral DNA breakthrough. Gastroenterology *130*, 1144–1152.

Lee, W.M., Grindle, K., Pappas, T., Marshall, D.J., Moser, M.J., Beaty, E.L., Shult, P.A., Prudent, J.R., and Gern, J.E. (2007). High-throughput, sensitive, and accurate multiplex PCR-microsphere flow cytometry system for large-scale comprehensive detection of respiratory viruses. J. Clin. Microbiol. *45*, 2626–2634.

Lee, S.H., Kim, S.W., Kang, J.Y., and Ahn, C.H. (2008). A polymer lab-on-a-chip for reverse transcription (RT)-PCR based point-of-care clinical diagnostics. Lab. Chip *8*, 2121–2127.

Li, Y., Cu, Y.T., and Luo, D. (2005). Multiplexed detection of pathogen DNA with DNA-based fluorescence nanobarcodes. Nat. Biotechnol. *23*, 885–889.

Li, H., McCormac, M.A., Estes, R.W., Sefers, S.E., Dare, R.K., Chappell, J.D., Erdman, D.D., Wright, P.F., and Tang, Y.W. (2007). Simultaneous detection and high-throughput identification of a panel of RNA viruses causing respiratory tract infections. J. Clin. Microbiol. *45*, 2105–2109.

Li, L., Victoria, J., Kapoor, A., Naeem, A., Shaukat, S., Sharif, S., Alam, M.M., Angez, M., Zaidi, S.Z., and Delwart, E. (2009). Genomic characterization of novel human parechovirus type. Emerg. Infect. Dis. *15*, 288–291.

Liu, R.H., Lodes, M.J., Nguyen, T., Siuda, T., Slota, M., Fuji, H.S., and McShea, A. (2006). Validation of a fully integrated microfluidic array device for influenza A subtype identification and sequencing. Anal. Chem. *78*, 4184–4193.

McNamara, D.T., Thomson, J.M., Kasehagen, L.J., and Zimmerman, P.A. (2004). Development of a multiplex PCR-ligase detection reaction assay for diagnosis of infection by the four parasite species causing malaria in humans. J. Clin. Microbiol. *42*, 2403–2410.

Mahony, J., Chong, S., Merante, F., Yaghoubian, S., Sinha, T., Lisle, C., and Janeczko, R. (2007). Development of a respiratory virus panel test for detection of twenty human respiratory viruses by use of multiplex PCR and a fluid microbead-based assay. J. Clin. Microbiol. *45*, 2965–2970.

Marchildon, P.A., Ciota, L.M., Zamaniyan, F.Z., Peacock, J.S., and Graham, D.Y. (1996). Evaluation of three commercial enzyme immunoassays compared with the 13C urea breath test for detection of *Helicobacter pylori* infection. J. Clin. Microbiol. *34*, 1147–1152.

Margeridon-Thermet, S., Shulman, N.S., Ahmed, A., Shahriar, R., Liu, T., Wang, C., Holmes, S.P., Babrzadeh, F., Gharizadeh, B., Hanczaruk, B., *et al.* (2009). Ultra-deep pyrosequencing of hepatitis B virus quasispecies from nucleoside and nucleotide reverse-transcriptase inhibitor (NRTI)-treated patients and NRTI-naive patients. J. Infect. Dis. *199*, 1275–1285.

Matthey, S., Nicholson, D., Ruhs, S., Alden, B., Knock, M., Schultz, K., and Schmuecker, A. (1992). Rapid detection of respiratory viruses by shell vial culture and direct staining by using pooled and individual monoclonal antibodies. J. Clin. Microbiol. *30*, 540–544.

Mellmann, A., Cloud, J., Maier, T., Keckevoet, U., Ramminger, I., Iwen, P., Dunn, J., Hall, G., Wilson, D., Lasala, P., *et al.* (2008). Evaluation of matrix-assisted laser desorption ionization-time-of-flight mass spectrometry in comparison to 16S rRNA gene sequencing for species identification of nonfermenting bacteria. J. Clin. Microbiol. *46*, 1946–1954.

Mellmann, A., Bimet, F., Bizet, C., Borovskaya, A.D., Drake, R.R., Eigner, U., Fahr, A.M., He, Y., Ilina, E.N., Kostrzewa, M., *et al.* (2009). High interlaboratory reproducibility of matrix-assisted laser desorption ionization-time of flight mass spectrometry-based species identification of nonfermenting bacteria. J. Clin. Microbiol. *47*, 3732–3734.

Mellors, J.W., Munoz, A., Giorgi, J.V., Margolick, J.B., Tassoni, C.J., Gupta, P., Kingsley, L.A., Todd, J.A., Saah, A.J., Detels, R., *et al.* (1997). Plasma viral load and CD4+ lymphocytes as prognostic markers of HIV-1 infection. Ann. Intern. Med. *126*, 946–954.

Merante, F., Yaghoubian, S., and Janeczko, R. (2007). Principles of the xTAG respiratory viral panel assay (RVP Assay). J Clin Virol *40 (Suppl 1)*, S31–35.

Merlino, J., Leroi, M., Bradbury, R., Veal, D., and Harbour, C. (2000). New chromogenic identification and detection of *Staphylococcus aureus* and methicillin-resistant S. aureus. J. Clin. Microbiol. 38, 2378–2380.

Michalet, X., Pinaud, F.F., Bentolila, L.A., Tsay, J.M., Doose, S., Li, J.J., Sundaresan, G., Wu, A.M., Gambhir, S.S., and Weiss, S. (2005). Quantum dots for live cells, *in vivo* imaging, and diagnostics. Science *307*, 538–544.

Miller, M.B., and Tang, Y.W. (2009). Basic concepts of microarrays and potential applications in clinical microbiology. Clin. Microbiol. Rev. *22*, 611–633.

Mokkapati, V.K., Sam Niedbala, R., Kardos, K., Perez, R.J., Guo, M., Tanke, H.J., and Corstjens, P.L. (2007). Evaluation of UPlink-RSV: prototype rapid antigen test for detection of respiratory syncytial virus infection. Ann. N. Y. Acad. Sci. *1098*, 476–485.

Morozova, O., and Marra, M.A. (2008). Applications of next-generation sequencing technologies in functional genomics. Genomics 92, 255–264.

Muddiman, D.C., Anderson, G.A., Hofstadler, S.A., and Smith, R.D. (1997). Length and base composition of PCR-amplified nucleic acids using mass measurements from electrospray ionization mass spectrometry. Anal. Chem. *69*, 1543–1549.

Nolte, F.S., Marshall, D.J., Rasberry, C., Schievelbein, S., Banks, G.G., Storch, G.A., Arens, M.Q., Buller, R.S., and Prudent, J.R. (2007). MultiCode-PLx System for multiplexed detection of seventeen respiratory viruses. J. Clin. Microbiol. *45*, 2779–2786.

Norrby, S.R., Nord, C.E., and Finch, R. (2005). Lack of development of new antimicrobial drugs: a potential serious threat to public health. Lancet Infect. Dis. *5*, 115–119.

O'Brien, W.A., Hartigan, P.M., Martin, D., Esinhart, J., Hill, A., Benoit, S., Rubin, M., Simberkoff, M.S., and Hamilton, J.D. (1996). Changes in plasma HIV-1 RNA and CD4+ lymphocyte counts and the risk of progression to AIDS. Veterans Affairs Cooperative Study Group on AIDS. N. Engl. J. Med. *334*, 426–431.

Oliveira, K., Procop, G.W., Wilson, D., Coull, J., and Stender, H. (2002). Rapid identification of *Staphylococcus aureus* directly from blood cultures by fluorescence in situ hybridization with peptide nucleic acid probes. J. Clin. Microbiol. *40*, 247–251.

Oliveira, K., Brecher, S.M., Durbin, A., Shapiro, D.S., Schwartz, D.R., De Girolami, P.C., Dakos, J., Procop, G.W., Wilson, D., Hanna, C.S., *et al.* (2003). Direct identification of *Staphylococcus aureus* from positive blood culture bottles. J. Clin. Microbiol. *41*, 889–891.

Overturf, G.D. (2008). CLIA waived testing in infectious diseases. Pediatr. Infect. Dis. J. *27*, 1009–1012.

Pabbaraju, K., Tokaryk, K.L., Wong, S., and Fox, J.D. (2008). Comparison of the Luminex xTAG respiratory viral panel with in-house nucleic acid amplification tests for diagnosis of respiratory virus infections. J. Clin. Microbiol. *46*, 3056–3062.

Paddock, C.D., and Childs, J.E. (2003). *Ehrlichia chaffeensis*: a prototypical emerging pathogen. Clin. Microbiol. Rev. *16*, 37–64.

Pakyz, A.L., MacDougall, C., Oinonen, M., and Polk, R.E. (2008). Trends in antibacterial use in US academic health centers: 2002 to 2006. Arch. Intern. Med. *168*, 2254–2260.

Patolsky, F., Zheng, G., Hayden, O., Lakadamyali, M., Zhuang, X., and Lieber, C.M. (2004). Electrical detection of single viruses. Proc. Natl. Acad. Sci. U.S.A. *101*, 14017–14022.

Payne, D.J., Gwynn, M.N., Holmes, D.J., and Pompliano, D.L. (2007). Drugs for bad bugs: confronting the challenges of antibacterial discovery. Nat. Rev. Drug Discov. *6*, 29–40.

Pejchalova, K., Zakovska, A., Mejzlikova, M., Halouzka, J., and Dendis, M. (2007). Isolation, cultivation and identification of *Borrelia burgdorferi* genospecies from *Ixodes ricinus* ticks from the city of Brno, Czech Republic. Ann. Agric. Environ. Med. *14*, 75–79.

Peterson, L.R. (2009). Bad bugs, no drugs: no ESCAPE revisited. Clin. Infect. Dis. *49*, 992–993.

Podzorski, R.P., Li, H., Han, J., and Tang, Y.W. (2008). MVPlex assay for direct detection of methicillin-resistant *Staphylococcus aureus* in nares and other swab specimens. J. Clin. Microbiol. *46*, 3107–3109.

Pusch, W., and Kostrzewa, M. (2005). Application of MALDI-TOF mass spectrometry in screening and diagnostic research. Curr. Pharm. Des. *11*, 2577–2591.

Qi, L., and Gao, X. (2008). Quantum dot-amphipol nanocomplex for intracellular delivery and real-time imaging of siRNA. ACS Nano *2*, 1403–1410.

Raja, S., Ching, J., Xi, L., Hughes, S.J., Chang, R., Wong, W., McMillan, W., Gooding, W.E., McCarty, K.S., Jr., Chestney, M., *et al.* (2005). Technology for automated, rapid, and quantitative PCR or reverse transcription-PCR clinical testing. Clin. Chem. *51*, 882–890.

Reisner, B.S., Gatson, A.M., and Woods, G.L. (1994). Use of Gen-Probe AccuProbes to identify *Mycobacterium avium* complex, *Mycobacterium tuberculosis* complex, *Mycobacterium kansasii*, and *Mycobacterium gordonae* directly from BACTEC TB broth cultures. J. Clin. Microbiol. *32*, 2995–2998.

Rice, L.B. (2008). Federal funding for the study of antimicrobial resistance in nosocomial pathogens: no ESKAPE. J. Infect. Dis. *197*, 1079–1081.

Ritchie, D.J., Alexander, B.T., and Finnegan, P.M. (2009). New antimicrobial agents for use in the intensive care unit. Infect. Dis. Clin. North Am. *23*, 665–681.

Roberts, K.J., Grusky, O., and Swanson, A.N. (2007). Outcomes of blood and oral fluid rapid HIV testing: a literature review, 2000–2006. AIDS Patient Care STDS *21*, 621–637.

Ronaghi, M. (2001). Pyrosequencing sheds light on DNA sequencing. Genome Res. *11*, 3–11.

Rossney, A.S., Herra, C.M., Brennan, G.I., Morgan, P.M., and O'Connell, B. (2008). Evaluation of the Xpert methicillin-resistant *Staphylococcus aureus* (MRSA) assay using the GeneXpert real-time PCR platform for rapid detection of MRSA from screening specimens. J. Clin. Microbiol. *46*, 3285–3290.

Saag, M.S., Holodniy, M., Kuritzkes, D.R., O'Brien, W.A., Coombs, R., Poscher, M.E., Jacobsen, D.M., Shaw, G.M., Richman, D.D., and Volberding, P.A. (1996). HIV viral load markers in clinical practice. Nat. Med. *2*, 625–629.

Schena, M., Shalon, D., Davis, R.W., and Brown, P.O. (1995). Quantitative monitoring of gene expression patterns with a complementary DNA microarray. Science *270*, 467–470.

Schmitt, M., Bravo, I.G., Snijders, P.J., Gissmann, L., Pawlita, M., and Waterboer, T. (2006). Bead-based multiplex genotyping of human papillomaviruses. J. Clin. Microbiol. *44*, 504–512.

Sefers, S.E., Raymer, A.K., Kilby, J.T., Persing, D.H., and Tang, Y.W. (2009). Prevalence and management of invalid GeneXpert enterovirus results obtained with cerebrospinal fluid samples: a 2-year study. J. Clin. Microbiol. *47*, 3008–3010.

Shepard, J.R., Addison, R.M., Alexander, B.D., Della-Latta, P., Gherna, M., Haase, G., Hall, G., Johnson, J.K., Merz, W.G., Peltroche-Llacsahuanga, H., *et al.* (2008). Multicenter evaluation of the *Candida albicans*/*Candida glabrata* peptide nucleic acid fluorescent in situ hybridization method for simultaneous dual-color identification of *C. albicans* and *C. glabrata* directly from blood culture bottles. J. Clin. Microbiol. *46*, 50–55.

Simen, B.B., Simons, J.F., Hullsiek, K.H., Novak, R.M., Macarthur, R.D., Baxter, J.D., Huang, C., Lubeski, C.,

Turenchalk, G.S., Braverman, M.S., *et al.* (2009). Low-abundance drug-resistant viral variants in chronically HIV-infected, antiretroviral treatment-naive patients significantly impact treatment outcomes. J. Infect. Dis. *199*, 693–701.

Sista, R., Hua, Z., Thwar, P., Sudarsan, A., Srinivasan, V., Eckhardt, A., Pollack, M., and Pamula, V. (2008). Development of a digital microfluidic platform for point of care testing. Lab. Chip. *8*, 2091–2104.

Sjoholm, M.I., Dillner, J., and Carlson, J. (2008). Multiplex detection of human herpesviruses from archival specimens by using matrix-assisted laser desorption ionization-time of flight mass spectrometry. J. Clin. Microbiol. *46*, 540–545.

Smalling, T.W., Sefers, S.E., Li, H., and Tang, Y.W. (2002). Molecular approaches to detecting herpes simplex virus and enteroviruses in the central nervous system. J. Clin. Microbiol. *40*, 2317–2322.

Smith, A.M., Heisler, L.E., Mellor, J., Kaper, F., Thompson, M.J., Chee, M., Roth, F.P., Giaever, G., and Nislow, C. (2009). Quantitative phenotyping via deep barcode sequencing. Genome Res. *19*, 1836–1842.

Soderlund-Strand, A., Dillner, J., and Carlson, J. (2008). High-throughput genotyping of oncogenic human papilloma viruses with MALDI-TOF mass spectrometry. Clin. Chem. *54*, 86–92.

Sogaard, M., Stender, H., and Schonheyder, H.C. (2005). Direct identification of major blood culture pathogens, including *Pseudomonas aeruginosa* and *Escherichia coli*, by a panel of fluorescence in situ hybridization assays using peptide nucleic acid probes. J. Clin. Microbiol. *43*, 1947–1949.

Spellberg, B., Guidos, R., Gilbert, D., Bradley, J., Boucher, H.W., Scheld, W.M., Bartlett, J.G., and Edwards, J., Jr. (2008). The epidemic of antibiotic-resistant infections: a call to action for the medical community from the Infectious Diseases Society of America. Clin. Infect. Dis. *46*, 155–164.

Sun, A.H., Fan, X.L., Li, L.W., Wang, L.F., Ans, W.Y., and Yan, J. (2009). Rapid detection of rpoB gene mutations in rif-resistant *M. tuberculosis* isolates by oligonucleotide microarray. Biomed. Environ. Sci. *22*, 253–258.

Talbot, G.H., Bradley, J., Edwards, J.E., Jr., Gilbert, D., Scheld, M., and Bartlett, J.G. (2006). Bad bugs need drugs: an update on the development pipeline from the Antimicrobial Availability Task Force of the Infectious Diseases Society of America. Clin. Infect. Dis. *42*, 657–668.

Tang, Y.W., and Persing, D.H. (2009). Diagnostic microbiology. In Encyclopedia of Microbiology, Schaechter, M., ed. (Oxford, Elsevier Press), pp. 308–320.

Tang, Y.W., Sefers, S.E., Li, H., Kohn, D.J., and Procop, G.W. (2005). Comparative evaluation of three commercial systems for nucleic acid extraction from urine specimens. J. Clin. Microbiol. *43*, 4830–4833.

Tang, Y.W., Kilic, A., Yang, Q., McAllister, S.K., Li, H., Miller, R.S., McCormac, M., Tracy, K.D., Stratton, C.W., Han, J., *et al.* (2007). StaphPlex system for rapid and simultaneous identification of antibiotic resistance determinants and Panton-Valentine leukocidin detection of staphylococci from positive blood cultures. J. Clin. Microbiol. *45*, 1867–1873.

Tang, W., Chow, W.H., Li, Y., Kong, H., Tang, Y.W., and Lemieux, B.K. (2010). Nucleic acid assay system for tier II labs and moderately complex clinics to detect HIV in low-resource settings. J. Infect. Dis. *201*, S46-S51.

Tanner, D.C., Weinstein, M.P., Fedorciw, B., Joho, K.L., Thorpe, J.J., and Reller, L. (1994). Comparison of commercial kits for detection of cryptococcal antigen. J. Clin. Microbiol. *32*, 1680–1684.

Tarr, C.L., Patel, J.S., Puhr, N.D., Sowers, E.G., Bopp, C.A., and Strockbine, N.A. (2007). Identification of *Vibrio* isolates by a multiplex PCR assay and rpoB sequence determination. J. Clin. Microbiol. *45*, 134–140.

Trevino, E.A., and Weissfeld, A.S. (2005). The case for point-of-care testing in infectious disease diagnosis. Clin. Microbiol. Newslet. *29*, 177.

Tripp, R.A., Alvarez, R., Anderson, B., Jones, L., Weeks, C., and Chen, W. (2007). Bioconjugated nanoparticle detection of respiratory syncytial virus infection. Int. J. Nanomedicine 2, 117–124.

Troesch, A., Nguyen, H., Miyada, C.G., Desvarenne, S., Gingeras, T.R., Kaplan, P.M., Cros, P., and Mabilat, C. (1999). *Mycobacterium* species identification and rifampin resistance testing with high-density DNA probe arrays. J. Clin. Microbiol. *37*, 49–55.

True, L.D., and Gao, X. (2007). Quantum dots for molecular pathology: their time has arrived. J. Mol. Diagn. *9*, 7–11.

Victoria, J.G., Kapoor, A., Dupuis, K., Schnurr, D.P., and Delwart, E.L. (2008). Rapid identification of known and new RNA viruses from animal tissues. PLoS Pathog. *4*, e100016.

Wang, D., Coscoy, L., Zylberberg, M., Avila, P.C., Boushey, H.A., Ganem, D., and DeRisi, J.L. (2002). Microarray-based detection and genotyping of viral pathogens. Proc. Natl. Acad. Sci. U.S.A. *99*, 15687–15692.

Wang, C., Mitsuya, Y., Gharizadeh, B., Ronaghi, M., and Shafer, R.W. (2007). Characterization of mutation spectra with ultra-deep pyrosequencing: application to HIV-1 drug resistance. Genome Res. *17*, 1195–1201.

Weber, B., Gurtler, L., Thorstensson, R., Michl, U., Muhlbacher, A., Burgisser, P., Villaescusa, R., Eiras, A., Gabriel, C., Stekel, H., *et al.* (2002). Multicenter evaluation of a new automated fourth-generation human immunodeficiency virus screening assay with a sensitive antigen detection module and high specificity. J. Clin. Microbiol. *40*, 1938–1946.

Wenzel, R.P. (2004). The antibiotic pipeline–challenges, costs, and values. N. Engl. J. Med. *351*, 523–526.

Wiedmann, M., Wilson, W.J., Czajka, J., Luo, J., Barany, F., and Batt, C.A. (1994). Ligase chain reaction (LCR)–overview and applications. PCR Methods Appl. 3, S51–64.

Wilson, W.J., Strout, C.L., DeSantis, T.Z., Stilwell, J.L., Carrano, A.V., and Andersen, G.L. (2002). Sequence-specific identification of 18 pathogenic microorganisms using microarray technology. Mol. Cell Probes *16*, 119–127.

Wittenbrink, M.M., Thiele, D., and Krauss, H. (1994). Comparison of dark-field microscopy, culture, and polymerase chain reaction (PCR) for detection of Borrelia burgdorferi in field-collected *Ixodes ricinus* ticks. Zentralbl Bakteriol. *281*, 183–191.

Wortmann, G., Weintrob, A., Barber, M., Scott, P., Zoll, S.T., Eshoo, M.W., Sampath, R., Ecker, D.J., and Massire, C. (2008). Genotypic evolution of *Acinetobacter baumannii* strains in an outbreak associated with war trauma. Infect. Control Hosp. Epidemiol. *29*, 553–555.

Wu, W., and Tang, Y.W. (2009). Emerging molecular assays for detection and characterization of respiratory viruses. Clin. Lab. Med. *29*, 673–693.

Xing, Y., Chaudry, Q., Shen, C., Kong, K.Y., Zhau, H.E., Chung, L.W., Petros, J.A., O'Regan, R.M., Yezhelyev, M.V., Simons, J.W., *et al.* (2007). Bioconjugated quantum dots for multiplexed and quantitative immunohistochemistry. Nat. Protoc. *2*, 1152–1165.

Zelada-Guillen, G.A., Riu, J., Duzgun, A., and Rius, F.X. (2009). Immediate detection of living bacteria at ultralow concentrations using a carbon nanotube based potentiometric aptasensor. Angew. Chem. Int. Ed. Engl. *48*, 7334–7337.

Zou, S., Han, J., Wen, L., Liu, Y., Cronin, K., Lum, S.H., Gao, L., Dong, J., Zhang, Y., Guo, Y., *et al.* (2007). Human influenza A virus (H5N1) detection by a novel multiplex PCR typing method. J. Clin. Microbiol. *45*, 1889–1892.

Changing the Therapeutic Paradigm in Antibacterial Drug Therapy and Discovery

20

Arturo Casadevall

Abstract

At the beginning of the twenty-first century the therapeutic paradigm for the treatment of infectious diseases can be summarized by three words: kill the bug. In other words, the overwhelming majority of therapeutic interventions against microbial diseases are designed to help the host by damaging the microbe directly and/or interfering with its ability to replicate in tissue (Casadevall, 2006). This strategy has been termed the second age of antimicrobial therapy and was preceded by the era of serum therapy, which differed in the fundamental manner that serum was primarily an immunotherapeutic agent than enhanced host defences (Casadevall, 2006). First and second age therapeutics differed in other ways including the chemistry of the therapeutic agent, their specificity and the form of manufacturing (Table 20.1). Second age therapeutics have been were tremendously successful and brought numerous drugs to the market that have saved countless lives. However, there are major trends at work that have significantly reduced the overall efficacy of second age therapeutics including widespread antimicrobial resistance, the emergence of new pathogenic microbes for which there are few drugs available and an epidemic of immunocompromised hosts where antimicrobial therapy is often less effective. Microbe-targeting strategies are limited in that they neglect the host; consequently, there are very few treatment strategies that aim to achieve a therapeutic outcome by enhancing host defences. Microbe-targeting strategies include both microbe-specific and -non-specific drugs, each of which can put tremendous selection pressure on microbes that often result in the emergence of resistance. Non-specific microbe-targeting strategies have the additional problem that they can select for resistance in non-targeted microbes and their effects on host flora can have a variety of unintended deleterious consequences on host homeostasis. This chapter will consider these strategies in light of their historical development and analyse the advantages and disadvantages of specific and non-specific antimicrobial strategies.

Immune serum provided the first effective antimicrobial strategy

The demonstration that administration of immune sera protected against toxins and bacterial infections (Behring and Kitasato, 1890; Klemperer and Klemperer, 1891) led to the development of serum therapy (Casadevall and Scharff, 1994, 1995). Serum therapy provided the first effective antimicrobial therapy and was rapidly developed against a broad range of infectious diseases including pneumococcal pneumonia, meningococcal meningitis, diphtheria, erysipelas, anthrax, and several viral diseases. The heyday of serum therapy was during the decades of the 1920s and 1930s when an extraordinary amount of scientific and clinical effort was devoted towards developing and improving serum therapy (Bullowa 1937). Although there was wide consensus that serum therapy was effective for certain infectious diseases such as pneumococcal pneumonia, there was considerable uncertainty whether serum was effective for others, such as tuberculosis

Table 20.1 The ages of antibacterial therapy

		Age I	Age II
Date		1890–1940s	1935-present
Agent		Serum (specific antibody)	Small molecule drug
Agent type		Immunoglobulin	Small molecule
Agent mass		>150,000 Daltons	About 1 kDa
Activity	Antiviral	Specific	Specific
	Antibacterial	Specific	Non-specific
	Antifungal	Not available	Non-specific
	Antimycobacterial	Not available	Specific
Mode of action		Opsonic, toxin neutralization, viral neutralization, complement activation	Interference with replication and/or directly microbicidal
Need for specific diagnosis prior to therapy		Yes	No
Efficacy		High	High
Toxicity		High	Low
Complexity		High	Low

(Glatman-Freedman and Casadevall, 1998). In general, serum therapy was found to be most effective when administered early after the onset of symptoms, and significant progress was made towards standardizing doses, at least for some diseases such as pneumococcal pneumonia (Felton, 1930). Serum was generated from two major sources: hyperimmunized large mammals such as horses and convalescent humans. Animal sera were used primarily for bacterial diseases when it was possible to immunize the host with the agent or part thereof. Human sera were used primarily for viral diseases that affected only humans, since in the absence of tissue culture techniques it was not possible to generate sufficient virus for hyperimmunizing animals. Nevertheless, it is noteworthy that in current times the use of convalescent serum for viral diseases such as influenza is again being considered since modern medicine has infrastructure in the form of blood banks and serological tests that could support such an approach (Luke *et al.*, 2010).

Serum therapy had considerable drawbacks (for recent reviews see (Casadevall and Scharff, 1994, 1995; Buchwald and Pirofski, 2003). First, it was complex, cumbersome and always required systemic administration. In the days before continuous intravenous infusions this required a physician to perform venepuncture and intravenous injection, usually remaining at the bedside during administration of therapy. Second, serum preparations were easily perishable and required constant refrigeration. Third, there was considerable lot-to-lot variation such that experience with one lot was not necessarily applicable to another. Fourth, it was expensive and costly to produce, since it required an immune animal or human host as its source. Fifth, and most importantly, serum therapy was accompanied by complications ranging from immediate hypersensitivity to a syndrome of arthralgias, fever, and rash that occurred in the days following the infusion known as 'serum sickness' (Feinberg, 1936). Serum sickness is thought to be the result of antigen–antibody complexes resulting from the induction of an antibody response to animal proteins. Although not known at the time it is likely that serum therapy with human sera also resulted in the inadvertent spread of certain infectious diseases such as hepatitis.

The discovery of sulfonamides in the mid-1930s, followed by the introduction of penicillin and its derivatives several years later, led to the abandonment of serum therapy. The transition from serum therapy to antibiotics and antimicrobial chemotherapy was rapid such that by 1950 the use of serum for the treatment of bacterial diseases had virtually stopped. Serum therapy was abandoned not because of lack of efficacy but because it could not compete with antibiotics and antimicrobial chemotherapy in ease of use, consistency of preparation, cost, and toxicity. In contrast to serum therapy, the new drugs were often available in oral formulation, came with standardized dosages and were much simpler and less costly to use. The toxicity of antibiotics and antimicrobial chemotherapy was significantly lower than serum therapy. A historical paradox in this momentous change in therapy was the fact that by the late 1930s research into protein chemistry was beginning to yield much purer and homogenous immunoglobulin preparations that promised much lower toxicity than conventional serum preparations. For example, the Cohn fractionation developed in the early 1940s produced relatively pure immunoglobulin preparations from serum (Berger, 2002). Hence, it is conceivable that if either immunoglobulin purification had been discovered earlier, or antibiotics later, immunoglobulin therapy would have maintained a place in the antimicrobial arsenal. In fact, there was considerable evidence at the time, and in subsequent studies, that the combination of specific antibody with antibiotics or antimicrobial chemotherapy was additive or synergistic (Casadevall and Scharff, 1995). However, such combination therapy was not pursued because its benefits were outweighed by the toxicity and complexity of using serum therapy. By the 1950s the situation with antimicrobial therapy was considered to be highly satisfactory, although there were already some clouds in the horizon with *Candida* superinfections in antibiotic-treated patients (Jawetz, 1956).

Despite the abandonment of serum therapy for the treatment of bacterial diseases, this mode retained a niche as an anti-toxin and anti-venom therapy (Blake *et al.*, 1976; Chippaux and Goyffon, 1983). To this day specific antibody remains the only effective therapeutic agent against toxin- and venom-mediated diseases. Beginning in the 1960s, human immunoglobulin preparations were developed that were active against specific viruses such as hepatitis, cytomegalovirus, and rabies. The discovery of hybridoma technology and monoclonal antibodies in 1975 promised the development of new antibody-based therapies for infectious diseases (Kohler and Milstein, 1975), but to date only one mAb has been licensed for the prevention of an infectious disease, that caused by respiratory syncytial virus. The paucity of monoclonal antibody therapeutics reflects in part their exquisite specificity, which in turns creates difficulties in drug development (Saylor *et al.*, 2009).

The emergence and dominance of non-specific therapies for bacterial and fungal diseases

The shift from serum therapy to antibiotics and antimicrobial chemotherapy also represented a fundamental shift in the approach to infectious diseases. Serum therapy was pathogen specific and consequently required a microbial diagnosis before it could be used effectively. For some antigenically variable pathogens such as the pneumococci, it was necessary to identify not only the offending pathogen but also its particular serotype. This was accomplished by the development of rapid tests such as the Quellung reaction with type specific sera (Sabin, 1930, 1933; Bullowa 1935). In contrast, antibiotics and antimicrobial chemotherapy had activity against a broad subset of medically important bacterial species and these drugs could be used empirically without making a specific diagnosis. The ability to treat an infectious disease without making a specific diagnosis had immediate advantages for the patient, for it meant that therapy was effective earlier that was undoubtedly life saving to countless individuals. However, non-specific therapy also had certain untoward consequences. The ability to treat effectively without a microbial diagnosis meant that diagnostic microbiology became less important and consequently it was underdeveloped. For example, a positive culture remained the gold standard for microbial diagnosis during most of the twentieth century despite the fact that recovery of microbes in pure culture was often a slow process requiring days. This in turn ushered in a culture of empiricism focused on the selection of appropriate antimicrobial therapy coverage rather than making a specific diagnosis. At the level of the individual the use of non-specific antibiotic and antimicrobial chemotherapy was less toxic than the older serum therapy, but by the 1950s it was clear that such therapy could disrupt host flora with the side-effect of candidiasis. Later in the century this therapy would be associated with other infectious diseases problems such as antibiotic-associated diarrhoea and the emergence of resistance among non-targeted microbes.

For antibacterial therapy, the change from serum therapy to antibiotics and antimicrobial chemotherapy was a clear shift from specific to non-specific therapy. It is noteworthy that other promising antibacterial therapies that were also abandoned at the time, such as phage therapy, were also pathogen-specific. However, this shift did not affect all areas of antimicrobial therapy equally. For example, no effective serum therapy was ever developed against fungal pathogens, and the first effective therapy drug, amphotericin B, had broad activity against fungi. Hence, for fungal diseases, therapy has always been non-specific. For viral diseases, most antiviral drugs are virus-specific and consequently antiviral therapy has always been largely pathogen-specific. For mycobacteria, the majority of drugs used are employed only in mycobacterial diseases treatment, and consequently, tuberculosis therapy is primarily pathogen-specific. Hence, shifts in the specificity of antimicrobial therapy have impacted primarily antibacterial therapy.

The current therapeutic paradigm

Current antimicrobial therapy is designed to help the host by killing and/or preventing the replication of the microbe in tissue. This is a fundamentally different approach than that of serum therapy which was designed to help the host by providing immunoglobulins that enhanced host immune function through opsonization and/or neutralizing microbial virulence factors. Antiviral therapy remains largely virus specific since viruses have few genes and their inherent specialization for a host means that they share few or no common molecular targets with other viruses for drug discovery. However, bacteria and fungi each has complex metabolic processes that differ from host metabolism, and antimicrobial drug discovery has focused on identifying compounds that kill, or prevent growth of broad classes of pathogenic species. The availability of relatively non-toxic antibacterial and antifungal therapy allows physicians to treat a variety of conditions in a variety of clinical settings and thus allows one drug discovery effort to address several clinical needs. Hence, the current therapeutic paradigm targets microbes with non-specific therapies for bacteria and fungi and with specific therapies for viruses and mycobacteria. Despite the preponderance of microbe-targeting therapies there are a few antimicrobial therapies that aim to achieve a therapeutic effect by enhancing host defences. Examples of host-targeted antimicrobial therapies include the use of interferons for chronic hepatitis and chronic granulomatous disorders, colony stimulating factors to hasten recovery from neutropenia, and intravenous immunoglobulin replacement therapies (Pirofski and Casadevall, 2006). However, with the field overwhelmingly focused on targeting microbes, host-targeted immune therapies are largely underdeveloped providing tremendous opportunities for future development. In fact, despite the molecular biology revolution that has seen many cytokines, chemokines and host proteins cloned and theoretically available for clinical use, few of these agents have been evaluated clinically. Similarly, antibody-based therapies remain underdeveloped for infectious diseases despite great technical strides in the development and availability of human antibody preparations that would have none of the disadvantages associated with the heterologous immune sera used in the era of serum therapy.

The advantages and disadvantages of specific and non-specific therapies

Both pathogen-specific and -non-specific therapies are effective clinically. This is apparent in the success of such pathogen-specific drugs such as isoniazid, ethambutol, and acyclovir and pathogen-non-specific drugs such as penicillins and tetracyclines. Hence, consideration of the advantages and disadvantages of pathogen-specific and -non-specific therapies should not include potential efficacy. Non-specific therapy has a clear advantage in mixed infections where the drug is likely to be effective against multiple organisms, and in situations where diagnostic information on pathogen identity is delayed or incomplete. In both situations, the administration of non-specific therapy has a much greater likelihood of providing effective therapy than using pathogen-specific drugs. Without the availability of diagnostic methods that provide rapid pathogen identification it is difficult to use pathogen-specific drugs effectively. The few pathogen-specific drugs used to today are effective because they are used to treat chronic diseases such as tuberculosis where the initiation of therapy can wait for a microbial diagnosis, diseases that are relatively easy to diagnose on clinical grounds such as herpes-related diseases, or those for which there are well developed diagnostic techniques such as HIV and other viruses. In fact, one can argue that pathogen-specific drugs cannot be successfully introduced into clinical practice without conditions that allow their effective use, and the current absence of rapid diagnostic techniques for many infectious diseases is a significant hindrance to their development.

The major advantage of pathogen-specific therapy is that it targets only the microbe causing disease and consequently has no, or only minimal, effects on non-targeted microbes. Although this advantage was not appreciated in the early years of the antibiotic era, there are various developments that could make this advantage increasingly attractive. These developments arise from the realization that non-pathogen specific drugs have profound effects on the host microflora, and possibly environmental microbes, which could have significant potential negative consequences on both the individual and society. Although few studies have analysed effects of antibiotic usage on the flora, the available data from deep sequencing experiments show profound disturbances in taxonomic composition of bacterial communities (Dethlefsen *et al.*, 2008). For example, a study of individuals who received ciprofloxacin revealed long term pervasive effects on gut flora with considerable individual variation, and the disturbance included the observation that some bacterial taxa did not recover after several weeks of observation (Dethlefsen *et al.*, 2008). In recent decades, several normally commensal microbes, such as *Enterococcus* spp., have emerged as pathogenic microbes in individuals treated with broad antibacterial therapy, and these strains often manifest considerable drug resistance. Broad spectrum therapy has also been associated with an epidemic of colitis caused by *Clostridium difficile*, presumably as a consequence of disruption of the normal host intestinal microflora. Mucocutaneous candidiasis is another example of a microbial disease that follows antibiotic-induced effects on the normal host-associated microflora. In addition to antimicrobial drug resistance and microbial diseases arising from disturbances in the host flora, there are increasing reports that alterations in the drug flora may contribute to the pathogenesis of certain diseases that historically have not been associated with microbes. For example, several studies have associated certain forms of cancer (Velicer *et al.*, 2004), asthma (Kozyrskyj *et al.*, 2007; Wickens *et al.*, 2008), and atopic disease (Kusel *et al.*, 2008) with prior antimicrobial drug use. Although none of these associations have risen to the level that causality has been proven, they do suggest the existence of potential links. In the case of breast cancer, the mechanisms that have been suggested include alterations in oestrogen metabolism and micronutrient absorption as a consequence of changes in the microbial flora (Velicer *et al.*, 2003). The association between previous antibiotic use and breast cancer remains controversial as various studies have found evidence for and against an association (Velicer *et al.*, 2006; Tamim *et al.*, 2008). For asthma and atopy, the association could reflect immunological alterations following disruptions in the microbial flora (Cornell *et al.*, 2008; Verhulst *et al.*, 2008). If any of these current tentative associations are established to be cause and effect relationships, then pathogen non-specific drugs will be at a significant disadvantage relative to pathogen-specific drugs with respect to certain patient populations.

Pathogen-specific therapies have clear advantages over non-specific theories assuming that one has rapid, accurate, and precise diagnostic methods and that the infectious disease is not caused by a mixed infection. However, we lack precise information on the relative proportions of infectious diseases caused by single organisms versus mixed infectious. Even in clinical situations where a single microbe is recovered from a sterile body site, one cannot be certain that the recovered microbe is the only cause of the disease. Although this uncertainty brings with it the conundrum of having to prove a negative, it is worthwhile to remember that our understanding of the association of certain microbes with disease is based largely on culture methods which provide a positive result only for culturable microbes. In recent years, nucleic acid analysis and deep sequencing methods have shown us that a large proportion, if not the majority, of microorganisms are not culturable by routine medical microbiological techniques. Hence, one cannot exclude the possibility that some infectious diseases currently associated with a single microbe are in fact polymicrobial infections, and that the efficacy of broad spectrum therapy reflects effects on other microbes. Although this scenario would at first raise intellectual questions on pathogen specific approaches, it is worth remembering that serum therapy was effective against numerous infectious diseases, and its historical efficacy despite this narrow specificity suggests that this scenario is probably not relevant for the majority of current infectious diseases. Nevertheless, the possibility of cryptic mixed infections must be considered during the development and use of pathogen-specific therapies, and the failure of such drugs would demand a re-consideration of the causal associations of certain microbes with certain infectious diseases.

Challenges for the development of pathogen-specific drugs

Given that pathogen-specific drugs make sense for many infectious diseases, the next issue is to discern the challenges facing the development of such therapies (Casadevall, 2009). Before even considering the problems associated with the development of pathogen-specific drugs, it important to acknowledge that in the past decade the entire area of antimicrobial drug discovery has been in crisis, with relatively few drugs in the pipeline (Projan, 2003; Projan and Shlaes, 2004; Spellberg *et al.*, 2004). The causes for these global trends are extremely complex and range from unfavourable economics for antimicrobial drugs relative to drugs for chronic diseases to the emergence of resistance to basic problems in identifying drug targets that would yield profitable, low-toxicity drugs. In this environment, perhaps the greatest challenge hindering the development of pathogen-specific drugs is economic. For antimicrobial drugs, the size of the market is proportional to the breadth of their antimicrobial activity and the prevalence of the infectious disease(s) for which the drug is effective. Since the process of drug discovery and development is enormously costly, and market size determines profit and the extent of reimbursement for development costs, pathogen-specific drugs are necessarily at a great economic disadvantage relative to non-pathogen specific therapies (Nathan, 2004). However, this hard law of antimicrobial economics does have some room for the development of certain pathogen-specific drugs. For example, consider the case of methicillin-resistant *Staphylococcus aureus* (MRSA) which has emerged a major pathogen in recent decades. The emergence of MRSA has severely limited the effective antimicrobial arsenal for staphylococcal disease, and currently vancomycin is first line therapy against this organism. However, vancomycin has broad activity against Gram-positive bacteria and its use has been associated with subsequent infections by other drug-resistant microbes such as *Enterococcus* spp. (Atta *et al.*, 2001; Patterson, 2001). In this situation where there is a high prevalence of disease cause by one microbe and where the therapeutic alternatives are few and have some untoward consequences, the development of an MRSA-specific drug could be economically viable. In fact, emergence of resistance for a microbe associated with a high prevalence of disease could make the economics of drug development against that pathogen favourable. This notion that resistance creates a market for pathogen-specific drugs could hasten the development of antimicrobial agents specific for particularly troublesome microbes that manifest multidrug resistance.

The second major challenge for the development of pathogen-specific therapies has already been alluded to: the underdevelopment of diagnostic techniques. Currently, diagnosing most bacterial diseases requires recovery of the microbe in pure culture followed by biochemical, and sometimes serological, identification. This approach takes time and obtaining often requires days. Given that the outcome of therapy is often a function of its timing, and that prompt therapy is associated with best outcomes, pathogen-specific therapy is at a severe disadvantage relative to non-specific therapy for the treatment of diseases where a microbial aetiology is suspected but a microbial identification is not available. Hence, successful development of pathogen-specific therapy would require the parallel development of rapid diagnostic methods for that pathogen that would support its proper use. In this regard the development of specific therapy of against influenza virus provides an encouraging precedent in the sense that rapid diagnostic tests were developed simultaneously with drug therapy, allowing physicians to discriminate influenza from other seasonal maladies and thus use specific antiviral therapy effectively.

The development of pathogen-specific therapy would involve a major rethinking of the drug discovery process. Currently, the approach is to look for common targets among pathogenic microbes that would allow the development of broad spectrum drugs. In contrast, pathogen-specific therapy would focus on differences between microbes to generate drugs that are active only against a particular type of microbe. Antibody-based therapeutics are already largely pathogen-specific but there are few examples of successful specific small molecule therapeutics, such as isoniazid. Consequently, there is scientific uncertainty whether drug development can identify enough unique pathways for each microbe that result in antimicrobial effects when interfered with by small molecules. One strategy here would be to focus on interfering with known virulence characteristics, such as capsule synthesis, toxin production, secretion, etc. (Clatworthy *et al.*, 2007; Lynch and Wiener-Kronish, 2008). Given that virulence factors often function in pathogenesis by disabling host immune mechanisms, a focus on interfering with their action would have the advantage of leaving the microbe vulnerable to killing by host defence mechanisms and could translate into less selection for drug resistance.

Another challenge for the development of pathogen-specific therapy is the culture of empiricism that pervades the current approach to the treatment of patients with presumed infectious diseases. In general, physicians tend to approach the problem of infectious diseases with a stochastic mind set where the decision of whether to institute antimicrobial therapy and the type of drugs to use is determined by estimating the probability of infection, the probable causative organisms, and their presumed drug susceptibility. Hence, many serious infectious diseases are treated empirically, with culture data serving to confirm or modify the suitability of the initial drug choices made. Consequently, the introduction of pathogen specific therapy would require a fundamental shift in the approach to the patient with the emphasis changing from empiricism to determinism, where the initial procedure would be to make a microbiologic diagnosis, provided that suitable diagnostic tools were available, and then selecting therapy. Although diagnosis followed by the selection of therapy is the usual sequence in much of clinical medicine, the availability of antimicrobial drugs with broad activity and low toxicity in combination with underdeveloped diagnostic methods has created in the field of infectious diseases a situation where therapy often precedes diagnosis. In fairness to physicians treating infectious diseases, this approach has merit in the sense that microbial diseases can progress very rapidly and, unlike other maladies, one often does not have the luxury of time. In fact, for chronic infectious diseases where a short delay in the initiation of therapy does not pose a significant risk to the long term outcome, such as pulmonary tuberculosis, physicians do try to make a diagnosis before therapy, although even there empiric therapy is often used while waiting for the results of sputum cultures.

The near and far horizons

Depending on the microbe, pathogen-specific therapy is either already available, in the near horizon or distant horizon, or way beyond our reach. Pathogen-specific therapy is used and successful against numerous viruses, *M. tuberculosis*, and several parasitic diseases like filariasis. In considering parasitic diseases, we are faced with the issue that the concept of 'specificity' is itself relative depending on the type of pathogen considered and its host. For example, ivermectin has activity against numerous helminths, yet in the case of humans with filariasis it is 'specific' in the sense that worms are not thought to be normal inhabitants of the commensal flora, and hence this therapy is specific for the pathogen despite its broad activity against worm species. Hence, the criterion of specificity can be considered in light of targeting one microbe among many, or targeting a microbe in a host where the microbe in question is sufficiently different from the host and its associated microflora such that only the pathogen is affected. In either case there is overwhelming evidence that pathogen-specific therapy is feasible, can be developed, and is effective clinically.

The microbes for which pathogen-specific drugs may be developed in the near horizon fall into three non-exclusive categories. First, some microbes have become so prevalent that by themselves they constitute a market of sufficient size to make a pathogen-specific drug economically viable. For example, in the case of *Staphylococcus aureus* it is possible and probable that a specific or very narrow spectrum drug would be developed if such a compound would be identified because there is a sufficiently large market to support the investment. Second, pathogen-specific drugs are likely to be developed for those few pathogenic microbes for which we have few or no therapeutic options. Here an example would be highly resistant organisms where existing therapies are unsatisfactory, such as certain *Acinetobacter* spp. Third, pathogen-specific drugs would be attractive for common drug-susceptible microbes where the current use of broad-spectrum therapy has been associated with untoward consequences. For example, the rise of *C. difficile* colitis as a major complication of broad spectrum therapy suggests a need for therapeutic options that spare the microbiota. In a world where many pathogen specific drugs are available it might be possible combine them for empiric therapy until a diagnosis is made, and afterwards those that are not needed can be discontinued.

Perhaps the greatest uncertainty as to the future development of pathogen-specific therapy is whether on-going studies associating broad-spectrum antimicrobial use with certain maladies such as asthma, cancer, and atopy are able to make causal associations between antibiotic use and subsequent chronic diseases. Strong evidence for the tremendous effects of non-specific therapy on host microbial flora is emerging from microbiome studies (Dethlefsen *et al.*, 2008). Given that host microbial flora has been increasingly implicated in the normal development of the immune system and that microbial products affect many aspects of metabolism, one can imagine how antimicrobial drug-induced alterations on the flora could result in disturbances in host homeostasis. If a single causal association is made, there would be tremendous public pressure for designing effective antimicrobial drugs that spare the host flora that would probably foster the development of more pathogen-specific drugs.

Assuming that the near and far horizons include more pathogen-specific drugs in the pharmacological arsenal, it is worthwhile to consider what such therapies will and will not accomplish. Pathogen-specific therapies should not select for resistance among non-targeted microbes but would remain vulnerable to the emergence of resistance for the targeted microbe. Broad spectrum therapies will not disappear since they will continue to have a role in the treatment of infectious diseases of polymicrobial aetiology. However, even in polymicrobial diseases there are often single microbial species that cause the majority of damage, and drugs targeting those microbes could make inroads into the therapy of mixed infections. The use of pathogen-specific therapies would bring greater complexity to the field of infectious diseases. An example of this is apparent from the fact that the care of HIV patients has become a virtual subspecialty within the field of infectious diseases as the advancements in HIV therapy have brought with them numerous drugs and diagnostic tests that in turn requires much specialized knowledge. Hence, the widespread use of pathogen-specific drugs would necessitate knowledge of more drugs, side-effects, diagnostic tests and microbial pathogenesis than is currently required by most physicians. Consequently, a future with more pathogen-specific drugs would be a more complex future with regards to selecting antimicrobial drugs. This in turn could lead to a rise in the importance of infectious disease specialists and a decrease in the use of antimicrobial therapy by non-specialist physicians.

Given the unfavourable economics for developing most pathogen-specific drugs, the introduction of numerous drugs would require new models for drug development. Since the economics of antimicrobial drugs are such that market size is directly proportional to activity and the prevalence of the susceptible microbes as pathogens, the development of many pathogen-specific drugs is not commercially viable unless special incentives are created to spur this activity. Here government and regulatory agencies may be able to design incentives. For example, linking the length of patent protection to the activity of the agent such that patent length is inversely proportional to activity, thereby conferring very long protection to pathogen-specific drugs, could impact the economics of antimicrobial development to favour narrower spectrum and single specificity drugs. Furthermore, there are some trends in place that could make the climate for pathogen-specific drugs significantly friendlier. These include an increasing focus on reducing medical errors that include nosocomial infections. Since some nosocomial infections such as *C. difficile* colitis, enterococcal bacteraemia, and candidiasis are often associated with prior broad spectrum antimicrobial drug use, one could envisage that pressures to reduce medical errors and complications could incentivize the use of antimicrobial drugs with narrower specificity, and hence smaller effects on host flora. Finally, any association of broad spectrum therapy with chronic non-infectious disorders such as asthma and cancer could bring on the cloud of future litigation, which in turn would exert tremendous pressures on drug development and physician selection.

Conclusion

In summary, pathogen-specific therapy is effective but largely unavailable owing to a variety of complex reasons that include unfavourable economics, underdeveloped diagnostics, and an emphasis in drug development on targeting microbial molecular targets common to many microbes rather than specific differences. Despite these enormous hurdles there are powerful trends in place that could catalyse the development of new pathogen-specific drugs. If any association is made between antimicrobial therapy, damage to the microflora and any of the chronic diseases that plague our society, the economic equation will change and there will be tremendous momentum for the development of drugs that spare the normal flora. This would inevitably force a reassessment of the cost–benefit and therapeutic index of non-specific therapies, which in turn could catalyse a shift back to pathogen-specific therapies. In the absence of such a detrimental association, antimicrobial therapy is likely to remain specific for viruses and non-specific for bacteria and fungi, with the steady development of narrower spectrum drugs that target microbes with high prevalence that are resistant to many broad spectrum drugs, like MRSA, and possibly Gram-negative bacteria (Theuretzbacher, 2009).

References

Atta, M.G., Eustace, J.A., Song, X., Perl, T.M., and Scheel, P.J., Jr. (2001). Outpatient vancomycin use and vancomycin-resistant enterococcal colonization in maintenance dialysis patients. Kidney Int. *59*, 718–724.

Behring, E.A. and Kitasato, S. (1890). Ueber das Zustandekommen der Diptherie-Immunität und der Tetanus-Immunität bei Thieren. Deutch. Med. Woch. *49*, 1113–1114.

Berger, M. (2002). A history of immune globulin therapy, from the Harvard crash program to monoclonal antibodies. Curr. Allergy Asthma Rep. *2*, 368–378.

Blake, P.A., Feldman, R.A., Buchanan, T.M., Brooks, G.F., and Bennett, J.V. (1976). Serologic therapy of tetanus in the United States. JAMA *235*, 42–44.

Buchwald, U.K. and Pirofski, L. (2003). Immune therapy for infectious diseases at the dawn of the 21st century: the past, present and future role of antibody therapy, therapeutic vaccination and biological response modifiers. Curr. Pharm. Des *9*, 945–968.

Bullowa, J.G.M. (1935). The reliability of sputum typing and its relation to serum therapy. J. Am. Med. Assoc. *105*, 1512–1518.

Bullowa, J.G.M. (1937). Serum therapy. In The Management of the Pneumonias (Oxford University Press, New York), pp. 283–362.

Casadevall, A. (2006). The third age of antimicrobial therapy. Clin. Infect. Dis. *42*, 1414–1416.

Casadevall, A. (2009). The case for pathogen-specific therapy. Expert. Opin. Pharmacother. *10*, 1699–1703.

Casadevall, A., and Scharff, M.D. (1994). 'Serum Therapy' revisited: Animal models of infection and the development of passive antibody therapy. Antimicrob. Agents Chemotherap. *38*, 1695–1702.

Casadevall, A., and Scharff, M.D. (1995). Return to the past: the case for antibody-based therapies in infectious diseases. Clin. Infect. Dis. *21*, 150–161.

Chippaux, J.P., and Goyffon, M. (1983). Producers of antivenomous sera. Toxicon *21*, 739–752.

Clatworthy, A.E., Pierson, E., and Hung, D.T. (2007). Targeting virulence: a new paradigm for antimicrobial therapy. Nat. Chem. Biol. *3*, 541–548.

Cornell, A., Shaker, M., and Woodmansee, D.P. (2008). Update on the pathogenesis and management of childhood asthma. Curr. Opin. Pediatr. *20*, 597–604.

Dethlefsen, L., Huse, S., Sogin, M.L., and Relman, D.A. (2008). The pervasive effects of an antibiotic on the human gut microbiota, as revealed by deep 16S rRNA sequencing. PLoS. Biol. *6*, e280.

Feinberg, S.M. (1936). The therapy of (Horse) serum reactions. General rules in the administration of therapeutic serums. JAMA *107*, 1717–1719.

Felton, L.D. (1930). The concentration of anti-pneumococcus serum. JAMA *94*, 1893–1896.

Glatman-Freedman, A., and Casadevall, A. (1998). Serum therapy for tuberculosis revisited: a reappraisal of the role of antibody-mediated immunity against *Mycobacterium tuberculosis*. Clin. Microbiol. Rev. *11*, 514–532.

Jawetz, E. (1956). Antimicrobial therapy. Ann. Rev. Microbiol. *10*, 85–114.

Klemperer, G., and Klemperer, F. (1891). Versuche uber immunisirung und heilung bei der pneumokokkeninfection. Berl. klin. Wchnschr. *28*, 833–835.

Kohler, G., and Milstein, C. (1975). Continuous cultures of fused cells secreting antibody of predefined specificity. Nature *256*, 495–497.

Kozyrskyj, A.L., Ernst, P., and Becker, A.B. (2007). Increased risk of childhood asthma from antibiotic use in early life. Chest *131*, 1753–1759.

Kusel, M.M., de, K.N., Holt, P.G., and Sly, P.D. (2008). Antibiotic use in the first year of life and risk of atopic disease in early childhood. Clin. Exp. Allergy *38*, 1921–1928.

Luke, T.C., Casadevall, A., Watowich, S.J., Hoffman, S.L., Beigel, J.H., and Burgess, T.H. (2010). Hark back: Passive immunotherapy for influenza and other serious infections. Crit. Care. Med. *38*, e66–73.

Lynch, S.V. and Wiener-Kronish, J.P. (2008). Novel strategies to combat bacterial virulence. Curr. Opin. Crit. Care *14*, 593–599.

Nathan, C. (2004). Antibiotics at the crossroads. Nature *431*, 899–902.

Patterson, J.E. (2001). Antibiotic utilization: is there an effect on antimicrobial resistance? Chest *119*, 426S-430S.

Pirofski, L.A. and Casadevall, A. (2006). Immunomodulators as an antimicrobial tool. Curr. Opin. Microbiol. *9*, 489–495.

Projan, S.J. (2003). Why is big Pharma getting out of antibacterial drug discovery? Curr. Opin. Microbiol. *6*, 427–430.

Projan, S.J. and Shlaes, D.M. (2004). Antibacterial drug discovery: is it all downhill from here? Clin. Microbiol. Infect. *10* (Suppl 4), 18–22.

Sabin, A.B. (1930). The microscopic agglutination test in pneumonia. Its application to rapid typing and control of serum therapy. J. Infect. Dis. *46*, 469–484.

Sabin, A.B. (1933). Immediate pneumococcus typing directly from sputum by the Neufeld reaction. JAMA *100*, 1584–1586.

Saylor, C., Dadachova, E., and Casadevall, A. (2009). Monoclonal antibody-based therapies for microbial diseases. Vaccine *27* (Suppl 6), G38–G46.

Spellberg, B., Powers, J.H., Brass, E.P., Miller, L.G., and Edwards, J.E., Jr. (2004). Trends in antimicrobial drug development: implications for the future. Clin. Infect. Dis. *38*, 1279–1286.

Tamim, H.M., Hanley, J.A., Hajeer, A.H., Boivin, J.F., and Collet, J.P. (2008). Risk of breast cancer in relation to antibiotic use. Pharmacoepidemiol. Drug Saf. *17*, 144–150.

Theuretzbacher, U. (2009). Future antibiotics scenarios: is the tide starting to turn? Int. J. Antimicrob. Agents *34*, 15–20.

Velicer, C.M., Lampe, J.W., Heckbert, S.R., Potter, J.D., and Taplin, S.H. (2003). Hypothesis: is antibiotic use associated with breast cancer? Cancer Causes Control *14*, 739–747.

Velicer, C.M., Heckbert, S.R., Lampe, J.W., Potter, J.D., Robertson, C.A., and Taplin, S.H. (2004). Antibiotic use in relation to the risk of breast cancer. JAMA *291*, 827–835.

Velicer, C.M., Heckbert, S.R., Rutter, C., Lampe, J.W., and Malone, K. (2006). Association between antibiotic use prior to breast cancer diagnosis and breast tumour characteristics (United States). Cancer Causes Control *17*, 307–313.

Verhulst, S.L., Vael, C., Beunckens, C., Nelen, V., Goossens, H., and Desager, K. (2008). A longitudinal analysis on the association between antibiotic use, intestinal microflora, and wheezing during the first year of life. J. Asthma *45*, 828–832.

Wickens, K., Ingham, T., Epton, M., Pattemore, P., Town, I., Fishwick, D., and Crane, J. (2008). The association of early life exposure to antibiotics and the development of asthma, eczema and atopy in a birth cohort: confounding or causality? Clin. Exp. Allergy *38*, 1318–1324.